DESCRIPTIVE PHYSICAL OCEANOGRAPHY:

AN INTRODUCTION

SIXTH EDITION

DESCRIPTIVE PHYSICAL OCEANOGRAPHY: AN INTRODUCTION

SIXTH EDITION

Lynne D. Talley

George L. Pickard

William J. Emery

James H. Swift

AMSTERDAM • BOSTON • HEIDELBERG • LONDON
NEW YORK • OXFORD • PARIS • SAN DIEGO
SAN FRANCISCO • SINGAPORE • SYDNEY • TOKYO

Academic Press is an imprint of Elsevier

Academic Press is an imprint of Elsevier
32 Jamestown Road, London NW1 7BY, UK
30 Corporate Drive, Suite 400, Burlington, MA 01803, USA
525 B Street, Suite 1800, San Diego, CA 92101-4495, USA

Sixth edition 2011

Notice

No responsibility is assumed by the publisher for any injury and/or damage to persons or property as a
matter of products liability, negligence or otherwise, or from any use or operation of any methods, products,
instructions or ideas contained in the material herein. Because of rapid advances in the medical sciences, in
particular, independent verification of diagnoses and drug dosages should be made.

British Library Cataloguing-in-Publication Data
A catalogue record for this book is available from the British Library

Library of Congress Cataloging-in-Publication Data
A catalog record for this book is available from the Library of Congress

ISBN: 978-0-7506-4552-2

For information on all Academic Press publications visit
our website at www.elsevierdirect.com

Typeset by TNQ Books and Journals Pvt Ltd

Contents

Supplementary Materials

http://booksite.academicpress.com/DPO

Supplementary Figures for many chapters

Java OceanAtlas Exercises

Preface

This new edition of *Descriptive Physical Oceanography: An Introduction* is dedicated to the memory of George L. Pickard (July 15, 1913–May 1, 2007), who was a physical oceanographer at the University of British Columbia. George was part of University of British Columbia's oceanography department from its inception. His training was in low temperature physics, with a Doctor of Philosophy from Oxford in 1937. After service in WW II in the Royal Aircraft Establishment, during which he survived a crash in the English Channel, he was appointed to the physics department at UBC. As a young member of the department, he was sent to Scripps Institution of Oceanography for "a year's training" in oceanography as part of the lobbying effort by John Tully for formation of Canada's first academic program in oceanography at UBC; the program was established in 1949 (Mills, 1994). George was director of the UBC Institute of Oceanography from 1958 to 1978. He retired from teaching in 1982.

George wrote the first and subsequent editions of this book as part of his teaching of physical oceanography, bringing in Bill Emery for the fourth and fifth editions as the material was updated and enlarged. He also co-authored with Stephen Pond the text *Introductory Dynamical Oceanography*. In 1950, George initiated time series measurements in many of the inlets along the British Columbia coastline, partially as training exercises for the UBC students; these observations continue to the present and constitute a tremendous source of climate-related information (http://www.pac.dfo-mpo.gc.ca/science/oceans/BCinlets/index-eng.htm, 2010). His research expanded to include coral reefs as well as fjords. A full CV is included on the website that accompanies this edition.

When George Pickard published the original DPO text in 1964, computers were just barely beginning to be introduced, and courses were taught at the blackboard and illustrated with slides. This sixth edition of DPO stands at the brink of fully electronic publishing and full online support for teaching. We therefore provide a website that includes all of the illustrations from the print text, many in color even if not reproduced in color herein. There are also many additional illustrations and supporting text on the website. Several chapters appear on the website and not in the print text, in order to keep the cost of the print text accessible: the full-length version of Chapter 7 on ocean dynamics; the final sections of Chapter 8 concerning estuaries, coral reefs, and adjacent seas; Chapter S15 on **Climate and the oceans**; and Chapter S16 on **Instruments and methods**.

Secondly, with this edition we introduce a digital set of tools and tutorials for working with and displaying ocean property data, using **Java OceanAtlas**. The software and representative data sets are also provided online, along with a step-by-step guide to using them and examples associated with most chapters of the print text. We strongly encourage students and lecturers to make use of these web-based materials.

This edition of DPO is also dedicated by LDT and JHS to our teachers, among them Joe Pedlosky, Mike McCartney, Val Worthington, Knut Aagaard and Eddy Carmack, and to our senior colleague, Joe Reid at SIO whose work is central to many chapters of this book, and who preceded LDT in teaching SIO course 210. The students of SIO 210 and colleagues who have team taught the course — M. Hendershott and P. Robbins — have provided annual motivation for recalling the essentials of large-scale descriptive oceanography. A number of colleagues and students provided invaluable feedback on parts of the text, including J. Reid, D. Sandwell, P. Robbins, J. Holte, S. Hautala, L. Rosenfeld, T. Chereskin, Y. Firing, S. Gille and the students from her data analysis laboratory course, B. Fox-Kemper, K. Aagaard, A. Orsi, I. Cerovecki, M. Hendershott, F. Feddersen, M. Mazloff, S. Jayne, and J. Severinghaus, as well as numerous comments from SIO 210 students. LDT gratefully acknowledges sabbatical support from the SIO department, hosted by Woods Hole Oceanographic Institution (T. Joyce and the Academic Programs Office) and the Université Joseph Fourier in Grenoble (B. Barnier and the Observatoire des Sciences de l'Univers de Grenoble).

L.D.T. and J.H.S.
Scripps Institution of Oceanography,
La Jolla, CA

1

Introduction to Descriptive Physical Oceanography

Oceanography is the general name given to the scientific study of the oceans. It is historically divided into physical, biological, chemical, and geological oceanography. This book is concerned primarily with the physics of the ocean, approached mainly, but not exclusively, from observations, and focusing mainly, but not exclusively, on the larger space and timescales of the open ocean rather than on the near-coastal and shoreline regions.

Descriptive physical oceanography approaches the ocean through both observations and complex numerical model output used to describe the fluid motions as quantitatively as possible. *Dynamical* physical oceanography seeks to understand the processes that govern the fluid motions in the ocean mainly through theoretical studies and process-based numerical model experiments. This book is mainly concerned with description based in observations (similar to previous editions of this text); however, in this edition we include some of the concepts of dynamical physical oceanography as an important context for the description. A full treatment of dynamical oceanography is contained in other texts. *Thermodynamics* also clearly enters into our discussion of the ocean through the processes that govern its heat and salt content, and therefore its density distribution.

Chapter 2 describes the ocean basins and their topography. The next three chapters introduce the physical (and some chemical) properties of freshwater and seawater (Chapter 3), an overview of the distribution of water characteristics (Chapter 4), and the sources and sinks of heat and freshwater (Chapter 5). The next three chapters cover data collection and analysis techniques (Chapter 6 and supplemental material listed as Chapter S6 on the textbook Web site http://booksite.academicpress.com/DPO/; "S" denotes supplemental material.), an introduction to geophysical fluid dynamics for graduate students who have varying mathematics backgrounds (Chapter 7), and then basic waves and tides with an introduction to coastal oceanography (Chapter 8). The last six chapters of the book introduce the circulation and water properties of each of the individual oceans (Chapters 9 through 13) ending with a summary of the global ocean in Chapter 14.

Accompanying the text is the Web site mentioned in the previous paragraph. It has four aspects:

1. Textbook chapters on climate variability and oceanographic instrumentation that do not appear in the print version
2. Expanded material and additional figures for many other chapters

1

3. A full set of tutorials for descriptive oceanography with data and sample scripts provided based on the Java Ocean Atlas (Osborne & Swift, 2009)
4. All figures from the text, with many more in color than in the text, for lectures and presentations.

1.1. OVERVIEW

There are many reasons for developing our knowledge of the ocean. Near-shore currents and waves affect navigation and construction of piers, breakwaters, and other coastal structures. The large heat capacity of the oceans exerts a significant and in some cases a controlling effect on the earth's climate. The ocean and atmosphere interact on short to long timescales; for example, the El Niño–Southern Oscillation (ENSO) phenomenon that, although driven locally in the tropical Pacific, affects climate on timescales of several years over much of the world. To understand these interactions it is necessary to understand the coupled ocean-atmosphere system. To understand the coupled system, it is first necessary to have a solid base of knowledge about both the ocean and atmosphere separately.

In these and many other applications, knowledge of the ocean's motion and water properties is essential. This includes the major ocean currents that circulate continuously but with fluctuating velocity and position, the variable coastal currents, the rise and fall of the tides, and the waves generated by winds or earthquakes. Temperature and salt content determine density and hence vertical movement. They also affect horizontal movement as the density affects the horizontal pressure distribution. Sea ice has its own full set of processes and is important for navigation, ocean circulation, and climate. Other dissolved substances such as oxygen, nutrients, and other chemical species, and even some of the biological aspects such as chlorophyll content, are used in the study of ocean physics.

Our present knowledge in physical oceanography rests on an accumulation of data, most of which were gathered during the past 150 years, with a large increase of in situ data collection (within the actual water) starting in the 1950s and an order of magnitude growth in available data as satellites began making ocean measurements (starting in the 1970s).

A brief history of physical oceanography with illustrations is provided as supplemental material on the textbook Web site (Chapter 1 supplement is listed as Chapter S1 on the Web site). Historically, sailors have always been concerned with how ocean currents affect their ships' courses as well as changes in ocean temperature and surface condition. Many of the earlier navigators, such as Cook and Vancouver, made valuable scientific observations during their voyages in the late 1700s, but it is generally considered that Mathew Fontaine Maury (1855) started the systematic large-scale collection of ocean current data using ship's navigation logs as his source of information. The first major expedition designed expressly to study all scientific aspects of the oceans was carried out on the British *H.M.S. Challenger* that circumnavigated the globe from 1872 to 1876. The first large-scale expedition organized primarily to gather physical oceanographic data was carried out on the German *FS Meteor,* which studied the Atlantic Ocean from 1925 to 1927 (Spiess, 1928). A number of photos from that expedition are reproduced on the accompanying Web site. Some of the earliest theoretical studies of the sea included the surface tides by Newton, Laplace, and Legendre (e.g., Wilson, 2002) and waves by Gerstner and Stokes (e.g., Craik, 2005). Around 1896, some of the Scandinavian meteorologists started to turn their attention to the ocean, because dynamical meteorology and dynamical oceanography have many common characteristics. Current knowledge of dynamical oceanography owes its progress to the work of Bjerknes, Bjerknes, Solberg, and Bergeron (1933), Ekman (1905, 1923), Helland-Hansen (1934), and others.

The post-war 1940s through 1960s began to produce much of the data and especially theoretical understanding for large-scale ocean circulation. With the advent of moored and satellite instrumentation in the 1960s and 1970s, the smaller scale, energetically varying part of the ocean circulation — the mesoscale — began to be studied in earnest. Platforms expanded from research and merchant ships to global satellite and autonomous instrument sampling. Future decades should take global description and modeling to even smaller scales (submesoscale) as satellite observations and numerical modeling resolution continue to evolve, and different types of autonomous sampling within the water column become routine. Physical oceanography has retained an aspect of individual exploration but large, multi-investigator, multinational programs have increasingly provided many of the new data sets and understanding. Current research efforts in physical oceanography are focused on developing an understanding of the variability of the ocean and its relation to the atmosphere and climate as well as continuing to describe its steady-state conditions.

1.2. SPACE AND TIMESCALES OF PHYSICAL OCEANOGRAPHIC PHENOMENA

The ocean is a fluid in constant motion with a very large range of spatial and temporal scales. The complexity of this fluid is nicely represented in the sea surface temperature image of the Gulf Stream captured from a satellite shown in Figure 1.1a. The Gulf Stream is the western boundary current of the permanent, large-scale clockwise *gyre* circulation of the subtropical North Atlantic. The Gulf Stream has a width of 100 km, and its gyre has a spatial scale of thousands of kilometers. The narrow, warm core of the Gulf Stream (red in Figure 1.1a) carries warm subtropical water northward from the Caribbean, loops through the Gulf of Mexico

around Florida and northward along the east coast of North America, leaves the coast at Cape Hatteras, and moves out to sea. Its strength and temperature contrast decay eastward. Its large meanders and rings, with spatial scales of approximately 100 km, are considered *mesoscale (eddy)* variability evolving on timescales of weeks. The satellite image also shows the general decrease in surface temperature toward the north and a large amount of small-scale eddy variability. The permanence of the Gulf Stream is apparent when currents and temperatures are averaged in time. Averaging makes the Gulf Stream appear wider, especially after the separation at Cape Hatteras where the wide envelope of meanders creates a wide, weak average eastward flow.

The Gulf Stream has been known and charted for centuries, beginning with the Spanish expeditions in the sixteenth century (e.g., Peterson, Stramma, & Kortum 1996). It was first mapped accurately in 1769 by Benjamin Franklin

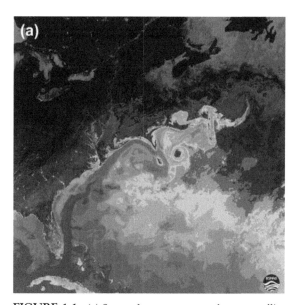

FIGURE 1.1 (a) Sea surface temperature from a satellite advanced very high resolution radiometer (AVHRR) instrument (Otis Brown, personal communication, 2009). This figure can also be found in the color insert.

(b)

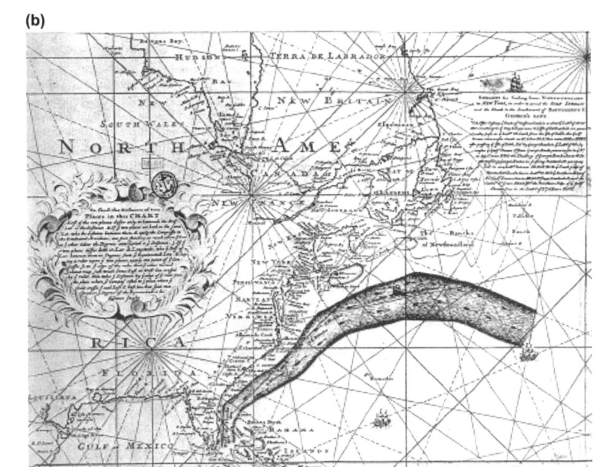

FIGURE 1.1   (b) Franklin-Folger map of the Gulf Stream. *Source: From Richardson (1980a).*

working together with whaling captain Timothy Folger (Figure 1.1b; from Richardson, 1980a).[1] The narrow current along the coast of the United States is remarkably accurate. The widening envelope of the Franklin/Folger current after separation from Cape Hatteras is an accurate depiction of the envelope of meandering apparent in the satellite image.

When time-mean averages of the Gulf Stream based on modern measurements are constructed, they look remarkably similar to this Franklin/Folger map.

The space and timescales of many of the important physical oceanography processes are shown schematically in Figure 1.2. At the smallest scale is molecular mixing. At small,

[1] Franklin noted on his frequent trips between the United States and Europe that some trips were considerably quicker than others. He decided that this was due to a strong ocean current flowing from the west to the east. He observed marked changes in surface conditions and reasoned that this ocean current might be marked by a change in sea surface temperature. Franklin began making measurements of the ocean surface temperature during his travels. Using a simple mercury-in-glass thermometer, he was able to determine the position of the current.

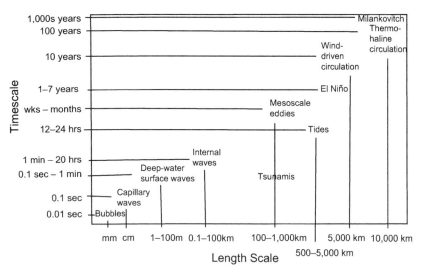

FIGURE 1.2 Time and space scales of physical oceanographic phenomena from bubbles and capillary waves to changes in ocean circulation associated with Earth's orbit variations.

macroscopic scales of centimeters, microstructure (vertical layering at the centimeter level) and capillary waves occur. At the slightly larger scale of meters surface waves are found, which have rapid timescales and somewhat long-lived vertical layers (fine structure). At scales of tens of meters are the internal waves with timescales of up to a day. Tides have the same timescales as internal waves, but much larger spatial scales of hundreds to thousands of kilometers. Surface waves, internal waves, and tides are described in Chapter 8.

Mesoscale eddies and strong ocean currents such as the Gulf Stream are found at spatial scales of tens of kilometers to several hundred kilometers and timescales of weeks to years (Figure 1.1a,b). The large-scale ocean circulation has a spatial scale the size of ocean basins up to the global ocean and a timescale ranging from seasonal to permanent, which is the timescale of plate tectonics that rearranges the ocean boundaries (Chapter 2). The timescales for wind-driven and thermohaline circulation in Figure 1.2 are actually the same for circulation of the flow through those systems (ten years around the gyre, hundreds of years through the full ocean); these are time-mean features of the

ocean and have much longer timescales. Climate variability affects the ocean, represented in Figure 1.2 by the El Niño, which has an interannual timescale (several years; Chapter 10); decadal and longer timescales of variability of the ocean circulation and properties are also important and described in each of the ocean basin and global circulation chapters.

We see in Figure 1.2 that short spatial scales generally have short timescales, and long spatial scales generally have long timescales. There are some exceptions to this, most notably in the tides and tsunamis as well as in some fine-structure phenomena with longer timescales than might be expected from their short spatial scales.

In Chapter 7, where ocean dynamics are discussed, some formal *non-dimensional parameters* incorporating the approximate space and timescales for these different types of phenomena are introduced (see also Pedlosky, 1987). A non-dimensional parameter is the ratio of dimensional parameters with identical dimensional scales, such as time, length, mass, etc., which are intrinsic properties of the flow phenomenon being described or modeled. Of special importance is whether the timescale of an ocean motion is greater than or less than about a day, which is the timescale for the earth's rotation. Earth's rotation has an enormous effect on how the ocean water moves in response to a force; if the force and motion are sustained for days or longer, then the motion is strongly influenced by the rotation. Therefore, an especially useful parameter is the ratio of the timescale of Earth's rotation to the timescale of the motion. This ratio is called the *Rossby number*. For the very small, fast motions in Figure 1.2, this ratio is large and rotation is not important. For the slow, large-scale part of the spectrum, the Rossby number is small and Earth's rotation is fundamental. A second very important non-dimensional parameter is the ratio of the vertical length scale (height) to the horizontal length scale; this is called the *aspect ratio*. For large-scale flows, this ratio is very small since the vertical scale can be no larger than the ocean depth. For surface and internal gravity waves, the *aspect ratio* is order 1. We will also see that dissipation is very weak in the sense that the timescale for dissipation to act is long compared with both the timescale of Earth's rotation and the timescale for the circulation to move water from one place to another; the relevant non-dimensional parameters are the Ekman number and Reynolds number, respectively. Understanding how the small Rossby number, small aspect ratio, and nearly frictionless fluid ocean behaves has depended on observations of the circulation and water properties made over the past century. These are the principal subjects of this text.

Ocean Dimensions, Shapes, and Bottom Materials

2.1. DIMENSIONS

The oceans are basins in the surface of the solid earth containing salt water. This chapter introduces some nomenclature and directs attention to features of the basins that have a close connection with the ocean's circulation and dynamical processes that are of importance to the physical oceanographer. More detailed descriptions of the geology and geophysics of the ocean basins are given in Seibold and Berger (1982), Kennett (1982), Garrison (2001), and Thurman and Trujillo (2002), among others. Updated data sets, maps, and information are available from Web sites of the National Geophysical Data Center (NGDC) of the National Oceanic and Atmospheric Administration (NOAA) and from the U.S. Geological Survey (USGS).

The major ocean areas are the Atlantic Ocean, the Pacific Ocean, the Indian Ocean, the Arctic Ocean, and the Southern Ocean (Figure 2.1). The first four are clearly divided from each other by land masses, but the divisions between the Southern Ocean and oceans to its north are determined only by the characteristics of the ocean waters and their circulations. The geographical peculiarities of each ocean are described in Section 2.11.

The shape, depth, and geographic location of an ocean affect the general characteristics of its circulation. Smaller scale features, such as locations of deep sills and fracture zones, seamounts, and bottom roughness, affect often important details of the circulation and of mixing processes that are essential to forcing and water properties. The Atlantic has a very marked "S" shape while the Pacific has a more oval shape. The Atlantic and Indian Oceans are roughly half the east-west width of the Pacific Ocean, which impacts the way that each ocean's circulation adjusts to changes in forcing. The Indian Ocean has no high northern latitudes, and therefore no possibility of cold, dense water formation. The edges of the Pacific are ringed with trenches, volcanoes, and earthquakes that signal the gradual descent of the ocean bottom crustal "plates" under the surrounding continental plates. In contrast, the Atlantic is the site of dynamic seafloor spreading as material added in the center of the Mid-Atlantic Ridge (MAR) pushes the plates apart, enlarging the Atlantic Ocean by a few centimeters each year.

Marginal seas are fairly large basins of salt water that are connected to the open ocean by one or more fairly narrow channels. Those that are connected by very few channels are sometimes called *mediterranean seas* after the

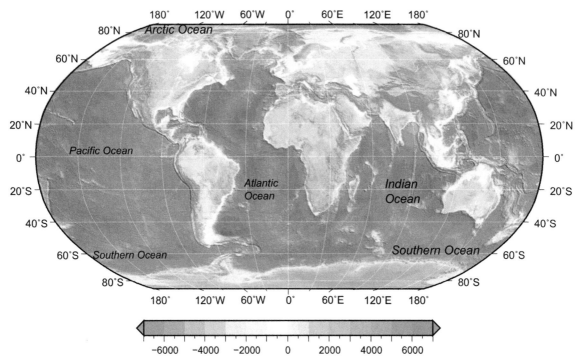

FIGURE 2.1 Map of the world based on ship soundings and satellite altimeter derived gravity at 30 arc-second resolution. Data from Smith & Sandwell (1997); Becker et al. (2009); and SIO (2008).

prototype, the (European) Mediterranean Sea. The Mediterranean provides an example of a negative water balance in a sea with less inflow (river runoff and precipitation) than evaporation. An excellent example of a positive water balance marginal sea (with net precipitation) is found in the Black Sea, which connects with the Mediterranean Sea. Both of these seas are discussed further in Chapters 5 and 9. Other examples of marginal seas that are separated from the open ocean by multiple straits or island chains are the Caribbean Sea, the Sea of Japan, the Bering Sea, the North Sea, the Baltic Sea, and so forth.

The term *sea* is also used for a portion of an ocean that is not divided off by land but has local distinguishing oceanographic characteristics; for example the Norwegian Sea, the Labrador Sea, the Sargasso Sea, and the Tasman Sea.

More of the earth's surface is covered by sea than by land, about 71% sea to 29% land. (The most recent elevation data for the earth's surface, used to construct Figure 2.2, show that 70.96% of the earth is ocean; see Becker et al., 2009.) Furthermore, the proportion of water to land in the Southern Hemisphere is much greater (4:1) than in the Northern Hemisphere (1.5:1). In area, the Pacific Ocean is about as large as the Atlantic and Indian Oceans combined. If the neighboring sectors of the Southern Ocean are included with the three main oceans north of it, the Pacific Ocean occupies about 46% of the total world ocean area, the Atlantic Ocean about 23%, the Indian Ocean about 20%, and the rest, combined, about 11%.

The average depth of the oceans is close to 4000 m while the marginal seas are generally about 1200 m deep or less. Relative to sea level,

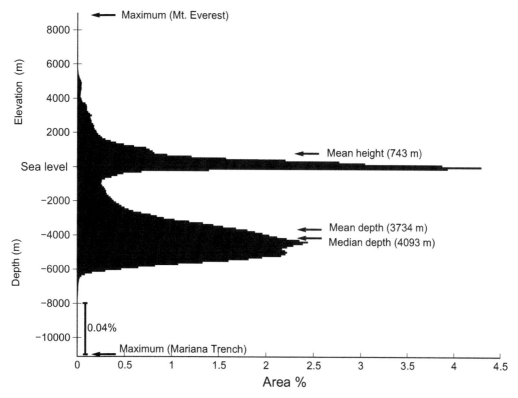

FIGURE 2.2 Areas of Earth's surface above and below sea level as a percentage of the total area of Earth (in 100 m intervals). Data from Becker et al. (2009).

the oceans are much deeper than the land is high. While only 11% of the land surface of the earth is more than 2000 m above sea level, 84% of the sea bottom is more than 2000 m deep. However, the maxima are similar: the height of Mt. Everest is about 8848 m, while the maximum depth in the oceans is 11,034 m in the Mariana Trench in the western North Pacific. Figure 2.2 shows the distributions of land elevations and of sea depths relative to sea level in 100 m intervals as the percentage of the total area of the earth's surface. This figure is based on the most recent global elevation and ocean bathymetry data from D. Sandwell (Becker et al., 2009). (It is similar to Figure 2.2 using 1000 m bins that appeared in previous editions of this text, based on data from Kossina, 1921 and Menard & Smith, 1966,

but the 100 m bins allow much more differentiation of topographic forms.)

Although the average depth of the oceans, 4 km, is a considerable distance, it is small compared with the horizontal dimensions of the oceans, which are 5000 to 15,000 km. Relative to the major dimensions of the earth, the oceans are a thin skin, but between the sea surface and the bottom of the ocean there is a great deal of detail and structure.

2.2. PLATE TECTONICS AND DEEP-SEA TOPOGRAPHY

The most important geophysical process affecting the shape and topography of the ocean basins is the movement of the earth's tectonic plates, described thoroughly in texts such as Thurman and Trujillo (2002). The plate boundaries are shown in Figure 2.3. *Seafloor spreading* creates new seafloor as the earth's plates spread apart. This creates the mid-ocean ridge system; the mid-ocean ridges of Figure 2.1 correspond to plate boundaries. The ocean plates spread apart at rates of about 2 cm/year (Atlantic) to 16 cm/year (Pacific), causing extrusion of magma into the surface at the centers of the ridges. Over geologic time the orientation of the earth's magnetic field has reversed, causing the ferromagnetic components in the molten new surface material to reverse. Spreading at the mid-ocean ridge was proven by observing the reversals in the magnetic orientations in the surface material. These reversals permit dating of the seafloor (Figure 2.3). The recurrence interval for magnetic reversals is approximately 500,000 to 1,000,000 years.

The 14,000 km long MAR is a tectonic *spreading center*. It is connected to the global mid-ocean ridge, which at more than 40,000 km long, is the most extensive feature of the earth's topography. Starting in the Arctic Ocean, the mid-ocean ridge extends through Iceland down the middle of the Atlantic, wraps around the tip of Africa, and then winds through the Indian and Pacific Oceans, ending in the Gulf of California. In all oceans, the mid-ocean ridge and other deep ridges separate the bottom waters, as can be seen from different water properties east and west of the ridge.

Deep and bottom waters can leak across the ridges through narrow gaps, called *fracture zones*, which are lateral jogs in the spreading center. The fracture zones are roughly vertical planes, perpendicular to the ridge, on either side of which the crust has moved in opposite directions perpendicular to the ridge. There are many fracture zones in the mid-ocean

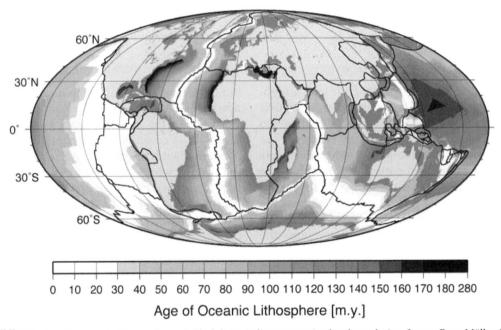

Age of Oceanic Lithosphere [m.y.]

FIGURE 2.3 Sea floor age (millions of years). Black lines indicate tectonic plate boundaries. *Source: From Müller, Sdrolias, Gaina, and Roest (2008).*

ridges. One example that is important as a pathway for abyssal circulation in the Atlantic is the Romanche Fracture Zone through the MAR close to the equator. Another example is the pair of fracture zones in the South Pacific (Eltanin and Udintsev Fracture Zones, Figure 2.12) that steer the Antarctic Circumpolar Current (ACC).

At some edges of the tectonic plates, one plate *subducts* (moves under) another. Subduction is accompanied on its landward side by volcanoes and earthquakes. Subduction creates deep *trenches* that are narrow relative to their length and have depths to 11,000 m. The deepest parts of the oceans are in these trenches. The majority of the deep trenches are in the Pacific: the Aleutian, Kurile, Tonga, Philippine, and Mariana. There are a few in other oceans such as the Puerto Rico and the South Sandwich Trenches in the Atlantic and the Sunda Trench in the Indian Ocean. Trenches are often shaped like an arc of a circle with an *island arc* on one side. Examples of island arcs are the Aleutian Islands (Pacific), the Lesser Antilles (Atlantic), and the Sunda Arc (Indian). The landward side of a trench extends as much as 10,000 m from the trench bottom to the sea surface, while the other side is only half as high, terminating at the ocean depth of about 5000 m.

Trenches can steer or impact boundary currents that are in deep water (Deep Western Boundary Currents) or upper ocean boundary currents that are energetic enough to extend to the ocean bottom, such as western boundary currents of the wind-driven circulation. Examples of trenches that impact ocean circulation are the deep trench system along the western and northern boundary of the Pacific and the deep trench east of the Caribbean Sea in the Atlantic.

Younger parts of the ocean bottom are shallower than older parts. As the new seafloor created at seafloor spreading centers ages, it cools by losing heat into the seawater above and becomes denser and contracts, which causes it to be deeper (Sclater, Parsons, & Jaupart, 1981). Ocean bottom depths range from 2 to 3 km for the newest parts of the mid-ocean ridges to greater than 5 km for the oldest, as can be seen by comparing the maps of seafloor age and bathymetry (Figures 2.1 and 2.3).

The rate of seafloor spreading is so slow that it has no impact on the climate variability that we experience over decades to millennia, nor does it affect anthropogenic climate change. However, over many millions of years, the geographic layout of Earth has changed. The paleocirculation patterns of "deep time," when the continents were at different locations, differed from the present patterns; reconstruction of these patterns is an aspect of paleoclimate modeling. By studying and understanding present-day circulation, we can begin to credibly model the paleocirculation, which had the same physical processes (such as those associated with the earth's rotation, wind and thermohaline forcing, boundaries, open east-west channels, equatorial regions, etc.), but with different ocean basin shapes and bottom topography.

Ocean bottom roughness affects ocean mixing rates (Sections 7.2.4 and 7.3.2). The overall roughness varies by a factor of 10. Roughness is a function of spreading rates and sedimentation rates. New seafloor is rougher than old seafloor. Slow-spreading centers produce rougher topography than fast spreading centers. Thus the slow-spreading MAR is rougher than the faster spreading East Pacific Rise (EPR; Figure 2.4). Slow-spreading ridges also have rift valleys at the spreading center, whereas fast-spreading ridges have an elevated ridge at the spreading center. Much of the roughness on the ridges can be categorized as *abyssal hills*, which are the most common landform on Earth. Abyssal hills are evident in Figures 2.4 and 2.5b, all along the wide flanks of the mid-ocean ridge.

Individual mountains (*seamounts*) are widely distributed in the oceans. Seamounts stand out clearly above the background bathymetry. In

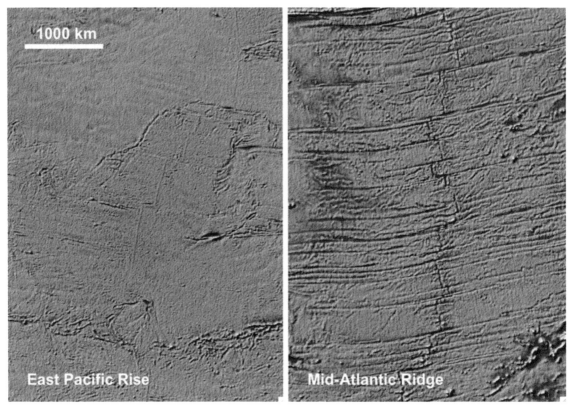

FIGURE 2.4 Seafloor topography for a portion of (a) the fast-spreading EPR and (b) the slow-spreading MAR. Note the ridge at the EPR spreading center and rift valley at the MAR spreading center. This figure can also be found in the color insert. (Sandwell, personal communication, 2009.)

the maps in Figure 2.4b there are some seamounts on the upper right side of the panel. In the vertical cross-section in Figure 2.5b, seamounts are distinguished by their greater height compared with the abyssal hills. The average height of seamounts is 2 km. Seamounts that reach the sea surface form islands. A *guyot* is a seamount that reached the surface, was worn flat, and then sank again below the surface. Many seamounts and islands were created by volcanic *hotspots* beneath the tectonic plates. The hotspots are relatively stationary in contrast to the plates and as the plates move across the hotspots, chains of seamounts are formed. Examples include the Hawaiian Islands/Emperor Seamounts chain, Polynesian island chains, the Walvis Ridge, and the Ninetyeast Ridge in the Indian Ocean.

Seamounts affect the circulation, especially when they appear in groups as they do in many regions; for instance, the Gulf Stream passes through the New England Seamounts, which affect the Gulf Stream's position and variability (Section 9.3). Seamount chains also refract tsunamis, which are ocean waves generated by submarine earthquakes that react to the ocean bottom as they propagate long distances from the earthquake source (Section 8.3.5).

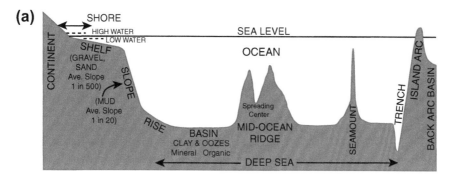

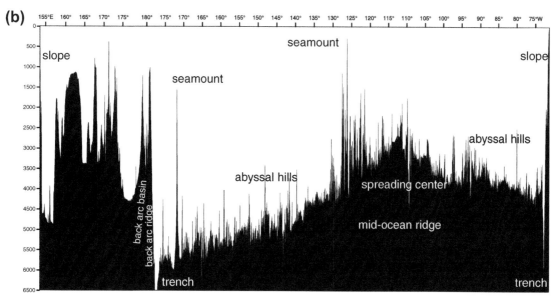

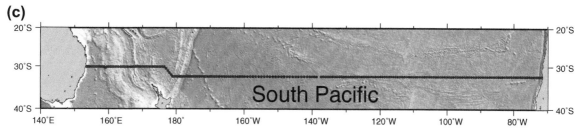

FIGURE 2.5 (a) Schematic section through ocean floor to show principal features. (b) Sample of bathymetry, measured along the South Pacific ship track shown in (c).

2.3. SEAFLOOR FEATURES

The continents form the major lateral boundaries to the oceans. The detailed features of the shoreline and of the sea bottom are important because of the way they affect circulation. Starting from the land, the main divisions recognized are the shore, the continental shelf, the continental slope and rise, and the deep-sea bottom, part of which is the abyssal plain (Figure 2.5a, b). Some of the major features of the seafloor, including mid-ocean ridges, trenches, island arcs, and seamounts, are the result of plate tectonics and undersea volcanism (Section 2.2 and Figure 2.3).

In some of the large basins the seafloor is very smooth, possibly more so than the plains areas on land. Sedimentation, which is mostly due to the incessant rain of organic matter from the upper ocean, covers the rough bottom and produces large regions of very smooth topography. Stretches of the *abyssal plain* in the western North Atlantic have been measured to be smooth within 2 m over distances of 100 km. The ocean bottom in the northeast Indian Ocean/Bay of Bengal slopes very smoothly from 2000 m down to more than 5000 m over 3000 km. This smoothness is due to sedimentation from the Ganges and Brahmaputra Rivers that drain the Himalayas. Bottom sediments can be moved around by deep currents; formation of undersea dunes and canyons is common. Erosional features in deep sediments have sometimes alerted scientists to the presence of deep currents.

Bottom topography often plays an important role in the distribution of water masses and the location of currents. For instance, bottom water coming from the Weddell Sea (Antarctica) is unable to fill the eastern part of the Atlantic basin directly due to the height of the Walvis Ridge (South Atlantic Ocean). Instead, the bottom water travels to the north along the western boundary of the South Atlantic, finds a deep passage in the MAR, and then flows south to fill the basin east of the ridge. At shallower depths the *sills* (shallowest part of a channel) defining the marginal seas strongly influences both the mid-level currents and the distribution of water masses associated with the sea. Coastal upwelling is a direct consequence of the shape of the coast and its related bottom topography. Alongshore currents are often determined by the coastal bottom topography and the instabilities in this system can depend on the horizontal scales of the bottom topography. Near the shore bottom topography dictates the breaking of surface gravity waves and also directly influences the local tidal expressions.

Much of the mixing in the ocean occurs near the boundaries (including the bottom). Microstructure observations in numerous regions, and intensive experiments focused on detection of mixing and its genesis, suggest that flow of internal tides over steep bottom slopes in the deep ocean is a major mechanism for dissipating the ocean's energy. Ocean bottom slopes computed from bathymetry collected along ship tracks show that the largest slopes tend to occur on the flanks of the fastest spreading mid-ocean ridges. With bathymetric slopes computed from the most recent bathymetric data (Figure 2.6) and information about the ocean's deep stratification, it appears the flanks of the mid-ocean ridges of the Atlantic, Southern Ocean, and Indian Ocean could be the most vigorous dissipation sites of ocean energy (Becker & Sandwell, 2008).

2.4. SPATIAL SCALES

Very often some of the characteristics of the ocean are presented by drawing a vertical cross-section of a part of the oceans, such as the schematic depiction of ocean floor features in Figure 2.5a. An illustration to true scale would have the relative dimensions of the edge of a sheet of paper and would be either

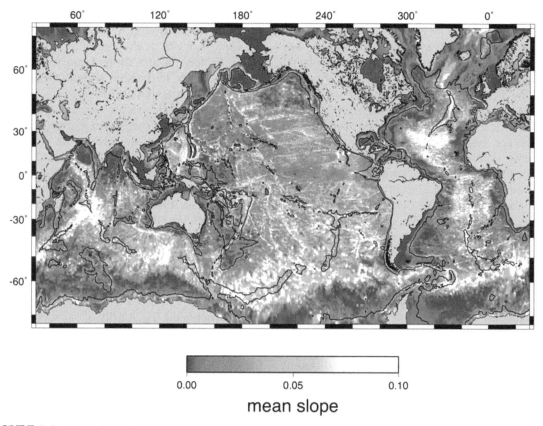

FIGURE 2.6 Mean slope of the ocean bottom, calculated from shipboard bathymetry and interpolated to a 0.5 degree grid. *Source: From Becker and Sandwell (2008).*

too thin to show details or too long to be convenient. Therefore, we usually distort our cross-section by making the vertical scale much larger than the horizontal one. For instance, we might use a scale of 1 cm to represent 100 km horizontally while depths might be on a scale of 1 cm to represent 100 m (i.e., 0.1 km). In this case the vertical dimensions on our drawing would be magnified 1000 times compared with the horizontal ones (a *vertical exaggeration* of 1000:1). This gives us room to show the detail, but it also exaggerates the slope of the sea bottom or of contours of constant water properties (*isopleths*) drawn on the cross-section (Figure 2.5b). In reality, such slopes are far less

than they appear on the cross-section drawings; for instance, a line of constant temperature (*isotherm*) with a real slope of 1 in 10,000 would appear as a slope of 1 in 10 on the plot.

2.5. SHORE, COAST, AND BEACH

The *shore* is defined as that part of the land-mass, close to the sea, that has been modified by the action of the sea. The words shore and *coast* have the same meaning. Shorelines (coasts) shift over time because of motion of the land over geologic time, changes in sea level, and erosion and deposition. The sedimentary record shows a series of marine intrusions and retreats corresponding to layers that reflect periods when the surface was above and below sea level. Variations in sea level between glacial and interglacial periods have been as much as 120 m. The ability of the coast to resist the erosional forces of the ocean depends directly on the type of material that makes up the coast. Sands are easily redistributed by the ocean currents whereas granitic coasts are slow to erode. Often sea level changes are combined with the hydrologic forces of an estuary, which dramatically change the dynamical relationship between the ocean and the solid surface.

The *beach* is a zone of unconsolidated particles at the seaward limit of the shore and extends roughly from the highest to the lowest tide levels. The landward limit of the beach might be vegetation, permanent sand dunes, or human construction. The seaward limit of a beach, where the sediment movement on- and offshore ceases, is about 10 m deep at low tide.

Coasts can be classified in many different ways. In terms of long timescales (such as those of plate tectonics; Section 2.2), coasts and continental margins can be classified as *active* or *passive*. Active margins, with active volcanism, faulting, and folding, are like those in much of the Pacific and are rising. Passive margins, like those of the Atlantic, are being pushed in front of spreading seafloor, are accumulating thick wedges of sediment, and are generally falling. Coasts can be referred to as *erosional* or *depositional* depending on whether materials are removed or added. At shorter timescales, waves and tides cause erosion and deposition. At millennial timescales, changes in mean sea level cause materials to be removed or added. Erosional coasts are attacked by waves and currents, both of which carry fine material that abrades the coast. The waves create *alongshore* and *rip currents* (Section 8.3) that carry the abraded material from the coastline along and out to sea. This eroded material can be joined by sediments discharged from rivers and form deltas. This type of erosion is fastest on *high-energy* coasts with large waves, and slowest on *low-energy* coasts with generally weak wave fields. Erosion occurs more rapidly in weaker materials than in harder components. These variations in materials allow erosive forces to carve characteristic features on coastlines such as sea cliffs and sea caves, and to create an alternation between bays and headlands.

Beaches result when sediment, usually sand, is transported to places suitable for continued deposition. Again these are often the quiet bays between headlands and other areas of low surf activity. Often a beach is in equilibrium; new sand is deposited to replace sand that is scoured away. Evidence for this process can be seen by how sand accumulates against new structures built on the shore, or by how it is removed from a beach when a breakwater is built that cuts off the supply of sand beyond it. On some beaches, the sand may be removed by currents associated with high waves at one season of the year and replaced by different currents associated with lower waves at another season. These currents are influenced by seasonal and interannual wind variations.

Sea level, which strongly affects coasts, is affected by the total amount of water in the ocean, changes in the containment volume of the world's ocean, and changes in the temperature/salinity characteristics of the ocean that alter its density and hence cause the water to expand or contract. Changes in the total amount of water are due primarily to changes in the volume of landfast ice, which is contained in ice sheets and glaciers. (Because sea ice floats in water, changes in sea ice volume, such as that in the Arctic or Antarctic, do not affect sea level.) Changes in containment volume are due to tectonics, the slow rebound of continents (continuing into the present) after the melt of landfast ice after the last deglaciation, and rebound due to the continuing melt of glacial ice. Changes in heat content cause seawater to expand (heating) or contract (cooling).

Sea level rose 20 cm from 1870 to 2003, including 3 cm in just the last 10 years (1993−2003). Because good global observations are available for that last 10 years, it is possible to ascribe 1.6 cm to thermal expansion, 0.4 cm to Greenland and Antarctic ice sheet melt, and 0.8 cm to other glacial melt with a residual of 0.3 cm. Sea level is projected to rise 30 ± 10 cm in the next 100 years mainly due to warming of the oceans, which absorb most of the anthropogenic heat increase in the earth's climate system. (See Bindoff et al., 2007 in the 4th assessment report of the Intergovernmental Panel on Climate Change.)

2.6. CONTINENTAL SHELF, SLOPE, AND RISE

The *continental shelf* extends seaward from the shore with an average gradient of 1 in 500. Its outer limit (the shelf break) is set where the gradient increases to about 1 in 20 (on average) to form the continental slope down to the deep sea bottom. The shelf has an average width of 65 km. In places it is much narrower than this, while in others, as in the northeastern Bering Sea or the Arctic shelf off Siberia, it is as much as ten times this width. The bottom material is dominantly sand with less common rock or mud. The shelf break is usually clearly evident in a vertical cross-section of the sea bottom from the shore outward. The average depth at the shelf break is about 130 m. Most of the world's fisheries are located on the continental shelves for a multitude of reasons including proximity of estuaries, depth of penetration of sunlight compared with bottom depth, and upwelling of nutrient-rich waters onto some shelves, particularly those off western coasts.

The *continental slope* averages about 4000 m vertically from the shelf to the deep-sea bottom, but in places extends as much as 9000 m vertically in a relatively short horizontal distance. In general, the continental slope is considerably steeper than the slopes from lowland to highland on land. The material of the slope is predominantly mud with some rock outcrops. The shelf and slope typically include submarine canyons, which are of worldwide occurrence. These are valleys in the slope, either V-shaped or with vertical sides, and are usually found off coasts with rivers. Some, usually in hard granitic rock, were originally carved as rivers and then submerged, such as around the Mediterranean and southern Baja, California. Others, commonly in softer sedimentary rock, are formed by turbidity currents described in the next paragraph. The lower part of the slope, where it grades into the deep-sea bottom, is referred to as the *continental rise*.

Turbidity currents (Figure 2.7) are common on continental slopes. These episodic events carry a mixture of water and sediment and are driven by the unstable sediments rather than by forces within the water. In these events, material builds up on the slope until it is no longer stable and the force of gravity wins out. Large amounts of sediment and bottom material crash down the slope at speeds up to 100 km/h. These events can snap underwater cables. The precise conditions that dictate when a turbidity current occurs vary with the slope of the valley and the nature of the material in the valley. Turbidity currents carve many of the submarine canyons found on the slopes. Some giant rivers, such as the Congo, carry such a dense load of suspended material that they form continuous density flows of turbid water down their canyons.

2.7. DEEP OCEAN

From the bottom of the continental slope, the bathymetric gradient decreases down the continental rise to the *deep-sea bottom*, the last and most extensive area. Depths of 3000−6000 m are found over 74% of the ocean basins with 1% deeper. Perhaps the most characteristic

FIGURE 2.7 Turbidity current evidence south of Newfoundland resulting from an earthquake in 1929. *Source: From Heezen, Ericson, and Ewing (1954).*

aspect of the deep-sea bottom is the variety of its topography. Before any significant deep ocean soundings were available, the sea bottom was regarded as uniformly smooth. When detailed sounding started in connection with cable laying, it became clear that this was not the case and there was a swing to regarding the sea bottom as predominantly rugged. Neither

view is exclusively correct, for we now know that there are mountains, valleys, and plains on the ocean bottom, just as on land. With the advent of satellite altimetry for mapping ocean topography, we now have an excellent global view of the distribution of all of these features (e.g., Figure 2.1; Smith & Sandwell, 1997), and can relate many of the features to plate tectonic processes (Section 2.2) and sedimentation sources and processes.

2.8. SILLS, STRAITS, AND PASSAGES

Sills, straits, and passages connect separate ocean regions. A *sill* is a ridge, above the average bottom level in a region, which separates one basin from another or, in the case of a fjord (Section 5.1), separates the landward basin from the sea outside. The *sill depth* is the depth from the sea surface to the deepest part of the ridge; that is, the maximum depth at which direct flow across the sill is possible. An oceanic sill is like a topographic saddle with the sill depth analogous to the saddlepoint. In the deep ocean, sills connect deep basins. The sill depth controls the density of waters that can flow over the ridge.

Straits, passages, and channels are horizontal constrictions. It is most common to refer to a strait when considering landforms, such as the Strait of Gibraltar that connects the Mediterranean Sea and the Atlantic Ocean, or the Bering Strait that connects the Bering Sea and the Arctic Ocean. Passages and channels can also refer to submarine topography, such as in fracture zones that connect deep basins. Straits and sills can occur together, as in both of these examples. The minimum width of the strait and the maximum depth of the sill can hydraulically control the flow passing through the constriction.

2.9. METHODS FOR MAPPING BOTTOM TOPOGRAPHY

Our present knowledge of the shape of the ocean floor results from an accumulation of sounding measurements (most of which have been made within the last century) and, more recently, using the gravity field measured by satellites (Smith & Sandwell, 1997). The early measurements were made by lowering a weight on a measured line until the weight touched bottom, as discussed in Chapter S1, Section S1.1 located on the textbook Web site http://booksite.academicpress.com/DPO/; "S" denotes supplemental material. This method was slow; in deep water it was uncertain because it was difficult to tell when the weight touched the bottom and if the line was vertical.

Since 1920 most depth measurements have been made with *echo sounders*, which measure the time taken for a pulse of sound to travel from the ship to the bottom and reflect back to the ship. One half of this time is multiplied by the average speed of sound in the seawater under the ship to give the depth. With present-day equipment, the time can be measured very accurately and the main uncertainty over a flat bottom is in the value used for the speed of sound. This varies with water temperature and salinity (see Section 3.7), and if these are not measured at the time of sounding an average value must be used. Research and military ships are generally outfitted with echo sounders and routinely report their bathymetric data to data centers that compile the information for bathymetric mapping. The bathymetry along the research ship track in Figure 2.5b was measured using this acoustic method.

The modern extension of these single echo sounders is a multi-beam array, in which many sounders are mounted along the bottom of the ship; these provide two-dimensional "swath" mapping of the seafloor beneath the ship.

Great detail has been added to our knowledge of the seafloor topography by satellite measurements. These satellites measure the earth's gravity field, which depends on the local mass of material. These measurements allow mapping of many hitherto unknown features, such as fracture zones and seamounts in regions remote from intensive echo sounder measurements, and provide much more information about these features even where they had been mapped (Smith & Sandwell, 1997). Echo sounder measurements are still needed to verify the gravity-based measurements since the material on the ocean bottom is not uniform; for instance, a bottom shaped by extensive sediment cover might not be detected from the gravity field. The bathymetry shown in Figure 2.5c, as well as in the global and basin maps of Figures 2.1 and Figures 2.8–2.12, is a blended product of all available shipboard measurements and satellite-based measurements.

2.10. BOTTOM MATERIAL

On the continental shelf and slope most of the bottom material comes directly from the land, either brought down by rivers or blown by the wind. The material of the deep-sea bottom is often more fine-grained than that on the shelf or slope. Much of it is pelagic in character, that is, it has been formed in the open ocean. The two major deep ocean sediments are "red" clay and the biogenic "oozes." The former has less than 30% biogenic material and is mainly mineral in content. It consists of fine material from the land (which may have traveled great distances through the air before finally settling into the ocean), volcanic material, and meteoric remains. The oozes are over 30% biogenic and originate from the remains of living organisms (plankton). The calcareous oozes have a high percentage of calcium carbonate from the shells of animal plankton, while the siliceous oozes have a high proportion of silica from the shells of silica-secreting planktonic plants and animals. The siliceous oozes are found mainly in the Southern Ocean and in the equatorial Pacific. The relative distribution of calcareous and siliceous oozes is clearly related to the nutrient content of the surface waters, with calcareous oozes common in low nutrient regions and siliceous oozes in high nutrient regions.

Except when turbidity currents deposit their loads on the ocean bed, the average rate of deposition of the sediments is from 0.1 to 10 mm per 1000 years, and a large amount of information on the past history of the oceans is stored in them. Samples of bottom material are obtained with a "corer," which is a 2–30 m long steel pipe that is lowered vertically and forced to penetrate into the sediments by the heavy weight at its upper end. The "core" of sediment retained in the pipe may represent material deposited from 1000 to 10 million years per meter of length. Sometimes the material is layered, indicating stages of sedimentation of different materials. In some places, layers of volcanic ash are related to historical records of eruptions; in others, organisms characteristic of cold or warm waters are found in different layers and suggest changes in temperature of the overlying water during the period represented by the core. In some places gradations from coarse to fine sediments in the upward direction suggest the occurrence of turbidity currents bringing material to the region with the coarser material settling out first and the finer later.

Large sediment depositions from rivers create a sloping, smooth ocean bottom for thousands of miles from the mouths of the rivers. This is called a deep-sea sediment *fan*. The largest, the Bengal Fan, is in the northeastern Indian Ocean and is created by the outflow from many rivers including the Ganges and Brahmaputra. Other examples of fans are at the mouths of the Yangtze, Amazon, and Columbia Rivers.

Physical oceanographers use sediments to help trace movement of the water at the ocean floor. Some photographs of the deep-sea bottom show ripples similar to a sand beach after the tide has gone out. Such ripples are only found on the beach where the water speed is high, such as in the backwash from waves. We conclude from the ripples on the deep-ocean bottom that currents of similar speed occur there. This discovery helped to dispel the earlier notion that all deep-sea currents are very slow.

Sediments can affect the properties of seawater in contact with them; for instance, silicate and carbonate are dissolved from sediments into the overlying seawater. Organic carbon, mainly from fecal pellets, is biologically decomposed (*remineralized*) into inorganic carbon dioxide in the sediments with oxygen consumed in the process. The carbon dioxide-rich, oxygen-poor pore waters in the sediments are released back into the seawater, affecting its composition. Organic nitrogen and phosphorus are also remineralized in the sediments, providing an important source of inorganic nutrients for seawater. In regions where all oxygen is consumed, methane forms from bacterial action. This methane is often stored in solid form called a methane hydrate. Vast quantities (about 10^{19} g) of methane hydrate have accumulated in marine sediments over the earth's history. They can spontaneously turn from solid to gaseous form, causing submarine landslides and releasing methane into the water, affecting its chemistry.

2.11. OCEAN BASINS

The *Pacific Ocean* (Figure 2.8) is the world's largest ocean basin. To the north there is a physical boundary broken only by the Bering Strait, which is quite shallow (about 50 m) and 82 km wide. There is a small net northward flow from the Pacific to the Arctic through this strait. At the equator, the Pacific is very wide so that tropical phenomena that propagate east-west take much longer to cross the Pacific than across the other oceans. The Pacific is rimmed in the west and north with trenches and ridges. This area, because of the associated volcanoes, is called the "ring of fire." The EPR, a major topographic feature of the tropical and South Pacific, is a spreading ridge that separates the deep waters of the southeast from the rest of the Pacific; it is part of the global mid-ocean ridge (Section 2.2). Fracture zones allow some communication of deep waters across the ridge. Where the major eastward current of the

Southern Ocean, the ACC (Chapter 13), encounters the ridge, the current is deflected.

The Pacific has more islands than any other ocean. Most of them are located in the western tropical regions. The Hawaiian Islands and their extension northwestward into the Emperor Seamounts were created by motion of the Pacific oceanic plate across the hotspot that is now located just east of the big island of Hawaii.

The Pacific Ocean has numerous marginal seas, mostly along its western side. In the North Pacific these are the Bering, Okhotsk, Japan, Yellow, East China, and South China Seas in the west and the Gulf of California in the east. In the South Pacific the marginal seas are the Coral and Tasman Seas and many smaller distinct regions that are named, such as the Solomon Sea (not shown). In the southern South Pacific is the Ross Sea, which contributes to the bottom waters of the world ocean.

The *Atlantic Ocean* has an "S" shape (Figure 2.9). The MAR, a spreading ridge down the center of the ocean, dominates its topography. Deep trenches are found just east of the Lesser Antilles in the eastern Caribbean and east of the South Sandwich Islands. The Atlantic is open both at the north and the south connecting to the Arctic and Southern Oceans. The northern North Atlantic is one of the two sources of the world's deep water (Chapter 9). One of the Atlantic's marginal seas, the Mediterranean, is evaporative and contributes high salinity, warm water to the mid-depth ocean. At the southern boundary, the Weddell Sea is a major formation site for the bottom water found in the oceans (Chapter 13). Other marginal seas connecting to the Atlantic are the Norwegian, Greenland, and Iceland Seas (sometimes known collectively as the Nordic Seas), the North Sea, the Baltic Sea, the Black Sea, and the Caribbean. The Irminger Sea is the region southeast of Greenland, the Labrador Sea is the region between Labrador and Greenland, and the Sargasso Sea is the open ocean region surrounding Bermuda. Fresh outflow

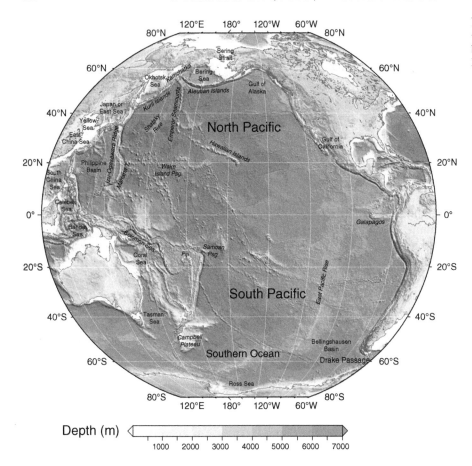

FIGURE 2.8 Map of the Pacific Ocean. Etopo2 bathymetry data from NOAA NGDC (2008).

from large rivers such as the Amazon, Congo, and Orinoco Rivers form marked low-salinity tongues at the sea surface.

The *Indian Ocean* (Figure 2.10) is closed off by land just north of the tropics. The topography of the Indian Ocean is very rough because of the ridges left behind as the Indian plate moved northward into the Asian continent creating the Himalayas. The Central Indian Ridge and Southwest Indian Ridge are two of the slowest spreading ridges on Earth. (As discussed previously, seafloor roughness from abyssal hills and fracture zones is highest at slower spreading rates, which is necessary in understanding the spatial distribution of deep mixing in the global ocean.) The only trench is the Sunda Trench

where the Indian plate subducts beneath Indonesia. The eastern boundary of the Indian Ocean is porous and connected to the Pacific Ocean through the Indonesian archipelago. Marginal seas for the Indian Ocean include the Andaman Sea, the Red Sea, and the Persian Gulf. The open ocean region west of India is called the Arabian Sea and the region east of India is called the Bay of Bengal.

The differential heating of land and ocean in the tropics results in the creation of the monsoon weather system. Monsoons occur in many places, but the most dramatic and best described monsoon is in the northern Indian Ocean (Chapter 11). From October to May the Northeast Monsoon sends cool, dry winds from the

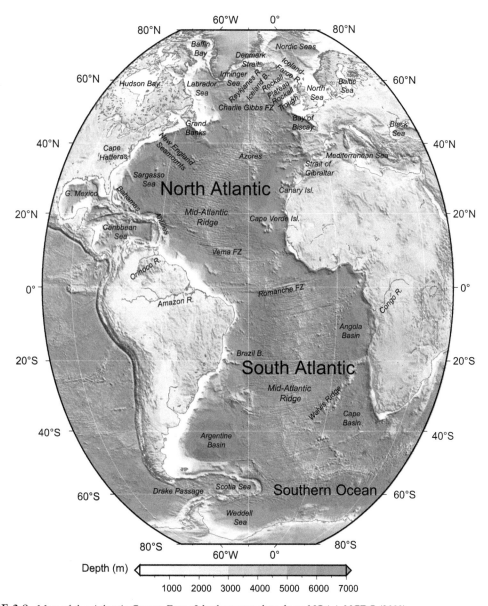

FIGURE 2.9 Map of the Atlantic Ocean. Etopo2 bathymetry data from NOAA NGDC (2008).

continental land masses in the northeast over the Indian subcontinent to the ocean. Starting in June and lasting until September, the system shifts to the southwest monsoon, which brings warm, wet rains from the western tropical ocean to the Indian subcontinent. While these monsoon conditions are best known in India, they also dominate the climate in the western tropical and South Pacific.

Most of the rivers that drain southward from the Himalayas — including the Ganges, Brahmaputra, and Irawaddy — flow out into the

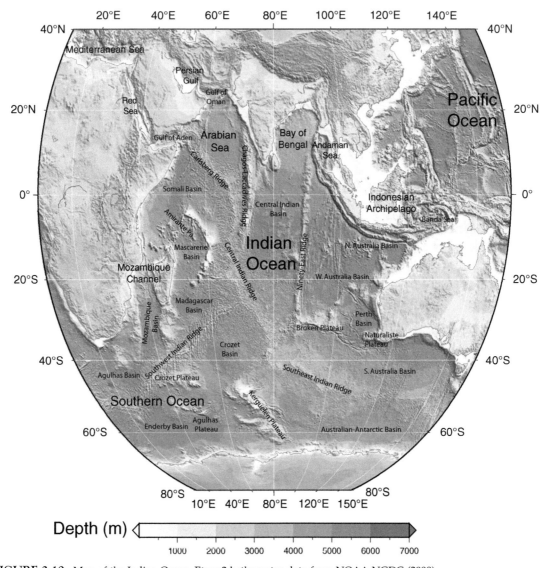

FIGURE 2.10 Map of the Indian Ocean. Etopo2 bathymetry data from NOAA NGDC (2008).

Bay of Bengal, east of India rather than into the Arabian Sea, west of India. This causes the surface water of the Bay of Bengal to be quite fresh. The enormous amount of silt carried by these rivers from the eroding Himalayan Mountains into the Bay of Bengal creates the subsurface geological feature, the Bengal Fan, which slopes smoothly downward for thousands of kilometers. West of India, the Arabian Sea, Red Sea, and Persian Gulf are very salty due to the dry climate and subsequent high evaporation. Similar to the Mediterranean, the saline Red Sea water is sufficiently dense to sink to mid-depth in the Indian Ocean and affects water properties over a large part of the Arabian Sea and western Indian Ocean.

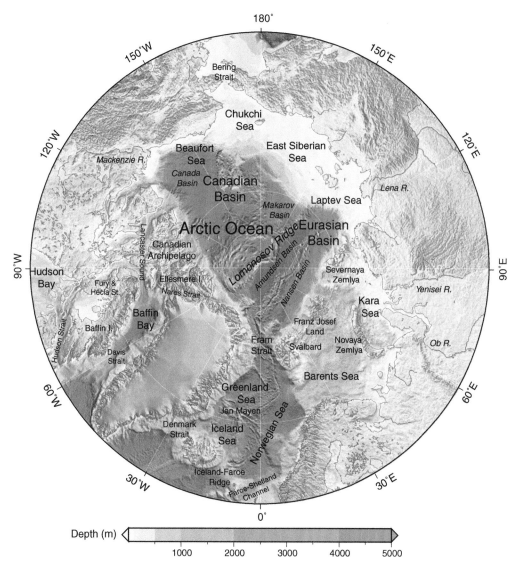

FIGURE 2.11 The Arctic Ocean. Etopo2 bathymetry data from NOAA NGDC (2008).

The *Arctic Ocean* (Figure 2.11) is sometimes not regarded as an ocean, but rather as a mediterranean sea connected to the Atlantic Ocean. It is characterized by very broad continental shelves surrounding a deeper region, which is split down the center by the Lomonosov Ridge. These shelf areas around the Arctic are called the Beaufort, Chukchi, East Siberian, Laptev, Kara, and Barents Seas. The Arctic is connected to the North Pacific through the shallow Bering Strait. It is connected to the Nordic Seas (Norwegian and Greenland) through passages on either side of Svalbard, including Fram Strait between Svalbard and Greenland. The Nordic Seas are separated from the Atlantic Ocean by the submarine ridge between Greenland, Iceland,

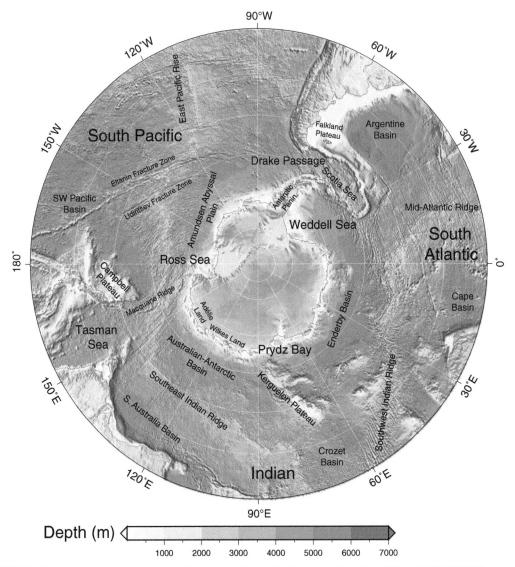

FIGURE 2.12 The Southern Ocean around Antarctica. Etopo2 bathymetry data from NOAA NGDC (2008).

and the UK, with a maximum sill depth of about 620 m in the Denmark Strait, between Greenland and Iceland. Dense water formed in the Nordic Seas spills into the Atlantic over this ridge. The central area of the Arctic Ocean is perennially covered with sea ice.

The *Southern Ocean* (Figure 2.12) is not geographically distinct from the Atlantic,

Indian, and Pacific Oceans, but is often considered separately since it is the only region outside the Arctic where there is a path for eastward flow all the way around the globe. This occurs at the latitude of Drake Passage between South America and Antarctica and allows the three major oceans to be connected. The absence of a meridional (north-south) boundary in Drake Passage changes the dynamics of the flow at these latitudes completely in comparison with the rest of the ocean, which has meridional boundaries. Drake Passage also serves to constrict the width of the flow of the ACC, which must pass in its entirety through the passage. The South Sandwich Islands and trench east of Drake Passage partially block the open circumpolar flow. Another major constriction is the broad Pacific-Antarctic rise, which is the seafloor spreading ridge between the Pacific and Antarctic plates. This fast-spreading ridge has few deep fracture zones, so the ACC must deflect northward before finding the only two deep channels, the Udintsev and Eltanin Fracture Zones.

The ocean around Antarctica includes permanent ice shelves as well as seasonal sea ice (Figures 13.11 and 13.19). Unlike the Arctic there is no perennial long-term pack ice; except for some limited ice shelves and all of the first-year ice melts and forms each year. The densest bottom waters of the world ocean are formed in the Southern Ocean, primarily in the Weddell and Ross Seas as well as in other areas distributed along the Antarctic coast between the Ross Sea and Prydz Bay.

Physical Properties of Seawater

3.1. MOLECULAR PROPERTIES OF WATER

Many of the unique characteristics of the ocean can be ascribed to the nature of water itself. Consisting of two positively charged hydrogen ions and a single negatively charged oxygen ion, water is arranged as a polar molecule having positive and negative sides. This molecular polarity leads to water's high dielectric constant (ability to withstand or balance an electric field). Water is able to dissolve many substances because the polar water molecules align to shield each ion, resisting the recombination of the ions. The ocean's salty character is due to the abundance of dissolved ions.

The polar nature of the water molecule causes it to form polymer-like chains of up to eight molecules. Approximately 90% of the water molecules are found in these chains. Energy is required to produce these chains, which is related to water's heat capacity. Water has the highest heat capacity of all liquids except ammonia. This high heat capacity is the primary reason the ocean is so important in the world climate system. Unlike the land and atmosphere, the ocean stores large amounts of heat energy it receives from the sun. This heat is carried by ocean currents, exporting or importing heat to various regions. Approximately 90% of the anthropogenic heating associated with global climate change is stored

in the oceans, because water is such an effective heat reservoir (see Section S15.6 located on the textbook Web site http://booksite.academic press.com/DPO/; "S" denotes supplemental material).

As seawater is heated, molecular activity increases and thermal expansion occurs, reducing the density. In freshwater, as temperature increases from the freezing point up to about 4°C, the added heat energy forms molecular chains whose alignment causes the water to shrink, increasing the density. As temperature increases above this point, the chains break down and thermal expansion takes over; this explains why fresh water has a density maximum at about 4°C rather than at its freezing point. In seawater, these molecular effects are combined with the influence of salt, which inhibits the formation of the chains. For the normal range of salinity in the ocean, the maximum density occurs at the freezing point, which is depressed to well below 0°C (Figure 3.1).

Water has a very high heat of evaporation (or heat of vaporization) and a very high heat of fusion. The heat of vaporization is the amount of energy required to change water from a liquid to a gas; the heat of fusion is the amount of energy required to change water from a solid to a liquid. These quantities are relevant for our climate as water changes state from a liquid in the ocean to water vapor in the atmosphere and to ice at polar latitudes. The heat energy

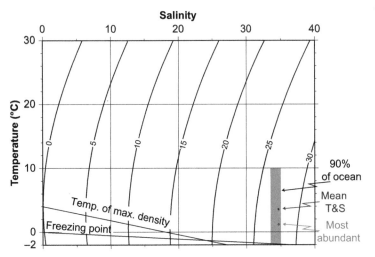

FIGURE 3.1 Values of density σ_t (curved lines) and the loci of maximum density and freezing point (at atmospheric pressure) for seawater as functions of temperature and salinity. The full density ρ is $1000 + \sigma_t$ with units of kg/m^3.

involved in these state changes is a factor in weather and in the global climate system.

Water's chain-like molecular structure also produces its high surface tension. The chains resist shear, giving water a high viscosity for its atomic weight. This high viscosity permits formation of surface *capillary waves*, with wavelengths on the order of centimeters; the restoring forces for these waves include surface tension as well as gravity. Despite their small size, capillary waves are important in determining the frictional stress between wind and water. This stress generates larger waves and propels the frictionally driven circulation of the ocean's surface layer.

3.2. PRESSURE

Pressure is the normal force per unit area exerted by water (or air in the atmosphere) on both sides of the unit area. The units of force are (mass × length/time²). The units of pressure are (force/length²) or (mass/[length × time²]). Pressure units in centimeters-gram-second (cgs) are dynes/cm² and in meter-kilogram-second (mks) they are Newtons/m². A special unit for pressure is the Pascal, where $1 \text{ Pa} = 1 \text{ N/m}^2$. Atmospheric pressure is usually measured in bars where $1 \text{ bar} = 10^6 \text{ dynes/cm}^2 = 10^5 \text{ Pa}$. Ocean pressure is usually reported in decibars where $1 \text{ dbar} = 0.1 \text{ bar} = 10^5 \text{ dyne/cm}^2 = 10^4 \text{ Pa}$.

The force due to pressure arises when there is a difference in pressure between two points. The force is directed from high to low pressure. Hence we say the force is oriented "down the pressure gradient" since the gradient is directed from low to high pressure. In the ocean, the downward force of gravity is mostly balanced by an upward pressure gradient force; that is, the water is not accelerating downward. Instead, it is kept from collapsing by the upward pressure gradient force. Therefore pressure increases with increasing depth. This balance of downward gravity force and upward pressure gradient force, with no motion, is called *hydrostatic balance* (Section 7.6.1).

The pressure at a given depth depends on the mass of water lying above that depth. A pressure change of 1 dbar occurs over a depth change of slightly less than 1 m (Figure 3.2 and Table 3.1). Pressure in the ocean thus varies from near zero (surface) to 10,000 dbar (deepest). Pressure

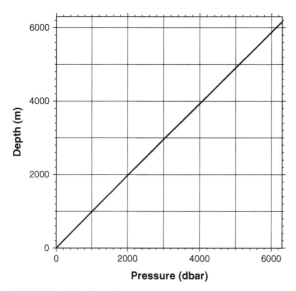

FIGURE 3.2 The relation between depth and pressure, using a station in the northwest Pacific at 41° 53'N, 146° 18'W.

TABLE 3.1 Comparison of Pressure (dbar) and Depth (m) at Standard Oceanographic Depths Using the UNESCO (1983) Algorithms

Pressure (dbar)	Depth (m)	Difference (%)
0	0	0
100	99	1
200	198	1
300	297	1
500	495	1
1000	990	1
1500	1453	1.1
2000	1975	1.3
3000	2956	1.5
4000	3932	1.7
5000	4904	1.9
6000	5872	2.1

Percent difference = (pressure − depth)/pressure × 100%.

is usually measured in conjunction with other seawater properties such as temperature, salinity, and current speeds. The properties are often presented as a function of pressure rather than depth.

Horizontal pressure gradients drive the horizontal flows in the ocean. For large-scale currents (of horizontal scale greater than a kilometer), the horizontal flows are much stronger than their associated vertical flows and are usually geostrophic (Chapter 7). The horizontal pressure differences that drive the ocean currents are on the order of one decibar over hundreds or thousands of kilometers. This is much smaller than the vertical pressure gradient, but the latter is balanced by the downward force of gravity and does not drive a flow. Horizontal variations in mass distribution create the horizontal variation in pressure in the ocean. The pressure is greater where the water column above a given depth is heavier either because it is higher density or because it is thicker or both.

Pressure is usually measured with an electronic instrument called a transducer. The accuracy and precision of pressure measurements is high enough that other properties such as temperature, salinity, current speeds, and so forth can be displayed as a function of pressure. However, the accuracy, about 3 dbar at depth, is not sufficient to measure the horizontal pressure gradients. Therefore other methods, such as the geostrophic method, or direct velocity measurements, must be used to determine the actual flow. Prior to the 1960s and 1970s, pressure was measured using a pair of mercury thermometers, one of which was in a vacuum ("protected" by a glass case) and not affected by pressure while the other was exposed to the water ("unprotected") and affected by pressure, as described in the following section. More information about these instruments and methods is provided in Section S6.3 of the supplementary materials on the textbook Web site.

3.3. THERMAL PROPERTIES OF SEAWATER: TEMPERATURE, HEAT, AND POTENTIAL TEMPERATURE

One of the most important physical characteristics of seawater is its temperature. Temperature was one of the first ocean parameters to be measured and remains the most widely observed. In most of the ocean, temperature is the primary determinant of density; salinity is of primary importance mainly in high latitude regions of excess rainfall or sea ice processes (Section 5.4). In the mid-latitude upper ocean (between the surface and 500 m), temperature is the primary parameter determining sound speed. (Temperature measurement techniques are described in Section S6.4.2 of the supplementary materials on the textbook Web site.)

The relation between temperature and heat content is described in Section 3.3.2. As a parcel of water is compressed or expanded, its temperature changes. The concept of "potential temperature" (Section 3.3.3) takes these pressure effects into account.

3.3.1. Temperature

Temperature is a thermodynamic property of a fluid, due to the activity or energy of molecules and atoms in the fluid. Temperature is higher for higher energy or heat content. Heat and temperature are related through the specific heat (Section 3.3.2).

Temperature (T) in oceanography is usually expressed using the Celsius scale ($°C$), except in calculations of heat content, when temperature is expressed in degrees Kelvin (K). When the heat content is zero (no molecular activity), the temperature is absolute zero on the Kelvin scale. (The usual convention for meteorology is degrees Kelvin, except in weather reporting, since atmospheric temperature decreases to very low values in the stratosphere and above.)

A change of $1°C$ is the same as a change of 1 K. A temperature of $0°C$ is equal to 273.16 K. The range of temperature in the ocean is from the freezing point, which is around $-1.7°C$ (depending on salinity), to a maximum of around $30°C$ in the tropical oceans. This range is considerably smaller than the range of air temperatures. As for all other physical properties, the temperature scale has been refined by international agreement. The temperature scale used most often is the International Practical Temperature Scale of 1968 (IPTS-68). It has been superseded by the 1990 International Temperature Scale (ITS-90). Temperatures should be reported in ITS-90, but all of the computer algorithms related to the equation of state that date from 1980 predate ITS-90. Therefore, ITS-90 temperatures should be converted to IPTS-68 by multiplying ITS-90 by 0.99976 before using the 1980 equation of state subroutines.

The ease with which temperature can be measured has led to a wide variety of oceanic and satellite instrumentation to measure ocean temperatures (see supplementary material in Section S6.4.2 on the textbook Web site). Mercury thermometers were in common use from the late 1700s through the 1980s. Reversing (mercury) thermometers, invented by Negretti and Zamba in 1874, were used on water sample bottles through the mid-1980s. These thermometers have ingenious glasswork that cuts off the mercury column when the thermometers are flipped upside down by the shipboard observer, thus recording the temperature at depth. The accuracy and precision of reversing thermometers is 0.004 and $0.002°C$. Thermistors are now used for most in situ measurements. The best thermistors used most often in oceanographic instruments have an accuracy of $0.002°C$ and precision of $0.0005-0.001°C$.

Satellites detect thermal infrared electromagnetic radiation from the sea surface; this radiation is related to temperature. Satellite sea surface temperature (SST) accuracy is about 0.5–0.8 K, plus an additional error due to the presence or absence of a very thin (10 μm) skin layer that can reduce the desired bulk (1–2 m) observation of SST by about 0.3 K.

3.3.2. Heat

The heat content of seawater is its thermodynamic energy. It is calculated using the measured temperature, measured density, and the specific heat of seawater. The specific heat is a thermodynamic property of seawater expressing how heat content changes with temperature. Specific heat depends on temperature, pressure, and salinity. It is obtained from formulas that were derived from laboratory measurements of seawater. Tables of values or computer subroutines supplied by UNESCO (1983) are available for calculating specific heat. The heat content per unit volume, Q, is computed from the measured temperature using

$$Q = \rho c_p T \qquad (3.1)$$

where T is temperature in degrees Kelvin, ρ is the seawater density, and c_p is the specific heat of seawater. The mks units of heat are Joules, that is, units of energy. The rate of time change of heat is expressed in Watts, where 1 W = 1 J/sec. The classical determinations of the specific heat of seawater were reported by Thoulet and Chevallier (1889). In 1959, Cox and Smith (1959) reported new measurements estimated to be accurate to 0.05%, with values 1 to 2% higher than the old ones. A further study (Millero, Perron, & Desnoyers, 1973) yielded values in close agreement with those of Cox and Smith.

The flux of heat through a surface is defined as the amount of energy that goes through the surface per unit time, so the mks units of heat flux are W/m^2. The heat flux between the atmosphere and ocean depends in part on the temperature of the ocean and atmosphere.

Maps of heat flux are based on measurements of the conditions that cause heat exchange (Section 5.4). As a simple example, what heat loss from a 100 m thick layer of the ocean is needed to change the temperature by 1°C in 30 days? The required heat flux is $\rho c_p \Delta T \, V/\Delta t$. Typical values of seawater density and specific heat are about 1025 kg/m^3 and 3850 J/(kg °C). V is the volume of the 100 m thick layer, which is 1 m^2 across, and Δt is the amount of time (sec). The calculated heat change is 152 W. The heat flux through the surface area of 1 m^2 is thus about 152 W/m^2. In Chapter 5 all of the components of ocean heat flux and their geographic distributions are described.

3.3.3. Potential Temperature

Seawater is almost, but not quite, incompressible. A pressure increase causes a water parcel to compress slightly. This increases the temperature in the water parcel if it occurs without exchange of heat with the surrounding water (*adiabatic* compression). Conversely if a water parcel is moved from a higher to a lower pressure, it expands and its temperature decreases. These changes in temperature are unrelated to surface or deep sources of heat. It is often desirable to compare the temperatures of two parcels of water that are found at different pressures. *Potential temperature* is defined as the temperature that a water parcel would have if moved adiabatically to another pressure. This effect has to be considered when water parcels change depth.

The *adiabatic lapse rate* or *adiabatic temperature gradient* is the change in temperature per unit change in pressure for an adiabatic displacement of a water parcel. The expression for the lapse rate is

$$\Gamma(S, T, p) = \left.\frac{\partial T}{\partial p}\right|_{heat} \qquad (3.2)$$

where S, T, and p are the measured salinity, temperature, and pressure and the derivative is taken holding heat content constant. Note that both the compressibility and the adiabatic lapse rate of seawater are functions of temperature, salinity, and pressure. The adiabatic lapse rate was determined for seawater through laboratory measurements. Since the full equation of state of seawater is a complicated function of these quantities, the adiabatic lapse rate is also a complicated polynomial function of temperature, salinity, and pressure. In contrast, the lapse rate for ideal gases can be derived from basic physical principles; in a dry atmosphere the lapse rate is approximately 9.8°C/km. The lapse rate in the ocean, about 0.1 to 0.2°C/km, is much smaller since seawater is much less compressible than air. The lapse rate is calculated using computer subroutines based on UNESCO (1983).

The potential temperature is (Fofonoff, 1985):

$$\theta(S, T, p) = T + \int_p^{pr} \Gamma(S, T, p) dp \qquad (3.3)$$

where S, T, and p are the measured (in situ) salinity, temperature, and pressure, Γ is the adiabatic lapse rate, and θ is the temperature that a water parcel of properties (S, T, p) would have if moved adiabatically and without change of salinity from an initial pressure p to a reference pressure p_r where p_r may be greater or less than p. The integration above can be carried out in a single step (Fofonoff, 1977). An algorithm for calculating θ is given by UNESCO (1983), using the UNESCO adiabatic lapse rate (Eq. 3.2); computer subroutines in a variety of different programming languages are readily available. The usual convention for oceanographic studies is to reference potential temperature to the sea surface. When defined relative to the sea surface, potential temperature is always lower than the actual measured temperature, and only equal to temperature at the sea surface. (On the other hand, when calculating potential density referenced to a pressure other than sea surface, pressure, potential temperature must also be referenced to the same pressure; see Section 3.5.)

As an example, if a water parcel of temperature 5°C and salinity 35.00 were lowered adiabatically from the surface to a depth of 4000 m, its temperature would increase to 5.45°C due to compression. The potential temperature relative to the sea surface of this parcel is always 5°C, while its measured, or in situ, temperature at 4000 m is 5.45°C. Conversely, if its temperature was 5°C at 4000 m depth and it was raised adiabatically to the surface, its temperature would change to 4.56°C due to expansion. The potential temperature of this parcel relative to the sea surface is thus 4.56°C. Temperature and potential temperature referenced to the sea surface from a profile in the northeastern North Pacific are shown in Figure 3.3. Compressibility itself depends on temperature (and salinity), as discussed in Section 3.5.4.

3.4. SALINITY AND CONDUCTIVITY

Seawater is a complicated solution containing the majority of the known elements. Some of the more abundant components, as percent of total mass of dissolved material, are chlorine ion (55.0%), sulfate ion (7.7%), sodium ion (30.7%), magnesium ion (3.6%), calcium ion (1.2%), and potassium ion (1.1%) (Millero, Feistel, Wright, & McDougall, 2008). While the total concentration of dissolved matter varies from place to place, the ratios of the more abundant components remain almost constant. This "law" of constant proportions was first proposed by Dittmar (1884), based on 77 samples of seawater collected from around the world during the Challenger Expedition (see Chapter S1, Section S1.2, on the textbook Web site), confirming a hypothesis from Forchhammer (1865).

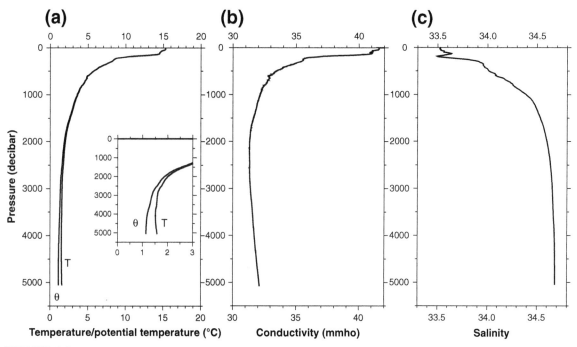

FIGURE 3.3 (a) Potential temperature (θ) and temperature (T) (°C), (b) conductivity (mmho), and (c) salinity in the northeastern North Pacific (36° 30′N, 135°W).

The dominant source of the salts in the ocean is river runoff from weathering of the continents (see Section 5.2). Weathering occurs very slowly over millions of years, and so the dissolved elements become equally distributed in the ocean as a result of mixing. (The total time for water to circulate through the oceans is, at most, thousands of years, which is much shorter than the geologic weathering time.) However, there are significant differences in total concentration of the dissolved salts from place to place. These differences result from evaporation and from dilution by freshwater from rain and river runoff. Evaporation and dilution processes occur only at the sea surface.

Salinity was originally defined as the mass in grams of solid material in a kilogram of seawater after evaporating the water away; this is the *absolute salinity* as described in Millero et al. (2008). For example, the average

salinity of ocean water is about 35 grams of salts per kilogram of seawater (g/kg), written as "S = 35 ‰" or as "S = 35 ppt" and read as "thirty-five parts per thousand." Because evaporation measurements are cumbersome, this definition was quickly superseded in practice. In the late 1800s, Forch, Knudsen, and Sorensen (1902) introduced a more chemically based definition: "Salinity is the total amount of solid materials in grams contained in one kilogram of seawater when all the carbonate has been converted to oxide, the bromine and iodine replaced by chlorine, and all organic matter completely oxidized."

This chemical determination of salinity was also difficult to carry out routinely. The method used throughout most of the twentieth century was to determine the amount of chlorine ion (plus the chlorine equivalent of the bromine and iodine) referred to as *chlorinity*, by titration with silver nitrate, and then to calculate salinity by a relation based on the measured ratio of chlorinity to total dissolved substances. (See Wallace, 1974, Wilson, 1975, or Millero et al., 2008 for a full account.) The current definition of salinity, denoted by S ‰, is "the mass of silver required to precipitate completely the halogens in 0.3285234 kg of the seawater sample." The current relation between salinity and chlorinity was determined in the early 1960s:

$$\text{Salinity} = 1.80655 \times \text{Chlorinity} \qquad (3.4)$$

These definitions of salinity based on chemical analyses were replaced by a definition based on seawater's electrical conductivity, which depends on salinity and temperature (see Lewis & Perkin, 1978; Lewis & Fofonoff, 1979; Figure 3.3). This conductivity-based quantity is called *practical salinity*, sometimes using the symbol psu for *practical salinity units*, although the preferred international convention has been to use no units for salinity. Salinity is now written as, say, S = 35.00 or S = 35.00 psu. The algorithm that is widely used to calculate salinity from conductivity and temperature is called the practical salinity scale 1978 (PSS 78). Electrical conductivity methods were first introduced in the 1930s (see Sverdrup, Johnson, & Fleming, 1942 for a review). Electrical conductivity depends strongly on temperature, but with a small residual due to the ion content or salinity. Therefore temperature must be controlled or measured very accurately during the conductivity measurement to determine the practical salinity. Advances in the electrical circuits and sensor systems permitted accurate compensation for temperature, making conductivity-based salinity measurements feasible (see supplemental materials in Chapter S6, Section S6.4.3 on the textbook Web site).

Standard seawater solutions of accurately known salinity and conductivity are required for accurate salinity measurement. The practical salinity (S_P) of a seawater sample is now given in terms of the ratio of the electrical conductivity of the sample at 15°C and a pressure of one standard atmosphere to that of a potassium chloride solution in which the mass fraction of KCl is 32.4356×10^{-3} at the same temperature and pressure. The potassium chloride solutions used as standards are now prepared in a single laboratory in the UK. PSS 78 is valid for the range S = 2 to 42, T = −2.0 to 35.0°C and pressures equivalent to depths from 0 to 10,000 m.

The accuracy of salinity determined from conductivity is ±0.001 if temperature is very accurately measured and standard seawater is used for calibration. This is a major improvement on the accuracy of the older titration method, which was about ±0.02. In archived data sets, salinities that are reported to three decimal places of accuracy are derived from conductivity, while those reported to two places are from titration and usually predate 1960.

The conversion from conductivity ratio to practical salinity is carried out using a computer subroutine based on the formula from Lewis (1980). The subroutine is part of the UNESCO (1983) routines for seawater calculations.

In the 1960s, the pairing of conductivity sensors with accurate thermistors made it possible to collect continuous profiles of salinity in the ocean. Because the geometry of the conductivity sensors used on these instruments change with pressure and temperature, calibration with water samples collected at the same time is required to achieve the highest possible accuracies of 0.001.

An example of the relationship between conductivity, temperature, and salinity profiles in the northeastern North Pacific is shown in Figure 3.3. Deriving salinity from conductivity requires accurate temperature measurement because the conductivity profile closely tracks temperature.

The concept of salinity assumes negligible variations in the composition of seawater. However, a study of chlorinity, density relative to pure water, and conductivity of seawater carried out in England on samples from the world oceans (Cox, McCartney, & Culkin, 1970) revealed that the ionic composition of seawater does exhibit small variations from place to place and from the surface to deep water. It was found that the relationship between density and conductivity was a little closer than between density and chlorinity. This means that the proportion of one ion to another may change; that is, the chemical composition may change, but as long as the total weight of dissolved substances is the same, the conductivity and the density will be unchanged.

Moreover, there are geographic variations in the dissolved substances not measured by the conductivity method that affect seawater density and hence should be included in absolute salinity. The geostrophic currents computed locally from density (Section 7.6.2), based on the use of salinity PSS 78, are highly accurate. However, it is common practice to map properties on surfaces of constant potential density or related surfaces that are closest to isentropic (Section 3.5). On a global scale, these dissolved constituents can affect the definition of these surfaces.

The definition of salinity is therefore undergoing another change equivalent to that of 1978. The *absolute salinity* recommended by the IOC, SCOR, and IAPSO (2010) is a return to the original definition of "salinity," which is required for the most accurate calculation of density; that is, the ratio of the mass of all dissolved substances in seawater to the mass of the seawater, expressed in either kg/kg or g/kg (Millero et al., 2008). The new estimate for absolute salinity incorporates two corrections over PSS 78: (1) representation of improved information about the composition of the Atlantic surface seawater used to define PSS 78 and incorporation of 2005 atomic weights, and (2) corrections for the geographic dependence of the dissolved matter that is not sensed by conductivity. To maintain a consistent global salinity data set, the IOC, SCOR, and IAPSO

(2010) manual strongly recommends that observations continue to be made based on conductivity and PSS 78, and reported to national archives in those practical salinity units. For calculations involving salinity, the manual indicates two corrections for calculating the absolute salinity S_A from the practical salinity S_P:

$$S_A = S_R + \delta S_A = (35.16504 \text{gkg}^{-1}/35)S_P + \delta S_A$$
$$(3.5)$$

The factor multiplying S_P yields the "reference salinity" S_R, which is presently the most accurate estimate of the absolute salinity of reference Atlantic surface seawater. A geographically dependent anomaly, δS_A, is then added that corrects for the dissolved substances that do not affect conductivity; this correction, as currently implemented, depends on dissolved silica, nitrate, and alkalinity. The mean absolute value of the correction globally is 0.0107 g/kg, and it ranges up to 0.025 g/kg in the northern North Pacific, so it is significant. If nutrients and carbon parameters are not measured along with salinity (which is by far the most common circumstance), then a geographic lookup table based on archived measurements is used to estimate the anomaly (McDougall, Jackett, & Millero, 2010). It is understood that the estimate (Eq. 3.5) of absolute salinity could evolve as additional measurements are made.

All of the work that appears in this book predates the adoption of the new salinity scale, and all salinities are reported as PSS 78 and all densities are calculated according to the 1980 equation of state using PSS 78.

3.5. DENSITY OF SEAWATER

Seawater density is important because it determines the depth to which a water parcel will settle in equilibrium — the least dense on top and the densest at the bottom. The distribution of density is also related to the large-scale circulation of the oceans through the geostrophic/thermal wind relationship (see Chapter 7). Mixing is most efficient between waters of the same density because adiabatic stirring, which precedes mixing, conserves potential temperature and salinity and consequently, density. More energy is required to mix through stratification. Thus, property distributions in the ocean are effectively depicted by maps on density (*isopycnal*) surfaces, when properly constructed to be closest to isentropic. (See the discussion of potential and neutral density in Section 3.5.4.)

Density, usually denoted ρ, is the amount of mass per unit volume and is expressed in kilograms per cubic meter (kg/m^3). A directly related quantity is the specific volume anomaly, usually denoted α, where $\alpha = 1/\rho$. The density of pure water, with no salt, at 0°C, is $1000\,kg/m^3$ at atmospheric pressure. In the open ocean, density ranges from about $1021\,kg/m^3$ (at the sea surface) to about $1070\,kg/m^3$ (at a pressure of 10,000 dbar). As a matter of convenience, it is usual in oceanography to leave out the first two digits and use the quantity

$$\sigma_{stp} = \rho(S,T,p) - 1000\,kg/m^3 \qquad (3.6)$$

where S = salinity, T = temperature (°C), and p = pressure. This is referred to as the in situ density. In earlier literature, $\sigma_{s,t,0}$ was commonly used, abbreviated as σ_t. σ_t is the density of the water sample when the total pressure on it has been reduced to atmospheric (i.e., the water pressure p = 0 dbar) but the salinity and temperature are as measured. Unless the analysis is limited to the sea surface, σ_t is not the best quantity to calculate. If there is range of pressures, the effects of adiabatic compression should be included when comparing water parcels. A more appropriate quantity is *potential density*, which is the same as σ_t but with temperature replaced by potential temperature and pressure replaced by a single reference pressure that is not necessarily 0 dbar. Potential density is described in Section 3.5.2.

The relationship between the density of seawater and temperature, salinity, and pressure is the equation of state for seawater. The equation of state

$$\rho(S,T,p) = \rho(S,T,0)/[1 - p/K(S,T,p)] \qquad (3.7)$$

was determined through meticulous laboratory measurements at atmospheric pressure. The polynomial expressions for the equation of state $\rho(S,T,0)$ and the bulk modulus $K(S,T,p)$ contain 15 and 27 terms, respectively. The pressure dependence enters through the bulk modulus. The largest terms are those that are linear in S, T, and p, with smaller terms that are proportional to all of the different products of these. Thus, the equation of state is weakly nonlinear.

Today, the most common version of Eq. (3.7) is "EOS 80" (Millero & Poisson, 1980; Fofonoff, 1985). EOS 80 uses the practical salinity scale PSS 78 (Section 3.4). The formulae may be found in UNESCO (1983), which provides practical computer subroutines and are included in various texts such as Pond and Pickard (1983) and Gill (1982). EOS 80 is valid for T = −2 to 40°C, S = 0 to 40, and pressures from 0 to 10,000 dbar, and is accurate to $9 \times 10^{-3}\,kg/m^3$ or better. A new version of the equation of state has been introduced (IOC, SCOR, and IAPSO, 2010), based on a new definition of salinity and is termed TEOS-10. Only EOS 80 is used in this book.

Historically, density was calculated from tables giving the dependence of the density on salinity, temperature, and pressure. Earlier determinations of density were based on measurements by Forch, Jacobsen, Knudsen, and Sorensen and were presented in the Hydrographical Tables (Knudsen, 1901). Cox et al. (1970) found that the σ_0 values (at T = 0°C) in "Knudsen's Tables" were low by about 0.01 (on average) in the salinity range from 15 − 40, and by up to 0.06 at lower salinities and temperatures.

To determine seawater density over a range of salinities in the laboratory, Millero (1967) used a magnetic float densimeter. A Pyrex glass float containing a permanent magnet floats in a

250 ml cell that contains the seawater and is surrounded by a solenoid, with the entire apparatus sitting in a constant temperature bath. The float is slightly less dense than the densest seawater and is loaded with small platinum weights until it just sinks to the bottom of the cell. A current through the solenoid is then slowly increased until the float just lifts off the bottom of the cell. The density of the seawater is then related to the current through the solenoid. The relation between current and density is determined by carrying out a similar experiment with pure water in the cell. The accuracy of the relative density determined this way is claimed to be $\pm 2 \times 10^{-6}$ (at atmospheric pressure). But as the absolute density of pure water is known to be only $\pm 4 \times 10^{-6}$, the actual accuracy of seawater density is more limited. The influence of pressure was determined using a high pressure version of the previously mentioned densimeter to measure the bulk modulus (K). K has also been determined from measurements of sound speed in seawater because sound speed depends on the bulk modulus and seawater compressibility.

The following subsections explore how seawater density depends on temperature, salinity, and pressure, and discusses concepts (such as potential and neutral density) that reduce, as much as possible, the effects of compressibility on a given analysis.

3.5.1. Effects of Temperature and Salinity on Density

Density values evaluated at the ocean's surface pressure are shown in Figure 3.1 (curved contours) for the whole range of salinities and temperatures found anywhere in the oceans. The shaded bar in the figure shows that most of the ocean lies within a relatively narrow salinity range. More extreme values occur only at or near the sea surface, with fresher waters outside this range (mainly in areas of runoff or ice melt) and the most saline waters in relatively confined areas of high evaporation (such as marginal seas). The ocean's temperature range produces more of the ocean's density variation than does its salinity range. In other words, temperature dominates oceanic density variations for the most part. (As noted previously, an important exception is where surface waters are relatively fresh due to large precipitation or ice melt; that is, at high latitudes and also in the tropics beneath the rainy Intertropical Convergence Zone of the atmosphere.) The curvature of the density contours in Figure 3.1 is due to the nonlinearity of the equation of state. The curvature means that the density change for a given temperature or salinity change is different at different temperatures or salinities.

To emphasize this point, Table 3.2 shows the change of density ($\Delta \sigma_t$) for a temperature change (ΔT) of $+1$ K (left columns) and the value of $\Delta \sigma_t$ for a salinity change (ΔS) of $+0.5$ (right columns). These are arbitrary choices for changes in temperature and salinity. The most important thing to notice in the table is how density varies at different temperatures and salinities for given changes in each. At high temperatures, σ_t varies significantly with T at all salinities. As temperature decreases, the rate of variation with T decreases, particularly at low salinities (as found at high latitudes or in estuaries). The change of σ_t with ΔS is about the same at all temperatures and salinities, but is slightly greater at low temperature.

3.5.2. Effect of Pressure on Density: Potential Density

Seawater is compressible, although not nearly as compressible as a gas. As a water parcel is compressed, the molecules are pushed closer together and the density increases. At the same time, and for a completely different physical reason, adiabatic compression causes the temperature to increase, which slightly offsets the density increase due to compression. (See discussion of potential temperature in Section 3.3.)

TABLE 3.2   Variation of Density ($\Delta\sigma_t$) with Variations of Temperature (ΔT) and of Salinity (ΔS) as Functions of Temperature and Salinity

Salinity	0	20	35	40	0	20	35	40
Temperature (°C)		$\Delta\sigma_t$ for $\Delta T = +1°C$				$\Delta\sigma_t$ for $\Delta S = +0.5$		
30	−0.31	−0.33	−0.34	−0.35	0.38	0.37	0.37	0.38
20	−0.21	−0.24	−0.27	−0.27	0.38	0.38	0.38	0.38
10	−0.09	−0.14	−0.18	−0.18	0.39	0.39	0.39	0.39
0	+0.06	−0.01	−0.06	−0.07	0.41	0.40	0.40	0.40

Density is primarily a function of pressure (Figure 3.4) because of this compressibility. Pressure effects on density have little to do with the initial temperature and salinity of the water parcel. To trace a water parcel from one place to another, the dependence of density on pressure should be removed. An early attempt was to use σ_t, defined earlier, in which the pressure effect was removed from density but not from temperature. It is now standard practice to use *potential density*, in which density is calculated using potential temperature instead of temperature. (The measured salinity is used.) Potential density is the density that a parcel would have if it were moved adiabatically to a chosen reference pressure. If the reference pressure is the sea surface, then we first compute the potential temperature of the parcel relative to surface pressure, then evaluate the density at pressure 0 dbar.[1] We refer to potential density referenced to the sea surface (0 dbar) as σ_θ, which signifies that potential temperature and surface pressure have been used.

The reference pressure for potential density can be any pressure, not just the pressure at the sea surface. For these potential densities, potential temperature is calculated relative to the chosen reference pressure and then the potential density is calculated relative to the same reference pressure. It is common to refer to potential density referenced to 1000 dbar as σ_1, referenced to 2000 dbar as σ_2, to 3000 dbar as σ_3 and so on, following Lynn and Reid (1968).

3.5.3. Specific Volume and Specific Volume Anomaly

The specific volume (α) is the reciprocal of density so it has units of m^3/kg. For some purposes it is more useful than density. The in situ specific volume is written as $\alpha_{s,t,p}$. The

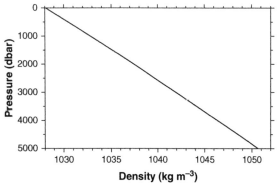

FIGURE 3.4 Increase in density with pressure for a water parcel of temperature 0°C and salinity 35.0 at the sea surface.

---

[1] The actual pressure at the sea surface is the atmospheric pressure, but we do not include atmospheric pressure in many applications since pressure ranges within the ocean are so much larger.

specific volume anomaly (δ) is also sometimes convenient. It is defined as:

$$\delta = \alpha_{s,t,p} - \alpha_{35,0,p} \qquad (3.8)$$

The anomaly is calculated relative to $\alpha_{35,0,p}$, which is the specific volume of seawater of salinity 35 and temperature $0°C$ at pressure p. With this standard δ is usually positive. The equation of state relates α (and δ) to salinity, temperature, and pressure. Originally all calculations of geostrophic currents from the distribution of mass were done by hand using tabulations of the component terms of δ, described in previous editions of this book. With modern computer methods, tabulations are not necessary. The computer algorithms for dynamic calculations (Section 7.5.1) still use specific volume anomaly δ, computed using subroutines, rather than the actual density ρ, to increase the calculation precision.

3.5.4. Effect of Temperature and Salinity on Compressibility: Isentropic Surfaces and Neutral Density

Cold water is more compressible than warm water; it is easier to deform a cold parcel than a warm parcel. When two water parcels with the same density but different temperature and salinity characteristics (one warm/salty, the other cold/fresh) are submerged to the same pressure, the colder parcel will be denser. If there were no salt in seawater, so that density depended only on temperature and pressure, then potential density as defined earlier, using any single pressure for a reference, would be adequate for defining a unique *isentropic surface*. An isentropic surface is one along which water parcels can move adiabatically, that is, without external input of heat or salt.

When analyzing properties within the ocean to determine where water parcels originate, it is assumed that motion and mixing is mostly along a quasi-isentropic surface and that mixing

across such a surface (quasi-vertical mixing) is much less important (Montgomery, 1938). However, because seawater density depends on both salinity and temperature, the actual surface that a water parcel moves along in the absence of external sources of heat or freshwater depends on how the parcel mixes along that surface since its temperature and salinity will be altered as it mixes with adjacent water parcels on that surface. This quasi-lateral mixing alters the temperature (and salinity) and therefore, the compressibility of the mixture. As a result, when it moves laterally, the parcel will equilibrate at a different pressure than if there had been no mixing. This means that there are no closed, unique isentropic surfaces in the ocean, since if our water parcel were to return to its original latitude and longitude, it will have moved to a different density and hence pressure because its temperature and salinity will have changed due to mixing along that surface. Note that these effects are important even without diapycnal mixing between water parcels on different isentropic surfaces (quasi-vertical mixing), which also can change temperature, salinity, and compressibility.

The density differences associated with these differences in compressibility can be substantial (Figure 3.5). For instance, water spilling out of the Mediterranean Sea through the Strait of Gibraltar is saline and rather warm compared with water spilling into the Atlantic from the Nordic Seas over the Greenland-Iceland ridge (Chapter 9). The Mediterranean Water (MW) density is actually higher than the Nordic Sea Overflow Water (NSOW) density where they flow over their respective sills, which are at about the same depth. However, the warm, saline MW ($13.4°C$, 37.8 psu) is not as compressible as the much colder NSOW (about $1°C$, 34.9 psu; Price & Baringer, 1994). The potential density relative to 4000 dbar of MW is lower than that of the more compressible NSOW. The NSOW reaches the bottom of the North Atlantic, while the MW does not. (As both types of water plunge

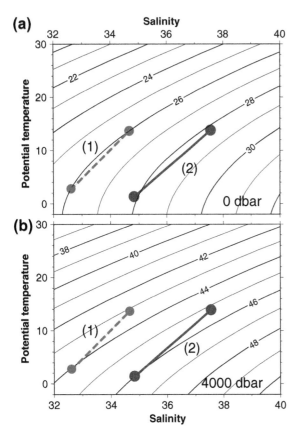

FIGURE 3.5 Potential density relative to (a) 0 dbar and (b) 4000 dbar as a function of potential temperature (relative to 0 dbar) and salinity. Parcels labeled 1 have the same density at the sea surface. The parcels labeled 2 represents Mediterranean (saltier) and Nordic Seas (fresher) source waters at their sills.

cold parcel is denser than the warm one at higher pressure (lower panel). The pair labeled 2 illustrates the MW (warm, salty) and NSOW (cold, fresh) pair with their properties at the sills where they enter the North Atlantic. At the sea surface, which neither parcel ever reaches, the Mediterranean parcel would actually be denser than the Nordic Seas parcel. Near the ocean bottom, represented by 4000 dbar (Figure 3.5b), the colder Nordic Seas parcel is markedly denser than the Mediterranean parcel. Therefore, if both parcels dropped to the ocean bottom from their respective sills, without any mixing, the Nordic Seas parcel would lie under the Mediterranean parcel. (In actuality, as already mentioned, there is a large amount of entrainment mixing as these parcels drop down into the North Atlantic.)

The surfaces that we use to map and trace water parcels should approximate isentropic surfaces. Early choices, that were an improvement over constant depth surfaces, included sigma-t surfaces (Montgomery, 1938) and even potential temperature surfaces (Worthington and Wright, 1970). A method, introduced by Lynn and Reid (1968), that produces surfaces that are closer to isentropic uses isopycnals with a reference pressure for the potential density that is within 500 m of the pressure of interest. Therefore when working in the top 500 m, a surface reference pressure is used. When working at 500 to 1500 m, a reference pressure of 1000 dbar is used, and so forth. Experience has shown this pressure discretization is sufficient to remove most of the problems associated with the effect of pressure on density. When isopycnals mapped in this fashion move into a different pressure range, they must be patched onto densities at the reference pressure in the new range. Reid (1989, 1994, 1997, 2003) followed this practice in his monographs on Pacific, Atlantic, and Indian Ocean circulations.

It is less complicated to use a continuously varying surface rather than one patched from different reference pressures, although in practice there is little difference between them.

downward, they entrain or mix with the waters that they pass through. This also has an effect on how deep they fall, so the difference in compressibility is not the only cause for different outcomes.)

Restating this more generally, changing the reference pressure for potential density alters the density difference between two water parcels (Figure 3.5). For the pair labeled 1, the densities are the same at the sea surface (upper panel). Because the cold parcel compresses more than the warm one with increasing pressure, the

"Neutral surfaces," introduced by Ivers (1975), a student working with J.L. Reid, use a nearly continuously varying reference pressure. If a parcel is followed along its path from one observation station to the next, assuming the path is known, then it is possible to track its pressure and adjust its reference pressure and density at each station. McDougall (1987a) refined this neutral surface concept and introduced it widely. Jackett and McDougall (1997) created a computer program for computing their version of this *neutral density*, based on a standard climatology (average temperature and salinity on a grid for the whole globe, derived from all available observations; Section 6.6.2), marching away from a single location in the middle of the Pacific. The Jackett and McDougall neutral density is denoted γ^N with numerical values that are similar to those of potential density (with units of kg/m^3). Neutral density depends on latitude, longitude, and pressure, and is defined only for ranges of temperature and salinity that occur in the open ocean. This differs from potential density, which is defined for all values of temperature and salinity through a well-defined equation of state that has been determined in the laboratory and is independent of location. Neutral density cannot be contoured as a function of potential temperature and salinity analogously to Figure 3.5 for density or potential density.

The advantage of neutral density for mapping quasi-isentropic surfaces is that it removes the need to continuously vary the reference pressure along surfaces that have depth variation (since this is already done in an approximate manner within the provided software and database). Neutral density is a convenient tool. Both potential and neutral density surfaces are approximations to isentropic surfaces. Ideas and literature on how to best approximate isentropic surfaces continue to be developed; neutral density is currently the most popular and commonly used approximation for mapping isentropes over large distances

that include vertical excursions of more than several hundred meters.

3.5.5. Linearity and Nonlinearity in the Equation of State

As described above, the equation of state (3.6) is somewhat nonlinear in temperature, salinity and pressure. That is, the equation of state (EOS 80) includes products of salinity, temperature and pressure. Sometimes, for practical purposes, in theoretical and simple numerical models the equation of state is approximated as linear and its pressure dependence is ignored:

$$\rho \approx \rho_0 - \alpha\rho\,(T - T_0) + \beta\rho\,(S - S_0);$$
$$\alpha = -(1/\rho)\,\partial\rho/\partial T \text{ and } \beta = (1/\rho)\,\partial\rho/\partial S \tag{3.9}$$

where ρ_0, T_0 and S_0 are arbitrary constant values of ρ, T and S; they are usually chosen as the mean values for the region being modeled. Here α is the *thermal expansion coefficient*, which expresses the change in density for a given change in temperature (and should not be confused with specific volume, defined with the same symbol in Section 3.5.3) and β is the *haline contraction coefficient*, also called the *saline contraction coefficient*, which is the change in density for a given change in salinity. The terms α and β are nonlinear functions of salinity, temperature, and pressure; their mean values are chosen for linear models. Full tables of values are given in UNESCO (1987). The value of α (at the sea surface and at a salinity of 35 g kg^{-1}) ranges from 53×10^{-6} K^{-1} at a temperature of 0°C to 257×10^{-6} K^{-1} at a temperature of 20°C. The value of β (at the sea surface and at a salinity of 35 g kg^{-1}) ranges from 785×10^{-6} kg g^{-1} (at a temperature of 0°C) to 744×10^{-6} kg g^{-1} (at a temperature of 20°C).

Nonlinearity in the equation of state leads to the curvature of the density contours in Figures 3.1 and 3.5. Mixing between two water parcels must occur along straight lines in the temperature/salinity planes of Figures 3.1 and 3.5.

contours, when two parcels of the same density but different temperature and salinity are mixed together, the mixture has higher density than the original water parcels. Thus the concavity of the density contours means that there is a contraction in volume as water parcels mix. This effect is called *cabbeling* (Witte, 1902). In practice, cabbeling may be of limited importance, having demonstrable importance only where water parcels of very different initial properties mix together. Examples of problems where cabbeling has been a factor are in the formation of dense water in the Antarctic (Foster, 1972) and in the modification of intermediate water in the North Pacific (Talley & Yun, 2001).

There are two other important mixing effects associated with the physical properties of seawater: thermobaricity and double diffusion. *Thermobaricity* (McDougall, 1987b) is best explained by the rotation with depth of potential density contours in the potential temperature—salinity plane (Section 3.5.4). As in Figure 3.5, consider two water parcels of different potential temperature and salinity in which the warmer, saltier parcel is slightly denser than the colder, fresher one. (This is a common occurrence in subpolar regions such as the Arctic and the Antarctic.) If these two water parcels are suddenly brought to a greater pressure, it is possible for them to reverse their relative stratification, with the colder, fresher one compressing more than the warmer one, and therefore becoming the denser of the two parcels. The parcels would now be vertically stable if the colder, fresher one were beneath the warmer, saltier one. Thermobaricity is an important effect in the Arctic, defining the relative vertical juxtaposition of the Canadian and Eurasian Basin Deep Waters (Section 12.2).

Double diffusion results from a difference in diffusivities for heat and salt, therefore, it is not a matter of linearity or nonlinearity. At the molecular level, these diffusivities clearly differ.

Because double diffusive effects are apparent in the ocean's temperature—salinity properties, the difference in diffusivities scales up in some way to the eddy diffusivity. Diffusivity and mixing are discussed in Chapter 7, and double diffusion in Section 7.4.3.2.

3.5.6. Static Stability and Brunt-Väisälä Frequency

Static stability, denoted by E, is a formal measure of the tendency of a water column to overturn. It is related to the density stratification, with higher stability where the water column is more stratified. A water column is statically stable if a parcel of water that is moved adiabatically (with no heat or salt exchange) up or down a short distance returns to its original position. The vigor with which the parcel returns to its original position depends on the density difference between the parcel and the surrounding water column at the displaced position. Therefore the rate of change of density of the water column with depth determines a water column's static stability. The actual density of the parcel increases or decreases as it is moved down or up because the pressure on it increases or decreases, respectively. This adiabatic change in density must be accounted for in the definition of static stability.

The mathematical derivation of the static stability of a water column is presented in detail in Pond and Pickard (1983) and other texts. The full expression for E is complicated. For very small vertical displacements, static stability might be approximated as

$$E \approx -(1/\rho)(\partial\rho/\partial z) \qquad (3.10a)$$

where ρ is in situ density. The water column is stable, neutral, or unstable depending on whether E is positive, zero, or negative, respectively. Thus, if the density gradient is positive downwards, the water column is stable and there is no tendency for vertical overturn.

For larger vertical displacements, a much better approximation uses local potential density, σ_n:

$$E = -(1/\rho)(\partial\sigma_n/\partial z) \qquad (3.10b)$$

Here the potential density anomaly σ_n is computed relative to the pressure at the center of the interval used to compute the vertical gradient. This local pressure reference approximately removes the adiabatic pressure effect. Many computer subroutines for seawater properties use this standard definition. An equivalent expression for stability is

$$E = -(1/\rho)(\partial\rho/\partial z) - (g/C^2) \qquad (3.10c)$$

where ρ is in situ density, g = acceleration due to gravity, and C = in situ sound speed. The addition of the term g/C^2 allows for the compressibility of seawater. (Sound waves are compression waves; Section 3.7.)

A typical density profile from top to bottom of the ocean has a surface mixed layer with low stratification, an upper ocean layer with an intermediate amount of stratification, an intermediate layer of high stratification (*pycnocline*), and a deep layer of low stratification (Section 4.2). The water in the pycnocline is very stable; it takes much more energy to displace a particle of water up or down than in a region of lesser stability. Therefore turbulence, which causes most of the mixing between different water bodies, is less able to penetrate through the stable pycnocline than through less stable layers. Consequently, the pycnocline is a barrier to the vertical transport of water and water properties. The stability of these layers is measured by E. In the upper 1000 m in the open ocean, values of E range from 1000×10^{-8} m^{-1} to 100×10^{-8} m^{-1}, with larger values in the pycnocline. Below 1000 m, E decreases; in abyssal trenches E may be as low as 1×10^{-8} m^{-1}.

Static instabilities may be found near the interfaces between different waters in the process of mixing. Because these instabilities occur at a small

vertical scale, on the order of meters, they require continuous profilers for detection. Unstable conditions with vertical extents greater than tens of meters are uncommon below the surface layer.

The *buoyancy* (*Brunt-Väisälä*) *frequency* associated with internal gravity waves (Chapter 8) is an intrinsic frequency associated with static stability. If a water parcel is displaced upward in a statically stable water column, it will sink and overshoot the original position. The denser water beneath its original position will force it back up into lighter water, and it will continue oscillating. The frequency of the oscillation depends on the static stability: the more stratified the water column, the higher the static stability and the higher the buoyancy frequency. The Brunt-Väisälä frequency, N, is an intrinsic frequency of internal waves:

$$N^2 = gE \approx g[-(1/\rho)(\partial\sigma_n/\partial z)] \qquad (3.11)$$

The frequency in cycles/sec (hertz) is $f = N/2\pi$ and the period is $\tau = 2\pi/N$. In the upper ocean, where E typically ranges from 1000×10^{-8} to 100×10^{-8} m^{-1}, periods are $\tau = 10$ to 33 min (Figure 3.6). For the deep ocean, $E = 1 \times 10^{-8}$ m^{-1} and $\tau \approx 6$ h.

The final quantity that we define based on vertical density stratification is the "stretching" part of the *potential vorticity* (Section 7.6). Potential vorticity is a dynamical property of a fluid analogous to angular momentum. Potential vorticity has three parts: rotation due to Earth's rotation (planetary vorticity), rotation due to relative motions in the fluid (relative vorticity, for instance, in an eddy), and a stretching component proportional to the vertical change in density, which is analogous to layer thickness (Eq. 7.41). In regions where currents are weak, relative vorticity is small and the potential vorticity can be approximated as

$$Q \approx -(f/\rho)(\partial\rho/\partial z) \qquad (3.12a)$$

This is sometimes called "isopycnic potential vorticity." The vertical density derivative is

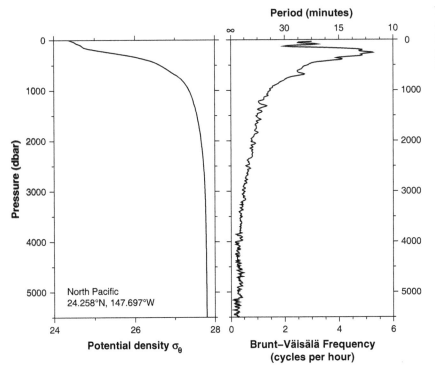

FIGURE 3.6 (a) Potential density and (b) Brunt-Väisälä frequency (cycles/h) and period (minutes) for a profile in the western North Pacific.

calculated from locally referenced potential density, so it can be expressed in terms of Brunt-Väisälä frequency:

$$Q = (f/g)N^2 \qquad (3.12b)$$

3.5.7. Freezing Point of Seawater

The salt in seawater depresses the freezing point below 0°C (Figure 3.1). An algorithm for calculating the freezing point of seawater is given by Millero (1978). Depression of the freezing point is why a mixture of salt water and ice is used to make ice cream; as the ice melts, it cools the water (and ice cream) below 0°C. At low salinities, below the salinity of most seawater, cooling water reaches its maximum density before freezing and sinks while still fluid. The water column then overturns and mixes until the whole water column reaches the temperature of maximum density.

On further cooling the surface water becomes lighter and the overturning stops. The water column freezes from the surface down, with the deeper water remaining unfrozen. However, at salinities greater than 24.7 psu, maximum density is achieved at the freezing point. Therefore more of the water column must be cooled before freezing can begin, so freezing is delayed compared with the freshwater case.

3.6. TRACERS

Dissolved matter in seawater can help in tracing specific water masses and pathways of flow. Some of these properties can be used for dating seawater (determine the length of time since the water was last at the sea surface; Section 4.7). Most of these constituents occur in such small concentrations that their variations do not significantly affect density variations or the relationship between chlorinity, salinity, and conductivity. (See Section 3.5 for comments on this.) These additional properties of seawater can be: conservative or non-conservative; natural or anthropogenic (man-made); stable or radioactive; transient or non-transient. The text by Broecker and Peng (1982) describes the sources and chemistry of many tracers in detail.

For a tracer to be *conservative* there are no significant processes other than mixing by which the tracer is changed below the surface. Even salinity, potential temperature, and hence density, can be used as conservative tracers since they have extremely weak sources within the ocean. This near absence of in situ sources and sinks means that the spreading of water masses in the ocean can be approximately traced from their origin at the sea surface by their characteristic temperature/salinity values. Near the surface, evaporation, precipitation, runoff, and ice processes change salinity, and many surface heat-transfer processes change the temperature (Section 5.4). Absolute salinity can be changed only very slightly within the ocean due to changes in dissolved nutrients and carbon (end of Section 3.4). Temperature can be raised very slightly by geothermal heating at the ocean bottom. Even though water coming out of bottom vents at some mid-ocean ridges can be extremely hot (up to 400°C), the total amount of water streaming out of the vents is tiny, and the high temperature quickly mixes away, leaving a miniscule large-scale temperature increase.

Non-conservative properties are changed by chemical reactions or biological processes within the water column. Dissolved oxygen is an example. Oxygen enters the ocean from the atmosphere at the sea surface. It is also produced through photosynthesis by phytoplankton in the sunlit upper ocean (photic zone or euphotic zone) and consumed by respiration by zooplankton, bacteria, and other creatures. Equilibration with the atmosphere keeps ocean mixed layer waters at close to 100% saturation. Below the surface layer, oxygen content drops rapidly. This is not a function of the temperature of the water, which generally is lower at depth, since cold water can hold *more* dissolved oxygen than warm water. (For example, for a salinity of 35: at 30°C, 100% oxygen saturation occurs at 190 μmol/kg; at 10°C it is 275 μmol/kg; and at 0°C it is 350 μmol/kg.) The drop in oxygen content and saturation with depth is due to respiration within the water column, mainly by bacteria feeding on organic matter (mostly dead plankton and fecal pellets) sinking from the photic zone. Since there is no source of oxygen below the mixed layer and photic zone, oxygen decreases with increasing age of the subsurface water parcels. Oxygen is also used by nitrifying bacteria, which convert the nitrogen in ammonium (NH_4) to nitrate (NO_3).

The rate at which oxygen is consumed is called the *oxygen utilization rate*. This rate depends on local biological productivity so it is not uniform in space. Therefore the decrease in oxygen from a saturated surface value is not a perfect indication of age of the water parcel, especially in the biologically active upper ocean and continental shelves. However, below the thermocline, the utilization rate is more uniform and changes in oxygen following a water parcel correspond relatively well to age.

Nutrients are another set of natural, non-conservative, commonly observed properties. These include dissolved silica, phosphate, and the nitrogen compounds (ammonium, nitrite, and nitrate). Nutrients are essential to ocean life so they are consumed in the ocean's surface layer where life is abundant; consequently, concentrations there are low. Nutrient content increases with depth and age, as almost a mirror image of the oxygen decrease. Silica is used by some organisms to form protective shells. Silica re-enters the water column when the hard parts of these organisms dissolve as they fall to the ocean floor. Some of this material reaches the seafloor and accumulates, creating a silica source on the ocean bottom as well. Some silica also enters the water column through venting at mid-ocean ridges. The other nutrients (nitrate, nitrite, ammonium, and phosphate) re-enter the water column as biological (bacterial) activity decays the soft parts of the falling detritus. Ammonium and phosphate are immediate products of the decay. Nitrifying bacteria, which are present through the water column, then convert ammonium to nitrite and finally nitrate; this process also, in addition to respiration, consumes oxygen. Because oxygen is consumed and nutrients are produced, the ratios of nitrate to oxygen and of phosphate to oxygen are nearly constant throughout the oceans. These proportions are known as "Redfield ratios," after Redfield (1934) who demonstrated the near-constancy of these proportions. Nutrients are discussed further in Section 4.6.

Other non-conservative properties related to the ocean's carbon system, including dissolved inorganic carbon, dissolved organic carbon, alkalinity, and pH, have been widely measured over the past several decades. These have both natural and anthropogenic sources and are useful tracers of water masses.

Isotopes that occur in trace quantities are also useful. Two have been widely measured: ^{14}C and ^{3}He. ^{14}C is radioactive and non-conservative. ^{3}He is conservative. Both have predominantly natural sources but both also have anthropogenic sources in the upper ocean. Isotope concentrations are usually measured and reported in terms of ratios to the more abundant isotopes. For ^{14}C, the reported unit is based on the ratio of ^{14}C to ^{12}C. For ^{3}He, the reported unit is based on the ratio of ^{3}He to ^{4}He. Moreover, the values are often reported in terms of the normalized difference between this ratio and a standard value, usually taken to be the average atmospheric value (see Broecker & Peng, 1982).

Most of the ^{14}C in the ocean is natural. It is created continuously in the atmosphere by cosmic ray bombardment of nitrogen, and enters the ocean through gas exchange. "Bomb" radiocarbon is an anthropogenic tracer that entered the upper ocean as a result of atomic bomb tests between 1945 and 1963 (Key, 2001). In the ocean, ^{14}C and ^{12}C are incorporated by phytoplankton in nearly the same ratio as they appear in the atmosphere. After the organic material dies and leaves the photic zone, the ^{14}C decays radioactively, with a half-life of 5730 years. The ratio of ^{14}C to ^{12}C decreases. Since values are reported as anomalies, as the difference from the atmospheric ratio, the reported oceanic quantities are generally negative (Section 4.7 and Figure 4.24). The more negative the anomaly, the older the water. Positive anomalies throughout the upper ocean originated from the anthropogenic bomb release of ^{14}C.

The natural, conservative isotope ^{3}He originates in Earth's mantle and is outgassed at vents in the ocean floor. It is usually reported in terms of its ratio to the much more abundant ^{4}He compared with this ratio in the atmosphere. It is an excellent tracer of mid-depth circulation, since its sources tend to be the tops of mid-ocean ridges, which occur at about 2000 m. The anthropogenic component of ^{3}He is described in the last paragraph of this Section.

Another conservative isotope that is often measured in seawater is the stable (heavy) isotope of oxygen, ^{18}O. Measurements are again reported relative to the most common isotope ^{16}O. Rainwater is depleted in this heavy isotope of oxygen (compared with seawater) because it is easier for the lighter, more common isotope of oxygen, ^{16}O, to evaporate from the sea and land. A second step of reduction of ^{18}O in atmospheric water vapor relative to seawater occurs when rain first forms, mostly at warmer atmospheric temperatures, since the heavier isotope falls out preferentially. Thus rainwater is depleted in ^{18}O relative to seawater, and rain formed at lower temperatures is more depleted than at higher temperatures. For physical oceanographers, ^{18}O content can be a useful indicator in a high latitude region of whether the source of freshwater at the sea surface is rain/runoff/ glacial melt (lower ^{18}O content), or melted sea ice (higher content). In paleoclimate records, it reflects the temperature of the precipitation (higher ^{18}O in warmer rain); ice formed during the (cold) glacial periods is more depleted in ^{18}O than ice formed in the warm interglacials and hence ^{18}O content is an indicator of relative global temperature.

Transient tracers are chemicals that have been introduced by human activity; hence they are anthropogenic. They are gradually invading the ocean, marking the progress of water from the surface to depth. They can be either stable or radioactive. They can be either conservative or non-conservative. Commonly measured transient tracers include chlorofluorocarbons, tritium, and much of the upper ocean ^{3}He and ^{14}C. Chlorofluorocarbons (CFCs) were introduced as refrigerants and for industrial use. They are extremely stable (conservative) in seawater. Their usage peaked in 1994, when recognition of their role in expanding the ozone hole in the atmosphere finally led to international conventions to phase out their use. Because different types of CFCs were used over the years, the ratio of different types in a water parcel can yield approximate dates for when the water was at the sea surface. Tritium is a radioactive isotope of hydrogen that has also been measured globally; it was released into the atmosphere through atomic bomb testing in the 1960s and then entered the ocean, primarily in the Northern Hemisphere. Tritium decays to ^{3}He with a half-life of 12.4 years, which is comparable to the circulation time of the upper ocean gyres. When ^{3}He is measured along with tritium, the time since the water was at the sea surface can be estimated (Jenkins, 1998).

3.7. SOUND IN THE SEA

In the atmosphere, we receive much of our information about the material world by means of wave energy, either electromagnetic (light) or mechanical (sound). In the atmosphere, light in the visible part of the spectrum is attenuated less than sound; we can see much farther away than we can hear. In the sea the reverse is true. In clear ocean water, sunlight may be detectable (with instruments) down to 1000 m, but the range at which humans can see details of objects is rarely more than 50 m, and usually less. On the other hand, sound waves can be detected over vast distances and are a much better vehicle for undersea information than light.

The ratio of the speed of sound in air to that in water is small (about 1:4.5), so only a small amount of sound energy starting in one medium can penetrate into the other. This contrasts with the relatively efficient passage of light energy through the air/water interface (speed ratio only about 1.33:1). This is why a person standing on the shore can see into the water but cannot hear any noises in the sea. Likewise, divers cannot converse underwater because their sounds are generated in the air in the throat and little of the sound energy is transmitted into the water. Sound sources used in the sea generate sound energy in solid bodies (transducers), for example, electromagnetically, in which the speed of sound is similar to that in water. Thus the two are acoustically "matched" and the transducer energy is transmitted efficiently into the sea.

Sound is a wave. All waves are characterized by amplitude, frequency, and wavelength (Section 8.2). Sound speed (C), frequency (n), and wavelength (λ) are connected by the wave equation $C = n\lambda$. The speed does not depend on frequency, so the wavelength depends on sound speed and frequency. The frequencies of sounds range from 1 Hz or less (1 Hz = 1 vibration per second) to thousands of kilohertz (1 kHz = 1000 cycles/sec). The wavelengths of sounds in the sea cover a vast range, from about 1500 m for $n = 1$ Hz to 7 cm for $n = 200$ kHz. Most underwater sound instruments use a more restricted range from 10 to 100 kHz, for which the wavelengths are 14 to 1.4 cm.

There are many sources of sound in the sea. A hydrophone listening to the ambient sound in the sea will record a wide range of frequencies and types of sounds, from low rumbles to high-frequency hisses. Some sources of undersea sounds are microseisms ($10 - 100$ Hz); ships ($50 - 1500$ Hz); the action of wind, waves, and rain at the surface ($1 - 20$ kHz); cavitation of air bubbles and animal noises ($10 - 400$ Hz); and fish and crustaceans ($1 - 10$ kHz). Noises associated with sea ice range from $1 - 10$ kHz.

Sound is a compressional wave; water molecules move closer together and farther apart as the wave passes. Therefore sound speed depends on the medium's compressibility. The more compressible a medium is for a given density, the slower the wave since more activity is required to move the molecules. The speed of sound waves in the sea, C, is given by

$$C = (\beta\rho)^{-1/2} \text{ where } \beta = \rho^{-1}(\partial\rho/\partial p)_{\theta,S.}$$
(3.13)

β is the adiabatic compressibility of seawater (with potential temperature and salinity constant), ρ is the density, p is the pressure, θ is the potential temperature, and S is the salinity. Since β and ρ depend (nonlinearly) on temperature and pressure, and to a lesser extent, salinity, so does the speed of sound waves. There are various formulae for the dependence of Eq. (3.13) on T, S, and p; all derived from experimental measurements. The two most accepted are those of Del Grosso (1974) and of Chen and Millero (1977); Del Grosso's equation is apparently more accurate, based on results from acoustic tomography and inverted echo sounder experiments (e.g., Meinen & Watts, 1997). Both are long and nonlinear polynomials, as is the equation of state. We present a simpler formula, which itself is simplified from Mackenzie (1981) and

is similar to Del Grosso (1974), to illustrate features of the relationship:

$$C = 1448.96 + 4.59T - 0.053T^2$$
$$+ 1.34(S - 35) + 0.016p$$
(3.14)

in which T, S, and p are temperature, salinity, and depth, and the constants have the correct units to yield C in m/s. The sound speed is 1449 m/s at T = 0°C, S = 35, and p = 0. The sound speed increases by 4.5 m/s for $\Delta T = + 1$ K, by 1.3 m/s for $\Delta S = + 1$, and by 16 m/s for $\Delta p = 1000$ dbar.

Sound speed is higher where the medium is less compressible. Seawater is less compressible when it is warm, as noted in the previous potential density discussion and apparent from the simplified equation (3.14). Seawater is also less compressible at high pressure, where the fluid is effectively more rigid because the molecules are pushed together. Salinity variations have a negligible effect in most locations. In the upper layers, where temperature is high, sound speed is high, and decreases downward with decreasing temperature (Figure 3.7). However, pressure increases with depth, so that at mid-depth, the decrease in sound speed due to cooler water is overcome by an increase in sound speed due to higher pressure. In most areas of the ocean, the warm water at the surface and the high pressure at the bottom produce maximum sound speeds at the surface and bottom and a minimum in between. The sound-speed minimum is referred to as the SOund Fixing And Ranging (SOFAR) channel. In Figure 3.6, the sound-speed minimum is at about 700 m depth. In regions where temperature is low near the sea surface, for instance at high latitudes, there is no surface maximum in sound speed, and the sound channel is found at the sea surface.

Sound propagation can be represented in terms of rays that trace the path of the sound (Figure 3.8). In the SOFAR channel, at about 1100 m in Figure 3.8, sound waves directed at

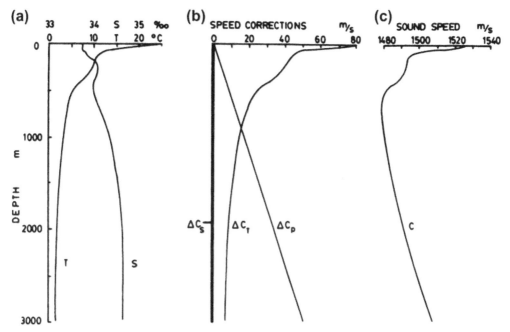

FIGURE 3.7 For station Papa in the Pacific Ocean at 39°N, 146°W, August, 1959: (a) temperature (°C) and salinity (psu) profiles, (b) corrections to sound speed due to salinity, temperature, and pressure, (c) resultant in situ sound-speed profile showing sound-speed minimum (SOFAR channel).

moderate angles above the horizontal are refracted downward, across the depth of the sound-speed minimum, and then refracted upward; they continue to oscillate about the sound-speed minimum depth. (Rays that travel steeply up or down from the source will not be channeled but may travel to the surface or bottom and be reflected there.) Low frequency sound waves (hundreds of hertz) can travel considerable distances (thousands of kilometers) along the SOFAR channel. This permits detection of submarines at long ranges and has been used for locating lifeboats at sea. Using the SOFAR channel to track drifting subsurface floats to determine deep currents is described in Chapter S6, Section S6.5.2 of the supplemental materials located on the textbook Web site.

The deep SOFAR channel of Figure 3.8b is characteristic of middle and low latitudes, where the temperature decreases substantially as depth increases. At high latitudes where the temperatures near the surface may be constant or even decrease toward the surface, the sound speed can have a surface minimum (Figure 3.8a). The much shallower sound channel, called a surface duct, may even be in the surface layer. In this case, downward directed sound rays from a shallow source are refracted upward while upward rays from the subsurface source are reflected downward from the surface and then refracted upward again. In this situation, detection of deep submarines from a surface ship using sonar equipment mounted in the hull may not be possible and deep-towed sonar equipment may be needed. In shallow water (e.g., bottom depth <200 m), reflection can occur both from the surface and from the bottom.

A pulse transmitted from a source near the SOFAR channel axis does not appear to

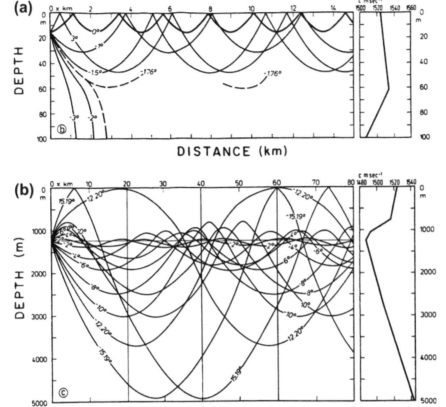

FIGURE 3.8 Sound ray diagrams: (a) from a shallow source for a sound-speed profile initially increasing with depth in upper mixed layer to a shallow minimum and then decreasing, and (b) from a sound source near the speed minimum in the sound channel for a typical open ocean sound-speed profile.

distant receivers as a sharp pulse but as a drawn-out signal rising slowly to a peak followed by a sharp cutoff. The peak before the cutoff is the arrival of the sound energy along the sound channel axis (direct signal), while the earlier arrivals are from sound that traveled along the refracted ray routes. It might appear in Figure 3.8b that the refracted rays have to travel a greater distance than the direct ray and would thus be delayed, but this is an illusion. Figure 3.8b is drawn with gross vertical exaggeration to enable the rays to be shown clearly, but the differences in distances traveled by refracted rays and the direct rays are very small; the greater speed

in the refracted ray paths compensates for the greater distance they travel, so the direct ray arrives last.

Sound is used widely to locate and observe solid objects in the water. Echo sounders are used to measure bottom depths to the ocean's maximum depth of more than 11,000 m. SONAR (**SO**und **N**avigation **A**nd **R**anging) can determine the direction and distance to a submarine at ranges of hundreds of meters and to schools of fish at somewhat lesser ranges. Sidescan sonars determine the structure of the ocean bottom and can be used to locate shipwrecks. Acoustically tracked Swallow floats (see Chapter S6, Section S6.5.2 of the supplemental material on the textbook Web site) provided some of the first direct observations of deep currents. Current speeds (or the speed of a ship relative to the water) are often measured using the reflection of sound waves from small particles moving with the water, applying the principle of Doppler shift. Because temperature and density affect the sound velocity, sound can be used to infer ocean water characteristics and their variations. Sound is used to measure surface processes such as precipitation, a measurement that is otherwise nearly impossible to determine.

In echo sounding, short pulses of sound energy are directed vertically downward where they reflect off of the bottom and return to the ship. (Echo sounders are also used to detect shoals of fish, whose air bladders are good reflectors of sound energy. Modern "fish finders" are simply low-cost echo sounders designed to respond to the fish beneath the vessel.) The acoustic travel time, t, yields the depth $D = C_o t/2$, where C_o is the mean sound speed between the surface and the bottom. Transducers in ordinary echo sounders are not much larger than the wavelength of the sound, so the angular width of the sound beam is large. Wide beams cannot distinguish the details of bottom topography. For special sounding applications, much larger sound sources that form a narrower beam are used. It is also possible to improve the resolution by using higher frequencies (up to 100 kHz or even 200 kHz), but the absorption of sound energy by seawater increases roughly as the square of the frequency, so higher frequency echo sounders cannot penetrate as deeply.

Inhomogeneities distort an initial sharp sound pulse so that the signal received at a hydrophone is likely to have an irregular tail of later arrival sounds. This is referred to as *reverberation*. One source of reverberation is the "deep scattering layer," which is biological in nature. This layer is characterized by diel (day/night) vertical migrations of several hundreds of meters; the organisms migrate toward the sea surface at dusk to feed and back down at dawn. This layer was first identified because of the scattering produced by the plankton and (gas-filled) fish bladders.

Sound is used to determine the speed of ocean currents and of ships, using a technique called acoustic Doppler profiling (see Chapter S6, Section S6.5.5.1 of the supplemental material located on the textbook Web site). Sound is transmitted from a source and reflects off the particles (mainly plankton) in the water and returns back to a receiver. If the source is moving relative to the particles, then the received sound wave has a different frequency from the transmitted wave, a phenomenon called Doppler shifting. Doppler shift is familiar to anyone who has listened to the sound of a siren when an emergency vehicle first approaches (Doppler shifting the sound to a higher frequency and, therefore, a higher pitch) and then retreats away (Doppler shifting the sound to a lower frequency and lower pitch). Acoustic Doppler speed logs are common on ships and give a relatively accurate measure of the speed of the ship through the water. If the ship's speed is tracked very precisely using, for instance, GPS navigation, then the ship speed can be subtracted from the speed of the ship relative to the water to yield the speed of the water relative to the GPS navigation, providing a measure of current speeds. Acoustic Doppler current profilers are also moored in the ocean to provide long-term records of current speeds.

Sound can be used to map the ocean's temperature structure and its changes, through a technique called acoustic tomography (see Chapter S6, Section S6.6.1 of the supplemental material located on the textbook Web site). Since sound speed depends on temperature, temperature changes along a ray path result in travel time changes. With extremely accurate clocks, these changes can be detected. If multiple ray paths crisscross a region, the travel time changes can be combined using sophisticated data analysis techniques to map temperature changes in the region. This technique has been especially useful in studying the three-dimensional structure of deep convection in regions and seasons that are virtually impossible to study from research ships. Similar techniques have been applied to very long distance monitoring of basin-average ocean temperature, which is possible because of the lack of attenuation of sound waves over extremely long distances (Munk & Wunsch, 1982). However, large-scale monitoring of ocean temperature changes using sound has been eclipsed by the global temperature–salinity profiling float array, Argo, which provides local as well as basin-average information.

Much more information about ocean acoustics can be found in textbooks such as Urick (1983).

3.8. LIGHT AND THE SEA

This is a very brief introduction to a complex subject. Full treatments are available in various sources; some suggestions are Mobley (1995) and Robinson (2004).

Sunlight with a range of wavelengths enters the sea after passing through the atmosphere. Within the upper layer of the ocean, up to 100 m depth or more, the visible light interacts with the water molecules and the substances that are dissolved or suspended in the water. The light provides energy for photosynthesis and also heats the upper layer. Processes of backscattering, absorption, and re-emission result in the visible light (ocean color) that emerges back from the ocean surface into the atmosphere. This emerging radiation is then measured with instruments above the sea surface, including satellites. For satellite observations, the atmosphere again affects the signal from the sea. Observations of ocean color by satellites can then be related to the processes within the ocean that affect the emerging light, including an abundance of phytoplankton, particulate organic carbon, suspended sediment, and so forth.

Absorption (attenuation) of the sun's energy in the upper layer depends on the materials within the water; therefore these materials affect how heating is distributed in the surface layer and affects mixed layer processes. General circulation models that are run with observed forcing sometimes use information about light attenuation, affecting mixed layer formation and, consequently, sea surface temperature in the model.

Section 3.8.1 describes the optical properties of seawater and Section 3.8.2 describes the quantity that is observed as ocean color. Examples of observations are shown in Chapter 4.

3.8.1. Optical Properties

The sun irradiates the earth with a peak in the visible spectrum (wavelengths from about 400 to 700 nm, from violet to red, where 1 nm = 10^{-9} m). Sunlight behaves differently in water and air. The ocean absorbs light in much shorter distances than the atmosphere. When this shortwave energy penetrates the sea, some of it is scattered, but much is absorbed, almost all within the top 100 m. The energy is attenuated approximately exponentially. This is the photic (euphotic) zone, where photosynthesis occurs. This penetration of solar energy into the ocean's upper layer is also important in the ocean's heat budget (Chapter 5).

A schematic overview of the ocean's optical processes is shown in Figure 3.9, after Mobley (1995), who provides much greater detail and precise expressions for each of the quantities in the diagram. Each of the quantities can be observed, with greater or lesser difficulty. At the top of the diagram, the external environmental quantities that determine the amount of radiation entering the ocean are the sun's radiance distribution, which depends on its position and on sky conditions; the sea state, since this determines how much radiation is reflected without entering the sea; and the ocean bottom, if it is shallow enough to intercept the

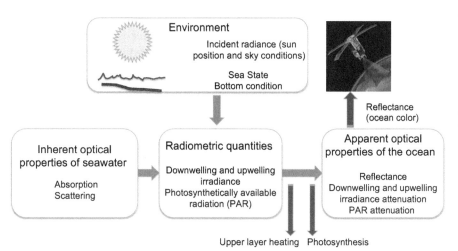

FIGURE 3.9 Schematic of optical processes in seawater. Adapted and simplified from Mobley (1995), with added indicators of seawater heating and photosynthesis, as well as satellite observation of ocean color.

light. The *inherent optical properties* of the seawater determine how it absorbs and scatters radiation, as a function of wavelength; this depends on the matter that is dissolved, suspended, or active (in the case of phytoplankton).

The environmental conditions and inherent optical properties work together through a radiative transfer equation to set the radiometric quantities of the medium. Here it is useful to provide some definitions of the radiometric quantities listed in the middle box of Figure 3.9. First, we note that light from a source, which could be at any point in a medium in which light is diffused or scattered, illuminates a complete sphere around the source. Therefore the solid angle, measured in "steradians" (sr), is a useful measure, similar to area. Next, the flux of energy from the light is measured in Watts (J/sec). The *radiance* is the flux of energy per unit area and per unit steradian; it is measured in units of $W/(sr\ m^2)$. If the radiance is measured as a function of wavelength of the light (i.e., spectral radiance), then its units are $W/(sr\ m^2\ nm)$ if wavelength is measured in nanometers.

The *irradiance* is the total amount of radiance that reaches a given point (i.e., where your optical measurement is made), so it is the sum of radiance coming in from all directions to the observation point; therefore it is the integral of radiance over all solid angles, and has units of W/m^2 for total irradiance, or $W/(m^2\ nm)$ for spectral irradiance (which is a function of wavelength). Next, *upwelling* irradiance is defined as the irradiance from all solid angles below the observation point; downwelling irradiance would come from all angles above that point.

Reflectance is the ratio of upwelling irradiance to downwelling irradiance, defined at a point; reflectance defined this way has no units. It is not the same as actual reflected light from the sea surface. Rather, reflectance is the light emerging from the ocean. For remote sensing, in which the radiation from the ocean's surface is being measured from a specified location, rather than from all directions, reflectance can be defined alternatively as the ratio of upwelling radiance to downwelling irradiance; in this case, reflectance has units of (sr^{-1}).

Finally, the amount of radiation available for photosynthesis (photosynthetically available radiation; PAR) is measured in photons $s^{-1}\ m^{-2}$.

Returning to Figure 3.9, the rightmost bottom box lists the *apparent optical properties* of the seawater. These include the rate at which light is attenuated within the water column, and how much light returns back out through the sea surface (indicated as reflectance). The irradiance and PAR are attenuated with increasing depth as the radiation is absorbed, scattered, and used for photosynthesis by phytoplankton. Attenuation is often approximately exponential. If attenuation were exactly exponential, of the form $I(z) = I_o e^{-Kz}$, where I_o is the radiation intensity at the sea surface, I the intensity at a depth z meters below the surface, and K the vertical attenuation coefficient of the water, then the apparent optical properties would be expressed in terms of the e-folding depth, K. The actual attenuation is not exponential, so the attenuation coefficient, K, is proportional to the depth derivative of the radiation intensity (and would be equal to the e-folding depth if the dependence were exponential).

The effects of depth and constant attenuation coefficient on light intensity are illustrated in Table 3.3, from Jerlov (1976). The coefficient K depends mainly on factors affecting absorption of light in the water and to a lesser extent on scattering. The last two columns of Table 3.3 indicate the range of penetrations found in actual seawater.

The smallest attenuation coefficient in Table 3.3 ($K = 0.02$ m^{-1}) represents the clearest ocean water and deepest penetration of light energy. Energy penetrates coastal waters less readily because of the extra attenuation due to suspended particulate matter and dissolved materials. The largest attenuation coefficient listed in the table, $K = 2$ m^{-1}, represents very turbid water with many suspended particles.

In seawater, the attenuation coefficient K also varies considerably with wavelength. Figure 3.10b shows the relative amounts of energy penetrating clear ocean water to 1, 10, and 50 m as a function of wavelength (solid curves). Light with blue wavelengths penetrates deepest; penetration by yellow and red is much less. That is, blue light, with wavelength of about 450 nm, has the least attenuation in clear ocean water. At shorter and longer wavelengths (in the ultraviolet and red), the attenuation is much greater. The increased attenuation in the ultraviolet is not important to the ocean's heat budget, because the amount of energy reaching sea level at such short wavelengths is small. Much more solar energy is contained in and beyond the red end of the spectrum. Virtually all of the energy at wavelengths shorter than the visible is absorbed in the top meter of water, while the energy at long wavelengths (1500 nm or greater) is absorbed in the top few centimeters.

All wavelengths are attenuated more in turbid water than in clear water. In clear ocean

TABLE 3.3 Amount of Light Penetrating to Specified Depths in Seawater as a Percentage of that Entering Through the Surface

Depth (m)	Vertical Attenuation Coefficient K (m^{-1})			Clearest Ocean Water	Turbid Coastal Water
	$K = 0.02$	$K = 0.2$	$K = 2$		
0	100%	100%	100%	100%	100%
1	98	82	14	45	18
2	96	67	2	39	8
10	82	14	0	22	0
50	37	0	0	5	0
100	14	0	0	0.5	0

Jerlov, 1976.

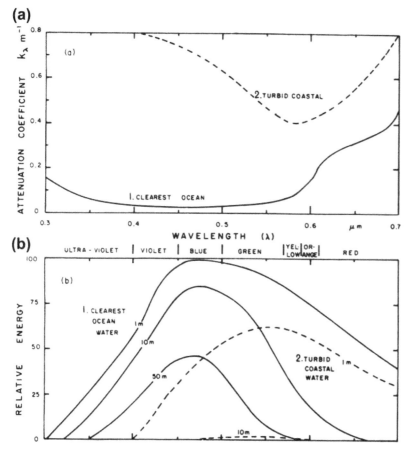

(a)

(b)

FIGURE **3.10** (a) Attenuation coefficient k_λ, as a function of wavelength λ (µm) for clearest ocean water (solid line) and turbid coastal water (dashed line). (b) Relative energy reaching 1, 10, and 50 m depth for clearest ocean water and reaching 1 and 10 m for turbid coastal waters.

water, there is enough light at 50 to 100 m to permit a diver to work, but in turbid coastal waters almost all of the energy may have been absorbed by a depth of 10 m. K is larger for turbid water than clear (Figure 3.10a), and the least attenuation is in the yellow part of the spectrum. In turbid water, less energy penetrates to 1 m and 10 m, and the maximum penetration is shifted to the yellow (Figure 3.10b). (The energy reaching 50 m in this turbid water is too small to show on the scale of this graph.)

In clear ocean water, the superior penetration of blue and green light is evident both visually when diving and also in color photographs taken underwater by natural light.

Red or yellow objects appear darker in color or even black as they are viewed at increasing depths because the light at the red end of the spectrum has been absorbed in the upper layers and little is left to be reflected by the objects. Blue or green objects retain their colors to greater depths.

The presence of plankton in seawater also changes the penetration depth of solar radiation, and hence the depth at which the sun's heat is absorbed. This changes the way the surface mixed layer develops, which can in turn impact the plankton, leading to a feedback. There are significant and permanent geographical variations in this vertical distribution of absorption, since some regions of the world ocean have much higher biological productivity than others.

3.8.2. Ocean Color

To the eye, the color of the sea ranges from deep blue to green to greenish yellow (Jerlov, 1976). Broadly speaking, deep or indigo blue color is characteristic of tropical and equatorial seas, particularly where there is little biological production. At higher latitudes, the color changes through green-blue to green in polar regions. Coastal waters are generally greenish.

Two factors contribute to the blue color of open ocean waters at low latitudes. Because water molecules scatter the short-wave (blue) light much more than the long-wave (red) light, the color seen is selectively blue. In addition, because the red and yellow components of sunlight are rapidly absorbed in the upper few meters, the only light remaining to be scattered by the bulk of the water is blue. Looking at the sea from above, sky light reflected from the surface is added to the blue light scattered from the body of the water. If the sky is blue, the sea will still appear deep blue, but if there are clouds, the white light reflected from the sea surface dilutes the blue scattered light from the water and the sea appears less intensely blue.

If there are phytoplankton in the water, their chlorophyll absorbs blue light and also red light, which shifts the water color to green. (This is also why plants are green.) The organic products from plants may also add yellow dyes to the water; these will absorb blue and shift the apparent color toward the green. These shifts in color generally occur in the more productive high-latitude and coastal waters. In some coastal regions, rivers carry dissolved organic substances that emphasize the yellowish green color. The red color that occurs sporadically in some coastal areas, the so-called red tide, is caused by blooms of reddish brown phytoplankton. Mud, silt, and other finely divided inorganic materials carried into the ocean by rivers can impart their own color to the water. In some fjords, the low-salinity surface layer may be milky white from the finely divided "rock flour" produced by abrasion in the glaciers and carried down by the melt water. The sediment can be kept in suspension by turbulence in the upper layer for a time, but when it sinks into the saline water, it flocculates (forms lumps) and sinks more rapidly. When diving in such a region the diver may be able to see only a fraction of a meter in the upper layer but be able to see several meters in the saline water below.

The color of seawater and depth of penetration of light were traditionally judged using a white Secchi disk (see Chapter S6, Section S6.8 of the supplemental materials located on the textbook Web site) lowered from the ship. This method has been superseded by a suite of instruments that measure light penetration at different wavelengths, transparency of the water at various wavelengths, and fluorescence. Most important, color observations are now made continuously and globally by color sensors on satellites.

Ocean color is a well-defined quantity, related to reflectance (Figure 3.9 and definition in Section 3.8.1). Reflectance, or ocean color, can be measured directly above the ocean's surface. Observations of ocean color since the 1980s have been made from satellites, and must be corrected for changes as the light passes upward through the atmosphere. Ocean color observations are then converted, through complex algorithms, to physically useful quantities such as the amount of chlorophyll present, or the amount of particulate organic carbon, or the amount of "yellow substance" (gelbstoff) that is created by decaying vegetation. With global satellite coverage, these quantities can be observed nearly continuously and in all regions.

Robinson (2004) provided a complete treatment of the optical pathways involved in ocean color remote sensing, starting with consideration of the total radiance observed by the satellite sensor. Many pathways contribute to the observed radiance. These can be grouped into an atmospheric path radiance (L_p), a "water-leaving radiance" from just below the sea surface (L_w), and a radiance due to all surface reflections (L_r) within the instantaneous field of view of the satellite sensor. The radiance L_s received at the satellite sensor is

$$L_s = L_p + TL_w + TL_r \qquad (3.15)$$

T is the transmittance, which gives the proportion of radiance that reaches the sensor without being scattered out of the field of view.

The water-leaving radiance provides the information about ocean color, so it is the desired observed quantity. It is closely related to the reflectance; the ratio of water-leaving radiance just above the sea surface to downwelling irradiance incident on the sea surface is the "remote sensing reflectance," or "normalized water-leaving radiance." The three net radiance terms depend on the wavelength and on the turbidity of the seawater. The largest

contribution is from the atmospheric pathway L_p. The water-leaving and reflected radiances are much smaller. Because of the weak signal strength for the ocean pathways (water-leaving radiance), ocean color remote sensing requires very precise atmospheric correction of the visible light sensed by the satellite. Complex radiative transfer models are invoked to carry out this correction and often the accuracy of the chlorophyll estimates depends critically on this atmospheric correction. After correction, the resultant radiances are analyzed for various components related to biological activity, particularly chlorophyll.

The biggest effect of chlorophyll on the spectrum of reflectance (normalized water-leaving radiance) is to reduce the energy at the blue end of the spectrum compared with the spectrum for clear water. This is demonstrated in Figure 3.11 (H. Gordon, personal communication, 2009). Here the spectrum of radiance is shown with and without correction for the atmosphere. When the atmosphere is not removed, there is virtually no difference between the spectra for low and high chlorophyll waters. When the atmospheric signal is

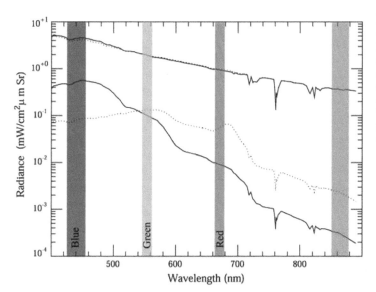

FIGURE 3.11 Example of observations of water-leaving radiance observed by the Multi-angle Imaging SpectroRadiometer (MISR), with bands observed by satellite color sensors indicated. Solid curves: low chlorophyll water (0.01 mg/m^3). Dotted curves: high chlorophyll water (10.0 mg/m^3). The two lower curves have the atmospheric signal removed. (H. Gordon, personal communication, 2009.)

removed, the desired difference emerges. The high chlorophyll spectrum is depressed at the blue end of the spectrum and elevated at the green and red.

Ocean color observations from satellites can also be used as a proxy for the attenuation properties of seawater, which can be used in mixed layer models that are set up to be run with observed atmospheric forcing. Absorption of solar radiation in the ocean heats the upper layer. The time and space distribution of absorption is directly related to the substances in the water column, and this affects ocean color. In practice, at present most mixed layer models in ocean general circulation models use a sum of two exponentially decaying functions that are proxies for attenuation of red light (quickly absorbed) and blue-green light (penetrating much deeper), with coefficients based on assumption of a particular Jerlov (1976) water type (Paulson & Simpson, 1977). However, explicit incorporation of biological effects on attenuation (incorporation of ocean color observations and of bio-optical models along with mixed layer models) is being tested widely and is a likely direction for the future, since the effects on modeled mixed layer temperature are clear (Wu, Tang, Sathyendranath, & Platt, 2007) and can in fact affect the temperature of the overlying atmosphere (Shell, Frouin, Nakamoto, & Somerville, 2003).

3.9. ICE IN THE SEA

Ice in the sea has two origins: the freezing of seawater and the ice broken off from glaciers. The majority of ice comes from the first of these sources and is referred to as *sea ice*; the glaciers supply "pinnacle" icebergs in the Northern Hemisphere and flat "tabular" icebergs in the Southern Hemisphere. Sea ice alters the heat and momentum transfers between the atmosphere and the ocean, is a thermal insulator, damps surface waves, changes the temperature

and salinity structure in the upper layer by melting and freezing, and is a major hindrance to navigation. Ice cover is an important part of Earth's climate feedbacks because of its high reflectivity, that is, its high *albedo* (Section 5.4.3). The ice—albedo feedback, which affects climate, is described in Section 5.4.5, and is especially important in the Arctic (Section 12.8).

3.9.1. Freezing Process

When water loses sufficient heat (by radiation, conduction to the atmosphere, convection, or evaporation) it freezes to ice, in other words, it changes to the solid state. Initial freezing occurs at the surface and then the ice thickens by freezing at its lower surface as heat is conducted away from the underlying water through the ice to the air.

The initial freezing process is different for fresh and low-salinity water than for more saline water because the temperature at which water reaches its maximum density varies with salinity. Table 3.4 gives the values of the freezing point and temperature of maximum density for water of various salinities. (Note that the values are for freezing, etc., at atmospheric pressure. Increased pressure lowers the freezing point, which decreases by about 0.08 K per 100 m increase in depth in the sea.)

To contrast the freezing process for freshwater and seawater, first imagine a freshwater lake where the temperature initially decreases from about 10°C at the surface to about 5°C

TABLE 3.4 Temperatures of the Freezing Point (t_f) and of Maximum Density ($t_{\rho max}$) for Fresh and Salt Water

S	0	10	20	24.7	30	35 psu
t_f	0	−0.5	−1.08	−1.33	−1.63	−1.91°C
$t_{\rho max}$	+3.98	+1.83	−0.32	−1.33	—	−°C

Note that the values for freezing and so forth are at atmospheric pressure. Increased pressure lowers the freezing point, which decreases by about 0.08 K per 100 m increase in depth in the sea.

at about 30 m depth. As heat is lost through the surface, the density of the water increases and vertical convective mixing (overturn) occurs with the temperature of the surface water layer gradually decreasing. This continues until the upper mixed layer cools to 3.98°C and then further cooling of the surface water causes its density to decrease and it stays near the top. The result is a rapid loss of heat from a thin surface layer, which soon freezes. For seawater of salinity = 35 psu of the same initial temperature distribution, surface cooling first results in a density increase and vertical mixing by convention currents occurs through a gradually increasing depth, but it is not until the whole column reaches −1.91°C that freezing commences. As a much greater volume of water has to be cooled through a greater temperature range than in the freshwater case, it takes longer for freezing to start in salt water than in freshwater. A simple calculation for a column of freshwater of 100 cm depth and 1 cm^2 cross-section initially at 10°C shows that it takes a heat loss of l63 J to freeze the top 1 cm layer, whereas for a similar column of seawater of S = 35 psu it takes a loss of 305 J to freeze the top 1 cm because the whole column has to be cooled to −1.91°C rather than just the top 1 cm to 0°C for the freshwater.

Note that as seawater of salinity <24.7 psu has a higher temperature of maximum density than its freezing point, it will behave in a manner similar to freshwater, although with a lower freezing point. For seawater of salinity >24.7 psu, (in high latitudes) the salinity generally increases with depth, and the stability of the water column usually limits the depth of convection currents to 30−50 m. Therefore ice starts to form at the surface before the deep water reaches the freezing point.

Generally, sea ice forms first in shallow water near the coast, particularly where the salinity is reduced by river runoff and where currents are minimal. When fully formed, this sea ice connected to the shore is known as "fast ice." The first process is the formation of needle-like crystals of pure ice, which impart an "oily" appearance to the sea surface (grease or frazil ice). The crystals increase in number and form a slush, which then thickens and breaks up into pancakes of approximately one meter across. With continued cooling, these pancakes grow in thickness and lateral extent, eventually forming a continuous sheet of floe or sheet ice.

Once ice has formed at the sea surface, when the air is colder than the water below, freezing continues at the lower surface of the ice and the rate of increase of ice thickness depends on the rate of heat loss upward through the ice (and any snow cover). This loss is directly proportional to the temperature difference between top and bottom surfaces and inversely proportional to the thickness of the ice and snow cover.

With very cold air, a sheet of sea ice of up to 10 cm in thickness can form in 24 hours, the rate of growth then decreasing with increasing ice thickness. Snow on the top surface insulates it and reduces the heat loss markedly, depending on its degree of compaction. For instance, 5 cm of new powder snow may have insulation equivalent to 250−350 cm of ice, while 5 cm of settled snow can be equivalent to only 60−100 cm of ice, and 5 cm of hard-packed snow can be equivalent to only 20−30 cm of ice.

As an example of the annual cycle of the development of an ice sheet at a location in the Canadian Arctic, ice was observed to start to form in September, was about 0.5 m thick in October, 1 m in December, 1.5 m in February, and reached its maximum thickness of 2 m in May—after which it started to melt.

3.9.2. Brine Rejection

In the initial stage of ice-crystal formation, salt is rejected and increases the density of the neighboring seawater, some of which then tends to sink and some of which is trapped among the ice crystals forming pockets called "brine cells."

The faster the freezing, the more brine is trapped. Sea ice in bulk is therefore not pure water-ice but has a salinity of as much as 15 psu for new ice (and less for old ice because gravity causes the brine cells to migrate downward in time). With continued freezing, more ice freezes out within the brine cells leaving the brine more saline, in a process called *brine rejection*. Some of the salts may even crystallize out.

Because of brine rejection, the salinity of first-year ice is generally 4−10 psu, for second-year ice (ice that has remained frozen beyond the first year) salinity decreases to 1−3 psu, and for multiyear ice salinity may be less than 1 psu. If sea ice is lifted above sea level, as happens when ice becomes thicker or rafting occurs, the brine gradually trickles down through it and eventually leaves almost salt-free, clear old ice. Such ice may be melted and used for drinking whereas melted new ice is not potable. Sea ice, therefore, is considered a material of variable composition and properties that depends very much on its history. (For more detail see Doronin & Kheisin, 1975.)

As a result of brine rejection, the salinity of the unfrozen waters beneath the forming sea ice increases. When this occurs in shallow regions, such as over continental shelves, the increase in salinity can be marked and can result in formation of very dense waters. This is the dominant mechanism for formation of the deep and bottom waters in the Antarctic (Chapter 13), and for formation of the densest part of the North Pacific Intermediate Water in the Pacific (Chapter 10). Brine rejection is a central process for modification of water masses in the Arctic as well (Chapter 12).

3.9.3. Density and Thermodynamics of Sea Ice

The density of pure water at $0°C$ is 999.9 kg/m^3 and that of pure ice is 916.8 kg/m^3. However the density of sea ice may be greater than this last figure (if brine is trapped among the ice crystals) or less (if the brine has escaped and gas bubbles are present.) Values from 924 to 857 kg/m^3 were recorded on the Norwegian Maud Expedition (Malmgren, 1927).

The amount of heat required to melt sea ice varies considerably with its salinity. For $S = 0$ psu (freshwater ice) it requires 19.3 kJ/kg from $-2°C$ and 21.4 kJ/kg from $-20°C$, while for sea ice of $S = 15$ psu, it requires only 11.2 kJ/kg from $-2°C$ but 20.0 kJ/kg from $-20°C$. The small difference of heat (2.1 kJ/kg) needed to raise the temperature of freshwater ice from $-20°C$ to $-2°C$ is because no melting takes place; that is, it is a true measure of the specific heat of pure ice. However, for sea ice of $S = 15$ psu, more heat (8.8 kJ/kg) is required to raise its temperature through the same range, because some ice near brine cells melts and thus requires latent heat of melting as well as heat to raise its temperature. Note also that less heat is needed to melt new ice ($S = 15$ psu) than old ice, which has a lower salinity.

3.9.4. Mechanical Properties of Sea Ice

Because of the spongy nature of first-year sea ice (crystals + brine cells) it has much less strength than freshwater ice. Also, as fast freezing results in more brine cells, the strength of ice formed this way is less than when freezing occurs slowly; in other words, sea ice formed in very cold weather is initially weaker than ice formed in less cold weather. As the temperature of ice decreases, its hardness and strength increase, and ice becomes stronger with age as the brine cells migrate downward. When ice forms in calm water, the crystals tend to line up in a pattern and such ice tends to fracture along cleavage planes more easily than ice formed in rough water where the crystals are more randomly arranged and cleavage planes are not formed.

The mechanical behavior of sea ice is complex when temperature changes. As the ice temperature decreases below its freezing point, the ice expands initially, reaches a maximum expansion, and then contracts. For instance, an ice floe of S = 4 psu will expand by 1 m per 1 km length between −2 and −3°C, reaches its maximum expansion at −10°C, and thereafter contracts slightly. Ice of S = 10 psu expands 4 m per 1 km length from −2 to −3°C, and reaches its maximum expansion at −18°C. The expansion on cooling can cause an ice sheet to buckle and "pressure ridges" to form, while contraction on further cooling after maximum expansion results in cracks, sometimes wide, in the ice sheet.

Pressure ridges can also develop as a result of wind stress on the surface driving ice sheets together. The ridges on top are accompanied by a thickening of the lower surface of the ice by four to five times the height of the surface ridges. Sea ice generally floats with about five-sixths of its thickness below the surface and one-sixth above, so relatively small surface ridges can be accompanied by deep ridges underneath — depths of 25 to 50 m below the sea surface have been recorded. Thickening of an ice sheet may also result from rafting when wind or tide forces one ice sheet on top of another or when two sheets, in compression, crumble and pile up ice at their contact. Old ridges, including piled up snow, are referred to as *hummocks*. As they are less saline than newer pressure ridges, they are stronger and more of an impediment to surface travels than the younger ridges.

3.9.5. Types of Sea Ice and its Motion

Sea ice can be categorized as fast ice (attached to the shore), pack ice (seasonal to multiyear ice with few gaps), and cap ice (thick, mostly multi-year ice), as described in Section 12.7.1. Several forces determine the motion of sea ice if it is not landfast:

(a) Wind stress at the top surface (the magnitude depending on the wind speed and the roughness of the ice surface as ridges increase the wind stress). Typical ice speeds are 1 to 2% of the wind speed.

(b) Frictional drag on the bottom of an ice sheet moving over still water tends to slow it down, while water currents (ocean and tidal) exert a force on the bottom of the ice in the direction of the current. Because current speeds generally decrease with increase in depth, the net force on deep ice and icebergs will be less than on thin ice, and pack ice will move past icebergs when there is significant wind stress.

(c) In the cases of (a) and (b), the effect of the Coriolis force (Section 7.2.3) is to divert the ice motion by 15−20 degrees to the right of the wind or current stress in the Northern Hemisphere (to the left in the Southern Hemisphere). (It was the observation of the relation between wind direction and ice movement by Nansen, and communicated by him to Ekman, that caused the latter to develop his well-known theory of wind-driven currents.) It is convenient to note that as surface friction causes the surface wind to blow at about 15 degrees to the left of the surface isobars, the direction of the latter is approximately that in which the ice is likely to drift (Northern Hemisphere).

(d) If the ice sheet is not continuous, collisions between individual floes may occur with a transfer of momentum (i.e., decrease of speed of the faster floe and increase of speed of the slower). Energy may go into ice deformation and building up ridges at impact. This is referred to as internal ice resistance and increases with ice concentration, that is, the proportion of area covered by ice. The effect of upper surface roughness (R on a scale of 1 to 9) and ice concentration (C on a scale of 1 to 9) on the speed of the ice V (expressed as a percentage of the wind speed) is given by: $V = R(1 - 0.08°C)$ (taken to only one decimal place), so that the speed of the ice increases with roughness but decreases with increased ice concentration. Note that for very close pack ice, stresses of wind or current are integrated over quite large areas and the local motion may not relate well to the local wind.

3.9.6. Polynyas and Leads

Regions of nearly open water within the ice pack are often found where one might expect to find ice. These open water areas are critical for air–sea heat exchange, since ice is a relatively good insulator. Small breaks between ice floes are called *leads*; these are created by motion of the ice pack and have random locations. Larger recurrent open water areas are called *polynyas*. There are two types of polynyas, depending on the mechanism maintaining the open water (Figure 3.12; see also Barber & Massom, 2007):

1. *Latent heat polynyas* are forced open by winds, often along coasts or ice shelf edges. New ice soon forms; latent heat from the forming sea ice is discharged to the atmosphere at a rate of as much as 200–500 W/m^2.

2. *Sensible heat polynyas* result from relatively warm water upwelling to the surface and melting the ice there. Another term often encountered is *flaw polynya*, which means that the polynya occurs at the boundary between fast ice and pack ice. Because most polynyas include a mixture of these forcings, nomenclature is tending toward being more specific about the forcing (mechanical–wind; convective–melting: Williams, Carmack, & Ingram, 2007).

Wind-forced polynyas are usually near coastlines or the edges of ice shelves or fast ice, where winds can be very strong, often forced by strong land–sea temperature differences (katabatic winds). The open water is often continually freezing in these polynyas since they are kept open through mechanical forcing. These wind-forced polynyas act as ice factories, producing larger amounts of new ice than regions where the ice is thicker and air–sea fluxes are minimized by the ice cover. If these polynyas occur over shallow continental shelves, the brine rejected in the ongoing sea ice formation process, together with temperatures at the

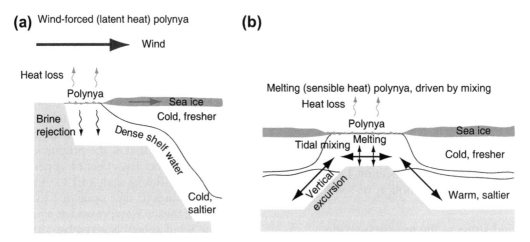

FIGURE 3.12 Schematics of polynya formation: (a) latent heat polynya kept open by winds and (b) sensible heat polynya kept open by tidal mixing with warmer subsurface waters (after Hannah et al., 2009).

freezing point, can produce especially dense shelf waters. This is one of the major mechanisms for creating very dense waters in the global ocean (Section 7.11), particularly around the coastlines of Antarctica (Chapter 13) and the Arctic (Chapter 12), as well as the densest (intermediate) water formed in the North Pacific in the Okhotsk Sea (Chapter 10).

Polynyas that are maintained by melting within the ice pack result from mixing of the cold, fresh surface layer with underlying warmer, saltier water. These polynyas might also produce sea ice along their periphery since the air—sea fluxes will be larger than through the ice cover, but the upward heat flux from the underlying warmer water means that they produce new ice much less efficiently than wind-forced polynyas. The vertical mixing can result from convection within the polynya, which can occur in deep water formation sites such as the Odden-Nordbukta in the Greenland Sea (Section 12.2.3). In shallow regions, the mixing process can be greatly enhanced by tides moving the waters over undersea banks (Figure 3.12b). A number of the well-known polynyas in the Canadian Archipelago in the Arctic Ocean are tidally maintained (Figure 12.23 from Hannah, Dupont, & Dunphy,

2009), as is a recurrent polynya over Kashevarov Bank in the Okhotsk Sea (Figure 10.29).

3.9.7. Ice Break-up

Ice *break-up* is caused by wave action, tidal currents, and melting. Melting of ice occurs when it gains enough heat by absorption of solar radiation and by conduction from the air and from nearby seawater to raise its temperature above the melting point. The absorption of radiation depends on the albedo of the surface (proportion of radiation reflected), which varies considerably; for example, the albedo for seawater is from 0.05 to 0.10 (it is a very good absorber of radiation), for snow-free sea ice it is from 0.3 to 0.4, while for fresh snow it is 0.8 to 0.9. Dark materials, like dirt and dust, have a low albedo of 0.1 to 0.25 and absorb radiation well. Such material on ice can form a center for the absorption of radiation and consequent melting of ice around it, so puddles can form. These can absorb heat because of the low albedo of water and may even melt right through an ice sheet. When any open water forms, it absorbs heat and causes rapid melting of ice floating in it.

Typical Distributions of Water Characteristics

4.1. INTRODUCTION

In this chapter, we describe the typical distributions of water properties such as temperature, salinity, oxygen, and nutrients. The properties were introduced in Chapter 3. Here we highlight distributions that are common, for instance, to the Atlantic, Pacific, and Indian Oceans, or to all subtropical regions, or to all three equatorial regions, and so on. The overview provides an essential framework for the heat and freshwater budgets of Chapter 5 and for the detailed descriptions of properties and circulation in each ocean basin presented in later chapters. Summaries of some of the large-scale water masses are included in Chapter 14.

Several central concepts are useful for studying large-scale water properties. First, most water properties are initially set at the sea surface and are then modified within the ocean through a process called *ventilation*. Ventilation is the connection between the surface and the ocean interior (similar to breathing). Second, the ocean is vertically stratified in density, and flow within the ocean interior is primarily along isentropic (isopycnal) surfaces rather than across them. That is, flow within the ocean interior is nearly adiabatic (without internal sources of heat and freshwater). Third, as a result, water properties are helpful for identifying flow paths from the surface into the interior, and for identifying forcing and mixing processes and locations. This is related to the usefulness of the concept of *water masses*, defined in the next section.

Most water characteristics have large and typical variations in the vertical direction, which encompasses an average of 5 km in the deep ocean, whereas variations of similar magnitude in the horizontal occur over vastly greater distances. For instance, near the equator, the temperature of the water may drop from 25°C at the surface to 5°C at a depth of 1 km, but it may be necessary to go 5000 km north or south from the equator to reach a latitude where the surface temperature has fallen to 5°C. The average vertical temperature gradient (change of temperature per unit distance) in this case is about 5000 times the horizontal one. However, the more gradual horizontal variations are important: the horizontal density differences are associated with horizontal pressure differences that drive the horizontal circulation, which is much stronger than the vertical circulation. To illustrate the three-dimensional distributions of water properties and velocities, we use a number of one- and two-dimensional representations, such as profiles, vertical sections, and horizontal maps.

Much of the geographic variation in properties in the oceans and atmosphere occurs in the north-south (*meridional*) direction. Properties are often much more uniform in the east-west (*zonal*) direction. A principal exception to the latter is the important zonal variation near boundaries, especially on the west sides of ocean basins. In addition to the major ocean basins, we also refer to general regions that are mainly distinguished by latitude ranges.

The *equatorial* region refers to the zone within several degrees of the equator, while *tropical* refers to zones within the tropics (23°N or °S of the equator). In the equatorial region, the effect of the earth's rotation on currents is minimal, leading to very distinctive currents and water property distributions compared with other regions. Within the tropics, there is net heating at the sea surface. The distinction between equatorial and tropical is often significant, but when the two are to be lumped together, they are referred to as the *low latitudes*. In contrast, the regions near the poles, north and south, are called the high latitudes. *Subtropical* refers to mid-latitude zones poleward of the tropics, characterized by atmospheric high pressure centers. *Polar* is used for the Arctic and Antarctic regions, where there is net cooling and usually sea ice formation. *Subpolar* refers to the region between the strictly polar conditions and those of the temperate mid-latitudes. The most marked seasonal changes take place in the temperate zones (approximately 30−60°N or °S).

Throughout this chapter and in subsequent chapters we refer to the concept of a *water mass*, which is a body of water that has had its properties set by a single identifiable process. This process imprints properties that identify the water mass as it is advected and mixed through the ocean. Most water masses are formed at the sea surface where their identifying characteristics are directly related to surface forcing, but some water masses acquire their characteristics (e.g., an oxygen minimum) through subsurface processes that might be biogeochemical as well as physical. Some water

masses are nearly global in extent while other water masses are confined to a region, such as a gyre in a specific ocean basin. Water masses have been given names that are usually capitalized. Some water masses have several names, simply because of the history of their study. *Water type* and *source water type* are useful related concepts; a water type is a point in property space, usually defined by temperature and salinity, and a source water type is the water type at the source of the water mass (e.g., Tomczak & Godfrey, 2003).

Each water mass is introduced in terms of (1) its identifying characteristic(s) and (2) the ocean process that creates that specific characteristic. Descriptive physical oceanographers often first identify an extremum or interesting central characteristic. They then seek to find the process that created that characteristic. Once the process is identified, additional information about the process is used to refine that water mass's definition, for example, the full density range might be assigned to the water mass. Information about the process and water mass distribution assists in studying the circulation.

The Mediterranean Water (MW) (Chapter 9) is an example of a water mass with a simple identifying characteristic. MW is a salinity maximum layer in the North Atlantic at mid-depth (1000−2000 m) and a lateral salinity maximum on any quasi-horizontal surface cutting through the layer (e.g., Figure 6.4). Its source is the saline outflow of water from the Mediterranean through the Strait of Gibraltar. Its high salinity results from excess evaporation and internal dense water formation within the Mediterranean Sea (see the textbook Web site, which contains supplementary materials, http://booksite.academicpress.com/DPO/ to view Section S8.10.2; "S" denotes supplemental material). The MW density range within the North Atlantic is a function of both its high density at the Strait of Gibraltar and also intense mixing with ambient (stratified) North

Atlantic water as it plunges down the continental slope after it exits the Strait of Gibraltar.

Subtropical Mode Water (STMW) is another example of a water mass with a simple vertical extremum; in this case its thickness (vertical homogeneity) is compared with waters above and below it. A type of STMW is found in each ocean's subtropical gyre (Sections 9.8.2, 10.9.1, and 11.8.1). STMW originates in a thick surface winter mixed layer that is then advected down along isopycnals into the ocean interior. STMW retains its signature of relative thickness, just as the MW retains its signature of high salinity. Slow mixing within the ocean interior eventually erodes these extrema, but they persist far enough from their sources to be useful tracers of flow.

Many other major world water masses are introduced in this chapter. Detailed descriptions of them and of their formation processes are provided in the ocean basin chapters (9 through 13), with a final summary in Chapter 14.

Taking into account the whole set of ocean properties and information about water masses, it is useful to think of the vertical structure in terms of four layers: upper, intermediate, deep, and bottom. The upper layer contains a surface mixed layer, thermocline and/or halocline, pycnocline, and other structures embedded in these (see descriptions with respect to temperature and density in Sections 4.2 and 4.4). The upper layer is in contact with the atmosphere, either directly or through broad flow (relatively directly) into the upper ocean through the subduction process described in Sections 4.4.1 and 7.8.5. The intermediate, deep, and bottom layers are all below the pycnocline, or at most, embedded within the bottom of it. These layers are identified by water masses that indicate surface origins, with respect to location and formation processes, and relative age.

Before describing some typical distributions of each of the water properties, the following information on ocean water temperatures and salinities is given for orientation (see Figure 3.1):

1. 75% of the total volume of the ocean water has a temperature between 0 and 6°C and salinity between 34 and 35 psu,
2. 50% of the total volume of the oceans has properties between 1.3 and 3.8°C and between 34.6 and 34.7 psu,
3. The mean temperature of the world ocean is 3.5°C and the mean salinity is 34.6 psu.

4.2. TEMPERATURE DISTRIBUTION OF THE OCEANS

The ocean and atmosphere interact at the sea surface. Surface forcing from the atmosphere and sun sets the overall pattern of sea surface temperature (SST) (Figure 4.1). High SST in the tropics is due to net heating, and low SST at high latitudes is due to net cooling. Beyond this simple meridional variation, the more complex features of SST result from ocean circulation and spatial variations in atmospheric forcing. The ocean's surface, which could include sea ice, provides the forcing at the bottom of the atmosphere through various kinds of heat forcing and as a source of water vapor.

SST ranges from slightly more than 29°C in the warmest regions of the tropics, to freezing temperature (about −1.8°C; Figure 3.1) in ice-forming regions, with seasonal variations especially apparent at middle to high latitudes.

Below the sea surface, we refer only to potential temperature so that the pressure effect on temperature is removed (Section 3.3 and Figure 3.3). The vertical potential temperature structure can usually be divided into three major zones (Figure 4.2): (1) the mixed layer, (2) the thermocline, and (3) the abyssal layer. This structure is typical of low and mid-latitudes with high SST. Relative to the four-layer structure introduced in Section 4.1, the first two zones are within the upper layer and

(a)

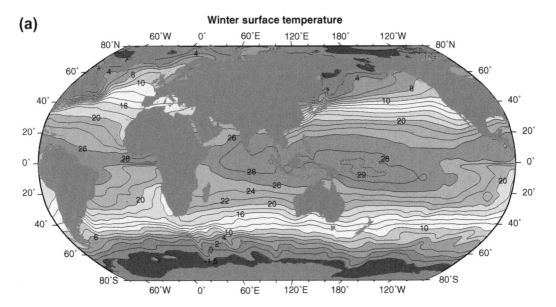

(b)

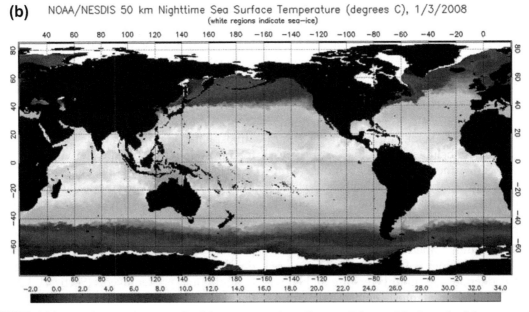

FIGURE 4.1   (a) Surface temperature (°C) of the oceans in winter (January, February, March north of the equator; July, August, September south of the equator) based on averaged (climatological) data from Levitus and Boyer (1994). (b) Satellite infrared sea surface temperature (°C; nighttime only), averaged to 50 km and 1 week, for January 3, 2008. White is sea ice. (See Figure S4.1 from the online supplementary material for this image and an image from July 3, 2008, both in color). *Source: From NOAA NESDIS (2009).*

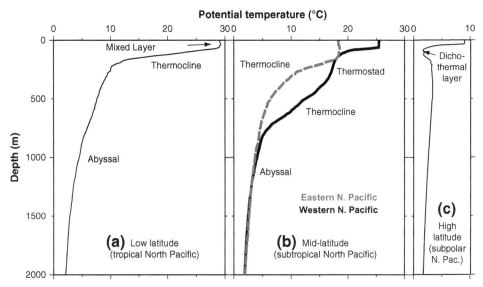

FIGURE 4.2 Typical potential temperature (°C)/depth (m) profiles for the open ocean in (a) the tropical western North Pacific (5°N), (b) the western and eastern subtropical North Pacific (24°N), and (c) the western subpolar North Pacific (47°N). Corresponding salinity profiles are shown in Figure 4.16.

the third temperature zone contains the intermediate, deep, and bottom layers.

In high latitudes where SST is low, this structure differs, and can have a mixed layer, a vertical temperature minimum and underlying maximum near the sea surface, and then the thermocline and abyssal layer.

The *mixed layer* (Section 4.2.2) is a surface layer of relatively well-mixed properties. In summer in low latitudes, it can be very thin or non-existent. In winter at middle to high latitudes, it can be hundreds of meters thick, and in isolated deep convection regions, the mixed layer can be up to 2000 m thick. Mixed layers are mixed by both wind and surface buoyancy forcing (air-sea fluxes). The *thermocline* (Sections 4.2.3 and 4.2.4) is a vertical zone of rapid temperature decrease with a depth of roughly 1000 m. In the abyssal layer, between the thermocline and ocean bottom, potential temperature decreases slowly. At high latitudes, a near-surface temperature minimum (*dichothermal* layer) is often found, a holdover from a cold

winter mixed layer that is "capped" with warmer waters in other seasons (Figure 4.2c); the underlying temperature maximum (*mesothermal* layer) results from advection of waters from somewhat warmer locations. This temperature structure is stable because there is strong salinity stratification, with fresher water in the surface layer.

Typical temperatures at subtropical latitudes are 20°C at the surface, 8°C at 500 m, 5°C at 1000 m, and 1−2°C at 4000 m. All of these values and the actual shape of the temperature profile are a function of latitude, as shown by the three different profiles in Figure 4.2.

There are some notable additions to this basic three-layered structure. In all regions, spring and summer warming produces a thin warm layer overlying the winter's mixed layer. In the western subtropical regions as well as other regions, there are often two thermoclines with a less stratified (more isothermal) layer (*thermostad*) between them, all within the upper 1000 m (Figure 4.2b). In some regions another mixed layer is found at the very bottom ("bottom boundary layer") and can be up to 100 m thick.

In many parts of the ocean, density is a strong function of temperature (Chapter 3), and has the same layered structure as temperature; that is, an upper layer, a *pycnocline* with rapidly increasing density, and an abyssal zone. Salinity usually has a more complicated vertical structure (Section 4.3). In regions of high precipitation and/or runoff (such as subpolar and high latitude regions and parts of the tropics), salinity may be more important than temperature in setting the vertical density structure, especially in the upper layer, since the water column must be vertically stable on average. A typical vertical salinity profile in these regions includes a relatively fresh surface layer with a *halocline* separating the surface layer from the higher salinity water below. The higher underlying salinity is an indication of a sea-surface source of water in a less rainy area. On the other hand, in the subtropics where the sea surface salinity is dominated by evaporation, surface water is usually more saline than the underlying water. Here temperature clearly dominates the vertical stability.

This three-layered structure is simpler than our simplest description of overall water mass structure, for which at least four layers are usually required (Section 4.1). The abyssal layer, in terms of temperature, usually includes at least two or possibly three separate water mass layers: intermediate, deep, and bottom waters. However, potential temperature is relatively low in all of these water mass-based layers, declining toward the bottom, and is not a useful indicator of these water mass layers.

4.2.1. Surface Temperature

The temperature distribution at the surface of the open ocean is approximately zonal, with the curves of constant temperature (isotherms)

running roughly east-west (Figure 4.1). Near the coast where the currents are diverted by the boundaries, the isotherms may swing more nearly north and south. Also, along the eastern boundaries of the oceans, surface temperatures are often lower due to upwelling of subsurface cool water, for example, along the west coast of North America in summer, causing the isotherms to trend equatorward. Upwelling also causes lower surface temperatures in the eastern equatorial Pacific and Atlantic.

The open ocean SST, averaged over all longitudes and displayed as a function of latitude (Figure 4.3), decreases from as high as 28°C just north of the equator to nearly −1.8°C near sea ice at high latitudes. This distribution corresponds closely with the input of short-wave radiation (mainly from the sun), which is highest in the tropics and lowest at high latitudes (Section 5.4.3). The corresponding mean zonal surface salinity and density are also shown. Salinity and density are discussed in Sections 4.3 and 4.4. Density is dominated by temperature. Salinity has subtropical maxima in both the Northern and Southern Hemispheres and a minimum just north of the equator.

Because many satellites observe SST and SST-related quantities, many different SST products are available, providing daily and longer term average maps with higher spatial and temporal resolution than the climatology based on in situ data shown in Figure 4.1a. Global SST based on infrared imagery for one week in January (boreal winter, austral summer) is shown in Figure 4.1b. (The equivalent image for July is included in the online supplementary materials as Figure S4.1.) The structures of ocean currents, fronts, upwelling regions, eddies, and meanders are more apparent in these nearly synoptic SST images.

Non-zonal features of global SSTs that are most apparent and important to note in Figure 4.1 include the *warm pool* and the *cold tongue*. The warm pool is the warmest SST region, located in the western tropical Pacific, through

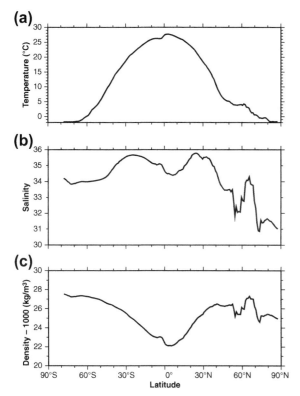

FIGURE 4.3 Variation with latitude of surface (a) temperature, (b) salinity, and (c) density averaged for all oceans for winter. North of the equator: January, February, and March. South of the equator: July, August, and September. Based on averaged (climatological) data from Levitus and Boyer (1994) and Levitus et al. (1994b).

the Indonesian passages, and into the tropical Indian Ocean. The cold tongue is the narrow tongue of colder water along the equator in both the eastern Pacific and Atlantic. This forms due to upwelling of thermocline water along the equator. Because the thermocline is shallower in the eastern Pacific and Atlantic than in the west, upwelling brings up colder water in the east.

In each ocean, warm regions are centered in the west, off the equator. Cooler waters cycle equatorward in the central and eastern parts of each ocean. These SST patterns reflect the anticyclonic circulation of the subtropical gyres (clockwise in the Northern Hemisphere, counterclockwise in the Southern Hemisphere), which advects warm water away from the tropics and cooler water toward the equator. There are also regions of warmer water in the eastern tropical North Pacific and North Atlantic. These are found east of the subtropical circulation and north of the cold tongue; high temperatures are not suppressed by either the anticyclonic circulation or equatorial upwelling.

In the subpolar North Pacific and North Atlantic, there is, again, evidence of the circulation in the SST pattern. Here the gyres are cyclonic (counterclockwise in the Northern Hemisphere). Warmer waters are advected northward in the eastern parts of these circulations (along the coast of British Columbia and along northern Europe). Warmer water extends far to the north in the Atlantic toward the Arctic, along the Norwegian coast. Cold waters are found in the western parts of these circulations, along the Kamchatka/Kuril region in the Pacific and Labrador/Newfoundland region in the Atlantic.

In the Southern Ocean, SST is not exactly zonal. This reflects excursions in the Antarctic Circumpolar Current (ACC), which is also not zonal. Colder waters are farther north in the Atlantic and Indian Oceans and pushed southward in the Pacific (Section 13.4).

In the satellite SST maps (Figures 4.1b and S4.1 from the online supplementary material), eddy-scale (100–500 km) features are apparent even with global maps, particularly where the color scaling provides large contrasts. Especially visible are the large wavelike structures in the equatorial regions; length scales of tropical waves are longer than at higher latitudes so they are better resolved in this map. The waves around the Pacific's equatorial cold tongue are the Tropical Instability Waves (TIWs), with timescales of about a month (Section 10.7.6).

4.2.2. Upper Layer Temperature and Mixed Layer

Within the ocean's near-surface layer, properties are sometimes very well mixed vertically, particularly at the end of the night (diurnal cycle) and in the cooling seasons (seasonal cycle). This is called the *mixed layer*. This layer is mixed by the wind and by buoyancy loss due to net cooling or evaporation at the sea surface. It is unmixed by warming and precipitation at the sea surface and by circulations within the mixed layer that move adjacent mixed waters of different properties over each other. Processes that create and destroy the mixed layer are described in much greater detail in Section 7.4.1. Here we focus on the observed structure and distribution of mixed layers.

As a rule of thumb, wind-stirred mixed layers do not extend much deeper than 100 or 150 m and can reach this depth only at the end of winter. On the other hand, infrequent vigorous cooling or evaporation at the sea surface can cause the mixed layer to deepen locally to several hundred meters, or even briefly in late winter to more than 1000 m in isolated deep convection locations. Mixed layers in summer may be as thin as 1 or 2 m, overlying a set of remnant thin mixed layers from previous days with storms, and thicker remnant mixed layers from winter. Because the mixed layer is the surface layer that connects the ocean and atmosphere, and because sea-surface temperature is the main way the ocean forces the atmosphere, observations of the mixed layer and understanding how it develops seasonally and on climate timescales is important for modeling and understanding climate.

A given vertical profile will not usually exhibit a thick, completely mixed layer of uniform temperature, salinity, and density. Most often, there will be small steps, nearly discontinuities, in the profiles due to daily restratification and remixing with layers sliding in from nearby. For a careful study of the mixed layer, the investigator assigns the mixed layer depth based on examination of every vertical profile. However, for general use (e.g., with the growing profiling float data set, or for use in upper ocean property mapping for fisheries, climate prediction, or navigational use), it is not feasible to examine each profile, and it is important to have consistent criteria for assigning the mixed layer depth. Functional definitions of mixed layer depth have been developed, mostly based on finding a set temperature or density difference between the surface observation and deeper observations; this is the so-called "threshold method." In tropical and mid-latitudes, temperature-based definitions are adequate, but at higher latitudes, it is common to find a subsurface temperature maximum lying underneath a low salinity surface layer. Currently, the most commonly used criterion is a density difference of $\sigma_\theta = 0.03 \text{ kg/m}^3$ or temperature difference of 0.2°C, as used in the mixed layer maps shown in Figure 4.4a,b (deBoyer Montégut et al., 2004). Other treatments have employed larger thresholds (e.g., 0.8°C in Kara, Rochford, & Hurlburt, 2003) or more detailed criteria that fit the observed vertical profiles rather than relying on a threshold (Holte & Talley, 2009). A global map of the maximum mixed layer depth, using the latter method, is shown in Figure 4.4c (Holte, Gilson, Talley, & Roemmich, 2011).

In all regions, winter mixed layers are much thicker than summer mixed layers. The main features of the global winter mixed layer maps are the thick mixed layers in the northern North Atlantic and in a nearly zonal band in the Southern Ocean. These regions correspond to maxima in anthropogenic carbon uptake (Sabine et al., 2004), so they have practical implications for global climate. These thick winter mixed layers are the main source of Mode Waters, which are identified as relatively thick layers in the upper ocean (Section 4.2.3).

Mixed layer development is affected by the amount of turbulence in the surface layer. This turbulence is generated by breaking surface and internal waves generated by the wind, decreasing with increasing depth. Mixed layer development can also be affected by Langmuir cells, which are transient helical circulations (in the vertical plane) aligned parallel to the wind (Section 7.5.2). These create the "wind rows" sometimes seen at the sea surface under the wind, where the water is pushed together, or converges, in the Langmuir cells, which reach to about 50 m depth and 50 m width, and can create turbulence that affects mixing in the mixed layer.

Another dynamical phenomenon present in the near-surface layer is the Ekman response to wind forcing, which forces flow in the ocean's surface layer off to the right of the wind in the Northern Hemisphere (and to the left in the Southern Hemisphere), because of the Coriolis force (Section 7.5.3). Turbulence in the surface layer acts like friction. In the Northern Hemisphere, each thin layer within the surface layer pushes the one below it a little more to the right, and with a little smaller

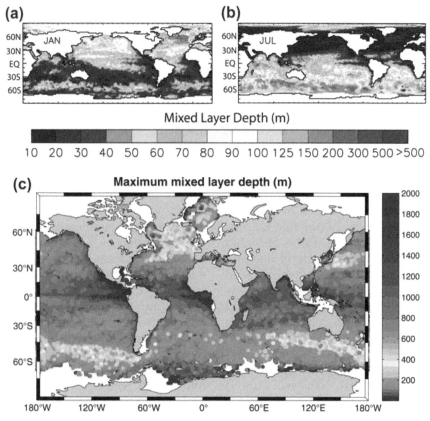

FIGURE 4.4 Mixed layer depth in (a) January and (b) July, based on a temperature difference of 0.2°C from the near-surface temperature. *Source: From deBoyer Montégut et al. (2004).* (c) Averaged maximum mixed layer depth, using the 5 deepest mixed layers in 1° × 1° bins from the Argo profiling float data set (2000–2009) and fitting the mixed layer structure as in Holte and Talley (2009). This figure can also be found in the color insert.

velocity than the layer above. This creates an "Ekman spiral" of decreasing velocities with increasing depth. The whole spiral occurs within the top 50 m of the ocean. If all of the velocities are added together to calculate the total transport in the Ekman layer, the net effect is that this Ekman transport moves at exactly right angles to the wind direction — to the right in the Northern Hemisphere and to the left in the Southern Hemisphere. Ekman velocities are small and do not generate turbulence. Thus they have no direct effect on mixed layer development and are affected by the upper layer turbulence but not by the mixed layer stratification. The Ekman response is crucial, however, for conveying the effect of the wind to the ocean, for development of the large-scale and long timescale ocean circulation, as also described in Chapter 7.

4.2.3. Thermocline, Halocline, and Pycnocline

Below the surface layer, which can be well mixed or can include messy remnants of local mixing and unmixing, temperature begins to decrease rapidly with depth. This rapid decrease ceases after several hundred meters, with only small vertical changes in temperature in the deep or abyssal layer that extends on down to the bottom. The region of higher vertical temperature gradient (rate of decrease of temperature with increasing depth) is called the *thermocline*. The thermocline is usually a *pycnocline* (high vertical density gradient). It is often hard to precisely define the depth limits, particularly the lower limit, of the thermocline. However, in low and middle latitudes, a thermocline is always present at depths between 200 and 1000 m. This is referred to as the main or permanent thermocline. In polar and subpolar waters, where the surface waters may be colder than the deep waters, there is often no permanent thermocline, but there is usually a permanent *halocline*

(high vertical salinity gradient) and associated permanent pycnocline.

The continued existence of the thermocline and pycnocline requires explanation. There are two complementary concepts, one based on vertical processes only, and the other based on horizontal circulation of the waters that form the thermocline away from where they outcrop as mixed layers in winter. Both concepts are important and work together.

The vertical processes that affect the thermocline are downward transfer of heat from the sea surface and either upwelling or downwelling (these depend on the location in the ocean and on what creates the vertical motion). One might expect that as the upper waters are warmest, heat would be transferred downward by diffusion despite the inhibiting effect of the stability in the pycnocline/thermocline, and that the temperature difference between the upper and lower layers would eventually disappear. However, the deeper cold waters are fed continuously from the sea surface at higher latitudes (deep and bottom water formation regions, mainly in the northernmost North Atlantic and Greenland Sea and in various regions around Antarctica). These deep inflows maintain the temperature difference between the warm surface waters and cold deep waters. The deep waters upwell and warm up through downward diffusion of heat. If upwelling from the bottommost layers to near the surface occurs through the whole ocean, the upward speed would be 0.5–3.0 cm/day. Unfortunately these speeds are too small to accurately measure with current instruments, so we are unable to test the hypothesis directly. The result of the downward vertical diffusion of heat balanced by this persistent upwelling of the deepest cold waters results in an exponential vertical profile of temperature (Munk, 1966), which approximates the shape of the permanent thermocline.

This simplified vertical model of the thermocline is depicted in Figure 4.5, which shows

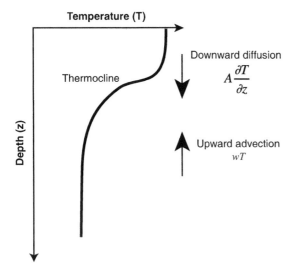

Temperature (T)

Thermocline

Downward diffusion

$$A\frac{\partial T}{\partial z}$$

Upward advection

wT

Depth (z)

FIGURE 4.5 Vertical processes that can maintain the thermocline in a simplified one-dimensional model.

a typical vertical temperature profile in the upper ocean containing the thermocline. The result of downward diffusion of heat is labeled as $A\frac{\partial T}{\partial z}$ and the result of upward vertical advection of colder, deeper water is labeled "wT". (Equation (7.46) shows these two terms are the vertical integrals of the vertical diffusion and vertical advection terms, assuming constant eddy diffusivity A and constant vertical velocity w. In this simplest of thermocline models, it is assumed that downward diffusion of heat is entirely balanced by upward advection.) If we assume that the difference between these two terms is a constant, we have an equation with an exponential solution for temperature T, which in many cases approximates the shape of the thermocline. We can use similar arguments relative to the vertical distribution of tracers like dissolved oxygen except that such tracers can have both sources and sinks within the water column, ultimately resulting in subsurface maxima or minima.

A second, more horizontal, adiabatic and complementary process for maintaining the thermocline/pycnocline was suggested by Iselin (1939) and further developed by Luyten,

Pedlosky, and Stommel (1983; Section 7.8.5). Iselin observed that the surface temperature-salinity relation along a long north-south swath in the North Atlantic strongly resembled the T-S relation in the vertical (Figure 4.6). He hypothesized that the waters in the subtropical thermocline therefore originate as surface waters farther to the north. As they move south, the colder surface waters *subduct* beneath the warmer surface waters to the south (using the term from Luyten et al., borrowed from plate tectonics). Subduction of many layers builds up the temperature, salinity, and density structure of the main pycnocline (thermocline) in the subtropical gyre. This process is adiabatic, not requiring any mixing or upwelling across isopycnals. Such one-dimensional diapycnal processes would then modify the thermocline structure, smoothing it out.

Double diffusion (Section 7.4.3) is another vertical mixing process that might affect the thermocline (pycnocline). This process might modify the relation between temperature and salinity within the pycnocline, smoothing the profile that results from adiabatic subduction (Schmitt, 1981).

The main thermoclines/pycnoclines of the world's subtropical gyres are permanent features. The temperature-salinity relation in the thermocline of each subtropical gyre is shown in Figure 4.7. The main thermoclines are identifiable in temperature/salinity relations, and they have a common formation history that is some combination of subduction and vertical upwelling/diffusion. Therefore, the waters in the thermocline can be identified as a water mass. This is the first water mass that we introduce systematically, rather than as an example. The thermocline water mass is *Central Water*. Central Water differs from typical water masses because it has a large range of temperature, salinity and density.

So far, we have referred to the "main," or permanent, thermocline. There are also permanent, double thermoclines in some large but

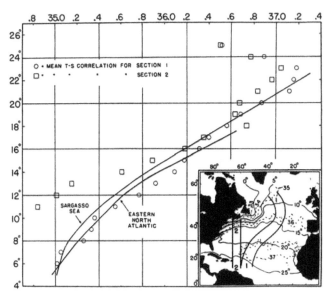

FIGURE 4.6 Temperature-salinity along surface swaths in the North Atlantic (dots and squares), and in the vertical (solid curves) at stations in the western North Atlantic (Sargasso Sea) and eastern North Atlantic. *Source: From Iselin (1939).*

geographically restricted regions. For instance, two thermoclines are found in the Sargasso Sea just south of the Gulf Stream. A layer of lower vertical stratification separates the two thermoclines. The layer of lower stratification is called a *thermostad* (or *pycnostad* for the equivalent density layer).

The thermostad/pycnostad is often given the water mass name, *Mode Water*. This is the second water mass that we introduce. Mode Water is

FIGURE 4.7 Potential temperature-salinity relation in the thermocline of each subtropical gyre. These are the Central Waters. R is the best fit of a parameter associated with double diffusive mixing (Section 7.4.3). *Source: From Schmitt (1981).*

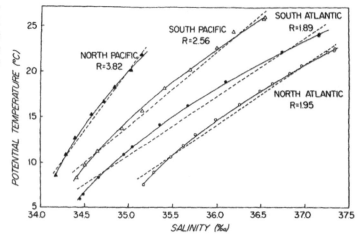

considered a water mass because it is identified by a particular characteristic (a vertical extremum in layer thickness), and because it has a specific formation process (subduction of thick mixed layers). The name "Mode Water" was introduced by Masuzawa (1969). Volumetrically there is more water in a particular temperature/salinity range than in the thermoclines above and below it, so Mode Water appears as a mode in the distribution of volume in temperature/salinity space.

In the region where the Mode Water outcrops as a thick mixed layer, the overlying thermocline is actually a seasonal thermocline that disappears in late winter. After Mode Water subducts, its thermostad is embedded in the permanent thermocline, creating a double thermocline.

4.2.4. Temporal Variations of Temperature in the Upper Layer and Thermocline

The temperature in the upper zone and into the thermocline varies seasonally, particularly in mid-latitudes. In winter the surface temperature is low, waves are large, and the mixed layer is deep and may extend to the main thermocline. In summer the surface temperature rises, the water becomes more stable, and a seasonal thermocline often develops in the upper layer.

The growth and decay of the seasonal thermocline is illustrated in Figure 4.8a using monthly mean temperature profiles from March 1956 to January 1957 taken at Ocean Weather Station P ("Papa") in the northeastern (subpolar) North Pacific. From March to August, the temperature gradually increases due to absorption of solar energy. A mixed layer from the surface down to 30 m is evident all the time. After August there is a net loss of heat and continued wind mixing; these erode away the seasonal thermocline until the isothermal condition of March is approached again. Note that March does not have the maximum heat

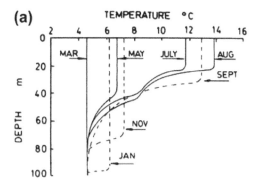

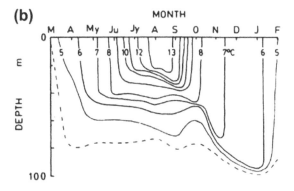

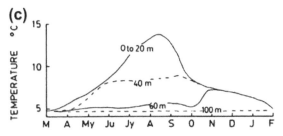

FIGURE 4.8 Growth and decay of the seasonal thermocline at 50°N, 145°W in the eastern North Pacific as (a) vertical temperature profiles, (b) time series of isothermal contours, and (c) a time series of temperatures at depths shown.

loss; rather, it is the last month of cooling before seasonal heating begins. Therefore total heat content is lowest in March. In tropical and subtropical locations, the summer mixed layer may be even thinner.

These same data may be presented in alternative forms; for instance, as a time series showing the depths of the isotherms during the year (Figure 4.8b). (The original data include the alternate months, which were omitted from Figure 4.8a to avoid crowding.) In Figure 4.8c the temperatures are plotted at selected depths. The different forms in which the thermocline appears in these three presentations should be noted. In Figure 4.8a, the permanent thermocline appears as a maximum gradient region in the temperature/depth profiles. In Figure 4.8b, the thermocline appears as a crowding of the isotherms, which rises from about 50 m in May to 30 m in August and then descends to 100 m in January. In Figure 4.8c, the thermocline appears as a wide separation of the 20- and 60-m isobaths between May and October, and between the 60- and 100-m isobaths after that as the thermocline descends.

At the highest latitudes, the surface temperatures are much lower than at lower latitudes, while the deep-water temperatures are little different. As a consequence, the main

thermocline might not be present at high latitudes, and only a seasonal thermocline might occur. In high northern latitudes, there is often a layer of cold water at 50–100 m (Figure 4.2c), with temperatures as low as −1.6°C, sandwiched between the warmer surface and deeper layers. As described at the beginning of Section 4.2, this cold layer is referred to as the dichothermal layer. The warmer surface water is often just seasonal, and the thermocline overlying the dichothermal layer is therefore seasonal.

Figure 4.9 shows the annual range of surface temperature over the globe. Annual variations at the surface rise from 1–2°C at the equator to between 5 and 10°C at 40° latitude in the open ocean, then decrease toward the polar regions (due to the heat required in the melting or freezing processes where sea ice occurs). Near the coast, larger annual variations (10–20°C) occur in sheltered areas and in the western subtropical regions of the Northern

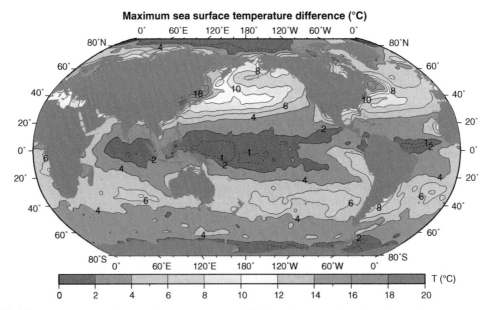

FIGURE 4.9 Annual range of sea surface temperature (°C), based on monthly climatological temperatures from the World Ocean Atlas (WOA05) (NODC, 2005a, 2009).

Hemisphere, where the Kuroshio and Gulf Stream are located and where surface heat loss is highest (Section 5.5, Figure 5.12). These annual variations in temperature decrease with depth and are rarely perceptible below 100–300 m. The maximum temperature at the surface occurs at the end of the warming season, in August/September in the Northern Hemisphere, and the minimum at the end of the cooling season, during February/March. Below the surface, the times of occurrence of the maxima and minima are delayed by as much as two months relative to the surface.

Diurnal variations of SST had been thought to be small (<0.4°C) prior to satellite observations. Such measurements, verified by in situ observations from a moored buoy in the Sargasso Sea over a period of two years (Stramma et al., 1986), have shown that diurnal variations to 1°C are common with occasionally higher values, up to 3–4°C. The larger diurnal variations of 1°C or more are observed in conditions of high insolation (solar radiation) and low wind speed, and are generally limited to the upper few meters of water. Similar diurnal variation has been observed elsewhere in the North Atlantic and in the Indian Ocean. In sheltered and shallow waters along the coast, values of 2–3°C are common.

4.2.5. Deep-Water Temperature and Potential Temperature

Below the thermocline, the temperature slowly decreases with increasing depth. (This vertical temperature change is much smaller than through the thermocline.) In the deepest waters, temperature can rise toward the bottom, almost entirely because the high pressure that compresses the water and raises its temperature adiabatically (Section 3.3.3, Figure 3.3). To interpret variations in temperature, even in shallow waters over a continental shelf as well as from the surface to thousands of meters, potential temperature (θ) should always be used. Potential temperature reflects the original temperature of the water when it was near the sea surface.

An example of this difference between in situ and potential temperature is shown in Table 4.1 and in Figure 4.10 using data collected in 1976 by the R/V T. Washington from the Mariana Trench (the deepest trench in the world ocean). While temperature (T) reaches a minimum at about 4500 m and thereafter increases toward the bottom, potential temperature is almost uniform. (Salinity also is almost uniform between 4500 m and the deepest observation as are potential densities relative to any reference pressure.) Uniform properties from

TABLE 4.1 Comparison of in situ and Potential Temperatures and Potential Densities Relative to the Sea Surface (σ_θ), 4000 dbar (σ_4) and 10,000 dbar (σ_{10}) in the Mariana Trench in the Western North Pacific

Depth (m)	Salinity (psu)	Temperature (°C)	θ (°C)	σ_θ (kg m^{-3})	σ_4 (kg m^{-3})	σ_{10} (kg m^{-3})
1487	34.597	2.800	2.695	27.591	45.514	69.495
2590	34.660	1.730	1.544	27.734	45.777	69.903
3488	34.680	1.500	1.230	27.773	45.849	70.015
4685	34.697	1.431	1.028	27.800	45.898	70.090
5585	34.699	1.526	1.004	27.803	45.904	70.099
6484	34.599	1.658	1.005	27.803	45.904	70.099
9940	34.700	2.266	1.007	27.804	45.904	70.099

Data from R/V T. Washington, 1976.

FIGURE 4.10 Mariana Trench: (a) in situ temperature, T, and potential temperature, θ (°C); (b) salinity (psu); (c) potential density σ_θ (kg m^{-3}) relative to the sea surface; and (d) potential density σ_{10} (kg m^{-3}) relative to 10,000 dbar.

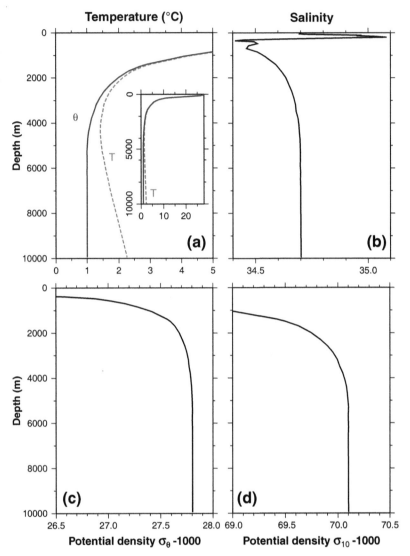

4500–9940 m imply that the trench is filled with water that passes over the sill into the trench, and that there is no other source of water. The slight increase in potential temperature with depth might be due to the weak geothermal heating acting on this nearly stagnant thick layer.

It is not necessary to go to the deepest part of the ocean to see the important differences between in situ temperature and potential temperature. Through most of the deep ocean, there is a temperature minimum well above the ocean bottom, with higher temperature at the bottom. However, potential temperature decreases to the ocean bottom almost everywhere. This is because the densest waters that fill the oceans are also the coldest, since salinity variations are mostly too weak to control the density stratification in the deep waters. There are some limited exceptions to the monotonic decrease with depth: in localized regions of densest water formation, at some mid-ocean ridges where geothermal heating slightly warms waters right on the ridges, and in the central South Atlantic where there is significant vertical salinity variation at mid-depth (Figure 4.11b).

A global map of potential temperature at the ocean bottom in the deep ocean (>3500 m depth) is shown in Chapter 14 (Figure 14.14b). The bottom temperature distribution is mostly set by the two sources of bottom water, from the Antarctic and the Nordic Seas. (Mid-ocean ridges also result in bottom temperature variations since they jut upward into warmer waters.) Bottom waters of Antarctic origin are the coldest; bottom temperatures are near the freezing point near Antarctica, with tongues of water colder than 0°C extending northward into the deep basins of the Southern Hemisphere. Bottom waters of northern Atlantic origin (which arise from overflows from the Nordic Seas) are considerably warmer with temperatures around 2°C.

4.2.6. Vertical Sections of Potential Temperature

We now view potential temperature using meridional cross-sections through each of the three oceans (Figures 4.11a, 4.12a, and 4.13a) to identify common and typical features. Salinity and potential density sections are also shown to keep the vertical sections from each ocean together. The salinity and density distributions are described in Sections 4.3 and 4.4.

In all oceans, the warmest water is in the upper ocean with the highest temperatures in the tropics. In the subtropics, the warm water fills bowl-shaped regions. These bowls define the upper ocean circulations, with westward flow on the equatorward side of the bowls and eastward flow on the poleward side of the bowls. Potential temperature decreases

downward through the thermocline into much more uniform, colder temperatures at depth. The coldest water is found at the surface at high latitudes (and is vertically stable because of low salinity surface water). The coldest water in these sections is in the Antarctic, since the northern ends of the sections do not extend into the Arctic. In the Antarctic, the cold isotherms slope steeply downward between 60 and 50°S. This marks the eastward flow of the ACC (Chapter 13).

There are distinct differences in potential temperature distributions between the Northern and Southern Hemispheres. The cold surface waters are much more extensive in the south. Even the two bowls of higher temperature are not symmetric; the southern bowl is more extensive than its northern counterpart. In the deep part of the Atlantic, Pacific, and Indian Oceans, the coldest waters are in the south (in the Antarctic) and the potential temperatures are slightly higher in the north.

4.3. SALINITY DISTRIBUTION

The mean salinity of the world ocean is 34.6 psu, based on integrating the climatological data in Java Ocean Atlas (Osborne & Swift, 2009; see the online supplementary materials located on the Web site for this text). There are significant differences between the ocean basins. The Atlantic, and especially the North Atlantic, is the saltiest ocean and the Pacific is the freshest (excluding the Arctic and Southern Ocean, which are both fresher than the Pacific). These basin differences are illustrated in Figure 4.14, which shows the mean salinity along well-sampled hydrographic sections, averaged zonally, and from top to bottom of the ocean.

Salinity sections from south to north in each ocean are included in Figures 4.11, 4.12, and 4.13. The following descriptions refer back to these sections. It is apparent after comparing salinity, potential temperature, and potential

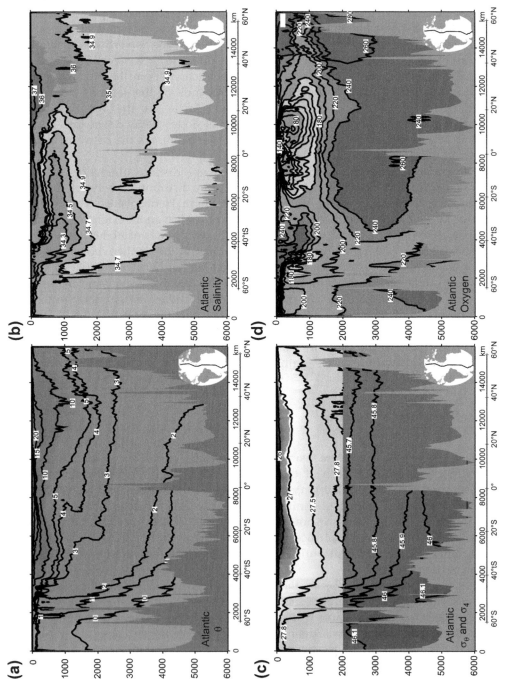

FIGURE 4.11 (a) Potential temperature (°C), (b) salinity (psu), (c) potential density σ_θ (top) and potential density σ_4 (bottom) and (d) oxygen (μmol/kg) in the Atlantic Ocean at longitude 20° to 25°W. Data from the World Ocean Circulation Experiment. This figure can also be found in the color insert.

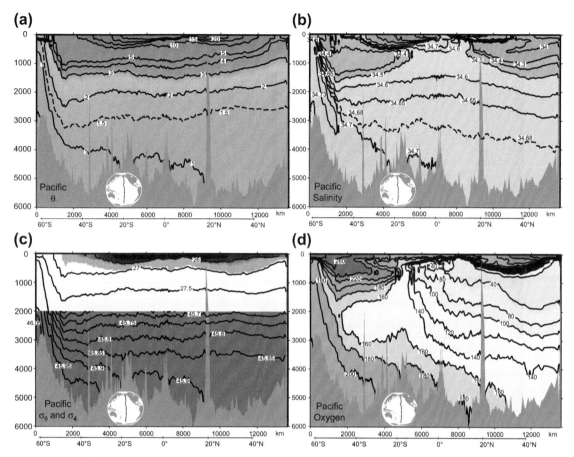

FIGURE 4.12 (a) Potential temperature (°C), (b) salinity (psu), (c) potential density σ_θ (top) and potential density σ_4 (bottom; kg m^{-3}), and (d) oxygen (μmol/kg) in the Pacific Ocean at longitude 150°W. Data from the World Ocean Circulation Experiment. This figure can also be found in the color insert.

density sections for each ocean that the salinity distribution is more complex than temperature and density. While potential temperature decreases monotonically to the bottom in most places, salinity has marked vertical structure; from the simplicity of the density field, it is apparent that it is dominated by potential temperature. Salinity therefore functions in part as a tracer of waters, even as it affects density in a small way.

More detailed depictions of the global salinity distribution and seasonal changes are available in the climatological (seasonally averaged) data set from Levitus, Burgett, and Boyer (1994b). They also showed the data used as the basis for the climatologies. There are far more observations (~90%) in the Northern Hemisphere than in the Southern Hemisphere (~10%), and far more observations in summer than in winter (e.g., Figure 6.13). (This is also true of temperature observations.) This sampling bias is rapidly being corrected in the upper 1800 m by the global profiling float program (Argo) that began in the 2000s.

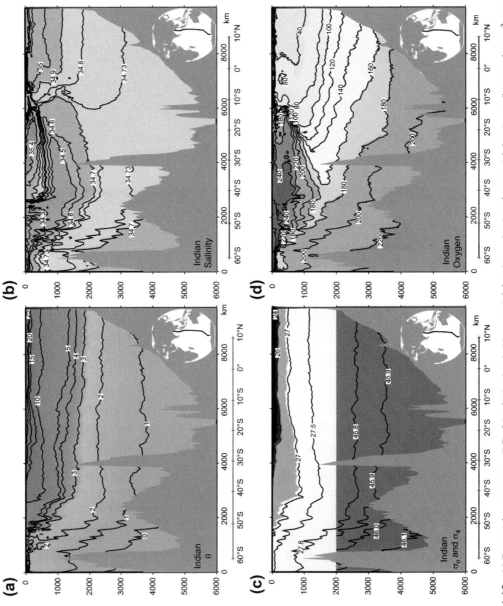

FIGURE 4.13 (a) Potential temperature (°C), (b) salinity (psu), (c) potential density σ_θ (top) and potential density σ_4 (bottom; kg m^{-3}), and (d) oxygen (μmol/kg) in the Indian Ocean at longitude 95°E. Data from the World Ocean Circulation Experiment. This figure can also be found in the color insert.

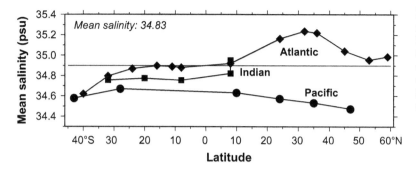

FIGURE 4.14 Mean salinity, zonally averaged and from top to bottom, based on hydrographic section data. The overall mean salinity is for just these sections and does not include the Arctic, Southern Ocean, or marginal seas. *Source: From Talley (2008).*

4.3.1. Surface Salinity

Surface salinity in the open ocean ranges from 33 to 37. Lower values occur locally near coasts where large rivers empty and in the polar regions where the ice melts. Higher values occur in regions of high evaporation, such as the eastern Mediterranean (salinity of 39) and the Red Sea (salinity of 41). On average, the North Atlantic is the most saline ocean at the surface (35.5 psu), the South Atlantic and South Pacific are less so (about 35.2 psu), and the North Pacific is the least saline (34.2 psu), which reflects the ocean basin differences in salinity over the whole ocean depth (Figure 4.14).

The salinity distribution at the ocean's surface is relatively zonal (Figure 4.15), although not as strongly zonal as sea-surface temperature. Unlike SST, which has a tropical maximum and polar

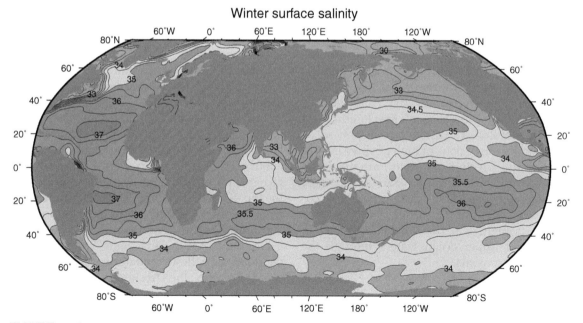

FIGURE 4.15 Surface salinity (psu) in winter (January, February, and March north of the equator; July, August, and September south of the equator) based on averaged (climatological) data from Levitus et al. (1994b).

minima, salinity has a double-lobed structure, with maxima in the subtropics in both hemispheres and minima in the tropics and subpolar regions. This meridional variation is also apparent in the global zonal average of surface salinity (Figure 4.3b). In that figure, the salinity maximum just north of 60°N (with corresponding density deviation) results from dominance of the North Atlantic waters over North Pacific at these latitudes. This is a combination of geography and the higher overall salinity of the North Atlantic; as the North Pacific closes off at these latitudes, the zonal average mainly includes more saline North Atlantic waters, even though internally the subpolar North Atlantic waters are fresher than its subtropical waters.

Surface salinity is set climatologically by the opposing effects of evaporation (increasing it) and precipitation, runoff, and ice melt (all decreasing it), mostly captured by the map of evaporation minus precipitation (Figure 5.4a). The meridional salinity maxima of Figures 4.3 and 4.15 are in the trade wind regions and subtropical high pressure regions where the annual evaporation (E) exceeds precipitation (P), so that (E−P) is positive. On the other hand, the surface temperature maximum is near the equator because the balance of energy into the sea has a single maximum there. Just north of the equator, precipitation is high and surface salinity is lower because of the Intertropical Convergence Zone (ITCZ) in the atmosphere.

Generally the regions of high positive evaporation minus precipitation (E−P) are displaced to the east of the subtropical salinity maxima. This lateral displacement results from the circulation (advection) of the surface waters, so that salinity is highest at the downstream end of the flow of upper ocean waters through the evaporation maxima.

4.3.2. Upper Layer Salinity

The vertical salinity distribution (Figure 4.16 and sections in Figures 4.11, 4.12, and 4.13) is more complicated than the temperature distribution. In the upper ocean, in the tropics, and subtropics and parts of the subpolar regions, temperature dominates the vertical stability (density profile). In the deep ocean, beneath the pycnocline, temperature also dominates over salinity. Therefore, warmer water (lower density) is generally found in the upper layers and cooler water (higher density) in the deeper layers. Salinity can have much more vertical structure, ranging from low to high, without creating vertical overturn. (In subpolar and high latitudes, where surface waters are quite fresh and also cold, salinity does dominate the vertical stability.) As a consequence of its less important role in dictating the density structure, salinity is a more passive tracer than temperature. Thus, salinity can often be used as a marker of the flow directions of water masses (minima or maxima).

In the subtropics, salinity is high near the sea surface due to subtropical net evaporation. Salinity decreases downward to a minimum in the vertical at 600−1000 m. Below this, salinity increases to a maximum, with the exact depths of the vertical minimum and maximum depending on the ocean. In the Atlantic and Indian Oceans, the salinity maximum is at depths of 1500−2000 m. In the Pacific, the maximum salinity is at the bottom.

In the tropics and southernmost part of the subtropical gyres, salinity is often slightly lower at the sea surface than in the main part of the subtropics. Salinity increases to a sharp subsurface maximum at depths of 100−200 m, close to the top of the thermocline. This maximum arises from the high salinity surface water in each subtropical gyre (Figures 4.7, 4.11b, 4.12b, 4.13b, and 4.15). This high salinity water subducts and flows equatorward and downward beneath the fresher, warmer tropical surface water, thus forming a salinity maximum layer. This shallow salinity maximum is found in the equatorward part of every subtropical gyre, merging into the tropics. Because it has an

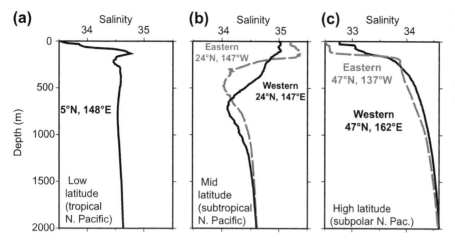

FIGURE 4.16 Typical salinity (psu) profiles for the tropical, subtropical, and subpolar regions of the North Pacific. Corresponding temperature profiles are shown in Figure 4.2.

identifiable characteristic (salinity maximum) and common formation history (subduction from the high salinity surface water at mid-latitudes), it has acquired status as a water mass. Several names are used for this water mass. Our preference is *Subtropical Underwater*, following Worthington (1976). It is also referred to as "salinity maximum water."

Low salinity layers also result from subduction, in this case from the fresher but denser northern outcrops of the subtropical gyres. Advection of these waters southward results in subduction and a low salinity layer that is found around the eastern and into the southern side of the anticyclonic gyre. In the North and South Pacific, these are extensive features called the *Shallow Salinity Minimum* in each ocean (Reid, 1973). In the subpolar North Atlantic, there is a much less-extensive shallow salinity minimum associated with the subarctic front (part of the North Atlantic Current); it is called *Subarctic Intermediate Water*.

In subpolar and high-latitude regions, with high precipitation, runoff, and seasonal ice melt, there is generally low salinity at the sea surface. The halocline, with a rapid downward increase of salinity, lies between the surface low-salinity layer and the deeper, saltier water. In such regions, the pycnocline is often determined by the salinity distribution rather than by temperature, which remains relatively cold throughout the year, and may have only a weak thermocline or even none at all. This condition, associated with runoff and precipitation, occurs throughout the subpolar North Pacific. In the Arctic and Antarctic and other regions of sea ice formation, ice melt in spring creates a similarly freshened surface layer.

This low salinity surface layer in regions like the subpolar North Pacific and around Antarctica permits a vertical temperature minimum near the sea surface, with a warmer layer below (the dichothermal and mesothermal layers, described in Section 4.2).

4.3.3. Intermediate Depth Salinity

At intermediate depths (around 1000−1500 m) in many regions of the world, there are horizontally extensive, vertically broad layers of either low salinity or high salinity. These are easily identified in Figures 4.11, 4.12, and 4.13 because of their vertical salinity extrema. In the North Pacific and Southern Hemisphere, the salinity minimum layer is at about a depth of

1000 m. The subpolar North Atlantic salinity minimum is at about a depth of 1500 m. The low salinity layers are located near the base of the pycnocline, with temperatures of 3–6°C. The two major intermediate-depth salinity maximum layers are in the North Atlantic and northern Indian Ocean (not to be confused with the deeper salinity maximum associated with North Atlantic Deep Water; NADW). They are considerably warmer than the low salinity intermediate waters. The vertical salinity extrema reflect specific formation processes, described briefly here and in more detail in later chapters. These layers are therefore labeled as water masses and called the "intermediate waters."

A map of the locations of the major intermediate water masses is provided in Chapter 14 (Figure 14.13). Their low salinity and their temperature ranges indicate that they originate at the sea surface at subpolar latitudes where surface waters are relatively fresh, but where surface waters are warmer than freezing. The *North Pacific Intermediate Water* (NPIW) originates in the northwest Pacific and is found throughout the North Pacific. *Labrador Sea Water* (LSW) originates in the northwest Atlantic and is found through the North Atlantic. LSW is also marked by high oxygen and chlorofluorocarbons, and retains these signatures even as it loses its salinity minimum as it becomes part of the NADW in the tropical and South Atlantic. *Antarctic Intermediate Water* (AAIW) originates in the Southern Ocean near South America and is found throughout the Southern Hemisphere and tropics. In these three ventilation regions, surface salinity is lower but density is higher than the upper ocean and thermocline waters in the subtropics and tropics. The ventilated intermediate waters spread equatorward and carry their low salinity signature with them.

The two major salinity maximum intermediate waters result from high salinity outflows from the Mediterranean and Red Seas. The source of these high salinity waters is surface inflow into these seas; high evaporation within the seas increases the salinity and cooling reduces their temperature, thus dense water is formed. When these saline, dense waters flow back into the open ocean, they are dense enough to sink to mid-depths.

Other, more local, intermediate waters are also identified by vertical salinity extrema. For instance, in the tropical Indian Ocean, a mid-depth salinity minimum originates from fresher Pacific Ocean water that flows through the Indonesian Passages (Chapter 11). This intermediate salinity minimum has been called Indonesian Intermediate Water or Banda Sea Intermediate Water (Rochford, 1961; Emery & Meincke, 1986; Talley & Sprintall, 2005).

Each of these intermediate waters is discussed in greater detail in the relevant ocean basin chapter (9–13).

4.3.4. Deep-Water Salinity

The deep waters of the oceans exhibit salinity variations that mark their origin. The North Atlantic is the saltiest of all of the oceans at the sea surface, so dense waters formed in the North Atlantic carry a signature of high salinity as they move southward into the Southern Hemisphere and then eastward and northward into the Indian and Pacific Oceans. This overall water mass is referred to as *North Atlantic Deep Water*. Dense waters formed in the Antarctic are colder and denser than North Atlantic dense waters, so they are found beneath waters of North Atlantic origin. The dense Antarctic waters are also fresher than North Atlantic waters; their progress northward into the Atlantic can be tracked through their lower salinity, where they are referred to as *Antarctic Bottom Water* (AABW). The vertical juxtaposition of the salty NADW and fresher AABW is apparent in the Atlantic vertical salinity section (Figure 4.11b). This NADW/AABW structure is also apparent in the southern Indian Ocean since both NADW and AABW enter the Indian Ocean from the south (Figure 4.13b).

The northern Indian Ocean is tropical so no dense waters are formed there, but the high salinity from the intermediate waters of the Red Sea penetrates and mixes quite deep, making northern Indian Ocean deep waters relatively saline (Figure 4.13b). The North Pacific does not form dense, abyssal waters because the sea surface in the subpolar North Pacific is too fresh to allow formation of waters as dense as those from the Antarctic and North Atlantic. Therefore, the salinity structure in the deep North Pacific is determined by the inflow of the mixture of Antarctic and North Atlantic deep waters from the south; this mixture is more saline than the local North Pacific waters so salinity increases monotonically to the bottom in the North Pacific (Figure 4.12b).

A global map of bottom salinity is shown in Chapter 14 (Figure 14.14c). Globally, the salinity variation in the deep waters is relatively small, with a range from 34.65 to 35.0 psu. Like bottom temperatures, the bottom salinities reflect the Antarctic and Nordic Seas origins of the waters. The Antarctic bottom waters are freshest, with salinities lower than 34.7 psu. The bottom waters of Nordic Sea origin are the saltiest, with salinity up to 35.0 psu. Full interpretation of the bottom salinity map also requires consideration of the varying bottom depth — as ridges cut up into overlying deep waters — and of downward diffusion of properties from the overlying deep waters, which are beyond the scope of this book.

Thus both the deep water temperature and deep water salinity have small ranges. The deep water environment is relatively uniform in character compared with the upper ocean and thermocline and even the intermediate layer. This relative uniformity is the result of the small number of distinct sources of dense waters, and the great distance and time that these waters travel, subjected to mixing with each other and to downward diffusion from layers above them.

4.3.5. Temporal Variations of Salinity

Salinity variations at all timescales are less well documented than temperature variations, because temperature is more easily measured. Annual variations of surface salinity in the open ocean are less than 0.5 psu. In regions of marked annual variation in precipitation and runoff, such as the eastern North Pacific and the Bay of Bengal and near sea ice, there are large seasonal salinity variations. These variations are confined to the surface layers because in such regions the effect of reduced salinity overrides the effect of temperature in reducing the seawater density. This keeps the low salinity water in the surface layer. Diurnal variations of salinity appear to be small, but again this is a conclusion based on very few observations. Local rainstorms produce fresh surface waters even in the open ocean that mix into the surrounding waters after several weeks.

Temporal salinity variations at a given location can be large at large-scale fronts between waters of different properties. These fronts are sometimes termed *water mass boundaries*. Temperature variations can also be quite large across these fronts. The fronts move about their mean location, on weekly, seasonal, and longer timescales. Meandering of the fronts and creation of eddies of the different types of waters can cause large salinity and temperature variations at a given location.

Interannual and long-term changes in large-scale salinity are observed and are part of the documentation of climate change. With the advent of the global profiling float array, it is becoming possible to document salinity changes in all regions of the non-ice-covered ocean; significant patterns of surface salinity change have already been detected (Hosoda, Suga, Shikama, & Mizuno, 2009; Durack &

Wijffels, 2010). Salinity variations in the North Atlantic and Nordic Seas are associated with changes in mixed layer convection and with changes in water mass formation in the Labrador and Greenland Seas (Chapters 9 and 12). LSW has dramatic decadal salinity variations that correspond with changes in its formation. See Figure S15.4 (Yashayaev, 2007) from the online supplementary material. Decades long freshening of the subpolar North Atlantic and Nordic Seas (see Chapter S15 from the online supplementary material), followed by salinification in recent years, has caused much interest in terms of NADW production rates. Large-scale, coherent salinity changes over several decades have been documented (Boyer, Antonov, Levitus, & Locarnini, 2005; Durack & Wijffels, 2010) and can be associated with large-scale changes in precipitation and evaporation that might be related to overall warming of the atmosphere (Bindoff et al., 2007).

4.3.6. Volumetric Distribution of Potential Temperature and Salinity

A classic (and typical) approach to looking at the water mass structure is to display various properties as a function of each other; a more modern statistical approach describes the water masses in terms of all of their properties (see Section 6.7). Potential temperature-salinity diagrams are used throughout the basin chapters (9–13) to illustrate the water masses. A volumetric θ-S diagram from Worthington (1981) is introduced as our first global summary of water properties (Figure 4.17). The method is described in Section 6.7.2.

The underlying θ-S in the upper panel (Figure 4.17a) shows three separate branches stretching from low θ-S to higher θ-S; these are the Central Waters of the pycnocline (as in Figure 4.7). The saltiest branch is the North Atlantic; the freshest branch is the North Pacific. The intermediate branch, with larger volumes,

is the three Southern Hemisphere basins (South Atlantic, South Pacific, and Indian). The importance of the Southern Ocean connection between these latter three basins is immediately apparent, as the three have properties that are more similar than the two Northern Hemisphere basins.

In the deep water (Figure 4.17b), the largest peak is the *Pacific Deep Water* (or Common Water); the large volume in a single θ-S class indicates how well mixed this water mass is, which is a direct result of its great age (Section 4.7). The coldest waters are the AABW, with the single ridge again indicating Southern Ocean circumpolar connectivity. Above about 0°C, the diagram splits into three branches — the Pacific Ocean, Indian/Southern Ocean, and Atlantic Ocean, from freshest to saltiest. The salty Atlantic ridge has a long portion of high volume, without a huge, single peak such as is found in the Pacific. This reflects the multiple sources of NADW and its relatively young, unmixed character.

4.4. DENSITY DISTRIBUTION

Potential density must increase with depth in a system in equilibrium. To be more precise, the water column must be statically stable, using the definition of static stability (Eq. 3.9) in Section 3.5.6. This means that potential density, using a local reference pressure, must increase with depth. While potential temperature and salinity together determine density, individually they can have maxima and minima in the vertical, as long as the water density increases with depth. The only exceptions to the monotonic increase occur at very short timescales, on the order of hours or less, which is the timescale for overturn. As soon as denser water flows over lighter water, or surface layer density is increased above that of the underlying water, the water column becomes unstable and will overturn and mix, removing the instability.

(a)

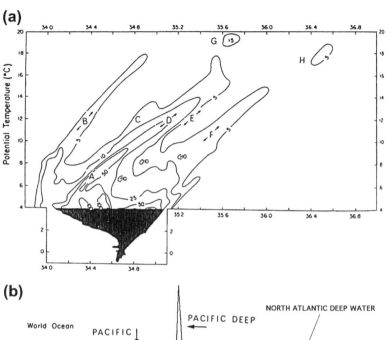

(b)

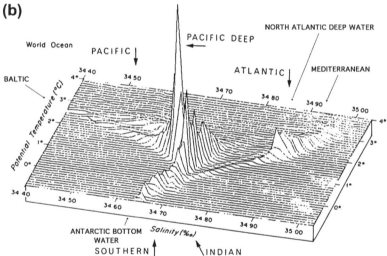

FIGURE 4.17 Potential tempera-ture-salinity-volume (θ-S-V) dia-grams for (a) the whole water column and (b) for waters colder than 4°C. The shaded region in (a) corresponds to the figure in (b). *Source: From Worthington (1981).*

In Chapter 3, we discussed the use of different reference pressures for reporting potential density, or equivalently for use of an empirically defined type of density such as neutral density (Section 3.5.4). The potential density that is used should best approximate the local vertical stability and isentropic surfaces. Profiles of poten-tial density relative to both the sea surface and 4000 dbar are used in constructing the potential density sections of Figures 4.11, 4.12, and 4.13. When spatial variability in temperature and salinity is very small, any type of potential density will increase monotonically with depth; an example is the potential density relative to both the sea surface and 10,000 dbar in the Mariana Trench (Figure 4.10). The North Pacific has little variation in temperature and salinity below the pycnocline, which is the vertical region of large

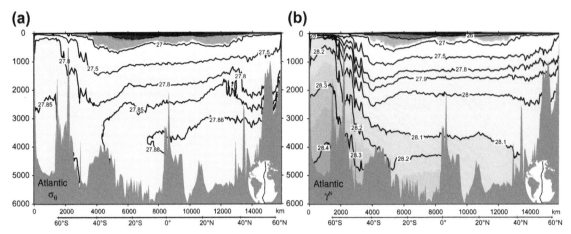

FIGURE 4.18   (a) Potential density σ_θ (kg m^{-3}) and (b) neutral density γ^N in the Atlantic Ocean at longitude 20° to 25°W. Compare with Figure 4.12c. Data from the World Ocean Circulation Experiment.

density change. Therefore all potential density choices yield stable-appearing vertical profiles (Figure 4.20).

In the South Atlantic, on the other hand, there are large-scale salinity inversions where the saline NADW is layered between the fresher AAIW and AABW (Figure 4.11b). Here the differences in compressibility of warmer and cooler waters begin to matter. Figure 4.11c emphasizes the local potential density structure, which is decidedly stable in the vertical. To illustrate the main drawback of using a surface reference pressure for deep-water density, a vertical section through the Atlantic Ocean of potential density relative to the sea surface, σ_θ, for the full water column is shown in Figure 4.18a. There is a large-scale inversion of σ_θ in the South Atlantic, most pronounced just south of the equator at a depth of about 3700 m. This is the base of the high salinity NADW layer. Potential temperature contours are compressed below the NADW (Figure 4.11a). Potential density referenced to 4000 dbar, σ_4, hence locally referenced, has no inversion (Figure 4.11c).

Neutral density γ^N (Section 3.5.4; Jackett & McDougall, 1997) is commonly used to represent the stable increase of "potential" density with depth.[1] Like choosing appropriate locally referenced potential densities, neutral density eliminates the apparent density inversions of Figure 4.18a and also removes the need to use multiple pressure reference levels such as in the use of 0 and 4000 dbar references in Figures 4.11c, 4.12c, and 4.13c. The neutral density γ^N section for the Atlantic is clearly monotonic, with γ^N increasing from top to bottom. The deep contours resemble those of σ_4 (Figure 4.11c), and the distortions of σ_θ in the region of the Mediterranean salinity maxima at about 2000 m in the North Atlantic are removed.

4.4.1. Density at the Sea Surface and in the Upper Layer

The density of seawater at the ocean surface increases from about $\sigma_\theta = 22$ kg/m^3 near the equator to $\sigma_\theta = 26 - 28$ kg/m^3 at 50–60° latitude, and beyond this it decreases slightly (Figures 4.3

---

[1] There continues to be energetic discussion of the most appropriate variable for density for constructing the most isentropic surfaces in the sense of the direction of motion of water parcels and the directions of along-isopycnal and diapycnal mixing.

Winter surface density

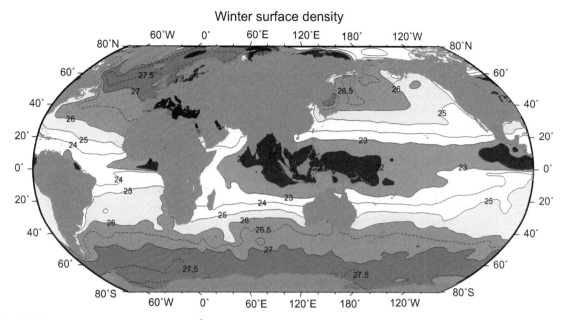

FIGURE 4.19 Surface density σ_θ (kg m^{-3}) in winter (January, February, and March north of the equator; July, August, and September south of the equator) based on averaged (climatological) data from Levitus and Boyer (1994) and Levitus et al. (1994b).

and 4.19), due to lower salinity at higher latitudes. Surface densities at high latitudes in the Antarctic and North Atlantic are higher than in the North Pacific even at the freezing point. North Pacific surface water must be less dense since its surface water is fresher.

In Figure 4.3, we see that the surface density averaged for all oceans follows surface temperature rather than surface salinity in the tropics and mid-latitudes. At the highest northern and southern latitudes, poleward of 50°, surface density follows salinity more than temperature, because temperature is close to the freezing point there, with little variation in latitude.

Surface density and the vertical stratification determine the depth to which surface waters will sink as they move away from their ventilation ("outcrop") region. The combination of surface temperature and salinity for a given density also affects the sinking because of their effect on compressibility, with warmer, saltier

water compressing less than colder, fresher water at the same surface density. Thus the colder, fresher parcel will become more dense and, consequently, deeper than the warmer, saltier parcel as they move into the ocean even though they start with the same surface density. See Section 3.5.4.

In late winter, surface waters reach their local density maximum as the cooling season draws to a close. (Cooling in many regions is also associated with evaporation, so both temperature and salinity may change together to create dense water, depending on the local amount of precipitation.) Late winter density is associated with the deepest mixed layers. As the warming season begins (March in the Northern Hemisphere, September in the Southern Hemisphere), the dense winter mixed layer is "capped" by warmer water at the surface. The capped winter waters move (advect) away from the winter ventilation region. If they move into a region where the winter surface waters are less dense, they sink beneath the local surface layers, and will not be reopened to the atmosphere during the next winter. This subduction process is a primary mechanism for moving surface waters into the ocean interior (Luyten et al., 1983; Woods, 1985 and Sections 4.2.3 and 7.8.5).

Longer timescale variations in surface density can affect the amounts of intermediate and deep waters that form and the overall size of the regions impacted by them. During major climate changes associated with glacial/inter-glacial periods, surface density distributions must have been strongly altered, resulting in very different deep water distributions.

Winter mixed layer depths vary from tens of meters to hundreds of meters, depending on the region (Figure 4.4). Because they have usually been detected using temperature criteria, we discussed mixed layers in some detail in Section 4.2.2. In the tropics, winter mixed layer depths may be less than 50 m. Winter mixed layer depths are greatest in the subpolar North Atlantic, reaching more than

1000 m in the Labrador Sea, and in the Southern Hemisphere around the northern edge of the major current that circles Antarctica at a latitude of about 50°S, reaching up to about 500 m thickness.

4.4.2. Pycnocline

Like potential temperature, potential density does not increase uniformly with depth (Figure 4.20). The vertical structure of density is similar to that of potential temperature. There is usually a shallow upper layer of nearly uniform density, then a layer where the density increases rapidly with depth, called the pycno-cline, analogous to the thermocline (Section 4.2.3). Below this is the deep zone where the density increases more slowly with depth (Figures 4.10, 4.11, 4.12, 4.18, and 4.20). There is much smaller variation with latitude of the deep-water density compared with upper ocean density. As a consequence, in high latitudes, where the surface density rises to $\sigma_\theta = 27$ kg/m³ or more, there is a smaller increase of density

FIGURE 4.20 Typical density/depth profiles for low and high latitudes (North Pacific).

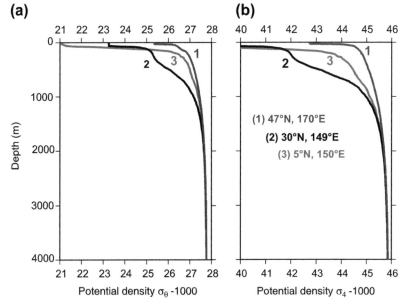

(a)

(b)

(1) 47°N, 170°E

(2) 30°N, 149°E

(3) 5°N, 150°E

Depth (m)

Potential density σ_θ -1000

Potential density σ_4 -1000

with depth than in the low latitudes, and the pycnocline is much weaker.

The double thermocline structure that occurs in some broad regions (described in Section 4.2), is mirrored in density because of the strong dependence of ocean density on temperature. Layers of lower vertical density gradients are called *pycnostads*.

In all regions, there is a seasonal pycnocline in the warm seasons. This results from seasonal warming and/or ice melt, overlying the remnant of the winter mixed layer, which forms a pycnostad in non-winter seasons. A permanent double pycnocline, with a pycnostad lying between the pycnoclines, is a common feature of subtropical regions. Mode Waters (Section 4.2.3) are pycnostads, and are best identified in terms of density stratification rather than temperature stratification; that is, a minimum in vertical stability is the best identifier of a Mode Water on a given vertical profile. Often Mode Waters and other water masses are tracked in terms of their potential vorticity (Eq. 3.11 and Section 7.7); the dominant contribution to potential vorticity in most of the ocean (except in strong currents) is proportional to the vertical stability. Potential vorticity is a useful tracer because it is a conserved dynamical quantity in the absence of mixing.

4.4.3. Depth Distribution of Potential Density

Potential density structure is simpler than potential temperature and salinity simply because the water column must be vertically stable. Potential density, appropriately defined, must increase downward. Below the pycnocline, vertical potential density variations are much smaller, similar to potential temperature structure. There are no large-scale inversions in density if the appropriate reference pressures are used, as described in Section 3.6 and as seen in the vertical section through the Atlantic Ocean (Figure 4.18 compared with Figure 4.11c

in Section 4.2.6). Horizontal variations in potential density are associated with horizontal pressure gradients and therefore with large-scale currents (Section 7.6).

Potential density structure is displayed along vertical sections through the north-south length of each ocean (Figures 4.11, 4.12, and 4.13). The main features are downward bowls in the upper to intermediate ocean in the subtropics, and a large upward slope toward the southern (Antarctic) end of the sections. Below about 2000 m, the total range of potential density is small, from about $\sigma_\theta = 27.6$ to 27.9 kg/m^3 (or $\sigma_4 = 45.6$ to 46.2 kg/m^3, which is potential density relative to 4000 dbar).

Because mixing is greatly inhibited by vertical stratification, there is a strong preference for flow and mixing to occur along isentropic surfaces, which are approximately isopycnals (surfaces of constant potential density). In the upper ocean, surfaces of constant σ_θ are useful. For instance, the processes that give ocean waters their particular properties act almost exclusively at the surface, and the origin of even the deepest water can be traced back to a region of formation at the surface somewhere. Because deep ocean water is of high density, it must form at high latitudes where cold, dense water is found at the surface. After formation, it spreads down almost isopycnally (reference pressures should be adjusted to account for temperature dependence of the compressibility). The sinking is combined with horizontal motion so that the water actually moves in a direction only slightly inclined to the horizontal. Even in the regions of largest isopycnal slopes, for instance in the southern part of Figures 4.11c, 4.12c, and 4.13c, the slopes are no more than several kilometers down over several hundred kilometers horizontally.

Even though there is large-scale structure in the deep ocean's salinity field (e.g., in the Atlantic in Figure 4.11b), temperature dominates the density structure in the deep ocean. Salinity is important for the density structure near the sea surface at high latitudes where precipitation or ice melt creates a low salinity surface layer, such as in the Arctic, in the region next to Antarctica, and in the subpolar North Pacific and coastal subpolar North Atlantic. In shallow coastal waters, fjords, and estuaries, salinity is often the controlling factor in determining density at all depths, while the temperature variations are of secondary importance (e.g., Table 3.1).

In much of the ocean, the density profile with increasing depth appears nearly exponential, asymptoting to a nearly constant value in the deep ocean (Figure 4.20). However, in some regions, where deep waters from very different sources are juxtaposed, there is a weak pycnocline (higher density variation) between them. An obvious example is between the NADW and AABW in the South Atlantic, which is where we illustrated the necessity for local referencing of the potential density. Such regions are most common in the subtropical Southern Hemisphere where dense Antarctic waters flow northward in a thick layer under slightly less dense deep waters flowing southward from the North Atlantic, Pacific, and Indian Oceans (Figures 4.11 and 4.18).

4.5. DISSOLVED OXYGEN

Seawater contains dissolved gases, including oxygen and carbon dioxide. Some of the transient tracers are dissolved gases (Section 4.6). The ocean is an important part of the global (atmosphere/land/ocean) cycle of carbon dioxide, which is a greenhouse gas. However, because of its complexity, we do not describe the ocean's carbon chemistry in this book, and instead refer readers to texts on biogeochemical cycles (e.g., Broecker & Peng, 1982; Libes, 2009).

Dissolved oxygen content is used as an important tracer of ocean circulation and an indicator of the time passed since a water parcel was at the sea surface (ventilated) (Section 3.6). The range of oxygen values found in the sea is from 0 to 350 μmol/kg (0 to 8 ml/L), but a large proportion of values fall within the more limited range from 40 to 260 μmol/kg (0 to 6 ml/L). The atmosphere is the main source of oxygen dissolved in seawater. At the sea surface, the water is usually close to saturation. Sometimes, in the upper 10−20 m, the water is supersaturated with oxygen, a by-product of photosynthesis by marine plants. Supersaturation also occurs near the sea surface if the water warms up through solar radiation that penetrates tens of meters into the ocean. (If the actual sea surface becomes warmer, it will lose its excess oxygen to the atmosphere, so supersaturation is not found right at the sea surface.) Sometimes surface waters are undersaturated; this occurs rarely in winter if mixing of the surface layer is especially intense, entraining underlying older, less saturated waters. (The equilibration time of surface waters — time required to restore the waters to 100% saturation — is a few days to a few weeks and is a function of wind speed and temperature.) Below the surface layers, the oxygen saturation is less than 100% because oxygen is consumed by living organisms and by the bacterial oxidation of detritus. Low values of dissolved oxygen in the sea are often taken to indicate that the water has been away from the surface for a long time, the oxygen having been depleted by the biological and detrital chemical demands.

Figure 4.21 shows typical dissolved oxygen profiles for the Atlantic and the Pacific for three latitude zones. Figures 4.11d, 4.12d, and 4.13d show oxygen along a south-north section for each ocean. Common features of the Atlantic and Pacific are (1) high oxygen close to the surface, (2) an oxygen minimum at 500−2500 m, (3) relatively high values below 1500 m in the Atlantic (NADW), (4) low values in the North Pacific beneath the surface layer, and (5) more similar subsurface distributions in the southern latitudes in both oceans. Distributions in the Indian Ocean are similar to those in the Pacific (south and tropics). The lower values in the deep water

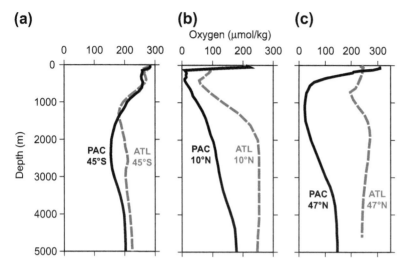

FIGURE 4.21 Profiles of dissolved oxygen (μmol/kg) from the Atlantic (gray) and Pacific (black) Oceans. (a) 45°S, (b) 10°N, (c) 47°N. Data from the World Ocean Circulation Experiment.

of the Pacific compared with the Atlantic indicate that the Pacific water has been away from the surface for a much longer time. In some regions of extremely low oxygen, such as the Black Sea and the bottom of the Cariaco Trench (off Venezuela in the Caribbean), hydrogen sulfide is present, created from the reduction of sulfate ion by bacteria. This indicates that the water has been stagnant there for a long time.

The oxygen minimum through the world oceans at mid-depth, overlying higher oxygen at the bottom, results from at least several mechanisms. One is that minimal circulation and mixing do not replace the oxygen consumed. A second is that the increase of density with depth (stability) allows biological detritus to accumulate in this region, which increases the oxidation rate. A third is that the bottom waters in each of the oceans are relatively high in oxygen because of their surface source in the Antarctic. In the Pacific and Indian Oceans, a three-layer structure is obtained, with high oxygen at the surface decreasing through the pycnocline, an oxygen minimum layer in the intermediate and deep water, and higher oxygen in the abyss. The Atlantic has a four-layer structure because of the juxtaposition of high oxygen content in the

NADW onto this three-layer structure (the thick layer of higher oxygen between 2000 and 4000 dbar in Figure 4.11d corresponds with high salinity in Figure 4.11b).

A pronounced vertical oxygen minimum is found in the upper ocean in the tropical Atlantic (Figure 4.11d), eastern tropical Pacific (Figure 4.12d), and in the northern Indian Ocean (Figure 4.13d). These shallow oxygen minima result from very high biological productivity in the surface waters in these regions. Bacterial consumption of the large amount of sinking detritus from these surface waters is large and consumes almost all of the dissolved oxygen within the upper 300–400 m of the ocean.

The production and utilization of oxygen in the sea are essentially biogeochemical matters (Section 3.6). Oxygen is a useful tracer, broadly indicative of a water parcel's age, but since it is non-conservative, it must be used carefully.

4.6. NUTRIENTS AND OTHER TRACERS

Other common water properties used as flow tracers or for identifying water masses include the nutrients (phosphate, nitrate, nitrite, silicic acid, and ammonium), dissolved gases other than oxygen and carbon dioxide, and plankton, which are small organisms (both plant and animal) that drift with the water. These water mass characteristics must be used with caution because, like oxygen, they are non-conservative; in other words, they may be produced or consumed within the water column. Other chemical and radioactive tracers are now measured widely (e.g. Broecker and Peng, 1982). Here we concentrate on the main features of the nutrient distributions, which are of interest to marine biologists as well as physical and chemical oceanographers.

Nutrient values are low in the upper few hundred meters with higher values in the deeper water (Figure 4.22). In the North Pacific, these deeper distributions are in the form of mid-depth maxima, extending from north to south, with highest values at 1000/2000 m for nitrate (NO_3) and phosphate (PO_4), and at 2000/3000 m for silicate. Additionally, there are maxima in the south in the dense water formed in the Antarctic. In the Atlantic, the mid-depth low nutrient tongues extending from north to south are associated with the NADW (Section 9.8). Maximum values are found in the south and along the bottom in the dense waters formed in the Antarctic (AABW).

The low values of nutrients in the upper layers result from utilization by phytoplankton in the surface layer (*euphotic zone*, exposed to sunlight), while the increase in deeper waters is because of their release back to solution by biological processes (respiration and nitrification, mostly microbial) during the decay of detrital material sinking from the upper layers. Therefore, nutrient distributions are approximately mirror images of the oxygen distribution. Phosphate and nitrate have similar distributions because they are governed by almost the same biological cycle (see discussion of Redfield ratios in Section 3.6). (Therefore only nitrate sections are included in

Figure 4.22.) The dissolved silica (silicic acids) distribution is not as closely similar. Silica has an additional source at the ocean bottom, as it can be dissolved into the seawater from the sediments or injected by hydrothermal sources.

Nutrient replenishment in the surface layers is strongly influenced by the physical processes of vertical diffusion, overturning, and upwelling. These bring nutrients from below the euphotic zone up to the surface. The impact of upwelling on surface nutrients is illustrated by the nitrate distribution at the sea surface (Figure 4.23). Nitrate is nearly zero in the subtropical regions where surface waters downwell (Chapters 9–11). Surface nitrate is non-zero (although small) where there is upwelling from just below the euphotic zone, which occurs in subtropical eastern boundary regions (Section 7.9.1), along the equator, and in the subpolar regions. These are regions of high biological productivity because of the nutrient supply to the euphotic zone (see map of depth of the euphotic zone in Figure 4.29).

At mid-depth, in the nutrient vertical maxima, the Pacific nutrients are higher than Atlantic values by a factor of about two for phosphate and nitrate, and by a factor of three to ten for silicate. This is due to the much greater age of the mid-depth and deep waters in the Pacific than in the North Atlantic. The lower dissolved oxygen values in the Pacific than in the Atlantic are attributed to the same cause.

Taken together, the oxygen and nutrient distributions, along with salinity, provide our principal identification for water masses below the pycnocline. The high oxygen, low nutrient, high salinity deep layer in the Atlantic Ocean is the NADW. The low oxygen, high nutrient layer in the Pacific Ocean is the Pacific Deep Water, and the same layer in the Indian Ocean is the *Indian Deep Water*. The higher oxygen, lower nutrient, very cold bottom layer in all oceans is the AABW. When considering more carefully the east-west distributions of

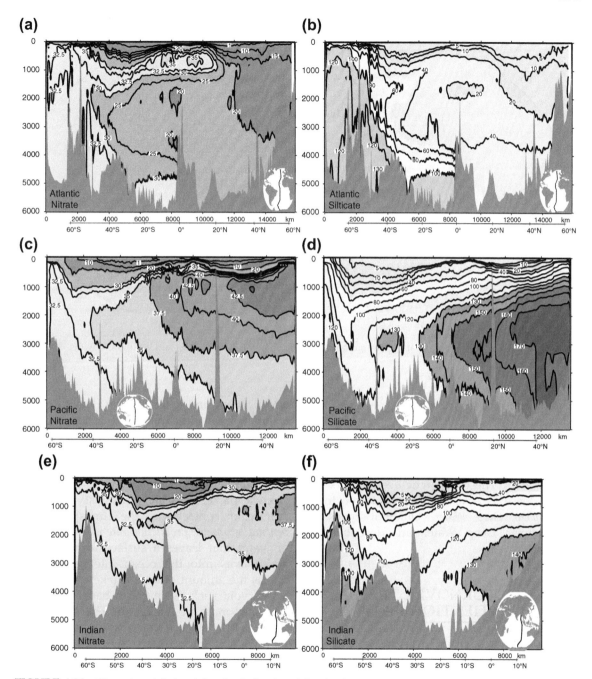

FIGURE 4.22 Nitrate (μmol/kg) and dissolved silica (μmol/kg) for the Atlantic Ocean (a, b), the Pacific Ocean (c, d), and the Indian Ocean (e, f). Note that the horizontal axes for each ocean differ. Data from the World Ocean Circulation Experiment. This figure can also be found in the color insert.

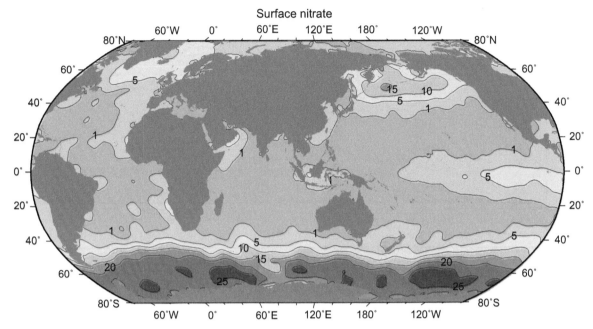

FIGURE 4.23 Nitrate (μmol/L) at the sea surface, from the climatological data set of Conkright, Levitus, and Boyer (1994).

properties in the Southern Hemisphere subtropics, we can also distinguish between higher oxygen, lower nutrient waters that come from the Antarctic compared with the low oxygen, high nutrient Pacific and Indian Deep Waters. These Antarctic deep waters are not as dense as AABW and are referred to as *Circumpolar Deep Water*. Different types of Circumpolar Deep Water are described in Chapter 13.

4.7. AGE, TURNOVER TIME, AND VENTILATION RATE

Estimates of age and overturning rates of ocean waters assist in understanding the overall distribution of temperature and salinity in the ocean, the replenishment rate of nutrients in the upper layers, and the exchange of gases between the atmosphere and ocean. The effectiveness and safety of the deep ocean as a dump for noxious materials depends on the turnover time of the deep waters. For these and many other applications, it is useful to estimate timescales for ocean ventilation.

Age as applied to ocean water is the time since a parcel of water was last at the sea surface, in contact with the atmosphere. The *ventilation rate* or *production rate* is the transport of water that leaves a surface formation site and moves into the ocean interior. *Turnover time* is the amount of time it takes to replenish a reservoir, such as an ocean basin or a layer or water mass in the ocean. The "reservoir" can also be construed in terms of a tracer rather than water particles (e.g., molecules of CO_2, or zooplankton, etc.). *Residence time* is the time a particle spends in a reservoir.

The *ages* of waters can be estimated using tracers (Section 3.6). Tracers that are biologically inert are more straightforward than those that are biologically active. Anthropogenic transient tracers that have measured histories in the atmosphere are useful in the upper ocean and well-ventilated parts of the deep ocean. The penetration of chlorofluorocarbons (CFCs; Pacific section shown in Figure 4.24a) and tritium (Pacific map in Figure 4.25b) is evidence of recent ventilation; absence of these tracers is clear evidence of age that is greater than 50–60 years.

Pairs of tracers whose concentration ratio changes with time can be used to estimate age, including pairs of CFCs with different atmospheric time histories that result in a changing ratio in surface source waters (Figure 4.25a). Similarly, because tritium (^{3}H) decays to ^{3}He with a half-life of about 12 years, the ^{3}H/^{3}He pair reflects age (ignoring the smaller amounts of natural ^{3}He injected in the deep ocean at the mid-ocean ridges). This tracer ratio method is straightforward only if the surrounding waters are free of the tracers because mixing between waters with different ratios complicates the age calculation. The tropical Pacific and North Pacific are ventilated only in the upper ocean, with no deep water sources except far to the south, so the CFC and tritium/^{3}He pair "ages" are especially useful for estimating water age there.

For the deep ocean, where water is too old to be dated using anthropogenic tracers, and also as an alternate method of estimating age in the better ventilated parts of the ocean, natural tracers such as oxygen, nutrients, and ^{14}C (Figure 4.24b) are useful. Biological activity reduces oxygen and increases nutrient content once the water moves away from the sea surface. If the oxygen consumption rate or nutrient remineralization rates are known as a function of location and temperature, then the age of a water mass as it moves away from its source can be estimated. As with anthropogenic tracers, simplifying assumptions about mixing with waters of different oxygen and nutrient content are required.

Radiocarbon can be used for dating just as with terrestrial organic matter (Section 3.6).

^{14}C is created in the atmosphere by cosmic rays and quickly becomes part of the atmospheric CO_2. It enters the ocean with the CO_2 that dissolves in surface water. When the surface water is subducted or incorporated in deeper waters, the decay of ^{14}C at a rate of 1% every 83 years results in increasingly negative values (deficits) along the pathways into the deep ocean. The largest deficits globally are found in the deep North Pacific, reflecting the great age of the waters there (Figure 4.24b). Use of ^{14}C deficits to precisely date ocean water is subject to caveats about mixing and also complications due to local sinking of organic matter from the surface and other sources of ^{14}C including nuclear testing. The gross estimate of ages of deep waters based directly on ^{14}C deficits is 275 years for the Atlantic, 250 years for the Indian, and 510 years for the Pacific. It is easy to see these age estimates are biased since the oldest waters always mix with younger waters and vice versa. Thus the age of the deep northern Pacific waters is likely higher than their ^{14}C age (around 1000 years), while the age of the deep northern Atlantic waters is lower, as evidenced by invasion of CFCs to the bottom (Broecker et al., 2004).

The *ventilation rate* (*production rate*) of a water mass or layer can be defined in several different ways, which can lead to somewhat different quantitative estimates. In all cases, the objective is to estimate the rate of injection of new water into a reservoir. One approach is to estimate production rate from the volume of the reservoir divided by its age:

$$R_P = \text{Volume}/\text{Age} \qquad (4.1a)$$

which has units of transport (m^3/sec). This is a straightforward concept, but difficult to implement since the ocean is not composed of simple boxes filled with waters of uniform age; therefore somewhat complex calculations and simple models are used to obtain ventilation rates from the continuous distribution of ages. If Eq. (4.1a)

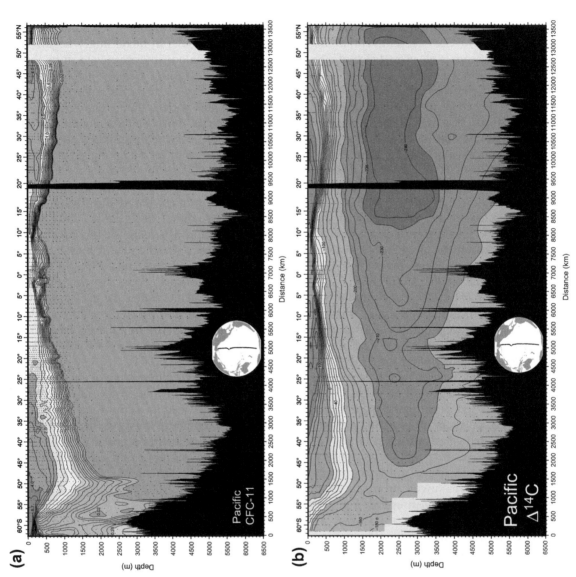

FIGURE 4.24 (a) Chlorofluorocarbon content (CFC-11; pmol/kg) and (b) $\Delta^{14}C$ (/mille) in the Pacific Ocean at 150°W. White areas in (a) indicate undetectable CFC-11. From the WOCE Pacific Ocean Atlas. *Source: From Talley (2007).*

(a)

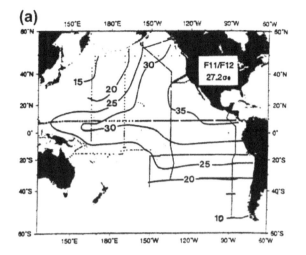

(b)

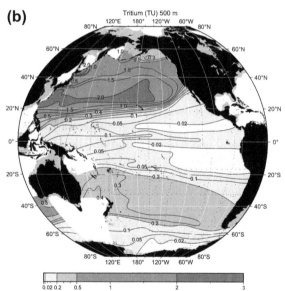

FIGURE 4.25 (a) Age (years) of Pacific Ocean waters on the isopycnal surface 27.2 σ_θ, using the ratio of chlorofluorocarbon-11 to chlorofluorocarbon-12. *Source: From Fine, Maillet, Sullivan, and Willey (2001).* (b) Tritium concentration at 500 m in the Pacific Ocean from the WOCE Pacific Ocean Atlas. *Source: From Talley (2007).*

is written in terms of turnover time (Eq. 4.2 below) instead of age, the rate that is obtained could differ since age and turnover time are usually not identical.

A related approach to estimating a ventilation rate using transient tracers starts with the total amount (inventory) of the tracer and concentration of the tracer at its source (sea surface). For instance, using CFC's with an inventory I_{CFC} (units of moles) and surface concentration C_{source} (units of moles/kg), the ventilation rate is given by (Smethie & Fine, 2001):

$$I_{CFC} = \sum R_P \, C_{source} \Delta t \qquad (4.1b)$$

Since both the source concentration and inventory vary with time, this ventilation rate is obtained iteratively.

Ventilation rates, R_P, are also estimated from observations of the transport of newly ventilated waters very close to the source of the water mass. Farther from the source, quantitative water mass identification techniques (Section 6.7) can be used to estimate the portion of observed transport that can be attributed to the source versus the portion due to mixing with other waters. Indirect estimates are also frequently used, based on measuring the buoyancy forcing that results in ventilation with simple or complex models to compute the ventilation rate; an approach using air–sea fluxes of heat and freshwater within surface outcropping regions of isopycnal layers was introduced by Walin (1982) and has been employed in numerous calculations.

Turnover time is the time it takes to replenish a reservoir. If in reference to water rather than a tracer, it is equal to the volume V of the water mass or layer, in units of m^3, divided by its outflow transport R_{out} in m^3/sec. If in reference to a tracer, it is the inventory of the tracer, in moles, divided by the transport of the tracer out of the reservoir in mole/sec. Turnover time, which has been defined generally for use in biogeochemistry, is written in terms of the exit flow because reservoirs are usually well-mixed, unlike the inflow sources, resulting in a simpler (proportional) relation between outflow volume transport and turnover time.

Turnover times for volume and for a tracer are given by:

$$T_{turnover} = \frac{V}{R_{out}} = \frac{\int\int\int dV}{\int\int v_{out} dA} \rightarrow \frac{\int\int\int dV}{\int\int v_{in} dA} \quad (4.2a)$$

$$T_{Cturnover} = \frac{\int\int\int \rho C dV}{\int\int \rho C v_{out} dA} \quad (4.2b)$$

where v_{out} and v_{in} are velocities out of and in to the reservoir, C is the concentration of a tracer (e.g., in μmol/kg) and ρ is density. Eq. (4.2a) can also be written in terms of mass rather than volume transport by including ρ in both the numerator and denominator. The rightmost term with inflow velocity in Eq. (4.2a) yields turnover time if the system is in steady state. In a steady state, Eq. (4.2b) could also be written in terms of the inflow.

Residence time is the time an individual water parcel spends in a reservoir. The average residence time is obtained by averaging over all of the water parcels passing through the reservoir. The average residence time is equal to the turnover time if the system is in steady state. The average residence time is twice the age if water is moving steadily through the reservoir, since the age is the average of newest to oldest waters in the reservoir.

4.8. OPTICAL PROPERTIES OF SEAWATER

The transparency of the ocean depends on how much suspended or living material is contained in it, as described in Section 3.8. If the water is very transparent, then solar radiation penetrates to greater depth than if there is much suspended material. Therefore optical properties of the surface waters affect surface layer heating, thereby affecting surface temperature and hence ocean-atmosphere interaction. Ocean color depends on suspended materials,

especially including chlorophyll-producing phytoplankton, so large-scale observations of color and other optical properties can be used to study biological productivity. Optical observations of ocean color using satellite remote sensing are used routinely to quantify the amount of chlorophyll-a (green pigment; McClain, Hooker, Feldmand, & Bontempi, 2006), and, more recently, particulate organic carbon (POC; Gardner, Mishonov, & Richardson, 2006; Stramski et al., 2008), and euphotic zone depth (Lee et al., 2007).

Prior to invention of electronic optical devices, transparency was measured using a Secchi disk (see Section S16.8 in the supplemental materials on the text Web site for information about this instrument). This was done by visually observing when the specially painted disk could no longer be seen from the ship's deck. An enormous number of Secchi disk depths (>120,000) were collected and are archived at the U.S. National Oceanographic Data Center (Lewis, Kuring, & Yentsch, 1988). The majority of the values were for the northern oceans and taken in the summer. There are large areas of the Southern Hemisphere open ocean where there are no values at all, but coastal areas were generally well sampled. Large Secchi depths are found in the open oceans at low and middle latitudes with lower values in higher latitudes and along most coasts. The latitudinal variation is apparent in Figure 4.26, which shows averages of Secchi depths along $180° \pm 20°$W for the Pacific, and along $35° \pm 10°$W for the Atlantic. Lewis et al. (1988) concluded that the prime source of variability in the open ocean is attenuating material in the water. The smaller Secchi depths correspond to higher chlorophyll-a values. The most marked feature in Figure 4.26 is the sharp decrease in Secchi depths beyond about 30° latitude, corresponding to higher productivity in the higher latitudes. The large Secchi depths in the Atlantic are in the Sargasso Sea, a region of notably low biological productivity. In a polynya in the

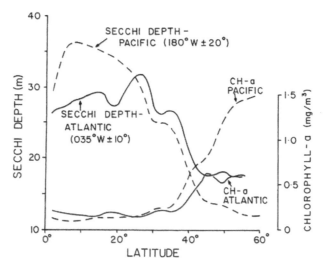

FIGURE 4.26 Mean Secchi disk depths as functions of latitude in the Pacific and Atlantic Oceans. *Source: From Lewis et al. (1988).*

Weddell Sea in 1986, a Secchi disk was visible to four observers at 79 m and disappeared at 80 m. This was claimed as a record: the Secchi depth calculated for distilled water is 80 m, so a greater depth is not possible. In coastal waters, values of 2–10 m are common, and in silty waters near rivers and in estuaries, values of less than 1 m are observed.

Modern in situ optical observations are made with instruments that measure many different aspects of seawater optical properties, which are affected by suspended materials, including sediments and plankton (Section 3.8; Figure 3.9). Fluorescence provides a measure of chlorophyll concentration and therefore, phytoplankton. Within the water column, light transmission, beam attenuation, and fluorescence, among other properties, are measured at different wavelengths to quantify the amount and type of suspended material (Gardner, 2009). As an example, beam attenuation measured with a transmissometer, at a visible wavelength (660 nm), is shown for the equatorial Pacific and the eastern subpolar North Pacific (Ocean Weather Station P or Papa; Figure 4.27). This instrument provides its own light as it is lowered through the water column, so the observation is related to the local amount of scattering and absorption by particles and absorption by water, and not to the actual penetration of sunlight. This particular beam attenuation can then be related to the amount of POC, which is measured from actual samples of seawater. The high beam attenuation in the uppermost layer indicates high POC.

Using ocean color remote sensing and in situ observations of chlorophyll-a and POC, algorithms have been developed to map the latter quantities. Chlorophyll-a maps are now standard remote sensing products. Seasonal maps of chlorophyll from remote sensing are shown in Figure 4.28. Notable features of the northern summer chlorophyll distribution include very low values throughout the subtropical gyres, high values in the equatorial regions and along parts of the ACC, very high values in the high northern latitudes and Arctic, and high values in coastal regions. In austral summer, the high latitude patterns reverse somewhat, with increased chlorophyll along the margins of

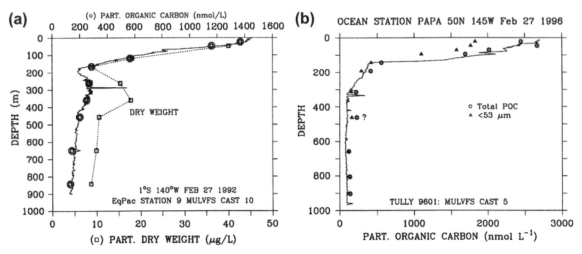

FIGURE 4.27 Profile of beam attenuation coefficient at 660 nm, from a transmissometer, converted to POC (solid line) and in situ measurements of POC (circles): (a) equatorial Pacific and (b) northeast Pacific at OWS Papa. *Source: From Bishop (1999).*

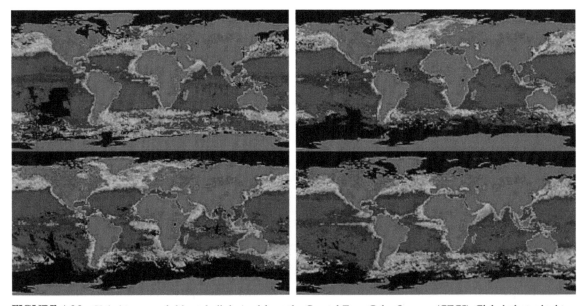

FIGURE 4.28 Global images of chlorophyll derived from the Coastal Zone Color Scanner (CZCS). Global phytoplankton concentrations change seasonally, as revealed by these three-month "climatological" composites for all months between November 1978–June 1986 during which the CZCS collected data: January–March (upper left), April–June (upper right), July–September (lower left), and October–December (lower right). Note the "blooming" of phytoplankton over the entire North Atlantic with the advent of Northern Hemisphere spring, and seasonal increases in equatorial phytoplankton concentrations in both Atlantic and Pacific Oceans and off the western coasts of Africa and Peru. Figure 4.28 will also be found in the color insert. See Figure S4.2 from the online supplementary material for maps showing the similarity between particulate organic carbon (POC) and chlorophyll. *Source: From NASA (2009a).*

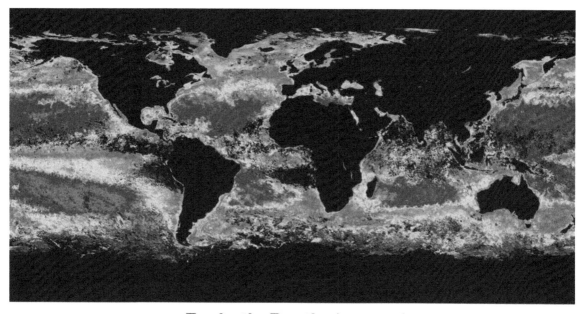

Euphotic Depth (meters)

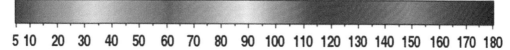

5 10   20   30   40   50   60   70   80   90   100 110 120 130 140 150 160 170 180

FIGURE 4.29 Euphotic zone depth (m) from the Aqua MODIS satellite, 9 km resolution, monthly composite for September 2007. (Black over oceans is cloud cover that could not be removed in the monthly composite.) See Figure S4.3 from the online supplementary material for the related map of photosynthetically available radiation (PAR). This figure can also be seen in the color insert. *Source: From NASA (2009b).*

Antarctica (now ice free) and reduced chlorophyll in the high northern latitudes. The POC distribution derived from ocean color is closely related to the chlorophyll-a distribution (Gardner et al., 2006; see Figure S4.2 in the online supplement).

The solar radiation that affects the upper ocean is quantified as photosynthetically available radiation (PAR; Section 3.8.1), and is mapped routinely from ocean color sensors (NASA, 2009b). An example (Figure S4.3) is included in the online supplementary materials. (In the NASA images, 1 Einstein = 1 mole of photons.) The reader is encouraged to visit the NASA Web site where images are continually posted and where the large seasonal variability is readily apparent.

The euphotic zone depth (Figure 4.29), which is defined as the depth of 1% light penetration, is also mapped from satellite color information using algorithms based on in situ observations (Lee et al., 2007). The euphotic zone depth is related to the historical Secchi disk depths (Figure 4.26 and Section S16.8 of the supplementary online materials); the features that were described previously for the zonally averaged Secchi depths are apparent in the satellite-based map.

Ocean color and derived products are mapped at a resolution of 4–9 km (as in Figure 4.29).

Color then complements remotely sensed SST data of a similar spatial resolution. Chlorophyll-a is somewhat independent of SST, so the two products provide powerful information about local circulation (advection) and upwelling (Simpson et al., 1986). The two fields are used extensively in studies of regional circulation and ecosystems. Examples of ocean color maps to illustrate regional circulation are included throughout later chapters.

Mass, Salt, and Heat Budgets and Wind Forcing

Conservation principles, such as the conservation of energy, of mass, of momentum, and so forth, are important in all of the sciences because these simple principles have very far-reaching results and valuable applications. This chapter discusses conservation of volume (or mass; Section 5.1), conservation of salt and freshwater (Sections 5.2 and 5.3), and conservation of heat energy (Section 5.4), as applied to the oceans. The ocean's heat budget and heat transports are described in Sections 5.5 and 5.6. Because heat and freshwater fluxes combine to make the buoyancy fluxes that are applied to the surface ocean, air—sea buoyancy fluxes are presented in Section 5.7. To complete the presentation of the principal drivers of the ocean circulation (prior to chapters describing the dynamics and circulation), wind forcing is included in Section 5.8.

5.1. CONSERVATION OF VOLUME AND MASS

The conservation of volume principle (or, as it is often called, the equation of continuity) is based on the fact that the compressibility of water is small. If water is flowing into a closed, full container at a certain rate, it must be flowing out somewhere else at the same rate or the level in the container must increase. "Containers" (such as bays, fjords, etc.) in the oceans are not closed in the sense of having lids (except when frozen over), but if, say, the mean observed sea level in a bay remains constant (after averaging out the waves and tides), then the bay is equivalent to a closed container.

For example, many of the fjords of Norway, western Canada, and Chile have large rivers flowing into their inland ends, but the mean sea level in them remains constant. We conclude from the continuity of volume that there must be outflow elsewhere, since evaporation is very unlikely to be large enough to balance the inflow. The only likely place for outflow is at the seaward end; if we measure the currents in fjords, we usually find a net outflow of the surface layer. However, when we actually measure the outflow, we might find a much greater volume flowing out to the sea than is coming in from the river. Because volume must be conserved, there must be another inflow; fjord current measurements usually show inflow in a subsurface layer. The river water, being fresh and therefore less dense than the seawater of the fjord, stays in the surface layers as it flows toward the sea. The subsurface inflow is freshened by mixing with the river water, and upwells into the surface layer where it flows out with the river water. (This is *estuarine* circulation; see

Chapter S8, Section S8.8 in the online supplementary material located at http://booksite.academicpress.com/DPO/; "S" denotes supplemental material.)

5.1.1. Conservation of Volume in a Closed Box

If we represent flows in and out of a closed region such as a fjord (Figure 5.1), and add precipitation (P) and river runoff (R), and subtract evaporation (E) from the water surface, conservation of volume may be stated as:

$$V_i + R + AP = V_o + AE \qquad (5.1)$$

or, rearranged slightly, as

$$V_o - V_i = (R + AP) - AE \equiv F. \qquad (5.2)$$

Here V is the *volume transport*, which expresses flow in terms of volume per second (m^3/sec) rather than as a speed (m/sec). The subscripts o and i represent outward and inward transports, respectively. The symbol R represents the river runoff as a volume transport that adds water into the basin. The symbols P and E are the precipitation and evaporation for each point and are therefore expressed in m/sec or an equivalent, such as cm/year. To calculate the total volume transport into the box due to precipitation and evaporation, they must be integrated over the surface area of the box. For the simple, illustrative approximation of Eqs. (5.1) and (5.2), if P and E have the same values at all points in the box, they can just be

multiplied by the surface area, A, of the box. F (for freshwater) is defined by the right side of Eq. (5.2), which is why the symbol $\equiv$ is used. The left side of Eq. (5.1) is the volume transport into the fjord. The right side of Eq. (5.1) is the volume transport out. The second equation simply says that the net volume flow of salt water balances the net volume flow of freshwater (when averaged over a suitable time period). This is an example of a *steady-state* situation in which some or all of the parts of a system may be in action, but at no point is there any change of motion (or of a property) with time.

(To be more precise, the principle expressed in Eqs. (5.1) and (5.2) should also include the density of water, and becomes a statement of mass conservation rather than of volume conservation. This is because simple heating of water will expand it slightly without adding any mass, so the true conservation principle is for mass. However, for most ocean applications, seawater density has such a small range of values that we can usually assume the density is uniform.)

Even though this conservation principle was discussed using an example of a nearly closed region, such as a fjord, it also applies just as well to any other closed "box" that might be drawn in the ocean. If our closed box includes the sea surface, then it will include P and E. If it has a coastline, then it will include R. If it has sea ice flowing into it and melting, or vice versa, then it includes yet another term for the water volume in the ice. Also, our box could be completely within the ocean somewhere, in which case the flows into the box must balance flows out of the box, as described next.

5.1.2. Open Ocean Continuity

Thinking about flow into and out of a closed box can be extended into the open ocean. Here we think of a hypothetical closed box, with sides, a top, and a bottom (Figure 5.2). We

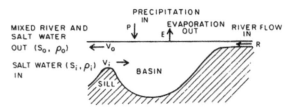

FIGURE 5.1 Schematic diagram of basin inflows and outflows for conservation of volume discussion.

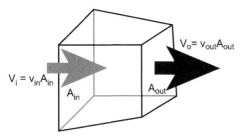

FIGURE 5.2 Continuity of mass for a small volume of fluid. By continuity, $V_o = V_i$.

then apply the same balance (Eq. 5.1) to this box. If none of the sides are next to the coast, then the runoff term R is zero. If the top of the box is inside the ocean and is not the sea surface, the precipitation and evaporation terms are also zero. Then the volume balance for the box becomes:

$$V_o - V_i = 0 \qquad (5.3)$$

This says that the transport into the box must equal the transport out of the box. (The full "continuity equation," expressed in partial derivative form, is given in Section 7.2.) In practice, in all open ocean areas, the volume transports into and out of boxes of interest are usually much larger than any precipitation or evaporation flux across the sea surface, so we use an approximate version of Eq. (5.3) even for boxes that include the sea surface (Section 5.1.3).

This principle of continuity is so fundamental that it might not seem very interesting, but it is the one law that applies in all situations, regardless of how complex the system becomes.

5.1.3. Radiation, Flux, and Diffusion

Before we can talk further about conservation of volume and salt, we need to understand how heat, water, salt, and other dissolved materials move around within the ocean and how they can be changed by physical processes (as opposed to chemical or biological processes).

There are three ways to physically change things inside the ocean: radiation, advection, and diffusion.

Radiation is how electromagnetic waves — heat and light — move. Radiation is most important in the atmosphere and less important in the ocean, since water is not very transparent. However, light does penetrate the ocean's surface layer ("euphotic zone"; Section 3.8), which is how the sun actually heats the ocean (Section 5.4). The ocean also radiates heat (infrared electromagnetic waves) out to the atmosphere (Section 5.4.2).

Advection is how the movement of a fluid "parcel" carries properties such as heat and salt. Sometimes we use the word *convection* when referring to vertical motion. As already introduced in Section 5.1.1, the basic concept here is *velocity*, which has units of length divided by time (m/sec) and a direction. A fluid is made up of countless molecules that move more or less together. If we draw some sort of surface in our minds through a part of the ocean, the surface will have an area (A; Figure 5.2). The *volume transport* of water through the area is equal to the velocity of the fluid (v) through the surface multiplied by the area A, or vA. The units of volume transport are m^3/sec. We can talk about transport of mass as well. Water, including seawater, has density ρ, which has units of mass/volume. *Mass transport* though our area is then density multiplied by velocity and area ($\rho v A$) and has units of kg/sec.

Seawater has dissolved matter in it, which has a concentration (C) of mass or molecules of matter per unit mass of seawater. (Recall our definition of salinity in Chapter 3.) We can talk about any dissolved matter, including salts. The transport of the dissolved matter becomes this concentration times density times velocity times area ($C \rho v A$) and has units of (mass of matter)/time or molecules/time, depending on how you write the concentration. For *salt transport*, we use salinity written as units of grams of salt per kilogram of seawater, and so salt transport has units of g/sec, which can also be written in terms of kg/sec by multiplying by 1000.

For specific dissolved substances, the concentration can be written in terms of moles per unit mass of seawater (mol/kg). Because concentrations of many common dissolved materials (such as oxygen and nutrients) are on the order of 10^{-6} mol/kg, the concentration unit of µmol/kg (micromoles per kilogram) is often used. Transports of these substances are then expressed as µmol/sec.

Heat transport uses the definition of heat Q in Eq. (3.1), which has units of energy (Joules). In place of "concentration," we use the specific heat multiplied by temperature (on the absolute temperature scale, Kelvin), and the transport is (Q v A), with units of J/sec. The unit for J/sec is a Watt, where 1 W = 1 J/sec.

Flux is directly related to transport. Flux is the transport per unit area. It can be thought of as "stuff" per unit time per unit area. For instance, heat flux is expressed as W/m^2, which is Joules per unit time per unit area.

Flux and transport are important when there is a difference between the transport of a property into a closed volume and the transport out of the same closed volume (Figure 5.2). The change in transport has to be related to a change in properties within the closed box. For example, if there is a larger flux of heat into a box than out of the box, then the water coming out of the box is cooler and must have been cooled inside the box. This could happen through a loss of heat out of the sea surface if one side of the box is at the sea surface. As another example, if there is a higher transport of oxygen into the box than out of the box, then oxygen could have been consumed (usually by bacteria) within the box. This change in transport through a box is called transport *divergence* (if more comes out than goes in) or transport *convergence* (if less comes out than goes in).

Advection is similar to flux, but advection occurs at a point rather than through the side of a volume. When a property is carried along by the flow, it is "advected." The equations in fluid mechanics that describe the change in a property at a point in a fluid include "advective terms," which indicate how the divergence or convergence of the flux of the property changes the property at that location.

Diffusion is the third way properties can change in a fluid. Diffusion is like flux convergence and divergence, but it happens at extremely small spatial scales. Molecules or tiny parcels of water bump around randomly (turbulence) and carry their properties with them. If there is a difference in heat or salt from one side of a region to the other side, then the random jostling will gradually smooth out the difference (or "gradient," which is the property difference divided by the separation distance, if you consider this distance as becoming very small).

Fick's law of diffusion says that the diffusive flux of "stuff" is proportional to its concentration gradient. Therefore diffusion will move stuff down a gradient (from high concentration to low concentration). If there is no gradient (no variation in concentration), there is no flux and hence no impact of diffusion. If the gradient is constant (meaning that the property difference centered at one location is the same as at another location), diffusion also has no effect on the property distribution because there must be a flux divergence or convergence for the property to change. In mathematical terms, the second spatial derivative of the concentration must be non-zero for diffusion to cause a change in concentration. More simply said, if there is more stuff at one location than at another, it will flux toward the lesser concentration. But the concentration will only change if there is a flux convergence or divergence. Therefore diffusion only acts if the property concentration gradient varies.

In turbulent flows, such as water in the ocean or air, we are sometimes more interested in properties and velocity changes over scales of many meters to many kilometers, or even thousands of kilometers, than over scales of centimeters or less. It is almost impossible to consider all ranges of motion at once, even though a single, limited set of fluid dynamics equations describes all of them. Therefore fluid dynamicists and modelers almost always make simplifying assumptions about the scales of motion (spatial and temporal) they are interested in, and often treat smaller scales (subgrid scale) as if they obey random, molecular motions.

Fluids such as water and air are highly turbulent, meaning that they are not very viscous. Turbulence at small scales is often considered to act on the larger scales of interest in a way that is analogous to random, molecular, microscopic motions. Thus, fluid dynamicists introduce the concept of *eddy diffusivity*, in which turbulent "eddies" at smaller scales accomplish the diffusion. Eddy diffusivity is much higher than molecular diffusivity since turbulent eddies carry properties much farther than molecular motions. Eddy diffusivity (and eddy viscosity) are discussed again in Section 5.4.7 and more formally and in more depth in Section 7.2.4.

5.2. CONSERVATION OF SALT

The principle of conservation of salt is based on the nearly accurate assumption that the total mass of dissolved salts in the ocean is constant. Rivers contribute a total of about 3×10^{12} kg of dissolved solids per year to the sea, which may sound like a lot, but it has negligible effect on salinity. The ocean is very large. The total amount of dissolved salt in the ocean is 5×10^{19} kg. Therefore the amount of salt brought into the ocean each year by the world's rivers increases the average ocean salinity by about one part in 17 million per year. We can only

measure salinity to an accuracy of about ± 0.001, or about 500 parts in 17 million if we assume the mean ocean salinity to be about 35. In other words, the oceans would increase in salinity each year by an amount which is only $1/500$ of our best accuracy of measurement if we neglect removal of salt for the moment. Salt is actually removed from the ocean in the form of evaporites, so salinity actually increases even more slowly, if at all, over geological time. For all practical purposes, it is assumed that the average salt content of the oceans is constant, at least over periods of tens or even hundreds of years.

Salinity, which is the dilution of the salts (Section 3.4), could vary at a barely measurable level with change in the total amount of water in the ocean, which depends on how much water is locked up in ice, especially in ice sheets (Greenland and Antarctica). If 1 m of water were added to the ocean from melting ice sheets and glaciers, the salinity change would be $1/4000 = 0.0003$, since the mean depth of the oceans is about 4000 m (Section 2.1). The maximum sea level change that might be expected from complete melting of Greenland (which is not an unreasonable possibility) is 7 m, leading to a mean salinity decrease of 0.002, which would be barely observable.

The early Greek philosophers were confused about the salinity of the ocean as compared to the freshwater of the rivers feeding into them. Rather than realizing that the very long-term accumulation of salts from these rivers caused the ocean's salinity, they postulated "salt fountains" at the bottom of the ocean. As recently as the nineteenth century, Maury (1855) believed that salt had been present in the oceans since "creation," in contrast to what he refers to as the "Darwin theory" (which is now understood to be mostly correct), that the salt is washed in by rivers.

The conservation of salt is usually applied to bodies of water that are smaller than the world ocean. Salt conservation in smaller bodies, such as the Mediterranean Sea, is a reasonable hypothesis. Salinity, however, can vary when the freshwater balance changes the dilution, mainly through changes in precipitation and evaporation patterns and amplitude. While the mean salinity of the ocean does not measurably change, the salinity of one region can increase while that of another region decreases, due to redistribution of freshwater. The salt conservation principle may be expressed symbolically as:

$$V_i \bullet \rho_i \bullet S_i = V_o \bullet \rho_o \bullet S_o \qquad (5.4)$$

where V_i and V_o are the volume transports of the inflowing and outflowing seawater for an enclosed region, S_i and S_o are their salinities, and ρ_i and ρ_o their densities. This is the equation for *salt transport*; it states that no salt is gained or lost inside the region. (For complete accuracy, Eq. 5.4 should be applied pointwise, using velocity rather than transport, and then added up over the whole area surrounding the volume being considered.) The left side of Eq. (5.4) is the rate at which salt is transported into the box, and the right side is the rate of transport out. Because the two densities will be the same within 3% (the difference between seawater and freshwater) the ρ's nearly cancel, leaving:

$$V_i \bullet S_i = V_o \bullet S_o \qquad (5.5)$$

This equation can be combined with the equation for conservation of volume (5.2) to give Knudsen's relations (Knudsen, 1901):

$$V_i = F \bullet S_o/(S_i - S_o) \text{ and } V_o = F \bullet S_i/(S_i - S_o) \qquad (5.6)$$

where F is due to runoff, precipitation, and evaporation (F = R + P − E), integrated over the full area. F is the freshwater volume input measured in m^3/sec. Equation (5.6) is useful for calculating volume transport if we know F and measure the salinities.

Conversely, if we know the transports and salinities around the perimeter of the region through measurements, we can calculate F:

$$F = V_i \bullet (S_i/S_o - 1) \text{ or } F = V_o \bullet (1 - S_o/S_i) \qquad (5.7)$$

This is the equation for *freshwater transport* and expresses how much freshwater is gained or lost inside the box. Equations (5.6) and (5.7) can be applied to any region, especially including marginal seas, estuaries, and fjords where inflow and outflow salinities are easily assigned. If F is positive (more runoff and precipitation than evaporation), then the marginal sea is considered to be "positive." If F is negative (net evaporation), then the marginal sea is called "negative."

Qualitative conclusions can be drawn from Eqs. (5.6) and (5.7). If both S_o and S_i are large, they must be similar because there is an upper limit to S in the ocean. Therefore $(S_i - S_o)$ must be small and both $S_o/(S_i - S_o)$ and $S_i/(S_i - S_o)$ must be large. Therefore V_i and V_o must be large compared with F, the excess of freshwater inflow over evaporation. That is, for large volume exchanges (large flushing rate), the salinity change will be small for a given amount of evaporation or precipitation. On the other hand, if the salinity difference between inflow and outflow is large (S_o much less or much more than S_i), then the exchange rate (V_i and V_o) is small for the same size F. Thus a body of water with large volume exchange will be better flushed and less likely to be stagnant than one with small volume exchange.

For the open ocean, where salinity and velocity vary continuously, it is more useful and accurate to calculate salt and freshwater transports as integrals of vS and v (1 − S/S_o) around the whole region being considered, where v and S are point observations of velocity and salinity (Wijffels, Schmitt, Bryden, & Stigebrandt, 1992; Wijffels, 2001; Talley, 2008). (The integration is in depth and horizontal distance around the region.) S_o is an arbitrary constant. The net volume transport F into the whole region should balance the amount gained and lost by runoff, precipitation, and evaporation, hence be very small.

5.3. THREE EXAMPLES OF THE TWO CONSERVATION PRINCIPLES

5.3.1. The Mediterranean Sea: An Example of Negative Water Balance

The Mediterranean Sea has a negative water balance — evaporation exceeds precipitation plus river runoff. There is a small net loss of volume due to net evaporation (i.e., for the volume transport Eq. 5.2, $E > (R + P)$ and F is negative). Because salt is conserved, the salinity increases. The saltier water is denser and sinks within the Mediterranean. This denser water flows out of the Mediterranean at the bottom of the sill at the Strait of Gibraltar, injecting this saltier water into the North Atlantic at depth (Section S8.10.2 in Chapter S8 located on the textbook Web site). The outflow, with salinity 38.4 psu, is balanced by inflow of less salty (36.1 psu) water from the North Atlantic in the upper layer (Figure 5.3a). The ratios of salinities in Eq. (5.6) both have values of about 16, which means that the inflow and outflow volume transports V_i and V_o are both greater by this factor than the air–sea freshwater loss, F.

Direct measurements of the upper layer inflow at the Strait of Gibraltar (Section S8.10.2 on the textbook Web site) give an average inflow transport of $V_i = 0.72$ Sv, where 1 Sv = 1×10^6 m³/sec. Then, from Eq. (5.6), $V_o = 0.68$ Sv and

$F = (R + AP) - AE = -0.04$ Sv; in other words, total evaporation exceeds freshwater input by 0.04×10^6 m³/sec. The units for inflow V_i can be converted to 2.3×10^4 km³/year. At this rate it would take about 165 years to fill the Mediterranean, which has a volume of 3.8×10^6 km³. (The Mediterranean does not "fill" since outflow balances inflow.) This "filling rate" is a measure of the mean *turnover time*, that is, the time required for replacement of all the Mediterranean water (sometimes called *flushing time* or *residence time*) (Section 4.7).

The deep salinity within the Mediterranean Sea is between 38 and 39 psu (Section S8.10.2 on the textbook Web site). The outflow salinity at the Strait of Gibraltar is lower than this because the outflow entrains (mixes with) lower salinity inflow as it passes through the strait into the North Atlantic.

5.3.2. The Black Sea: An Example of Positive Water Balance

Even though it is adjacent to the Mediterranean, the Black Sea (Section S8.10.3 located on the textbook Web site) is a "positive" basin, in which there is a net gain of freshwater from the atmosphere and runoff (Figure 5.3b). The salinity of the inflow (bottom layer) is approximately 35 psu. The salinity of the outflow (upper layer) is much lower at 17 psu. The ratios of salinities in Eq. (5.6) are 1 and 2, respectively, indicating that the transports V_i and V_o, which are the Black Sea's exchange with the Mediterranean, are of the same order as the air–sea

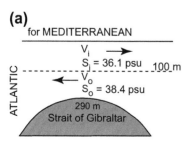

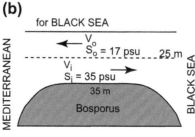

FIGURE 5.3 Schematic diagrams of inflow and outflow characteristics for (a) Mediterranean Sea (negative water balance; net evaporation), (b) Black Sea (positive water balance; net runoff/precipitation).

freshwater balance flux F. Measured values are approximately $V_i = 9.5 \times 10^3$ m^3/s (300 km^3 yr^{-1} of saline water) and $V_o = 19 \times 10^3$ m^3/s (600 km^3 yr^{-1} of fresher water), giving F = (R + P) − E = 9.5×10^3 m^3/sec (Oguz et al., 2006). That is, there is a net flux of freshwater into the Black Sea from an excess of runoff and precipitation compared with evaporation. (The mean salinity of the deep Black Sea is about 22.4 psu and the surface layer is much fresher, which means there is a net gain of freshwater.) Using the Black Sea volume of 0.6×10^6 km^3, a turnover time of 1000–2000 years can be calculated. These turnover, or flushing-time calculations, are very rough, but the contrast with the 165 years turnover time for the Mediterranean is notable given that these seas are connected.

Independent oceanographic measurements support the contrast in turnover time between the Mediterranean and Black Seas, as the bulk of the Mediterranean water has an oxygen content of over 160 µmol/kg (>4 ml/L) whereas the Black Sea water below 200 m has no dissolved oxygen but a large amount of hydrogen sulfide (over 6 ml/L), indicative of great age. The Mediterranean is described as well flushed or well ventilated, whereas the Black Sea is stagnant below 95 m. As described in Chapter 9, the physical reason for the ventilation of the Mediterranean is that deep water is formed by winter evaporation and cooling at the surface in the north. In the Black Sea, precipitation and river runoff decrease the salinity and density so much that even severe winter cooling cannot make the water dense enough to sink. Thus, regional climate dictates turnover time.

5.3.3. Salt and Freshwater Transports in the Open Ocean

The concepts of salt and freshwater transports are important for global water balances. It rains more in some regions than in others, and there is more evaporation from the sea surface in some regions than in others; yet, on the whole, the salinity distribution of the world oceans is mostly in steady state. The ocean does not become saltier over time in evaporation regions or fresher in net precipitation regions. (This is not to say that there are no small daily or seasonal changes, or small and perhaps important climate changes over the course of years to decades. Rather, the general distribution observed in the 1990s, described in Chapter 4, applies as well to several hundred years ago and perhaps even several hundred years hence.)

Evaporation, precipitation, and runoff (see map in Figure 5.4a) affect only the total water content (freshwater) and not the total amount of salt. Salt remains, by and large, in the ocean. (The amounts flung into the air, where they might become important condensation nuclei for clouds, are infinitesimal, and have no impact on ocean salt budgets; the input rates from weathering are also miniscule.) However, evaporation, precipitation, and runoff do change the concentration of salt, that is, the salinity. Globally there is net evaporation (red regions in Figure 5.4a) reaching more than 150 cm/year in each of the southeastern subtropical regions. Net precipitation (blue regions) is high in the tropics beneath the ascending air of the atmosphere's Hadley circulation (Intertropical Convergence Zone). Net precipitation is also found in the subpolar regions of both hemispheres, in the Antarctic and Arctic.

For the steady-state salinity distribution in the ocean, freshwater must be transported within the ocean from regions of net precipitation to regions of net evaporation. (The rest of the freshwater cycle is completed through the atmosphere, which must transport moisture from regions of net evaporation to those of net precipitation. The net freshwater transport into each area of the ocean must exactly balance the net freshwater transport in the atmosphere over the same area.) The total volume transports associated with open ocean freshwater transport

(a)

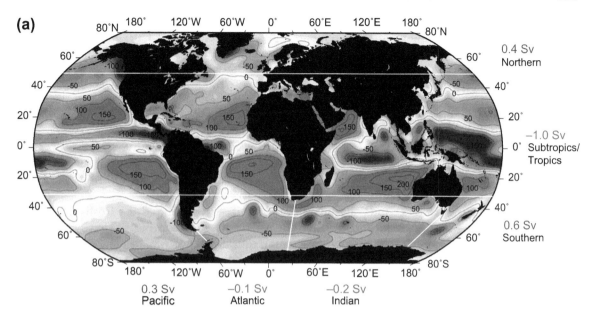

0.4 Sv
Northern

−1.0 Sv
Subtropics/
Tropics

0.6 Sv
Southern

0.3 Sv   −0.1 Sv   −0.2 Sv
Pacific   Atlantic   Indian

(b)

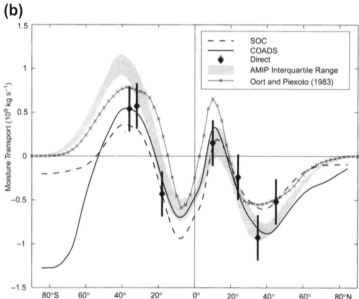

FIGURE 5.4 (a) Net evaporation and precipitation (E−P) (cm/yr) based on climatological annual mean data (1979–2005) from the National Center for Environmental Prediction. Net precipitation is negative (blue), net evaporation is positive (red). Overlain: freshwater transport divergences (Sverdrups or 1×10^9 kg/sec) based on ocean velocity and salinity observations. This figure can also be found in the color insert. *After Talley (2008)*. (b) Meridional (south to north) freshwater mass transport (Sverdrups), positive northward, based on ocean velocity and salinity observations (direct) and based on atmospheric analyses (continuous curves). *Source: From Wijffels (2001)*.

are very small compared with the net volume transports. That is, while the circulation regularly transports volume at rates of 10×10^6 to 100×10^6 m^3/sec from one location to another, freshwater transport into large ocean regions (gain or loss to the atmosphere in that region) is on the order of 0.1 to 1.0×10^6 m^3/sec. As an example, consider the total freshwater transport into the central North Pacific between latitudes 25°N and 35°N. This region has a surface area of 16.2×10^{12} m^2. This is a region of net evaporation, resulting in higher surface salinities than in the tropics and subpolar region. The net freshwater flux F into the ocean is 0.11×10^6 m^3/sec, based on climatology (Figure 5.4a). The circulation in this region is dominated by a strong western boundary current called the Kuroshio, which flows northward along the western boundary in a band that is about 100 km wide (Section 10.3.1). Most of this water turns around and flows back southward across the width of the North Pacific, mainly within the upper ocean. If we apply Eq. (5.6) to this situation, using an inflow Kuroshio volume transport for the upper ocean of 25×10^6 m^3/sec and a freshwater gain F of 0.11×10^6 m^3/sec, we calculate that the salinity of the southward flow should be about 0.15 psu lower than the salinity of the northward Kuroshio. If we go to the actual data for the upper 1000 m of the ocean, we find that the average salinity of the Kuroshio is 34.73 psu, and the average salinity of the southward return flow is 34.60 psu, which substantiates our estimate.

Global estimates of meridional (south to north) ocean freshwater transports (Figure 5.4b) have been constructed from the total distribution of evaporation/precipitation/runoff (as in Figure 5.4a). These freshwater transports are all less than 1 Sverdrup (1 Sv = 1×10^6 m^3/sec, which is equivalent to the units of 1×10^9 kg/sec in Figure 5.4b). Even the weakest ocean currents transport much more total water volume than this. The freshwater transport of

Eq. (5.6) is the excess amount of freshwater at one location compared with another. Thus, the freshwater transport is the amount of dilution or evaporation required to change the salinity in a given region. In other words, what we are really calculating (and what we can really compare with the precipitation, evaporation, and runoff) is the divergence or convergence of freshwater transport into a given region. When calculating these transports over complete ocean basins, an arbitrary reference salinity is chosen such that all other salinities are compared with it and the freshwater transports calculated accordingly. That is, the arbitrary constant salinity S_0 in the denominator in Eq. (5.6) must be a single number for the whole global calculation.

The freshwater divergences (net freshwater transport into the indicated areas) labeled in Figure 5.4a show more graphically the pattern of these differences in freshwater transports with latitude. Where the freshwater transport increases toward the north, freshwater is being added to the ocean. This occurs in the rainy belts from 80°S to about 40°S, from 10°S to 10°N, and from 40°N to 80°N (also see Figure 5.4b). Where the freshwater transport decreases toward the north, freshwater is being removed. These are the evaporation regions of the subtropics, from 40°S to 10°S and from 10°N to 40°N.

The total freshwater transport for the globe must balance to nearly zero when averaged over several years, given that the ocean's mean salinity is constant (Section 5.2). Thus the freshwater transport curves of Figure 5.4b should start at zero in the south at Antarctica and end at zero at the North Pole. The "indirect" estimates of freshwater transports, based on precipitation and evaporation (Comprehensive Ocean Atmosphere Data Set, or COADS and from the National Oceanography Centre, Southampton or NOCS), do not balance because they are based on surface observations of rainfall and evaporation, which have large errors, especially in the Southern Ocean. The "direct" estimates, which are calculated from ocean velocity and salinity observations, fall along the curves from the indirect estimates. This indicates that both estimates are detecting a similar signal. Both panels of Figure 5.4 show net precipitation at high southern and northern latitudes, and net evaporation in the subtropics. Net precipitation at the equatorial latitudes is evident in Figure 5.4b. The map (Figure 5.4a) shows, additionally, that the Atlantic and Indian Oceans are net evaporative, while the Pacific has net precipitation. This accounts for the relative saltiness of the Atlantic and Indian Oceans compared with the Pacific. The Southern Ocean, south of 30°S, is fresher than all of these.

Higher evaporation in the Atlantic compared with the Pacific is associated with the trade winds. In the Atlantic, they originate from the dry continent (Mideast and northern Africa), whereas in the Pacific they have only the narrow Central American landmass to cross; that is, there is a zonal atmospheric transport of moisture from the Atlantic to Pacific (Zaucker & Broecker, 1992).

5.4. CONSERVATION OF HEAT ENERGY; THE HEAT BUDGET

5.4.1. Heat Budget Terms

The spatial and temporal variations of ocean temperatures are indications of heat transfer by currents, absorption of solar energy, loss by evaporation, and so forth. The size and character of the temperature variations depend on the net rate of heat flow (transport) into or out of a water body. Heat budgets quantify these balances. In the following list, the symbol Q represents the rate of heat flow measured in Joules per second (Watts) per square meter, W/m^2. Subscripts are used to distinguish the different components of the heat budget. These components include:

Q_s = rate of inflow of solar energy through the sea surface (shortwave radiation)
Q_b = net rate of heat loss by the sea as longwave radiation to the atmosphere and space (back radiation)
Q_h = rate of heat loss/gain through the sea surface by conduction (the sensible heat flux)
Q_e = rate of heat loss/gain by evaporation/condensation (the latent heat flux)
Q_v = rate of heat loss/gain by a water body due to currents (the advective term)

Other sources of heat inflow, such as that from the earth's interior, change of kinetic energy of waves into heat in the surf, heat from chemical or nuclear reactions, and so forth, are all small and can mostly be neglected relative to the previously listed terms. The heat budget for a particular body of water is:

$$Q_T = Q_s + Q_b + Q_h + Q_e + Q_v \qquad (5.8)$$

where Q_T is the total rate of gain or loss of heat of the body of water (T refers to total). A schematic of average values of these terms is shown in Figure 5.5. Q_v, the advective heat flux, is not shown in Figure 5.5. The advective heat flux, which is internal to the ocean and is the product of velocity and temperature gradient (Section 7.3.1), can range from 1 to over 20 units on the scale of Figure 5.5.

When Eq. (5.8) is used for heat-budget calculations, numerical values have a positive sign if the water gains heat and a negative sign if they represent a heat loss from the sea. Solar heat flux Q_s values are always positive (heat gain) and longwave back radiation Q_b values are almost always negative (heat loss). Latent heat fluxes, Q_e, are almost always negative. Sensible heat flux, Q_h, can be negative or positive depending on the sign of the temperature difference between the air and water. Advective heat flux, Q_v, depends on the difference in temperature between the water flowing into the region and water flowing out of the region. These volume transports are equal by the Conservation

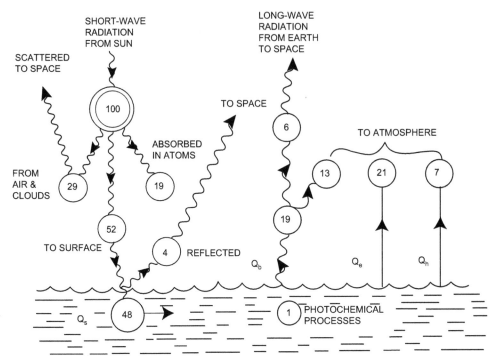

FIGURE 5.5 Distribution of 100 units of incoming shortwave radiation from the sun to Earth's atmosphere and surface: long-term world averages.

of Volume (Eq. 5.3), and in actuality differ only by the very small freshwater gain or loss within a region. Therefore Q_v may be positive (inflow of warmer water and outflow of colder water) or negative (opposite case).

Observations of solar radiation, back radiation, and latent and sensible heat flux are at points on the sea surface, with units of W/m^2. To obtain their total impact on the heat content in a body of water in Watts, they must be multiplied by the sea surface area (m^2) of the body. (For continuously varying values, this is actually a sum of the heat fluxes at each unit area of the sea surface, or equivalently, an area integral.) Advection through the sides of the water body must likewise be calculated at all points on the sides and summed for each unit area (equivalent to an area integral).

If the temperature of a body of water is not changing with time, this does not mean that there is no heat exchange. It simply means that the algebraic sum of the terms on the right side of the heat-budget equation (5.8) is zero—net heat inflow equals net heat outflow, an example of a steady-state condition. If we apply the heat-budget equation to the world ocean as a whole, Q_v must be zero because then all the advection is internal and must add up to zero. Also, if we average over a whole year or number of years then the seasonal changes average out and Q_t becomes zero. The equation for the oceans in this case simplifies to

$$Q_s + Q_b + Q_h + Q_e = Q_{sfc} = 0. \qquad (5.9)$$

The global distribution of each of the four components is examined next. Typical relative values in Figure 5.5 are only intended as an indication of the general range and must not be used for specific calculations. The largest component is the shortwave radiation Q_s and it is always positive (input of heat into ocean). The other three components usually represent a loss of heat from the ocean. The sensible heat flux Q_h varies with time and place, having maximum values in the north-western North Atlantic and North Pacific, but is generally the smallest term. Latent heat flux Q_e is the second largest term in the heat-balance equation and has large seasonal variations. Longwave radiation Q_b has the smallest variations.

The following sections explain how each of these heat flux components is calculated. The observed quantities are temperature, humidity, wind speed, cloud cover, and surface reflectivity. These are measured from routine observation stations, ships, ocean buoys, and, increasingly, from satellites. The heat fluxes are calculated from these observations, based on empirical approximations called "bulk formulae," with basic physical principles only loosely at the core. While there has been modest progress made in our understanding of the physical principles of turbulent heat exchanges, this progress has not transitioned into a more formal analytical description of the individual heat flux terms. The only alternatives to these bulk estimates are precise observations of the individual heat fluxes. Such observations are sufficiently complex that they cannot be routinely made.

Local experiments have been carried out at island stations, moorings, and research ships, which have provided time series of accurate measurements of heat exchange, including diurnal components, and also provided data to improve the bulk estimates. A long-term goal is to improve air–sea heat exchange estimates to have errors of less than 10 W/m^2. Hopefully, at some point, satellite measurements will provide global, accurate coverage of each component of the heat exchange.

Our discussion of the commonly used bulk estimates closely follows Josey, Kent, and Taylor (1999), with a summary of satellite techniques given by Liu and Katsaros (2001).

Maps of annual (and seasonal) averages of each of the heat flux components as well as descriptions of their patterns are provided in Section 5.5 (and in the online supplement to this chapter).

5.4.2. Shortwave and Longwave Radiation: Elements of Radiation Theory

Before discussing the shortwave and long-wave radiation terms, Q_s and Q_b, certain aspects of electromagnetic radiation theory must first be reviewed. First, *Stefan's Law* states that all bodies radiate energy at a rate proportional to the fourth power of their absolute temperature T (expressed in Kelvin as K = °C + 273.15). This energy is in the form of electromagnetic radiation over a range or spectrum of wavelengths. Second, the concentration of energy is not the same at all wavelengths but has a marked peak at a wavelength of l_m given by *Wien's Law*: $l_m \bullet T = 2897$ μm K, where T is again the absolute temperature (in degrees Kelvin) of the radiating body. Therefore bodies at higher temperatures radiate energy at preferentially shorter wavelengths and vice versa.

The sun has a surface temperature T of approximately 6000 K and radiates energy in all directions at a rate proportional to $T^4 = 6000^4$. According to Wien's Law this energy is concentrated around a wavelength of 0.5 μm (1 μm $= 10^{-6}$ m); 50% of this energy is in the visible part of the electromagnetic spectrum (about 0.35 to 0.7 μm), whereas 99% is of wavelengths shorter than 4 μm. This energy is referred to as *shortwave* radiation and is the source of the Q_s term in the heat budget. The shortwave energy that reaches the ocean (after passing through the atmosphere and clouds, but not including the portion that is reflected) is absorbed by the water where it is converted into heat energy. This increases the temperature of the water, consistent with its specific heat, which relates temperature to heat energy (Section 3.3.2).

The *longwave* radiation term Q_b represents the electromagnetic energy that is radiated outward by the earth (land and sea) at a rate depending on the absolute temperature of the local surface. Taking an average temperature of $17°C = 290$ K for the sea, it is radiating energy at a rate proportional to 290^4. This is a much smaller rate than for the sun. Wien's Law then says that the earth's peak radiation wavelength is longer since the temperature is lower. The wavelength at which the sea radiation reaches its maximum is about 10 μm (i.e., in the thermal infrared). About 90% of the sea radiation is in the wavelength range from 3 to 80 μm.

5.4.3. Shortwave Radiation (Q_s)

The sun is the dominant source of energy for Earth. Most of the sun's energy is in the visible (short) wavelength part of the electromagnetic spectrum. Because of absorption and scattering in the atmosphere and reflection, only 50% or less of this radiation reaches the earth's surface. In Figure 5.5, this loss of shortwave radiation is represented by the 29 units lost to space by scattering from the atmosphere and clouds, 19 units absorbed in the atmosphere and clouds, and 4 units reflected from the sea surface. The remaining 48 units enter the sea as the Q_s term of the heat budget. Of these 48 units, about 29 units reach the sea as *direct* radiation from the sun and 19 units as *indirect* scattered radiation from the atmosphere (*sky* radiation). This distribution represents a long-term world-area average; instantaneous values vary diurnally, seasonally, and with locality and cloud cover.

Shortwave radiation input to the sea is typically calculated using two different methods: from bulk formulae using in situ observations

and from a suite of satellite observations. Direct measurements of the energy arriving at the sea surface are made with a pyranometer (see Section S6.8 located on the textbook Web site), but it is not practical to do this over large areas, or for prediction. Such direct observations are used to derive the bulk formulae and develop the satellite algorithms and calibrations.

The following bulk formula is in general use for the shortwave radiation flux penetrating the ocean's surface, using traditional surface-based observations of cloud cover:

$$Q_s = (1 - \alpha)Q_c(1 - 0.62C + 0.0019\theta_N), \quad (5.10)$$

in which Q_c is the incoming clear-sky solar radiation (measured above the atmosphere in units of W/m^2, and often referred to as the "solar constant," even though the value is not constant in time or space), C is the monthly mean fractional cloud cover, α is the *albedo* (fraction of radiation that is reflected), and θ_N is the noon solar elevation in degrees (Taylor, 2000). In practical calculations, Q_s is not allowed to exceed Q_c. The terms in Eq. (5.10) are explained in the next subsection. Absorption of the radiation by the sea is discussed in Section 5.4.3.2. Annual mean values for Q_s are shown below in Section 5.5 and Figure 5.11.

Satellite-based shortwave radiation calculations include observing the incident solar radiation at the top of the atmosphere, composition of the atmosphere including water vapor content and clouds, and information on surface conditions including atmospheric reflectivity. A major effort has been put into observing cloud conditions from satellites (International Satellite Cloud Climatology Project or ISCCP, and the Atmospheric Radiation Monitoring or ARM program). The top of the atmosphere radiation is measured through the Earth Radiation Budget Experiment (ERBE). These products are combined in the Surface Radiation Budget Program at NASA. An example of a map of the shortwave radiation from ERBE is shown in Figure S5.1 located on the textbook Web site. These shortwave radiation estimates are still bulk estimates since they involve observations of the external parameters that affect radiation rather than being direct measurements of the radiation penetrating the ocean's surface.

5.4.3.1. Factors Affecting Shortwave Radiation Reaching Earth's Surface

In the expression for shortwave radiation (Eq. 5.10), the central quantity is the incoming clear-sky solar radiation Q_c. The rate at which energy reaches the outside of the atmosphere from the sun is called the *solar constant* and, as obtained from satellite measurements above most of the earth's atmosphere, is about 1365–1372 W/m^2, perpendicular to the sun's rays. In Figure 5.5 this penetration of shortwave radiation is represented at the top left as 100 units of incoming shortwave radiation. In addition to direct sunlight, the sea also receives a significant amount of energy from the sky, such as sunlight scattered or absorbed and re-radiated by the atmosphere, clouds, and so forth. The skylight component increases in importance at high latitudes. For instance, at Stockholm (59°N), for a clear sky in July, about 80% of Q_s will be direct sunlight and only 20% skylight. In December, only 13% will be direct sunlight and 87% skylight. However, the total amount of energy reaching the ground will be less in December than in July, so 87% of skylight in December represents a smaller energy flow than the 20% in July.

The incoming shortwave radiation is partially reflected upward both from the atmosphere (clouds and water vapor) and from Earth's surface. The albedo, α, in Eq. (5.10) is the ratio of the radiation reflected from the surface to the incoming radiation, expressed in percent. The albedo is also called "reflectance," and depends on the properties of the reflecting surface. The albedo of water (most of the ocean) is about 10–12% but can be higher or lower depending on the suspended matter and sea state. The albedo of sea ice can be much higher but depends strongly on ice type and whether it has snow cover. The albedo of new ice can be as low as 5–20% (see Section 3.9 for sea ice formation stages), while the albedo for first year ice can be as high as 60%. The albedo for multiyear ice without snow cover is about 70%. The albedo of snow is between 60 and 90%. Land surface albedo depends on vegetation and ranges between 0.5 and 30%. Clouds also reflect solar radiation and contribute greatly to the albedo of the whole earth system.

Some values for albedo, extracted from Payne's (1972) tables, are given in Table 5.1, assuming complete transmittance through the atmosphere (no clouds) and an average sea state (neither flat nor extremely rough). The smoother the sea state, the higher the reflection, therefore the albedo has a (small) wind speed dependence. The albedo also depends on the sun's elevation since direct sunlight is reflected more.

The reflection characterized by albedo is diffuse. It is not the same as reflection from a mirror, which is called "specular reflection." Specular reflection from the ocean surface is known as *sun glint*. Sun glint patterns (Figure 5.6) are likely caused by variations in the specular reflection of sunlight from the ocean's surface due to variations in the surface waves caused by variations in the winds.

The bulk formula (5.10) also depends on cloud cover C, which is the fraction of the sky covered by clouds. Part of the incoming radiation is reflected, absorbed, or scattered by

TABLE 5.1 Reflection Coefficient (Albedo) and Transmission Coefficient ($\times$100) for Sea Water

Sun's Elevation:	90°	60°	30°	20°	10°	5°
Amount reflected (%):	2	3	6	12	35	40
Amount transmitted into water (%):	98	97	94	88	65	60

Payne, 1972

partially absorbs the radiation. A beam of radiation of one square meter cross-section covers an area of one square meter of calm sea surface when the sun is vertically overhead. At lower elevations, the beam strikes the sea surface obliquely and is distributed over a larger area. The energy density, or amount per square meter of sea surface, therefore decreases as the sun moves further from the vertical. (This explains why the equatorial regions are warm and the polar regions are cold, and explains why the seasonal radiation changes are greatest at mid-latitudes.) The absorption is due to the combined effect of gas molecules, dust in the atmosphere, water vapor, and so forth. When the sun is directly overhead ($\theta_N = 90°$), the radiation passes through the atmosphere by the shortest possible path and the absorption is at a minimum. When the sun elevation is less than 90 degrees, the radiation path is longer and the absorption is greater.

5.4.3.2. *Absorption of Shortwave Radiation in the Sea*

Shortwave radiation is not absorbed in the ocean's surface skin layer (approximately 10 μm), but instead penetrates to 1−100 m, depending on wind stirring and incident shortwave flux magnitude. The absorption decreases exponentially with depth (Section 3.8). The shortwave penetration affects the way the mixed layer restratifies after being mixed by wind or cooling. Shortwave radiation also penetrates below the mixed layer in many regions, particularly at low latitudes. The solar energy penetration allows for growth of phytoplankton, the ocean's chlorophyll-producing plants, in the near-surface euphotic zone.

The penetration depth of absorption depends on both the wavelength of the light and the optical properties of the water. The water's optical properties and attenuation of solar radiation also depend on particle concentration in the water, which can be composed of sediment (near-coastal) and plankton (everywhere). In

FIGURE 5.6 Sun glint in the Mediterranean Sea. *Source: From NASA Visible Earth (2006a).*

clouds. The factor, 0.062, multiplying C was worked out empirically from direct observations of shortwave radiation reaching the surface. One of the biggest sources of error in computing the shortwave radiation, and in the whole air−sea heat exchange budget, is the cloud cover estimate. Cloud cover estimates prior to the 1980s were mainly subjective. Automated techniques for measuring cloud cover include satellite observations and radar observations from land-based networks for weather prediction and satellite observations; both methods were introduced in the 1980s.

Finally, the solar radiation reaching the sea surface in Eq. (5.10) depends on the sun's elevation, θ_N, for two reasons: (1) dependence of the sea surface area of intersection of a "beam" of sunlight and (2) dependence of the path length of the beam through the atmosphere, which clear water, the e-folding depth for attenuation of light is about 50 m (Table 3.2, Figure 5.7). In water with a heavy load of sediments or biological particles, for instance during major plankton blooms, the radiation is absorbed much closer to the sea surface with an e-folding scale of less than 5 m.

When more of the solar radiation is absorbed close to the surface, the surface temperature increases faster than where the water is clear. The heating rate can differ by a factor of 100.

5.4.4. Longwave Radiation (Q_b)

The radiation term, Q_b, in the heat budget (Eq. 5.9) is the amount of energy lost or gained by the sea as longwave (thermal infrared) radiation. The back radiation is the difference between the energy radiated outward from the sea surface and the longwave radiation received by the sea from the atmosphere. Both the sea surface and the atmosphere radiate approximately as "black bodies," at a rate proportional to the fourth power of their absolute temperature, according to Stefan's Law (Section 5.4.2). The outward radiation from the sea at these wavelengths is generally greater than the inward longwave thermal radiation from the atmosphere, so Q_b usually presents a loss of energy from the sea (hence the subscript "b" for back radiation). Q_b is expressed through the following empirical bulk formula, evaluated by Josey et al. (1999) as the most accurate of several differing formulations:

$$Q_b = \varepsilon\sigma_{SB}T_w^4(0.39 - 0.05e^{1/2})(1 - kC^2)$$
$$+ 4\varepsilon\sigma_{SB}T_w^3(T_w - T_A). \qquad (5.11)$$

Here ε is the emittance of the sea surface (0.98); σ_{SB} is the Stefan-Boltzmann constant (5.67×10^{-8} W m^{-2} K^{-4}); T_w is the surface water temperature in Kelvin; T_A is the air temperature in Kelvin, which is usually measured on a ship near the bridge that may be as much as 8–10 m above the sea surface; e is the water vapor pressure; k is a cloud cover coefficient that is determined empirically and that increases with latitude; and C is the fractional cloud cover (as in Section 5.4.3).

Maps of longwave radiation from the ocean's surface are shown below in Section 5.5 (mean in Figure 5.11 in this chapter and seasonal variations in Figure S5.4 of the online supplement seen on the textbook Web site).

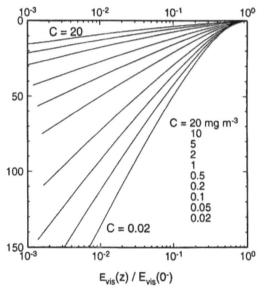

FIGURE 5.7 Absorption of shortwave radiation as a function of depth (m) and chlorophyll concentration, C (mg m^{-3}). The vertical axis is depth (m). The horizontal axis is the ratio of the amount of radiation at depth z to the amount of radiation just below the sea surface, at depth "0." Note that the horizontal axis is a log axis, on which exponential decay would appear as a straight line. ©American Meteorological Society. Reprinted with permission. *Source: From Morel and Antoine (1994).*

5.4.4.1. Factors Affecting Longwave Radiation

The first term in the longwave heat flux (Eq. 5.11) is essentially the black body radiation term (product of the Stefan-Boltzmann constant and the fourth power of the water temperature). Pure Stefan's Law assumes a perfect black body. Each actual object has its own "gray-body" emittance ε. Emittance is always a fraction less than 1, depending on the molecular structure of the body. Water has a relatively high emittance, hence the value of 0.98 given earlier.

While Q_b basically follows the fourth power of temperature, there are several parts to Eq. (5.11) that must be determined empirically. It is difficult to directly measure Q_b over large areas, although it can be measured locally with a radiometer (see the description in Chapter S6, Section S6.8 located on the textbook Web site). Early studies estimated the heat loss using data published by Ångström (1920). He showed that the net loss depends upon the absolute temperature of the sea surface and upon the water-vapor content of the atmosphere immediately above it. The surface temperature, T_W, determines the rate of outward flow of energy. The water vapor pressure e effectively determines the inward flow from the atmosphere because the water vapor in the atmosphere is the main source of its longwave radiation.

The two empirical terms multiplying the T_W^4 term include water vapor pressure e and cloud cover C. Water vapor in the atmosphere radiates longwave energy back to the sea, thus reducing the net longwave radiation from the sea. Clouds reduce the longwave back radiation from the sea to space. This effect of cloud cover is familiar on land where the frost that results from radiative cooling is more frequent on clear nights than on cloudy ones. With the sky completely covered with substantial cloud (C = 1), in a region where the factor k = 0.2, this cloud exponent factor is 0.2. The reason for the big difference between clear and cloudy conditions is that the atmosphere, particularly its water-vapor content, is relatively transparent to radiation in the range from about 8 to 13 μm, which includes the peak of the radiation spectrum for a body at the temperature of the sea. In clear weather, energy at 8−13 μm passes through the more transparent "wavelength window" in the atmosphere and out into space where it is lost from the earth system.

Global cloud cover is represented by an image from NASA's MODIS instrument (Figure 5.8). (Climatological cloud cover for the four seasons is also available as Figure S5.3 from the online supplement located at the textbook Web site.) Over the oceans, cloud cover is high in the polar regions and in zonal stripes in the Intertropical Convergence Zone. Cloud cover is low in the subtropical regions, where

FIGURE 5.8 Cloud fraction (monthly average for August, 2010) from MODIS on NASA's Terra satellite. Gray scale ranges from black (no clouds) to white (totally cloudy). *Source: From NASA Earth Observatory (2010).*

we will see that evaporation greatly dominates precipitation.

Returning to the expression for longwave radiation (Eq. 5.11), the second term is proportional to the difference between the water temperature and the air temperature just above the water. This represents the atmospheric feedback to the longwave radiation radiated at the sea surface due to atmospheric moisture (Thompson & Warren, 1982). The temperature difference is generally small, so the correction is only important in a few special regions. An example is where warm surface ocean currents flow under a cold overlying atmosphere, such as in the western boundary currents in the North Pacific and North Atlantic.

5.4.4.2. Sea Surface Temperature and Penetration Depth of Longwave Radiation

Longwave radiation depends mainly on sea surface temperature (SST). But what is the appropriate measure of SST? From what depth ranges is the sea surface emitting longwave radiation? Water is nearly opaque to longwave radiation. The incoming longwave radiation from the atmosphere is absorbed in the top millimeters, unlike incoming shortwave radiation that penetrates much deeper (Section 5.4.3.2). Thus, the outward longwave radiation is determined by the temperature of the literal surface or skin temperature of the sea, which is less than one millimeter thick.

The bulk surface temperature, characterizing the upper few meters of the ocean and measured with in situ instruments (such as thermistors on buoys or in engine intake water) is not the skin temperature. Instead it represents the temperature about 0.5 to 1 m beneath the surface. Skin and bulk temperatures are equivalent only when the bulk layer is well mixed vertically, as in the presence of breaking surface waves and strong surface winds. Models of skin layer physics (Castro, Wick, & Emery, 2003; Wick, Emery, Kantha, & Schluessel,

1996; Wick, 1996) suggest that the difference between the skin and bulk temperatures is proportional to the wind speed, which affects surface waves, and net air–sea heat flux, which affects mixing.

Regardless of the actual physical process, the empirical bulk formula (5.11) was developed to be used with the traditional bulk SST and not with the skin layer temperature.

5.4.4.3. Outgoing Longwave Radiation (OLR)

"Outgoing longwave radiation" (OLR) is the total infrared radiation at wavelengths of 5 to 100 μm that escapes from the top of Earth's atmosphere back into space. Most of this longwave energy is emitted from the surface of the ocean while some of it is emitted by the land and the atmosphere. The OLR can be computed from infrared satellite data and is generated as a standard product by the National Oceanic and Atmospheric Administration (NOAA) from their polar-orbiting satellites. It was also a product of the ERBE program (Section 5.4.3).

An example of satellite-derived OLR is shown in Figure 5.9. Maximum OLR dominates the mid-latitudes in all ocean regions, where cloudiness is low, broken only by equatorial minima that are related to the cloudy regions of the atmosphere's Intertropical Convergence Zone along 5°N to 10°N and to the Walker circulation (Section 7.9.2). Associated with the latter, an equatorial minimum dominates the Indonesian archipelago and extends into the western Pacific.

5.4.5. Effect of Ice and Snow Cover on the Radiation Budget

When the sea surface becomes covered with a layer of ice, and especially if snow covers the ice, there is a marked change in the heat-radiation budget as described in the section on shortwave radiation. First of all, ice cover significantly reduces heat exchange between the ocean and atmosphere — 1 meter of ice will almost totally

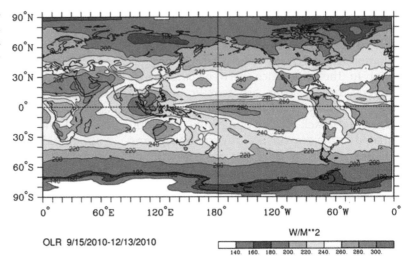

OLR 9/15/2010-12/13/2010

W/M**2

140. 160. 180. 200. 220. 240. 260. 280. 300.

insulate the ocean. However, sea ice is always moving and is full of leads (breaks that expose open water). Ocean heat loss in leads is intense, and new ice forms quickly. Therefore heat budgets in ice-covered areas must take into account ice type, thickness, and concentration (percent coverage). Where there is ice cover, the heat flux into the water column becomes

$$Q = k(T_w - T_s)/h \qquad (5.12)$$

where k is the ice conductivity, T_w is the water temperature just below the ice, T_s is the temperature at the upper surface of the ice, and h is the ice thickness. The water temperature is assumed to be at the freezing point. The surface temperature of the ice and its thickness are more difficult to determine without local measurements. The availability of satellite-based observations of ice cover using microwave imagery has greatly improved knowledge and regular mapping of ice cover. The type and thickness of ice is harder to estimate from satellite observations, but there are various approaches for obtaining information about the distribution of new, first-year, and multiyear ice that are then translated to estimates of ice thickness. Remote sensing of ice thickness remains one of the major hurdles in air–sea flux estimation at high latitudes.

A second important effect of ice is that it is highly reflective (high albedo), much more so than water (lower albedo). This mostly impacts the incoming shortwave radiation (Section 5.4.3). Sea ice and snow also reflect most incident solar radiation, so they have a high albedo in comparison with open water (Figure 5.10). However, the back radiation (Q_b) heat loss is much the same for ice as for water (due to the relative similarity of surface temperature). Therefore there is a smaller net gain ($Q_s - Q_b$) by ice and snow surfaces than by water. Thus as sea ice melts back, more solar radiation is absorbed by the water, which then warms and causes more ice to melt. This is a positive feedback, which is called *ice-albedo feedback*.

The ice balance in a region such as the Arctic is somewhat delicate (Section 12.7). If the sea ice were melted all the way at a given time, the increased net heat gain ($Q_s - Q_b$) could maintain an ice-free Arctic Ocean. On the other hand, this could increase evaporation, which would increase precipitation in the high northern latitudes, increasing snow cover and high latitude albedo, which would have a cooling effect. The present situation of ever-decreasing Arctic Ocean ice cover (Section 12.8) suggests that the ice-albedo feedback effect dominates.

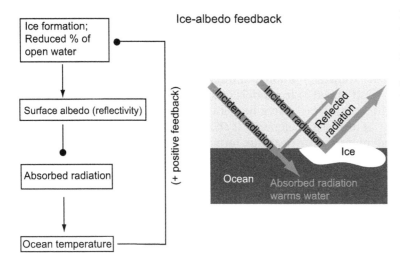

Ice-albedo feedback

FIGURE 5.10 Ice-albedo feedback. In the feedback diagram, arrowheads (closed circles) indicate that an increase in one parameter results in an increase (decrease) in the second parameter. The net result is a positive feedback, in which increased sea ice cover results in ocean cooling that then increases the ice cover still more.

5.4.6. Evaporative or Latent Heat Flux (Q_e)

Evaporation requires a supply of heat from an outside source or from the remaining liquid. (This is why one feels cold when wet after swimming, as evaporation of the water requires heat.) Therefore evaporation, besides implying loss of water volume, also implies loss of heat. The rate of heat loss is

$$Q_e = F_e \cdot L \qquad (5.13)$$

where F_e is the rate of evaporation of water in kg $\sec^{-1} m^{-2}$ and L is the latent heat of evaporation (vaporization) in kilojoules ($1 \text{ kJ} = 10^3 \text{ J}$). For pure water, L depends on the temperature of the water T in °C: $L = (2494 - 2.2\,T)$ kJ/kg. At 10°C, the latent heat is about 2472 kJ/kg, which is larger than its value of 2274 kJ/kg (540 cal/gm) at the boiling point. While one can see steam after the boiling point is reached, more volume is being evaporated at temperatures well below boiling.

The average amount of evaporation F_e from the sea surface is about 120 cm/yr, in other words, the equivalent of the sea surface sinking more than 1 m per year. Local values range from an annual minimum of as little as 30 to 40 cm/yr in high latitudes to maxima of 200 cm/yr in the tropics associated with the trade winds. This decreases to about 130 cm/yr at the equator where the mean wind speeds are lower.

How is the evaporation rate F_e determined? The rate of water loss from a pan of water can be measured, but this has serious practical difficulties. For large area estimates and for prediction, a formula using easily measured parameters is desirable. Evaporation is basically a diffusive process that depends on how water vapor concentration changes with height above the sea surface and on the processes that cause diffusion. In Section 5.1.3, we discussed eddy diffusion, which is analogous to molecular diffusion, except that the turbulence in the air or water is the process that diffuses properties rather than movement of individual molecules. Air turbulence controls the diffusion that creates evaporation. Air turbulence depends on wind speed, therefore we expect the evaporation rate to depend on wind speed. (This explains why we feel cooler when the wind is blowing.)

A semi-empirical ("bulk") formula for evaporation that depends on wind speed and the vertical change in water vapor content is frequently used:

$$F_e = \rho\, C_e\, u(q_s - q_a). \qquad (5.14)$$

Here ρ is the density of air, C_e is the transfer coefficient for latent heat, u is the wind speed in meters per second at 10 m height, q_s is 98% of the saturated specific humidity at the sea surface temperature, and q_a is the measured specific humidity. The factor of 98% for saturated humidity over seawater compensates for the salinity. The saturated specific humidity over distilled water may be obtained from tables of physical or meteorological constants.

Specific humidity is the mass of water vapor per unit mass of air, in g/kg. Saturated specific humidity is the maximum weight of water vapor that the air can hold for a given temperature. Relative humidity is the amount of water vapor divided by the saturated water vapor expressed in percent. Therefore it is equal to the specific humidity divided by the saturated specific humidity.

The empirical ("bulk") formula for heat flux due to evaporation (latent heat flux) is therefore, in W/m^2,

$$Q_e = F_e \bullet L = \rho\, C_e\, u(q_s - q_a)L. \qquad (5.15)$$

In most regions of the ocean, the saturated specific humidity q_s is greater than the actual specific humidity q_a. Since all of the other terms in the practical formula (5.15) are positive, evaporative heat loss occurs in these regions. As long as the sea temperature is more than about 0.3 K greater than the air temperature, there will be a loss of heat from the sea due to evaporation. Only in a few regions is the reverse the case, when the air temperature is higher than the sea temperature and the humidity is sufficient to cause condensation of water vapor from the air into the sea. This results in a loss of heat from the air into the sea. The Grand Banks off Newfoundland and the coastal seas off northern California are examples of regions where the latent heat flux Q_e is *into* the sea (numerically

positive). The fog in these regions is the result of cooling the atmosphere.

The heat loss due to evaporation occurs from the topmost layer of the sea, like longwave radiation and unlike shortwave radiation heat gain. In models of air–sea heat exchange, the evaporative heat loss is applied to the ocean surface element. It should also be noted that this latent heat flux term is usually the largest of the heat flux terms other than the incoming shortwave radiation. Unlike the short- and longwave terms discussed previously, there is no straightforward estimate of latent heat exchange from satellite observations. Therefore the latent flux is best estimated from in situ measurements, which will be presented later.

5.4.7. Heat Conduction or Sensible Heat Flux (Q_h)

The last process that we discuss for heat exchange between the sea surface and atmosphere, sensible heat flux, arises from a vertical difference (gradient) in temperature in the air just above the sea. This is perhaps the simplest of the heat flux terms in Eq. (5.9) to understand. If temperature decreases upward, heat will be conducted away from the sea, resulting in an ocean heat loss. If the air temperature increases upward, heat will be conducted into the sea. The rate of loss or gain of heat is proportional to the air's vertical temperature gradient, and to the heat conductivity (for which we use an eddy diffusivity or conductivity, A_h):

$$Q_h = -A_h\, c_p\, dT/dz \qquad (5.16)$$

As described for eddy diffusion of water vapor, the eddy conductivity A_h depends on wind speed. The vertical gradient of temperature is measured as a difference between the sea surface temperature and the air temperature. The bulk formula for sensible heat flux is written, with units of W/m^2, as:

$$Q_h = \rho\, c_p\, C_h\, u(T_s - (T_a + \gamma z)) \qquad (5.17)$$

where ρ is the air density; C_h is the transfer coefficient for sensible heat (derived from the eddy conductivity); u is the wind speed in meters per second at 10 m height; T_s is the surface temperature of the ocean (assumed to be equal to the air temperature immediately above the ocean surface); T_a is the air temperature; z is the height where T_a is measured; and γ is the adiabatic lapse rate of the air, which accounts for changes in air temperature due to simple changes in height and pressure.

The sensible heat flux cannot be estimated from satellite measurements and must be estimated from in situ data using these bulk formulae. Global maps of such computations are presented in Section 5.5.

5.4.8. Dependence of the Latent and Sensible Heat Transfer Coefficients on Stability and Wind Speed

Latent and sensible heat transfers are computed using bulk formulae like Eqs. (5.15) and (5.17) using in situ observations. Values for the transfer coefficient for sensible heat for various air–sea temperature differences and different wind speeds are given in Table 5.2 (from Smith, 1988). The transfer coefficients in the two expressions, C_e and C_h, depend on whether the ocean is warmer or colder than the atmosphere, and whether the atmosphere is undergoing deep or vigorous convection. If the sea is warmer than the air above it, there will be a loss of heat from the sea because of the direction of the temperature gradient. However, larger scale atmospheric convection will increase the heat transfer away from the sea surface. Convection occurs because the air near the warm sea is heated, expands, and rises, carrying heat away rapidly. In the opposite case, where the sea is cooler than the air, convection does not occur. Therefore, for the same temperature difference between sea and air, the rate of heat

TABLE 5.2 Some Values for the Sensible Heat Transfer Coefficient, C_h, as Functions of $(T_s - T_a)$ and Wind Speed u

| | Wind Speed u in m/sec | | | |
$(T_s - T_a)$ (K)	2	5	10	20
−10	—	—	0.75	0.96
−3	—	0.62	0.93	0.99
−1	0.34	0.87	0.98	1.00
+1	1.30	1.10	1.02	1.00
+3	1.50	1.19	1.06	1.01
+10	1.87	1.35	1.13	1.03

Smith, 1988

loss when the sea is warmer is greater than the rate of gain when the sea is cooler.

For example, for $(T_s - T_a) = -1$ K, that is, for the sea cooler than the air $(T_s < T_a)$, the stability in the air is positive. When the sea is warmer than the air, for example, $(T_s - T_a) = +1$ K, the air is unstable and heat conduction away from the sea is promoted, so the transfer coefficient is larger than 1. The blank areas in the table are for highly stable conditions (unusual) where Smith's analysis breaks down.

For the transfer coefficient for evaporation, Smith commented that measurements in open sea conditions are relatively rare, particularly for high wind speeds. After reviewing the available data, he recommended $C_e = 1.20\ C_h$. That is, the physical process causing the transfer coefficient is similar for both evaporation and heat conduction.

5.5. GEOGRAPHIC DISTRIBUTION AND TEMPORAL VARIATION OF THE HEAT-BUDGET TERMS

Maps and description of the four components of the surface heat flux are given in this section. A monthly climatology[1] of fluxes from the

---

[1] A monthly climatology is the average of values from a given month over all the years of analysis (see Section 6.6.2).

National Oceanography Center, Southampton (NOCS; Grist & Josey, 2003) is used here, but we could use other available climatologies for description of the basic patterns and magnitudes as well. The NOCS fluxes are based on carefully quality-controlled ship observations covering more than a century, from the COADS database. The annual mean heat flux components are shown in the text; four monthly maps representing the seasonal variations for each component and the net heat flux are provided in the online supplement as Figures S5.2–S5.7 along with seasonal cloud cover over the oceans (see Figure S5.3 based on data from da Silva, Young, & Levitus, 1994) since it has a large impact on the shortwave and longwave radiation.

Large and Yeager (2009) created a product in which input fields from reanalysis and satellite observations are systematically adjusted and combined to balance heat and freshwater budgets. Their fluxes are used for the mean buoyancy flux map in Figure 5.15. The mean heat and freshwater air–sea flux maps that are combined for the buoyancy flux are shown in Figure S5.8 seen on the textbook Web site.

Other commonly used global air–sea flux products are from weather prediction models that have been systematically "reanalyzed" to create consistent data sets over many years of runs. The two major reanalyses are from the National Centers for Environmental Prediction (NCEP; Kalnay et al., 1996) and from the European Centre for Medium-range Weather Forecasts (ECMWF).

5.5.1. Annual Mean Values of the Components of the Heat Budget

The four components of the air–sea heat flux are shown in Figure 5.11 and their sum, the net heat flux, in Figure 5.12. Basically heat is added to the ocean through shortwave radiation (incoming sunlight) and mostly lost from the ocean through the other three components. The shortwave radiation, Q_s, (Figure 5.11a)

depends mainly on latitude. It adds 50 to 150 W/m^2 of heat to the ocean in subpolar latitudes, and 150 to almost 250 W/m^2 in the subtropics and tropics (Figures 5.11a and Figure S5.2 located on the textbook Web site). Shortwave radiation does vary from exact dependence on latitude. The highest shortwave flux, of almost 250 W/m^2, is in patches in the eastern tropical Pacific and the western tropical Indian Ocean along the Arabian Peninsula. Lower tropical shortwave heat gain is found in wide regions in the eastern parts of the oceans. For instance, in the eastern Pacific, the 200 W/m^2 contours bulge toward the equator in both the North and South Pacific. These variations in shortwave radiation are due to spatially varying cloud cover (Figure 5.8), which partially blocks incoming shortwave radiation.

Longwave radiation, Q_b, (Figure 5.11b) results in net heat loss from the ocean, even though there is some longwave radiation into the ocean from the atmosphere. The radiation heat loss centers around 50 W/m^2 over much of the earth. Longwave radiation does not have a large range of values because it depends on the absolute temperature (Kelvin and not Celsius). The relative changes in temperature are just a small fraction of the total temperature. The relative humidity also does not change much over the sea. For instance, a seasonal change of sea temperature from 10 to 20°C changes the outward radiation proportional to the ratio $293^4/283^4$ or about 1.15, only a 15% increase. At the same time the atmospheric radiation inward would increase and reduce the net rate of loss below this figure. The small seasonal and geographic changes of Q_b contrast with the large seasonal changes of Q_s. Variations in longwave radiation with latitude follow cloud cover rather than surface temperature. Longwave radiative heat loss is highest in the subtropics (>50 W/m^2) where the cloud cover is smaller than in the equatorial and subpolar regions.

Latent heat flux, Q_e, (Figure 5.11c) is the largest heat loss component at all latitudes. It

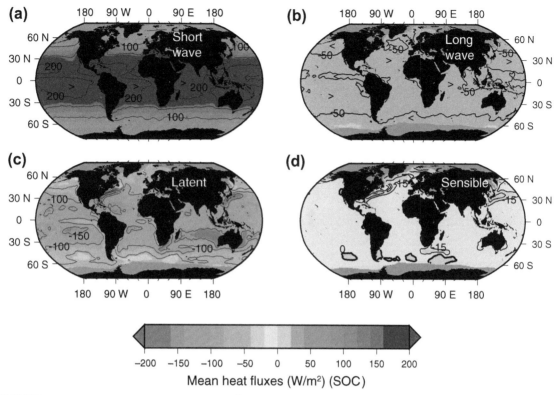

FIGURE 5.11 Annual average heat fluxes (W/m²). (a) Shortwave heat flux Q_s. (b) Longwave (back radiation) heat flux Q_b. (c) Evaporative (latent) heat flux Q_e. (d) Sensible heat flux Q_h. Positive (yellows and reds): heat gain by the sea. Negative (blues): heat loss by the sea. Contour intervals are 50 W/m² in (a) and (c), 25 W/m² in (b), and 15 W/m² in (d). Data are from the National Oceanography Centre, Southampton (NOCS) climatology (Grist and Josey, 2003). This figure can also be found in the color insert.

is strongest (more than 100 W/m² heat loss) in the subtropical regions of low cloud cover, where dry air descends from aloft onto the oceans. Variations in Q_e from west to east in the stormy regions of the western North Atlantic and western North Pacific are associated with dry winds blowing off the continents, creating greater latent heat flux. Latent heat loss also depends mildly on temperature since warmer water evaporates more easily than colder water.

Sensible heat flux, Q_h, (Figure 5.11d) is usually the smallest of all of the components over most of the ocean (−15 to 0 W/m²). It is slightly larger in the western North Atlantic

and western North Pacific, where latent heat loss is also large. This is because the air–sea temperature contrast is large in these regions, where cold air blows off the continent over the warm western boundary currents. Sensible heat loss is much larger in winter in some regions than is apparent from these maps of the mean components. A small amount of heat gain from sensible heat exchange is shown in the Antarctic, but the reader should understand that these data are especially poor.

The total air–sea heat flux based on the NOCS climatology (Figures 5.12 and 5.13) is the sum of the four components of Figure 5.11. (Total air–sea heat flux from a different climatology is

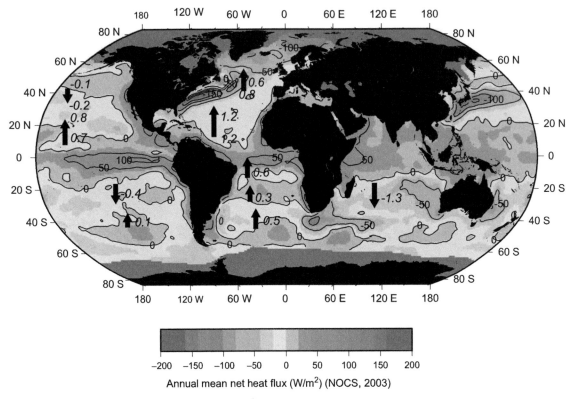

FIGURE 5.12 Annual average net heat flux (W/m²). Positive: heat gain by the sea. Negative: heat loss by the sea. Data are from the NOCS climatology (Grist and Josey, 2003). Superimposed numbers and arrows are the meridional heat transports (PW) calculated from ocean velocities and temperatures, based on Bryden and Imawaki (2001) and Talley (2003). Positive transports are northward. The online supplement to Chapter 5 (Figure S5.8) includes another version of the annual mean heat flux, from Large and Yeager (2009). This figure can also be found in the color insert.

shown in Figure S5.8 on the textbook Web site, and is combined with freshwater fluxes to produce the total buoyancy flux seen in Figure 5.15 in this chapter. Comparison of these two maps provides a useful indication of uncertainty in the total flux.) The ocean gains heat in the tropics and loses heat at higher latitudes. The most heat is gained along the equator, especially in the eastern Pacific. Regions of net heat gain spread away from the equator on the eastern sides of the oceans, in the regions where colder water is upwelled to the surface. Patches of heat gain are found in the Antarctic, corresponding to regions where the sensible heat flux is into the ocean in Figure 5.11d, but where

there are also almost no winter data to balance observations of summer heat gains.

The greatest mean annual heat losses occur in the Gulf Stream region of the North Atlantic, the Kuroshio of the North Pacific, and in the Nordic Seas north of Iceland and west of Norway. In the Southern Hemisphere, the Agulhas/Agulhas Return Current is the region of largest heat loss. The Brazil and East Australian Currents are marked by heat loss, as is the Leeuwin Current, which is the only southward-flowing eastern boundary current. Each of these regions is characterized by fast poleward flow of warm water that loses its heat locally rather than over a large region; the highest heat losses are where

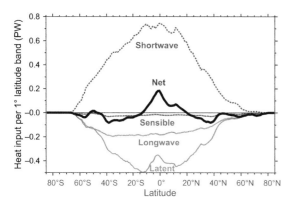

FIGURE 5.13 Heat input through the sea surface (where 1 PW = 10^{15} W) (world ocean) for 1° latitude bands for all components of heat flux. Data are from the NOCS climatology (Grist and Josey, 2003).

these warm waters encounter dry, cold continental air in winter (Gulf Stream and Kuroshio).

The local values of each term in the heat budget and the total heat flux can be summed all the way around the earth in each latitude band (Figure 5.13). The numbers used in this figure are the heat gain or loss in each 1° latitude/longitude square multiplied by the area of the square. Then all the heat gains or losses for a single latitude are added together to get the total heat gain or loss in Watts for each latitude band. The total (net) heat gain or loss in each latitude band (Figure 5.13) is the sum of these four. As is apparent in the maps in Figure 5.11, shortwave radiation heats the ocean while the other three components cool it. Latent heat loss is the largest of these three heat losses, but longwave radiation is also significant. The sensible heat contribution is very small.

All of the heat budget components are larger in the Southern Hemisphere than in the Northern Hemisphere. Part of the reason may be there is more ocean area in the Southern Hemisphere. Shortwave radiation is also slightly skewed because Earth is closer to the sun in January, which is the summertime in the Southern Hemisphere (Section 5.5.2). The net heat exchange is positive (heating) in the low latitudes and negative (cooling) at higher latitudes. The net heat flux is also skewed, with slightly more heating in the low-latitude Southern Hemisphere. The net heat flux distribution requires a transport of excess heat from low to high latitudes in order to maintain a climatologically steady state (Section 5.6).

5.5.2. Seasonal Variations in the Components of the Heat Budget

Each component of the heat budget varies in time. Components can vary on short (diurnal) to long (decades to millennia) timescales, but at mid-latitudes, seasonal variation has the largest amplitude and impact on weather. Seasonal maps for each of the air−sea heat flux components, and also cloud cover, are shown in Figures S5.2−S5.7 located on the textbook Web site. Only a short summary of salient results is presented here.

The march of the seasons from summer to winter is apparent in the shortwave radiation maps (see Figure S5.2 located on the textbook Web site), with much more shortwave radiation reaching the summertime hemisphere. Northern Hemisphere winter radiation is higher than Southern Hemisphere winter radiation because Earth is closer to the sun in January than in July, so the winter seasons are not identical. (This has paleoclimate ramifications, as it highlights the importance of the exact orbit of Earth, which varies slowly, changing the distribution of incoming radiation.)

For longwave radiation, seasonal variations are small (see Figure 5.4 of the online supplemental material), just as geographical variations are small (Section 5.5.1 of this chapter), because the radiation variations depend only weakly on temperature. Within these small variations, longwave radiation is larger in the winter hemisphere than in the summer hemisphere, mainly because of greater cloud cover in summer.

Latent heat loss through the seasons is strongest (most negative) in the winter (Figure S5.5

located on the textbook Web site). The Northern Hemisphere's western boundary currents (Gulf Stream and Kuroshio) are clearly marked by their large latent heat loss in fall and winter. Southern Hemisphere latent heat loss is less associated with the western boundary currents and more associated with the central subtropical gyres where evaporation is high.

Despite the relatively small contribution of sensible (conductive) heat flux to the annual mean net heat flux, its seasonal variations (Figure S5.6 located on the textbook Web site) are striking because of the sign change in the temperature difference between the ocean and overlying air. Sensible heat flux causes heat loss in winter when the overlying air is colder than the ocean. Significant heat loss is found in the western boundary current regions, more than 100 W/m^2 in the climatological maps, and much higher in individual storms. Sensible heat flux heats the ocean in higher latitudes in summer, when the air is warmer than the ocean.

5.6. MERIDIONAL HEAT TRANSPORT

The ocean gains heat in the tropics between 30°S and 30°N and loses heat at higher latitudes, when zonally averaged (along latitudes) and over the year (Figure 5.13). There is net radiative heat gain by the ocean at lower latitudes because the solar radiation Q_s is greater than the longwave radiation Q_b between the equator and about 30° to 40°N (Figures 5.11 and 5.13). At higher latitudes, longwave heat loss is greater than shortwave heat gain; evaporative heat loss is also higher, and therefore, overall, there is a net heat loss. These zonally averaged patterns of heat gain and loss also apply to the atmosphere.

Because the oceans as a whole are not warming or cooling (except for the very small rates that are indeed significant for climate studies),

we expect a nearly exact balance between heat gain and loss when summed over all of the ocean area. This requires a net flux of heat toward both poles, from the lower latitudes of net heat gain to the higher ones of net heat loss. This poleward heat flux is carried by both the ocean and atmosphere. Both transport warm water or air toward the pole and cooler water or air toward the equator, although not symmetrically in all oceans (see the following section). The atmosphere carries much more of this heat than the ocean (see the following section), but the ocean's role in heat transport is important, especially at low to mid-latitudes.

The meridional (north-south) heat transports by currents within the oceans are calculated in three separate and independent ways. The first two methods are *indirect*, in which the ocean's heat transport is inferred from heat balances rather than from measurements of ocean velocity and interior temperature.

The first indirect method uses the surface heat fluxes (as in Figures 5.12 and 5.13), which are summed within latitude bands. The ocean must then transport enough heat into or out of each latitude band to balance the heat lost or gained through the sea surface in that band.

The second indirect method starts with the heat exchange of the whole Earth's system with outer space; that is, at the top of the atmosphere (TOA). Then heat transports are calculated for the atmosphere from meteorological data. The ocean's heat transport is the TOA flux minus the atmosphere's flux. The first such estimates (Oort & Vonder Haar, 1976), based on 9 years of radiation measurements from satellites, showed the ocean heat transport to be a maximum of 60% of the total at 20°N, 25% at 40°N, and 9% at 60°N. However, as observations have improved, and especially with the addition of a special satellite mission to measure radiation at the TOA (ERBE), estimates of the total heat transport and atmospheric heat transport have become higher, leaving the ocean heat

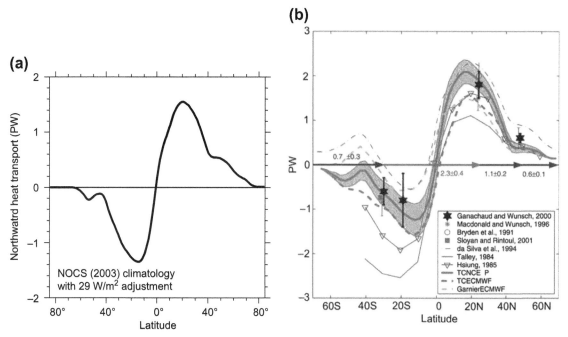

FIGURE 5.14 Poleward heat transport (W) for the world's oceans (annual mean). (a) Indirect estimate (light curve) summed from the net air–sea heat fluxes of Figures 5.12 and 5.13. Data are from the NOCS climatology, adjusted for net zero flux in the annual mean. Data from Grist and Josey (2003). A similar figure, based on the Large and Yeager (2009) heat fluxes is reproduced in the online supplement (Figure S5.9). (b) Summary of various direct estimates (points with error bars) and indirect estimates. The direct estimates are based on ocean velocity and temperature measurements. The range of estimates illustrates the overall uncertainty of heat transport calculations. ©American Meteorological Society. Reprinted with permission. *Source: From Ganachaud and Wunsch (2003).*

transport at about the same original values, but a smaller fraction of the whole (Trenberth & Caron, 2001).

The third type of ocean heat transport calculation is *direct*, based on measuring velocity and temperature across a whole cross-section of the ocean through which there is zero net mass transport. (If there is net transport, then additional cross-sections forming a "box" with balanced mass must be included.). The net heat transported through the section (or box) can then be calculated, and must balance the total heat gain or loss through the sea surface on either side of the section (or within the box). That is, if the section is located at 30°N in Figure 5.12, then there is net gain of heat to the south and net loss of heat to the north. The section's velocities and temperature should then show a net flow of warmer waters northward and colder waters southward.

Global annual mean ocean heat transport, calculated using all three methods, is from the heat-absorbing tropics to the cooling regions at mid to high latitudes (Figures 5.12 and 5.14). The maximum rate is 1 to 2×10^{15} W at about 20° to 30° latitude in each hemisphere. This is 20 to 30% of the total global energy transport of about 6×10^{15} W; the atmosphere transports more heat than the ocean everywhere (Trenberth & Caron, 2001). The Pacific Ocean exports heat poleward in both hemispheres. The Indian Ocean exports heat southward because of the absence of a high latitude Northern Hemisphere region.

As previously mentioned, the global heat budget must almost exactly balance; that is, there is almost no net heat gain or loss for the whole Earth, and the same for the ocean alone.[2] The meridional ocean heat transports based on Grist and Josey's (2003) fluxes are nearly, but not exactly, in balance. In Figure 5.14b we also show an "adjusted" meridional heat transport curve that balances exactly, with zero transport in both the north and south, achieved by adding 2.5 W/m^2 at every grid point. The adjusted curve is within the error of Grist and Josey's (2003) calculation. Discussion in the next paragraphs is in reference to this adjusted transport.

A counterintuitive result, found in every heat transport product, is that the heat transport throughout the Atlantic, including the South Atlantic, is northward (arrows and annotation in Figure 5.12). This is because there is so much heat loss in the subpolar North Atlantic and Nordic Seas. To feed this heat loss, there must be a net northward flow of upper ocean water throughout the length of the Atlantic, which is returned southward by deeper colder water. In all oceans, the subtropical gyre circulation in just the upper ocean carries heat poleward (Section 14.2.2) (Talley, 2003). This part of the heat transport in the South Atlantic is not strong enough to overcome the northward heat transport due to the top-to-bottom overturn that produces North Atlantic Deep Water. The Pacific Ocean does not have a top-to-bottom overturn with associated meridional heat transport, so there is an asymmetry between the Pacific and Atlantic heat transports.

Regarding the apparently odd northward heat transport throughout the Atlantic even while the Pacific and Indian Oceans follow the expected pattern of poleward heat transport in both hemispheres, Henry Stommel (personal communication) told an interesting story. One of the last studies by Georg Wüst (1957) was a study of north-south transports in the South Atlantic. Although he computed and published the transports of oxygen, salinity, nutrients, and so forth, he did not publish the meridional heat flux, which is the easiest to compute since only temperature profiles are required. Stommel suspected that Wüst computed the heat transport but found that it appeared to go in the wrong direction, namely from the south to north (toward the equator), as we see in Figure 5.12. This violated Wüst's intuition, which required the heat to flow from the tropical north to the colder polar south. To verify his suspicion, Stommel sought out Wüst's former students. He managed to locate a German Admiral Noodt who wrote to say that, yes, Professor Wüst did not publish the heat transport because it appeared that it "flew in the wrong direction" (sic). This view did not change until the 1970s when new studies (Bennett, 1976) clearly displayed that the meridional heat transport in the South Atlantic is northward.

5.7. BUOYANCY FLUXES

Buoyancy forcing changes the density of seawater. External forcing is due to heat fluxes (heating and cooling) and freshwater fluxes (evaporation and precipitation plus runoff from land, see the preceding sections). Almost all of these forcings are from (or through) the atmosphere, with only a very minor component from Earth's crust below.[3] Brine rejection due to

---

[2] Major climate change on the order of 1 to 5°C, such as global warming or a shift into an ice age, would be associated with a net ocean heat gain or loss on the order of 1 to 10 W/m^2, calculated for a 1000 m thick layer of water over 100 or 10 years, respectively. It is also well known that global warming associated with a doubling of CO_2 in the atmosphere corresponds to a net change in heat flux of 4 W/m^2.

[3] Geothermal heat flux is typically 0.05 W/m^2, in comparison to typical solar heating of 250 W/m^2, weaker by a factor of 5000.

sea ice formation is an effective direct means of fractionating (redistributing) the water column density, by freshening the sea water locked up in sea ice and increasing the seawater density below the ice by release of the salt into the water column. Sea ice maps and brine rejection are described in the Arctic and Southern Ocean chapters (Chapters 12 and 13).

A global air—sea buoyancy flux map (annual mean) is shown in Figure 5.15, based on Large and Yeager (2009). It is the sum of the mean air—sea heat flux and mean freshwater flux (evaporation minus precipitation/runoff). While the units of buoyancy are inverse density, or m^3/kg, the mapped flux is converted to heat flux units (W/m^2); this is simply because most present-day depictions of air—sea fluxes are in terms of heat, and thus intuition is largely based on heat. The air—sea heat and freshwater fluxes from Large and Yeager (2009) used for Figure 5.15 are shown in the online supplement (Figure 5S.8). In terms of a textbook description,

these fluxes differ only slightly from the NOCS fluxes shown earlier.

The buoyancy flux map strongly resembles the heat flux map because freshwater forcing, while essential to the salinity balance, is weak. Buoyancy loss (density gain) is most vigorous in subtropical western boundary current separation regions, where heat loss is large (Section 5.5.1). The other region of large buoyancy loss, due to heat loss, is the subpolar North Atlantic and Nordic Seas. The associated northward transport of heat and hence buoyancy in the Atlantic is related to the meridional overturning circulation (Section 14.2).

Buoyancy gain is largest in the tropics, particularly over the cool surface waters in the eastern equatorial Pacific (Section 10.7.2). In this "cold tongue," the sea surface temperature is persistently lower than the air temperature, leading to heat gain. The equatorial Atlantic is also a region of high buoyancy gain, in part due to freshwater contributions from the

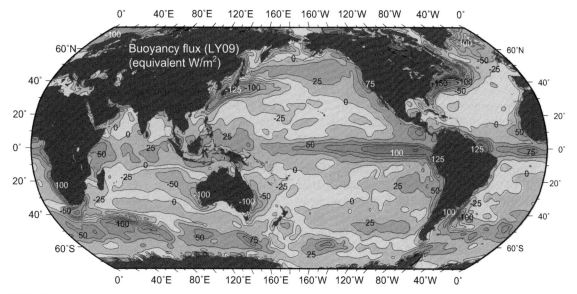

FIGURE 5.15 Annual mean air—sea buoyancy flux converted to equivalent heat fluxes (W/m^2), based on Large and Yeager (2009) air—sea fluxes. Positive values indicate that the ocean is becoming less dense. Contour interval is 25 W/m^2. The heat and freshwater flux maps used to construct this map are in the online supplement to Chapter 5 (Figure S5.8).

Amazon, Orinoco, Congo, and Niger River outflows. The equatorward eastern boundary currents (Peru-Chile, Benguela, California, and Canary) are also regions of buoyancy gain, associated with heating, and dynamically associated with upwelling.

Perhaps most counterintuitive are the high latitude regions where the seawater actually becomes less dense due to air-sea fluxes, rather than being cooled and becoming denser. The two large regions that stand out are the subpolar North Pacific and the Southern Ocean within and south of the Antarctic Circumpolar Current. Both are open ocean upwelling regions, and are also regions of equatorward Ekman transport. These two processes supply cooler water to the sea surface, which is apparently met with net heat gain.

Although freshwater fluxes generally contribute much less than heat fluxes to the total buoyancy flux, freshwater fluxes in the Southern Ocean and subpolar North Pacific tip the balance toward stronger and broader regions of buoyancy gain by the ocean. Freshwater fluxes also make a difference where heat fluxes are small, such as throughout the subtropical gyres outside the western boundary currents, where contributions from both heat and freshwater fluxes (net evaporation) are on the order of 10 W/m².

Intense heat losses in ice formation regions are not represented well in Figure 5.15. Brine rejection is the main agent for producing very dense water, but it is not included at all as it only redistributes buoyancy internally. The heat losses in coastal polynyas where large amounts of sea ice are formed have small spatial scales. For instance, the Weddell and Ross Seas both appear as regions of net buoyancy gain, whereas this is where very dense water is formed through cooling and brine rejection; cooling in cavities under the ice shelves is also a factor here and cannot be represented in these air–sea flux maps (Chapter 13). In the North Pacific, the brine rejection region of the Okhotsk Sea does not appear in this map, which instead

shows a net buoyancy gain driven by runoff from the Amur River.

5.8. WIND FORCING

Surface wind stress is the principal means for forcing the ocean circulation, through a near-surface frictional (turbulent) layer and the mass convergences and divergences in that layer (see Chapter 7). The convergences/divergences are directly related to the wind stress curl. Global wind stress and wind stress curl are shown in Figure 5.16. Seasonal variation is important, especially in monsoonal regions, so mean August and February winds are also shown.

The largest scale annual mean wind patterns are the easterly trade winds in the tropics and westerly winds poleward of 30° latitude in both hemispheres. Annual mean winds and wind stress are strongest in the westerly wind belt of the Southern Hemisphere (40°S to 60°S). In the summer hemispheres in the tropics, the summer monsoon with winds blowing from the ocean to the continent is apparent in all three ocean basins, but is most pronounced in the northwestern Indian Ocean. The opposing monsoon is also evident in the winter hemispheres (represented by February in the Northern and August in the Southern Hemisphere). In the Northern Hemisphere winter, the westerly winds are strongly developed around low pressure centers in the North Pacific (Aleutian Low) and North Atlantic (Iceland Low). In the Southern Hemisphere winter, strong southerlies are apparent around the Antarctic; these are the wintertime katabatic winds (gravity-driven flow down the sloping ice sheet). Similar winds are apparent off the Greenland ice cap in Northern Hemisphere winter.

Global mean wind stress curl (Figure 5.16d) from the QuikSCAT satellite is an extraordinary recent result (Chelton, Schlax, Freilich, & Milliff, 2004), with important detail that is not resolved in the coarser resolution NCEP winds displayed in the other panels of Figure 5.16. Wind stress curl is related to ocean circulation because the curl indicates Ekman convergence/divergence that then drives interior equatorward/poleward Sverdrup transport (Section 7.8 and maps in Figure 5.17). Ekman downwelling is present throughout the subtropical regions. Ekman upwelling is present in the subpolar regions and Antarctic, and in long zonal bands in the tropics. These features are evident in any map of mean wind stress curl, including those with much coarser spatial resolution as shown in the basin chapters.

With the high resolution winds, persistent smaller scale features in wind stress curl are

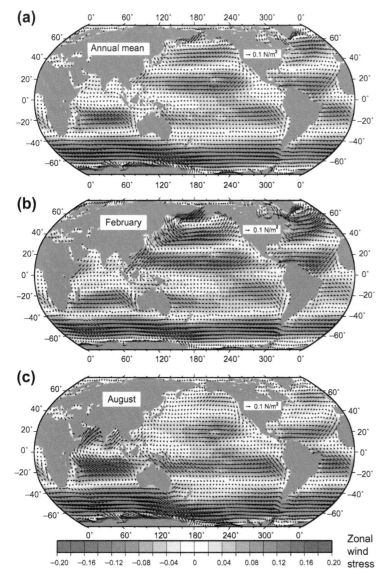

FIGURE 5.16 Mean wind stress (arrows) and zonal wind stress (color shading) (N/m^2): (a) annual mean, (b) February, and (c) August, from the NCEP reanalysis 1968–1996 (Kalnay et al., 1996). (d) Mean wind stress curl based on 25 km resolution QuikSCAT satellite winds (1999–2003). Downward Ekman pumping (Chapter 7) is negative (blues) in the Northern Hemisphere and positive (reds) in the Southern Hemisphere. *Source: From Chelton et al. (2004).* This figure can also be found in the color insert.

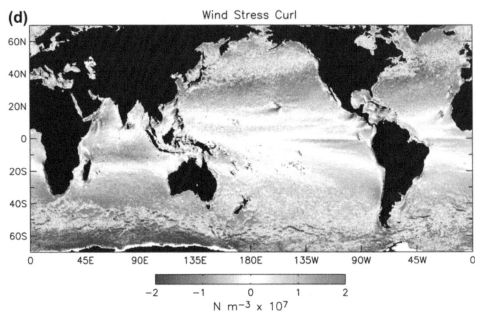

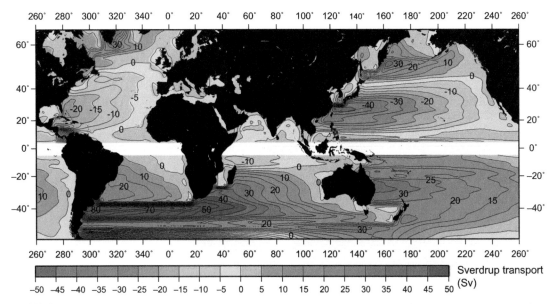

FIGURE 5.16   (*Continued*).

FIGURE 5.17   Sverdrup transport (Sv), where blue is clockwise and positive is counterclockwise circulation. Wind stress data are from the NCEP reanalysis 1968–1996 (Kalnay et al., 1996). The mean annual wind stress and wind stress curl used in this Sverdrup transport calculation are shown in Figure 5.16a and in the online supplement, Figure S5.10.

apparent in the lee of large islands and mountain gaps; examples include the Hawaiian Islands among many others, and west of Central America, where strong winds force eddy generation in the Gulf of Tehuantepec (Chapter 10). Also apparent in the altimetric product (and not in the coarser reanalysis products) are wind stress curl patterns that follow the major western boundary currents such as the Gulf Stream, Kuroshio, and Agulhas, suggesting that these ocean fronts affect the position of the winds, constituting a feedback (Chelton et al., 2004).

The general circulation of the upper ocean is mainly driven by wind stress through the Sverdrup balance (Section 7.8). The Sverdrup transport computed from NCEP reanalysis wind stress curl is shown in Figure 5.17. (The NCEP wind stress curl that is the basis for Figure 5.17 is shown in the online supplement, Figure S5.10). Its pattern and magnitude are similar to that calculated from the mean QuikSCAT winds shown in Figure 5.16d (Risien & Chelton, 2008). This global map is mainly described in the later basin chapters as context for the circulation. The Sverdrup transport is computed as the zonal integral of the wind stress curl, integrated westward from the eastern meridional boundary in a given basin. For the Southern Ocean, the eastern boundary is the Chilean coast of South America and the integration extends westward across all three oceans until reaching the Argentine and Brazil coast of South America. Sverdrup transport is not computed at the latitudes of the Drake Passage because there is no meridional boundary there. It is also not shown for the equatorial region because the dynamics there are more complex.

6

Data Analysis Concepts
and Observational Methods

Our basic information about the oceans comes from observations and, increasingly, from numerical model output. To assist with reading later chapters and other observational oceanography literature, this chapter provides an overview of commonly used methods for analyzing observations. The basis of the statistical methods that are commonly used for oceanographic observations and numerical model output is not fully and mathematically described here (to a point where linear algebra is required), as is necessary in many modern approaches. There are many good starting points for a full course on modern data analysis; some useful texts are Bendat and Piersol (1986), von Storch and Zwiers (1999), Chatfield (2004), Emery and Thomson (2001), Bevington and Robinson (2003), Wunsch (1996), and so on. The Wolfram (2009) Web site provides demonstrations of basic statistical concepts. Press, Flannery, Teukolsky, and Vetterline's (1986) *Numerical Recipes* is useful for moving from concepts to the practice of data analysis.

In data analysis, we begin with *observations* or *determinations* of the value of a variable, such as pressure, time, temperature, conductivity, oxygen content, and so forth. These are collected using oceanographic instruments at particular times and locations that are chosen through

a sampling strategy (Section 6.1). From these imperfect observations, containing both instrumental and sampling error (Section 6.2), we *estimate* the true field and its statistical properties as a function of time and/or space (Sections 6.3 through 6.7). Sources of error are crucial to identify and are expressed in terms of statistical quantities. Errors arise from the accuracy of the instrumental measurements and from sampling that is discretized in time or space and finite in duration.

A large amount of supplementary material for this chapter appears on the textbook Web site http://booksite.academicpress.com/DPO/ as Chapter S16 ("S" denotes supplemental material). This supplement consists of an extended, fully illustrated description of instrumentation and methods for collecting information about the ocean, including accuracies and sources of error. Chapter S16 includes some of the sampling issues for physical oceanography (Section S16.1), platforms for observations (research and merchant ships; Section S16.2), instruments for in situ (within the water column) observations (Sections S16.3–S16.8), an overview of satellite remote sensing (Section S16.9), and oceanographic archives (Section S16.10).

As discussed in the supplementary historical materials for Chapter 1 (located on the textbook

Web site as Chapter S1) and the supplementary materials on instrumentation for this chapter (Chapter S16), observations in large-scale physical oceanography through the 1950s were designed to resolve the mean or long-term, large-scale structure of the ocean. Observations of time-variable phenomena were focused on smaller spatial scales and higher frequencies, including waves and tides. Because of the relatively small volume of data prior to the era of electronic sampling, satellite instruments, and computer analysis, individual observations were given greater consideration than they often are today. Sparse sampling still characterizes parameters that are difficult to measure, for instance, some of the chemical properties that require seawater samples and specialized laboratories for analysis. Error detection for such observations relies on good laboratory practice, including use of standards; comparisons with previous, possibly sparse, observations; and careful review of sample collection logs and analysis procedures.

In contrast, modern instruments that measure nearly continuous vertical profiles — underway sampling systems such as expendable bathythermographs (XBTs) and acoustic Doppler current profilers (ADCPs), moored current meters, autonomous drifting and guided systems, and satellites — can generate large volumes of digital data. These large data sets can be treated statistically to identify data errors, to map fields, to generate statistical information such as means and trends, and to detect embedded time and space patterns and correlations among different observed parameters (Section 6.4). Some of the basic concepts of time series analysis, including brief introductions to spectral and empirical orthogonal function methods, are included at a rudimentary level (Sections 6.5 and 6.6).

Oceanographic data are, by nature, three-dimensional in space and have time variation. Spatial sampling is almost always irregular, making good statistical techniques for mapping and analysis beneficial. Objective mapping is one common approach and is based on minimizing the difference between the mapped field and the observations in a least squares sense (Section 6.4).

Least squares methods are central to many common data analysis techniques, including those that seek to estimate the absolute velocity field from vertical profiles of temperature and salinity (see Section 7.6 for information on the derivation of dynamic height). Geostrophic velocity calculations require an accurate estimate of a quantity that is not measured: the geostrophic velocity at just one depth (any depth). Estimating this unknown "reference" velocity is important because transport calculations and budgets for all parameters depend on having an absolute velocity field. "Inverse methods" based on least squares estimation have been developed over the past several decades to yield the optimal estimate of these unknowns (Wunsch, 1996). Large-scale observed velocity fields from drifters or floats with climatological hydrographic data have also been merged using least squares techniques. Most recently, oceanographic data assimilation (or state estimation) methods, in which observations are incorporated in computer ocean models, are moving toward providing these absolute velocity fields since the ocean is now becoming sampled enough for practical data assimilation. Inverse and data assimilation methods are not introduced here, but we do provide some acquaintance with the principles of least squares methods on which they are based (Section 6.3.4).

Methods for the presentation and analysis of hydrographic (water property) data are also provided (Section 6.7). This traditional study has evolved to include statistical analysis as well. We introduce a reasonably common recent method, optimum multiparameter analysis (OMP), to illustrate the possibilities for modern water mass analysis. OMP is also based on least squares methods.

Many terms and concepts are introduced in this chapter, so a glossary of terms with short definitions of many of the concepts is found at the end.

The most important rule in data analysis is that there are no absolutely fixed methods. Data should be plotted, played with, replotted, combined with other data, and so forth, until the objective is achieved, which for scientists usually means discovering something new about the ocean. On the other hand, common understanding and application of data analysis techniques, and especially those that involve estimating error, are absolutely essential when combining and comparing results from different instruments, different properties, and different scientists.

6.1. OCEANOGRAPHIC SAMPLING

Modern physical oceanographic data are collected using many different platforms and instruments (see Chapter S16 located at the textbook Web site). Research ships have provided a long historical data set and continue to provide important modern observations. Analysis of these data sets often requires dealing with irregular temporal sampling and inhomogeneous spatial sampling (much higher resolution in some spatial directions than others). Instrumentation has also evolved and sampling philosophies have changed over time; in order to combine historical and modern data, changes in instrumentation, measurement error, and sampling have to be considered. Modern in situ data sets increasingly include large data sets from autonomous samplers such as floats and drifters and of vertical profiles collected from merchant ships.

Satellites (and sometimes aircraft) collect remotely sensed measurements of surface parameters such as sea surface height (SSH) and sea surface temperature (SST). Relative to ship and buoy measurements, satellites sample the ocean's surface so quickly that the observations can be regarded as almost *synoptic*. The word "synoptic" comes from meteorology, where it refers to weather in both space and timescales. In practical physical oceanography, the word is usually applied to observations taken at nearly a single time relative to the timescale of interest, similar to a "snapshot," and that are interpreted to contain only spatial information. For the ocean, the synoptic timescale is about two weeks, that is, the timescale of evolution of an eddy. For the non-seasonal part of large-scale ocean circulation, which varies on interannual and longer timescales, synoptic could be as long as a season or even a year or two.

All observations are collected with finite sampling intervals and over finite lengths of time. Observations that might be considered synoptic or nearly simultaneous for a given phenomenon might not resolve faster-evolving motions very well. For instance, sampling at "eddy timescales" demonstrably misses faster motions, such as tides or barotropic waves, and will contaminate interpretation of the desired timescale. This contamination is called *aliasing*, in which the actual underlying fast motion, being badly sampled in time, takes on the appearance of a much longer timescale (Section 6.5.3).

Moored instruments collect temporally continuous information on currents, float-tracks, temperature, salinity, and other chemical quantities; these can be treated using more straightforward time series methods than the ship- and satellite-based data sets. Now equipped with satellite transmission systems, these buoys report data in near real time for processing and incorporation into model studies. Because of the cost of individual moorings, these data sets tend to be spatially isolated; experiments and observation systems that incorporate multiple moorings have to be designed carefully based on whether it is advantageous to have observations at adjacent moorings correlated (or not) over the timescales of interest.

Sampling and analysis strategies for each of these various systems are based on the space and timescale of the phenomenon observed. Is the observing system looking at capillary and surface gravity waves or at changes in the North Atlantic's meridional overturning circulation? Sampling resolution must be sufficient to measure variations at the space and timescales of interest. This means that samples must be sufficiently frequent in space or time to resolve the highest frequency of variability, and the entire record must be long enough in space or time to contain a minimum number of cycles of the important fluctuations of the variable of interest.

For instance, in vertical profiling, the vertical variation in most properties is larger through the pycnocline than through the abyssal waters. Therefore vertical sample spacing or resolution through the pycnocline and in the upper ocean should be much closer than in the abyss (unless all are oversampled, as with a nearly continuous profiling instrument such as a conductivity-temperature-depth profiler; CTD). Another example is horizontal sampling in regions that contain a mixture of spatial scales; for example, the western North Atlantic contains both the Gulf Stream and the ocean interior offshore of the Gulf Stream. Good sampling strategy would include better spatial resolution across the narrow currents than in the broader flow regimes. (The best sampling might be considered to be the highest possible horizontal spacing throughout, but ship time and cost are also important factors, so sampling schemes are based on prior knowledge of the time and space scales of interest.)

For time or space coverage from an instrument that records nearly continuously, these data are averaged and/or filtered to yield the time/space scale of interest. Thus, the data processing for, say, a surface wave experiment is very different from the data processing for seasonal variability.

6.2. OBSERVATIONAL ERROR

Errors in observations, due to imperfections in the actual measured values, can be characterized in terms of *accuracy* and *precision*. In Chapter 3 and in the supplementary Web site materials Chapter S16, we report the accuracy and precision of different instruments measuring common ocean properties. *Accuracy* is how well the observation reproduces a well-defined standard that is usually set by an international group. High accuracy means that the difference between the observation and the international standard is small. Every instrument or observational technique has to be calibrated to match these international standards to the level required by the user and, therefore, the manufacturer or engineer. Accuracy of a measurement or data set is reported in terms of offset and standard deviation of the offset. Bias error (see following text) is directly affected by the accuracy of an observation.

Precision is the repeatability of an observation using a given instrument or observing system. An instrument or system could be highly inaccurate (e.g., due to lack of calibration), but highly precise, meaning that its variations due to instrument noise are very small. Precision is related to random error (see the next paragraphs). When observations are reported, the level of precision affects the number of significant places in the report; for instance, highly precise ocean temperature observations are listed out to four decimal places ($10^{-4\circ}C$).

There are two basic types of error in all data sets: systematic or bias error and random error. *Bias errors* are an offset of the measured values from the true values. Such errors can result from poor sampling strategy, failure of the sensor, error in the recording system, measurement inaccuracy, or insufficient record length for a time series that is averaged. For example, sampling choices can inadvertently create a "fair weather" bias. Many more ship-based observations are made in summer than in winter, especially at high latitudes; this biases the cumulative historical data set toward warm conditions. Another example is infrared satellite sensing of SST, which requires clear sky (cloud-free) conditions, thus biasing these observations.

The bias error associated with the measurements is separate from the bias errors that can be introduced by the statistical methods used to analyze the data set. Statistical methods (e.g., how an average is weighted) can introduce or offset bias. It is usually desirable to use statistical estimates that are unbiased. The estimators given in Section 6.3 are unbiased.

Random error or noise arises from variations at different time or space scales than the process of interest for the particular experiment. These can be both intrinsic to the observed variable (hence the desired true statistical property) or due to instrument or sampling error. The root-mean-square (rms) standard deviation (Section 6.3) is the calculated quantity associated with this noise. To minimize noise, data are averaged or filtered. For time series, this means collecting a record that is long enough to cover many cycles of the process of interest. These cycles are then averaged together. Figure 9.6 shows an example with snapshots of the path of the Gulf Stream and the mean value of this path. The noise or variance is a measure of the envelope of all of the meanders around this mean.

Most time or space series analysis techniques have been developed for long and continuous data sets. Many oceanographic data sets are sampled irregularly, which leads to problems with analyzing the variability in the data sets. A source of irregular sampling in time may be the failure of an instrument or its replacement by another instrument with clearly different sampling characteristics. Irregular sampling is also a consequence of combining historical data for a particular analysis. While individual cruises or experiments might have been organized for a specific task, a combination of these different sampling programs will not have regular temporal spacing. The resulting gaps in the time or space series mean that temporal variability is often not well resolved.

Even if sampling is at regular intervals, the true field cannot be continuously sampled. The discrete time interval (or distance) between samples leads to error in estimating processes that have short time (or space) scales. The *Nyquist frequency* (Section 6.5.3) is the highest frequency that can be resolved with a given sampling interval. Anything happening at a frequency higher than the Nyquist frequency is then very badly sampled, but is still in the record. These higher frequency signals appear as much lower frequency signals; this is called aliasing. It is highly desirable to design the observing strategy, specifically the sampling interval, to minimize aliasing.

6.3. BASIC STATISTICAL CONCEPTS

Every variable (such as temperature, salinity, pressure, velocity, etc.) has a set of true statistical behaviors; every set of observations of the variable is an imperfect representation of these statistical behaviors. In-depth data analysis courses and textbooks carefully cover the differences and similarities between the true statistics and estimation of these statistics and the associated error that arises simply because of the always imperfect sampling. The estimated statistics are called "sample statistics." Here, the sample statistics (mean, variance, standard deviation, etc.), rather than the true ones, are presented. We assert but do not derive the important relations that show that these expressions provide "unbiased estimates" of the true mean, variance, covariance, and so forth. An unbiased estimate is one that preserves the true value without introducing bias through the estimation method. Sometimes biased estimators can be more useful or more practical.

6.3.1. Mean, Variance, Standard Deviation, and Standard Error

The abundance of data in modern physical oceanography means that most analyses use averages of data values rather than individual samples. Modern instruments such as CTDs, current meters, satellite instruments, and so forth, collect many samples per second. Data processing usually starts with averages over many samples, for example, over 1 second or some other time interval. The sample mean $\bar{x}$ of a data set x that has been measured N equally spaced times is

$$\bar{x} = \frac{1}{N}\sum_{i=1}^{N} x_i \qquad (6.1)$$

With an increasing number of observations, Eq. (6.1) approaches the true mean if there is no external source of bias error. (That is, in the absence of instrument calibration problems, Eq. 6.1 is an unbiased estimate of the true mean.)

An *anomaly* is the difference between a measured value and the mean value (6.1):

$$x' = x - \bar{x}. \qquad (6.2)$$

The quantity x' is also often referred to as a *deviation* from the mean; it is also referred to as an anomaly. Because many observational studies are concerned with time or space variation, calculation and display of anomalies is common in oceanography, meteorology, and climate science. Depending on the study, it might also be common to remove a seasonal cycle from the original data set by computing monthly or seasonal means rather than the overall mean, and then displaying the anomalies relative to the monthly or seasonal means. For instance, a time series of surface pressure in the North Atlantic can be averaged over its approximately 50-year record, the average (mean) removed, and the anomaly time series analyzed to search for signals like the North Atlantic Oscillation or El Niño-Southern Oscillation (ENSO) influence.

The *variance* of the data set x is the mean value of the squared deviations of each measurement from the mean. The variance gives the inherent variability of the data set including variability of the true field and variability due to random sampling or instrument error. For sampled values, the best (unbiased) estimate of variance is the sum of the squared deviations, divided by (N−1), rather than N:

$$\sigma^2 = \frac{1}{N-1}\sum_{i=1}^{N}(x_i - \bar{x})^2$$

$$= \frac{1}{N-1}\left[\sum_{i=1}^{N}(x_i)^2 - \frac{1}{N}\left(\sum_{i=1}^{N} x_i\right)^2\right] \qquad (6.3)$$

The last expression provides a computationally efficient way to compute the variance, using only one pass through the data. The square root of Eq. (6.3) is the *standard deviation*, σ. Variance and standard deviation are intrinsic properties of the variable; Eq. (6.3) approaches the true variance with an increasing number of observations.

The *rms error* or *standard error* s_ε of the observed data set is the square root of the mean value of the difference between the true mean and the sample mean, averaged over many realizations of the sample mean. The standard error is related to the standard deviation from the mean as

$$s_\varepsilon = \frac{\sigma}{\sqrt{N}} \qquad (6.4)$$

Thus, the rms error of the mean, $\bar{x}$, is smaller by $\frac{1}{\sqrt{N}}$ than the standard deviation, σ, of an individual measurement x. The standard error decreases with increasing numbers of

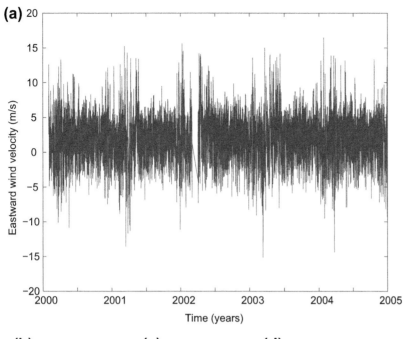

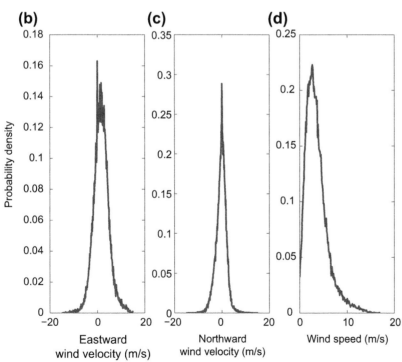

FIGURE 6.1 Example of time series and probability density functions (pdfs). (a) Eastward wind speed (m/sec) from an ocean buoy in Santa Monica Basin. (b) pdf of eastward wind velocity. (c) pdf of northward wind velocity. (d) pdf of wind speed. (*Constructed from Gille, 2005*).

observations; it is not an inherent property of the field that is being measured, but is a property of the sampling and instrumentation.

The profound difference between standard deviation (which is property of the true field) and standard error (which is a property of the sampling) is illustrated in the salinity climatology in Figure 6.13, in which the standard error is large in undersampled regions, while the standard deviation is large in regions of high oceanographic variability.

6.3.2. Probability Density Function

The *probability density function* (pdf) is the most basic building block for statistical description of a variable. Although the reader is much more likely to encounter spectral analysis or empirical orthogonal functions (see the following sections) in various publications, it is best to introduce pdfs first in order to develop intuition about estimates and confidence intervals.

The pdf of the true field is a measure of how likely the variable is to have a certain value. The probability of falling somewhere in the entire range of possible values is 1. The observed pdf is basically a histogram, that is, counts of the number of occurrences of a value in a given range. The histogram is then "normalized" to produce the pdf by dividing by the total number of observations and the bin widths. The more observations there are, the closer the histogram comes to the pdf of the true field, assuming that observational bias error is low (accuracy of observations is high).

Probability distribution functions can have many different shapes, depending on the variable and on the physical processes. As an example, from Gille (2005), a time series of wind velocity from an ocean buoy off the coast of southern California is shown in Figure 6.1. The data are hourly samples for four years. To compute the pdfs, the number of samples of velocities/speeds in each 0.1 m/sec bin was counted to create a histogram; for the pdf, the

values in each bin were normalized by dividing by the total number of hourly samples (43,797) and by the bin width (0.1). The two wind velocity pdfs are somewhat symmetric about 0, but they are not quite Gaussian (bell-shaped, Eq. 6.5). The wind speed pdf cannot be centered at 0 since wind speeds can only be positive; this pdf resembles a Rayleigh distribution, which has positive values only, a steep rise to a maximum and then a more gradual fall toward higher values.

A pdf with a uniform distribution would have equal likelihoods of any value within a given range. The pdf would look like a "block." Random numbers generated by a random number generator, for instance, could have a uniform distribution (the same number of occurrences for each value).

One special form of pdf has a "bell shape" around the mean value of the variable. Such a pdf is called a Gaussian distribution or a *normal distribution*. Expressed mathematically, a pdf of the variable x with a normal distribution is

$$pdf = \frac{1}{\sigma\sqrt{2\pi}}e^{-(x-\bar{x})^2/2\sigma^2} \qquad (6.5)$$

where the mean $\bar{x}$ is defined in Eq. (6.1) and the standard deviation σ in Eq. (6.2). A field that is the sum of random numbers has a normal distribution. The pdf associated with calculating the mean value has a normal distribution. Thus if we measure a large number of sample means of the same variable, the distribution of these mean values would be normal.

The pdf associated with a sum of squared random variables is called a *chi-squared distribution*. Squared variables show up in basic statistics in the variance (6.2), so the chi-squared distribution is important for estimates of variance. Gaussian and chi-squared distributions have a special place in statistical analysis, especially in assessing the quality of an estimate (confidence intervals), as described at the end of the next section.

6.3.3. Covariance, Auto-Covariance, Integral Timescale, Degrees of Freedom, and Confidence Intervals

If two or more variables are measured, it is useful to quantify how closely they depend on each other. For instance, we might want to know how closely temperature and velocity are correlated with each other. The sample *covariance* is the statistical relation between two observed variables, for example x and y, each sampled N equally spaced times:

$$\text{cov}(x, y) = \frac{1}{N-1} \sum_{i=1}^{N} (x_i - \bar{x})(y_i - \bar{y}) \quad (6.6)$$

With an infinite number of samples, this approaches the true covariance. The sample *correlation* is covariance divided by the sample standard deviations:

$$\rho_{x,y} = \frac{\text{cov}(x, y)}{\sigma_x \sigma_y} \quad (6.7)$$

in which the standard deviations for both x and y are defined as in Eq. (6.3).

The autocovariance and autocorrelation are the same expressions as (6.6) and (6.7), but with the two variables replaced by a single variable measured at different times. In this case the sum is over all pairs separated by the same time difference within the time series. For instance, if velocity (indicated here by the variable x) is measured every hour for four years (e.g., Figure 6.1), then time lags, denoted by τ, of 1 hour up to 4 years are available. If the record length is T, with a total of N samples at a sampling interval of Δt (so $T = N\Delta t$), the autocorrelation for a given lag $\tau = n\Delta t$ is

$$\rho_{x,x}(\tau) = \frac{1}{\sigma_x^2} \frac{1}{M} \sum_{i=1}^{(N-n)\Delta t} x'(t_i - n\Delta t)x'(t_i) \quad (6.8)$$

The anomaly x' was defined in Eq. (6.2). The autocorrelation is typically calculated for all time lags. The value of M can be either N (total number of samples), or N−n (total number of

pairs at lag τ). If the total number of pairs is chosen, then Eq. (6.8) is an unbiased estimate of the autocorrelation, because the estimated autocorrelation is not offset from the true autocorrelation. However, this unbiased estimate becomes very large at large lags, where N−n becomes very small. If the total number of samples N is chosen, then Eq. (6.8) is a biased estimate of the autocorrelation, but it has good behavior at large lags. The unbiased estimate is best for looking at behavior at small lags, such as for finding decorrelation timescales (see the next paragraph).

Using a simulated temperature record for a Pacific island, the unbiased and biased autocorrelation functions are calculated and plotted as a function of lag (Figure 6.2; Gille, 2005). The autocorrelation is 1 at zero lag, as it should be, since the values should be perfectly correlated with themselves. The unbiased estimate blows up at large time lag, but is well-behaved at small lag. The biased estimate (which is the default in the Matlab software package used by many oceanographers) is well-behaved at large lag. At small lags (Figure 6.2d), the autocorrelation decreases to a zero crossing at about 6 months. The time lag for the zero crossing is one measure of the "decorrelation timescale" for the variable, that is, the time lag for which samples become uncorrelated. Since the autocorrelation hovers around zero for several months in Figure 6.2d, the decorrelation timescale is somewhat ambiguous, but is in the range 6−14 months.

An *integral timescale* T_{int} for the observed variable is defined as the time integral of the autocorrelation (e.g., Gille, 2005; Rudnick, 2008). The integral timescale is another measure of the decorrelation timescale. For the sample autocorrelation in Eq. (6.8), T_{int} is the sum of the autocorrelations multiplied by the time lag bin width; this is the area under the autocorrelation function in Figure 6.2b and c. In practice, the sum is computed starting with just a small total time interval; this is increased

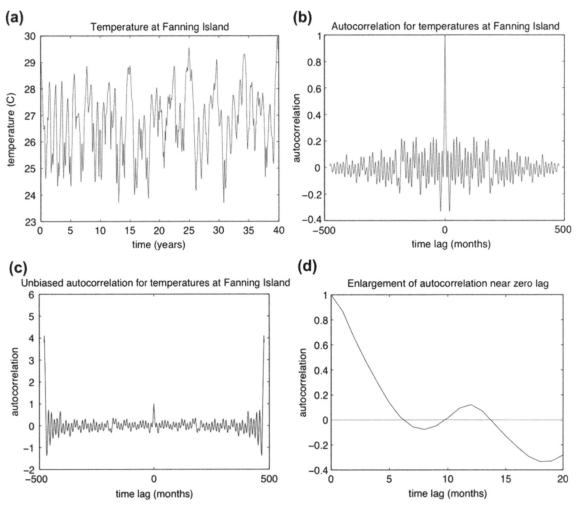

FIGURE 6.2   (a) Time series of temperature at Fanning Island (Pacific Ocean) from the NCAR Community Ocean Model. (b) Autocorrelation normalized to a maximum value of 1 (biased estimate with averages divided by N). (c and d) Auto-correlation (unbiased estimate with averages divided by N−n). *Source: From Gille (2005).*

incrementally until reaching the entire length of the record. There will usually be a maximum value for the integral at one of the intermediate integration limits. This maximum value is the integral timescale. (Thus the misbehavior of the unbiased estimate at large lags can be ignored, so either the biased or unbiased auto-correlation function can be used.) For the time series in Figure 6.2, the computed integral time-scale is 9.3 months, which can be compared with

the crude decorrelation timescale of >6 months estimated from the first zero crossing.

The effective *degrees of freedom* of an estimate are related to the integral timescale. The N measurements are not necessarily independent of each other. How many independent samples do we really have, for instance, in a nearly continuously sampled time series? The number of independent samples, which is the same as the "degrees of freedom," is the total length of the time series divided by the integral timescale:

$$N_{dof} = T/T_{int} \qquad (6.9)$$

How many degrees of freedom are desirable for a good estimate? There are textbook answers to this question, but the real answer lies in how well you want or need to know the answer; that is, how large are the errors? If they are too large, little can be learned from the data set. The "error bars" are formally the *confidence intervals* calculated from the data set. Confidence intervals are central to most data-oriented analyses.

Confidence intervals depend on the number of degrees of freedom, and they also depend on the standard error, hence on the standard deviation of the time series. (Again, a full textbook description and derivation is recommended; see for instance Bendat & Piersol, 1986.) Suppose we have a set of averaged, observed values X of the variable x. We have already found the standard deviation and the number of degrees of freedom. Therefore we already know the standard error $s_\varepsilon = \sigma/\sqrt{N_{dof}}$. Suppose we are looking for the probability that the true average $\overline{X}$ exists within a given interval. What is that interval? The statement of probability P is

$$P[X - s_\varepsilon t_{Ndof}(\alpha/2) \leq \overline{X} \leq$$
$$X + s_\varepsilon t_{Ndof}(\alpha/2)] = 1 - \alpha \qquad (6.10)$$

If, for instance, we wish to find 95% confidence intervals, then $(1-\alpha) = 95\%$ and $\alpha = 5\%$. The statement (Eq. 6.10) is then read as "there is a 95% probability that the true mean $\overline{X}$ lies within the interval from $X - s_\varepsilon t_N$ to $X + s_\varepsilon t_N$ where X is the sample mean." The factors t_N are the "Student t variables" with N degrees of freedom, and which depend on the choice of confidence interval. Here the N is the calculated degrees of freedom N_{dof}. (The Student t-test is appropriate for a variable that has a Gaussian distribution; because we craftily started out with a variable X that was already an average, we can be pretty certain that the distribution is

Gaussian. This is a consequence of the Central Limit Theorem. (See Bendat &Piersol, 1986 for a discussion of this important theorem.) Once you have estimated the degrees of freedom and chosen an α, the t-variables are found from a lookup table, available in most statistics textbooks or online; they can also be obtained through functions in Matlab or Mathematica.

For a 95% confidence interval and with 10 degrees of freedom, the t variable is 2.23. For 10 degrees of freedom and a 90% confidence interval, the t variable is 1.81, whereas for a 99% confidence interval, it is 3.16. As the number of degrees of freedom increases, the t variables become smaller and the confidence interval shrinks.

An example of a plot with confidence intervals, in this case at the 90% level, is shown in Figure 6.3. This is a graph of global ocean heat content in the upper 700 m since the 1950s, constructed from all available temperature profile data at the time of the analysis. Because there are confidence intervals on the plot, it is possible to conclude that the upper ocean has warmed since the 1950s, and that the warming is "significant" in a formal sense. However, there is an important limitation to the use of confidence intervals, for which it is assumed that error is random. When there is also a problem of accuracy (formally, bias error), then confidence intervals are simply not adequate. The graph in Figure 6.3 should be compared with the more recent version of global ocean heat content in Figure S15.15 on the textbook Web site, from Domingues et al. (2008). There was a protracted episode of low quality temperature data in the 1970s, due to error in assigning depths to falling XBT profilers (see instrument description in Section S16.4.2.5 in the online supplementary materials); this led to artificially high temperatures in the 1970s in Figure 6.3. Domingues et al. (2008) recognized and corrected for this accuracy problem; the improved estimate of heat content change is much more monotonic over the full record from the 1950s to the 2000s.

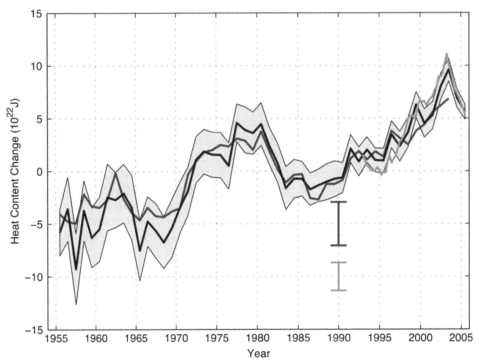

FIGURE 6.3 Example of time series with confidence intervals. Global ocean heat content (10^{22} J) for the 0 to 700 m layer, based on Levitus et al. (2005a; black curve), Ishii et al. (2006; full record gray curve and larger error bar), and Willis et al. (2004; darker gray after 1993 and shorter error bar). Shading and error bars denote the 90% confidence interval. Compare with Figure S15.15 seen on the textbook Web site from Domingues et al. (2008) which uses improved observations. *Source: From the IPCC AR4, Bindoff et al., 2007; Climate Change 2007: The Physical Science Basis. Working Group I Contribution to the Fourth Assessment Report of the Intergovernmental Panel on Climate Change, Figure 5.1. Cambridge University Press.)*

This discussion about confidence intervals is very limited because this text is not primarily about statistics. The subject is large and subtle. Most actual properties are not normally distributed, although they might be close. Techniques for choosing confidence intervals can therefore vary. Several other important methods are used for assessing uncertainty, including bootstrap and Monte Carlo methods. Those who wish to pursue serious analysis of data sets are urged to study this subject in much greater depth.

6.3.4. Least Squares Analysis

When two or more variables are measured, we can use correlation to see how closely they are related (Section 6.3.3). This gives us a single number. Beyond correlation, we can make a guess at the functional dependence of one variable on the other. To determine parameters in the relation between the variables, we perform some kind of fit based on assumptions or a model of how the variables are related. One of the most prominent methods of this type is *least squares analysis*. Wunsch (1996) provided an excellent, deep introduction to least squares and applications to inverse modeling and related topics in physical oceanography.

As the simplest common example of least squares, consider two time series, $x(t)$ and $y(t)$. Suppose that we believe that one depends linearly on the other, for whatever reason, so we have an equation that we assume is a pretty good relation between them, but with parameters that we do not yet know. That is, suppose:

$$y(t_i) = a\, x(t_i) + b \qquad (6.11)$$

We then determine the parameters a and b by minimizing the difference between the two time series in a least squares sense:

$$\varepsilon = \sum_{i=1}^{N} (y_i - ax_i - b)^2 \qquad (6.12)$$

where a and b are unknown parameters to be determined and ε, which is the sum of the squared differences between the two expressions, is the squared "misfit" we seek to minimize. (ε is also referred to as a "cost function," which leads into the topics of inverse modeling and data assimilation based on least squares.) To find the best values for a and b, take the partial derivatives of ε with respect to a and separately with respect to b, set the derivatives to zero, and solve for a and b. Just to show how this works with an even simpler model, in which b = 0, the solution for parameter a comes from solving

$$\frac{\partial \varepsilon}{\partial a} = \frac{\partial}{\partial a} \sum_{i=1}^{N} \left(-2ay_i x_i + a^2 x_i^2 \right) = 0$$
$$a = \left(\sum_{i=1}^{N} y_i x_i \right) \bigg/ \left(\sum_{i=1}^{N} x_i^2 \right) \qquad (6.13)$$

As the "model" becomes more complex, but still linear, solving for the parameters becomes much simpler if basic linear algebra is employed. That is, the parameters in a matrix "**A**" that relate vectors **x** and **y** might be expressed as, **y** = **Ax**, where the bold type indicates vectors and matrices. Solution of even the simplest of these problems is beyond this chapter. However, it is straightforward to carry through and to actually calculate the parameters in a problem like Eq. (6.12) using software packages like Matlab that include linear algebra and matrix operations.

Simple least squares fits are often employed in calibrations. For instance, for CTD conductivity calibration, a number of highly accurate salinity values might have been obtained externally using bottle samples. These can be converted to conductivity (see Chapter 3), and then the measured CTD conductivity can be compared with the sample conductivities at the same locations. The differences between the two data sets (bottle and CTD) would be expressed as a sum of squared differences, and the CTD conductivity fits, using least squares, to the bottle samples. The fits can be linear, quadratic, cubic, and so forth, to provide the best possible calibration, which depends on the underlying physical response of the conductivity sensor (it usually also has pressure and temperature dependence, adding more complexity to the fitting process).

In more general circumstances, beyond the simple time series observations used in the previous paragraphs, the size of the vectors **y** and **x** can differ so much that **A** is not a square matrix. Their meanings can also differ, where **x** can be a set of observations, and **y** the field being sought. These become "overdetermined" problems if there are fewer unknowns than equations, and "underdetermined" problems if there are more unknowns than equations. It is not necessary to use the squared difference as the ideal "norm" for minimization; a more educated choice would depend on the statistics of the differences being minimized. For least squares, it is assumed that the differences have a Gaussian distribution.

More advanced linear least squares analysis can also add external constraints to the assumed functional relationship between two data sets. This enters the realm of *inverse models* as applied to estimation of the unknown reference velocity for a geostrophic velocity profile that is first calculated from vertical density profiles using dynamic or steric height (Wunsch, 1996). It also enters the realm of *data assimilation*, in which the proper dynamical relations between different fields are presumed and all of the fields are adjusted, in a dynamically consistent manner, to most closely fit the observations using least squares. But the basic assumptions and practices are the same: the assumption that there are linear relationships, and the practice of minimizing the differences between two functions based on some assumed norm.

But the real question is whether even complicated linear models are valid. In many cases they are. The assumption of a simple linear relationship is like the origin of spectral analysis in the late nineteenth century, when the assumption was that a given time series could be fit or explained in terms of just one or two sinusoidal functions. Spectral analysis moved far beyond this by generalizing the valid representation of a given time series in terms of a very large number of sinusoidal functions that can completely represent the data (Section 6.5). Spectral analysis is not always ideal since the underlying processes may not be sinusoidal, in which case a spectrum can be obtained but might not be as easy to interpret as an analysis using more appropriate orthogonal basis functions. Similarly, in many cases we know that linear relationships are valid, because we know independently what the dynamical relationship between two data sets is expected to be, at least after components that have little to do with the dynamical relationship of interest are filtered out. If the relationship is linear (and the differences are Gaussian), then linear least squares methods are a valuable place to start.

6.4. VARIATION IN SPACE: PROFILES, VERTICAL SECTIONS, AND HORIZONTAL MAPS

Observations that sample the ocean spatially are often combined to produce vertical cross-sections and quasi-horizontal maps. These data are almost never collected on a regular grid.

Temporal coverage within spatially sampled data sets also varies greatly.

Beyond simply plotting the data as profiles or values along a float track, for example, a frequent step in working with spatially distributed data is to map it to a regular grid. The most simplistic method is to average the data within a given grid "box" defined by latitude and longitude. This "bin-average" method produces a field that can be described and can be adequate for a given study, especially if the data set is very large. But a bin-averaged field is often not the optimal field for quantifying and studying the dynamics.

Objective mapping is a more complex and common method for mapping randomly spaced data to a specified set of locations. Objective mapping for oceanography was introduced by Bretherton, Davis, and Fandry (1976) for analysis of eddy-scale observations in the Gulf Stream region; their paper remains the definitive basic treatment. Objective mapping provides the "least square error linear estimate" of the field (Bretherton et al., 1976) and, importantly, an error field that depends on sampling locations. Objective mapping methods are most useful if the estimates are unbiased, and there are different approaches to achieving this (e.g., Le Traon, 1990). With objective mapping techniques, external constraints on the fields can be incorporated (e.g., the mapped velocity field is geostrophic), and different types of data can be combined.

We do not present any details of the objective mapping method, as linear algebra is not included in this text. Objective maps are basically weighted averages of the data in the neighborhood of the grid points. But the weighting requires information on the horizontal shape and scale of smoothing as a function of distance from the grid point. The more influence given to data from farther away, the smoother and larger scale the mapped field. This weighting information should come from the spatial covariance, but this is not always (or usually) known since it is also obtained from the actual observations. In practice, simplified functions for weighting are often chosen *a priori*; these are often either exponentials or Gaussians (squared exponentials). The weighting can be anisotropic, with different horizontal decay scales in different directions. This is useful for studies of frontal or coastal regions where correlations are larger in one direction than in the other. Anisotropic scales are also useful for mapping data onto vertical sections since the vertical and horizontal scales differ enormously.

6.4.1. Variation in the Vertical Direction

6.4.1.1. Sampling

Many ocean observations are collected as vertical profiles. Because the ocean varies more strongly in the vertical than in the horizontal, sampling strategies usually provide far more vertical resolution than horizontal resolution. In addition, the upper ocean is more strongly stratified than the deeper ocean in most places, and most properties and currents reflect the stratification. So sampling strategies often include higher vertical resolution in and above the pycnocline than in the deep ocean.

Instruments that are typically used for vertical profiling, like a CTD, XBT, profiling float, or lowered ADCP (LADCP), measure "continuously." For instance, commonly used CTDs sample at 24 Hz (24 samples per second). Profile processing usually involves averaging over short pieces of the continuous sample to produce a series of data at regularly spaced pressures (e.g., 1 or 2 dbar) or times (e.g., 1 second). The averaging reduces the profile noise resulting from smaller scale processes, such as microstructure on scales of centimeters. These smoothed data series then resolve most of the phenomena of interest for a large-scale or mesoscale study, since ocean layering is often on scales of 10 m and more. (Obviously a different sampling and averaging approach is necessary for studying the much smaller vertical scales of the microstructure and fine structure associated with mixing processes.)

Bottle sampling requires a choice of observation depths, using prior knowledge of the basic field and nearby or concurrent CTD profiles, which provide temperature/salinity information that can assist choice of bottle-sampling depths. In the past, it was common to sample at standard depths. This is no longer considered good practice for hydrographic sampling since we now focus on mapping the three-dimensional property distributions, or properties on surfaces that cut through the ocean, for example, isopycnal surfaces. A hydrographic property field is best mapped if there is some randomness in the vertical sampling from station to station, but always with enough samples to define the vertical gradients in the property.

6.4.1.2. Vertical Profiles

The distribution of properties with depth is illustrated with temperature/depth, salinity/depth, and oxygen/depth profiles (e.g., Figure 4.2, etc.), and is usually the first step in examining hydrographic data. It is useful to plot the data as soon as possible so that problems with equipment, sampling, or laboratory analyses can be identified and remedied before many more samples or stations are collected. Data from multiple stations, geographical positions, or times can be displayed together, to differentiate between true ocean structure and biased or noisy data. When analysts work with large data sets, collected over many years, rather than with a data set that is newly acquired, they often average the data after interpolation to either standard depths or standard densities, and examine or reject data points that are outliers because they are outside the range of the standard deviation of the data set. (This involves some iteration if additional data continue to be collected.)

With nearly continuous and regularly sampled vertical profile data, such as from CTDs, XBTs, LADCPs, and so forth, it is easy to interpolate to any desired vertical level (depth, pressure, density, temperature, etc.). With sampling intervals on the order of 10 dbar or less, simple linear interpolation can be adequate. Some properties are more sparsely sampled, such as chemical samples from a rosette water sampler, where sampling intervals can range up to several hundred decibars. Vertical interpolation using smoother methods, such as a cubic spline, can then be advantageous; care should be used to choose procedures that do not introduce spurious (unmeasured) maxima or minima in the profiles. The Akima cubic spline has been found to be especially good in this respect.

6.4.1.3. *Vertical Sections*

To examine the geographical distribution of properties across a basin and in the vertical direction, cross-sections through the ocean ("vertical sections") are produced from a set of stations, often located along a substantially straight path (e.g., Figures 4.11 through 4.13 and many other examples through the text). Such sections are an invaluable step in data quality control. For quality control and to communicate the accuracy of a given sampled field, it is useful to include station locations and also sample locations if the property is sampled with discrete bottles (e.g. vertical sections of chlorofluorocarbon and $\Delta^{14}C$ in Figure 4.24).

To construct vertical sections from profile or water sample data, interpolation methods are usually required. Objective mapping (Section 6.4.1) is a commonly used method. Roemmich (1983) introduced a useful method for mapping vertical section data that relies on the station separation to set horizontal decorrelation scales, and very crude information about vertical stratification to set vertical decorrelation scales. The assumptions are that the scientists collecting the data will know to sample more closely across strong dynamical features such as western boundary currents, the equator, and fronts, and that this information should not be lost in the process of objectively mapping the data. Almost all of the vertical sections presented in this text were objectively mapped using Roemmich's method.

Since flow is mostly along isopycnal surfaces, with only a small diapycnal component, it can be useful to display data as a function of density rather than pressure or depth. An Atlantic salinity section is plotted in Figure 9.17 as a function of both depth and neutral density (Section 3.5.4). (The profile/sample data were first interpolated to a large number of neutral densities using cubic splines and then objectively mapped.) The contours are much "flatter" in the isopycnal coordinate than they are in the depth coordinate. This suggests that flow is more along isopycnal surfaces than along surfaces of constant depth and helps to justify analysis of large-scale properties along isopycnal surfaces (Section 6.4.2).

6.4.2. Variation in the Horizontal Direction

Some types of measurements lend themselves exclusively to horizontal mapping. Examples include temperature, velocity, surface height, and air—sea flux fields from surface drifter and satellite observations. Within the ocean, because of strong vertical stratification, flow mostly follows isopycnal surfaces, which are substantially horizontal. Therefore velocity and water property data from within the ocean are often mapped on quasi-horizontal surfaces, including constant depth or isopycnal surfaces. Examples are found throughout the chapters of this text.

Horizontal maps are usually created by choosing the desired surface, interpolating the data in the vertical to the surface, and then mapping the vertically interpolated data. Various methods are used for vertical interpolation (Section 6.4.1.2). Horizontal mapping of the data to a latitude/longitude or distance/distance grid is also an interpolation exercise. It is often carried out using objective mapping (see beginning of Section 6.4), or some other procedure that chooses data within a given radius of the grid point and then creates a weighted mean of the data depending on distance from the grid point.

Maps that illustrate the influence of the high salinity Mediterranean Water (MW; Section 9.8.3.2) are shown in Figure 6.4. Three kinds of maps are shown: at a constant depth, on an isopycnal surface, and at a "core layer." All three have their uses. The isopycnal surface is expected to be the most representative of the actual flow. The choice of pressure reference for the isopycnal is important for creating a surface that best follows the flow; in this figure, a reference pressure at 1000 dbar was chosen since the core of the MW is around 1200 dbar. A neutral density surface would also be an effective choice (see Figure 6.4 caption). Core layers (surfaces defined by vertical extrema of properties like salinity) were introduced and used extensively in the 1930s by German oceanographers. The core layer in Figure 6.4c is the MW salinity maximum.

For all three maps, care must be taken in interpreting "tongues" of high or low salinity as indicating flow direction. If mixing is relatively strong, then a horizontal tongue may indicate flow direction. On the other hand, if there were no mixing, then flow would have to follow contours of the mapped property, and go around the tongue. Therefore the main usefulness of a core layer is to show the area of influence of a particular water mass.

The three maps in Figure 6.4 are complementary, and all show that the highest salinity in this depth range originates at the Strait of Gibraltar, where the high salinity MW exits into the North Atlantic. The idea of the core layer method is that the high salinity pool in the North Atlantic indicates movement of the water away from the Strait of Gibraltar. Because of mixing, a core gradually weakens along its length.

Horizontal velocity mapping is important for studying circulation. At horizontal length scales ranging from mesoscale (ten to hundreds of km) to global scale (thousands of km), the horizontal velocities are nearly geostrophic and therefore non-divergent.[1] The horizontal velocities can then be represented by a streamfunction (Section 7.6). Consequently, it is desirable to map continuous contours that align with the velocity vectors; error maps can be produced as the difference between the mapped non-divergent velocities and the original velocity data, as well as the usual error estimates due to the mapping procedure and due to measurement error and variance.

Large-scale velocity and streamfunction maps are shown in other chapters to illustrate the sea surface and 900 m circulation. These are based on surface drifter plus altimeter data (Niiler, Maximenko, & McWilliams, 2003) and subsurface float data (Davis, 2005). (These instruments are described in the supplementary online material in Chapter S16, Section S16.5.) These two treatments of "Lagrangian" data (Section 7.2) followed somewhat different routes to produce the non-divergent fields, and both illustrate the sensible creation and application of mapping techniques based on the desired product and the types of available data. Niiler et al. (2003) combined the surface drifter with satellite altimeter data, using a least squares procedure and dynamical constraints to produce the mean surface streamfunction. Davis (2005) produced mean velocity vector maps constrained to produce a non-divergent field; he then used objective mapping to produce the geostrophic streamfunction.

---

[1] Typical horizontal current speeds range from 1 cm/sec up to 200 cm/sec (about 200 km/day or about 2 knots) in the swift western boundary currents (Gulf Stream, Kuroshio), in the Antarctic Circumpolar Current and in the upper ocean equatorial currents, to a fraction of 1 cm/sec in much of the surface layer and in the deep waters. The vertical speeds associated with the large-scale circulation are much less, on the order of 10^{-5} cm/sec or 1 cm/day; these are essentially unmeasurable except with extremely good instruments and data filtering.

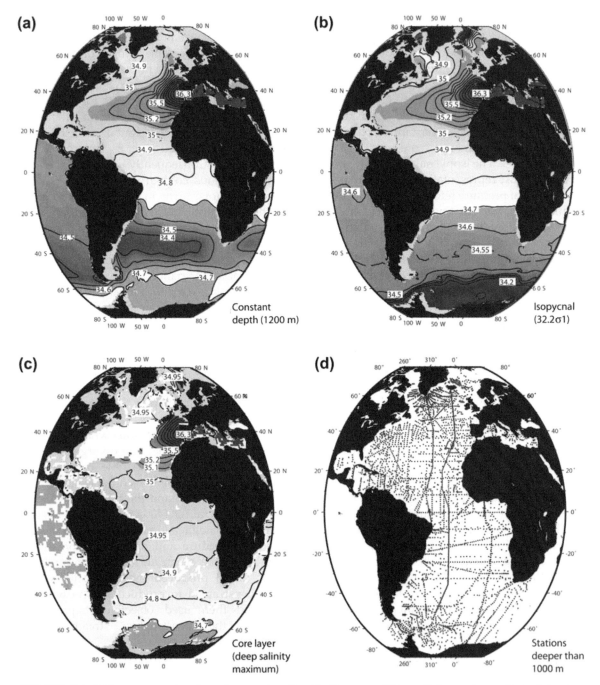

FIGURE 6.4 Different types of surfaces for mapping. The Mediterranean Water salinity maximum illustrated using: (a) a standard depth surface (1200 m); (b) an isopycnal surface (potential density $\sigma_1 = 32.2$ kg/m^3 relative to 1000 dbar, $\sigma_\theta \sim 26.62$ kg/m^3 relative to 0 dbar, and neutral density ~ 26.76 kg/m^3); (c) at the salinity maximum of the Mediterranean Water and North Atlantic Deep Water (white areas are where there is no deep salinity maximum); and (d) data locations used to construct these maps. This figure can also be found in the color insert.

Davis and also Gille (2003), who mapped the velocity field at 900 m in the Southern Ocean from subsurface floats, provided in-depth information about their mapping methods, the details of which are beyond this text.

Horizontal mapping of velocity fields has been carried out on regional scales using density profile and ADCP velocity data; the sampling strategies here have been quasi-grids with observations in three dimensions rather than just along a single section. The density information provides the vertical shear of the geostrophic velocity through the thermal wind balance. The ADCP data provide information for the geostrophic velocity referencing, but contain all timescales of motion, including ageostrophic velocity as well as geostrophic. An example from the California Current is shown in Figure 6.5 (Chereskin & Trunnell, 1996); other similar maps have been produced

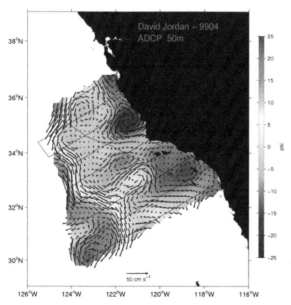

FIGURE 6.5 Objective mapping of velocity data, combining density and ADCP velocity measurements. California Current: absolute surface streamfunction and velocity vectors in April, 1999, using the method from Chereskin and Trunnell (1996). This figure can also be found in the color insert. *Source: From Calcofi ADCP (2008).*

for the Azores Front (Rudnick, 1996) and the Antarctic Circumpolar Current in Drake Passage (Figure 13.9 from Lenn, Chereskin, Sprintall, & Firing, 2008). These publications include extensive information on the mapping techniques created for these specific data sets.

6.5. VARIATION IN TIME

All ocean flows and properties vary in time. Here we introduce some basic ways of displaying and analyzing time series (data display, spectral methods), as well as some common but more advanced methods (empirical orthogonal functions).

6.5.1. Time Series Data Display

Examples of time series are shown in Figures 6.6 through 6.9. Others appear throughout the chapters of this book. The first step in working with a time series is usually a simple plot of the property versus time (e.g., Figure 6.6a). For data collected as profiles, overlays of all, or a subset of, the profiles are useful for seeing the variability and variance in the time series (Figure 6.6b). It can be useful to make a "waterfall" plot, with the profiles offset from each other by a fixed increment of the observed property, to see individual features and how they might propagate through the time series.

Profile data, such as from profiling floats, are often contoured like a vertical section with time and depth (or pressure or density) as the axes rather than distance and depth. Similarly, if time series data are collected from a number of locations along a repeated track, display of the data as a contour plot as a function of time and the spatial dimension can be useful. This type of plot is called a *Hovmöller diagram*. This type of display is used in this book to show the evolution of Arctic sea ice (Figure 12.22) and the westward propagation at mid-latitudes typical of Rossby wave behavior (Figure 14.18).

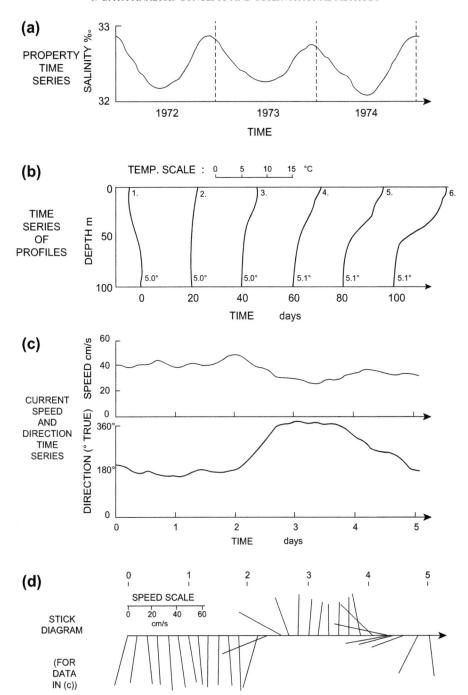

FIGURE 6.6 Examples of time series plots: (a) property/time, (b) time series of profiles, (c) current speed and direction, and (d) stick diagram for data of (c).

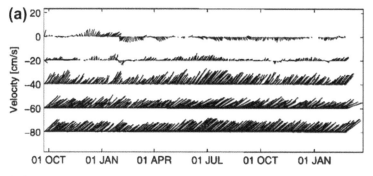

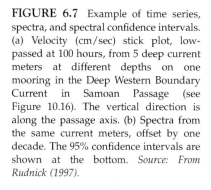

FIGURE 6.7 Example of time series, spectra, and spectral confidence intervals. (a) Velocity (cm/sec) stick plot, low-passed at 100 hours, from 5 deep current meters at different depths on one mooring in the Deep Western Boundary Current in Samoan Passage (see Figure 10.16). The vertical direction is along the passage axis. (b) Spectra from the same current meters, offset by one decade. The 95% confidence intervals are shown at the bottom. *Source: From Rudnick (1997).*

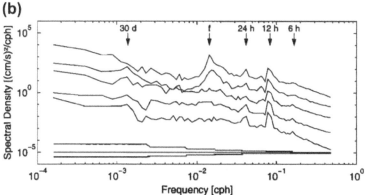

6.5.2. Velocity (Vector) Data Time-Series Analysis

Vector fields such as velocity are slightly more complicated to present than scalar time series because they include two quantities: magnitude (i.e., speed) and direction. The *stick plot* is a useful vector display method, representing speed and direction by a line drawn to scale out from a time axis at each observation time (Figures 6.6d and 6.7). Currents from a set of instruments on a mooring can be plotted above each other to give a visual idea of the correlation of the records at different depths (Figure 6.7).

Another alternative for displaying velocity data is a *progressive vector diagram* (see example in Figure S7.14a from Chereskin, 1995 on the textbook Web site), in which the displacements from each time step are added to produce an apparent particle track in space. This is not the

track followed by an actual particle, but it is a useful visual representation of the velocity time series.

Vector fields with geographic coverage (latitude and/or longitude) can be plotted as vectors on a map (Figure 6.5). A Hovmöller diagram with a time axis and a position axis can also be used, with vectors plotted as a function of time and distance.

6.5.3. Spectral Analysis

In Section 6.3, we introduced some concepts for quantifying basic properties of a time series. In Section 6.5.2 we described some simple approaches to viewing the data. More in-depth analysis of a time series, for instance using spectral analysis techniques, can yield much more information about ocean processes such as their timescales, repeatability, and evolution.

(a)

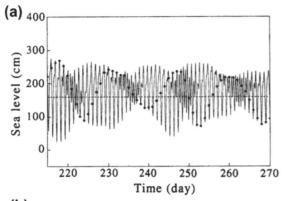

(b)

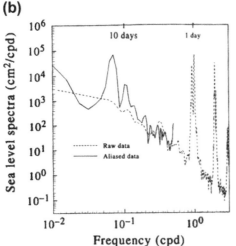

FIGURE 6.8 Example of a time series, spectrum, and spectral aliasing. (a) Tide record at Victoria, British Columbia (July 29 to September 27, 1975). The heavy dots are a once per day subsampling of the record. (b) Power spectrum of the complete tidal record (dashed) and the subsampled record (solid), showing how the diurnal and semi-diurnal tidal energy are aliased to periods of 10 days and longer. *Source: From Emery and Thompson (2001).*

Spectral analysis, or "Fourier analysis," is a straightforward approach to extracting more information from a time series. Spectral analysis is useful for determining tidal components in a time series of sea-surface height or currents (Figures 6.7a and 6.8a), or for deciding if there is significant variability at a seasonal or inter-annual frequency. "Significance" is a formal

concept for spectral analysis, based on an estimate of error for the energy at each frequency and related to the confidence intervals described in Section 6.3 for probability density functions. Here we briefly describe some of the basic concepts. Further information can be obtained from a number of textbooks and Web sites. Examples can also be found in Press et al. (1986), Emery and Thomson (2001), Chatfield (2004), von Storch and Zwiers (1999), and Wolfram (2009). The following paragraph draws partially on Emery and Thomson and notes from Gille (2005).

Each true time-varying process in the ocean can be represented as an integral (continuous sum) of an infinite number of orthogonal basis functions such as sines and cosines; that is, an infinite time series can be fit to specified functions using least squares. If the functions are orthogonal to each other and if there are enough of them (e.g., an infinite series), then the time series can be completely represented as a sum of these functions. Because many of the external forcings for the ocean recur regularly, (orthogonal) periodic functions that describe both the forcing and the ocean are a sensible place to start. For instance, the tidal record in Figure 6.8a clearly includes periodic components.

In spectral analysis, the orthogonal basis functions are sines and cosines for a range of frequencies. With actual data sets, we must also deal with the finiteness and discrete sampling of the observed time series and produce error estimates for the contribution of each frequency to the overall process.

Spectral analysis is often as useful in the spatial domain as in the time domain, that is, for yielding information about spatial scales (wavelengths). This is especially helpful when analyzing spatial-temporal data to study waves, which are characterized by wavelengths and frequencies. On the other hand, for large-scale oceanography, other techniques such as empirical orthogonal functions (Section 6.6) that do not presume

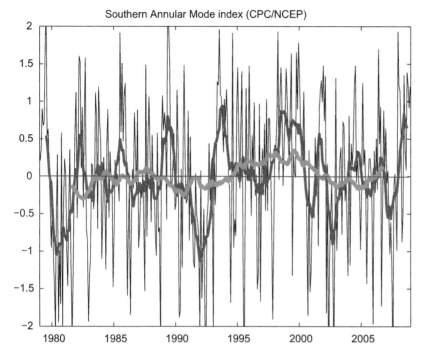

Southern Annular Mode index (CPC/NCEP)

FIGURE 6.9 Lowpass filtering by averaging the time series: Southern Annular Mode monthly index from the NCEP Climate Prediction Center (thin black) with 1- and 5-year running means (mid-weight and heavy, respectively), with uniform weighting. Data from Climate Prediction Center Internet Team (2006). *Source: From Roemmich et al. (2007).*

periodicity can be more useful than spectral analysis to study spatial structures.

The simplest approach to representing a time series in terms of sines and cosines projects the time series onto these functions, using a least squares fit or Fourier transform (and at first ignoring the important issues with discrete sampling and a finite length time series). The Fourier transform yields an amplitude for each frequency. Mathematically, the Fourier transform $X(f)$ of a sample time series $x(t)$, as a function of frequency f, and its spectral density $S(f)$ are

$$X(f_j) = \sum_{i=1}^{N} x(t_i) e^{-i2\pi f_j t_i} = \sum_{i=1}^{N} x_i e^{-i2\pi(j-1)(i-1)/N}$$

(6.14a)

$$S(f_j) = |X(f_j)|^2, j = 0, ..., N-1 \quad (6.14b)$$

The frequency distribution of the squared amplitudes (6.14b) is known as a "periodogram." The periodogram does not have statistical value (having large error at each frequency) because it includes no averaging. The "power spectrum" is calculated by averaging the periodogram, and is therefore a statistical quantity. The averaging is sometimes taken over multiple realizations of the periodogram, which can be calculated from multiple realizations of the time series. (In practice, this means taking a long time series and chopping it into shorter time series, calculating the periodogram for each piece, and averaging them.) Averaging to create the power spectrum can also be over a frequency range in the periodogram, thus reducing the frequency resolution. (Later we see that this is equivalent to reducing the length of the time series, so the two types of averaging are equivalent.)

It turns out that the spectrum obtained from the periodogram is identical to that obtained as the Fourier transform of the autocovariance function (which already includes the squared amplitudes); this latter is the Blackman-Tukey approach. Since it is now efficient to compute Fourier transforms using readily available fast Fourier transform (fft) software (Press et al., 1986; Matlab, Mathematica, etc.), the periodogram approach is now the most commonly used.

The total "energy" in the power spectrum is equal to the total variance in the time series (Parseval's theorem). Thus the area under the spectrum (in a graph of the spectral amplitude vs. frequency) is the total variance of the time series. If the spectrum is normalized by dividing by the variance, then the area under the spectrum is 1, and the spectral values give the fraction of the total variance at each frequency.

The power spectrum is often displayed divided by the frequency interval at each frequency. This is called the power spectral density ("density" since it is divided by the interval). The units of power spectral density are spectral energy/frequency. For instance, for a sea level spectrum, the units might then be m^2/cps (where cps is cycles/second). For a velocity record, the units would be $(m/s)^2$/cps. For a temperature record, the units would be $(^\circ C)^2$/cps.

There are a number of important details to be considered when working with real, discretely sampled data from a finite length time series. We never have an infinite time series. Turning on and then turning off the sampling of a continuing process means that a "box car" has been applied to the actual process (the box car has multiplied the true time series). The calculated spectrum then includes the box car; the sudden jump up and drop down in amplitude at the beginning and end of sampling has unfortunate, unwanted spectral properties, that is, the "ringing" of the Gibbs phenomenon,

which creates unphysical energy at high frequencies. To avoid this, sample time series are multiplied by a window that "tapers" smoothly to zero amplitude at the beginning and end of the time series. The references listed at the beginning of this section describe the commonly used windows.

The actual frequencies that can be analyzed for a sample time series depend on the total length of the time series and on the discrete sampling interval for the time series. (If the time series is sampled irregularly, there is an additional set of considerations that are not discussed here.) In Eq. (6.15), the length of the time series is T, the sampling interval is Δt, and the total length of the time series is $T = N \Delta t$, where N is the total number of samples.

The lowest frequency, f_0, that can be resolved by a given time series is called the "fundamental frequency":

$$f_0 = 1/T = 1/(N\Delta t) \qquad (6.15)$$

in units of Hertz (or $\omega = 2\pi/T$ if frequency ω is in terms of radians). However, such a low frequency relative to the record length is sampled only once in the record, so confidence in the amplitude estimate is low. Energy at frequencies lower than the fundamental frequency will appear as a trend in the record. It is common practice to first fit a linear trend to the time series and remove the trend from the time series prior to performing the Fourier analysis. (Recognition of problems with spectral estimation when very few cycles of an oscillation have been sampled is vitally important for large-scale oceanographic data analysis if timescales of interest are tens to hundreds of years, especially when looking at climate variability and change. Oceanographic time series are not long enough to resolve these low frequencies very well.)

In the spectrum, the fundamental frequency is also the difference in frequency between adjoining frequency components, f_1 and f_2. That is,

$$\Delta f = |f_2 - f_1| = 1/N\Delta t \qquad (6.16)$$

The two frequencies are well resolved for $\Delta f = 2/N\Delta t$ and $3/2N\Delta t$, just resolved for $\Delta f = 1/N\Delta t$, and not resolved for $\Delta f = 1/2N\Delta t$.

The highest frequency that can be observed depends on the sampling interval, Δt, because two samples are required to sample a given frequency (the "sampling theorem"). The maximum resolved frequency is

$$f_N = 1/2\Delta t. \qquad (6.17)$$

This is the *Nyquist frequency*. As with the fundamental frequency, the estimate of spectral amplitude at the Nyquist frequency is poor since the sampling does not resolve the sinusoidal character. Note that if f_N is the highest frequency we can measure and if f_0 is the limit of frequency resolution, then the Nyquist frequency also gives the maximum number of Fourier components that can be estimated in any analysis:

$$f_N/f_0 = (1/2\Delta t)/(1/N\Delta t) = N/2. \qquad (6.18)$$

What happens if there is energy in the time series at frequencies that are higher than the Nyquist frequency? Energy from the under-sampled higher frequencies appears in the spectrum at much lower frequencies. This is called aliasing, as mentioned in Sections 6.1 and 6.2. For any actual time series, there will always be higher frequencies that are not sampled; this presents a problem only if they have a significant amount of energy. An example of aliasing from Emery and Thompson (2001) is shown in Figure 6.8b. If the well-measured tidal record in the top panel is subsampled with just one observation per day, a much lower, erroneous, frequency appears in the spectrum. When the spectra of the original record and the sub-sampled records are computed, the correct spectrum has peaks at the well-known tidal frequencies, while the subsampled record produces a spectrum without these peaks, but also with the energy folded back (aliased) into lower frequencies. This erroneously boosts the spectral amplitude at lower frequencies.

Figure 6.8b also illustrates the sampling theorem: the highest frequency that is resolvable with sampling once per day is 1/(2 days). Thus the solid curve ends at this frequency.

As another example of aliasing, satellite altimeters measure SSH. Their orbits sample a given location every 10 days. However, at almost all locations in the ocean there is significant tidal energy at semi-diurnal and/or diurnal frequencies. There is also under-sampled, high-frequency SSH variability due to fluctuations in atmospheric pressure and barotropic motions in the ocean. (Barotropic variability is solely due to dynamical changes in surface height without compensation, or "baroclinicity," in the ocean interior. Barotropic variability has much shorter timescales than baroclinic; see Sections 7.6 and 7.7.) These energetic, higher frequency signals must be managed when analyzing altimetric spectra, which can be approached using models of tides and the barotropic variability and analyses of atmospheric pressure (Chelton et al., 2001; Stammer, Wunsch, & Ponte, 2000). Stammer and Wunsch (1999) showed a spectrum of SSH from the altimeter, with an aliased semi-diurnal tidal peak.

The significance of a spectral estimate is measured with a confidence interval. These are calculated similarly to those for the basic time series (Eq. 6.10). The spectral density at a given frequency is like an energy, meaning that it is the square of an amplitude. To obtain useful (significant) spectral estimates, there must be some averaging. This can be done by either averaging spectral estimates from many independent time series sampling the same process, or by averaging together spectral estimates for a range of adjoining frequencies. Since the spectral estimate is then a sum of squares, it has a chi-square probability density function. (See Emery and Thomson (2001) to learn about chi-square distributions.) The chosen confidence intervals are then determined by the chi-square distribution and knowledge of the effective number of degrees of freedom in each spectral estimate. Again this is very similar to Eq. (6.10), but instead of a Student t-test, the function used to evaluate is the chi-square distribution.

It is common practice to show the 95% confidence interval with a spectral estimate. An example from Rudnick (1997) is shown in Figure 6.7. In this spectrum, the 95% confidence interval varies with frequency because the averaging in the spectral estimate and therefore the number of degrees of freedom differ with the frequency.

In spectral analysis, it is assumed that the physical processes that have been sampled are *stationary*. In a stationary process, observations from one time period yield the same spectrum as observations from another time period. However, the underlying processes can change. Ocean currents that produce instabilities, and hence eddies, can change with time; for example, seasonally or in response to climate variability. The spectra of the eddies would then change. A more complicated method, called *wavelet analysis*, recognizes that the processes underlying the spectrum might be changing, and thus the spectrum might vary with time. Empirical orthogonal functions (Section 6.6) also do not require stationarity of the time series, so they can be generally more useful than spectra for some large-scale (spatial and time) oceanographic processes.

6.5.4. Filtering Data

We are often interested in isolating phenomena with a particular frequency in a data record, whether a time series or spatial data set. For instance, if the study is focused on tides, the lower and higher frequencies present in the data set might be removed. If the study is focused on decadal variation, the internal waves or tides might be removed. To do this, the data set is filtered to remove frequencies that are not of interest. Filters can be applied in either the time domain or the frequency domain. As for other aspects of data analysis presented in this chapter, only the rudimentary concepts are provided. Among the many treatments of filtering, Press et al. (1986) provided general, practical advice about filtering; Bendat and Piersol (1986) provided much more of the complete background and mathematics; and Emery and Thomson (2001) provided a thorough treatment as commonly practiced in physical oceanography. The Matlab signal processing toolbox includes many different filters as well as the capability to design filters.

In the time domain, the output from a filter at a given time is a weighted sum of the input data from a range of times. For example, to filter in the time domain,

$$y(t_j) = \sum_{i=1}^{N} w(t_i - t_j)x(t_i) \qquad j = 1, N \quad (6.19a)$$

where x is the original data, y is the output, and the w's are the weights, which depend on the difference in time between the data point and the output. What does a given filter (choice of weights) do to the frequencies in the time series? To answer this, the filter weights can be Fourier transformed to the frequency domain and plotted as a function of frequency. This is called the frequency response of the filter.

To filter in the frequency domain, the time series of data are first Fourier transformed to form a periodogram (unaveraged spectrum). Then the periodogram is filtered, with weights that depend on frequency, and the results are Fourier transformed back to the time domain. For example:

$$Y(f_i) = \sum_{i=1}^{N} W(f_i)X(f_i) \qquad i = 1, N \quad (6.19b)$$

where f is frequency, X is the Fourier transformed data, Y is the filtered spectrum, and W are the weights in the frequency domain. An equivalent method is to design a filter shape (the weights, W) in the frequency domain and Fourier transform this filter to the time domain; the resulting time series then becomes the time domain weights, w.

A "lowpass" filter removes high frequencies above a cut-off frequency, retaining only the low frequencies. A "highpass" filter is the opposite, retaining high frequencies only. A "bandpass" filter retains frequencies in the middle of the record, removing both the low and high frequencies. This filter is characterized by its central frequency and bandwidth.

Lowpass filtering is equivalent to smoothing a time series. This is the easiest type of filtering to understand within the time domain. Box car averaging is a simple lowpass method, in which a segment of the data record is averaged, with equal weight given to each point in the segment; the uniform weights resemble a box. The box car can be moved through the record, with overlapping segments; this is called a running mean. The weights in Eq. (6.19a) for a box car filter are all the same size up to the length of the segment (summing to 1, producing an average value of the data over the chosen time interval) and 0 for all other times. An example of one- and five-year running means, with box car weighting, for the climate index called the Southern Annular Mode (Section 13.8) is shown in Figure 6.9 (after Roemmich et al., 2007).

Box car averaging is often all that is necessary for a given purpose. However, the frequency response of a box car filter can be undesirable because of the Gibbs phenomenon: the sudden drop to 0 for the weights means that there is high-frequency ringing in the filtered data set. Weights chosen for a low pass filter can be tapered to zero at the ends of the filter, much like windowing in spectra.

Lowpass filtering can also be done spectrally in the frequency domain rather than the time domain. The time series can first be Fourier transformed. All undesired high frequencies can then be set to zero amplitude (or tapered to zero

amplitude). Then the filtered data record can be reconstituted using an inverse Fourier transform.

Bandpass and highpass filtering are conceptually easiest to understand in the frequency domain, since the objective is to remove certain frequencies. The crudest method is to take the Fourier transform of the time series, then set the amplitude of all undesired frequencies to zero, and then take the inverse Fourier transform to reconstitute the time series, which will now be missing all the undesired frequencies. However, such simple removal of undesired frequencies also creates problems in the inverse Fourier transform similar to the Gibbs phenomenon. Therefore, it is desirable to taper (window) when removing undesired frequencies.

Bandpass and highpass filtering are often carried out in the time domain. The crudest method of highpass filtering is to subtract the lowpass record from the original data record. Bandpass filtering can be produced by successive application of low and highpass filtering (Emery & Thomson, 2001). However, it is more desirable to design these filters to diminish the Gibbs phenomenon. The time domain filter can be constructed as the Fourier transform of a frequency domain filter. For bandpass filters, the narrower the desired band of frequencies, the longer the time series must be to produce the narrow band; this is readily understood from the wide shape of the Fourier transform of a very narrow signal.

There are many subtleties associated with filtering that are not described here.

6.6. MULTIDIMENSIONAL SAMPLING

The ocean is often sampled in time and in at least two dimensions in space. This is especially true in the present era of satellite programs, which collect data over large parts of the ocean surface at regular intervals. With regular data for the whole surface, observers often wish to extract signals that indicate processes. For instance, ENSO is a time-varying climate process with an underlying quasi-periodicity of 3 to 7 years (see Figure 10.28b). Over the course of several years, different parts of the sea surface in the tropical Pacific have changed in height and temperature at different times. There is also variability at the sea surface, in the atmosphere, and in vegetation far from the tropical Pacific; observers are interested in knowing how much of the variability can be traced to or participates in ENSO.

Covariance and correlation are the most basic calculations used to analyze observations of different properties in different locations and/ or at different times (Sections 6.1 and 6.3). These calculations show how temperature in the western Pacific varies with temperature in the eastern Pacific; or how the correlation of time series of some property such as surface temperature at all locations on the globe with the time series of, say, a climate index can provide useful displays of the spatial distribution of that climate variation (see the figures in Chapter S15 on the textbook Web site).

Beyond point-to-point correlation analysis, it is useful to look at the large-scale geographic "modes" of variability. This can be approached by extracting wavelike signals from the data sets, using spectral analysis, or by allowing the data set to define its own spatial patterns. The latter approach, which uses empirical orthogonal functions, can produce patterns that can clearly be identified as ENSO, the Southern Annular Mode, and so forth. The time series of the identified climate pattern can then be correlated with observed time series at different locations to begin to identify sources of local variability.

We first briefly describe common methods of dealing with both spatially and temporally sampled variability (spectral analysis and empirical orthogonal functions), and then common methods of displaying time-averaged data with spatial distribution (climatologies and atlases).

6.6.1. Multidimensional Time Series Data and Empirical Orthogonal Functions

Spectral analysis can be applied to spatial and temporal sampling. Wavenumber-frequency spectra (bispectra), which represent both the spatial and temporal aspects of the data, are very useful in studying waves that are typically described theoretically in terms of dispersion relations that relate frequency and wavenumber (Section 8.2). Wave fields are typically at least two-dimensional in space, so wavenumber-wavenumber spectra are also useful. Figure 6.10 shows two recent examples of frequency-wavenumber spectra: (a) for large-scale, much lower frequency equatorial waves in the Pacific Ocean (Shinoda, Kiladis, & Roundy, 2009) and (b) for very high frequency surface gravity waves (Herbers, Elgar, Sarap, & Guza, 2002). In both panels, theoretical dispersion relations are overlaid. In Figure 6.10b, the observed spectrum can then be used to determine if a given theory accounts for the observations.

A drawback of spectral analysis is its underlying assumption that the processes are periodic (in space or time, depending on how it is applied). While many ocean processes indeed satisfy this assumption — ocean surface waves, tides, and large-scale waves such as Kelvin and Rossby waves (Figure 6.10a) — many large-scale ocean and climate processes do not. This is especially true of the spatial patterns for large-scale ocean responses to changing forcing, where the geography begins to dominate the patterns. Therefore it is useful to move beyond spectral analysis to find basis functions that better represent the underlying ocean processes.

Empirical orthogonal functions (EOFs) are regularly used in oceanography, meteorology, and climate science for analyzing space-time data sets such as satellite or SST time series. EOF analysis was introduced for meteorology by Lorenz (1956). It is similar to principal component analysis used in other sciences (and

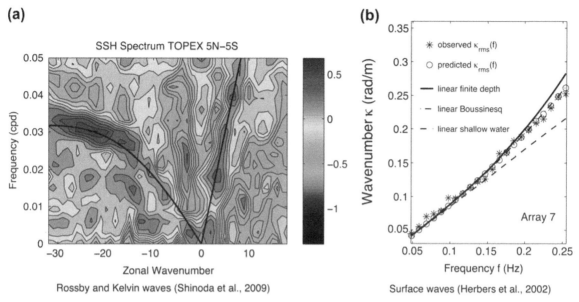

FIGURE 6.10 Examples of frequency-wavenumber spectra. (a) Equatorial waves (Kelvin and Rossby) from SSH anomalies, compared with theoretical dispersion relations (curves). Figure 6.10a can also be found in the color insert. *Source: From Shinoda et al. (2009).* (b) Surface gravity waves: observed two-dimensional spectrum (*) averaged over wavenumber at each frequency, and compared with several theoretical dispersion relations. *Source: From Herbers et al. (2002).*

occasionally in meteorology and oceanography). Unlike spectral analysis, in which the temporal and spatial dependence are represented as sines and cosines, the EOF procedure defines its own set of functions that can be used to describe the process most efficiently. Each EOF is "orthogonal" to the others, which means that each one represents something unique about the process. (In spectral analysis, each sine and cosine function is orthogonal to all the others.)

EOFs are determined through a linear least squares process, minimizing the difference between the observations and the EOFs; that is, the basic ideas presented in Section 6.3.4 for least squares apply to finding these much more complex functions. We do not present any of the method here, but refer to texts such as Wilks (2005) and von Storch and Zwiers (1999). Following the procedures in these texts and in basic publications on EOFs, calculation

of EOFs is straightforward because the useful linear algebra software is easily available (Matlab, Mathematica). Other multivariate approaches are also described in these texts, including canonical correlation analysis, in which the pattern with the highest possible correlation between two time series is sought.

EOFs are shown in this text in reference to modes of climate variability: the Pacific Decadal Oscillation (PDO), North Pacific Gyre Oscillation (NPGO), the Arctic Oscillation, the Southern Annular Mode, and so forth (see Chapter S15 on the textbook Web site). Other climate modes introduced throughout, such as the ENSO and the North Atlantic Oscillation, are also easily and often described in terms of EOFs.

EOFs are typically ordered by amplitude (percentage of variance of the observations explained by that EOF); that is, the first mode explains the most variance of the signal (by design because of the least squares approach), the second mode explains as much of the remaining variance as possible, and so on. Typically, only the two or three EOFs with the largest amplitudes are significant while the others might be below the noise level of the observations. The major climate modes listed in the previous paragraph tend to be first or second EOFs.

In a classic oceanographic application of EOF analysis, including the reasoning for an EOF approach and an extensive appendix describing the EOF method, Davis (1976) analyzed historical (1947–1974) SST and sea level pressure (SLP) anomalies in the North Pacific Ocean. Using data that had already been gridded to latitudes and longitudes for each month, he constructed monthly anomalies by removing the long-term monthly mean at each grid point, and then calculated the EOFs from the anomalies (Figures 6.11 and 6.12). These figures are typical of EOF displays, including the spatial

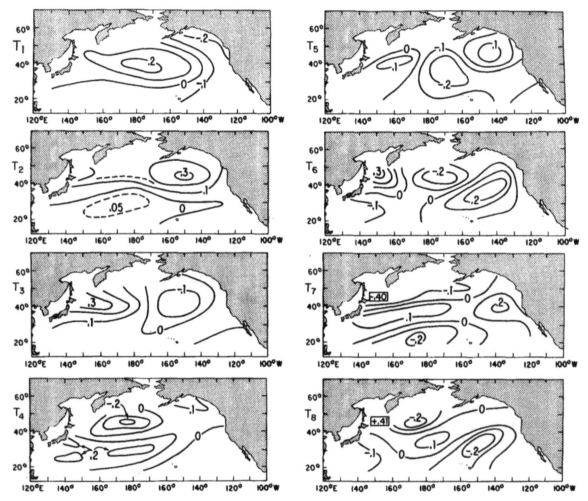

FIGURE 6.11 Example of empirical orthogonal functions (EOFs): the eight principal EOFs describing the sea surface temperature anomalies. ©American Meteorological Society. Reprinted with permission. *Source: From Davis (1976).*

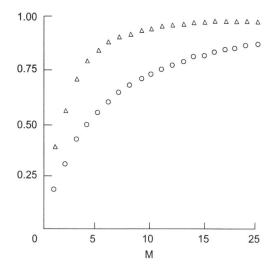

FIGURE 6.12 The cumulative fraction of total sea surface temperature (circles, o) and sea level pressure (triangles, △) anomaly variance accounted for by the first M empirical orthogonal functions. ©American Meteorological Society. Reprinted with permission. *Source: From Davis (1976).*

pattern of the principal EOFs, ranked in terms of variance of the overall signal explained by that EOF, and a plot of the variance for each mode. The spatial patterns in Figure 6.11 illustrate non-sinusoidal EOFs and how each successive mode has more spatial complexity and is visually orthogonal to the other modes.

Davis (1976) showed that the SLP variability could be explained by fewer EOFs than the SST variability (Figure 6.12), likely due to the inherently smoother SLP field. A second important result was the spatial distributions of the modes (Figure S6.3 on the textbook website). The first EOF of SLP looks like the Aleutian Low pattern and the first EOF of SST is the temperature variability that accompanies variations in the Aleutian Low strength. This first EOF is essentially the North Pacific part of the PDO, also called the North Pacific Index (see online materials that include Chapter S15). The second EOF is now associated with the North Pacific Gyre Oscillation (NPGO) (DiLorenzo et al., 2008).

EOF presentations often include the computed time series of the amplitude of the dominant EOF modes. These time series can then be analyzed using spectral analysis or some other approach, or simply employed to study correlations with other fields and other modes. Davis (1976) presented the frequency spectrum but not the actual time series. Time series of climate indices that are actually EOF amplitudes are commonly used; for instance, the Southern Annular Mode index seen in Figure 6.9.

6.6.2. Climatologies and Atlases

In oceanography and meteorology, it is useful to use mean values distributed geographically to produce a *climatology*. A climatology is generally understood to be the mean value over many years, usually including at least several decades. The mean can be over all months, but often the mean values are constructed for individual months or seasons, in which case they can be referred to as monthly or seasonal climatologies.

Climatologies are usually constructed from observations that are irregularly sampled in time. Therefore some sort of weighting of each data point is important so that, for instance, well-sampled summers and poorly-sampled winters do not bias the mean value toward summer. It can be useful to first construct short period mean values, such as daily averages, before constructing the monthly to annual mean. Data gaps are often filled by some interpolation or mapping procedure before averaging.

Observations used for climatologies are also usually irregularly sampled in space; the same weighting and data gap issues that apply for time are applied here. Objective mapping is a common method for producing a spatially gridded data set. Simple geographic binning is also used if there are many observations.

Construction of a climatology almost always involves a data quality step, which can be very extensive. Published climatologies, therefore, are often accompanied by carefully quality-controlled data sets.

Climatologies are essential for large-scale numerical modeling and also for data assimilation, which is based on numerical modeling. A good mean field is essential as a startup condition in these models since "spinup" from a state of rest without stratification would be hopelessly inefficient.

Commonly used ocean climatologies include the National Oceanographic Data Center's (NODC) World Ocean Atlas (WOA05; NODC, 2005a) and its antecedents, beginning with Levitus (1982), which are based on the archived, quality-controlled original data World Ocean Data (most recent version WOD05; NODC, 2005b). An example from WOA05 that illustrates the climatological salinity at 500 m with several types of available indications of error is shown in Figure 6.13; the online climatology from NODC includes many other properties and depths as well. Another hydrographic climatology in general use, along with its quality-controlled data set (Hydrobase), was produced for the Atlantic, Pacific, and Indian Oceans by Lozier, Owens, and Curry (1995), Macdonald,

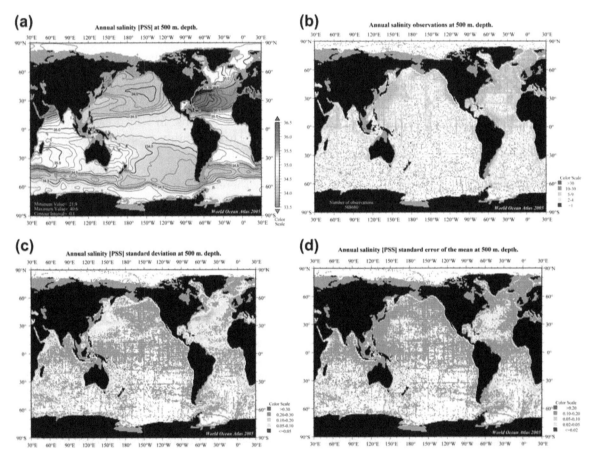

FIGURE 6.13   Illustration of a climatology: Salinity at 500 m. (a) Climatological annual mean, (b) data distribution, (c) standard deviation, and (d) standard error. Many other properties and depths are also available online (NODC, 2005a). *Source: From Antonov et al. (2006).*

Suga, and Curry (2001), and Kobayashi and Suga (2006), respectively. Hydrographic data in this climatology were averaged along isopycnal surfaces, producing useful improvements in regions of large isopycnal slopes (strong currents) over those averaged on constant pressure surfaces.

Atlases are more vaguely defined than climatologies. Traditionally atlases were books filled with maps — or for oceanography, vertical sections of ocean properties in addition to maps — for visualization of mostly the mean fields, or at most, seasonal fields. Atlases using data from the International Geophysical Year (1957–1958) were published in the 1960s (Fuglister, 1960; Worthington & Wright, 1970). Atlases of vertical sections from the Geosecs expeditions of the 1970s were also published and are widely used (Bainbridge et al., 1981 to 1987). The World Ocean Circulation Experiment atlases of vertical sections and maps are now published both in print and online (Orsi &Whitworth, 2005; Talley, 2007, 2011; Koltermann, Jancke, & Gouretski, 2011). Many figures from these atlases are reproduced in other chapters of this text.

Modern atlases include graphics based on averaged data (climatologies) and most are now digital. Perhaps the first of these, a print atlas based on all available NODC data, was published by Levitus (1982); this was produced specifically to provide climatological fields for general ocean circulation models. The latest version of the NODC atlases is the World Ocean Atlas 2005 (WOA05), which is exclusively digital (NODC, 2005a). Figure 6.13 is from this NODC atlas. Some modern digital atlases also include software for the user to produce their own graphics based on individual data. Java Ocean Atlas (JOA; Osborne & Swift, 2009) is the basis for the DVD distributed with this text. Ocean Data View (ODV, 2009), developed by R. Schlitzer in the 1990s, is a widely used display package and database that is easily adapted to optimum multiparameter analysis (Section 6.7.3).

6.7. WATER PROPERTY (WATER MASS) ANALYSES

Much of the descriptive oceanography in other chapters of this text is associated with the large-scale ocean circulation and water mass distributions. Data sets that describe this circulation extend back more than a century. Techniques to work with the data to discern sources and influence of water masses have been based on using several different characteristic properties of the water masses in addition to its density to trace them. This section describes some of these traditional techniques, which remain effective for studying water mass distributions and the associated ocean circulation (Sections 6.7.1 and 6.7.2). These methods are being replaced by more statistical techniques as the data sets grow. A technique that has been widely adopted in recent years is *optimum multiparameter analysis* (OMP), which is a least squares approach to estimating the fraction of a given source water (Section 6.7.3).

6.7.1. Analysis Using Two Characteristics

Observed properties such as potential temperature, salinity, dissolved oxygen, and so forth, may have important correlations with each other. (Density is not included because it is derived from temperature and salinity.) These dependencies may be regional or have time variations. The high correlations between properties arise because most ocean water masses (Section 4.1) acquire their characteristics at the surface of the sea in particular localities. The water properties are determined there by the local climate, and when the water sinks along density surfaces it carries these properties with it. The characteristic and unique combination of different water properties that arises from a given source or process in the ocean provides the definition of a given water mass (Section 4.1). Water masses are useful because they can be recognized by these combinations and hence the associated processes deduced from their distributions and modifications.

To show these combinations, characteristic diagrams of the water properties were first introduced by Helland-Hansen (1916), who plotted temperature against salinity (T-S plots) for individual oceanographic stations. Potential temperature rather than temperature is almost always used now to remove the adiabatic effect of pressure on temperature. (Use of θ is important even in water as shallow as 100 m, if one is attempting to identify the water with its surface source or follow it down into the ocean or deduce its mixing properties.) Potential temperature-salinity diagrams are used throughout this text to illustrate water mass distributions.

Each point on the θ-S diagram corresponds to a particular potential density, so potential density is often contoured (Figure 6.14). Different reference pressures for potential density can be used, as explored with Figure 3.5. (Neutral density is not contoured on θ-S plots since it is defined empirically based on the actual temperature and salinity properties of the water column, rather than based on the equation of state; therefore it is not defined for the full two-dimensional potential temperature/salinity plane. It could, however, be indicated for given observations.)

The θ-S diagram in Figure 6.14 illustrates water types, water masses and mixing diagrams (Section 4.1.1). Water types are points in property-property space, and represent source waters. As the water advects away from its source and

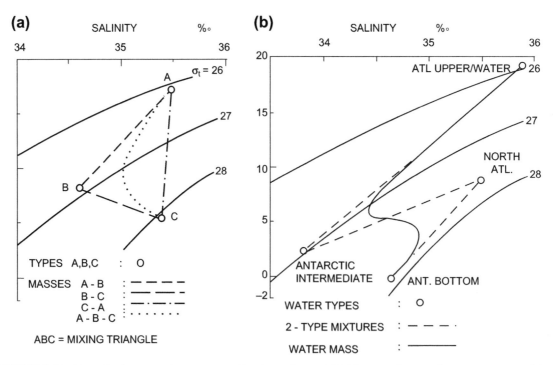

FIGURE 6.14   Example of a potential temperature (θ)-salinity diagram. (a) Schematic showing three water types and their mixing products. (b) θ-S diagram from the central North Atlantic with water masses labeled, illustrating how mixing connects the extrema. The contoured field on the diagrams is the density σₜ since this figure is reproduced from an earlier version of this text, although as indicated in Chapter 3, it is advisable to use a potential density parameter.

mixes with other waters from other sources, its identifying properties spread to a range of properties. If the water type originates as a vertical extremum such as a salinity minimum, then as it mixes downstream, the vertical extremum might remain, marking the influence of the original water type. The overall envelope of these gradually mixing properties identifies the water mass. However, the sources of waters below the pycnocline are so well separated in space and hence in properties that the water types and masses are relatively easily defined.

It has long been the practice to compute fractional mixing rates of end points (source waters) for a given water parcel, based on θ-S properties. This has been readily extended to other conservative and possibly non-conservative tracers. A good assumption for conservative tracers is that mixing occurs along straight lines; for mixing of θ and S, this is along straight lines in the θ-S plane (Section 3.5.5). Non-conservative tracers, such as oxygen and nutrients, are more problematic since they depend on water parcel age. Extended methods include use of conservative parameters such as "PO," "NO," and "N*" that take advantage of the linear mixing assumption (Broecker, 1974; Gruber & Sarmiento, 1997). Redfield ratios (Section 3.6) are based on fixed proportions of, say, phosphate ("P") and nitrate ("N") production while oxygen ("O") is consumed. As soon as additional properties are included, error estimates are desired, and it is understood that end points are not necessarily well defined, and a more quantitative approach becomes useful. This end point mixing practice has therefore evolved into the more statistical approach of OMP analysis (Section 6.7.3).

The additional independent information about source waters and mixing available from tracers other than potential temperature and salinity is evident in two-parameter plots for a multitude of tracers (Figure 6.15). When more than two parameters are available, one can think in terms of multi-dimensional space,

and the relationships between the parameters that reveal the source waters and mixing rapidly become more easily utilized through OMP.

Finally, property-property plots such as Figure 6.15 are useful visual tools for checking data quality. Since locally envelopes of profiles in property-property space can often be quite "tight," or have small variance, outliers can be identified and flagged for additional quality checking. In Figure 6.15, for instance, slightly low salinity values in the nitrate and silicate versus salinity plots suggest that each step in obtaining these values be checked, often by going back to original log sheets and laboratory notebooks to see if there were any issues or uncertainties in data collection or analysis procedures. (These particular values were found to be accurate and were therefore retained.)

6.7.2. Volumetric θ-S Characteristics of Ocean Waters

The volume of water with a given property or set of properties can be a useful diagnostic of relative quantity, or reservoir size, of the ocean's water masses. This technique for potential temperature-salinity was pioneered by Montgomery (1958) and subsequently reworked for the world oceans by Worthington (1981; see Figure 4.17 in this text). In principle, volumetric assays for any set of properties, not just potential temperature and salinity, can be produced and displayed. Column inventories of properties such as chlorofluorocarbons and CO_2 have emerged as important tools for understanding the ocean's role in the global carbon system.

The volumetric θ-S diagram in Figure 4.17 was produced by choosing a "bin" size for potential temperature and salinity (e.g., 0.1°C and 0.01 psu) and the volume was calculated for each bin from observed oceanographic data. When done originally without benefit of computer interpolation, this was an extremely tedious exercise. It can be done relatively easily now, and is a feature of the JOA package

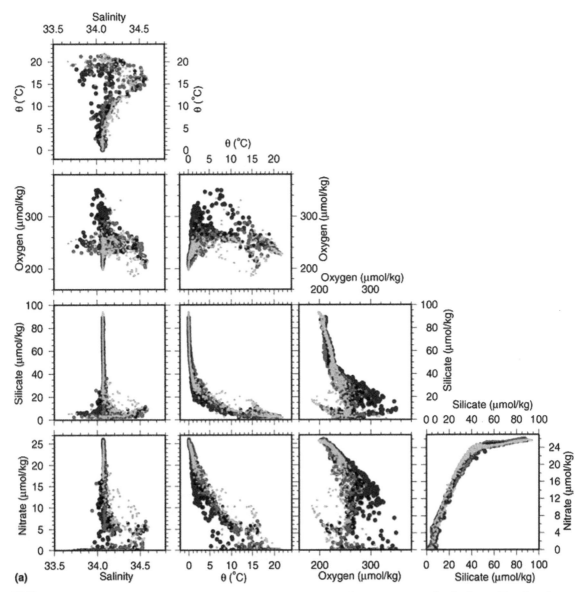

FIGURE 6.15 Example of property-property plots for a variety of different properties, for the Japan/East Sea. *Source: From Talley et al. (2004).*

distributed as part of this text, in the supplemental materials on the Web site.

The last form of traditional, three-dimensional characteristic diagram we show is the temperature-salinity-time (T-S-t) diagram. This is a compact way of showing the sequence of combinations of water properties with time. As an example, Figure 6.16 shows monthly mean values for three zones of the Australian Great Barrier Reef lagoon (Pickard, 1977). In

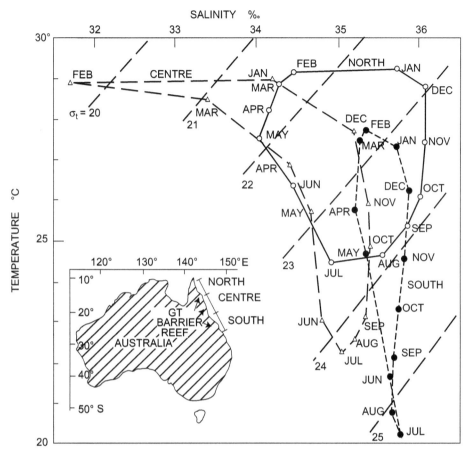

FIGURE 6.16 Temperature-salinity-time (T-S-t) diagrams for shallow lagoon waters inside the Great Barrier Reef. *Source: From Pickard (1977).*

the south, the annual variation is mainly in temperature. In the north there are large variations of both temperature and salinity, while in the center zone there is an extreme salinity variation. The reason for the differences is that the north and center zones are subject to heavy monsoonal rains in the austral summer (January to April) while the south zone escapes these. The very low salinity in the center is due to the rivers that drain much larger inland areas than the smaller rivers in the north. In situ temperatures used for these T-S-t diagrams, which is appropriate for these surface data.

6.7.3. Optimum Multiparameter Analysis

All waters in the interior of the ocean are understood to be a mixture of waters that have some well-defined source at the sea surface. Those sources have associated water properties that depend on location. All measured chemical properties can be used in some way to help define the source waters and the relative mixing. Mixing in the ocean is mostly linear and hence proportional. For a given water parcel with several measured properties, there are likely to be multiple source waters. If there are more properties than source waters, then formally the problem of determining the relative mixture of the source waters in the parcel is "over-determined," meaning there are more equations (one for each property) than unknowns (proportion of each source water).

A formal method for determining the relative proportions of the source waters was introduced by Tomczak (1981), and developed formally with a least squares approach by Mackas, Denman, and Bennett (1987) and others, including Tomczak. Software written in Matlab was developed for general use by M. Tomczak and has been further developed and provided for general use by Karstensen (2006), whose Web site also contains practical information about OMP and a bibliography. The output of OMP analysis for a given observed water parcel is the fraction of each assumed source water, which should add in total to 1.0. OMP thus first requires selection of at least two source waters, with as many observed parameters as possible. If, for instance, two source waters are assumed, with six parameters each (temperature, salinity, oxygen, nitrate, silicate, and potential vorticity), then there would be six equations with two unknowns for each water parcel; hence this is an overdetermined system. OMP finds the best least squares solution, that is, the best choice of fractions of the source waters for each observed water parcel.

Mathematically, for an example of three conservative parameters and two source waters, the linear equations expressing the fractions of each source water for a given water parcel are

$$
\begin{aligned}
x_1\theta_1 + x_2\theta_2 &= \theta_{obs} + R_\theta \\
x_1S_1 + x_2S_2 &= S_{obs} + R_S \\
x_1PV_1 + x_2PV_2 &= PV_{obs} + R_{PV} \\
x_1 + x_2 &= 1 + R_M
\end{aligned}
\tag{6.20}
$$

where the conservative parameters chosen here are potential temperature (θ), salinity (S), and

potential vorticity (PV). The fourth equation is conservation of mass. (It is not necessary to include PV; if included, the system is overdetermined. It is only necessary that the system not be underdetermined.) The x's are the mass fractions of the two source waters in the observed ("obs") water parcel. The R's on the right-hand side are residuals that permit solution in a least squares manner of this overdetermined system of four equations and two unknowns. Solution of the system proceeds by minimizing the squared residuals R (difference between the left-hand sides and the observed values) for each equation simultaneously. This linear algebra step is omitted here, since we are not introducing this level of mathematics in this textbook. A complete description of the remaining steps and procedures is available in the cited references (Mackas et al., 1987; Karstensen, 2006) as well as in numerous papers that use OMP (e.g., Tomczak & Large, 1989; Maamaatuaiahutapu et al., 1992; Poole & Tomczak, 1999).

OMP analysis often includes a constraint that the fractions of the source waters be non-negative (Mackas et al., 1987). A measured water parcel can fall outside the a priori range of the source water characteristics ending up with a non-physical negative fraction. (This does not necessarily mean that the source waters need to be redefined, unless so many observations yield negative fractions that the a priori choice of source waters is clearly inadequate.) For instance, in the Kuroshio-Oyashio region, the obvious source waters are "pure" Kuroshio and "pure" Oyashio water. However, there could be fresh, near-coastal water parcels that might lie outside the assumed ranges. If a nonnegativity constraint is enforced, then such a parcel would be assigned a 1.0 fraction of Oyashio and 0.0 fraction of Kuroshio water. It might sometimes be better not to impose a constraint, because information about whether water properties can be explained by chosen source waters can be lost.

We show an example of OMP application in the southwestern Atlantic, where numerous water masses meet in the Brazil-Malvinas confluence (Maamaatuaiahutapu et al., 1992). Their analysis used six properties (temperature, salinity, oxygen, phosphate, nitrate, and silicate) plus mass conservation, and seven source water types (point sources). This is an exactly determined rather than overdetermined system. Three of the water mass fraction sections are shown in Figure 6.17. The source waters here are well separated in properties, and also dominate different vertical layers. The result is quantitative information on the mount of each water mass at each location on the section, as opposed to simply subjective labeling, or a more traditional attempt at calculating water mass fractions using just temperature and salinity.

The global maps of North Atlantic Deep Water and Antarctic Bottom Water fractions shown in Chapter 14 (Figure 14.15), from Johnson (2008), are results of OMP analysis.

OMP analysis can be carried out along quasi-isopycnals if desired. If isopycnal mixing is assumed, then temperature and salinity are not independent. Since potential density, regardless of how it is referenced in pressure, is not conserved when two water parcels mix (because of cabbeling), OMP can reveal the extent of cabbeling (Yun & Talley, 2003).

Approaches to determining the distribution of different source waters through the ocean are continually being updated and improved. Thus this section is just an introduction to the general topic, and creative approaches are strongly encouraged.

GLOSSARY

The following list summarizes a number of the basic terms introduced in this chapter.

Accuracy Difference between an estimate and the "true" value. High accuracy means that this difference is small.

Aliasing Folding of spectral energy above the Nyquist frequency back into the frequencies below the Nyquist, creating higher spectral energy at these frequencies than is actually in the time series.

Anomaly Difference between an observation and the mean value, regardless of how the mean value is defined.

Climatology Time mean values of a geographically mapped field.

Correlation Normalized version of the covariance. It is equal to the covariance divided by the product of the standard deviations of the two variables. Correlation ranges from −1 to +1.

Covariance A measure of the covariability of two variables, computed as the averaged sum of the cross-product of the variations from the respective means of the two variables.

Determination or observation Actual direct measurement of a variable, e.g., the length of a piece of wood using a ruler. Synonyms include observation, measurement, or sample.

Empirical orthogonal functions (EOFs) Set of orthogonal basis functions that can completely describe (sum to) a given field. EOFs are often used to describe the spatial structure of a time-varying field in place of spectral analysis in the spatial domain.

Estimation Value for one variable derived from one or more determinations (either of the variable of interest or of other related variables), for example, the estimation of salinity from the determination of conductivity and

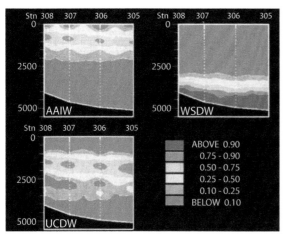

FIGURE 6.17 Example of optimum multiparameter (OMP) water mass analysis. Southwestern Atlantic about 36°S, showing the fraction of three different water masses. Antarctic Intermediate Water, AAIW; Upper Circumpolar Deep Water, UCDW; and Weddell Sea Deep Water, WSDW. This figure can also be found in the color insert. *Source: From Maamaatuaiahutapu et al. (1992).*

temperature. This also refers to the use of repeated "determinations" to define a statistical parameter such as the mean or standard deviation. Thus, we can speak of an "estimate of the mean".

Filter To output a data set as a weighted sum of the original data.

Gaussian population or distribution Probability density function (pdf) characterized by a symmetric bell curve defined by a mean and a variance (or standard deviation). It is also known as a "normal" population or distribution.

Inverse methods Ways to find the best estimate of an unknown quantity in an underdetermined system, usually using a least squares approach. In large-scale physical oceanography, this has most often been applied to estimating geostrophic reference velocities.

Least squares methods Ways to fit one function or data set to another function and/or data set by minimizing the sum of the squared differences between the two series.

Mean Average of a series of measurements over a fixed time interval such as a week, a month, a year, and so forth, or over a specific spatial interval (square kilometer, a 1 degree square, a five degree square, etc.).

Nyquist frequency Highest detectable frequency in a time series, equal to half the sampling frequency of the time series.

Objective mapping A statistically unbiased method of mapping irregularly spaced observations.

Precision Difference between one estimate and the mean of several obtained by the same method, that is, reproducibility (includes random errors only).

Probability density function (pdf) Sampling population from which the data are collected. This can be depicted by a histogram showing the frequency of occurrence of each data value.

Random error This results from basic limitations in the method, for example, the limit to one's ability to read the temperature of a thermometer. It is possible to determine a value for this type of error by statistical analysis of a sufficient number of measurements because it affects precision. Truly random errors have a Gaussian distribution with zero mean.

Standard deviation Square root of the variance.

Synoptic sampling A way of sampling the conditions as they exist at a given time over a broad area (a snapshot).

Systematic error or bias Error that results from a basic (but unrealized) fault in the method that causes values to be consistently different from the true value. Systematic error cannot be detected by statistical analysis of values obtained and affects accuracy.

Variance Mean square difference between a sample value and the sample mean.

Dynamical Processes for Descriptive Ocean Circulation

The complete version of this chapter (Chapter S7) appears on the textbook Web site http://booksite.academicpress.com/DPO/. The sections and equations are identical, but the explanatory text and figures are greatly truncated in this book. Figures, chapters, and sections that appear only on the Web site are denoted by "S" in their name such as Figure S7, Chapter S7, and so forth. Tables mentioned in this chapter appear only on the Web site.

7.1. INTRODUCTION: MECHANISMS

Ultimately, motion of water in the ocean is driven by the sun, the moon, or tectonic processes. The sun's energy is transferred to the ocean through buoyancy fluxes (heat fluxes and water vapor fluxes) and through the winds. Tides create internal waves that break, creating turbulence and mixing. Earthquakes and turbidity currents create random, irregular waves including tsunamis. Geothermal processes heat the water very gradually with little effect on circulation.

Earth's rotation profoundly affects almost all phenomena described in this text. Rotating fluids behave differently from non-rotating fluids in ways that might be counterintuitive.

In a non-rotating fluid, a pressure difference between two points in the fluid drives the fluid toward the low pressure. In a fluid dominated by rotation, the flow can be *geostrophic*, perpendicular to the pressure gradient force, circling around centers of high or low pressure due to the Coriolis effect.

Ocean circulation is often divided conceptually into *wind-driven* and *thermohaline* (or buoyancy-dominated) components. Wind causes waves, inertial currents, and Langmuir cells. At longer timescales, which involve the Coriolis effect, wind drives the near-surface frictional layer and, indirectly, the large-scale gyres and currents that are usually referred to as the wind-driven circulation. Thermohaline circulation is associated with heating and cooling ("thermo"), and evaporation, precipitation, runoff, and sea ice formation, all of which change salinity ("haline"). Thermohaline-dominated circulation is mostly weak and slow compared with wind-driven circulation. In discussing thermohaline effects, it is common to refer to the *overturning circulation*, which involves buoyancy changes. The energy source for thermohaline circulation importantly includes the wind and tides that produce the turbulence that is essential for the diffusive upwelling across isopycnals that closes the thermohaline overturning. Both the wind-driven and thermohaline circulations are almost completely in geostrophic balance, with the forcing that drives them occurring at higher order.

7.2. MOMENTUM BALANCE

Fluid flow in three dimensions is governed by three equations expressing how velocity (or momentum) changes, one for each of the three physical dimensions. Each of the three momentum equations includes an acceleration term (how velocity changes with time), an advection term (see Section 5.1.3), and forcing terms:

Density $\times$ (Acceleration + Advection)

$$= \text{Forces per unit volume} \qquad (7.1)$$

Forces per unit volume = Pressure

gradient force + Gravity + Friction   (7.2)

Expressions (7.1) and (7.2) are each three equations, one for each of the three directions (e.g., east, north, and up). The terms in Eqs. (7.1) and (7.2) are illustrated in Figure 7.1.

The inclusion of advection means that Eq. (7.1) is the expression of momentum change in a *Eulerian* framework, where the observer sits at a fixed location relative to Earth. Equation (7.1) can be written without the advection term, in a *Lagrangian* framework, where the observer drifts along with the fluid flow. (See also Section S16.5 in the online Supplement.)

For a rotating geophysical flow, we, as observers, sit within a rotating "frame of reference" attached to the rotating Earth. For this reference frame, the acceleration term on the left-hand side of Eq. (7.1) is rewritten to separate local acceleration due to an actual local force from the effects of rotation. The effects that are separated out are the *centrifugal* and *Coriolis accelerations* (Section 7.2.3).

The frictional force in Eq. (7.1) leads to *dissipation* of energy due to the fluid's *viscosity*.

7.2.1. Acceleration and Advection

Acceleration is the change in velocity with time. If the vector velocity is expressed in Cartesian coordinates as $\mathbf{u} = (u, v, w)$ where the bold $\mathbf{u}$ indicates a vector quantity, and u, v, and w are the positive eastward (x-direction), northward (y-direction) and positive upward (z-direction) velocities, then

$$\text{x-direction acceleration} = \partial u/\partial t \qquad (7.3a)$$

with similar expressions for the *y*- and *z*-directions.

Advection is defined in Section 5.1.3. Advection is how the flow moves properties (including scalars such as temperature or salinity) and vectors (such as the velocity). Advection can change the flow property if there is a gradient in the property through which the fluid moves. In the x-momentum equation, the advection term is

x-direction advection

$$= u\, \partial u/\partial x + v\, \partial u/\partial y + w\, \partial u/\partial z \qquad (7.3b)$$

The *substantial derivative* is the sum of the acceleration and advection terms:

$$Du/Dt = \partial u/\partial t + u\, \partial u/\partial x + v\, \partial u/\partial y$$
$$+ w\, \partial u/\partial z \qquad (7.4)$$

7.2.2. Pressure Gradient Force and Gravitational Force

Pressure is defined in Section 3.2. The flow of fluid due to spatial variations in pressure is also described. In mathematical form, the pressure gradient force is

x-direction pressure gradient force

$$= -\partial p/\partial x \qquad (7.5)$$

The *gravitational force* between Earth and the object or fluid parcel is directed toward the center of mass of Earth. Gravitational force is

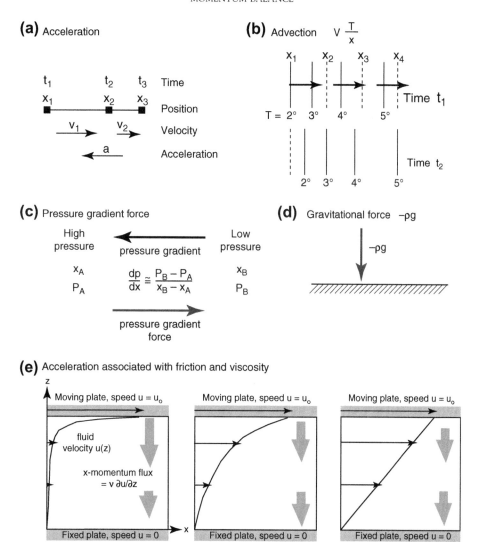

FIGURE 7.1 Forces and accelerations in a fluid: (a) acceleration, (b) advection, (c) pressure gradient force, (d) gravity, and (e) acceleration associated with viscosity υ.

mass of the object $\times$ gravitational acceleration g, equal to 9.780318 m²/sec (at the equator). The gravitational force per unit volume is

z-direction gravitational force per unit

volume $= -\rho g$ (7.6)

7.2.3. Rotation: Centrifugal and Coriolis Forces

Centrifugal force is the apparent outward force on a mass when it is rotated. Since Earth rotates around a fixed axis, the direction of centrifugal force is always outward away from the axis, opposite to the direction of gravity at the equator; at Earth's poles it is zero. (*Centripetal force* is the necessary inward force that keeps the mass from moving in a straight line; it is the same size as centrifugal force, with the opposite sign. Centripetal force is real; centrifugal force is just an apparent force.) The mathematical expression for centrifugal acceleration (force divided by density) is

$$\text{centrifugal acceleration} = \Omega^2 r \quad (7.7)$$

where Ω is the rotation rate of Earth, equal to $2\pi/T$ where T is the length of day, and r is Earth's radius. Because the centrifugal acceleration is nearly constant in time and points outward, away from Earth's axis of rotation, we usually combine it formally with the gravitational force, which points toward Earth's center. We replace g in Eq. (7.6) with an effective gravity g, which has a weak dependence on latitude. Hereafter, we do not refer separately to the centrifugal force. The surface perpendicular to this combined force is called the *geoid*. If the ocean were not moving relative to Earth, its surface would align with the geoid.

The second term in a rotating frame of reference included in the acceleration equation (7.1) is the *Coriolis force*. When a water parcel, air parcel, bullet, hockey puck, or any other body that has little friction moves, Earth spins out from under it. By Newton's Law, the body moves in a straight line if there is no other force acting on it. As observers attached to Earth, we see the body appear to move relative to our location. In the Northern Hemisphere, the Coriolis force causes a moving body to appear to move to the right of its direction of motion (Figure S7.2b). In the Southern Hemisphere, it moves to the left.

The Coriolis force is non-zero only if the body is in motion, and is important only if the body travels for a significant period of time. Coriolis force is larger for larger velocities as well. Mathematically, the Coriolis force is

$$\text{x-momentum equation}: -2\Omega\sin\varphi\, v \equiv -f\, v$$
$$(7.8a)$$

$$\text{y-momentum equation}: 2\Omega\sin\varphi\, u \equiv f\, u \quad (7.8b)$$

$$\text{Coriolis parameter}: f \equiv 2\Omega\sin\varphi \quad (7.8c)$$

where "$\equiv$" denotes a definition, Ω is the rotation rate, φ is latitude, u is velocity in the x-direction, v is velocity in the y-direction, and where the signs are appropriate for including these terms on the left-hand side of Eq. (7.1). The *Coriolis parameter*, f, is a function of latitude and changes sign at the equator, and it has units of $\sec^{-1}$. (The non-dimensional parameter called the *Rossby number* introduced in Section 1.2 is Ro $= 1/fT$ or Ro $= U/fL$, where U, L, and T are characteristic velocity, length, and timescales for the flow.)

7.2.4. Viscous Force or Dissipation

Fluids have viscous molecular processes that smooth out variations in velocity and slow down the overall flow. These molecular processes are very weak, so fluids can often be treated, theoretically, as "inviscid" rather than viscous. However, it is observed that turbulent fluids like the ocean and atmosphere actually act as if the effective viscosity were much larger than the molecular viscosity. *Eddy viscosity* is introduced to account for this more efficient mixing (Section 7.2.4.2).

7.2.4.1. Molecular Viscosity

We can think of molecular viscosity by considering two very different types of coexisting motion: the flow field of the fluid, and, due to their thermal energy, the random motion of molecules within the flow field. The random molecular motion carries (or advects) the larger scale velocity from one location to another, and then collisions with other molecules transfer their momentum to each other; this smoothes out the larger-scale velocity structure (Figure S7.3).

The viscous stress within a Newtonian fluid is proportional to the velocity shear. The proportionality constant is the *dynamic viscosity*, which has meter-kilogram-second (mks) units of kg/m-sec. The dynamic viscosity is the product of fluid density times a quantity called the *kinematic viscosity*, which has mks units of m^2/sec. For water, the kinematic viscosity is 1.8×10^{-6} m^2/sec at 0°C and 1.0×10^{-6} m^2/sec at 20°C (Table S7.1). Flow is accelerated or decelerated if there is a variation in viscous stress from one location to another (Figure 7.1e).

Formally, for a *Newtonian fluid*, which is defined to be a fluid in which stress is proportional to strain (velocity shear), and if viscosity has no spatial dependence, viscous stress enters the momentum equations as

x-momentum dissipation

$$= \upsilon(\partial^2 u/\partial x^2 + \partial^2 u/\partial y^2 + \partial^2 u/\partial z^2) \quad (7.9)$$

where υ is the molecular (kinematic) viscosity. (The dynamic viscosity is $\rho\upsilon$.) Molecular viscosity changes flow very slowly. Its effectiveness can be gauged by a non-dimensional parameter, the *Reynolds number*, which is the ratio of the dissipation timescale to the advective timescale: $Re = UL/\nu$. When the Reynolds number is large, the flow is nearly inviscid and most likely very turbulent; this is the case for flows governed by molecular viscosity. When Earth's rotation and hence the Coriolis term is important, the non-dimensional parameter of most interest for judging the effectiveness of dissipation is the Ekman number: $E = \upsilon/fH^2$. Nearly inviscid rotating flows have very small Ekman number. From matching observations and theory we know that the ocean currents dissipate energy much more quickly than we can predict using molecular viscosity. How this happens is described next.

7.2.4.2. Eddy Viscosity

Mixing at spatial scales larger than those quickly affected by molecular viscosity is generally a result of turbulence in the fluid. Turbulent motions stir the fluid, deforming and pulling it into elongated, narrow filaments. A stirred fluid mixes much faster than one that is calm and subjected only to molecular motion. We refer to the effect of this turbulent stirring/mixing on the fluid as *eddy viscosity*. For large-scale ocean circulation, the "turbulent" motions are mesoscale eddies, vertical fine structure, and so on, with spatial scales smaller than the larger scales of interest. Like molecular viscosity, eddy viscosity should be proportional to the product of turbulent speed and path length. Therefore, horizontal eddy viscosity is generally much larger than vertical eddy viscosity (Table S7.1).

To mathematically include eddy viscosity, the viscous terms in Eqs. (7.1) and (7.9) are replaced by the eddy viscosity terms:

x-momentum dissipation

$$= A_H(\partial^2 u/\partial x^2 + \partial^2 u/\partial y^2) + A_V(\partial^2 u/\partial z^2)$$
$$(7.10)$$

where A_H is the horizontal eddy viscosity and A_V is the vertical eddy viscosity. A_H and A_V have units of kinematic viscosity, m^2/sec in mks units. (Although we often use these Cartesian coordinates, the most relevant stirring/mixing directions are along isopycnals (isentropic surfaces) and across isopycnals (*diapycnal mixing*), so the coordinate system used in Eq. (7.10) is better modeled by rotating it to have the "vertical" direction perpendicular to isopycnal surfaces, and replace A_H and A_V with eddy viscosities that are along and perpendicular to those surfaces.)

Although eddy viscosity is much larger than molecular viscosity, the ocean is nevertheless nearly inviscid, in the sense that the Reynolds number is large and the Ekman number is small even when eddy viscosities are used.

7.2.5. Mathematical Expression of Momentum Balance

The full momentum balance with spatially varying eddy viscosity and rotation is:

$$Du/Dt - fv = \partial u/\partial t + u\,\partial u/\partial x$$
$$+ v\,\partial u/\partial y + w\,\partial u/\partial z - fv$$
$$= -(1/\rho)\partial p/\partial x + \partial/\partial x(A_H \partial u/\partial x)$$
$$+ \partial/\partial y(A_H \partial u/\partial y) + \partial/\partial z(A_V \partial u/\partial z)$$

$$(7.11a)$$

$$Dv/Dt + fu = \partial v/\partial t + u\,\partial v/\partial x + v\,\partial v/\partial y$$
$$+ w\,\partial v/\partial z + fu = -(1/\rho)\partial p/\partial y$$
$$+ \partial/\partial x(A_H \partial v/\partial x) + \partial/\partial y(A_H \partial v/\partial y)$$
$$+ \partial/\partial z(A_V \partial v/\partial z)$$

$$(7.11b)$$

$$Dw/Dt = \partial w/\partial t + u\,\partial w/\partial x + v\,\partial w/\partial y$$
$$+ w\,\partial w/\partial z = -(1/\rho)\partial p/\partial z - g$$
$$+ \partial/\partial x(A_H \partial w/\partial x) + \partial/\partial y(A_H \partial w/\partial y)$$
$$+ \partial/\partial z(A_V \partial w/\partial z)$$

$$(7.11c)$$

Here the standard notation "D/Dt" is the substantial derivative defined in Eq. (7.4).

The full set of equations describing the physical state of the ocean must also include the *continuity* (or *mass conservation*) equation (Section 5.1):

$$D\rho/Dt + \rho(\partial u/\partial x + \partial v/\partial y + \partial w/\partial z) = 0$$

$$(7.11d)$$

If density changes are small, Eq. 7.11d is approximated as

$$\partial u/\partial x + \partial v/\partial y + \partial w/\partial z = 0 \qquad (7.11e)$$

which is known as the *continuity* equation.

The set is completed by the equations governing changes in temperature, salinity, and density, which are presented in the following section.

7.3. TEMPERATURE, SALINITY, AND DENSITY EVOLUTION

Evolution equations for temperature and salinity — the equation of state that relates density to salinity, temperature, and pressure, and thus an evolution equation for density — complete the set of equations (7.11a–d) that describe fluid flow in the ocean.

7.3.1. Temperature, Salinity, and Density Equations

Temperature is changed by heating, cooling, and diffusion. Salinity is changed by addition or removal of freshwater, which alters the dilution of the salts. Density is then computed from temperature and salinity using the equation of state of seawater. The "word" equations for temperature, salinity, and density forcing include:

temperature change

+ temperature advection/convection

= heating/cooling term + diffusion

$$(7.12a)$$

salinity change + salinity advection/convection

= evaporation/precipitation/runoff

/brine rejection + diffusion   (7.12b)

equation of state(dependence of density on

salinity, temperature, and pressure)   (7.12c)

density change + density advection/convection

= density sources + diffusion   (7.12d)

Written in full, these are

$$DT/Dt = \partial T/\partial t + u\,\partial T/\partial x + v\,\partial T/\partial y$$
$$+ w\,\partial T/\partial z = Q_H/\rho c_p + \partial/\partial x(\kappa_H \partial T/\partial x)$$
$$+ \partial/\partial y(\kappa_H \partial T/\partial y) + \partial/\partial z(\kappa_V \partial T/\partial z)$$

$$(7.13a)$$

$$DS/Dt = \partial S/\partial t + u\, \partial S/\partial x + v\, \partial S/\partial y + w\, \partial S/\partial z$$
$$= Q_S + \partial/\partial x(\kappa_H \partial S/\partial x)$$
$$+ \partial/\partial y(\kappa_H \partial S/\partial y) + \partial/\partial z(\kappa_V \partial S/\partial z)$$

(7.13b)

$$\rho = \rho(S, T, p) \tag{7.13c}$$

$$D\rho/Dt = \partial\rho/\partial t + u\, \partial\rho/\partial x + v\, \partial\rho/\partial y$$
$$+ w\, \partial\rho/\partial z = (\partial\rho/\partial S)\, DS/Dt$$
$$+ (\partial\rho/\partial T)\, DT/Dt + (\partial\rho/\partial p)\, Dp/Dt$$

(7.13d)

where Q_H is the heat source (positive for heating, negative for cooling, applied mainly near the sea surface), c_p is the specific heat of seawater, and Q_S is the salinity "source" (positive for evaporation and brine rejection, negative for precipitation and runoff, applied at or near the sea surface). κ_H and κ_V are the horizontal and vertical eddy diffusivities, analogous to the horizontal and vertical eddy viscosities in the momentum equations (7.11a—d) (Table S7.1 located on the textbook Web site). The full equation of state appears in Eq. (7.13c), from which the evolution of density in terms of temperature and salinity change can be computed (Eq. 7.13d). The coefficients for the three terms in Eq. (7.13d) are the haline contraction coefficient, the thermal expansion coefficient, and the adiabatic compressibility, which is proportional to the inverse of sound speed (Chapter 3).

7.3.2. Molecular and Eddy Diffusivity

The molecular diffusivity κ for each substance depends on the substance and the fluid. The molecular diffusivity of salt in seawater is much smaller than that for heat (Table S7.1). This difference results in a process called "double diffusion" (Section 7.4.3). *Eddy diffusivity* is the equivalent of eddy viscosity for properties like heat and salt. A globally averaged vertical eddy diffusivity of $\kappa_V = 1 \times 10^{-4}$ m²/sec accounts for the observed average vertical density structure (Section 7.10.2; Munk, 1966). However, the directly observed vertical (or diapycnal) eddy diffusivity in most of the ocean is a factor of 10 lower: $\kappa_V \sim 1 \times 10^{-5}$ m²/sec, implying that there are regions of much higher diffusivity to reach the global average. Measurements show huge enhancements of diapycnal eddy diffusivity in bottom boundary regions, especially where topography is rough (Figure 7.2) (Polzin, Toole, Ledwell, & Schmitt, 1997; Kunze et al., 2006), and on continental shelves where tidal energy is focused (Lien & Gregg, 2001). In the surface layer, eddy diffusivities and eddy viscosities are also much greater than the Munk value (e.g., Large, McWilliams, & Doney, 1994) (Section S7.4.1). Horizontal eddy diffusivities κ_H are estimated to be between 10^3 and

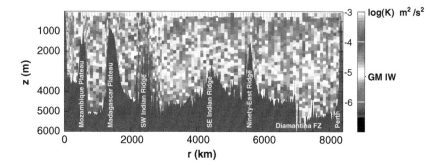

FIGURE 7.2 Observed diapycnal diffusivity (m²/s²) along 32°S in the Indian Ocean, which is representative of other ocean transects of diffusivity. See Figure S7.4 for diffusivity profiles. This figure can also be found in the color insert. ©American Meteorological Society. Reprinted with permission. Source: From Kunze et al. (2006).

10^4 m^2/sec, with large spatial variability (e.g., Figure 14.17). κ_H is much larger than κ_V.

7.4. MIXING LAYERS

Mixing occurs throughout the ocean. While weak, mixing is essential for maintaining the observed stratification and can regulate the strength of some parts of the circulation.

7.4.1. Surface Mixed Layer

The surface layer (Section 4.2.2) is forced directly by the atmosphere through surface wind stress and buoyancy (heat and freshwater) exchange. For a surface layer that is initially stably stratified (Figure 7.3a), sufficiently large wind stress will create turbulence that mixes and creates a substantially uniform density or mixed layer (Figure 7.3b). This typically results in a discontinuity in properties at the mixed layer base.

The upper layer can also be mixed by buoyancy loss through the sea surface, increasing the density of the top of the surface layer and causing it to overturn (*convect*) to a greater depth (Figure 7.3c–e).

This type of mixed layer typically has no discontinuity in density at its base. Heat or freshwater gain decreases the density of the top of the surface layer, resulting in a more stably stratified profile. If the wind then mixes it, the final mixed layer is shallower than the initial mixed layer (Figure 7.3f–h). (Mixed layer observations typically show much more vertical structure than might be expected from these simple ideas.)

The thickest mixed layers occur at the end of winter (Figure 4.5), after an accumulation of months of cooling that deepens the mixed layer and increases its density. For large-scale oceanographic studies, these end-of-winter mixed layers set the properties that are subducted into the ocean interior (Section 7.8.5).

7.4.2. Bottom Mixed Layers

Near the ocean bottom turbulence, and hence mixing, can be generated by currents or current shear caused by the interaction with the bottom. In shallow (e.g., coastal) waters, complete mixing of the water column occurs if the depth is shallow enough and the tidal currents are fast enough (see reviews in Simpson, 1998 and Brink, 2005 and more extended discussion in Section S7.4.2 on the textbook Web site). At longer timescales on the shelf, a bottom Ekman layer can develop in which frictional and Coriolis forces balance (Ekman, 1905 and Section 7.5.3), with the bottom slope also affecting the layer.

Enhanced turbulence in a bottom boundary layer can be created by movement of water across rough topography and by breaking of internal waves that reflect off the topography and result in higher eddy diffusivity values (Figure 7.2). This can create "steppy" vertical profiles near the bottom some distance from the mixing site (Figure S7.6a located on the textbook Web site).

Bottom currents due to density differences can also cause mixing. One example is a turbidity current down an underlying bottom slope (Section 2.6). Another example is the overflow of dense water across a sill, as seen at the Strait of Gibraltar (Chapter 9). The dense water flows down the continental slope as a *plume*, mixing vigorously with the lighter water around it (Figure S7.6b). This turbulent process is called *entrainment*. Density differences due to the injection of lighter water into the ocean also cause mixing and entrainment. An example is hot hydrothermal water injected at mid-ocean ridges and hotspots that entrain ambient waters as the plumes rise.

7.4.3. Internal Mixing Layers

In the interior of the ocean (i.e., away from boundaries), continuous profiling instruments have shown that vertical profiles of water properties — temperature and salinity, and hence density — are often not smooth (Figure 7.3i)

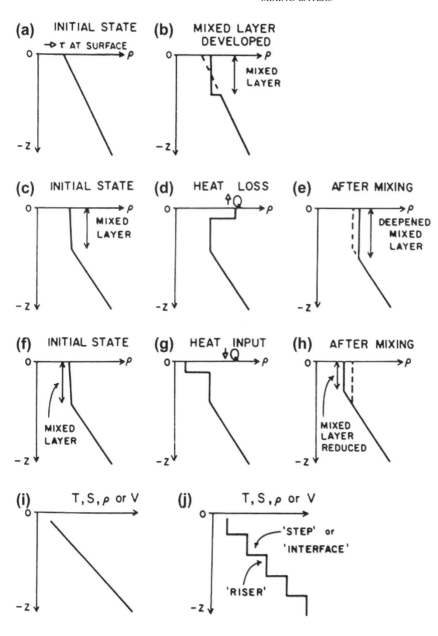

FIGURE 7.3 Mixed layer development. (a, b) An initially stratified layer mixed by turbulence created by wind stress; (c, d, e) an initial mixed layer subjected to heat loss at the surface which deepens the mixed layer; (f, g, h) an initial mixed layer subjected to heat gain and then to turbulent mixing presumably by the wind, resulting in a thinner mixed layer; (i, j) an initially stratified profile subjected to internal mixing, which creates a stepped profile. Notation: τ is wind stress, Q is heat (buoyancy).

but "stepped" (Figure 7.3j). Turbulence and/or double diffusion mix the water column internally and can create such steps.

Breaking internal waves (Chapter 8) can create internal mixing (Section 8.4; Rudnick et al., 2003). Vertical shear from other sources can also result in turbulence. On the other hand, vertical stratification stabilizes the mixing. One way to express this trade-off is through a non-dimensional parameter called the *Richardson number* (Ri):

$$Ri = N^2/(\partial u/\partial z)^2 \qquad (7.14)$$

$$N^2 = -g\,(\partial\rho/\partial z)/\rho_0 \qquad (7.15)$$

where N is the Brunt-Väisälä frequency (Section 3.5.6) and the vertical shear of the horizontal speed is $(\partial u/\partial z)$. If the Richardson number is small, the stratification is weak and the shear is large, so we expect mixing to be vigorous.

Stirring between two horizontally adjacent waters with strongly contrasting temperature and salinity results in *interleaving* or *fine structure*, with layering of one to tens of meters on both sides of the front. A much smaller scale of vertical structure, on the order of centimeters (*microstructure*), is associated with the actual mixing at the interfaces between the interleaving layers.

Heat diffuses about 100 times faster than salt (Table S7.1). *Double diffusion* arises from these differing molecular diffusivities, acting at scales of centimeters to meters, and can also create well-mixed internal layers. When warm, salty water lies above cold, fresh water, the saltier water becomes denser and tends to sink into the lower layer and vice versa (Figure S7.7a). The alternating columns are called *salt fingers*. Lateral diffusion occurs between the "fingers" and produces a uniform layer that may be meters to tens of meters thick in the ocean. When cold, fresh water lies above warm, salty water (Figure S7.7b), the fresher upper layer becomes warmer and rises within the upper layer; this is called the *diffusive* form of double diffusion. Salt fingering effects are observed in the ocean where there are strong contrasts in salinity, for instance, where salty Mediterranean Water enters the Atlantic (Figure S7.7c). Diffusive interfaces are observed in high latitude regions with dichothermal layers (Sections 4.2, 4.3.2 and Figure S7.7d).

7.5. RESPONSE TO WIND FORCING

The wind blows over the sea surface exerting stress and causing the water to move within the top 50 m. Initially the wind excites small capillary waves that propagate in the direction of the wind. Continued wind-driven momentum exchange excites a range of surface waves (Chapter 8). The net effect of this input of atmospheric momentum is a stress on the ocean (*wind stress*). For timescales of about a day and longer, Earth's rotation becomes important and the Coriolis effect enters in, as described in the following subsections.

7.5.1. Inertial Currents

At timescales of a day or so (after build-up of surface waves and possibly Langmuir circulation), the ocean responds to a wind stress impulse with transient motions known as "inertial currents." These are a balance of the Coriolis force and the time derivatives of the initial horizontal velocities caused by the wind stress. In the Northern Hemisphere, the water particles trace out clockwise circles (Figure S7.8a). In the Southern Hemisphere, inertial currents are counterclockwise.

(Mathematically, inertial currents are the solution of

$$\partial u/\partial t = fv \qquad (7.16a)$$

$$\partial v/\partial t = -fu \qquad (7.16b)$$

which is taken from Eq. (7.11a and b) assuming that advection, pressure gradient forces, and dissipation are very small.)

Inertial currents are often observed in surface drifter trajectories and surface velocity moorings in the wake of a storm (Figure 7.4). Inertial periods are often very close to tidal periods, so separating tidal and inertial effects in time series is sometimes difficult.

After the wind starts to blow impulsively, the current will initially oscillate around and then, after several days, settle frictionally to a steady flow at an angle to the wind (Figure S7.8b from Ekman, 1905). This becomes the surface Ekman velocity (Section 7.5.3).

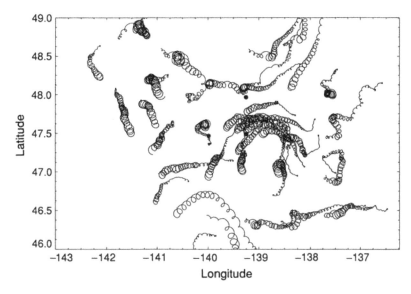

FIGURE 7.4 Observations of near-inertial currents. Surface drifter tracks during and after a storm. ©*American Meteorological Society. Reprinted with permission. Source: From d'Asaro et al. (1995).* See Figure S7.8 for schematics of inertial currents and Ekman's (1905) original hodograph.

7.5.2. Langmuir Circulation

"Langmuir circulation" (LC) is another transient response to impulsive wind forcing (Langmuir, 1938). A lengthier discussion with illustrations is provided in Section S7.5.2 in the online supplementary material. LCs are visually evident as numerous long parallel lines or streaks of flotsam ("windrows") that are mostly aligned with the wind (Figure S7.9). The streaks are formed by the convergence caused by helical vortices with a typical depth and horizontal spacing of 4–6 m and 10–50 m, but they can range up to several hundred meters horizontal separation and up to two to three times the mixed layer depth (Figure S7.10). Alternate cells rotate in opposite directions, causing convergence and divergence between alternate pairs of cells. The cells can be many kilometers long. Langmuir circulations generally occur only for wind speeds greater than 3 m/sec and appear within a few tens of minutes of wind onset. The mechanism for producing Langmuir circulation is still a matter of study and beyond the scope of this text. See Smith (2001) and Thorpe (2004) for further discussions.

7.5.3. Ekman Layers

Wind stress is communicated to the ocean surface layer through viscous (frictional) processes that extend several tens of meters into the ocean. For timescales longer than a day, the response is strongly affected by Coriolis acceleration. This wind-driven frictional layer is called the *Ekman layer*. The physical processes in an Ekman layer include only friction (eddy viscosity) and Coriolis acceleration. Velocity in the Ekman layer is strongest at the sea surface and decays exponentially downward, disappearing at a depth of about 50 m. It coexists with, but is not the same as, the mixed layer depth or euphotic zone depth.

The two most unusual characteristics of an Ekman layer (compared with a frictional flow that is not rotating) are (1) the horizontal velocity vector spirals with increasing depth (Figure 7.5) and (2) the net transport integrated through the Ekman layer is *exactly* to the right of the wind in the Northern Hemisphere (left in the Southern Hemisphere).

The surface water in an Ekman layer moves at an angle to the wind because of Coriolis acceleration. If eddy viscosity is independent

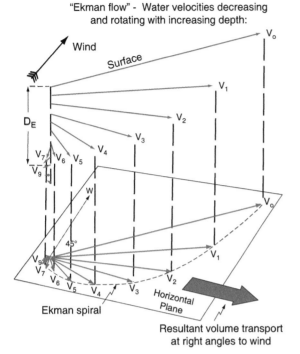

"Ekman flow" - Water velocities decreasing and rotating with increasing depth:

FIGURE 7.5 Ekman layer velocities (Northern Hemisphere). Water velocity as a function of depth (upper projection) and Ekman spiral (lower projection). The large open arrow shows the direction of the total Ekman transport, which is perpendicular to the wind.

of depth, the angle is 45 degrees to the right of the wind in the Northern Hemisphere; otherwise it differs somewhat from 45 degrees. Due to the frictional stress proportional to the eddy viscosity A_V, each layer from the surface on down accelerates the next layer below to the right (Northern Hemisphere) and has a weaker velocity than the layer above it. The complete structure is a decaying "spiral." If the velocity arrows are projected onto a horizontal plane, their tips form the Ekman spiral (Figure 7.5).

The Ekman layer depth is the e-folding depth of the decaying velocity:

$$D_E = (2A_v/f)^{1/2} \qquad (7.17)$$

Using a constant eddy viscosity of 0.05 m²/sec from within the observed range (Section 7.5.5), the Ekman layer depths at latitudes 10, 45, and 80 degrees are 63, 31, and 26 m, respectively. The vertically integrated horizontal velocity in the Ekman layer is called the *Ekman transport*:

$$U_E = \int u_E(z)\, dz \qquad (7.18a)$$

$$V_E = \int v_E(z)\, dz \qquad (7.18b)$$

where u_E and v_E are the eastward and northward velocities in the Ekman layer, and U_E and V_E are the associated Ekman transports. (Ekman "transport" has units of depth times velocity, hence m²/sec, rather than area times velocity.) Ekman transport in terms of the wind stress is derived from Eq. (7.11):

$$U_E = \tau^{(y)}/(\rho f) \qquad (7.19a)$$

$$V_E = -\tau^{(x)}/(\rho f) \qquad (7.19b)$$

where $\tau^{(x)}$ and $\tau^{(y)}$ are the wind stresses positive in the east and north directions, assuming no time acceleration, advection, or pressure gradient force, and setting the eddy friction stress at the sea surface equal to the wind stress. The Ekman transport is *exactly* perpendicular and to the right (left) of the wind in the Northern (Southern) Hemisphere (large arrow in Figure 7.5). For applications of Ekman layers to general circulation (Sections 7.8 and 7.9), only the Ekman transport matters. Thus, the actual eddy viscosity and Ekman layer thickness are unimportant.

Bottom Ekman layers that are 50 to 100 m thick can develop if there is a flow along the bottom. In shallow water, the top and bottom Ekman layers can overlap, so that the right-turning tendency in the top layer (Northern Hemisphere) will overlap the left-turning tendency in the bottom layer. If there is a wind stress at the top surface that would produce an Ekman layer of depth D_E in deep water, then in water of depth h, the approximate angle α between the wind and the surface flow is as listed in Table S7.2. As water depth decreases, the net flow is more in the direction of the wind.

7.5.4. Ekman Transport Convergence and Wind Stress Curl

When the wind stress varies with position so that Ekman transport varies with position, there can be a convergence or divergence of water within the Ekman layer. Convergence results in downwelling of water out of the Ekman layer. Divergence results in upwelling into the Ekman layer. This is the mechanism that connects the frictional forcing by wind of the surface layer to the interior, geostrophic ocean circulation (Section 7.8).

The vertical velocity w_E at the base of the Ekman layer is obtained from the divergence of the Ekman transport, by vertically integrating the continuity equation Eq. (7.11e) over the depth of the Ekman layer:

$$(\partial U_E/\partial x + \partial V_E/\partial y) = \nabla \cdot U_E$$
$$= -(w_{surface} - w_E) = w_E$$
$$(7.20)$$

where U_E is the horizontal vector Ekman transport (Eq. 7.18) and it is assumed that the vertical velocity at the sea surface, $w_{surface}$, is 0. When Eq. (7.20) is negative, the transport is convergent and there must be downwelling (w_E at the base of the Ekman layer is negative). The relation of Ekman transport divergence to the wind stress from Eq. (7.19a,b) is

$$\nabla \cdot U_E = \partial/\partial x(\tau^{(y)}/(\rho f)) - \partial/\partial y(\tau^{(x)}/(\rho f))$$
$$= k \cdot \nabla \times (\tau/\rho f) \qquad (7.21)$$

where τ is the vector wind stress and k is the unit vector in the vertical direction. Therefore, in the Northern Hemisphere ($f > 0$), upwelling into the Ekman layer results from positive wind stress curl, and downwelling results

from negative wind stress curl. Downwelling is referred to as *Ekman pumping*. Upwelling is sometimes referred to as Ekman suction.

A global map of wind stress curl was shown in Figure 5.16d, and is referred to frequently in subsequent chapters because of its importance for Ekman pumping/suction, although the mapped quantity should include the Coriolis parameter, f, to be related directly to upwelling and downwelling.

Equatorial upwelling due to Ekman transport results from the westward wind stress (trade winds). These cause northward Ekman transport north of the equator and southward Ekman transport south of the equator. This results in upwelling along the equator, even though the wind stress curl is small, more or less because of the Coriolis parameter dependence in Eq. (7.21).

The coastline is the other place where Ekman transport divergence or convergence can occur, and it is *not* included in Eq. (7.21) because this divergence is due to the boundary condition at the coast and not wind stress curl. If the wind blows along the coast, then Ekman transport is perpendicular to the coast, so there must be either downwelling or upwelling at the coast to feed the Ekman layer (Figure 7.6). This is one mechanism for creation of coastal upwelling and subtropical eastern boundary current systems (Section 7.9).

7.5.5. Observations of Ekman Response and Wind Forcing

The Ekman theory has major consequences for wind-driven ocean circulation. Thus it is important to confirm and refine Ekman's theory with ocean observations. Observations of Ekman response are difficult because of the time dependence of the wind. California Current observations produced an easily visible Ekman-like response because the wind direction was relatively steady (Figure 7.7 and Chereskin, 1995).

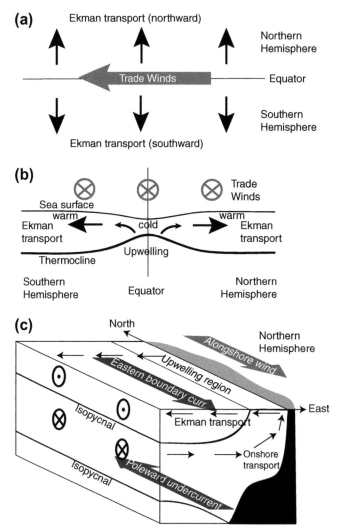

FIGURE 7.6   (a) Ekman transport divergence near the equator driven by easterly Trade Winds. (b) The effect of equatorial Ekman transport divergence on the surface height, thermocline, and surface temperature. (c) Coastal upwelling system due to an alongshore wind with offshore Ekman transport (Northern Hemisphere).

An Ekman response to the wind for a large part of the Pacific Ocean is apparent in the average 15 m velocity from surface drifters (Figure 7.8). Velocities are to the right of the wind stress in the Northern Hemisphere and to the left in the Southern Hemisphere.

7.6. GEOSTROPHIC BALANCE

7.6.1. Pressure Gradient Force and Coriolis Force Balance

Throughout most of the ocean at timescales longer than several days and at spatial scales

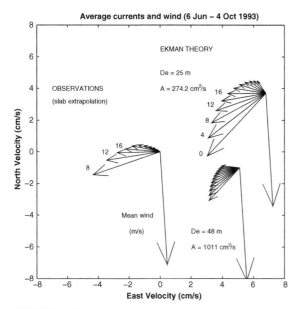

Average currents and wind (6 Jun – 4 Oct 1993)

EKMAN THEORY

De = 25 m

A = 274.2 cm²/s

OBSERVATIONS

(slab extrapolation)

Mean wind

(m/s)

De = 48 m

A = 1011 cm²/s

North Velocity (cm/s)

East Velocity (cm/s)

FIGURE 7.7 Observations of an Ekman-like response in the California Current region. Observed mean velocities (left) and two theoretical Ekman spirals (offset) using different eddy diffusivities (274 and 1011 cm²/s). The numbers on the arrows are depths. The large arrow is the mean wind. See Figure S7.14 for the progressive vector diagram. *Source: From Chereskin (1995).*

longer than several kilometers, the balance of forces in the horizontal is between the pressure gradient and the Coriolis force. This is called "geostrophic balance" or *geostrophy*.

In a "word" equation, geostrophic balance is

horizontal Coriolis acceleration

$$= \text{horizontal pressure gradient force} \quad (7.22)$$

This is illustrated in Figure 7.9. The pressure gradient force vector points from high pressure to low pressure. In a non-rotating flow, the water would then move from high to low pressure. However, with rotation, the Coriolis force exactly opposes the pressure gradient force, so that the net force is zero. Thus, the water parcel does not accelerate (relative to Earth). The parcel moves exactly perpendicular to both the pressure gradient force and the Coriolis force. A heuristic way to remember the direction of geostrophic flow is to think of the pressure gradient force pushing the water parcel from high to low pressure, but Coriolis force moves the parcel off to the right (Northern Hemisphere) or the left (Southern Hemisphere).

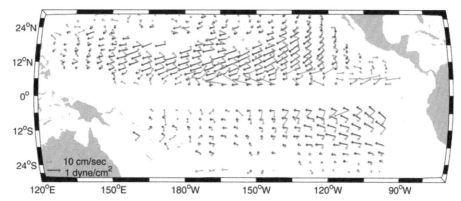

FIGURE 7.8 Ekman response. Average wind vectors (blue) and average ageostrophic current at 15 m depth (red). The current is calculated from 7 years of surface drifters drogued at 15 m, with the geostrophic current based on average density data from Levitus et al. (1994a) removed. (No arrows were plotted within 5 degrees of the equator because the Coriolis force is small there.) This figure can also be found in the color insert. ©*American Meteorological Society. Reprinted with permission. Source: From Ralph and Niiler (1999).*

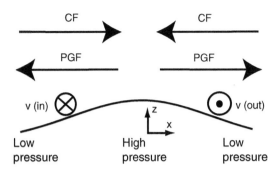

FIGURE 7.9 Geostrophic balance: horizontal forces and velocity. PGF = pressure gradient force. CF = Coriolis force. v = velocity (into and out of page). See also Figure S7.17.

The vertical force balance that goes with geostrophy is *hydrostatic balance* (Section 3.2). The vertical pressure gradient force, which points upward from high pressure to low pressure, is balanced by gravity, which points downward.

The mathematical expression of geostrophy and hydrostatic balance, from Eq. (7.11), is

$$-fv = -(1/\rho)\partial p/\partial x \qquad (7.23a)$$

$$fu = -(1/\rho)\partial p/\partial y \qquad (7.23b)$$

$$0 = -\partial p/\partial z - \rho g \qquad (7.23c)$$

An alternate form for Eq. (7.23c), used for dynamic height calculations (Section 7.6.3), is

$$0 = -\alpha\,\partial p/\partial z - g \qquad (7.23d)$$

where α is specific volume. From Eq. (7.23a and b), if the Coriolis parameter is approximately constant $(f = f_o)$, the geostrophic velocities are approximately non-divergent:

$$\partial u/\partial x + \partial v/\partial y = 0 \qquad (7.23e)$$

Formally in fluid dynamics, such a non-divergent velocity field can be written in terms of a *streamfunction* ψ:

$$u = -\partial\psi/\partial y \text{ and } v = \partial\psi/\partial x \qquad (7.23f)$$

From Eqs. (7.23a and b) the streamfunction for geostrophic flow is $\psi = p/(f_o\rho_o)$. Therefore, maps of pressure distribution (or its proxies like dynamic height, steric height, or geopotential anomaly; Section 7.6.2), are maps of the geostrophic streamfunction, and flow approximately follows the mapped contours.

Geostrophic balance is intuitively familiar to those with a general interest in weather reports. Weather maps show high and low pressure regions around which the winds blow. Low pressure regions in the atmosphere are called *cyclones*. Flow around low pressure regions is thus called *cyclonic* (counterclockwise in the Northern Hemisphere and clockwise in the Southern Hemisphere). Flow around high-pressure regions is called *anticyclonic*.

In the ocean, higher pressure is caused by a higher mass of water lying above the observation depth. At the "sea surface," pressure differences are due to an actual mounding of water relative to the geoid. Over the complete width of the Atlantic or Pacific Ocean anticyclonic gyres, the total contrast in sea surface height is about 1 m.

The geostrophic velocities at the sea surface could be calculated if the appropriately time-averaged sea surface height were known (as yet not possible for the time mean, but definitely possible from satellite altimetry for variations from the mean). The geostrophic velocity at the sea surface in terms of sea surface height η above a level surface is derived from Eqs. (7.23a and b):

$$-fv = -g\partial\eta/\partial x \qquad (7.24a)$$

$$fu = -g\partial\eta/\partial y \qquad (7.24b)$$

To calculate the horizontal pressure difference below the sea surface, we have to consider both the total height of the pile of water above our observation depth and also its density, since the total mass determines the actual pressure at our observation depth (Figure 7.10 in this

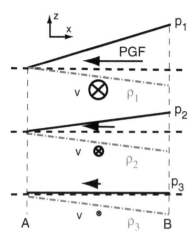

FIGURE 7.10 Geostrophic flow and thermal wind balance: schematic of change in pressure gradient force (PGF) with depth. The horizontal geostrophic velocity v is into the page for this direction of PGF and is strongest at the top, weakening with depth, as indicated by the circle sizes. Density (dash-dot) increases with depth, and isopycnals are tilted. With the sea surface at B higher than at A, the PGF at the sea surface (h_1) is to the left. The PGF decreases with increasing depth, as indicated by the flattening of the isobars p_2 and p_3.

A useful rule of thumb for geostrophic flows that are surface-intensified is that, when facing downstream in the Northern Hemisphere, the "light/warm" water is to your right. (In the Southern Hemisphere, the light water is to the left when facing downstream.) It can be useful to memorize the example for the Gulf Stream recalling that the current flows eastward with warm water to the south. Geostrophic flow with vertical shear, which requires sloping isopycnals, is often called baroclinic. Geostrophic flow without any vertical shear is often called barotropic. Barotropic flow is driven only by horizontal variations in sea surface height. Most oceanic geostrophic flows have both barotropic and baroclinic components.

Mathematically, the thermal wind relations are derived from the geostrophic and hydrostatic balance Eq. (7.23):

$$-f\partial v/\partial z = (g/\rho_0)\partial \rho/\partial x \qquad (7.25a)$$

$$f\partial u/\partial z = (g/\rho_0)\partial \rho/\partial y \qquad (7.25b)$$

(Here we have used the *Boussinesq approximation*, where ρ can be replaced by the constant ρ_0 in the x and y momentum equations, whereas the fully variable density ρ must be used in the hydrostatic balance equation.)

To calculate geostrophic velocity, we must know the absolute horizontal pressure difference between two locations. If we have only the density distribution, we can calculate only the geostrophic shear. To convert these relative currents into absolute currents, we must determine or estimate the absolute current or pressure gradient at some level (*reference level*). A common, but usually inaccurate, referencing approach has been to assume (without measuring) that the absolute current is zero at some depth (*level of no motion*). In the next subsection, we introduce the "dynamic" method widely used to calculate geostrophic velocities (shear), and continue the discussion of reference velocity choices.

chapter and Figure S7.19 on the textbook Web site). The variation in geostrophic flow with depth (the *geostrophic velocity shear*) is therefore proportional to the difference in density of the two water columns on either side of our observation location. The relation between the geostrophic velocity shear and the horizontal change (gradient) in density is called the *thermal wind relation*.

The thermal wind relation is illustrated in Figure 7.10. The sea surface is sloped, with surface pressure higher to the right. This creates a pressure gradient force to the left, which drives a surface geostrophic current into the page (Northern Hemisphere). The density ρ increases with depth, and the isopycnals are tilted. Therefore the geostrophic velocity changes with depth because the pressure gradient force changes with depth due to the tilted isopycnals.

7.6.2. Geopotential and Dynamic Height Anomalies and Reference Level Velocities

Historically and continuing to the present, it has been too difficult and too expensive to instrument the ocean to directly observe velocity everywhere at every depth. Density profiles, which are much more widely and cheaply collected, are an excellent data set for estimating geostrophic velocities using the thermal wind relations and estimates of a reference level velocity. To calculate geostrophic velocity shear from density profiles, oceanographers have created two closely related functions, *geopotential anomaly* and *dynamic height,* whose horizontal gradients represent the horizontal pressure gradient force. Another closely related concept, *steric height,* is used to study variations in sea level. The gradient of the *geopotential,* Φ, is in the direction of the local force due to gravity (modified to include centrifugal force). Geopotential is defined from hydrostatic balance (Eq. 7.23c) as

$$d\Phi = g\,dz = -\alpha\,dp \qquad (7.26a)$$

where α is specific volume. The units of geopotential are m^2/sec^2 or J/kg. For two isobaric surfaces p_2 (upper) and p_1 (lower), the geopotential is

$$\Phi = g\int dz = g(z_2 - z_1) = -\int \alpha dp \quad (7.26b)$$

Geopotential height is defined as

$$Z = (9.8 \text{ m s}^{-2})^{-1}\int g\,dz$$
$$= -(9.8 \text{ m s}^{-2})^{-1}\int \alpha\,dp \qquad (7.26c)$$

and is nearly equal to geometric height. This equation is in mks units; if centimeter-gram-second (cgs) units are used instead, the multiplicative constant would change from 9.8 m s^{-2} to 980 cm s^{-2}. Most practical calculations,

including common seawater computer subroutines, use the specific volume anomaly

$$\delta = \alpha(S, T, p) - \alpha(35, 0, p) \qquad (7.26d)$$

to compute the geopotential anomaly

$$\Delta\Phi = -\int \delta\,dp. \qquad (7.26e)$$

The geopotential height anomaly is then defined as

$$Z' = -(9.8 \text{ m s}^{-2})^{-1}\int \delta\,dp. \qquad (7.26f)$$

Geopotential height anomaly is effectively identical to *steric height anomaly,* which is defined by Gill and Niiler (1973) as

$$h' = -(1/\rho_o)\int \rho\,dz \qquad (7.27a)$$

in which the *density anomaly* $\rho' = \rho - \rho_o$. Using hydrostatic balance and defining ρ_o as $\rho(35,0,p)$, Eq. (7.27a) is equivalent to Tomczak and Godfrey's (1994) steric height (anomaly)

$$h' = \int \delta\,\rho_o\,dz \qquad (7.27b)$$

which can be further manipulated to yield

$$h' = (1/g)\int \delta\,dp. \qquad (7.27c)$$

This is nearly identical to the geopotential height anomaly in Eq. (7.26f), differing only in the appearance of a standard quantity for g. In SI units, steric height is in meters.

Dynamic height, D, is closely related to geopotential, Φ, differing only in sign and units of reporting. Many modern publications and common computer subroutines do not distinguish between dynamic height and geopotential anomaly. The unit traditionally used for dynamic height is the dynamic meter:

$$1 \text{ dyn m} = 10 \text{ m}^2/sec^2. \qquad (7.28a)$$

Therefore dynamic height reported in dynamic meters is related to geopotential anomaly as

$$\Delta D = -\Delta \Phi/10 = \int \delta \, dp/10. \qquad (7.28b)$$

Its relation to the geopotential height and steric height anomalies is

$$10 \, \Delta D = -9.8 \, Z' = gh'. \qquad (7.28c)$$

The quantities ΔD and Z' are often used interchangeably, differing only by 2%. With use of the dynamic meter, maps of dynamic topography are close to the actual geometric height of an isobaric surface relative to a level surface; a horizontal variation of, say, 1 dyn m, means that the isobaric surface has a horizontal depth variation of about 1 m. Note that the geopotential height anomaly more closely reflects the actual height variation, so a variation of 1 dyn m would be an actual height variation closer to 1.02 m.

 Geostrophic velocities at one depth relative to those at another depth are calculated using Eq. (7.25) with geopotential anomalies, steric height anomalies, or dynamic heights. In SI units, and using dynamic meters for dynamic height, the difference between the northward velocity v and eastward velocity u at the pressure surface p_2 relative to the pressure surface p_1 is

$$f(v_2 - v_1) = 10 \, \partial \, \Delta D/\partial x = -\partial \, \Delta \Phi/\partial x$$
$$= g \partial h'/\partial x \qquad (7.29a)$$

$$f(u_2 - u_1) = -10 \, \partial \, \Delta D/\partial y = \partial \, \Delta \Phi/\partial y$$
$$= -g \partial h'/\partial y \qquad (7.29b)$$

where the dynamic height or geopotential anomalies are integrated vertically from p_1 to p_2. The surface p_1 is the *reference level*. (Comparison of Eq. 7.29 with Eq. 7.23 shows that the dynamic height and geopotential anomalies are streamfunctions for the difference between geostrophic flows from one depth to another.)

 How is the velocity at the reference level chosen? Since the strength of ocean currents decreases from the surface downward in many (but not all) regions, for practical reasons, a deep level of no motion has often been presumed. A much better alternative is to use a "level of known motion" based on direct velocity observations. Satellite altimetry by itself is insufficient for the ocean's mean flow field because the spatial variations of Earth's geoid are vastly larger than the ocean's sea-surface height variations; the **GR**avity **a**nd **E**arth **C**limate **E**xperiment (GRACE) is helping to resolve this geoid problem. Modern practice requires that the flow field that is defined by many density profiles must satisfy overall constraints such as mass conservation. The constraints then help narrow the choices of reference level velocities, which can be done formally (see Wunsch, 1996). Ocean state estimation (data assimilation), which merges observations with an ocean model, is currently the focus of most activity for construction of velocity fields from density profiles.

 As an example of the geostrophic method, we calculate dynamic height and velocity profiles from density profiles across the Gulf Stream (Figure 7.11 in this chapter and Table S7.3 on the textbook Web site). The isopycnals sloping upward toward the north between 38° and 39°N mark the Gulf Stream (Figure 7.11a). The geostrophic velocity profile is calculated between stations "A" and "B" relative to an arbitrary level of no motion at 3000 m. Station A has lower specific volume (higher potential density) than station B (Figure 7.11b). The surface dynamic height at A is therefore lower than at B and the surface pressure gradient force is toward the north, from B to A. Therefore, the geostrophic velocity at the midpoint between the stations (Figure 7.11d) is eastward and is largest at the sea surface. This means that the sea surface must tilt downward from B to A. The vertical shear is largest in the upper 800 m where the difference in dynamic heights is largest.

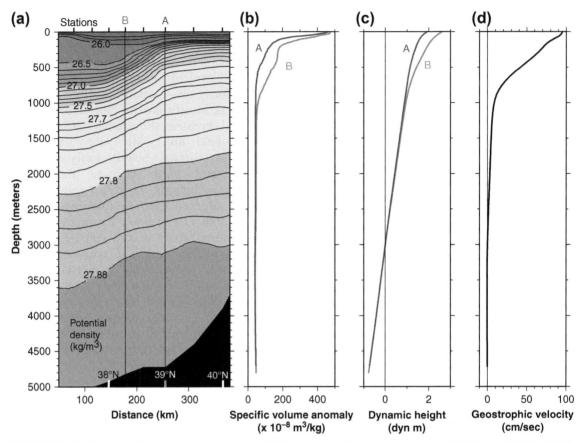

FIGURE 7.11 Geostrophic flow using observations. (a) Potential density section across the Gulf Stream (66°W in 1997). (b) Specific volume anomaly δ (×10^{-8} m^3/kg) at stations A and B. (c) Dynamic height (dyn m) profiles at stations A and B, integrated from 3000 m depth. (d) Eastward geostrophic velocity (cm/sec), assuming zero velocity at 3000 m. This figure is described in detail in Section S7.6.2 of the online supplement.

7.6.3. Dynamic Topography and Sea Surface Height Maps

Dynamic height at one surface relative to another is the streamfunction for the geostrophic flow at that surface relative to the other, as an extension of Eq. (7.23f). Flows are along the contours with the high "hills" to the right of the flow in the Northern Hemisphere (to the left in the Southern Hemisphere). The speed at any point is proportional to the steepness of the slope at that point, in other words, inversely proportional to the separation of the contours.

Dynamic topography maps (equivalently, steric height) are shown in Chapter 14 and throughout the ocean basin chapters (9–13) to depict the geostrophic flow field. At the sea surface, all five ocean basins have highest dynamic topography in the west in the subtropics. The anticyclonic flows around these highs are called the *subtropical gyres*. The Northern Hemisphere oceans have low dynamic topography around 50° to 60°N; the cyclonic flows around these lows are the *subpolar gyres*. Tightly spaced contours along the western boundaries indicate the swift western boundary currents for each of the gyres. Low values are found all the way around Antarctica; the band of tightly spaced contours to its north marks the eastward Antarctic Circumpolar Current. The contrast in dynamic height and sea-surface height from high to low in a given gyre is about 0.5 to 1 dynamic meters.

7.6.4. A Two-Layer Ocean

It is frequently convenient to think of the ocean as composed of two layers in the vertical, with upper layer of density ρ_1 and lower layer of density ρ_2 (Figure S7.21). The lower layer is assumed to be infinitely deep. The upper layer thickness is h + H, where h is the varying height of the layer above the sea level surface and H is the varying depth of the bottom of the layer. We sample the layers with stations at "A" and "B." Using the hydrostatic equation (7.23c), we compute the pressure at a depth Z at the stations:

$$p_A = \rho_1\, g(h + H) + \rho_2\, g(Z - H) \qquad (7.30a)$$

$$p_B = \rho_1\, g(h_B + H_B) + \rho_2\, g(Z - H_B) \qquad (7.30b)$$

Here Z represents a common depth for both stations, taken well below the interface. If we assume that $p_A = p_B$, which amounts to assuming a "level of no motion" at Z, we can compute a surface slope, which we cannot measure in terms of the observed density interface slope:

$$\frac{h_A - h_B}{\Delta x} = \frac{\rho_2 - \rho_1}{\rho_1} \frac{H_A - H_B}{\Delta x} \qquad (7.31a)$$

We then use Eq. (7.30a) to estimate the surface velocity v:

$$fv = g\frac{h_A - h_B}{\Delta x} = g\frac{\rho_2 - \rho_1}{\rho_1} \frac{H_A - H_B}{\Delta x} \qquad (7.31b)$$

7.7. VORTICITY, POTENTIAL VORTICITY, ROSSBY AND KELVIN WAVES, AND INSTABILITIES

An apparent "problem" with the geostrophic balance (Eq. 7.23a,−b) is that it does not include any of the external forces that make the ocean flow; it has only pressure gradient and Coriolis force. How do we insert external forces such as the wind? In formal geophysical fluid dynamics, we would show that these forces *are* in the momentum equations, but are so weak that we safely consider the flows to be geostrophic (to lowest order). To reinsert the external forces, we have to consider the "vorticity" equation, which is formally derived from the momentum equations by combining the equations in a way that eliminates the pressure gradient force terms. (It is straightforward to do.) The resulting equation gives the time change of the vorticity, rather than the velocities. It also includes dissipation, variation in Coriolis parameter with latitude, and vertical velocities, which can be set externally by Ekman pumping.

The text that follows in this section is a greatly truncated version of the full text found at the textbook Web site (Section S7.7), which includes numerous figures and examples. For a more thorough treatment, it is recommended that the full text be used.

7.7.1. Vorticity

Vorticity is twice the angular velocity at a point in a fluid. It is easiest to visualize by thinking of a small paddle wheel immersed in the fluid (Figure 7.12). If the fluid flow turns the paddle wheel, then it has vorticity. Vorticity is a vector, and points out of the plane in which the fluid turns. The *sign* of the vorticity is given by the "right-hand" rule. If you curl the fingers on your right hand in the direction of the turning paddle wheel and your thumb points upward, then the vorticity is positive. If your thumb points downward, the vorticity is negative.

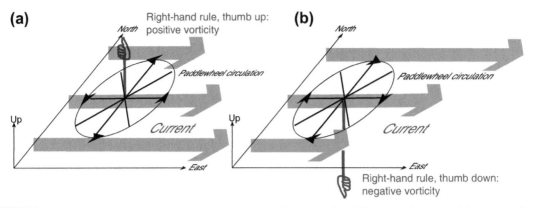

FIGURE 7.12   Vorticity. (a) Positive and (b) negative vorticity. The (right) hand shows the direction of the vorticity by the direction of the thumb (upward for positive, downward for negative).

The vorticity vector $\boldsymbol{\omega}$ is the curl of the velocity vector $\mathbf{v}$:

$$\boldsymbol{\omega} = \nabla \times \mathbf{v} = \mathbf{i}(\partial v/\partial z - \partial w/\partial y)$$
$$+ \mathbf{j}(\partial w/\partial x - \partial u/\partial z) + \mathbf{k}(\partial v/\partial x - \partial u/\partial y)$$
$$(7.32)$$

Vorticity, therefore, has units of inverse time, $(\text{sec})^{-1}$.

Fluids (and all objects) have vorticity simply because of Earth's rotation. This is called *planetary vorticity*. The vector planetary vorticity points upward, parallel to the rotation axis of Earth, which has an angular rotation rate of Ω:

$$\omega_{\text{planetary}} = 2\Omega \qquad (7.33)$$

where $\Omega = 2\pi/\text{day} = 2\pi/86160\ \text{sec} = 7.293 \times 10^{-5}\ \text{sec}^{-1}$, so $\omega_{\text{planetary}} = 1.4586 \times 10^{-4}\ \text{sec}^{-1}$.

The vorticity of the fluid motion relative to Earth's surface (Eq. 7.32) is called the *relative vorticity*. The total or absolute vorticity of a piece of fluid is the sum of the relative vorticity and planetary vorticity.

For large-scale oceanography, only the local vertical component of the total vorticity is used because the fluid layers are thin compared with Earth's radius, so flows are nearly horizontal. The local vertical component of the planetary vorticity is exactly equal to the Coriolis parameter f (Eq. 7.8c) and is therefore maximum and positive at the North Pole ($\varphi = 90°\text{N}$), maximum and negative at the South Pole ($\varphi = 90°\text{S}$), and 0 at the equator.

The local vertical component of the relative vorticity from Eq. (7.32) is

$$\varsigma = (\partial v/\partial x - \partial u/\partial y) = \text{curl}_z \mathbf{v} \qquad (7.34)$$

where $\mathbf{v}$ is the horizontal velocity vector. The local vertical component of the absolute vorticity is therefore $(\varsigma + f)$. The geostrophic velocities calculated from Eq. (7.23) (Section 7.6) are often used to calculate relative vorticity.

7.7.2. Potential Vorticity

Potential vorticity is a dynamically important quantity related to relative and planetary vorticity. Conservation of potential vorticity is one of the most important concepts in geophysical fluid dynamics, just as conservation of angular momentum is a central concept in solid body mechanics. Potential vorticity takes into account the height of a water column as well as its local spin (vorticity). If a column is shortened and flattened (preserving mass), then it must spin more slowly. On the other hand, if a column is stretched and thinned (preserving mass), it should spin more quickly similar to a spinning ice skater or diver who spreads his or her arms out and spins more slowly (due to conservation of angular momentum). Potential vorticity, when considering only the local vertical components, is

$$Q = (\zeta + f)/H \qquad (7.35)$$

where H is the depth, if the fluid is unstratified. When the fluid is stratified, the equivalent version of potential vorticity is

$$Q = -(\zeta + f)(1/\rho)(\partial\rho/\partial z). \qquad (7.36)$$

When there are no forces (other than gravity) on the fluid and no buoyancy sources that can change density, potential vorticity Q is conserved:

$$DQ/Dt = 0 \qquad (7.37)$$

where "D/Dt" is the substantial derivative (Eq. 7.4). Figures S7.24–S7.26 and text describing the trade-offs between the relative, planetary, and stretching vorticity are found on the textbook Web site. All that we note here is that f varies with latitude, with huge consequences for ocean currents and stratification. Therefore, a special symbol is introduced to denote the change in Coriolis parameter with northward distance y, or in terms of latitude φ and Earth's radius R_e:

$$\beta = df/dy = 2\Omega \cos\phi/R_e \qquad (7.38)$$

We often refer to the "β-effect" when talking about how changes in latitude affect currents, or the very large-scale, mainly horizontal Rossby waves for which the β-effect is the restoring force, described next.

7.7.3. Rossby Waves

The adjustment of any fluid to a change in forcing takes the form of waves that move out and leave behind a steady flow associated with the new forcing. We describe some general

properties of waves in Chapter 8. The large-scale, almost geostrophic circulation adjusts to changing winds and buoyancy forcing mainly through "planetary" or *Rossby waves* and *Kelvin waves* (Section 7.7.6). Pure Rossby (and Kelvin) waves are never found except in simplified models and lab experiments. However, much of the ocean's variability can be understood in terms of Rossby wave properties, particularly the tendency for westward propagation relative to the mean flow.

Most of the physical motivation for these waves, including illustrations (Figures S7.27–S7.29), are in the full online version located at the textbook Web site. Only the most basic facts are in the following list.

1. Rossby waves have wavelengths of tens to thousands of kilometers. Therefore particle motions in Rossby waves are almost completely transverse (horizontal, parallel to Earth's surface).
2. The restoring force for Rossby waves is the variation in Coriolis parameter f with latitude, so all dispersion information includes β (Eq. 7.38). As a water column is shoved off to a new latitude, its potential vorticity must be conserved (Eq. 7.37). Therefore, the water column height (long Rossby waves) or relative vorticity (short Rossby waves) begins to change. As with all waves, the column overshoots, and then has to be restored again, creating the wave.
3. All Rossby wave crests and troughs move only westward (relative to any mean flow, which could advect them to the east) in *both* the Northern and Southern Hemispheres; that is, the phase velocity is westward (plus a northward or southward component).
4. The group velocity of Rossby waves is westward for long wavelengths (more than about 50 km) and eastward for short wavelengths (even though the phase velocity is westward).

5. Velocities in Rossby waves are almost geostrophic. Therefore, they can be calculated from variations in pressure, for instance, as measured by a satellite altimeter, which observes the sea-surface height (e.g., Rossby wave-like behavior in Figures 14.18 and 14.19).

7.7.4. Rossby Deformation Radius and Rossby Wave Dispersion Relation

The length scale that separates long from short Rossby waves is called the *Rossby deformation radius*. It is the intrinsic horizontal length scale for geostrophic or nearly geostrophic flows, relative to which all length scales are compared. Again, see the full online version at http://booksite.academicpress.com/DPO/ Chapter S7 for more information.

The Rossby deformation radius in an unstratified ocean is

$$R_E = (gH)^{1/2}/f \qquad (7.39a)$$

where H is the ocean depth scale. R_E is called the *barotropic Rossby deformation radius* or "external" deformation radius. Barotropic deformation radii are on the order of thousands of kilometers.

The Rossby deformation radius associated with the ocean's stratification is

$$R_I = NH_s/f \qquad (7.39b)$$

where N is the Brunt-Väisälä frequency (7.15), and H_s is an intrinsic scale height for the flow. R_I is called the *baroclinic deformation radius* (or "internal" deformation radius). The vertical length scale H_s associated with the first baroclinic mode is about 1000 m, which is the typical pycnocline depth. R_I for the first baroclinic mode varies from more than 200 km in the tropics to around 10 km at high latitudes (Figure S7.28a) (Chelton et al., 1998).

The dispersion relation (Section 8.2) for first mode baroclinic Rossby waves is

$$\omega = \frac{-\beta k}{k^2 + l^2 + (1/R_I)^2} \qquad (7.40)$$

where ω is the wave frequency, k and l are the wavenumbers in the east-west (x) and north-south (y) directions, β is as in Eq. (7.38), and R_I is as given in Eq. (7.39b). Highest frequency (shortest period) occurs at the wavelength associated with the Rossby deformation radius (Figure S7.29). The shortest periods vary from less than 50 days in the tropics to more than 2 to 3 years at high latitudes (Figure S7.28b from Wunsch, 2009). Poleward of about 40 to 45 degrees latitude there is no first baroclinic mode at the annual cycle, so seasonal atmospheric forcing cannot force the first baroclinic mode at these higher latitudes.

7.7.5. Instability of Geostrophic Ocean Currents

Almost all water flows are unsteady. When gyre-scale flows break up, they do so into large eddies, on the order of tens to hundreds of kilometers in diameter or larger (see Section 14.5). The size of the eddies is often on the order of the Rossby deformation radius. The eddies usually move westward, like Rossby waves.

Instabilities of flows are often studied by considering a mean flow and then finding the small perturbations that can grow exponentially. This approach is called "linear stability theory"; it is linear because the perturbation is always assumed to be small relative to the mean flow, which hardly changes at all. When perturbations are allowed to grow to maturity, when they might be interacting with each other and affecting the mean flow, the study has become nonlinear.

We define three states: *stable, neutrally stable,* and *unstable*. A stable flow returns to its original state after it is perturbed. A neutrally stable flow remains as is. In an unstable flow, the perturbation grows.

The two sources of energy for instabilities are the *kinetic energy* and the *potential energy* of the mean flow. Recall from basic physics that kinetic energy is ½ mv² where m is mass and v is speed; for a fluid we replace the mass with density ρ, or just look at the quantity ½ v². Also recall from basic physics that potential energy comes from raising an object to a height; the work done in raising the object gives it its potential energy. In a stratified fluid like the ocean, there is no *available potential energy* if isopycnals are flat, which means that nothing has been moved and nothing can be released. For there to be usable or available potential energy, isopycnals must be tilted.

Barotropic instabilities feed on the kinetic energy in the horizontal shear of the flow. *Baroclinic instabilities* draw on the available potential energy of the flow. Baroclinic instability is peculiar to geostrophic flows, because Earth's rotation makes it possible to have a mean geostrophic flow with mean tilted isopycnals. On the other hand, barotropic instability is similar to instabilities of all sheared flows including those without Earth's rotation.

7.7.6. Kelvin Waves

Coastlines and the equator can support a special type of hybrid wave called a "Kelvin wave," which includes both gravity wave and Coriolis effects. Kelvin waves are "trapped" to the coastlines and trapped at the equator, which means that their amplitude is highest at the coast (or equator) and decays exponentially with offshore (or poleward) distance. Kelvin waves are of particular importance on eastern boundaries since they transfer information poleward from the equator. They are also central to how the equatorial ocean adjusts to changes in wind forcing, such as during an El Niño (Chapter 10).

Kelvin waves propagate with the coast to the right in the Northern Hemisphere and to the left in the Southern Hemisphere. At the equator, which acts like a boundary, Kelvin waves propagate only eastward. In their alongshore direction of propagation, Kelvin waves behave just like surface gravity waves and obey the gravity wave dispersion relation (Section 8.3). However, unlike surface gravity waves, Kelvin waves can propagate in only one direction. Kelvin wave wavelengths are also very long, on the order of tens to thousands of kilometers, compared with the usual surface gravity waves at the beach. Although the wave propagation speed is high, it can take days to weeks to see the transition from a Kelvin wave crest to a Kelvin wave trough at a given observation point.

In the across-shore direction, Kelvin waves differ entirely from surface gravity waves. Their amplitude is largest at the coast. The offshore decay scale is the Rossby deformation radius (Section 7.7.4).

Lastly, Kelvin wave water velocities in the direction perpendicular to the coast are exactly zero. The water velocities are therefore exactly parallel to the coast. Moreover, the alongshore velocities are geostrophic, so they are associated with pressure differences (pressure gradient force) in the across-shore direction.

7.8. WIND-DRIVEN CIRCULATION: SVERDRUP BALANCE AND WESTERN BOUNDARY CURRENTS

The large-scale circulation in the ocean basins is asymmetric, with swift, narrow currents along the western boundaries, and much gentler flow within the vast interior, away from the side boundaries. This asymmetry is known as *westward intensification* of the circulation; it occurs in both the Northern and Southern Hemispheres and in the subtropical and subpolar gyres. Sverdrup (1947) first explained the mid-ocean vorticity balance now called the "Sverdrup interior" solution. Stommel (1948) and Munk (1950) provided the first (frictional) explanations for the western boundary currents, and Fofonoff (1954) showed how very different the circulation would be without friction.

7.8.1. Sverdrup Balance

The gentle interior flow of the (non-equatorial) oceans can be described in terms of its meridional (north-south) direction. In the subtropical gyres, the interior flow is toward the equator in both the Northern and Southern Hemispheres. In the subpolar gyres, the interior flow is poleward in both hemispheres. These interior flow directions can be understood through a potential vorticity argument introduced by Sverdrup (1947), so we call the applicable physics the "Sverdrup balance."

Consider a schematic of the subtropical North Pacific (Figure 7.13). The winds at the sea surface are not spatially uniform (Figures 5.16a−c and S5.10a). South of about 30°N, the Pacific is dominated by easterly trade winds. North of this, it is dominated by the westerlies. This causes northward Ekman transport under the trade winds, and southward Ekman transport under the westerlies. As a result, there is Ekman convergence throughout the subtropical North Pacific (Figures 5.16d and S5.10a).

The convergent surface layer water in the subtropics must go somewhere so there is downward vertical velocity at the base of the (50 m thick) Ekman layer. At some level between the surface and ocean bottom, there is likely no vertical velocity. Therefore there is net "squashing" of the water columns in the subtropical region (also called *Ekman pumping*).

This squashing requires a decrease in either planetary or relative vorticity (Eq. 7.35). In the

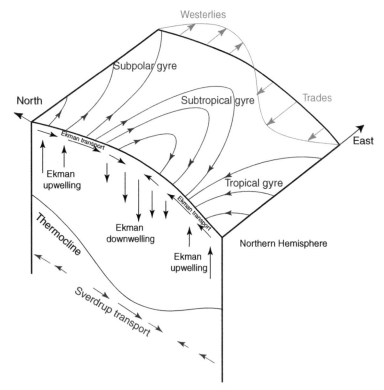

FIGURE 7.13 Sverdrup balance circulation (Northern Hemisphere). Westerly and trade winds force Ekman transport, creating Ekman pumping and suction and hence Sverdrup transport. See also Figure S7.12.

ocean interior, relative vorticity is small, so planetary vorticity must decrease, which results in the equatorward flow that characterizes the subtropical gyre (Figure S7.26).

The subpolar North Pacific lies north of the westerly wind maximum at about 40°N. Ekman transport is therefore southward, with a maximum at about 40°N and weaker at higher latitudes. Therefore there must be upwelling (*Ekman suction*) throughout the wide latitude band of the subpolar gyre. This upwelling stretches the water columns (Eq. 7.35), which then move poleward, creating the poleward flow of the subpolar gyre.

The *Sverdrup transport* is the net meridional transport diagnosed in both the subtropical and subpolar gyres, resulting from planetary vorticity changes that balance Ekman pumping or Ekman suction.

All of the meridional flow is returned in western boundary currents, for reasons described in the following sections. Therefore, subtropical gyres must be anticyclonic and subpolar gyres must be cyclonic.

Mathematically, the Sverdrup balance is derived from the geostrophic equations of motion with variable Coriolis parameter f (Eq. 7.23a,b). The x- and y-momentum equations are combined to form the vorticity equation:

$$f(\partial u/\partial x + \partial v/\partial y) + \beta v = 0 \qquad (7.41)$$

Using the continuity equation

$$\partial u/\partial x + \partial v/\partial y + \partial w/\partial z = 0 \qquad (7.42)$$

Eq. (7.41) becomes the potential vorticity balance

$$\beta v = f\,\partial w/\partial z. \qquad (7.43)$$

This important equation states that water column stretching in the presence of rotation is balanced by a change in latitude (Figure S7.26).

In Eq. (7.43), the vertical velocity w is due to Ekman pumping. From Eqs. (7.20) and (7.21):

$$w = \partial/\partial x(\tau^{(y)}/\rho f) - \partial/\partial y(\tau^{(x)}/\rho f) = \text{"curl }\tau\text{"} \qquad (7.44)$$

where τ is the vector wind stress, $\tau^{(x)}$ is the zonal wind stress, and $\tau^{(y)}$ is the meridional wind stress. Assuming that the vertical velocity w is zero at great depth, Eq. (7.43) can be vertically integrated to obtain the *Sverdrup balance*:

$$\beta(M^{(y)} - (\tau^{(x)}/f)) = \partial/\partial x(\tau^{(y)}) - \partial/\partial y(\tau^{(x)})$$
$$= \text{"curl }\tau\text{"}$$
$$(7.45)$$

where the meridional (south-north) mass transport $M^{(y)}$ is the vertical integral of the meridional velocity v times density ρ. The second term on the left side is the meridional Ekman transport. Thus, the meridional transport in the Sverdrup interior is proportional to the wind stress curl corrected for the Ekman transport.

The meridional transport $M^{(y)}$ is the *Sverdrup transport*. A global map of the Sverdrup transport integrated from the eastern to the western boundary is shown in Figure 5.17. The size of the integral at the western boundary gives the western boundary current transport since Sverdrup's model must be closed with a narrow boundary current that has at least one additional physical mechanism beyond those in the Sverdrup balance (a shift in latitude because of water column stretching driven by Ekman transport convergence). Physics of the boundary currents are discussed in the following sections.

7.8.2. Stommel's Solution: Westward Intensification and Western Boundary Currents

Because the Sverdrup balance applies to the whole ocean basin, the return flow must be in a narrow, swift meridional jet where the potential vorticity balance is different from the Sverdrup balance. Stommel (1948) included dissipation of potential vorticity Q on the right-hand side of Eq. (7.37), and showed that the returning flow must be along the western boundary (Figure S7.31). His potential vorticity balance is change in planetary vorticity balanced by bottom friction. Stommel's idealized circulation resembles the western-intensified Gulf Stream and Kuroshio subtropical gyres. Much more discussion is provided in Section S7.8.2 in the textbook Web site http://booksite.academicpress.com/DPO/.

7.8.3. Munk's Solution: Western Boundary Currents

Like Stommel, Munk (1950) also showed western intensification of the gyres, but used a more realistic type of dissipation that could work equally well with a stratified ocean. To the potential vorticity conservation (Eq. 7.37), Munk added friction between the currents and the side walls. A narrow, swift jet along the western boundary returns the Sverdrup interior flow to its original latitude (Figure S7.32).

We use Munk's model to understand why the returning jet must be on the western rather than the eastern boundary (Figure 7.14). In the Sverdrup interior of a subtropical gyre, Ekman pumping squashes the water columns, which then move equatorward to lower planetary vorticity. To return to a higher latitude (i.e., increase the planetary vorticity), higher vorticity must be put back into the fluid through either stretching or relative vorticity. Stretching through wind stress curl in very narrow regions does not occur where and when it would be needed. Therefore, there must be an input of relative vorticity. Relative vorticity in a narrow boundary current is high because the horizontal shear is high. At the side wall, the velocity parallel to the wall is zero; it increases offshore, so the boundary current has positive relative vorticity if it is on a western boundary (Figure 7.14a). This vorticity is injected into the fluid by the friction at the wall, which then allows the planetary vorticity to change and the fluid to return to its original latitude; such a circulation closure would not be possible if the boundary current were on the eastern boundary (Figure 7.14b). The online version of this text includes a more detailed explanation (see textbook Web site http://booksite.academicpress.com/DPO/).

7.8.4. Fofonoff's Solution: Large-Scale Inertial Currents

In addition to interior flows created from Ekman pumping or other external sources of vorticity, there are free, unforced modes of circulation, as shown by Fofonoff (1954), in a model with no wind input and no friction (Figure S7.34). This type of circulation is called an "inertial circulation". Without an external vorticity input, the interior flow is *exactly* zonal (east-west), because there is no way to change its planetary vorticity (due to the β-effect). Suppose there is westward flow across the middle of the ocean. When it reaches the western boundary, it gets back to the eastern boundary to feed back into the westward flow by moving either northward or southward along the western boundary in a very narrow current that can have as much relative vorticity as needed. Suppose it is northward. Then the relative vorticity of this frictionless current is positive, allowing it to move to a higher latitude. It then jets straight across the middle of the ocean, reaches the eastern boundary, and forms another narrow jet and moves southward, feeding into the westward flow in the interior.

Aspects of the Fofonoff inertial solution are found in highly energetic regions, such as near the separated Gulf Stream where its transport increases far above that predicted by the Sverdrup balance. The recirculation gyres associated with this energetic part of the Gulf Stream can be partially thought of in terms of Fofonoff gyres.

7.8.5. Wind-Driven Circulation in a Stratified Ocean

What happens to the wind-driven circulation theories in a stratified ocean? Water moves

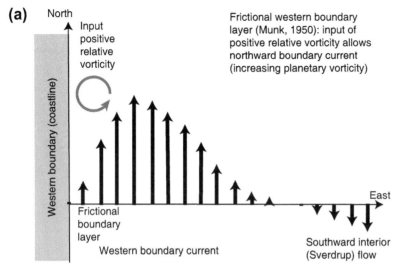

(a)

North

Input positive relative vorticity

Frictional western boundary layer (Munk, 1950): input of positive relative vorticity allows northward boundary current (increasing planetary vorticity)

East

Western boundary (coastline)

Frictional boundary layer

Western boundary current

Southward interior (Sverdrup) flow

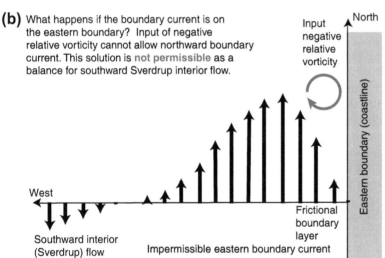

(b) What happens if the boundary current is on the eastern boundary? Input of negative relative vorticity cannot allow northward boundary current. This solution is not permissible as a balance for southward Sverdrup interior flow.

North

Input negative relative vorticity

Eastern boundary (coastline)

West

Southward interior (Sverdrup) flow

Frictional boundary layer

Impermissible eastern boundary current

FIGURE   7.14 (a)   Vorticity balance at a western boundary, with sidewall friction (Munk's model). (b) Hypothetical eastern boundary vorticity balance, showing that only western boundaries can input the positive relative vorticity required for the flow to move northward.

down into the ocean, mostly along very gradually sloping isopycnals. Where streamlines of flow are connected to the sea surface, we say the ocean is directly *ventilated* (Figure 7.15). Where there is Ekman pumping (negative wind stress curl), the Sverdrup interior flow is equatorward (Section 7.8.1). Water columns at the local mixed layer density move equatorward and encounter less dense water at the surface. They slide down into the subsurface along isopycnals, still moving equatorward. This process is called *subduction* (Luyten, Pedlosky, & Stommel, 1983), using a term borrowed from plate tectonics. The subducted waters then flow around the gyre and enter the western boundary current if they do not first enter the tropical circulation. The details are beyond the scope of this text.

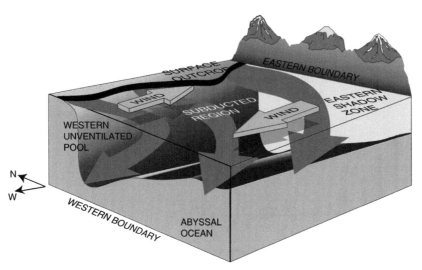

FIGURE 7.15 Subduction schematic (Northern Hemisphere). See Figure S7.35 for additional schematics, including obduction.

In each subducted layer, there can be three regions (Figure 7.15): (1) a *ventilated region* connected from the sea surface as just described, (2) a western unventilated pool with streamlines that enter and exit from the western boundary current without entering the surface layer, and (3) an eastern quiet (*shadow*) zone between the easternmost subducting streamline and the eastern boundary. A continuous range of surface densities is found in the subtropical gyre; the water column is directly ventilated over this full range, with waters at each density coming from a different seasurface location depending on the configuration of streamlines on that isopycnal. This is called the "*ventilated thermocline*"; in water mass terms, this process creates the Central Water. The maximum density of the ventilated thermocline is set by the maximum winter surface density in the subtropical gyre (Stommel, 1979).

The opposite of subduction is *obduction*, borrowed again from plate tectonics by Qiu and Huang (1995). In obducting regions, waters from subsurface isopycnals come up and into the surface layer. These are generally upwelling regions such as the cyclonic subpolar gyres and

the region within and south of the Antarctic Circumpolar Current.

Wind-driven circulation occurs in non-ventilated stratified regions as well. It is most vigorous in regions connected to the western boundary currents where water can enter and exit the western boundary. In these regions, the western boundary currents and their separated extensions usually reach to the ocean bottom. These dynamics are also beyond our scope.

7.9. WIND-DRIVEN CIRCULATION: EASTERN BOUNDARY CURRENTS AND EQUATORIAL CIRCULATION

7.9.1. Coastal Upwelling and Eastern Boundary Currents

The eastern boundary regions of the subtropical gyres have strong but shallow flow that is dynamically independent of the open ocean gyre regimes. Upper ocean eastern boundary circulation is driven by alongshore wind stress that creates onshore (or offshore) Ekman transport that creates upwelling (or downwelling; Section 7.5.4). Beneath or inshore of the equatorward eastern boundary currents there is a poleward undercurrent or counter-current. Coastal upwelling systems are not restricted to eastern boundaries; the southern coast of the Arabian peninsula has the same kind of system. These circulations are fundamentally different from western boundary currents, which are tied to potential vorticity dynamics (Section 7.8).

The classical explanation of eastern boundary currents is that equatorward winds force Ekman flow offshore, which drives a shallow upwelling (on the order of 200 m deep) in a very narrow region adjacent to the coast (on the order 10 km; Figure 7.6c). The upwelling speed is about 5–10 m/day. Because of stratification, the source of upwelled water is restricted to layers close to the sea surface, usually between 50 and 300 m.

The zone of coastal upwelling can be extended to more than 100 km offshore by an increase in longshore wind strength with distance offshore; this is observed in each eastern boundary upwelling system due to topographic steering of the winds by the ocean-land boundary. The offshore Ekman transport therefore increases with distance offshore, which requires upwelling through the whole band (Bakun & Nelson, 1991). The zone is identified by positive wind stress curl (Figure 5.16d).

Upwelled water is cooler than the original surface water. It originates from just below the euphotic zone and therefore is also rich in nutrients. Cool surface temperatures and enhanced biological productivity are clear in satellite images.

Upwelling is strongly seasonal, due to seasonality in the winds. Onset of upwelling can be within days of arrival of upwelling-favorable winds. In one example, off the coast of Oregon, the surface temperature dropped by 6°C in two days after a longshore wind started.

Coastal upwelling is accompanied by a rise in upper ocean isopycnals toward the coast (Figure 7.6). This creates an equatorward geostrophic surface flow, the *eastern boundary current*. These currents are narrow (<100 km width and near the coast), shallow (upper 100 m), strong (40 to 80 cm/sec), and strongly seasonal. The actual flow in an eastern boundary current system includes strong, meandering eddies and offshore jets/filaments of surface water, often associated with coastline features such as capes (Figure 10.6). Actual eastern boundary currents are some distance offshore, at the axis of the upwelling front created by the offshore Ekman transport.

Poleward undercurrents are observed at about 200 m depth beneath the equatorward surface currents. They are driven by a poleward pressure gradient force along the eastern boundary. When upwelling-favorable winds weaken or disappear, the equatorward flow also disappears and the poleward undercurrent extends up to the surface (there is no longer an undercurrent).

7.9.2. Near-Surface Equatorial Currents and Bjerknes Feedback

Circulation within about 2° latitude of the equator is very different from non-equatorial circulation because the Coriolis parameter f vanishes at the equator. Equatorial circulation is driven by easterly trade winds in the Pacific and Atlantic and by the seasonally reversing monsoonal winds in the Indian Ocean.

Since the Coriolis parameter vanishes and there is no frictional Ekman layer at the equator, the easterly trade winds drive equatorial surface flow due westward in a frictional surface layer ("normal panel" in Figure 10.27). The westward surface current is shallow (50 to 100 m) and of medium strength (10 to 20 cm/sec). In each of the three oceans, this westward surface flow is a part of the *South Equatorial Current*. The water piles up gently in the west (to about 0.5 m height) and leaves a depression in the east.

This creates an eastward pressure gradient force (from high pressure in the west to low pressure in the east). The pressure gradient force drives an eastward flow called the *Equatorial Undercurrent* (EUC). The EUC is centered at 100 to 200 m depth, just below the frictional surface layer. The EUC is only about 150 m thick. It is among the strongest ocean currents (>100 cm/sec). The pileup of waters in the western equatorial region also results in a deepened pycnocline there, called the *warm pool*, and a shoaling of the pycnocline in the eastern equatorial region (upwelling into the *cold tongue*). Coriolis effects set in a small distance from the equator, and off-equatorial Ekman transport enhances upwelling in the equatorial band. This enhances the cold tongue. The east-west contrast in temperature along the equator maintains the atmosphere's *Walker circulation*, which has ascending air over the warm pool and descending over the eastern colder area.

The Walker circulation is an important part of the trade winds that create the warm pool and cold tongue, so there can be a feedback between the ocean and atmosphere; this is called the *Bjerknes feedback* (Bjerknes, 1969) (Figure S7.36b located on the textbook Web site). If something weakens the trade winds, as happens at the beginning of an El Niño event (Chapter 10), the westward flow at the equator weakens and upwelling weakens or stops. Surface waters in the eastern regions therefore warm. Water in the deep warm pool in the west sloshes eastward along the equator, thinning the pool. The change in SST weakens the Walker circulation/ trade winds even more, which further exacerbates the ocean changes. This is an example of a *positive feedback*.

In the Indian Ocean, the prevailing equatorial winds are monsoonal, meaning that trade winds are only present for part of the year. This creates seasonally reversing equatorial currents and inhibits the formation of the warm pool/cold tongue structure. The Indian Ocean sea surface temperature is high at all longitudes.

7.10. BUOYANCY (THERMOHALINE) FORCING AND ABYSSAL CIRCULATION

Heating and cooling change the ocean's temperature distribution, while evaporation, precipitation, runoff, and ice formation change the ocean's salinity distribution (Chapters 4 and 5). Collectively, these are referred to as *buoyancy*, or *thermohaline*, forcing. Buoyancy processes are responsible for developing the ocean's stratification, including its abyssal properties, pycnocline, thermocline, halocline, and upper layer structure (other than in wind-stirred mixed layers).

Abyssal circulation refers to the general category of currents in the deep ocean. The *overturning circulation*, also called the *thermohaline circulation*, is the part of the circulation associated with buoyancy changes, and overlaps spatially with the wind-driven upper ocean circulation; it also includes shallow elements that are independent of the abyssal circulation. In the overturning circulation, cooling and/or salinification at the sea surface causes water to sink. This water must rise back to the warm surface, which requires diffusion of heat (buoyancy) downward from the sea surface. The source of eddy diffusion is primarily wind and tidal energy.

7.10.1. Buoyancy Loss Processes (Diapycnal Downwelling)

Water becomes denser through net cooling, net evaporation, and brine rejection during sea ice formation. We have already described brine rejection (Section 3.9); it is responsible for creating the densest bottom waters in the global ocean (Antarctic Bottom Water and parts of the Circumpolar Deep Water) and also in the regional basins where it is operative (Arctic Ocean, Japan Sea, etc.). Here we focus on *convection* created by net buoyancy loss in the open ocean, when surface water becomes denser than water below, and advects and mixes downward. Diurnal (daily) convection occurs at night in areas where the surface layer restratifies strongly during the day. During the annual cycle, cooling usually starts around the autumnal equinox and continues almost until the spring equinox. The resulting convection eats down into the surface layer, reaching maximum depth and density at the end of winter when the cumulative cooling reaches its maximum. A convective mixed layer can be hundreds of meters thick by the end of winter, whereas a wind-stirred mixed layer is limited to about 150 m by the depth of wind-driven turbulence.

Ocean convection is usually driven by surface cooling. Excess evaporation can also create convection, but the latent heat loss associated with evaporation is usually dominant. "Deep" convection is a loose term that usually refers to creation of a surface mixed layer that is thicker than about 1000 m. Deep convection has three phases: (1) preconditioning (reduction in stratification), (2) convection (violent mixing), and (3) sinking and spreading. (See Killworth, 1983 and Marshall & Schott, 1999.)

Convective regions have a typical structure (Figure S7.37). These include: (a) a *chimney*, which is a patch of 10 km to more than 100 km across within which preconditioning can allow convection and (b) convective *plumes* that are about 1 km or less across. The plumes are about the same size across as they are deep.

Deep convection occurs in only a very few special locations around the world: Greenland Sea, Labrador Sea, Mediterranean Sea, Weddell Sea, Ross Sea, and Japan (or East) Sea. These sites, with the exception of the isolated Japan Sea, ventilate most of the deep waters of the global ocean.

7.10.2. Diapycnal Upwelling (Buoyancy Gain)

The structure of the basin and global scale overturning circulations depends on both the amount of density increase in the convective source regions and the existence of a buoyancy (heat) source at lower latitudes that is at least as deep as the extent of the cooling (Sandström, 1908; Figure S7.38). Since there are no significant local deep heat sources in the world ocean, waters that fill the deep ocean can only return to the sea surface as a result of diapycnal eddy diffusion of buoyancy (heat and freshwater) downward from the sea surface (Sections 7.3.2 and 5.1.3).

Munk's (1966) diapycnal eddy diffusivity estimate of $\kappa_v = 1 \times 10^{-4}$ m^2/sec (Section 7.3) was based on the idea of isolated sources of deep water and widespread diffusive upwelling of this deep water back to the surface. From all of the terms in the temperature and salt equations (7.12) and (7.13), Munk assumed that most of the ocean is dominated by the balance

vertical advection = vertical diffusion (7.46a)

$$w \, \partial T/\partial z = \partial/\partial z(\kappa_v \partial T/\partial z) \qquad (7.46b)$$

Munk obtained his diffusivity estimate from an average temperature profile and an estimate of about 1 cm/day for the upwelling velocity w, which can be based on deep water formation rates and an assumption of upwelling over the whole ocean. The observed diapycnal eddy diffusivity in the open ocean away from boundaries is much smaller than Munk's estimate, which must be valid for the globally averaged ocean structure. This means that there must be much larger diffusivity in some regions of the ocean, now thought to be at the boundaries, at large seamount and island chains, and possibly the equator (Section 7.3).

7.10.3. Stommel and Arons' Solution: Abyssal Circulation and Deep Western Boundary Currents

Deep ocean circulation has been explained using potential vorticity concepts that are very familiar from Sverdrup balance (Section 7.8;

Stommel, 1958; Stommel & Arons, 1960a,b). The sources of deep water are very localized. The deep water fills the deep ocean layer, which would raise the upper interface of this layer if there were no downward eddy diffusion (Section 7.10.2). This upwelling stretches the deep ocean water columns. Stretching requires a poleward shift of the water columns to conserve potential vorticity (Eq. 7.36). The predicted interior flow is therefore counterintuitive — it runs toward the deepwater source! (Actual abyssal flow is strongly modified from this by the major topography that modifies the β-effect, by allowing stretched columns to move toward shallower depths rather than toward higher latitude.)

Deep Western Boundary Currents (DWBCs) connect the isolated deepwater sources and the interior poleward flows. Whereas unambiguous poleward flow is not observed in the deep ocean interior (possibly mostly because of topography), DWBCs are found where they are predicted to occur by the Stommel and Arons abyssal circulation theory (Figure 7.16; Warren, 1981). One such DWBC runs southward beneath the Gulf Stream, carrying dense waters from the Nordic Seas and Labrador Sea. Swallow and Worthington (1961) found this current after being convinced by Stommel to go search for it.

7.10.4. Thermohaline Oscillators: Stommel's Solution

An entirely different approach to understanding the meridional overturning circulation considers changes in overturn due to changes in dense water production by reducing the ocean to just a few boxes (Stommel, 1961). Such box models show how even the simplest model of climate change, for example, can lead to complex results. In this case, *multiple equilibria* result, that is, the system can jump suddenly between quite different equilibrium states. Stommel reduced the ocean to two connected boxes representing dense, cold, fresh high latitudes and light, warm, saltier low latitudes (Figure S7.41). In each box, the temperature and salinity are set by: (1) flux of water between the boxes (thermohaline circulation) that depends on the density difference between the boxes and (2) restoring temperature and salinity to a basic state over some set time period.

Stommel found that several different thermohaline circulation strengths exist for a given set

FIGURE 7.16 Global abyssal circulation model, assuming two deep water sources near Greenland and Antarctica (filled circles), filling a single abyssal layer. (These sources are actually at different densities.) *Source: From Stommel (1958).*

of choices of model parameters (externally imposed temperature and salinity, restoration timescales for temperature and salinity, and factor relating the flow rate to the density difference between the boxes). As the basic state was slowly changed, perhaps by reduction of the basic high-latitude salinity (which reduces its density), the flow rate slowly changed and then suddenly jumped to a different equilibrium rate. When the basic state salinity was then slowly increased, the system jumped back to a higher flow rate but at a very different basic salinity than during its decreasing phase. Thus this system exhibits *hysteresis*: it has different equilibrium states depending on whether the state is approached from a much higher salinity or a much lower salinity.

The coupled atmosphere-sea-ice-land-physics-biology-chemistry climate system is far more complex than the two simple boxes in this very simple Stommel oscillator model. Yet its multiple equilibria and hysteresis behavior have been useful in demonstrating the potential for abrupt and relatively large changes in climate and, more specifically, for interpretation of numerical models of the changes in overturning circulation that could result from changes in external forcing.

8

Gravity Waves, Tides, and Coastal Oceanography

8.1. INTRODUCTION

This chapter continues the dynamical discussion of Chapter 7, starting with an overview of the properties of waves (Section 8.2), and moving to surface and internal gravity waves and tides (Sections 8.3 to 8.6). These sections are truncated here, but appear in full in Chapter S8 of the online supplement located at the textbook Web site http://booksite.academicpress.com/DPO/; "S" denotes supplemental material.

This chapter then continues in supplementary form on the textbook Web site, covering several aspects of the coastal regime: coastal runoff, estuaries, and coral reefs (Sections S8.7 to S8.9). The supplement ends with descriptions of circulation and water properties in various adjacent seas of the Atlantic (Mediterranean, Black, Baltic and North Seas), Pacific (Bering, Okhotsk, Japan, Yellow, East and South China Seas, Gulf of California), and Indian Ocean (Red Sea and Persian Gulf) (Section S8.10).

Relevant advanced treatments of waves include Phillips (1977), Lighthill (1978), Pedlosky (2003) and Mei, Stiassnie, and Yue (2005). Some suggestions for coastal oceanography texts are Komar (1998), Van Dorn (1993), Open University (1999), Tomczak (2002), and Stewart (2008). Comprehensive reviews of coastal oceanography and adjacent seas are found in many volumes of *The Sea* (Brink & Robinson, 1998; Robinson & Brink, 1998, 2005, 2006; Bernard & Robinson, 2009).

8.2. GENERAL PROPERTIES OF WAVES

Waves are the displacement of parcels in a medium, such as water, that has a force that pushes the parcel back to its initial position, where it overshoots and is then restored back again, overshoots, and is then restored. For example, for *surface gravity waves* (Section 8.3), the medium is water (the air–sea interface) and the restoring force is gravity acting on parcels displaced vertically at the interface. All types of waves are generated by some external force that creates the initial displacement of particles away from their equilibrium position. For surface gravity waves, the most common external force is the wind, although undersea earthquakes can also generate them (tsunamis).

Waves are described in terms of their wavelength, period, amplitude, and direction (Figure 8.1). The *wavelength* (L) is the distance from one wave crest to the next or from one trough to the next. Another quantity used to describe the length of waves is the *wavenumber* (k), where $k = 2\pi/L$ and has units of radians

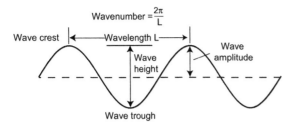

FIGURE 8.1 Schematic of a sinusoidal wave.

per unit length. The wave *period* (T) is the time between observing successive crests (or troughs) passing a fixed point. The wave *frequency* (ω) is $2\pi/T$ with units of radians per unit time. (Frequency given in hertz would be $1/T$.) The wave *amplitude* is conventionally one-half the height of the wave from crest to trough.

Two different types of velocity describe how all waves travel: phase velocity and group velocity. *Phase velocity* is the velocity of individual wave crests. The phase velocity c_p is

$$c_p = L/T = \omega/k \qquad (8.1)$$

where L is the wavelength, T is the wave period, ω is the frequency, and k is the wavenumber. Waves are *non-dispersive* if the phase velocity is a single constant for all wavelengths. If the phase velocity is not constant, waves are *dispersive* and separate from each other. The *dispersion relation* for a given type of wave (e.g., surface waves, internal waves, or acoustic waves) expresses the wave frequency in terms of the wavelength or wavenumber, that is, $\omega = \omega(k)$. For waves that move in several different directions, a wavenumber is defined for each direction. For the (x, y, z) directions, these wavenumbers are often called (k, l, m). The phase speed (Eq. 8.1) is defined in each of these directions.

Wave energy moves at a different speed from the wave crests for most types of waves. This speed is called the *group velocity*. In readily recognized examples (such as the waves in a boat wake), a wave group (packet) moves

out from the source, and individual waves propagate through the packet. The packet moves at the group velocity. In deep-water surface waves, the phase velocity is faster than the group velocity, so it looks like the waves just appear from one side of the packet, move through, and disappear out the other side. Formally, group velocity (c_g) is the derivative of frequency with respect to wavenumber. In one dimension, this is

$$c_g = \partial\omega/\partial k \qquad (8.2)$$

For two and three dimensions, the group velocity is a vector:

$$\mathbf{c}_g = (\partial\omega/\partial k, \partial\omega/\partial l, \partial\omega/\partial m) \qquad (8.3)$$

For non-dispersive waves, the group velocity must be the same constant as the phase velocity.

8.3. SURFACE GRAVITY WAVES

8.3.1. Definitions and Dispersion Relation

The restoring force for surface gravity waves is gravity, assisted by the large difference between the density of air and that of water, which acts against any disturbance of the free surface. Any external forcing that can momentarily mound up the water causes surface gravity waves: wind, a passing boat, slumping of the ocean bottom caused by an earthquake, and so forth.

Surface gravity waves with wavelengths that are much shorter than the water depth are referred to as *deep-water waves* or "short waves." Water particle motion in a deep-water wave is nearly circular in the vertical plane parallel to the direction of wave propagation. The diameter of the circles decreases exponentially with depth and the wave does not "feel" the bottom. At the other extreme are *shallow-water waves* or "long waves", whose wavelengths are greater than the water depth. The water particles move elliptically in the vertical plane rather than in circles.

The dispersion relation for an ideal (linear, sinusoidal), short (deep-water), surface gravity wave is

$$\omega = \sqrt{gk} \tag{8.4a}$$

Therefore the phase velocity (Eq. 8.1) of a short surface gravity wave is

$$c_p = \frac{\omega}{k} = \sqrt{\frac{g}{k}} \tag{8.4b}$$

and the group velocity (Eq. 8.2) is

$$c_g = \frac{\partial \omega}{\partial k} = \frac{1}{2}\sqrt{\frac{g}{k}} \tag{8.4c}$$

Therefore, short surface gravity waves (large k) move more slowly than longer surface gravity waves (smaller k), and energy propagates at a different, slower, speed than phase.

For shallow-water (long) gravity waves, in water of depth (d), the dispersion relation is

$$\omega = k\sqrt{gd} \tag{8.5a}$$

Their phase and group velocities are

$$c_p = \frac{\omega}{k} = \sqrt{gd} \tag{8.5b}$$

$$c_g = \frac{\partial \omega}{\partial k} = \sqrt{gd} \tag{8.5c}$$

When the group speed is a constant (and therefore equal to the phase speed), as in Eq. (8.5c), the waves are non-dispersive; that is, energy moves at the same speed for all wavelengths.

8.3.2. Wind-Forced Surface Gravity Waves

On an extremely calm day, the ocean surface appears glassy, with no visible short waves. As the wind starts to blow, small capillary waves form and the water surface begins to appear slightly rough. The wind-forced waves grow and change through differences in air pressure created by the wind between the front and backside of the waves. The pressure differences become larger as the surface gravity waves grow. Nonlinear interactions between the waves spread energy to longer wavelengths and lower frequencies.

The resulting wave state produced by local winds is called the *wind-sea*. These wind-forced surface gravity waves have periods and wavelengths that range from about 1 to 25 sec and about 1 to 1000 m. The amplitude and frequency/wavelength of waves generated locally by the wind depends on the wind *duration* (time over which the wind blows), *fetch* (distance over which the wind blows), and strength. In a storm, with wind gusts in many different directions, the rough sea surface becomes choppy with waves traveling in all directions (*confused sea*). *Whitecaps* appear when the wind strength exceeds about 10 knots (3 m/sec). A *fully developed sea* arises after the wind blows for many days with a very long fetch. The white caps and foam in the photograph in Figure 8.2a are characteristic of strong wind conditions, in this case, in the Gulf of Tehuantepec in the eastern Pacific (Section 10.7.6 and wind curl map in Figure 5.16d).

When the wind slows, the shorter waves are damped out, leaving behind longer, slower, smoother waves called *swell*. Swell wavelengths are tens of meters. Swell can propagate exceedingly long distances with little damping. Using the dispersion relation (8.4), swell with a period of 14 sec travels at a phase speed of 22 m/sec and group speed of 11 m/sec, taking approximately 5 days to propagate from the Gulf of Alaska to the north shore of Hawaii, a distance of about 4500 km. The swell arriving on distant beaches often is fairly narrow-band in frequency, so it consists of well-defined *sets* separated by relatively quiescent intervals.

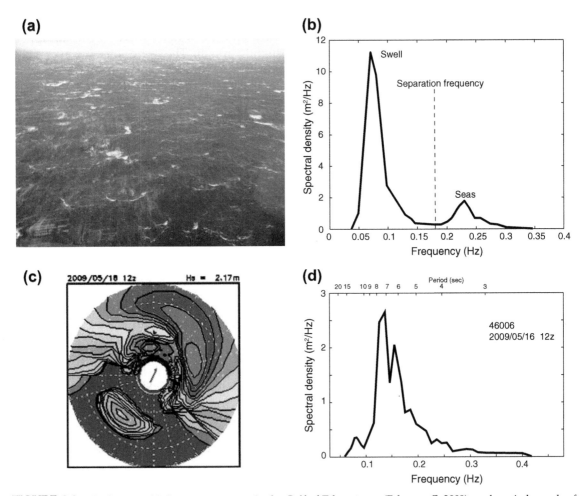

FIGURE 8.2 Wind waves. (a) Open ocean waves in the Gulf of Tehuantepec (February 7, 2009), under wind speeds of 20–25 m/sec, including actively breaking waves, old foam patches, and streaks of foam (K. Melville, personal communication, 2009). (b) Example of a surface wave spectrum (spectral density) in which ocean swell and wind-seas are well separated in frequency. *Source: From National Data Buoy Center (2006).* (c, d) Directional wave spectrum (spectral density) and spectrum, without clear separation between swell and the wind-sea, from the NE Pacific (station 46006, 40°53′ N 137°27′ W, May 16, 2009. In (c), wave periods are from about 25 sec at the center of the ring to 4 sec at the outer ring. Blue is low energy, purple is high. Direction of the waves is the same as direction relative to the center of the circle. Gray arrow in center indicates wind direction. "Hs" indicates significant wave height. Figure 8.2c can also be found in the color insert. *Source: part c is from NOAA Wavewatch III (2009) and part d is from National Data Buoy Center (2009).*

Since waves of many different frequencies and wavelengths are present in the open ocean at the same time, the surface gravity wave field is often described using spectral analysis (Section 6.5.3). The spectra can often have two separate peaks associated with the local wind-seas and with the swell (Figure 8.2b). However, many spectra show no clear separation between swell and wind seas (Figure 8.2d). Directional wave spectra that show energy as a function of frequency and the direction of the waves (Figure 8.2c) can clarify what is happening in the non-directional spectrum.

Description and forecast of the wave state in the open ocean is crucial for shipping. Wave observations from buoys and satellites are analyzed using global wave models (Figure 8.3). The *significant wave height* in the left panel of Figure 8.3 is the average height of the highest one-third of the waves. For the right panel, wave spectra at each location were used to determine the wave period with maximum (peak) energy; vectors show the direction of propagation of the peak energy waves.

8.3.3. Beaches, Breaking Waves and Associated Set-Up, and Near-Shore Currents

Surface gravity waves move from offshore generation regions to the near-shore region where they impact the beach and coastline. We distinguish between the *beach*, the *surf zone* (where waves break), and the *swash zone* (where water from the broken waves runs up the beach). Offshore sand bars and reefs are also important for how waves break and how currents are set up in the surf zone. Beaches exist in a delicate balance between variable

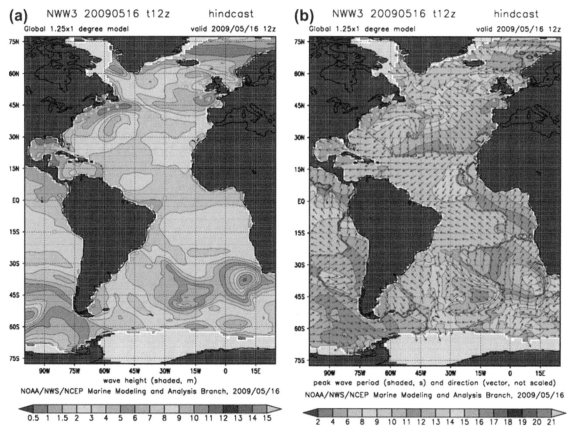

FIGURE 8.3 (a) Significant wave height (m) and (b) peak wave period (s) and direction (vectors) for one day (May 16, 2009). Figure 8.3a and 8.3b can also be found in the color insert. *Source: From NOAA Wavewatch III (2009).*

waves, tides, and near-shore currents. The waves and local currents often have strong seasonality with resulting variations in beach structure and composition (Yates, Guza, O'Reilly, & Seymour, 2009).

Based on their interaction with impinging waves, beaches can be classified as *dissipative* or *reflective*. Dissipative beaches remove much of the wave energy. Dissipation is enhanced by a mild bottom slope and rough material. Reflective beaches reflect much of the wave energy; these are more steeply sloped and/or composed of smooth material.

As a surface wave approaches the shore, it "feels" the bottom at a depth that depends upon its wavelength. The wave slows down, becoming shorter and steeper while retaining the same period. Its height increases to the point where it breaks. Observations suggest that waves typically break when $H/d = 0.8$, where H = wave height and d = depth to the bottom.

If the incident waves approach the beach at an angle, their direction changes to be more perpendicular as they shoal; this causes the wave crests to become more parallel to the shore. This is called *refraction*. Refraction occurs because the phase speed of the waves decreases as the depth decreases. The offshore part of the crest, in deeper water, moves more quickly toward shore and the whole crest pivots. (This is *Snell's law of refraction*.) Some of the energy of the incident waves also reflects from the shoaling bottom as they come ashore.

If the bottom depth has alongshore variation, as it does on almost every beach, then refraction and reflection of incoming waves vary alongshore. This can result in focusing of wave energy in some locations. This is illustrated in Figure 8.4a, which shows a large swell approaching the shore at La Jolla, California. A major underwater canyon is situated toward the front of the photograph, on the north side of the pier; this reflects much of the swell to the left, while some swell continues onshore (Thomson, Elgar, & Herbers, 2005).

Breaking waves are typically classified as: (a) *spilling breakers*, (b) *plunging breakers*, or (c) *surging breakers* (Figure 8.5). Spilling breakers occur on the mildest sloping beach, plunging breakers on a moderately sloping beach, and surging breakers on a steep beach where the wave reaches the beach before it has a chance

FIGURE 8.4 (a) Surf zone, looking toward the south at the Scripps Pier, La Jolla, CA. *Source: From CDIP (2009).* (b) Rip currents, complex pattern of swell, and alongshore flow near the head of a submarine canyon near La Jolla, CA, *Photo courtesy of Steve Elgar (2009).*

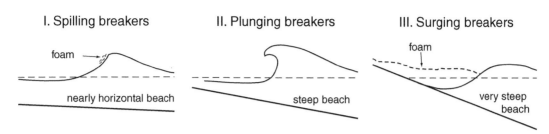

I. Spilling breakers   II. Plunging breakers   III. Surging breakers

foam

nearly horizontal beach   steep beach   very steep beach

FIGURE 8.5 Types of breaking waves: (I) spilling breaker, (II) plunging breaker, and (III) surging breaker. *Source: From Komar (1998).*

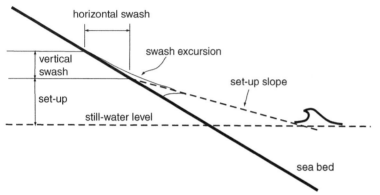

horizontal swash

swash excursion

vertical swash

set-up slope

set-up

still-water level

sea bed

FIGURE 8.6 Features of the surf zone. *Source: From Komar and Holman (1986).*

to break. The type of breaker also depends on wave steepness (ratio of wave height to wavelength). The greatest dissipation in the surf zone occurs for spilling breakers, and the greatest reflection of incident waves back to deep water occurs for surging breakers. A given surf zone may include a combination of these different types of breakers. Breaking wave heights are reported similarly to open ocean wave heights, in terms of significant breaker height (average height of the one-third highest breakers) and maximum breaker height.

Breaking waves transport momentum to the near shore region. This creates wave *set-up*, which is a rise in mean water level above the mean still water line (Figure 8.6). There is a complementary set-down as an incoming wave trough reaches the beach. Waves that are 3 m high offshore can produce a set-up of 50 cm (Guza & Thornton, 1982; Komar &

Holman, 1986). The total run-up on the beach is the sum of set-up, *swash* (landward flow of water) of individual larger waves, and swash due to longer period (>20 sec) surf beat. Swash on a reflective beach is more strongly affected by the incident waves. Swash on a dissipative beach is more affected by longer period *edge waves*. Edge waves are surface gravity waves with relatively long periods (>20 sec) that travel along, and are trapped to, the shoaling beach with amplitude decreasing offshore. They are forced by incident surface gravity waves.

The transport of mass onshore resulting from breaking waves must be compensated by offshore flow. There are two types of offshore flow: *undertow* and *rip currents*. Undertow balances the mass two-dimensionally: the onshore transport near the sea surface in the breaking wave zone is balanced by offshore transport in a layer at the bottom. Rip currents, on the other hand, return the mass back out to sea in horizontally limited jets (Figure 8.4b). Rip currents occur when alongshore flow is generated by alongshore variability in wave breaking. This leads to variability in set-up, which leads to an alongshore pressure gradient that drives alongshore flow. The location of rip currents can be controlled by bathymetry or shoreline shape, or can be transient, depending on instability of the alongshore current. Rip current intensity varies with incident wave amplitude, and is weak to absent in low wave conditions.

8.3.4. Storm Surge

Sea level is affected by local storm systems that drive water onshore. Storms have both very low atmospheric pressure and strong winds. The low pressure raises the sea surface locally within the storm. The winds create large waves, which can generate significant set-up at the coastline. The winds can also push water onshore. Both result in a rise in local sea level, called a *storm surge.*

The size of a storm surge depends on the strength of the storm and on the slope of the bottom. For gradually sloping shelves with shallow water far offshore, storm surges can be large, as in the North Sea. When the shelf depth increases quickly offshore, storm surge can be quite small, as on the west coast of North America, where storm surge is usually dwarfed by the tides. Many storm surges pass quickly and unremarkably, but when they coincide with maximum tidal height they can be disastrous; hurricane-force winds in conjunction with a high spring tide flooded low-lying areas of the North Sea in 1953.

Low-lying areas with tropical cyclones, hence extremely strong winds and low atmospheric pressure, are particularly susceptible to storm surges. In Bangladesh, storm surges in 1970 (Bhola cyclone) and 1991 (Bangladesh cyclone) reached 10 and 6 m, respectively, with enormous loss of life (World Meteorological Organization, 2005a,b). A storm surge of about 9 m in the Gulf of Mexico resulting from Hurricane Katrina (2005) created the most destructive natural disaster in U.S. history (Figure 8.7).

8.3.5. Tsunamis

Surface gravity waves can be forced by seismic shifts in submarine topography and other large, abrupt forcing events such as underwater landslides, meteorite impacts, and underwater volcanic eruptions. If there is a sudden submarine earthquake in which the bottom drops on one side of the fracture, the result is a displacement of seawater from top to bottom above the fracture of the same amplitude as the bottom shift (see Gonzalez, 1999). The sudden seawater displacement creates a surface gravity wave called a *tsunami*, which is the Japanese word for harbor wave.

Tsunami wavelengths are hundreds to thousands of kilometers. Since this is much greater than the ocean depth, the tsunami is a shallow water wave (Eq. 8.5). Therefore, the speed and time for a tsunami to propagate from one point to another in the ocean are set by the ocean depth. Frequencies are 10 minutes to about 2 hours (Mei et al., 2005). In the open ocean, where the depth is 4000 to 5000 m, the speed is 200 to 220 m/sec (17,280 km/day), so tsunamis take up to a day to cross a large ocean basin like the Pacific or Indian Ocean.

Tsunamis propagate with little decay across vast ocean expanses. Most of the energy is concentrated in the initial packet (Figure 8.7a and b). The initial arrival may be either a rise or a fall in sea level. The shape and separation of the peaks, as well as the dispersion, depend on the shape of the initial deformation due to the earthquake and on the bottom geometry.

All of the energy in an idealized tsunami in a flat-bottomed ocean is initially distributed around a circle centered on the earthquake. As the tsunami front moves out, the circle radius increases and the energy per unit length along

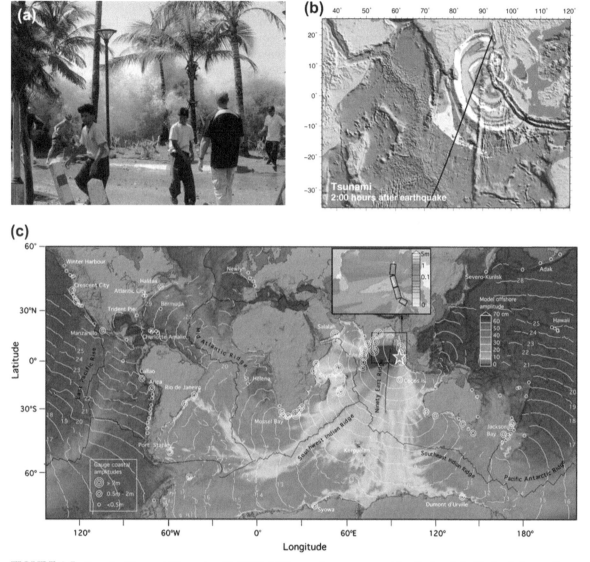

FIGURE 8.7 Sumatra Tsunami (December 26, 2004). (a) Tsunami wave approaching the beach in Thailand. *Source: From Rydevik (2004).* (b) Simulated surface height two hours after earthquake. *Source: From Smith et al. (2005).* (c) Global reach: simulated maximum sea-surface height and arrival time (hours after earthquake) of wave front. Figure 8.7c can also be found in the color insert. *Source: From Titov et al. (2005).*

the circumference of the circle decreases. The tsunami refracts and scatters as it crosses deep topographic features, resulting in less energy density in some regions and more in others. Mid-ocean ridges may act as waveguides for the tsunami waves (Figure 8.7c).

When a tsunami reaches the shoaling continental slope, its wave speed decreases and the wave refracts to approach the coastline in a more normal direction, just like any surface gravity wave. Part of its energy can reflect from the shelf and part can generate waves, which can be either coastally trapped or reflected. Because its wavelengths are very long, the wave steepness of a tsunami is small. Thus shoaling tsunamis behave as surging breakers, with little loss of energy through breaking until reaching the beach. Amplitudes of run-up from submarine earthquakes can reach 10 to 30 m. Because of their large run-up, tsunamis can flood large coastal regions in a short time (half the period of the wave, which is about half an hour or less).

Tsunami energy can be focused by mid-ocean features and also by the natural resonance of specific continental shelves and harbors. For instance, Crescent City, California, is particularly susceptible to large tsunamis due to a combination of the offshore Mendocino Fracture Zone that focuses tsunami energy as it crosses the ocean, and the natural resonances of the local continental shelf and the harbor (Horrillo, Knight, & Kowalik, 2008).

8.4. INTERNAL GRAVITY WAVES

This section is a brief introduction to the *internal gravity waves*, or *internal waves*, that ride the internal stratification in the ocean, and how they are affected by Earth's rotation. The ocean is stably stratified almost everywhere. Therefore a water parcel that is displaced, say, upward, encounters water of lower density and falls back downward and vice versa. This results in an oscillation, hence a wave. The restoring force is the buoyancy force, which is the product of gravity and the difference in density between the displaced water parcel and its neighbors at the same pressure. Internal gravity waves are similar in this respect to the surface gravity waves on the strong air–sea density interface. Because the stratification within the ocean is much weaker than that

between the air and water, the restoring action is weaker and the waves have much lower frequencies than surface waves of comparable wavelengths. For the same reason, water particles can travel large distances up and down in internal waves: amplitudes of tens of meters are common for internal waves.

Internal waves are mostly generated by tides, which interact with topography and generate internal tides (baroclinic tides), and by the wind, which stirs the mixed layer and generates internal waves with frequencies close to the inertial frequency (associated with Earth's rotation). Following Gill (1982), we take two approaches to considering internal waves: (a) waves on an interface between two layers of different density and (b) waves in a continuously stratified ocean. These two types of waves have quite different behaviors.

8.4.1. Interfacial Internal Gravity Waves

An interfacial internal wave is illustrated in Figure 8.8. This kind of internal wave is strikingly similar to a surface gravity wave. It propagates horizontally and involves heaving up and down of the sharp interface between the two layers, whose densities are ρ_1 and ρ_2. The principal modification from surface gravity

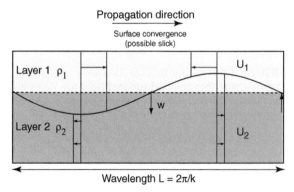

FIGURE 8.8 Schematic of a simple interfacial internal wave in a two-layer flow. *Source: After Gill (1982).*

waves is that the density difference, $\Delta\rho = \rho_1 - \rho_2$, between the two layers is much smaller than the density difference between air and water. The phase and group speeds of the interfacial internal wave are like those of shallow water surface waves (Eq. 8.5):

$$c_p^2 \approx g\frac{\Delta\rho}{\rho}H_1 \equiv gH_1 \qquad (8.6)$$

where H_1 is the mean thickness of the upper layer, ρ is the mean density and g' is called the "reduced gravity." It is assumed in Eq. (8.6) that the upper layer (1) is much shallower than the deeper layer (2); if they are of comparable depth, then the factor H_1 becomes a more complicated combination of both layer depths.

In Figure 8.8, the wave is propagating to the right. The water at the node in the center of the diagram (zero between the crest and the trough) is moving downward. The horizontal velocities are highest at the crests and troughs. There is a convergence at the node behind the crest; if the wave has very large amplitude, it can produce a surface slick (Figure 8.9c).

An example of internal waves that are nearly like interfacial waves is shown in Figure 8.9. The temperature fluctuations are due to internal waves heaving the thermocline up and down. The amplitude is up to 8 m, even in this very shallow water (15 m depth). Just to the north, surface slicks parallel to the coastline are often observed on calm days (Figure 8.9c); these are due to internal waves similar to those at the Mission Bay site. (Surface slicks are also caused by convergent surface flow in Langmuir circulations, Section 7.5.2, but these typically occur on windy days when surface wave activity is high.)

8.4.2. Internal Gravity Waves in a Continuously Stratified Ocean

Now consider waves within a continuously stratified ocean (or atmosphere), ignoring the upper and lower boundaries. Vertical stratification is the most important external ocean property for characterizing these waves. The Brunt-Väisälä (buoyancy) frequency, N, introduced in Section 3.5.6, is the maximum frequency for internal gravity waves. The maximum frequency is higher for higher stratification (higher N). The wave periods range from several minutes in the well-stratified upper ocean, to hours in the weakly stratified deep ocean. Waves at the Brunt-Väisälä frequency propagate entirely horizontally, with water particles moving exactly vertically with maximum exposure to the stratification (Figure 8.10).

Because internal waves can have periods on the order of hours, low frequency internal gravity waves are influenced by Earth's rotation (Eq. 7.8). The lowest frequency waves are pure inertial waves, whose frequency is equal to the Coriolis parameter, f. These have particle motions that are entirely in the horizontal plane, with no vertical component that can feel the vertical stratification. The full range of internal wave frequencies, ω, is

$$f \leq \omega \leq N \qquad (8.7)$$

Because f depends on latitude (0 at the equator and maximum at the poles), the allowable range of frequencies depends on latitude as well as on stratification.

The complete dispersion relation for internal waves in a continuously stratified flow is given here, without derivation, in terms of horizontal and vertical wavenumbers k, l, and m:

$$\omega^2 = \frac{(k^2 + l^2)N^2 + m^2 f^2}{k^2 + l^2 + m^2} \qquad (8.8)$$

This has been simplified by assuming that N has no variation and that f is constant (constant latitude). Even if more complicated stratification is included, Eq. (8.8) can still be a good approximation to the local behavior of the internal waves.

The internal wave frequencies from f to N are set entirely by the angle of the wave vector with

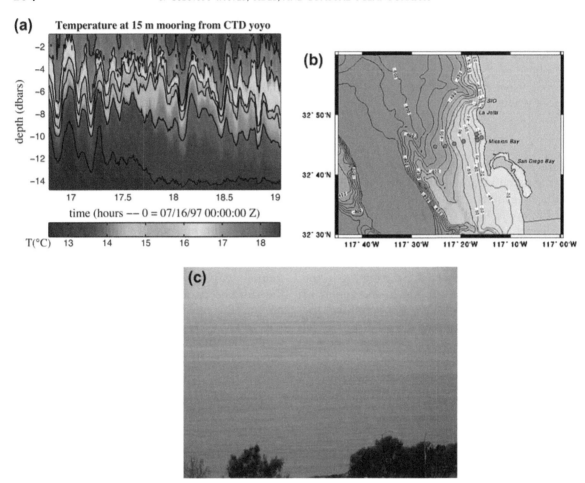

FIGURE 8.9 Internal wave observations. (a) Temperature as a function of time and depth on June 16, 1997 at location shown in (b) (Lerczak, personal communication, 2010). (b) Map of mooring location in water of 15 m depth west of Mission Bay, California. *Source: From Lerczak (2000).* (c) Ocean surface west of Scripps Institution of Oceanography (map in b) on a calm day; the bands are the surface expression of internal waves propagating toward shore. (Shaun Johnston, personal communication, 2010).

the vertical (θ in Figure 8.10). As the wave vector tilts from horizontal toward the vertical, the water particles feel less and less stratification, and the frequency decreases until finally reaching its lowest value, f. Manipulation of the dispersion relation (8.8) (not derived here) shows that the frequencies do not depend on the actual wavenumber, only on the angle of the wave vector with the horizontal. This differs

entirely from surface gravity waves and from interfacial waves (Section 8.4.1).

The group velocity (c_g) of internal gravity waves is exactly at right angles to the phase velocity (c_p). Thus the energy propagation direction, which is always given by the group velocity, is in the direction that the particles move. And finally, the group velocity for both the highest frequency (N) and lowest frequency

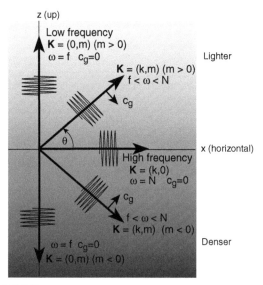

FIGURE 8.10 Schematic of properties of internal waves. The direction of phase propagation is given by the wavevector (k, m) (heavy arrows). The phase velocity (c_p) is in the direction of the wavevector. The group velocity (c_g) is exactly perpendicular to the wavevector (shorter, lighter arrows).

(f) internal waves is 0 in all directions (upward and horizontally).

At near-inertial frequencies (close to f), downward group velocity from the mixed layer is accompanied by *upward* phase velocity, and the particles move in clockwise ellipses that are almost circular. Because the Coriolis parameter, f, is 0 at the equator, internal waves of very low frequency can be found in the equatorial region, with periods of many days (10 days at 3 degrees latitude to infinite at the equator).

8.4.3. Internal Wave Generation and Observations

Internal waves within the water column (other than the interfacial waves described in Section 8.4.1) are primarily generated by winds that generate disturbances in the surface mixed layer and by the tides sloshing over bottom topography. Internal waves then propagate energy from the disturbances into the ocean interior (e.g., Polton, Smith, MacKinnon, & Tejada-Martinez, 2008). Nonlinear interactions between the internal waves generated at many different sites then spread the energy to internal waves at other frequencies.

Observed waves are usually analyzed by spectral analysis, including filters to remove frequencies that are not characteristic of internal waves (Section 6.5). Observed internal wave spectra are so similar from one place to another that it took several decades of work to begin to delineate variations in the spectrum due to local generation. The general form of the internal wave spectrum was introduced by Garrett and Munk (1972, 1975); their later modification is referred to as the Garrett-Munk 79 spectrum (Munk, 1981), and remains widely used. Much of what is now known about internal wave distributions and generation has arisen from understanding the reasons for the nearly universal (empirical) spectral shape and from describing differences from this shape.

The energetic tides, which have very specific frequencies dictated by the moon and sun orbits (Section 8.6), produce internal waves when they sweep over topography, if their frequency falls between f and N. This means that the propagation direction relative to the vertical of tidally generated internal waves can be precisely predicted because the direction is set, exactly, by the wave frequency. Tidally generated internal waves that propagate energy upward and outward have been observed from the Hawaiian ridge (Figure 8.11b).

Energy can pile up in internal waves, usually as the waves propagate toward shallow water near the coast, creating large localized disturbances called *solitary waves* or solitons. Internal solitons are associated with tides moving over banks or straits. Internal solitary waves have been observed in a number of locations; acoustic backscattering from wave-generated turbulence was used to produce the extraordinary images from the continental shelf off Oregon in Figure 8.11c (Moum et al., 2003).

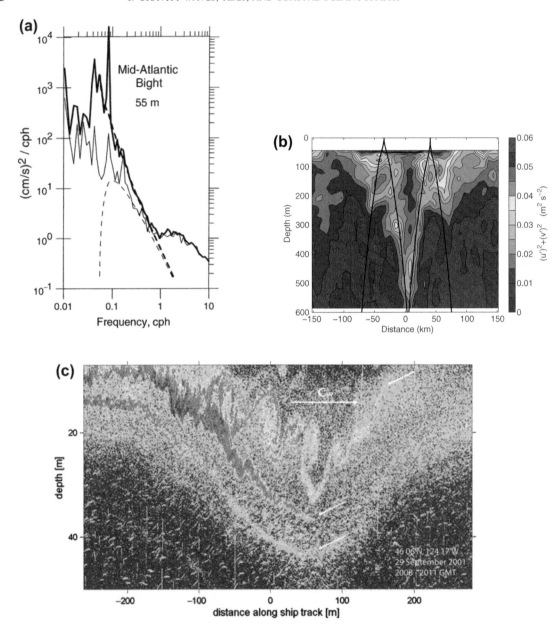

FIGURE 8.11 Internal wave observations. (a) Rotary spectra from a current meter at 55 m depth in the Mid-Atlantic bight: bold is clockwise and thin is counterclockwise; the dashed curve is the modified Garrett-Munk spectrum. *Source: From Levine (2002).* (b) Velocity variance (variability) observed along a section crossing the Hawaiian Ridge, which is located just below the bottom of the figure at 0 km; the black rays are the (group velocity) paths expected for an internal wave with frequency equal to the M_2 tide; distance (m) is from the center of the ridge. *Source: From Cole, Rudnick, Hodges, & Martin (2009).* This figure can also be found in the color insert. (c) Breaking internal solitary wave, over the continental shelf off Oregon. The image shows acoustic backscatter: reds indicate more scatter and are related to higher turbulence levels. Figure 8.11 can also be found in the color insert. ©American Meteorological Society. Reprinted with permission. *Source: From Moum et al. (2003).*

8.5. LARGE-SCALE CONTINENTAL SHELF AND COASTAL-TRAPPED WAVES

The physical boundary in the coastal ocean permits a particular class of large-scale surface gravity wave for which rotation is important. These *coastal-trapped waves* have their highest amplitude at or near the coast and decay away toward the open ocean (see review in Brink, 1991). They have large length scales (tens to hundreds of kilometers) and are *subinertial* (frequencies lower than the inertial frequency). The purest such coastal-trapped wave, for an ideal ocean with a flat bottom and vertical sides, is the *Kelvin wave* (Section 7.7.6).

The more general coastal-trapped (or topographic) waves behave like Kelvin waves, but are strongly modified by the side slope (continental slope), and are more like Rossby waves (Section 7.7.3). These *topographic Rossby waves* always propagate with shallow water to the right in the Northern Hemisphere (and to the left in the Southern Hemisphere). *Continental shelf waves* are similar to topographic Rossby waves, but solved with a bottom configuration that includes the continental shelf and slope and a flat deep ocean bottom offshore of the slope.

8.6. TIDES

The once or twice daily rise and fall of the tides and their long-term variations are the most predictable of all oceanographic phenomena. Water piles up against the coast during the *flood* tide, and falls away during the *ebb* tide. As the waves associated with internal tides in the deep ocean run into seamounts, ridges, or the ocean sides, they break and become turbulent, becoming the major source of dissipation in the deep ocean (Section 8.4). In this section we present only a brief introduction to tides. More complete pedagogical discussions of this important topic are available in many sources

such as Pugh (1987), Komar (1998), Open University (1999), Stewart (2008), and Garrison (2001). Hendershott's introductory lecture in Balmforth, Llewellyn-Smith, Hendershott, and Garrett (2005) is especially helpful.

8.6.1. The Equilibrium Tide

The moon and sun exert gravitational forces on Earth, including its thin shell of ocean. In 1687, Sir Isaac Newton published the expression for the gravitational attractive force between two bodies:

$$F = G \frac{mM}{r^2} \tag{8.9a}$$

Here F is the gravitational attractive force directed along the line separating the two bodies in Newtons; r is the distance between them in meters; m is the mass of one body (e.g., the moon); M is the mass of the other body (e.g., the Earth), and G is Newton's universal gravitational constant (6.67×10^{-11} Nm^2kg^{-2}).

The *equilibrium tide* is the shape that the ocean would take due to the gravitational attraction of the moon or sun on the water if Earth were a pure water-covered planet, with no continents and no topography. The tide-generating force on the ocean is the difference between (1) the gravitational attraction of the moon (or sun) at Earth's center of mass and (2) the gravitational attraction of the moon (and sun) on the ocean. As shown in Fig 8.12a, this is the difference between the force (F_C) between the moon and Earth centers, and the force between the moon and either the far side of Earth (F_A at the "antipodal point") or the near side of Earth (F_S at the "sublunar point"). (These statements can also refer to the sun.) In Figure 8.12a, the force differences are $T_A = F_A - F_C$ or $T_S = F_S - F_C$. We see right away, without even writing down the expressions for the forces, that T_A and T_S are the same size and pointed in opposite directions from each other. This results in a bulge of ocean toward the moon on the sublunar (near) side

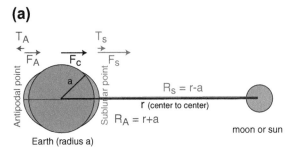

Equilibrium tide force balances

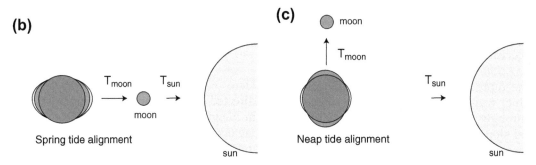

FIGURE 8.12 The equilibrium tide. (a) Tide-generating force due to the moon or sun. (b) Earth-moon-sun alignment during spring tide, which also includes the case when the moon is opposite the sun. *Source: After NOAA (2008).* (c) Alignment during neap tide. In (a), the F's are the net gravitational acceleration at the antipodal, center, and sublunar points, and the T's are the net tidal gravitational accelerations.

and a bulge away from the moon on the antipodal (far) side. As Earth rotates, there are therefore two bulges and hence two high tides per day.

Derivation of the shape of the equilibrium tide for every point on Earth is complicated (Komar, 1998; Open University, 1999; Stewart, 2008), beyond our scope, and is not included here. But we can derive the simpler expressions for the maximum tidal amplitudes at the sublunar and antipodal points. Writing down the expressions for the forces using Eq. (8.9a), the lunar gravitational acceleration (force per unit mass) at a point in the ocean that lies a distance R from the center of mass of the moon is toward the moon and of size Gm/R^2 where m is the mass of the moon. Meanwhile, at the center of Earth, the acceleration of the center of mass of Earth (in its orbit about the center of mass of the Earth-moon system) is

toward the moon and of size Gm/r^2 where r is the distance from Earth's center to the moon's center. The distance of the moon to the sublunar point is $R = R_s = r-a$, where a is Earth's radius. The tidal acceleration of a fluid parcel at the sublunar point is toward the moon:

$$T_S = \frac{Gm}{R_S{}^2} - \frac{Gm}{r^2} \sim \frac{2Gma}{r^3} \qquad (8.9b)$$

(A Taylor series expansion assuming a $\ll$ r yields this approximate result.) The distance of the moon to the antipodal point is $R = R_A = r + a$. The tidal acceleration of a fluid parcel at the antipodal point is

$$T_A = \frac{Gm}{R_A{}^2} - \frac{Gm}{r^2} \sim \frac{-2Gma}{r^3} \qquad (8.9c)$$

which is directed away from the moon. These accelerations are illustrated in Figure 8.12a.

Thus, the tide-generating force at Earth's surface has a component *toward* the moon on the side of Earth facing the moon, and also *away* from the moon on the other side of Earth. This is simply because the force at the ocean's surface on the side facing the moon is greater than at the center of Earth, while the force on the ocean on the side opposite the moon is less than at the center of Earth.[1]

Moreover, the tide-generating force decreases as the inverse third power of the distance to the tide-generating body, even though the Newtonian attraction decreases like the inverse square of the distance (Eq. 8.9a). This is because the differences taken in Eq. (8.9b, c) are between two large terms that nearly cancel.

Earth, and observers fixed to it, rotate under the equilibrium tidal potential so that an Earth-bound observer sees a high equilibrium tide when the tide-generating body is at its highest elevation above the horizon and another of equal magnitude when the tide-generating body is at its lowest elevation below the horizon. When the sun is the tide-generating body, the two maximum equilibrium tides are 12 hours apart. However, when the moon is the tide-generating body, the interval between them is about 12 hours, 25 minutes, because the moon orbits Earth in the same direction as Earth's rotation. Therefore, high and low tides due to the moon occur slightly less often than twice per day, and the time of high and low tide shifts with each day. In wave language, these hypothetical tides have frequencies of two cycles per solar and lunar day, respectively, and so are called *semidiurnal*.

When the tide-generating body is in Earth's equatorial plane, then the two high tides are of equal size. However, the sun is in Earth's equatorial plane only twice per year and the moon is in this plane only twice per month. As a result, the size of the two high tides each day at a given point on Earth differs. This is called the *daily inequality* (also called diurnal inequality). The solar daily inequality is greatest twice per year, at the solstices, when the sun is at its greatest distance from Earth's equatorial plane; the solar daily inequality vanishes at the intervening equinoxes. The lunar daily inequality varies similarly over the tropical month, which is defined by successive northward passages of the moon across Earth's equatorial plane. In wave language, the occurrence of a daily inequality may be viewed as the constructive and destructive interference of a semidiurnal tide (two cycles per day) with a diurnal tide (one cycle per day). Because the solar daily inequality vanishes twice per year, there are two solar diurnal tidal components that interfere destructively twice per year.

The lunar equilibrium tide amplitude is about 20 cm, which is much smaller than the actual tides observed along many coasts and harbors; the difference is due to the influence of coastal boundaries (Section 8.6.2). The sun is much farther away from Earth than the moon, so even though it has much greater mass than the moon, the solar tidal forcing is only about half that of the lunar tidal forcing. (However, the solar tide response for given locations can be larger than the lunar tide response.) When

---

[1] An equivalent derivation of the equilibrium tide-generating force is in terms of the centrifugal force associated with the rotation of Earth around the center of mass of the moon-Earth system (the *barycenter*, which is located about 4670 km from Earth's center hence inside the Earth). The gravitational acceleration between the moon and Earth centers, F_C, is balanced by the centrifugal acceleration of this rotation around the barycenter ($F_{cf} = -F_C$). The centrifugal acceleration around the barycenter is the same at every point on Earth because Earth is a rigid body (e.g. M. Hendershott's Lecture 1 in Balmforth et al., 2005). In an Earth-centered coordinate system, the tide-generating acceleration is then the sum of this invariant centrifugal acceleration and the gravitational acceleration of the ocean towards the moon, which depends on nearness to the moon. That is, $T_S = F_{cf} + F_S = -F_C + F_S$ and $T_A = F_{cf} + F_A = -F_C + F_A$, which are identical to the expressions given above.

the Earth, moon, and sun are aligned (Figure 8.12b), and also when the moon is exactly opposite the sun, the lunar and solar tides reinforce each other, producing very large high tides. (This alignment is called syzygy.) These are called *spring tides* (two per month). When the moon is perpendicular to the Earth-sun axis, the lunar and solar tides do not reinforce each other, and the two periods of smallest high tides of the month occur; these are called *neap tides*. This aspect of the semi-monthly variation in tidal amplitude is sometimes called the fortnightly tide (one fortnight equals two weeks).

The orbit of the moon around Earth is elliptical rather than circular. Therefore, once a month the moon is closest to Earth (*perigee*) and the lunar tidal range is highest. Once a month, the moon is farthest from Earth (*apogee*) and the lunar tidal range is smallest. Similarly, Earth's orbit around the sun is elliptical. When the sun is closest to Earth (perihelion, which occurs around January 2), the spring tide is largest (perigean spring tide). When the sun is farthest from Earth (aphelion, which occurs around July 2), the tidal range is reduced.

The plane of Earth's orbit around the sun is called the *ecliptic*. Earth's equatorial plane is tilted about $23°26'$ to the ecliptic. The plane of the moon's orbit is tilted about 5 degrees to the ecliptic; this tilt is referred to as the moon's *declination*. Thus the maximum declination of the moon is about $28° 26'$ and its minimum declination is about $18° 26'$. The moon's orbit precesses with a period of 18.6 years; during this period, the moon's declination shifts from its minimum to its maximum. Therefore the size of the lunar daily inequality varies with a period of 18.6 years.

8.6.2. Dynamic Tides

Because the Earth, sun, and moon motions are well known, the tide-generating force is very precisely known. Given the regularity and predictability of the forcing, why do tides at any given location on the coastline differ from those farther along the coast, why are actual tides at coastlines sometimes much larger than the equilibrium tide, and why are some locations dominated by semidiurnal tides while others are dominated by diurnal tides? The continents block the free propagation of the equilibrium tide westward as the Earth turns. The result is a complex pattern of tides that move around each of the ocean basins. Depending on how each basin responds to each particular frequency in the tide-generating force, the tide that results at any given location is unique, being a function of the lunar and solar tidal forcing and the basin and coastline geometry. The frequency of each component is determined astronomically. The relative amplitudes of the components depend on location.

The primary tidal frequencies are semidiurnal (twice a day due mainly to the lunar tide) and diurnal (once a day). In some locations there is almost no semidiurnal component, while in other locations there may be almost no diurnal. The tide is usually expressed in terms of tidal constituents. The principal constituents, in order of amplitude, are M_2 (lunar semidiurnal), K_1 (luni-solar declinational, diurnal), S_2 (solar semi-diurnal), O_1 (lunar diurnal, accounting for the moon's declination), N_2 (lunar elliptical, semidiurnal), P_1 (solar diurnal, accounting for the sun's declination), K_2 (luni-solar declinational, semidiurnal), and a number of other semidiurnal, diurnal, fortnightly, and longer period frequencies. Tables of the constituents, their equilibrium tide amplitude, and their periods are given in numerous textbooks (e.g., Defant, 1961; Komar, 1998; Stewart, 2008). Computer software is readily available for prediction of tides based on the tidal constituents and an observed tidal record. National agencies provide such predictions as a service (e.g., NOAA CO-OPS, 2010).

The spring and neap tides are illustrated in Figure 8.13 using a two-month record at

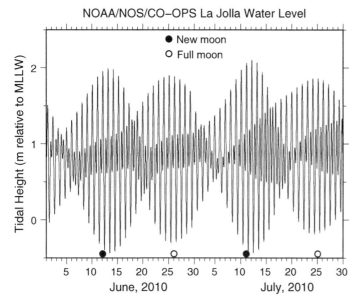

NOAA/NOS/CO–OPS La Jolla Water Level

- ● New moon
- ○ Full moon

FIGURE 8.13 Tides at La Jolla, California. Data from NOAA CO-OPS (2010).

Los Angeles. The small circles show the times of the full and new moons, which coincide with the spring (highest) tides. The neap tides are the lulls in between. The mixture of semidiurnal and diurnal tides produces the two separate envelopes of tides – the lower high tides near the center and the higher high tides that show that spring-neap cycles. Plots from other locations can look quite different depending on the relative amplitudes of the diurnal and semidiurnal components.

A global map of the M_2 tide is shown in Figure 8.14. The curves in Figure 8.14a are *cotidal lines*, which indicate the time of passage of high tide (measured in terms of phase from 0 to 360°). Where the cotidal lines intersect, the amplitude is zero (Figure 8.14b). These special points are called *amphidromes*.

In addition to possibly large changes once or twice daily in the volume of water at the coast, tides can also promote vertical mixing and break down the stratification of the water. Water moving in and out over a subsurface bottom feature is referred to as "tidal flushing." Georges Bank provides a nice example of tidal mixing effects. The M_2 tide impinging on the bank creates clockwise circulation, especially a jet along its northern side that can reach 100 cm/sec (Chen & Beardsley, 2002 and Figure 8.15a). A "tidal mixing front" appears at the edge of Georges Bank when the water column is stratified, separating well-mixed water over the shallow bank from stratified water offshore (Figure 8.15b). Tidal mixing moves colder, nutrient-rich water onto the bank from greater depths off the bank. The result is very high productivity over Georges Bank, as observed by satellite surface color images indicating high chlorophyll content (Figure 8.15c).

In ice-forming regions at high latitudes, there is often a subsurface temperature maximum beneath the much fresher, colder (freezing) surface layer. Tidal mixing over banks in such places can create polynyas (open water) by mixing the subsurface warm water to the surface where it can melt the sea ice (Section 3.9; Figures 3.12, 10.29, and 12.23).

In marginal seas, gulfs, and estuaries, the source of tidal forcing differs from the open

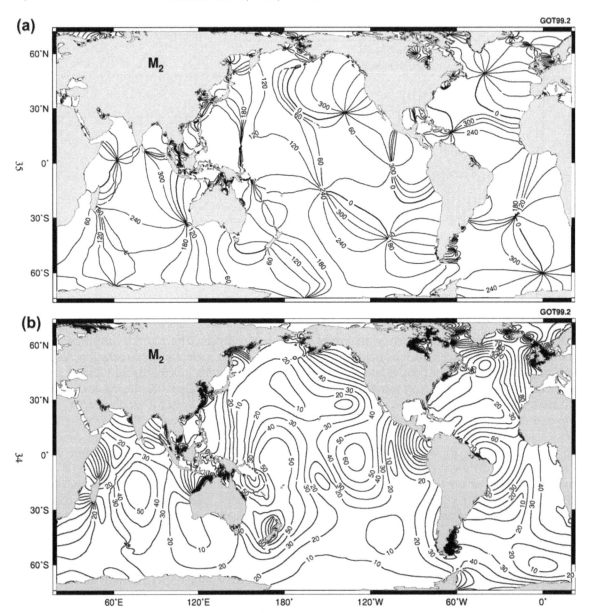

FIGURE 8.14 Maps of (a) cotidal (phase) lines (°) and (b) tidal amplitude (cm) for the M_2 tide (lunar semidiurnal). *Source: From Ray (1999).*

ocean. Tidal currents from the open ocean impinge on the coastal region, forcing motions in the coastal ocean and estuaries. These are called *co-oscillation tides* (Bowden, 1983). Co-

oscillation tides can have non-trivial amplitudes if there is a resonance between the open ocean tide and the natural frequency of the basin. If there is a resonance, the highest amplitude of

(a)

(c)

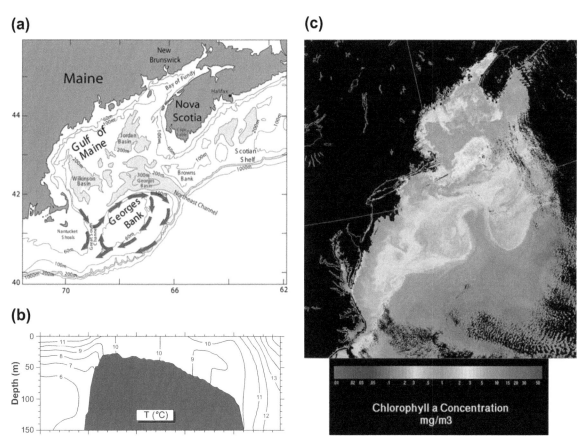

(b)

FIGURE 8.15   Tidal effects on Georges Bank. (a) Schematic circulation and (b) summer temperature (°C) structure. TMF = Tidal Mixing Front. SBF = Shelf Break Zone. *Source: From Hu et al. (2008).* (c) Chlorophyll a concentration (mg/m^3) on October 8, 1997, from the SeaWiFS satellite. Figure 8.15c can also be found in the color insert. *Source: From Sosik (2003).*

the tide is at the head of the estuary or gulf. The Gulf of Maine and Bay of Fundy (location in Figure 8.15a) have a maximum tidal amplitude of 15 m; this is a strong co-oscillation tide, in resonance with the M_2 tide at a period of 13.3 hours (Garrett, 1972).

8.7. WATER PROPERTIES IN COASTAL REGIONS: RIVER RUNOFF

In this brief section, which appears only in the online supplementary materials for the

textbook (Section S8.7), we discuss and illustrate the impact of river runoff on coastal conditions, and the importance of river runoff in global freshwater budgets for the ocean.

8.8. ESTUARIES

An estuary, in the strictest definition, is formed at the mouth of a river, where the river meets the sea (Dyer, 1997). The defining characteristic of estuarine circulation is that inflow is denser than outflow, which is diluted relative to the inflow. Estuaries are classified in terms of their shape and their stratification. Discussion of estuarine stratification, circulation, and flushing times appears only in the online supplementary materials for the textbook in Section S8.8.

8.9. CORAL REEFS

The physical oceanography of coral reefs was of particular interest to George Pickard, the original author of this text. He published several papers and a book on the Great Barrier Reef in 1977 (Pickard, Donguy, Henin, & Rougerie, 1977). We retain this section of the coastal oceanography chapter, but it has not been updated from the 5th edition. The material appears only in the online supplementary materials for the textbook in Section S8.9.

8.10. ADJACENT SEAS

This is a lengthy description of the circulation and water properties in a number of adjacent (marginal) seas of the Atlantic Ocean (Mediterranean, Black, Baltic, and North Seas), the Pacific Ocean (South China, East China, Yellow, and Japan or East Seas; Okhotsk, and Bering Seas), and the Indian Ocean (Red Sea and Persian Gulf). This material appears only in the online supplementary materials for the textbook in Section S8.10.

9

Atlantic Ocean

9.1. INTRODUCTION AND OVERVIEW

The Atlantic Ocean is a long, narrow ocean basin bisected by the Mid-Atlantic Ridge (MAR) (Figure 2.9). Wind-driven gyres and the wind-driven tropical circulation dominate transports in the upper ocean (Figure 9.1). The gyres and their western boundary currents include the anticyclonic subtropical gyres of the North Atlantic (Gulf Stream and North Atlantic Current) and South Atlantic (Brazil Current), and the cyclonic subpolar gyre of the northern North Atlantic (East Greenland Current and Labrador Current). The subtropical gyres include eastern boundary current upwelling systems: the Canary Current system in the North Atlantic and Benguela Current System (BCS) in the South Atlantic. The tropical circulation is predominantly zonal (east-west), including the North Equatorial Countercurrent and the South Equatorial Current, and has a low-latitude western boundary current (North Brazil Current; NBC).

Conversion of upper ocean waters to denser intermediate and deep waters (*meridional overturning circulation* or *thermohaline circulation*) in the northern North Atlantic is associated with a deep circulation, including Deep Western Boundary Currents (DWBCs; Section 7.10.3). Most of the final conversion from the surface

to the deeper layers occurs within the Labrador Sea and Nordic Seas (Chapter 12). This conversion also affects the Atlantic's upper ocean circulation: it increases the northward transport in the North Atlantic's Gulf Stream and North Atlantic Current by approximately 10% and provides a connection of tropical and subtropical waters to the subpolar North Atlantic. This overturning circulation results in net northward heat transport through all latitudes of the Atlantic, as it draws warm, saline surface waters north and sends dense, cold, fresher waters south at depth. In the South Atlantic, this reverses the usual poleward direction of subtropical heat transport found in all other subtropical regions (Section 5.6).

In the south, the Atlantic connects with the other oceans through the Southern Ocean (Chapter 13). As it enters the South Atlantic from the Drake Passage, the Subantarctic Front (SAF) of the Antarctic Circumpolar Current (ACC) makes an important northward excursion along the coast of South America as the Malvinas (or Falkland) Current and then loops partially back southward to begin a long, slow, southward drift as it moves eastward to the Indian Ocean and beyond to the Pacific. Warm surface water from the Indian Ocean enters the South Atlantic where the Agulhas Current rounds the southern tip of Africa. Most of the Agulhas retroflects back to the Indian Ocean

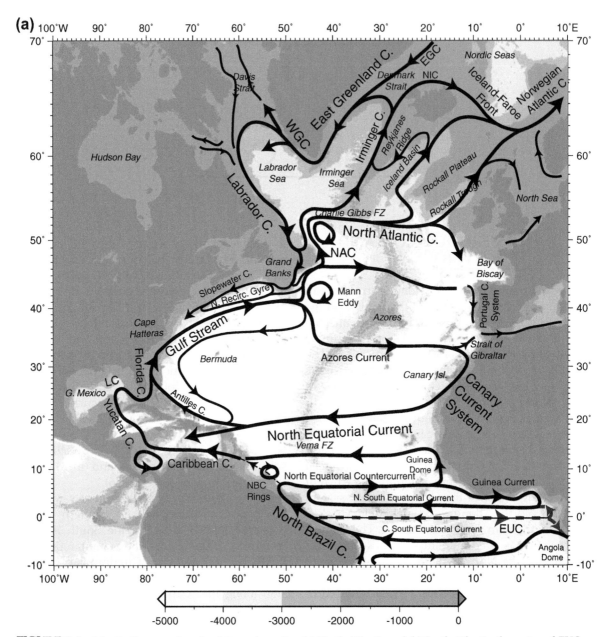

FIGURE 9.1 Atlantic Ocean surface circulation schematics. (a) North Atlantic and (b) South Atlantic; the eastward EUC along the equator just below the surface layer is also shown (gray dashed).

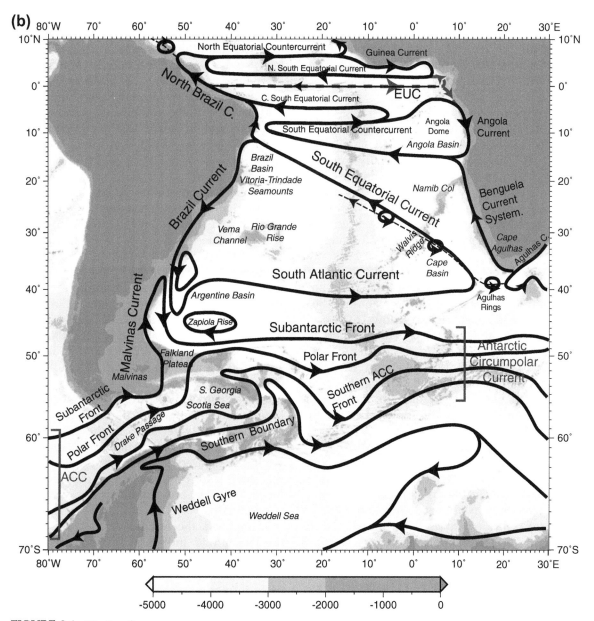

FIGURE 9.1 (*Continued*).

(Chapter 11), but the process sheds large eddies of Indian Ocean water that move northwestward into the Atlantic. A small portion of the Agulhas waters also enters the South Atlantic's Benguela Current. Dense bottom waters from the Antarctic enter the Atlantic from the Weddell Sea.

In the north, the Atlantic connects with the Nordic Seas and Arctic Ocean (Chapter 12), which are separated topographically from the North Atlantic by the ridge running from Greenland to Iceland and then from Iceland to the Faroe and Shetland Islands. Northward flow from the Atlantic to the Nordic Seas feeds into the Norwegian Atlantic Current along the coast of Norway. Southward flow back into the Atlantic occurs in the fresh surface layer in the East Greenland Current (EGC) and also through Davis Strait into the Labrador Sea, and as dense subsurface overflows over each of the three channels in the Greenland-Shetland ridge. These overflows form the dense deep waters of the North Atlantic and the deep part of the Atlantic's branch of the global overturning circulation. (The other branch is associated with dense water production in the Antarctic.)

The marginal seas of the North Atlantic also include important sites for water mass mixing and conversion. The subtropical western boundary current flows through the Intra-American Seas (Caribbean Sea and Gulf of Mexico) before emerging back into the North Atlantic. The Mediterranean Sea has a series of nearly separated sub-basins (each with characteristic water mass formation and circulation) and its own marginal sea, the Black Sea. Net evaporation in the Mediterranean contributes about one-third of the observed salinity difference between the Atlantic and Pacific Oceans. The Mediterranean's dense water re-enters the North Atlantic from the Strait of Gibraltar. In the northwest, the Labrador Sea (which is more of a large embayment than a marginal sea) is the site of intermediate water formation

that contributes to the meridional overturning circulation. Baffin Bay, to its north, connects the Atlantic to the Arctic Ocean west of Greenland, and has its own internal water mass formation process. In the northeast are the shallow intra-European shelf seas, the North Sea, and the Baltic Sea. These marginal seas are described in detail in Chapter S8 of the online supplement located at http://booksite. academicpress.com/DPO/; S denotes supplemental material.

Water masses of the upper ocean in the Atlantic are similar to those found in the wind-driven gyres of the other oceans, including those associated with thermocline ventilation (Central Waters and Subtropical Underwater; STUW), and those associated with the strong currents (Subtropical Mode Water; STMW). The North Atlantic Current of the subpolar North Atlantic has its own mode water — the Subpolar Mode Water (SPMW).

The northern North Atlantic and its adjacent seas produce new deep water (North Atlantic Deep Water; NADW) for the global ocean. The local, convective sources are the Labrador, Mediterranean, and Nordic Seas. Because of the local deep water sources, the deep waters of the northern North Atlantic are relatively young, measured in decades, in contrast to the deep and bottom waters of the North Pacific, which are hundreds of years old (Chapter 10).

In this text, the Atlantic is presented before the Pacific and Indian Oceans because it has been historically central for development of ideas about the general circulation and water mass formation. From a pedagogical point of view, it might be more advantageous to present the North Pacific circulation first (Chapter 10), since the wind-driven subtropical and subpolar circulations dominate in the North Pacific, while the North Atlantic's upper ocean circulation also includes a significant inflow to the deep overturn of the meridional overturning circulation (MOC). In both oceans the wind-driven subtropical circulation transports 30 to more than 140 Sv (depending on location). Whereas 15 to 20 Sv of water weave through the upper North Atlantic's circulation as part of the MOC to form NADW, less than 2 Sv traverse the North Pacific, ultimately forming the analogous North Pacific Intermediate Water (Chapters 10 and 14).

Climate variability in the Atlantic is vigorous (Section 9.9 and online supplementary Chapter S15). Much of the quasi-decadal variability in the North Atlantic is associated with the North Atlantic Oscillation (NAO), which is linked to the Arctic Oscillation. Effects of the NAO at the sea surface include variability in the westerly winds, air–sea buoyancy fluxes, and surface ocean properties. The NAO affects the subtropical and subpolar circulations and water mass formation rates and properties. Longer term variation in Atlantic Ocean circulation at centennial to millennial scales, called the Atlantic Multidecadal Oscillation (AMO), has been described, and is important in understanding possible anthropogenic climate change in the Atlantic Ocean. Tropical climate modes intrinsic to the Atlantic Ocean are also observed, and are separate from the El Niño-Southern Oscillation (ENSO) of the Pacific Ocean, whose effects also intrude into the Atlantic sector. In the south, the Southern Annular Mode (SAM) has a major center of action in the Weddell Sea and South Atlantic sector of the Southern Ocean.

9.2. FORCING

The long-term mean external forcing for the Atlantic's general circulation is described in this section. Seasonal effects are mostly not covered. Some of the interannual to decadal variations in forcing are discussed in the online supplementary chapter on climate variability (Chapter S15).

9.2.1. Wind Forcing

Wind stress drives ocean circulation via frictional Ekman transport in the surface layer.[1] The surface layer's convergences and divergences then drive interior ocean circulation (Chapter 7). The annual mean and seasonal mean wind stress is shown in Figure 5.16 (global) with supplementary Atlantic-only maps available online (Figure S9.3). The east-west (zonal) part of the wind field includes mainly westerly winds north of 30°N and south of 30°S, and easterly trade winds in the region between. Meridional components notably include equatorward winds along the eastern boundaries: along northern Africa from the Strait of Gibraltar to about 10°N, and along southern Africa up to about 10°S. These large-scale, longshore winds force the Canary and Benguela Current Systems.

The Ekman transport divergence (upwelling) and convergence (downwelling) are represented by the wind stress curl of Figures 5.16d and S9.3a through Eqs. (7.21) and (7.44). Downwelling regions fill the subtropics and upwelling regions occur in the northern North Atlantic and in the Southern Ocean south of about 50°S.

The Sverdrup transport (Section 7.8) is shown in Figures 5.17 and S9.3b. The subtropical gyres are the regions of equatorward interior flow closed by poleward western boundary currents. Based on the location of southward Sverdrup transport, the North Atlantic's subtropical gyre extends northward to 50°N to 52°N, the latitude of the UK. The northward western boundary currents for this gyre include the Gulf Stream System and the North Atlantic Current east of Newfoundland. The South Atlantic's subtropical gyre extends well south of Africa. The southward western boundary current for this circulation is the Brazil Current.

The maximum Sverdrup transport predicted for the Gulf Stream from these National Centers

---

[1] Except at the equator, where the frictional layer transport is directly downwind.

for Environmental Prediction (NCEP) mean winds is about 20 Sv. The NCEP winds are known to be too weak (Taylor, 2000); the Sverdrup transport is likely to be more like 30–50 Sv. This is much less than the maximum Gulf Stream transport of more than 140 Sv, arising from its recirculation gyres. In the South Atlantic, the Brazil Current Sverdrup transport at 30°S (where Africa forms an eastern boundary) is a comparable 25 Sv. Just south of the tip of Africa (Cape Agulhas), the Sverdrup transport at the South American coast jumps to more than 85 Sv because the west-east integration includes the full width of the Indian Ocean, from the coast of Australia/Tasmania westward to South America. The actual circulation does not include an interior zonal jet at 35°S across the South Atlantic; instead, the Agulhas, which would feed such a jet, retroflects (turns abruptly eastward) and creates eddies that propagate into the South Atlantic. The observed Brazil Current transport does jump to higher values south of 34°S, but this appears to be associated with local recirculation, as in the Gulf Stream (Section 9.5).

9.2.2. Buoyancy Forcing

Buoyancy forcing is the sum of heat and freshwater air–sea fluxes (Figures 5.4a, 5.12, 5.15 and online supplementary Figure S9.4). The Atlantic Ocean has two of the largest annual mean heat/buoyancy loss regions on the globe: in the Gulf Stream where it separates from the North American coast at 35–38°N ($>200 \text{ W/m}^2$) and in the Nordic Seas ($>100 \text{ W/m}^2$). Both are associated with poleward transport of warm water that is cooled by the atmosphere, including large latent (evaporative) heat loss. Similarly, in the South Atlantic, the Brazil Current and Agulhas retroflection are regions of heat loss ($>100 \text{ W/m}^2$). Net heat gain occurs in the tropics, with the highest gain (greater than 100 W/m^2) along the equator. Heat is also gained in narrow ribbons along the coasts, associated with the upwelling systems.

Net evaporation minus precipitation (E−P) minus runoff for the Atlantic shows the typical large subtropical net evaporation regions centered at 10–20° latitude on both sides of the equator, flanking the tropical net precipitation region associated with the Intertropical Convergence Zone (ITZC). Net precipitation/runoff is found in the subpolar North Atlantic, especially around the continental margins and in the adjacent seas (as runoff). E−P for the Atlantic is tipped toward net evaporation compared with the Pacific Ocean, so its mean salinity is higher than in the Pacific. The higher overall salinity of the Atlantic is due to larger evaporation throughout the subtropics.

Air−sea buoyancy flux is dominated by heat flux with a smaller contribution from freshwater flux (online supplementary Figures S5.8 and S9.4). Net evaporation in the subtropics enlarges the subtropical buoyancy loss regions to cover the full gyre in both the North and South Atlantic. Freshwater input from the Amazon, Congo, and Orinoco Rivers is greater than 0.4 Sv, on the order of the largest components of the global freshwater budget (Dai & Trenberth, 2002; Talley, 2008). Freshwater input in subpolar coastal regions is also evident (Newfoundland region, British Isles). Buoyancy loss, even within the Mediterranean Sea where evaporation greatly increases salinity, is nevertheless controlled mainly by heat loss. (Evaporation is accompanied by latent heat loss from the ocean.)

9.3. NORTH ATLANTIC CIRCULATION

The surface circulation of the North Atlantic (Figures 9.1 and 9.2a; Figure S9.1 and Tables S9.1 and S9.2 in the online supplement) includes an anticyclonic *subtropical gyre* and a cyclonic *subpolar gyre* that stretches northward into the Nordic Seas. Basics of the surface circulation have been well known since the nineteenth century (e.g., review in Peterson, Stramma, &

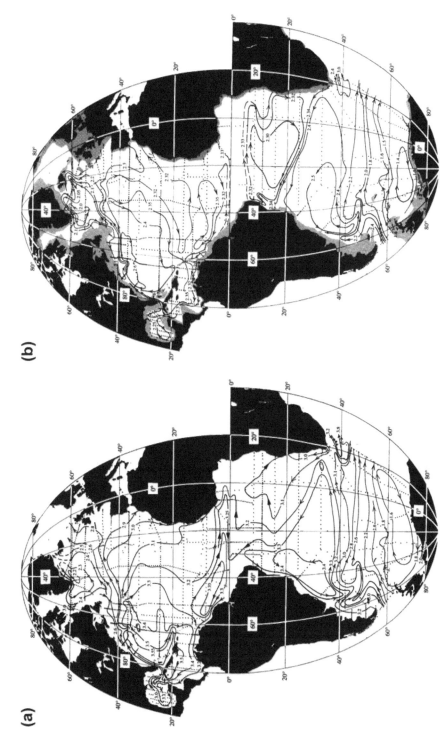

FIGURE 9.2 Steric height (10 m² s⁻²) at (a) 0 dbar and (b) 500 dbar, adjusted to estimate the absolute geostrophic circulation. *Source: From Reid (1994).*

Kortum, 1996). By the mid-twentieth century, volume transports had been estimated (Sverdrup, Johnson, & Fleming, 1942), and the modern picture of the surface circulation began to emerge, with depiction of intense, narrow western boundary currents and recirculations (e.g., Iselin, 1936; Defant, 1961; Dietrich, 1963).

The North Atlantic's *subtropical gyre*, like all subtropical gyres, is asymmetric, with strong, narrow western boundary currents and broad southward flow throughout the central and eastern subtropics. The subtropical western boundary current is composed of two connected portions: the *Gulf Stream System* south of about 40°N, and part of the *North Atlantic Current System* east of Newfoundland and north of 40°N. The eastern boundary upwelling system is called the *Canary and Portugal Current System*. The westward flow on the equatorward side of the gyre is the *North Equatorial Current*.

The cyclonic *subpolar gyre* is less asymmetric and more strongly controlled by topography.[2] It has swift, narrow western boundary currents along Greenland and Labrador (EGC and Labrador Current) that are connected by the *West Greenland Current* (WGC), which is on an eastern boundary. The North Atlantic Current (NAC) is the eastward flow on the southern side of the subpolar region; branches of the NAC flow northeastward toward the Nordic Seas. At the sea surface, the cyclonic subpolar gyre encompasses both the subpolar North Atlantic and the Nordic Seas (Chapter 12). Southward return flow from the Nordic Seas occurs in the EGC.

The subtropical and subpolar surface circulations are connected through the NAC, with net northward transfer of upper ocean water required by the MOC.

With increasing depth, the anticyclonic subtropical gyre shrinks westward and northward toward the Gulf Stream System. The cyclonic subpolar gyre becomes closed south of the Greenland-Faroe ridges. At depths below about 1500 m, the "abyssal" circulation becomes evident, with emergence of a DWBC that carries the newly formed intermediate and deep waters from the subpolar North Atlantic southward toward the equator (Section 9.6). Below the depth of the MAR, the circulation is confined to various abyssal basins, but on average transports the northern North Atlantic waters southward and the bottom waters from the Southern Ocean northward.

9.3.1. Subtropical Circulation

We start our detailed description of circulation with the subtropical gyre (Figure 9.1 and Figure S9.1 and Table S9.1, which are found in the online supplement). The subtropical western boundary current system consists of both the Gulf Stream System (Section 9.3.2) and the more northern NAC (Section 9.3.4). The Canary and Portugal Current Systems are the eastern boundary current system (Section 9.3.3).

9.3.2. Gulf Stream System

The Gulf Stream System consists of multiple segments with different names depending on location (and author). Nomenclature, therefore, can be confusing (Stommel, 1965).[3] We will follow Stommel's definition, in which the *Florida Current* refers to the western boundary current through the constriction between Florida and the Bahamas, and *Gulf Stream* refers to the continuation of this boundary current north of

---

[2] Topographic control is a greater factor in the subpolar region than in the subtropics, due to deep penetration of the currents resulting from weaker vertical stratification and a larger Coriolis parameter.

[3] "I often use the term Gulf Stream in a more general sense than that proposed by Iselin; and I do not speak of the Florida Current as extending to Cape Hatteras, but restrict the use of this term to mean the current actually within the Florida Straits. Unfortunately, the naming of things is more a matter of common usage than of good sense" (Stommel, 1965).

the Florida Straits and after it separates from the western boundary at Cape Hatteras and flows eastward out to sea. The phrase *Gulf Stream Extension* may also be used to describe the separated current, especially east of the New England Seamounts.

The subtropical Gulf Stream System begins where the North Equatorial Current, joined by the northward low latitude western boundary current, enters the Caribbean Sea through the complex of the Antilles islands (Figures 9.1 and 9.3 and Figure S9.5 in the online supplementary material). The maximum sill depth for currents entering the Caribbean Sea is 1815 m at Anegada Passage (Fratantoni, Zantopp, Johns, & Miller, 1997), reflected in nearly uniform properties below sill depth (see Figure 9.7). The exit sill depth through the Straits of Florida, described in the following text, is much shallower at 640 m, and limits the maximum density of waters that can flow completely through the Intra-American

Seas. (Denser waters can flow northward east of the Antilles.) Within the Caribbean, the upper ocean circulation consists of the westward *Caribbean Current* and a local wind-driven cyclonic circulation in the Colombia Basin.

Net transport into the Caribbean is estimated at 28.4 Sv (Johns, Townsend, Fratantoni, & Wilson, 2002). Below sill depth, there is a vigorous cyclonic circulation of about 15 Sv that simply moves the deep waters around in this isolated deep basin (Joyce, Hernandez-Guerra, & Smethie, 2001). The Caribbean Current forms into a western boundary current along the Honduran coast, called the *Cayman Current*, and then exits northward into the Gulf of Mexico through the Yucatan Channel as the *Yucatan Current*. Moored observations of the Yucatan Current from 1999–2001 showed a mean transport of 23 Sv and maximum surface velocity in excess of 130 cm/sec, sometimes reaching 300 cm/sec (Candela et al., 2003;

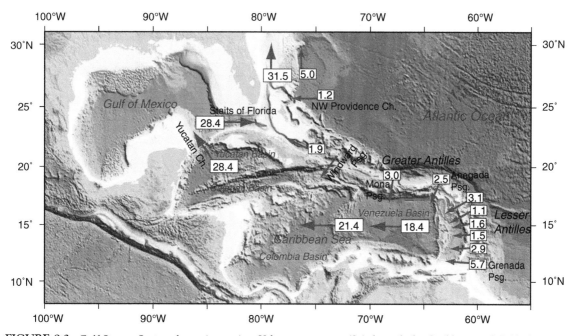

FIGURE 9.3 Gulf Stream System formation region. Volume transports (Sv) through the Caribbean and Gulf of Mexico. *After Johns et al. (2002).*

Cetina et al., 2006) (online supplementary Figure S9.6). The velocity structure is typical of strong currents restricted to a narrow channel, with a central core of flow and weak, flanking countercurrents (opposite direction).

After entering the Gulf of Mexico, the western boundary current, now named the *Loop Current*, flows northward to the middle of the Gulf and turns east toward the Straits of Florida. Loops, characterized by high sea surface temperature (SST), frequently pinch off, forming anticyclonic eddies that propagate westward, often ending their existence on the shelf of the eastern Texas coast (Figure 9.4a).

From the Gulf of Mexico, the western boundary current escapes into the North Atlantic. It turns northward along the coast of Florida and forms the *Florida Current* and the

Gulf Stream. A small part of the Gulf Stream originates in the *Antilles Current*, which is a highly variable, weak western boundary current in the open ocean east of the Antilles, Puerto Rico, Cuba, and the Bahamas (Rowe et al., 2010).

The Florida Current/Gulf Stream is a narrow, intense, northward flow. The Florida Current is well monitored in the confined strait between Florida and the Bahamas (Figure 9.5a and online supplementary Figures S9.7 and S9.8). Maximum surface velocities exceed 180 cm/sec, concentrated in a 20 km band in the western part of the channel over the continental slope. The mean transport at 27°N is 32 Sv with seasonal and interannual variability each of the order ±2 to 3 Sv; maximum seasonal transport occurs in summer (Baringer & Larsen, 2001).

(a)

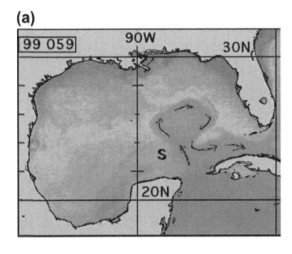

(b)

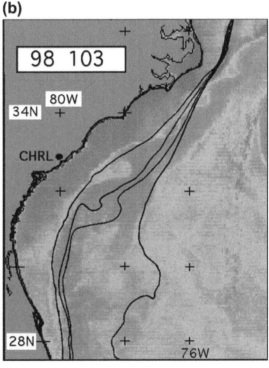

FIGURE 9.4 Sea surface temperature from the GOES satellite. (a) Gulf of Mexico showing the Loop Current beginning to form an eddy. (b) Gulf Stream, showing meander at the Charleston Bump and downstream shingling. Black contours are isobaths (100, 500, 700, 1000 m). This figure can also be found in the color insert. *Source: From Legeckis, Brown and Chang (2002).*

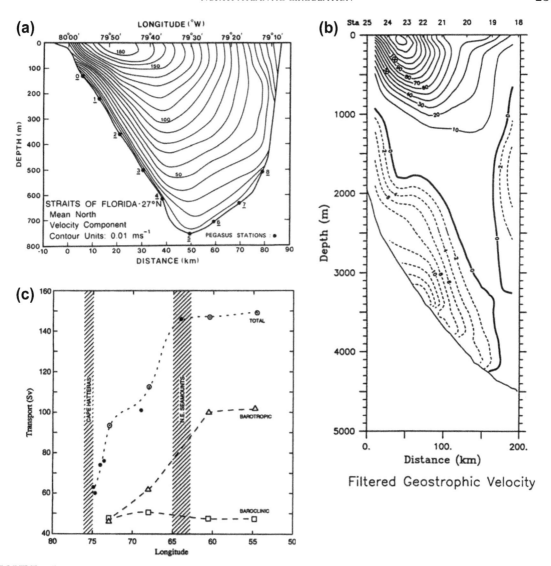

FIGURE 9.5 Gulf Stream velocity sections and transports. (a) Mean velocity of the Florida Current at the Straits of Florida at 27°N. *Source: From Leaman, Johns, and Rossby (1989).* (b) Smoothed geostropic velocity at Cape Hatteras. *Source: From Pickart and Smethie (1993).* (c) Gulf Stream transport (Sv) at different longitudes; Cape Hatteras and the New England Seamounts are indicated by hatching. Barotropic and baroclinic transports are indicated. *Source: From Johns et al. (1995).*

After the Gulf Stream emerges from the Florida Straits, it remains a western boundary current until leaving the coast at Cape Hatteras (about 35°N, 75° 30′W). This location is referred to as the *separation point*. An SST image (Figure 9.4b) shows the narrow boundary current, with a quasi-permanent meander at 32°N due to topography ("Charleston Bump"; Bane & Dewar, 1988) and time dependent "shingle" structures in which meanders peel backward on the inshore side of the current.

This segment of the Gulf Stream System is the prototype for western boundary currents, informing simple theoretical models of the "Gulf Stream" and other subtropical western boundary currents (e.g., Section 7.8). The current extends to the ocean bottom over the continental slope while its typical width remains <100 km; its volume transport increases to more than 90 Sv at the separation point (Leaman, Johns, & Rossby, 1989), fed by westward flow inflow from the Sargasso Sea, including the vigorous recirculation gyre. The mean velocity section at Cape Hatteras (Figure 9.5b) shows the concentrated Gulf Stream and the southward flow inshore and beneath it in the DWBC (Pickart & Smethie, 1993).

East of separation at Cape Hatteras, the Gulf Stream is one of the most powerful currents in the world's oceans in terms of volume transport (up to 140 Sv), maximum velocity (up to 250 cm/ sec), average velocity (about 150 cm/sec), and eddy variability. It reaches to the ocean bottom with bottom velocities exceeding 2 cm/sec. It remains a narrow (<120 km wide), but strongly meandering current for hundreds of kilometers, carrying a warm, saline core of surface water far eastward into the North Atlantic (Figure 1.1a). Its structure is asymmetric, with strongest surface flow on the northern (western) side of the current, shifting southward (eastward) with depth. The current decays quickly to the north of this core; temperature and salinity also change rapidly here (Figure 9.7). This sharp transition is often called the "cold wall" of the Gulf Stream.

The instantaneous Gulf Stream is far from steady. Its meanders often become large enough to pinch off into rings on both sides (Section 9.3.6). The envelope of its meandering paths is illustrated by the positions of its cold wall (Figure 9.6): it is narrowest at Cape Hatteras and then spreads to about 300 km in width downstream, which is 3 times wider than its instantaneous width. Between Cape Hatteras and about 69°W, the Gulf Stream envelope widens but follows sloping bottom topography,

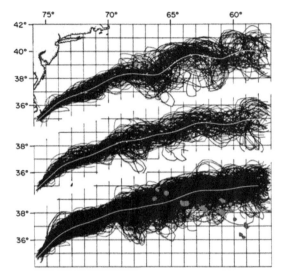

FIGURE 9.6 Gulf Stream northern edges every two days from infrared surface temperature for (top) April to December 1982, (middle) all of 1983, and (bottom) April 1982 to September 1984. The faint white curves are the mean tracks. *Source: From Cornillon (1986).*

which perhaps constrains its meandering. East of the New England Seamounts and 69°W, large-scale meandering sets in. The envelope width compares remarkably well with the historical Franklin and Folger map of the Gulf Stream location (Figure 1.1b from Richardson, 1980a).

The Gulf Stream transport increases rapidly downstream, from about 60 Sv at separation to more than 140 Sv at 65°W (Figure 9.5c). It then loses water to the south, much of it prior to reaching 50°W. Between its separation point and to at least 55°W, there is mean westward surface flow just south of the Gulf Stream, referred to as the Gulf Stream *recirculation*. With the Gulf Stream, this forms the *recirculation gyre* (sometimes called the "Worthington Gyre"). The total transport of the Gulf Stream is many times larger in this recirculation region than predicted by Sverdrup transport theory (Section 7.8). The recirculation is likely driven by the Gulf Stream's instability, which forces westward flow on its flanks, and inertial overshoot of the separated current.

North of the Gulf Stream, the westward flow of the Slope Water Current forms an elongated cyclonic gyre with the Gulf Stream, called the Northern Recirculation Gyre (Hogg, Pickart, Hendry, & Smethie, Jr., 1986). Here the wind stress curl drives upwelling. The westward current is partly supplied from the Labrador Current.

At the sea surface, the Worthington Gyre extends all along the Gulf Stream (Figure 9.1). Its southward flow offshore of the Florida Current turns eastward into the central western North Atlantic at about 22–25°N. This is called the *Subtropical Countercurrent*, and has an exact analog in the North Pacific's Kuroshio gyre circulation (Section 10.3.1). The eastward flow then bends back to join the westward flow North Equatorial Current. The entire recirculation and Subtropical Countercurrent form the so-called "C-shape" of the surface gyre.

Even though it is losing water to the south, part of the Gulf Stream continues eastward to the Grand Banks of Newfoundland at 50°W. Here a portion turns northward and re-forms as a western boundary current east of Flemish Cap, where it is called the *North Atlantic Current* (Section 9.3.4). The remainder of the Gulf Stream continues eastward and southward, splitting into two branches, one at 42–43°N and one farther south at 35°N, called the *Azores Current*. The branch at 42°N passes north of the Azores and weakens considerably to the east. The Azores Current extends eastward toward the Strait of Gibraltar where a small amount of surface water flows into the Mediterranean Sea. Other than this remarkably zonal jet, the subtropical gyre primarily turns southward in the central and eastern North Atlantic. These flows turn westward and feed the *North Equatorial Current*,[4] completing our anticyclonic circuit of the subtropical gyre.

The separated Gulf Stream is mostly in geostrophic balance and is vertically sheared, so its isopycnals and isotherms slope up toward the north, with a 300 to 500 m depth change across the 150 km wide current (Figure 9.7 and online supplementary Figure S9.9). It contains a warm and salty core close to the surface, due to advection from lower latitudes. The cold wall on the north side of the Gulf Stream (which was tracked in Figure 9.6) is a front that is less than 20 km wide.

9.3.3. Canary and Portugal Current Systems

The subtropical gyre has a classic eastern boundary upwelling regime: the Canary Current System south of the Strait of Gibraltar and the Portugal Current System north of the Strait. These are separated by the eastward Azores Current, which is associated with the Mediterranean inflow (New, Jia, Coulibaly, & Dengg, 2001). The eastern boundary currents are associated with large-scale alongshore winds that create offshore Ekman transport (Figure 5.16 and online supplementary Figure S9.3a; Sections 7.9 and 10.3.1).

The Canary Current System (Figure 9.8), along the North African coast, is the more energetic, better developed, and better studied of the two systems (Mittelstaedt, 1991). The Canary Current is the equatorward (southward) near-coastal current. It is present year-round, but its termination in the south is seasonally dependent. Between 20°N and 23°N, the Canary Current turns offshore to join the North Equatorial Current. The upwelling-favorable winds from Gibraltar to Cape Blanc are strongest in summer. The equatorward winds are strongest offshore, leading to positive wind stress curl in the Canary Current region, which augments the upwelling forcing. A poleward undercurrent flows along the continental shelf beneath the Canary Current, north of 25°N, centered at about 600 m depth, with a mean speed of about 5 cm/sec.

---

[4] The NEC is not actually "Equatorial," because it is separated from the equator by the vigorous eastward flow of the North Equatorial Countercurrent. Instead, the NEC is the equatorward side of the subtropical gyre.

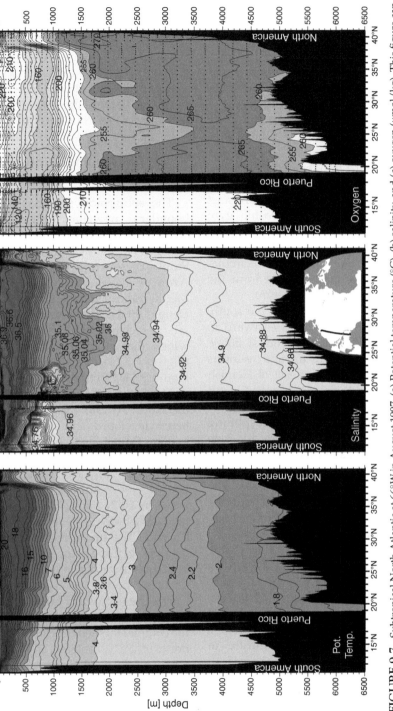

FIGURE 9.7 Subtropical North Atlantic at 66°W in August 1997. (a) Potential temperature (°C), (b) salinity, and (c) oxygen (μmol/kg). This figure can also be found in the color insert. (*World Ocean Circulation Experiment section A22.*)

(a)   **(b)**

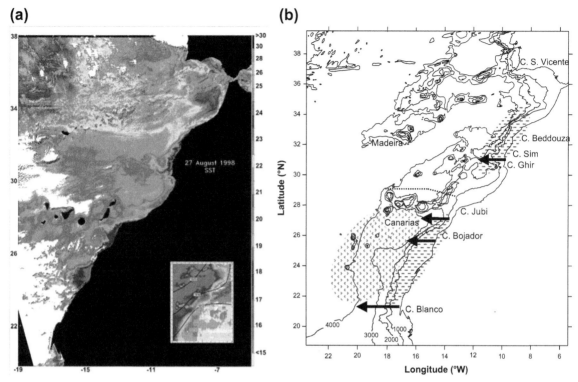

FIGURE 9.8 Canary Current System. (a) SST (satellite AVHRR image) on August 27, 1998. This figure can also be found in the color insert. (b) Schematic of upwelling (horizontal bars), eddy fields (dots), and preferred filaments (arrows). *Source: From Pelegrí et al. (2005).*

Like other eastern boundary current systems, the Canary Current is synoptically complex, containing large offshore jets of upwelled water associated with capes in the coastline (Figure 9.8). The Canary Islands at 28–29°N create especially vigorous filaments and an eddy field south of the islands.

The Portugal Current is part of the generally southward mean flow along the eastern boundary north of the Strait of Gibraltar. Inflow is from the branch of the NAC that lies about 45°N. Seasonality is marked, with southward winds in spring and summer that reverse to northward in fall and winter. This creates a reversal in the coastal surface flow from equatorward in summer (Portugal Current), to poleward in autumn and winter (Portugal Coastal

Countercurrent; Ambar & Fiuza, 1994). Spring and summer are thus the upwelling seasons. During upwelling season, there is a poleward undercurrent called the Portugal Coastal Undercurrent. The deep extension of the poleward undercurrent is an important conduit for northward flux of Mediterranean Water (MW) exiting from the Strait of Gibraltar (Section 9.8.3).

9.3.4. North Atlantic Current

The NAC begins as a northward western boundary current at about 40°N, 46°W, east of the Grand Banks of Newfoundland, fed by a branch of the Gulf Stream. At 51°N, the NAC separates from the boundary and turns abruptly eastward, in a feature referred to as the *Northwest Corner* (Rossby, 1996, 1999; Zhang & Hunke, 2001). The NAC then flows eastward as a free jet, steered by the Charlie Gibbs Fracture Zone at 52°N, and splits into multiple branches. The southward branches become part of the North Atlantic's anticyclonic subtropical circulation. The northward branches, which retain locally intense frontal structures, feed into the subpolar circulation and on northward into the Nordic Seas.

Where the NAC is a western boundary current, it has two roles. It is dynamically part of the wind-driven subtropical gyre circulation, responding to the Sverdrup forcing across the width of the North Atlantic. It also carries the 10 to 20 Sv of northward flow of the MOC. The northward flow eventually enters the Norwegian Sea and the Irminger Sea, where it is a source of the dense waters formed in the Nordic and Labrador Seas, respectively.

The eastward flow of the NAC has some aspects in common with the simpler North Pacific Current (Section 10.3.1), but connectivity of the subtropical and subpolar circulations differs. As the pathway for the upper ocean part of the Atlantic's MOC, the NAC includes more net northward transport than the North Pacific Current since the North Pacific's MOC is much weaker. As a western boundary current of the subtropical gyre circulation, the NAC has no counterpart in the North Pacific.

Formation of the NAC is complicated, illustrated schematically in the online supplementary Figure S9.10. A branch of the separated Gulf Stream turns north roughly along the 4000 m isobath east of the Grand Banks and is joined by colder water from the inshore Slopewater Jet and from a northward turn of the Labrador Current, which also lies inshore of the NAC (Figure 9.9a). By the time it reaches the southern flank of the Flemish Cap, the NAC can be considered a true western boundary current, extending well inshore of the 4000 m isobath (Figure 9.9b). Here the NAC's velocity structure is similar to the Gulf Stream's

velocity structure, with maximum mean surface velocity >60 cm/sec and northward flow extending to the ocean bottom (Meinen & Watts, 2000). The core of the current shifts offshore with increasing depth. The southward flow inshore of the NAC, intensified at the ocean bottom, is the DWBC (Section 9.6). The NAC transport at 42°30′N has been observed to exceed 140 Sv, of which about 50 Sv recirculates in a local, permanent eddy (Mann Eddy). The NAC's net northward transport is thus about 90 Sv (Meinen & Watts, 2000). Of this, 15 to 20 Sv can be considered to be part of the MOC.

As the NAC follows the deep isobaths along the western boundary east of Newfoundland, it reaches a latitude where both the integrated Sverdrup transport becomes zero and the isobaths turn offshore in the Northwest Corner, as mentioned previously. Waters on the subpolar side of the NAC are cold, fresh, and highly oxygenated to great depth. Waters on its warm side are nearly subtropical. The sharp front across the NAC is called the *Subarctic Front*. There is some subduction of fresher surface waters along this front, resulting in a shallow salinity minimum on the warm side of the front, called the Subarctic Intermediate Water. Discussion of the NAC further downstream is in the next section.

9.3.5. Subpolar Circulation

The *subpolar gyre* in the North Atlantic is the quasi-cyclonic circulation north of 50°N (Figures 9.1 and 9.2, and Figure S9.1 and Table S9.2 located in the online supplement). It is divided into western and eastern regimes on either side of the Reykjanes Ridge. The western part is a cyclonic gyre in the Labrador and Irminger Seas. The eastern part is northeastward surface flow in several topographically-controlled branches of the NAC that continue northward into the Nordic Seas. If we consider the subpolar North Atlantic together with the Nordic Seas, the surface flow makes a complete cyclonic gyre.

Below the depth of the Greenland-Iceland-Faroe Ridge, the subpolar North Atlantic flow is cyclonic throughout the region (Figure 9.2b).

The eastward NAC in the western North Atlantic forms the southern side of the subpolar circulation, as well as the northern side of the subtropical circulation. Its Subarctic Front is steered through the Charlie Gibbs Fracture Zone in the MAR. The NAC then splits into a part that turns southward to the subtropics,

(a)

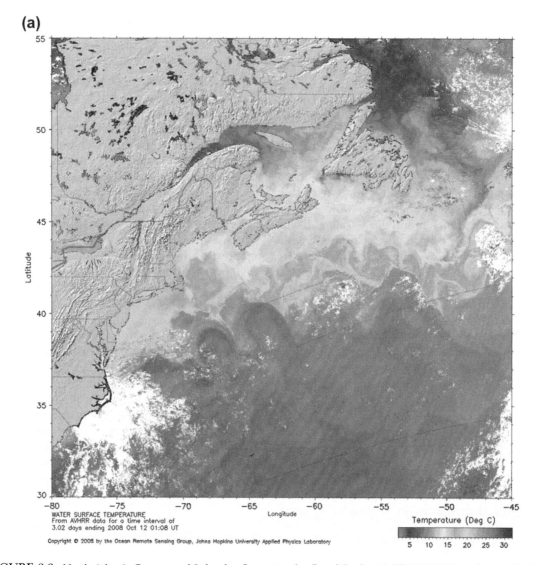

FIGURE 9.9 North Atlantic Current and Labrador Current at the Grand Banks. (a) SST (AVHRR) on October 12, 2008, showing cold Labrador Current moving southward along the edge of the Grand Banks. *Source: From Johns Hopkins APL Ocean Remote Sensing (1996).* This figure can also be found in the color insert. (b) North Atlantic Current and DWBC velocity section (solid contours and numbers) with temperature contours, from August 1993 to January 1994, from about 48°W to 41°W at about 42°N. Velocity contours are 10 cm/sec. *Source: From Meinen and Watts (2000).*

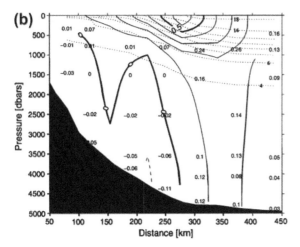

FIGURE 9.9 (*Continued*).

including the Portugal Current, and two north-eastward branches (Fratantoni, 2001; Flatau, Talley, & Niiler, 2003; Brambilla & Talley, 2008). The subtropical branch is associated with typical subtropical gyre subduction. The north-eastward branches are part of the subpolar circulation. The first turns northward into the Iceland Basin east of the Reykjanes Ridge and the second turns northward into Rockall Trough, close to the eastern boundary. As they reach the Iceland-Faroe Ridge, both branches join the *Iceland-Faroe Front* and move northward into the Norwegian Atlantic Current in the Nordic Seas (Section 12.2).

The western cyclonic gyre begins with a branch of the NAC that turns northward into the *Irminger Current* along the western flank of the Reykjanes Ridge. This turns west and south, joining the EGC coming out of the Nordic Seas, then the northward flow in the WGC and finally the southward flow in the Labrador Current along the Labrador coast. The Labrador Current also sweeps in waters from the Arctic through Baffin Bay and Davis Strait. The Labrador and EGCs are western boundary currents. The WGC is a more unstable eastern boundary current, with eddies shed at Cape Farewell at the southern tip of Greenland that move

westward into the Labrador Sea, creating enhanced eddy kinetic energy (EKE) there. Transport estimates for the EGC and WGC are 16 and 12 Sv, respectively, with the EGC eddies absorbing the loss (Holliday et al., 2007).

The subpolar circulation is so strongly steered by topography that the flow around the Labrador Sea is sometimes referred to as the "Rim Current." Within the Labrador Sea (and probably also the Irminger Sea) the cyclonic Rim Current has a weak offshore countercurrent, running clockwise around the sea (Lavender, Davis, & Owens, 2000). The countercurrent reflects an enhanced cyclonic dome near the Rim Current. This could localize the deep convection involved in Labrador Sea Water (LSW) production closer to the offshore side of the current than to the center of the Labrador Sea (Section 9.8.3; Pickart, Torres, & Clarke, 2002).

From the Labrador Sea, the Labrador Current continues southward to the Newfoundland region; in SST images it is cold (Figure 9.9a). Most of the current flows through Flemish Pass between Newfoundland and Flemish Cap and then southward along the continental shelf break to the Tail of the Grand Banks. Here the cold water evident in SST images disappears (Figure 9.9a). Part of the current turns back northward and joins the inshore side of the NAC. Part of the current continues westward following the continental slope toward Nova Scotia, well north of the Gulf Stream. At the sea surface this westward flow is called the *Slope Water Current*. The deeper southward and westward boundary flow (Figure 9.9b, below 1000 m) is the DWBC, carrying new, dense LSW and Nordic Seas Overflow Waters southward (Section 9.6).

9.3.6. North Atlantic Eddy Variability and Gulf Stream Rings

North Atlantic eddy variability is represented by the global EKE and coherent eddy maps of Figures 14.16 and 14.21 and in Fratantoni (2001) (Figure S9.11 in the online supplement). From south to north, the highest EKE is found associated with the western boundary currents: in the North Brazil Current, in the Gulf of Mexico in the Loop Current, along the Gulf Stream, and in the separated Gulf Stream with its large meanders and ring creation. The NAC continues the axis of higher EKE toward the north and east along 50°N. Within the subtropical gyre, there is also slightly enhanced EKE in the Azores Current near 35°N. In the subpolar region, there is high EKE in the EGC and in the eddy band in the Labrador Sea spawned at Cape Farewell.

The overall level of EKE is lower in the subpolar gyre than at lower latitudes. This is related to the weaker baroclinicity of the subpolar gyre; that is, the water column is less stratified, the energetic currents have less vertical shear, and isopycnals are less sloped.

Subsurface eddy variability is better observed in the Gulf Stream region of the North Atlantic than in any other part of any ocean, using acoustically tracked subsurface floats (Owens, 1991). High EKE occurs directly beneath the high surface EKE of the Gulf Stream, and decreases in amplitude with depth.

Gulf Stream rings are especially large, energetic, closed eddies formed when meanders of the Gulf Stream pinch off, forming anticyclonic warm-core rings to the north and cyclonic cold-core rings to the south (Figure 9.10). The Gulf Stream does not have strongly preferred meandering sites, unlike the other subtropical western boundary currents; ring formation occurs all along the front from 70°W to the Grand Banks. The surface temperature image in Figure 1.1a includes two obvious cold-core rings south of the Gulf Stream and one warm-core ring to the north. Gulf Stream rings can have surface speeds exceeding 150 cm/sec, be 150 to 300 km in diameter, be more than 2000 m deep, and can have lifetimes of more than a year. At any time, in the area west of 55°W and north of about 30°N, there may be 3 (anticyclonic) warm-core rings north of the Gulf Stream and 10 (cyclonic)

cold-core rings to the south (Richardson, 1983). Approximately five warm-core, and five to eight cold-core rings form per year.

In ring formation, a meander forms, closes up, and then separates from the Gulf Stream (Parker, 1971) (online supplementary Figure S9.12). The ring is in nearly solid-body rotation to about 60 km from the center, which differs from the form it would have if it were simply a closed loop of the Gulf Stream. Once formed, both cold- and warm-core rings propagate westward. Cold-core rings also move southward in the recirculation and are often found offshore of the Gulf Stream as far south as 28°S (Richardson, 1980c, 1983). Rings exchange biologically productive water from north of the Gulf Stream with much less productive Sargasso Sea water. Therefore, warm-core rings appear in ocean color images as areas of low chlorophyll while cold-core rings have high chlorophyll.

9.4. TROPICAL ATLANTIC CIRCULATION

We describe the circulation in the tropical Atlantic briefly, reserving a more complete description of typical features of equatorial circulation for the Pacific (Section 10.7). The principal near-surface currents for both the tropics and the South Atlantic are shown in Figures 9.1 and 9.11 and listed in Table S9.3 in the online supplement. At the equator, the Atlantic extends from 45°W to 10°E, a distance of about 6000 km. Because the equatorial Pacific is more than twice this wide, the wind-driven equatorial current systems differ in some respects, especially in strength. The tropical Atlantic is bisected by the MAR, which has a major east-west fracture zone — the Romanche Fracture Zone — close to the equator. In the east, the tropical region is limited to the north by the curve of the African coastline.

The tropical circulation responds strongly to the trade wind forcing, which has large seasonal

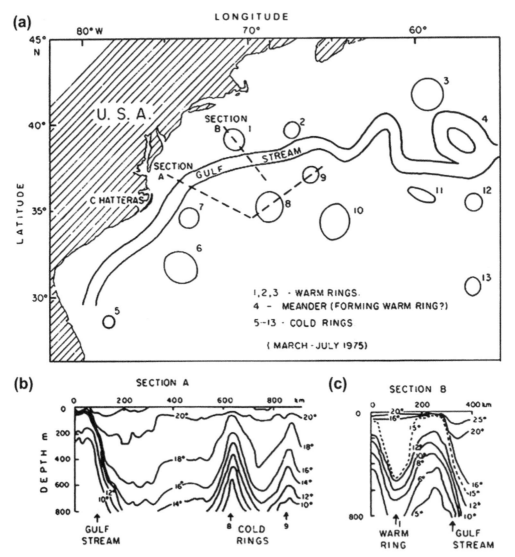

FIGURE 9.10 Gulf Stream rings. (a) Locations of Gulf Stream and of warm- and cold-core rings in March to July 1975. (b, c) Vertical temperature sections along lines A and B in (a) showing Gulf Stream and cold- and warm-core ring structures. *After Richardson et al. (1978).*

changes (Stramma & Schott, 1999) as well as interannual variability (Section 9.9 and online supplementary Chapter S15). Seasonal wind changes are related to shifts in strength and location of the ITCZ, which is most strongly developed in the summer hemisphere. The freshwater from the Amazon and Orinoco rivers empties into the western boundary region and spreads northwest into the Caribbean Sea and Gulf Stream System. The Congo River freshwater spreads southward in the Angola Current along the African boundary.

Circulation within 10° of the equator is nearly zonal at depths above the strong topography. The

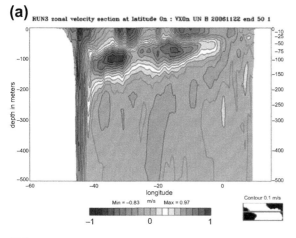

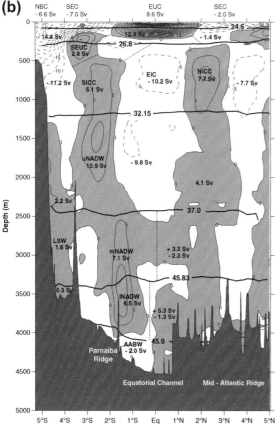

dominant surface flow is westward, in the *South Equatorial Current* (SEC). The "South SEC" is the westward flow in the northern part of the South Atlantic's subtropical gyre. When it reaches the South American coast, it splits into the southward Brazil Current and the northward North Brazil Current. The "Central SEC" and "North SEC" straddle the equator to about 5−7° latitude; directly on the equator there is also a weak westward flow, driven by the trade winds. This equatorial part of the SEC is bounded to the north by the vigorous eastward flow of the *North Equatorial Countercurrent* (NECC), which is associated with the ITCZ wind forcing. The NECC bounds the tropical circulation, separating it from the North Equatorial Current of the North Atlantic's subtropical gyre.

As the NECC flows eastward, it encounters Africa and splits into a northward flow toward Dakar and the eastward Guinea Current along the coast, with surface speeds in excess of 100 cm/sec (Richardson & Reverdin, 1987). The Guinea Current follows the coast and eventually turns south and joins the westward North SEC. The northward flow turns westward to join the North Equatorial Current (NEC). The eastern tropical region between the NEC and NECC, which forms a cyclone, is an upwelling region called the *Guinea Dome* (Siedler, Zanbenberg, Onken, & Morlière, 1992).

At about 7−8°S between the South SEC and the Central SEC, there is a quasi-permanent (seasonal) *South Equatorial Countercurrent* (SECC), associated with the southern hemisphere ITCZ. The SECC terminates at the coast of Africa, where it is joined by upwelling flow from the Equatorial Undercurrent (EUC). This turns southward along the coast, forming the Angola Current, and then westward into the South SEC, forming a cyclonic upwelling region called the *Angola Dome* (Wacongne & Piton, 1992). With increasing depth,

FIGURE 9.11 Tropical current structures. (a) Eastward velocity along the equator, from a data assimilation. This figure can also be found in the color insert. *Source: From Bourlès et al. (2008).* (b) Mean zonal transports (Sv) (gray eastward) and water masses at 35°W. *Source: From Schott et al. (2003).*

the cyclonic gyre enlarges and is more "gyre-like," as the eastward flow beneath the surface SECC is more pronounced and permanent (Gordon & Bosley, 1991).

Both the Angola and Guinea Domes are regions of upwelling and great biological productivity. This results in a large subsurface tropical oxygen minimum layer, with the lowest oxygen centers in each of the domes, hence on either side of the equator (Stramma, Johnson, Sprintall, & Mohrholz, 2008; Karstensen, Stramma, & Visbeck, 2008). This signal, centered at about 500 m depth, is an obvious feature of any vertical oxygen section in the equatorial region (e.g., Figure 4.11d).

Along the equator, just below the sea surface at 60 to 120 m depth, the EUC flows eastward, similar to the EUC in the Pacific (Figures 9.11a and 10.23c). An eastward pressure gradient force created by the easterly trade winds, which pile surface water up in the west, drives the EUC. These create a weak version of the equatorial Pacific's warm pool and cold tongue. The EUC core shoals from deeper than 100 m near the western boundary to about 30 m at the eastern boundary. Eastward currents in the EUC core can exceed 80 cm/sec and, occasionally, 100 cm/sec, but do not reach the much larger mean velocities of the Pacific's EUC (Wacongne, 1990; Giarolla, Nobre, Malaguti, & Pezzi, 2005).

The full suite of subsurface flows in the equatorial region (Figure 9.11b) continues the strong zonal character of the surface currents. The correspondence with the Pacific equatorial currents is remarkable (Figures 10.20a–10.21). On the equator beneath the EUC are found the westward *Equatorial Intermediate Current* and the "stacked jets" of alternating flow down to about 2000 m. On either side of the equator, at 2–4° latitude, there are eastward flows centered around 500 to 1000 m; in the Atlantic these are referred to as the South and North Equatorial Undercurrents and the deeper parts are referred to as the South and North Intermediate Countercurrents (SICC and NICC). The

transports of each of these currents are substantial, exceeding 5–10 Sv.

The strong zonal flows in the tropical Atlantic are unstable, and routinely form regular trains of planetary waves and eddies (Legeckis & Reverdin, 1987; Steger & Carton, 1991). These are the Atlantic Tropical Instability Waves (TIWs); they correspond with the TIWs in the tropical Pacific where they were discovered first (Section 10.7.6). TIWs form on the northern and southern edges of the equatorial cold tongue. They have a wavelength of about 900 km, which means there are typically about 4 to 5 waves across the width of the Atlantic, and they propagate westward (phase) at about 25 cm/sec. The cold tongue and hence the TIWs are seasonal features, appearing each summer. TIWs form within several weeks of the cold tongue's appearance, grow, and begin to break, similar to the Pacific TIWs. The energy source is mostly barotropic instability (Jochum et al., 2004). The TIWs on the northern and southern flanks appear to be independent of each other. The rolling up of the breaking waves is a dramatic feature in satellite images, and forms large anticyclonic eddies with diameters of about 500 km that last for more than a month before decaying away.

The tropical Atlantic's low latitude western boundary current is the NBC, which flows northward starting from the bifurcation of the SEC at the South American coast at around 10–15°S. It extends to intermediate depth (~800 m), carrying surface water down through the Antarctic Intermediate Water (AAIW) northward into the North Atlantic. The surface water also includes much of the 0.2 Sv of fresh water discharge from the Amazon. Part of the NBC turns east near the equator, joining the EUC. The remainder crosses the equator, and then splits into a portion that joins the eastward NECC and a portion that continues northward along the western boundary.

The NBC is surface-intensified (Figure 9.12b), with velocities exceeding 90 cm/sec at the

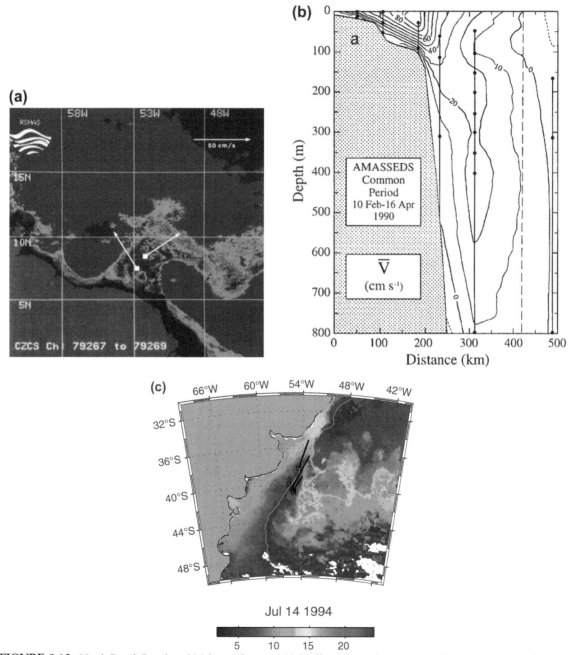

FIGURE 9.12 North Brazil, Brazil, and Malvinas Currents. (a) Satellite ocean color image (CZCS) of the NBC retroflection prior to ring formation. *Source: From Johns et al. (1990).* (b) Mean velocity (cm/sec) from current meters in the NBC at about 4°N in 1990. *Source: From Johns et al. (1998).* (c) Infrared satellite image of the Brazil-Malvinas confluence. Black lines are current vectors at moorings, at approximately 200 m depth. Light curve is the 1000 m isobath. This figure can also be found in the color insert. *Source: From Vivier and Provost (1999).* (d) Malvinas current mean velocities (cm/sec) at about 41°S, based on current meters (crosses and diamonds) and satellite altimetry. Positive velocities are northward. *Source: From Spadone and Provost (2009).*

(d)

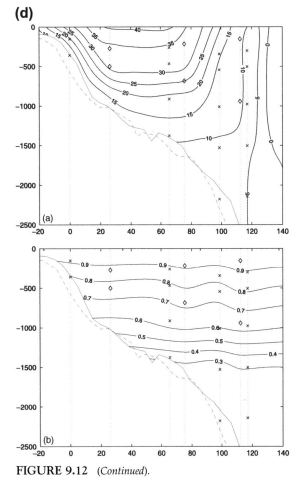

FIGURE 9.12   *(Continued)*.

The NBC is a site of notable mesoscale variability. It does not easily cross the equator. At about 5 to 7°N, it retroflects and spawns enormous anticyclonic rings (400 km diameter) called "North Brazil Current rings" (Figure 9.12a). Each carries about 1 Sv of NBC water northward. Three or more rings are formed each year. Surface velocities in the rings are 30 to 80 cm/sec, and the translational speed toward the northwest is about 10 cm/sec. The rings are deep-reaching, easily trapping floats at 900 m (Richardson, Hufford, Limeburner, & Brown, 1994).

The continuation of a northward current past the retroflection of the NBC is called the Guiana Current, but the region is highly variable as a result of the large rings. The flow and rings impinge on the southern islands of the Caribbean Sea, and join the westward flow into the sea as the Caribbean Current (Section 9.3).

Below the NBC, there are opposing DWBCs (Section 9.6.2). One approaches the equator from the north carrying NADW centered at 2000–3000 m depth. A deeper one comes from the south carrying Antarctic Bottom Water (AABW). Near the equator, there is a tendency for all layers of the western boundary current system to detrain or leak water eastward along the equator. The equatorial Romanche Fracture Zone channels the deepest equatorial flows into the deep eastern basins.

surface, decreasing to about 20 cm/sec at 200 m. The NBC's mean transport at 4°N is 26 Sv, based on moored observations (Johns et al., 1998). The transport has two sources: the wind-driven circulation and the Atlantic's MOC. Fratantoni, Johns, Townsend, and Hurlburt (2000) found that of the 14 Sv of MOC transport that travel northward through the system, 7 Sv are carried in the NBC and its continuation to the Guiana Current, 3 Sv are carried by NBC rings (see the previous section), and the remainder are carried into the upper ocean's interior circulation.

9.5. SOUTH ATLANTIC CIRCULATION

The South Atlantic surface circulation consists of the eastward Antarctic Circumpolar Current (ACC) in the south, an anticyclonic subtropical gyre that is partially contiguous with the Indian Ocean's subtropical gyre, and a cyclonic tropical circulation gyre (Figures 9.1, 9.2, and online supplementary Figure S9.1).

The subtropical gyre's western boundary current is the Brazil Current, flowing southward along the coast of South America. The eastward flow on the south side of the gyre is the *South Atlantic Current* (SAC). The eastern boundary upwelling system is the Benguela Current System (BCS). The broad westward flow on the north side of the subtropical gyre is the *South Equatorial Current* (SEC), which splits at the western boundary into the Brazil Current and the NBC.

South of the subtropical Brazil Current gyre, we enter the domain of the ACC, whose northernmost front is the SAF (Chapter 13). The SAF enters the South Atlantic from the Pacific along the northern boundary of the Drake Passage and immediately turns northward along the coast of South America as the Malvinas Current (or Falkland Current). The Malvinas can be thought of as the western boundary current of a cyclonic (subpolar-type) circulation that is forced by positive wind stress curl north of the latitude of the Drake Passage. The Malvinas and Brazil Currents encounter each other almost head-on at 36−38°S, in the *Brazil-Malvinas confluence*. Both separate from the coast and loop southward. They retain their identities as separate fronts and move eastward into the South Atlantic along separate paths east of 50°W. The Malvinas Current front is then again called the SAF, and heads eastward and slightly southward across the South Atlantic and Indian Oceans at about 50°S.

The South Atlantic's subtropical gyre is a conduit for northward flow of upper ocean waters to the North Atlantic, where they are ultimately transformed to dense deep waters in the Labrador and Nordic Seas as part of the global overturning circulation. Much of the South Atlantic's net northward flow originates from the Indian Ocean via the Agulhas. This enters the South Atlantic in the northwestward flow in the BCS and as large Agulhas rings. Somewhat denser near-surface flow enters from the Drake Passage and moves northward through the South Atlantic's subtropical gyre as Subantarctic Mode Water (SAMW) and AAIW. The net northward flow of surface through intermediate waters moves westward in the subtropical gyre with the SEC, reaches the western boundary, and flows northward as part of the transport of the NBC.

In the deep layers of the South Atlantic, the circulation is strongly modified by topography. The DWBCs carry NADW southward and abyssal AABW northward (Section 9.6.2). But even in the NADW-dominated layers, there is northward flow of Circumpolar Deep Water (CDW), in regions other than the DWBC.

9.5.1. Subtropical Gyre

The South Atlantic's anticyclonic subtropical gyre is forced by anticyclonic wind stress curl, which causes Ekman pumping and equatorward Sverdrup transport across the South Atlantic (Figures 5.16d, 5.17 and online supplementary S9.3). The Brazil Current at the western boundary is its narrow poleward return flow.

We emphasize two unique aspects of the South Atlantic's subtropical gyre compared with other oceans, both partly associated with the Agulhas, which is the western boundary current of the Indian Ocean's subtropical gyre (Chapter 11). The first is the throughput of upper ocean water as part of the Atlantic's MOC (Section 9.7). Low latitude Pacific and Indian Ocean waters enter the Atlantic via the Agulhas retroflection region and ultimately return southward in the deep ocean as NADW. The South Atlantic also imports water from the Pacific through the Drake Passage that joins the northward upper ocean flow to the NADW formation sites. Both of these warm water sources to the South Atlantic are part of the global overturning circulation (Section 14.3; Figure 14.11).

A second special aspect is the connection of the wind-driven subtropical gyre circulations of the South Atlantic and Indian Oceans. The subtropical wind forcing (Ekman convergence) extends south of the African continent to around 50°S, all the way from the South American coast eastward to Australia and New Zealand

(Figure 5.16d). Thus the South Atlantic and Indian subtropical gyres are connected from 34 to 50°S, with mostly eastward flow in the SAC plus westward flow in the Agulhas.

The eastward flow of the SAC is organized into two nearly zonal fronts at about 35°S and 40°S (see online supplementary Figure S9.13a from Juliano & Alvés, 2007). The SAC transport, including both fronts, is about 30 Sv. Maximum surface speeds in the fronts are around 20 cm/sec. The cores of the currents extend down to about 800 m. These are both part of the subtropical front first identified by Krummel (1882) and called the "subtropical convergence" by Deacon (1933). We suggest calling the two fronts the *North* and *South Subtropical Fronts* following Belkin and Gordon (1996) and Provost et al. (1999). (At the western boundary, the two fronts originate in the separated Brazil Current and are therefore called the Brazil Current Front and the Subtropical Front by Peterson, 1992 and Tsuchiya, Talley, & McCartney, 1994.)

The North Subtropical Front terminates in the east where it meets the Agulhas retroflection and Benguela Current just west of Africa. The front may turn continuously northward where it becomes the outer front of the BCS. The South Subtropical Front continues on eastward into the Indian Ocean, clearly separated from, and south of, the Agulhas Return Current (Chapter 11; Belkin & Gordon, 1996).

The SEC originates in the Benguela Current at 34°S at the tip of Africa and moves generally northwestward toward the broad western boundary region between 15 and 30°S (Figure 9.1). Agulhas rings generated at the retroflection move westward more zonally than the overall SEC, following the more zonal depth-integrated streamfunction (Biastoch, Böning, & Lutjeharms, 2008).

9.5.2. Brazil Current

The Brazil Current begins its southward flow in the surface layer at about 10−15°S

along the coast of Brazil. Deeper parts of the Brazil Current begin at successively higher latitudes (toward the south). An anticyclonic recirculation gyre on the offshore side of the Brazil Current, south of about 30°S, increases the Brazil Current's transport toward the south. The Brazil Current begins to separate from the coast at the Brazil-Malvinas confluence at about 36°S, where the cold Malvinas water intrudes inshore of the warm Brazil Current (Figure 9.12c). The main transport of the Brazil Current finally leaves the continental shelf somewhat farther south, around 38°S.

A quantitative overview of Brazil Current structure and transport appears to be lacking. Maamaatuaiahutapu, Garçon, Provost, and Mercier (1998) provided references and a partial summary of transport estimates. The Brazil-Malvinas confluence has been well-measured with moored velocity observations, but upstream locations have not. Large-scale inverse circulation models have generally focused on 32°S. Many studies have used shallow zero velocity reference levels for geostrophic velocities, thus underestimating the current strength. Considering only estimates that move beyond zero reference velocity assumptions, the southward Brazil Current transport is estimated at 2.5 Sv at its very beginnings at 12°S, with increasing southward transport of 4 Sv, 11 Sv, 17 Sv, 22 Sv, and 41 Sv at 15°S, 27°S, 31°S, 34°S, and 36°S, respectively (online supplementary Figure S9.14b from Zemba, 1991; Sloyan & Rintoul, 2001; Stramma, Ikeda, & Peterson, 1990). The Brazil Current has a large recirculation gyre that begins south of 30°S and greatly increases in transport south of 35°S. By 36°S, the total Brazil Current transport is 70 to 80 Sv, with about half participating in the recirculation gyre (Zemba, 1991; Peterson, 1992). About 30 Sv actually exits to the east into the South Atlantic gyre (Stramma & Peterson, 1990).

9.5.3. Malvinas Current and Subantarctic Front

The Malvinas Current originates from the SAF in the Drake Passage (Figure 9.1 and supplementary Figure S9.14a). It flows northward as a western boundary current after leaving the Drake Passage, following the South American shelf roughly along the 1000 m isobath, up to about 38°S. The Malvinas meets the Brazil Current head-on at this point, as seen in satellite SST images of the Brazil-Malvinas confluence (Figure 9.12c). Both currents separate from the coast and move offshore and southward. The expected location of the Brazil Current separation (based on wind stress curl and Sverdrup transport) is about 10° latitude south of the actual separation point (Figure 5.17 and supplementary Figure S9.3b); the powerful Malvinas appears to push the Brazil Current separation northward (Spadone & Provost, 2009).

Mean surface velocities in the Malvinas Current are about 40 cm/sec. The Malvinas is most intense inshore of the 1500 m isobath, and extends to the ocean bottom (Figure 9.12d). Transport from the moored array of Figure 9.12d is estimated to be 42 Sv with a variability of 12 Sv (Spadone & Provost, 2009), which is weaker than the maximum transport estimate of 70 Sv from Peterson (1992).

Where the Malvinas separates from the coast, it turns sharply southward in a feature referred to as the Malvinas, or Falkland, Loop. Eddies are generated, mixing with the much warmer and more saline Brazil Current water. Subsurface waters in the Malvinas may not turn as strongly southward, which injects some of the denser waters into the subtropical gyre. This is the primary mechanism by which the AAIW enters the SAC (Talley, 1996a).

After looping back to the south, the Malvinas reaches the northern escarpment of the Falkland Plateau and turns to the east along the bathymetry. Here it is again referred to as the Subantarctic Front. The SAF is distinct from the two subtropical fronts. It crosses the South Atlantic at around 50°S, which is approximately the latitude of zero wind stress curl (Figure 5.16d).

In the ACC, the Polar Front lies south of the SAF. After exiting the Drake Passage to the South Atlantic, the Polar Front remains in the Scotia Sea while the SAF executes the large northward Malvinas Current loop. The two fronts converge along the north side of the Falkland Plateau and then remain relatively close to each other, but distinct in water mass properties, across the rest of the South Atlantic (Figure 13.1).

9.5.4. Benguela Current System

The BCS is the eastern boundary current system for the subtropical South Atlantic. The BCS extends from 34°S at Cape Agulhas at the southern tip of Africa northward to 14°S. Details can be found in Shillington (1998) and Field and Shillington (2006). The BCS is unique among eastern boundary currents because of its role in the northward transport of warm waters in the global overturning circulation. Some portion of the Benguela originates in the warm, saline Agulhas Retroflection waters that round the southern tip of Africa. The warmth and higher salinity of waters on its poleward end distinguishes the BCS from the other eastern boundary current systems.

The BCS has the characteristics of a classic upwelling system (Section 7.9) with upwelling-favorable equatorward winds, offshore Ekman transport, an equatorward surface current (the Benguela Current), and a poleward undercurrent. The equatorward winds that force the BCS are strongest offshore, leading to positive wind stress curl close to shore (Figure 5.16d). As in the other eastern boundary current upwelling systems, the positive wind curl is associated with local upwelling in a wider band than just within the near-coastal strip.

Upwelling season in the southern Benguela is in the austral summer (December through February), but continues year-round in the central and northern Benguela, based on appearance of cold eastern boundary waters in satellite SST images (Figure 9.13). As in other eastern boundary current systems, the BCS is marked by offshore jets of cold water associated with the coastline. Preferred locations are northwest of Lüderitz (24–26°S), and at 28–30°S.

At the northern boundary of the BCS, the northward, cold Benguela Current meets the southward, warm Angola Current. (The Angola Current is part of the mid-ocean cyclonic gyre that occupies the tropical South Atlantic, described in Section 9.4.) The resulting *Angola-Benguela Front* is located near Cape Frio at about 16°S (Shillington, 1998). The front is evident in the SST in Figure 9.13a and b.

9.5.5. South Atlantic Eddy Variability and Agulhas Rings

Eddy variability in the South Atlantic is illustrated in the global maps of surface EKE and coherent eddies in Figures 14.16 and 14.21. The

(a)

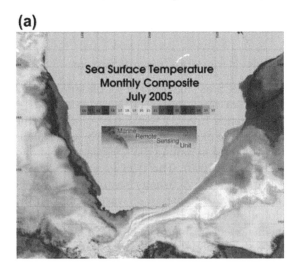

(b)

(c)

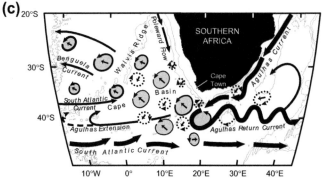

FIGURE 9.13 Benguela Current and Agulhas retroflection. (a, b) AVHRR SST monthly composite for July (winter) and December (summer) 2005. *Source: From UCT Oceanography Department (2009).* (c) Schematic of Agulhas retroflection and eddies, with flow directions in the intermediate water layer. Gray-shaded rings are the Agulhas anticyclones. Dashed rings are cyclones that are generated in the Agulhas. This figure can be found in the color insert. *Source: From Richardson (2007).*

highest EKE occurs in the equatorial region and North Brazil Current, and in the Malvinas/Brazil Current confluence and Agulhas retroflection. Each of these regions spawns anticyclonic rings.

Brazil Current warm-core rings form at the large southward meander of the Brazil Current Front after it separates from the coast, reaching 45°S near 50°W (supplementary Figure S9.14c from Lentini, Goni, & Olson, 2006). Approximately 6 rings with a diameter of about 100 km are shed at this meander per year. They drift southward at a mean speed of 10 km/day and have a lifetime on the order of 40 days. This meander and ring formation area encircle a topographic feature, the Zapiola Rise, above which there is much lower EKE (Figure 14.16), and around which there is a permanent anticyclonic flow, the "Zapiola Eddy," which is more well defined with increasing depth (Section 9.6).

Agulhas rings are anticyclonic, warm-core rings that form when the Agulhas protrudes westward south of Africa and retroflects back to the east, between 15°E and 20°E (Figure 9.13c and Section 11.4.2). The centers of the rings are warm and saline in contrast with the local South Atlantic waters. The rings are 100–400 km in diameter, with maximum speeds of more than 100 cm/sec at the surface, and up to 10 cm/sec even at 4000 m depth (supplementary Figure S9.15 from van Aken et al., 2003). Like Gulf Stream rings, Agulhas rings are in solid body rotation out to the locus of maximum surface speed.

About six Agulhas rings are produced each year and propagate westward into the South Atlantic, with about three reaching the South American coast and entering the North Brazil Current (Gordon, 2003) (supplementary Figure S9.13b). Each ring contributes a volume transport into the South Atlantic of 0.5 to 1.5 Sv (Richardson, Lutjeharms, & Boebel, 2003). The 6 rings per year with 3 to 9 Sv thus represent a significant fraction of the exchange from the Indian to the Atlantic, with the rest carried by connection to the Benguela Current. Part of this transport is simply part of the extended Atlantic/Indian anticyclonic circulation north of the ACC, and part of it contributes to transport of warm water in the global overturning circulation.

9.6. DEPTH DEPENDENCE OF THE ATLANTIC OCEAN CIRCULATION

The circulation in the upper 1000 to 1500 m of the Atlantic is mostly associated with wind forcing through Ekman pumping and subduction/obduction (Section 7.8). This circulation's depth dependence depends on regime (tropical, subtropical, subpolar). The vigorous subtropical western boundary currents and equatorial current systems associated with wind forcing extend weakly to the bottom, but their lateral extent is very limited. Outside the energetic wind-driven western boundary regimes, circulation below the subtropical and tropical pycnocline may be mostly associated with buoyancy forcing and overturning circulation. This includes weak interior ocean flows that are easily masked by the eddy field and the slightly more vigorous DWBCs (Section 7.10.3), which are observed at all latitudes in the Atlantic. In contrast, the wind-forced subpolar North Atlantic circulation, while most vigorous at the sea surface, extends to the ocean bottom where it is merged with the buoyancy-driven circulation; the whole complex mostly follows topographic contours.

9.6.1. Depth Dependence of the Wind-Driven Circulation

The subtropical and subpolar gyre circulations change with depth. The energetic wind-driven circulation of the upper ocean decreases in energy with depth and also changes shape laterally. The main points about depth dependence of the wind-driven, anticyclonic *subtropical circulation* apply to all subtropical gyres, including those of the North and South Atlantic:

1. The western boundary currents and their extensions penetrate to the ocean bottom, but are vertically sheared so that the highest velocities are in the upper ocean. The recirculation gyres, which are directly adjacent to and result from these strong currents, also penetrate to the ocean bottom.
2. The subtropical gyres shrink westward and poleward with increasing depth, becoming compressed into their western boundary currents and separated extensions.
3. The subtropical gyre circulation can be conceptualized as multiple layers in which streamlines begin at the sea surface and move downward along isopycnals into the interior ocean (*ventilation* through the process of *subduction*, Section 7.8.5), and deeper layers of anticyclonic circulation that are not connected to the sea surface (locally unventilated). Ventilated layers contain unventilated regions where the streamfunctions do not connect to the sea surface. The flow in each of the layers is rotated relative to that in the overlying and underlying layer, so that at any given location (latitude-longitude), the waters on different isopycnals forming the local vertical profile will have come from different geographic locations at the sea surface. This creates the subtropical pycnocline structure (Central Water).

The main points about the depth dependence of the wind-driven, cyclonic *subpolar circulation* are

1. The circulation is nearly "equivalent barotropic," meaning the surface current structure extends to the ocean bottom (barotropic), even though it diminishes in strength with depth (equivalent).
2. Ekman divergence in the surface layer drives upwelling, so there are no regions of subduction, hence interior ventilation, via wind-driven flow along streamlines.

(Ventilation in this region is due to the buoyancy-driven circulation, through convection or brine rejection.)
3. In the subpolar North Atlantic, the Greenland-Iceland-Shetland ridge strongly constrains the subpolar circulation. The flow above the sill depth of the ridge extends northward into the Nordic Seas, and is part of a much larger regional cyclonic circulation. Below sill depth, the subpolar circulation is constrained to follow the complicated isobaths. Thus in the North Atlantic, there is a shift in the shape of subpolar circulation above and below this ridge.
4. The coexisting overturning circulation also has deep currents that follow isobaths. It is not straightforward to distinguish wind-driven and thermohaline features in the North Atlantic's subpolar gyre. Bottom intensification is an indication that a given flow has a significant thermohaline component. (Examples are DWBCs and the plunging plumes that overflow from the Nordic Seas and Mediterranean, neither of which are wind-driven features.)

9.6.1.1. Depth Dependence of the Subtropical Gyre Circulation

The North Atlantic's subtropical circulation at the sea surface is anticyclonic, with its intense western boundary current, but the subtropical "gyre" is not quite closed (Figure 9.2a). On the equatorward side, the gyre's streamlines merge broadly with the equatorial circulation, into the eastward NECC. The subtropical Brazil Current gyre also smoothly merges into the SECC. In the North Atlantic, the region of highest steric height at the sea surface parallels the Florida Current and Gulf Stream, with a maximum offshore of the Antilles Current. In the South Atlantic, the region of highest steric height stretches from 15 to 40°S.

However, just 250 m below the sea surface and well represented by the 500 dbar map (Figure 9.2b), the subtropical gyres in both hemispheres are considerably more "gyre-like," with large areas of closed streamlines. For the Gulf Stream, the region of closed streamlines is shifted away from the eastern boundary and toward the separated Gulf Stream. The region of highest steric height shifts to north of about 30°N. The Brazil Current gyre likewise tightens toward the pole and the west, shifting to south of 30°S. At 500 dbar, both subtropical gyres have tightened further into their western and poleward corners.

At 1000 dbar in the North Atlantic (Reid, 1994), the Gulf Stream System is greatly reduced spatially, to the separated eastward flow of the Gulf Stream and its two recirculation gyres to the north and south. At 1000 dbar in the South Atlantic, the strongest part of the gyre shrinks toward the Brazil Current's southwest corner/separation point.

The Gulf Stream and NAC and their recirculation gyres penetrate to the ocean bottom (Figure 9.14 and supplementary Figure S9.16). Acoustically tracked floats at 2000 m show this penetration, and also the vanishing of statistically important mean flow in other regions (Owens, 1991). At greater depth, current meter observations along 55°W show the Gulf Stream and its flanking recirculations (Hogg, 1983). (See also Figure 9.5b.)

In the South Atlantic, the poleward shrinkage of the subtropical gyre has been observed with acoustically tracked floats at the bottom of the pycnocline and in the AAIW layer (Boebel et al., 1999; supplementary Figure S9.17). The westward return flow of the subtropical gyre is much more zonal and intense in these direct observations than in the Reid (1994) streamfunctions. The anticyclonic Zapiola Eddy, embedded in the eastward flow of the ACC and SAC in the central Argentine Basin (Section 9.5.5), is more of a closed circulation at depth than at the sea surface.

9.6.1.2. Depth Dependence of the North Atlantic's Subpolar Gyre

The subpolar gyre is divided at the sea surface into western and eastern domains (Section 9.3.5). The western subpolar domain, west of the Reykjanes Ridge, has an almost closed cyclonic surface circulation (the Rim Current introduced in Section 9.3.5). The eastern domain is the NAC, which flows eastward at about 50°N and then turns northeastward and crosses the Iceland-Faroe-Shetland ridge into the Norwegian Sea.

With depth, the Rim Current extends to the ocean bottom (Figures 9.2 and 9.14). By mid-depth, and down to the bottom, this circulation is filled with newly formed intermediate and deep waters (Section 9.8). By 700 m depth, the mean flow also includes a counterflow offshore of the boundary current, creating a cyclonic dome in the shape of a "donut" around the Labrador Sea and western Irminger Sea (Lavender et al., 2000; supplementary Figure S9.18). This donut is the preferred locale of the deeper convection that creates the densest SPMW and LSW (Section 9.8.2).

In the eastern subpolar domain, the Iceland-Faroe-Shetland ridge alters the northeastward NAC. Below sill depth, the flow must be closed to the north, and it becomes continuously cyclonic, nearly following bathymetric contours (Figures 9.2, 9.14 and supplementary Figure S9.19 from Bower et al., 2002). Vertical shear is required for this configuration and most likely occurs mainly on the eastern flanks of the Reykjanes Ridge and Rockall Plateau.

9.6.2. Deep Circulation and Deep Western Boundary Currents

This is a brief overview of the part of the weak lateral circulation below the main pycnocline that is mostly associated with density changes. The deep circulation is often described in terms of water masses (Section 9.8), since the direction of flow is often inferred

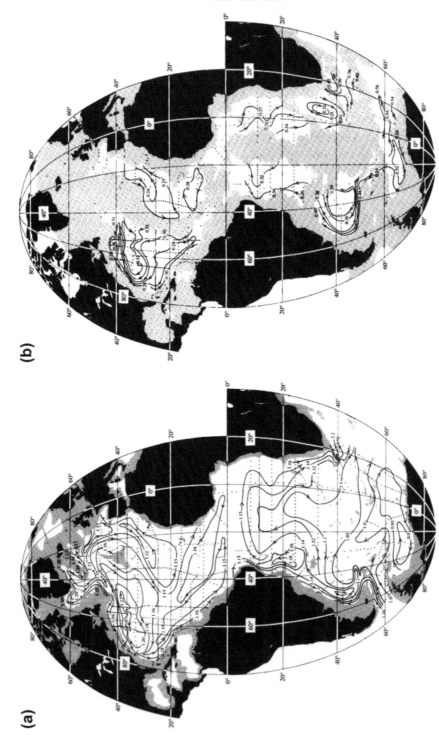

FIGURE 9.14 Steric height (10 m² s⁻²) at (a) 2500 dbar and (b) 4000 dbar, adjusted to estimate the absolute geostrophic circulation. *Source: From Reid (1994).*

from property distributions (because there are few direct velocity observations). The associated overturning circulation is described in Section 9.7.

9.6.2.1. Lateral Circulation and Basin Connections

Looking at the lateral circulation at 2500 and 4000 dbar (Figure 9.14), the Gulf Stream and its recirculation features are still present, as is the subpolar North Atlantic circulation and a residual of the Brazil-Malvinas confluence. Circulation in the South Atlantic may be broadly cyclonic, circling the MAR. Along the coast of Africa, beneath the Benguela Current, a deep-reaching poleward boundary current occurs at 2500 dbar, analogous to deep poleward flow in the South Pacific at about the same latitudes. This transports NADW out of the Atlantic and into the Indian Ocean. At 4000 dbar, flow in this Cape Basin, south of the Walvis Ridge, is likely cyclonic.

Abyssal flows are affected by the topography. Deep flows often follow topographic contours and mixing can be related to the structure of the topography. The mid-ocean ridges confine deep waters to the abyssal basins. Fracture zones in the ridges allow for limited exchange through sometimes vigorous, turbulent flow of waters from one deep basin to another. Bottom waters in the downstream basin tend to be relatively uniform with properties set where the basin was filled at the fracture zone. Principal fracture zones affecting Atlantic deep and bottom waters include the following, each of which has been studied locally: the Vema and Hunter Channels (northward flow of AABW from the Argentine to the Brazil Basin), the Namib Col in the Walvis Ridge (southeastward flow of AABW and NADW into the Cape Basin), the Romanche Fracture Zone in the MAR at the equator (eastward flow of AABW and NADW), the Vema Fracture Zone at 11°N in the MAR (eastward AABW flow into the eastern North Atlantic), and the Charlie Gibbs Fracture Zone

at 52°N in the MAR (eastward flow of the Denmark Strait Overflow Water and Labrador Sea Water).

9.6.2.2. Deep Western Boundary Currents

The dense water masses formed in the northern North Atlantic must, on average, spread southward while the dense waters formed in the Southern Ocean must, on average, spread northward. DWBCs that respond to spatially limited sources of dense water and net upwelling in the ocean interior are part of the circulation of these newly formed dense waters. (Dynamically, it is important to recall from Section 7.10.3 that the DWBCs do not necessarily flow away from their deep sources, but in the case of the Atlantic, they mostly do.)

Historically, Wüst (1935) showed preferential southward spreading of the North Atlantic's oxygenated, saline deep waters along the western boundary, foreshadowing later discovery of the DWBC there. In the 1950s, following H. Stommel's advice, Swallow and Worthington (1961) measured the southward DWBC beneath the Gulf Stream off the coast of South Carolina (Section 7.10.3). Through the 1960s and 1970s, DWBCs were traced worldwide (Warren, 1981). Work since then has refined estimates of transports, described exchange between DWBCs and the interior, considered the continuity of DWBCs, and studied local aspects of interaction of DWBCs with other strong circulation systems.

Along the western boundary of the Atlantic, DWBCs associated with both NADW and AABW are found. The northern DWBC originates in overflows from the Nordic Seas joined by mid-depth waters from the Labrador Sea (Figure 9.15a); direct velocity measurements east of Greenland show the plume of dense overflow water moving to the bottom of the northern North Atlantic (supplementary Figure S9.20 from Dickson and Brown, 1994). The AABW DWBC lies beneath and offshore of the NADW

DWBC (Figures 9.14, 9.15b and Figure 9.25). Existence of the DWBCs is illustrated by a series of velocity sections in the textbook Web site (Figures S9.21, S9.22, S9.23). Oxygen and chlorofluorocarbons (CFCs) are heightened in the DWBCs because they carry recently formed NADW (LSW and Nordic Seas Overflow; NSOW), which has elevated atmospheric gas concentrations, into the subtropical North Atlantic (Figures 9.22 and 9.7).

The DWBC that carries NADW southward is centered around 2500 m, but it extends up to at least 1500 m in the North Atlantic and tropics, and down to the bottom in the North Atlantic (Figures 9.5, 9.9, 9.11). The NADW's DWBC begins to form as soon as the NSOWs spill across the Greenland-Iceland-Shetland ridges into the deep subpolar North Atlantic, forming the cyclonic Rim Current at depth (Figure 9.15a). This abyssal current follows the boundary

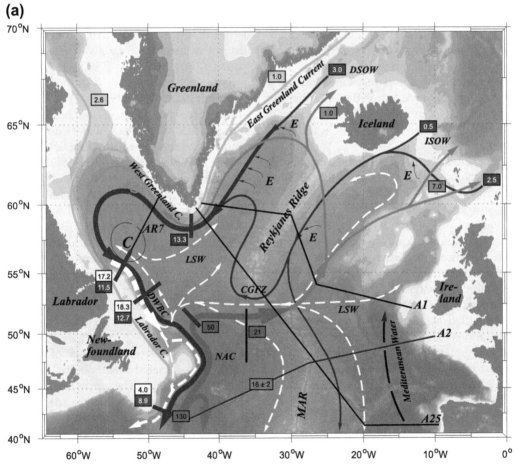

FIGURE 9.15 Schematics of deep circulation. (a) NSOW (blue), LSW (white dashed), and upper ocean (red, orange, and yellow) in the northern North Atlantic. *Source: From Schott and Brandt (2007).* (b) Deep circulation pathways emphasizing DWBCs (solid) and their recirculations (dashed). Red: NSOW. Brown: NADW. Blue: AABW. This figure can also be found in the color insert. *(M.S. McCartney, personal communication, 2009.)*

(b)

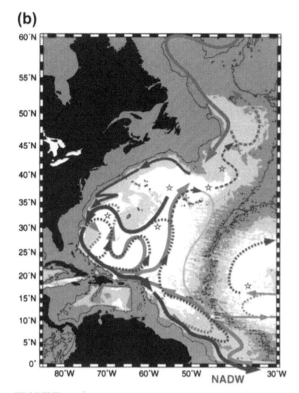

FIGURE 9.15   (*Continued*).

around Greenland and into the Labrador Sea, where it picks up LSW along the western boundary. The whole layered complex flows out of the Labrador Sea into the Northwest Corner of the NAC. Part of it joins the eastward NAC flowing through the Charlie Gibbs Fracture Zone, and part continues southward along the western boundary to the east of Flemish Cap, and then along the Grand Banks of Newfoundland (Figure 9.9b). This part of the DWBC moves southward along the western boundary under and inshore of the Gulf Stream, where it is seen at Cape Hatteras (Figure 9.5b). Interaction with the Gulf Stream is complex (Pickart & Smethie, 1993), as suggested by the brown zigzag in Figure 9.15b. The NADW's DWBC moves on to the equator, where part of the flow turns eastward along the equator

(Figure 9.11). Part continues into the South Atlantic, and leaves the western boundary at 25 to 40°S (Figure 9.14a).

Velocities in the NADW's DWBC are on the order of 5 to 20 cm/sec and more. Transports are on the order of 10 to 35 Sv, and depend on latitude since the DWBC has significant recirculations, indicated schematically in Figure 9.15b. The recirculations increase the local transport and also mix the DWBC waters with the interior waters, greatly increasing the transit time of water parcels down the western boundary as measured by transient tracers such as CFCs. At 26.5°N, for instance, the southward DWBC throughput is as much as 22 Sv, while the net southward transport is 35 Sv, of which 13 Sv are due to the deep recirculation gyre (Bryden, Johns, & Saunders, 2005a).

Now consider the northward flow of the AABW's DWBC, which becomes organized in the southwestern South Atlantic. The AABW DWBC moves northward along the coast of South America offshore of and deeper than the southward DWBC carrying NADW (Figure 9.25). Its northward transport is in the order of 7 Sv in the South Atlantic and into the North Atlantic. As the AABW approaches the equator, part of its transport turns eastward with the NADW (Figure 9.11), and the rest crosses the equator. At this point it crosses to the *eastern* boundary of the basin, riding along the western flank of the MAR, rather than remaining a DWBC (Figures 9.14b and 9.15b). Some of it passes into the deep eastern North Atlantic through the Vema Fracture Zone. The AABW loses its transport through upwelling into the overlying isopycnals, which contain NADW, and disappears.

9.6.2.3. Recirculations and Time Dependence

The DWBCs are the most energetic part of the deep circulation with velocities up to tens of centimeters per second. The simple, laminar boundary currents predicted by theory are unlikely to be either simple or laminar. One analogy might be the extent to which eastern boundary currents are modeled as simple laminar flows that arise in response to offshore Ekman transport whereas in actuality they are full of local jets and eddies. Another is the extent to which the actual Gulf Stream is predicted by simple Sverdrup balance/western boundary current theory. Both systems have much greater spatial and temporal variability than simple theories suggest, although the simple theories provide the most basic understanding of the existence of these systems.

DWBCs lie entirely in the deep ocean, so their spatial and temporal variability have been difficult to observe. Differences between the observed DWBCs and simplified theory include geographically localized detrainment of water from the DWBCs, and large-scale, permanent recirculation gyres (e.g., Figure 9.15b). High-resolution numerical modeling suggests ongoing creation of DWBC eddies along the South American coast (Dengler et al., 2004), while deep Lagrangian float observations show considerable eddy activity (Hogg & Owens, 1999).

Several detrainment locations and recirculation gyres for the DWBCs have been described. At each, the DWBC properties change significantly as water is exchanged with the interior. Starting from the north and with the NADW DWBC, there is detrainment at the exit to the Labrador Sea and along the Newfoundland/Flemish Cap region (Bower, Lozier, Gary, & Böning, 2009). A second is at the Gulf Stream separation point at Cape Hatteras (Pickart & Smethie, 1993). There is a tropical set, at about 20 and 5°N, and a large detrainment at the equator, all as part of the Guiana abyssal gyre (Kanzow, Send, & McCartney, 2008). In the South Atlantic, there is a change in DWBC character as it passes the easternmost point of South America (around 8°S) with either a recirculation gyre between 20 and 8°S (Reid, 1994;

Friedrichs, McCartney, & Hall, 1994) or a change to a more eddy-like character (Dengler et al., 2004). A major detrainment occurs at about 20°S upon encountering the Vitória-Trindade Seamounts, forming the southern boundary of a recirculation gyre (Tsuchiya et al., 1994; Hogg & Owens, 1999). The final detrainment is where the NADW DWBC, along with the Brazil Current, encounters the Malvinas Current/SAF.

The AABW DWBC, flowing northward through the South Atlantic, also has several major transitions. Large temporal variability is found where it has been directly observed. The first transformation occurs where this DWBC leaves the Argentine Basin and enters the Brazil Basin, at the Rio Grande Rise around 32°S, where its deepest flow is confined to the narrow Vema and Hunter Channels (Hogg, Siedler, & Zenk, 1999). The second is at the Vitória-Trindade Seamounts which interrupt and deflect the DWBC eastward at 20°S (Hogg & Owens, 1999). A third large change occurs at the equator, where the northward flow of AABW shifts over to the eastern side of the Guiana Basin (McCartney & Curry, 1993).

9.7. MERIDIONAL OVERTURNING CIRCULATION IN THE ATLANTIC

The MOC of the Atlantic, which is part of the global overturning circulation (Chapter 14), is a double cell consisting of (1) northward flow of upper ocean waters that become denser in the northern North Atlantic and flow out southward at depth, eventually becoming NADW, and (2) northward flow of dense AABW that upwells into the lower part of the NADW, disappearing by the mid-latitude North Atlantic.

The upper ocean waters that flow northward from the South Atlantic to feed the overturn in the North Atlantic originate as: (a) upper Indian Ocean waters from the Agulhas retroflection, and (b) the slightly more dense AAIW and Upper Circumpolar Deep Water (UCDW: Section 9.8.3). These are transformed into NADW components in the Labrador, Mediterranean, and Nordic Seas. The water masses are described in Section 9.8.

Superimposed above this in the surface layers are shallow overturning cells that move the warm, light tropical surface waters poleward in the subtropics and return them as cooler, denser subducted pycnocline waters. While these shallower cells might not grab our attention because they do not have global scale, they are responsible for most of the ocean's poleward heat redistribution (Chapter 5).

Returning to the full-depth MOC, meridional transports are computed in layers from zonal coast-to-coast sections (Section 14.2; see example in Figure 9.16 from Talley, 2008; compare with Bryden et al., 2005b and Ganachaud, 2003, whose results are included in supplementary Figures S9.24 and S9.25). The net southward transport of NADW is typically 15–25 Sv (depending on latitude), which is almost all carried by the DWBC (Section 9.6.2). Northward transports that feed this are divided into 3 to 7 Sv of bottom water with the remainder in the upper ocean layers (AAIW and pycnocline).

An overturning transport streamfunction can then be calculated for display; the method is described in Section 14.2.3. An Atlantic overturning streamfunction from a high-resolution global ocean model is shown in Figure 14.8 (Maltrud & McClean, 2005). This particular calculation shows a maximum NADW cell of 22 Sv centered at 40°N, with a typical NADW transport of 16 Sv for the length of the Atlantic. This particular model also has almost no AABW bottom cell, which is a common problem with ocean circulation models at this time; data-based transport estimates show a much more robust AABW influx (Figure 9.16).

The Atlantic's MOC is included in all schematics of global overturn (Figures 14.10 and 14.11 and the original sources upon which these were based). The NADW layer that exits the Atlantic upwells to the surface waters and downwells to the bottom waters that then feed back into the Atlantic. The upwelling occurs broadly through the Indian and Pacific Oceans and in the Southern Ocean within the ACC latitudes. The "downwelling" is formation of dense waters around Antarctica from upwelled surface waters that include NADW (Chapter 14).

The meridional heat and freshwater transports accompanying the overturning circulation are discussed in Section 14.3, in the context of the transports for all oceans. Briefly, the heat transport is northward throughout the length of the Atlantic, with a maximum in the subtropical North Atlantic just south of the Gulf Stream separation, where ocean heat loss is maximum. The northward sign found even in the South Atlantic is due to the additional heat loss region of the Nordic Seas, and is thus associated with the full-depth MOC. Freshwater transport is more complicated to discuss, but the most important result for the Atlantic's overturning circulation is that the NADW cell transports freshwater southward, because the northward-flowing upper ocean waters are saline, and the new NADW is fresher because it incorporates much of the Arctic and subpolar net precipitation and runoff. (It is dense enough to sink because it is cooler than the inflowing surface waters.)

9.8. ATLANTIC OCEAN WATER MASSES

The hydrographic structure and basic water masses of the Atlantic Ocean were introduced in Chapter 4 in terms of four layers in the vertical: surface through pycnocline, intermediate, deep, and abyssal. Upper ocean water mass structures and processes are similar in all oceans. However, the North Atlantic forms deep waters locally, unlike the North Pacific and Indian, resulting in a complete asymmetry

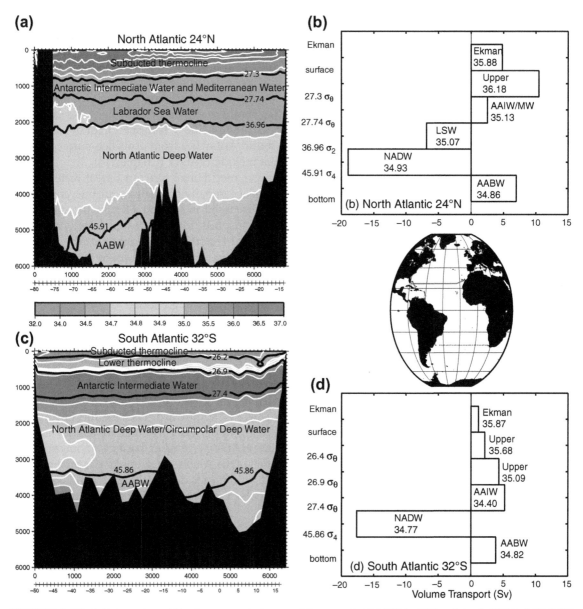

FIGURE 9.16 Salinity and meridional transport in isopycnal layers at (a, b) 24°N in 1981 and (c, d) 32°S in 1959/1972. The inset map shows section locations. The isopycnals (σ_θ, σ_2, σ_4) that define the layers are contoured in black on the salinity sections. Figures 9.16a, c can also be found in the color insert. See also online supplementary Figures S9.24 and S9.25 for examples from Bryden, Longworth, and Cunningham (2005b) and Ganachaud (2003). *After Talley (2008), based on Reid (1994) velocities.*

between the three oceans in their deep water ages and age-related properties (oxygen, nutrients, CFCs, etc.; Sections 4.5 and 4.6; Figures 4.11 and 4.22). Together with the supply of younger waters in the Antarctic, which is common to all three oceans, much of the interior Atlantic is affected by ventilation changes within decades. Therefore care must be taken in combining Atlantic data sets from different decades.[5]

Most of the main water masses of the Atlantic Ocean are shown in Figures 9.17 and 9.18 and listed in Table S9.4 in the online supplement. We start with summary potential temperature-salinity diagrams (Section 9.8.1) and follow with details about the water masses from shallow to deep, illustrated with vertical sections and maps of properties on isopycnals.

9.8.1. Potential Temperature vs. Salinity and Oxygen

Many water masses are identified by vertical maxima or minima of salinity or oxygen. The principal water masses are therefore first illustrated with salinity and oxygen versus potential temperature. Figure 9.18 includes thousands of bottle samples collected in the WOCE. A much older but useful schematic potential temperature-salinity (T-S) diagram from the 5th edition of this text, based on Sverdrup et al. (1942), is included in the online materials along with a display of T-S diagrams in each 5° latitude-longitude square (Figures S9.26 and S9.27). It is useful to consider these diagrams together with the surface property maps and vertical sections of Chapter 4. Each water mass introduced here is considered in more detail in subsequent sections.

Looking at the overall ranges of properties, the highest and lowest temperatures are the 29−30°C of tropical surface water and the negative temperatures of the bottom water from the Antarctic. (Water at the freezing point is found on coastal shelves in the North Atlantic, but is too fresh to appear here.) Highest salinities at highest temperatures are in the subtropical surface waters at 11−24°S and 20−30°N. In oxygen, the ridge that tilts from 200 μmol/kg at high temperature to 350 μmol/kg at low temperature is the locus of 100% saturation, hence surface water. Lowest oxygen occurs in the low latitude, upper ocean oxygen minimum zones, resulting from high biological productivity.

Below the warm tropical surface water, the nearly linear T-S relation in both the North Atlantic and South Atlantic is called the Central Water. This is the main pycnocline of the subtropical gyre of each ocean. Central Waters originate from surface waters that subduct from different locations and have a range of densities (Sections 4.2.3 and 7.8.5). The North Atlantic Central Water is saltier than the South Atlantic Central Water, and is, in fact, the saltiest Central Water of all five oceans (Figure 4.7).

---

[5] Water characteristics of the Atlantic have been surveyed numerous times, which is advantageous for in situ study of climate variability. The first basin-wide survey with temperature, salinity, and oxygen measurements from top to bottom was carried out on the German *Meteor* from 1925−1927 (Wüst, 1935). A second major survey was carried out in 1957−1958 as part of the International Geophysical Year, intentionally repeating many of the Meteor sections to obtain a direct comparison of the distribution of water properties after the interval of 30 years. Throughout the 1970s and 1980s, much of the Atlantic was surveyed for chemical tracers, as well as basic hydrographic properties, along with a number of newly eddy-resolving sections; all vertical sampling included conductivity, temperature, depth (CTD) profiling as well as bottle samples. In the 1990s, all of the Atlantic was surveyed again as part of the World Ocean Circulation Experiment (WOCE). These various experiments are summarized in numerous papers in Siedler, Church, and Gould (2001). Post-WOCE, hydrographic sampling continues at a high pace to observe the clear changes in deep water properties associated with surface changes and to follow anthropogenic carbon signals into the ocean.

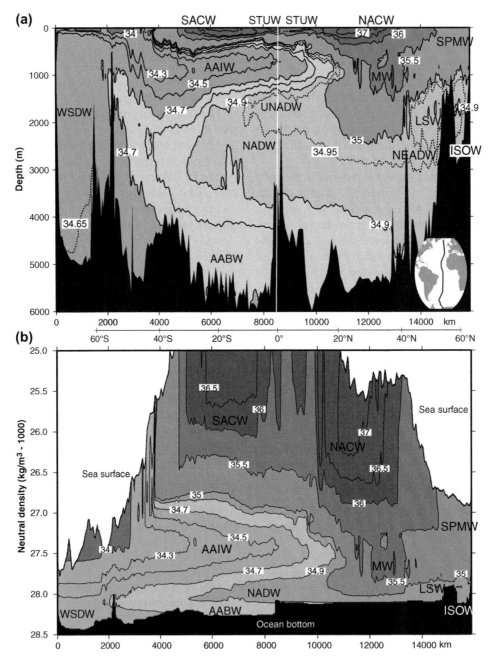

FIGURE 9.17 Location of most major Atlantic water masses using a meridional salinity section at 20–25°W, as a function of (a) depth and (b) neutral density (γ^N). (White areas at high density are the ocean bottom. White areas at low density (top of figure) are above the sea surface.) Inset map in (a) shows station locations. Acronyms are within the text and in Table S9.4 in the online supplement. (See also Figure 4.11b.) (World Ocean Circulation Experiment sections A16 and A23)

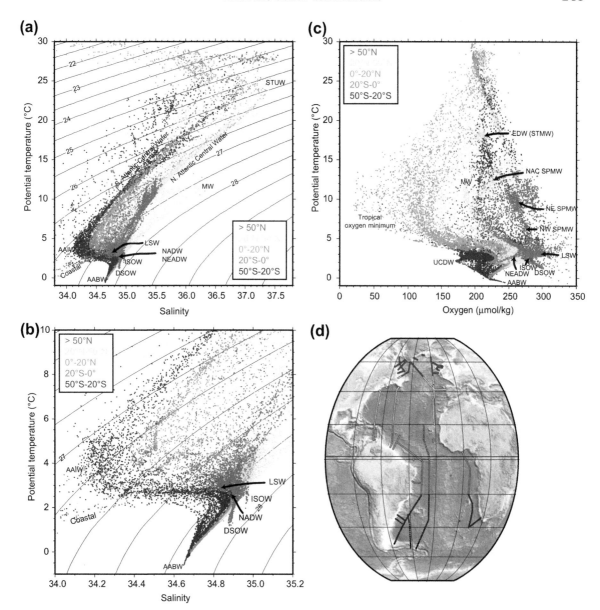

FIGURE 9.18 Potential temperature (°C) versus salinity for (a) full water column, and (b) water colder than 10°C. (c) Potential temperature versus oxygen for full water column. (d) Station location map. Colors indicate latitude range. Contours are potential density referenced to 0 dbar. Data are from the World Ocean Circulation Experiment (1988–1997). This figure can also be found in the color insert.

Within the Central Water, subducted high salinity surface water creates a near-surface salinity minimum called Subtropical Underwater (STUW). This is a noisy presence at around 25°C in this sample-based T-S relation, but is much clearer in the vertical sections shown in the following text. STMWs are also subducted, but they cannot be seen in T-S because they fit the Central Water T-S relation; they are more easily seen in oxygen (described later).

In the North Atlantic, proceeding downward, the high salinity MW is the salty protrusion in T-S at mid-latitudes (20–50°N band). Although warm, around 12°C, MW is so salty that it is almost as dense as the much colder NSOW.

The denser North Atlantic waters are best represented in T-S north of 50°N (red points in Figure 9.18a, b). The LSW is the main salinity minimum. The set of colder, fresher, less dense points that extend toward low salinity away from the LSW are Greenland and Labrador coastal waters. *Northeast Atlantic Deep Water* (NEADW) is the cluster of higher salinity waters colder than about 3–4°C. Protruding toward lower salinity from the NEADW are the two NSOWs: DSOW and *Iceland-Scotland Overflow Water* (ISOW). DSOW is fresher and more oxygenated than ISOW.

The low salinity AAIW (around 4°C and 34.1 psu) and the high salinity NADW (at 2–3°C) are found in the tropical and South Atlantic. At the bottom, we see the narrow tail of the AABW.

Oxygen (Figure 9.18c) provides independent water mass information. The tropical oxygen minimum was already mentioned. Clusters of points around 18°C, 11–14°C and 8–11°C indicate the high-volume mode waters (Hanawa & Talley, 2001). They are the *Eighteen Degree Water* (EDW) (identical to STMW), NAC SPMW, northeastern Atlantic SPMW (NE SPMW), and northwestern Atlantic SPMW (NW SPMW). All have high oxygen saturation.

Farther down, at lower temperature, the high oxygen LSW and DSOW appear. ISOW does not have an obvious oxygen extremum due to entrainment of lower oxygen water as it overflows into the North Atlantic. At the bottom (lowest temperature), the higher oxygen AABW is seen. In these deeper waters, two features that appear in oxygen but not well in salinity are the NEADW, which is an axis toward lower oxygen in the northern latitudes (red points) at about 3°C, and the low oxygen UCDW at about 2°C in the eastern South Atlantic.

9.8.2. Atlantic Ocean Upper Waters

9.8.2.1. *Surface Water and Mixed Layer*

Surface temperature (Figure 4.1) is highest, up to 30°C, in a band north of the equator, in the ITCZ. The equatorial cold tongue is evident in the east, separated from the South Atlantic's colder subtropics by a band of warmer water at 10°S. The Gulf Stream and Brazil Current subtropical gyres both include poleward intrusion of warmer waters along their western boundaries and equatorward swoops of cooler waters in the east. In the satellite SST image in Figure 1.1a, the separated Gulf Stream is evident as a narrow band of warm water, with the cooler recirculation just to its south. The NAC and its northwest corner are evident east of Newfoundland, and the warm waters of the NAC spread northward toward Iceland and Scotland. The coldest waters are in the EGC, and, in the Labrador Sea, in the Davis Strait and southward in the Labrador Current. In the southern South Atlantic, the cold Malvinas Current loops northward and the strongly eddying Subantarctic and Polar Fronts stretch eastward.

Atlantic sea-surface salinity (SSS), as in all oceans, is dominated by the alternation of regions of net precipitation/runoff with those of net evaporation (Figure 4.15, Section 4.3). Lower surface salinity is found in the tropics, especially beneath the ITZC; low surface salinity is evident due to runoff from the Amazon and

Congo Rivers as well. Net evaporation in the subtropics results in maximum surface salinity there. The northward swing of the fresh Malvinas loop in the Southern Ocean is apparent. In the North Atlantic's subpolar gyre, lowest salinities are along the coasts of Greenland and Labrador, and in the slope water region north of the Gulf Stream. On a global scale, the high Atlantic SSS, extending far into the Nordic Seas, reflects the overall higher salinity of the Atlantic, which facilitates deep water formation here and not in the much fresher North Pacific.

The basin-wide pattern of sea surface density (Figure 4.19, Section 4.4) mainly follows SST, with lowest density in the tropics and higher density at higher latitudes. Salinity affects the surface density in the subpolar North Atlantic compared with the North Pacific: at a given latitude, the saltier North Atlantic is significantly denser than in the North Pacific. Large river outflows (Amazon, Congo, and Orinoco) also cause the lowest surface densities in the tropics.

The winter surface mixed layer in the Atlantic (Figure 4.4a and c, Section 4.2.2) is markedly thick throughout the subpolar gyre and into the Labrador Sea; this thick layer is called Subpolar Mode Water (Figure 9.19b) (McCartney & Talley, 1982; Hanawa & Talley, 2001). A band of thick mixed layers is also found just south of the Gulf Stream (EDW or STMW; Figure 9.19a) (Worthington, 1959). In the South Atlantic, winter mixed layers are also thicker farther south near the ACC (SAMW), but

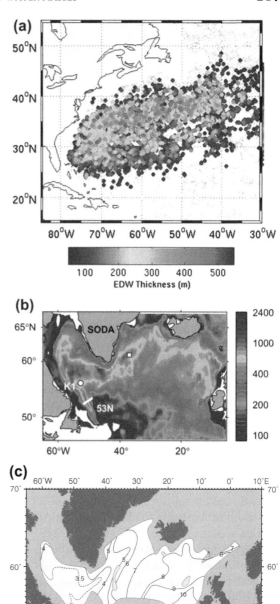

FIGURE 9.19 Mode waters. (a) EDW thickness from all Argo profiles from 1998–2008. The EDW is defined here by $17°C \leq T \leq 19°C$ and $dT/dZ \leq 0.006°C/m$. The small gray dots in the background indicate profiles without EDW. (*Young-Oh Kwon, personal communication, 2009.*) (b) March mixed layer depth from a data-assimilating model (SODA). *Source: From Schott et al. (2009).* (c) Potential temperature (°C) of the late winter mixed layer, shown only where the mixed layer is more than 200 m thick. This is the SPMW. *Source: From McCartney and Talley (1982).*

much thicker winter mixed layers are found at similar southern latitudes in the Pacific and Indian Oceans (McCartney, 1977).

9.8.2.2. Central Water and Subtropical Underwater

Central Water includes the water in the main pycnocline in each subtropical gyre (Section 9.8.1 and Section 4.2.3). The North and South Atlantic Central Waters extend to depths of 300 m on either side of the equator, deepen to 600 to 900 m at mid-latitudes, and are somewhat shallower at the poleward side of their gyres. The density range of Central Water is set by the winter surface density within the Ekman pumping region of the subtropical gyre. The highest such density in the North Atlantic is around $\sigma_\theta = 27.2$ kg/m^3, which outcrops in winter around 52°N. The southern edge of North Atlantic Central Water is the southern edge of the subtropical gyre, which is nicely marked by the onset of very low oxygen in the tropics at about 20°N (Figure 4.11d).

In the South Atlantic, defining the maximum density of subduction based on the region of Ekman downwelling is problematic because the subtropical gyre is connected to the Indian Ocean's subtropical gyre over a broad latitude range between Africa and the ACC. The maximum outcrop density in the combined South Atlantic-Indian subtropical gyre is $\sigma_\theta \sim 26.9$ kg/m^3, south of Australia. Within the South Atlantic proper, west of Africa, the maximum density of gyre-wide winter outcropping could be as low as $\sigma_\theta \sim 26.2$ kg/m^3. As in the North Atlantic, the tropical oxygen minimum marks the northern edge of the subtropical gyre and Central Water.

The STUWs are shallow vertical salinity maximum, within the upper 100 m. STUWs are relatively minor water masses in terms of areal extent, volume, and formation rates (~1−2 Sv; O'Connor, Fine, & Olson, 2005), but they nicely illustrate the subtropical subduction process (Figure S9.28 in the online supplementary material). They are embedded in Central Water

arising from equatorward subduction from the subtropical SSS maxima. In the South Atlantic, STUWs occur between about 13 and 6°S; in the North Atlantic, the range is 12°N to 20−25°N depending on longitude. North Atlantic STUW potential density is $\sigma_\theta \sim 25.5$ kg/m^3. South Atlantic STUW has a larger density range, centered at about $\sigma_\theta = 24$ to 24.5 kg/m^3 in the western South Atlantic but denser in the eastern South Atlantic.

The salinity contrast between the STUW salinity maximum and the underlying fresher water can be large, leading to favorable conditions for salt fingering (Section 7.4.3.2). Schmitt, Perkins, Boyd, and Stalcup (1987) observed multiple stepped layers of 5−30 m thickness, indicative of salt fingering, beneath the salinity maximum east of Barbados. The layers had remarkable coherence over hundreds of kilometers in the horizontal and remarkable persistence in time.

9.8.2.3. Mode Waters

Mode Waters are layers that, in terms of isopycnal spacing, are relatively thick compared with surrounding waters on the same isopycnals and in the vertical (Section 4.2). The North Atlantic has several STMWs and its SPMW.

The North Atlantic's principal STMW is found south of the Gulf Stream; it is also called Eighteen Degree Water because of its typical temperature. EDW is the archetype of all STMW (Worthington, 1959; Masuzawa, 1969). It can be seen on any vertical section crossing the Gulf Stream (Figure 9.7). EDW is a permanent feature, with observations dating back to the *Challenger* expedition in 1873 (Worthington, 1976). It has relatively homogeneous properties centered at about 18°C, 36.5 psu, and $\sigma_\theta = 26.5$ kg/m^3, with some spatial and temporal variability. EDW originates in thick winter mixed layers adjacent to the Gulf Stream and within the tight recirculation gyre (Section 9.3.2). The mixed layer thickness can reach to more than 500 m (Figure 9.19a). EDW subducts southward into the western subtropical gyre, creating a low stability, subsurface layer far from the Gulf Stream, throughout most of the Sargasso Sea. The estimated EDW formation rate is 2–5 Sv (Kwon & Riser, 2004). This formation is a conversion from warmer, lighter Gulf Stream waters to the characteristic 18°C of EDW.

Madeira Mode Water, on the southern flank of the Azores Current front, is another STMW and is clearly separate from, and weaker than, the EDW. It is somewhat cooler (16–18°C), saltier (36.5–36.8 psu), and denser ($\sigma_\theta = 26.5$–26.8 kg/m^3) than EDW (Siedler, Kuhl, & Zenk, 1987; New et al., 2001). Its formation rate and volume are much smaller than those of EDW. Whereas EDW is a year-round water mass, Madeira Mode Water is eliminated every year. This difference can be expressed in terms of their residence times: the EDW has a residence time of 3–5 years (which results in a permanent reservoir), whereas the Madeira Mode Water's residence time is about 6–9 months.

The South Atlantic's subtropical gyre has a number of different mode waters related to the complex frontal system associated with the Brazil and Malvinas Currents and the SAF (Tsuchiya et al., 1994). Provost et al. (1999) documented three STMWs in the western South Atlantic (Figure S9.29 in the online supplementary material). The coldest and densest (12–14°C, 35.1 psu, $\sigma_\theta = 26.7$ kg/m^3) is on the north side of the SAF. It is actually the warmest form of SAMW (Chapter 13), but it subducts into the South Atlantic's subtropical gyre like a typical STMW. The second mode water, at ~13.5°C, 35.3 psu, $\sigma_\theta = 26.6$ kg/m^3, is the principal STMW associated with the separated Brazil Current Front. The third STMW is lighter, warmer, and less extensive.

Returning to the North Atlantic, the most significant mode water in terms of volume and impact on internal ocean properties is the SPMW, which is found throughout the subpolar region (Figure 9.19b, c and Figure S9.30 in the online supplementary material). SPMW is an important part of the upper ocean water that

feeds into the NADW, in both the Nordic Seas and the Labrador Sea. SPMW (as depicted originally in McCartney & Talley, 1982) is a broad water mass arrayed around the cyclonic gyre, essentially identical with the winter surface mixed layer. It is generally more than 400 m thick, and is much thicker on the Iceland-Faroe Ridge and in the Irminger and Labrador Seas. The warmest, lightest SPMW (14°C, $\sigma_\theta = 26.9$ kg/m^3) is found east and south of the NAC. As the NAC moves eastward across the North Atlantic, its SPMW becomes progressively colder and denser, reaching about 11°C, $\sigma_\theta = 27.2$ kg/m^3 near the British Isles where the NAC bifurcates. Much of this lightest part of the SPMW subducts southward into the subtropical gyre, behaving as an STMW of the NAC.

The NAC turns northeastward, split into at least three permanent meandering fronts, each with its own progression of SPMWs on its eastern (warm) side (Brambilla & Talley, 2008). These SPMWs do not subduct, but instead continue in the surface layer, becoming progressively colder, fresher, and denser toward the north. The branches east of the Reykjanes Ridge (in the Iceland Basin and Rockall Trough) carry SPMW that cools to 8°C by the Iceland-Faroe Ridge. This SPMW enters the Nordic Seas via the Norwegian Atlantic Current as part of the Atlantic Water that eventually is transformed to NADW (Chapter 12).

The third NAC branch is the Irminger Current, west of the Reykjanes Ridge (Section 9.3.5). Its SPMW progresses toward even colder, fresher, and denser properties around the Irminger Sea, following the East and West Greenland Currents into the Labrador Sea. This SPMW is a source of the LSW (and Irminger Sea Water), which at about 3–3.5°C, form the upper part of the NADW (see the next section).

9.8.3. Intermediate Waters

Below the surface layer and pycnocline, at intermediate depths of about 500–2000 m, the

Atlantic Ocean includes three intermediate water masses, usually identified by vertical salinity extrema. These are the low salinity LSW in the north, the high salinity MW in the subtropical North Atlantic, and the low salinity AAIW in the South Atlantic and tropical Atlantic (summary map in Figure 14.13).

These intermediate water masses are characterized by geographically limited source regions, unlike the upper ocean water masses. LSW forms by deep convection in the central western Labrador Sea, one of the few sites in the global ocean of such convection (Marshall & Schott, 1999). AAIW enters the South Atlantic at the Brazil-Malvinas Current confluence; its low salinity source is the freshest SAMW of the southeastern Pacific (Section 13.5). MW enters the North Atlantic as a dense overflow through the Strait of Gibraltar (Section S8.10.2 in the online supplemental material).

9.8.3.1. Labrador Sea Water

LSW is the intermediate depth water mass of the subpolar and western subtropical North Atlantic. LSW is characterized by (1) a lateral and vertical salinity minimum in the subpolar North Atlantic; (2) a lateral and vertical minimum in potential vorticity (maximum in layer thickness) in the subpolar North Atlantic and subtropical western boundary region; and (3) a lateral and vertical extremum in dissolved gases that mark recent ventilation, such as oxygen and CFCs.

These LSW characteristics result from its convective formation process and young age relative to other waters at the same depth. LSW forms in the Labrador Sea, between Labrador and Greenland, where winter mixed layers exceed 800 m and can reach to 1,500 m depth (Figure 9.20a and supplementary Figure S9.31). (The deep and bottom waters of the Labrador Sea are denser NSOW and NADW, which are never penetrated by the Labrador Sea's deep convection.) The LSW source water is mostly SPMW entering the Labrador Sea from the

Irminger Sea, and includes fresh surface water from Baffin Bay through the Davis Strait. The deep winter mixed layers within the Labrador Sea are capped by lower density in spring, and the thick layers collapse somewhat thereafter forming the relatively uniform and thick layer of LSW. The resulting thick layer of cold, fresh, dense, oxygenated LSW appears in the leftmost panels of each property in Figure 9.20. During the year (1997) of these observations, the new LSW had properties of 2.9–3.0°C, 34.84 psu, and $\sigma_\theta = 27.78$ kg/m^3. LSW properties are variable (Chapter S15 in the online supplemental material); the temperature minimum (<2.8°C) at about 2000 m in the figure is remnant LSW from more vigorous convection at a historically low temperature several years prior.

The new layer of LSW moves southward out of the Labrador Sea following the Labrador Current (Figure 9.15a), as evident in salinity, oxygen, CFCs, and potential vorticity, all of which have extrema in the LSW (Figure 9.21 and also supplementary Figure S15.4). Upon reaching the northwest corner of the NAC, part of the LSW turns eastward with the NAC and part continues on southward past Flemish Cap. The LSW that moves eastward mostly turns northward into the Irminger Sea, while part moves on eastward through the Charlie Gibbs Fracture Zone into the eastern subpolar gyre and northward into the Iceland Basin and into the Rockall Trough. Because of the shorter path to the Irminger Sea, LSW is fresher and more oxygenated there than in the Iceland Basin and Rockall Trough (Figures 9.20 and 9.21).

The LSW in the Labrador Current turns westward as part of the DWBC into the slope water region north of the Gulf Stream. It then moves southward inshore of and beneath the Gulf Stream, which can be seen in the high CFCs at 1000–1700 m at the western boundary at 24°N; elevated oxygen and reduced salinity are also found here (Figure 9.22). Neither is straightforward: Lagrangian floats that should track the southward progression of LSW do not make the

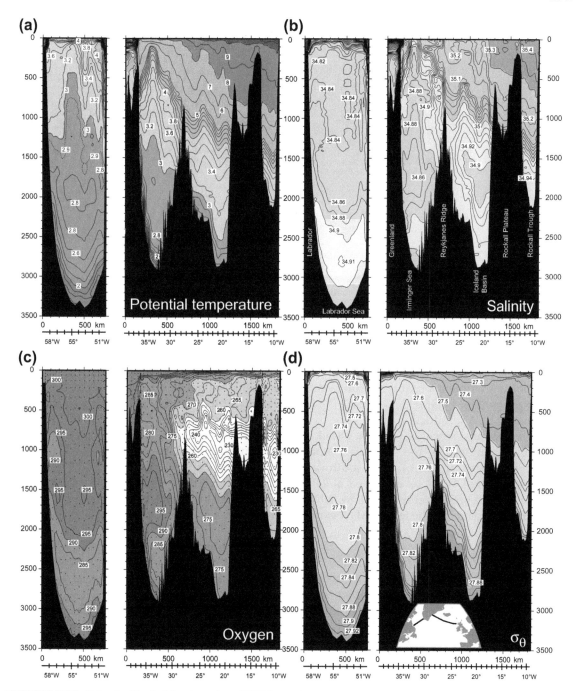

FIGURE 9.20 Subpolar North Atlantic at about 55°N from May to June, 1997. (a) Potential temperature (°C), (b) salinity, (c) oxygen, and (d) potential density (σ_θ) in the Labrador Sea (left side) and from Greenland to Ireland (right side). This figure can also be found in the color insert. (World Ocean Circulation Experiment sections AR7W and A24.)

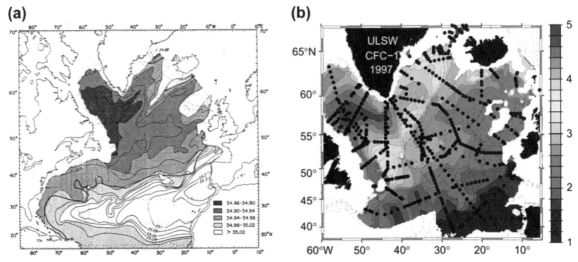

FIGURE 9.21 LSW. (a) Salinity at the LSW potential vorticity minimum. Dark curve is the limit of the PV minimum; salinity on an intersecting isopycnal is shown south and east of this limit. *Source: From Talley and McCartney (1982).* (b) Chlorofluorocarbon-11 (pmol/kg) in the upper LSW layer, at $\sigma_\theta \sim 27.71$ kg/m^3. Figure 9.21b can be found in the color insert. *Source: From Schott et al. (2009) and from Kieke et al. (2006).*

turn (Bower et al., 2009), while the interaction with the Gulf Stream is complex, resulting in significant entrainment of Gulf Stream waters (Pickart & Smethie, 1993).

Estimated LSW production rates vary from 2 to 11 Sv; the most recent is 3 to 9 Sv based on the CFC inventory in the subpolar North Atlantic (Kieke et al., 2006). The southward export rate in the DWBC, as estimated at 24°N (Figures 9.16 and 9.22), is around 6–8 Sv, which is a significant fraction of the total NADW export of 15–20 Sv.

9.8.3.2. Mediterranean Water

The North Atlantic also contains a high salinity water mass, the MW (also called Mediterranean Overflow Water), at about the same depth and density range as LSW. Salinity maps representing the MW were shown in Figure 6.4, at constant depth, on an isopycnal, and at the core of maximum salinity. MW enters the Atlantic as dense water at the Strait of Gibraltar (Figure 9.23a,b). The total outflow is about 0.7 Sv at 38.4 psu and $\sigma_\theta = 28.95$ kg/m^3 (Section

S8.10.2 in the online supplemental material). The overflow plunges downward, entraining ambient water that reduces its salinity and density. It follows the topography to the right, turning northward into the Gulf of Cadiz where it splits into two cores (Figure 9.23c). It reaches its neutral buoyancy and depth of 1000–1500 m by about Cape St. Vincent (Candela, 2001).

As the overflow encounters the sharp northward bend in topography at Cape St. Vincent and other topographic features along the Iberian peninsula, anticyclonic eddies of nearly pure MW are spun off (Bower, Armi, & Ambar, 1997; Richardson, Bower, & Zenk, 2000; Candela, 2001). These "Meddies" propagate southwestward and westward into the North Atlantic, retaining their coherence and high salinity for enormously long distances and over 2–3 years (Figure S9.32 in the online supplementary material). They are entirely subsurface. At formation, the Meddies are small — about 9 km diameter. After aging and propagation, their radii become 20–100km, with a thickness of about 650 m, and centered at about 1000 m depth. Approximately

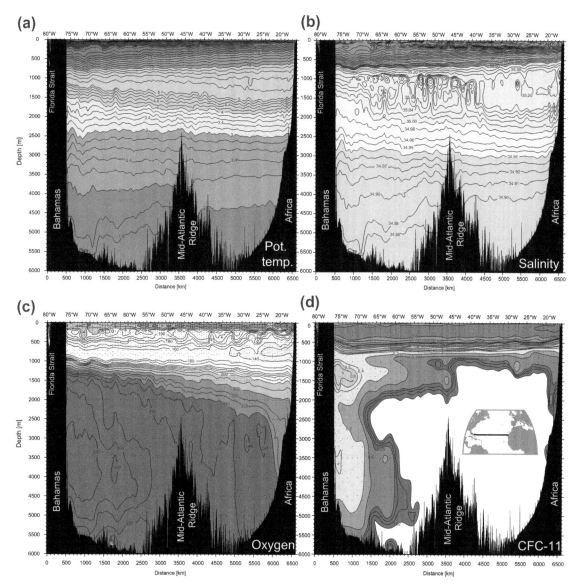

FIGURE 9.22 Subtropical North Atlantic at 24°N from July to August 1992. (a) Potential temperature (°C), (b) salinity, (c) oxygen (μmol/kg), and (d) CFC-11 (pmol/kg) at 24°N. This figure can be found in the color insert. (World Ocean Circulation Experiment section A05). *Adapted: From WOCE Atlantic Ocean Atlas; Jancke, Gouretski, and Koltermann (2011).*

15—20 may be formed each year (Figure S9.33 in the online supplementary material). They may carry up to 50% of the MW into the North Atlantic.

Advection of high salinity MW into the subtropical North Atlantic forms the characteristic "Mediterranean salt tongue" (Figure 6.4). Because the feature is so striking, it is tempting to jump to simple conclusions about the associated circulation. However, the salt tongue does not mirror the circulation,

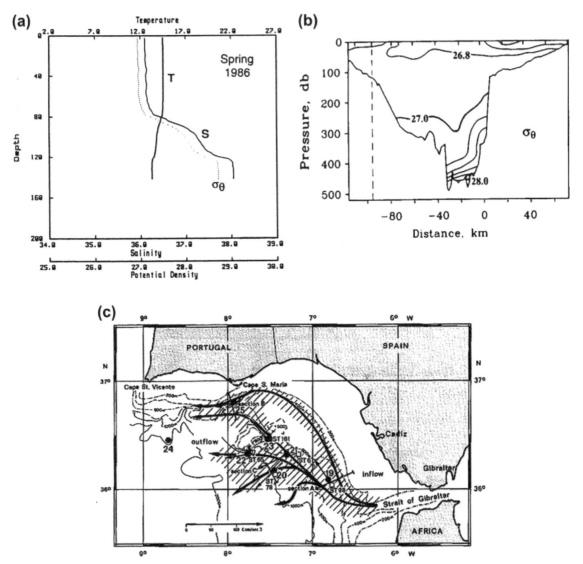

FIGURE 9.23 (a) Temperature, salinity, and potential density profiles near the strait sill in spring. *Source: From Bray, Ochoa, and Kinder (1995).* (b) Potential density at 6° 30′W just west of the Strait of Gibraltar, in spring. *Source: From Ochoa and Bray (1991).* (c) Outflow pathways of the MW. *Source: From Zenk (1975).*

which is very weak, but instead is associated with eddy diffusivity related to planetary wave and eddy-like motions (Richardson & Mooney, 1975; Spall, Richardson, & Price, 1993). The associated horizontal eddy diffusivities have reasonable magnitudes of 8 to 21×10^6 cm^2/sec.

MW injects high salinity down to intermediate depth in the North Atlantic, contributing to the characteristic high salinity of the NADW (Figure

9.17). On the other hand, open ocean subtropical evaporation in the North Atlantic dominates in establishing the North Atlantic as the saltiest ocean, with the Mediterranean contributing about 30% of this enhancement (Talley, 1996b).

9.8.3.3. Antarctic Intermediate Water

AAIW is the third major intermediate water of the Atlantic Ocean. It is the low salinity layer at about 1000 m in the South Atlantic and tropical Atlantic (Figures 9.17, 9.18, and 14.13). The AAIW's salinity minimum originates near Drake Passage where it is related to the densest, coldest, freshest SAMW (Chapter 13). The AAIW's northern boundary mostly coincides with the southern boundary of MW, at about 20°N (Figure 14.13 and Figure S9.34 and S9.35 in the online supplementary material). AAIW is fresher than LSW since its Southern Ocean source waters are fresher than the LSW's subpolar source waters in the more evaporative North Atlantic.

In the southern South Atlantic, the AAIW salinity minimum is slightly denser than $\sigma_\theta = 27.1$ kg/m^3 (4°C, 34.2 psu). In the tropics, the AAIW layer is eroded from above and is subject to diapycnal diffusion that increase its "core" density, potential temperature, and salinity to about $\sigma_\theta = 27.3$ kg/m^3, 5°C, and 34.5 psu (Talley, 1996a).

The net northward transport of AAIW into the Atlantic is estimated at 5–7 Sv (Figure 9.16). In the South Atlantic, AAIW is advected eastward away from the Malvinas-Brazil Current confluence and then northward and westward around the anticyclonic subtropical gyre. It returns to the South American coast and enters the North Brazil Current System. It is advected eastward near the equator as part of the zonally elongated equatorial current system. In the tropics, the AAIW joins vertically with UCDW (Tsuchiya, Talley, & McCartney, 1994). This complex moves northward into the Gulf Stream System and NAC, where the remnants of AAIW/UCDW are marked by elevated nutrients rather than low salinity (Tsuchiya, 1989).

9.8.4. Deep and Bottom Waters

The deep and bottom waters of the North Atlantic consist of the NADW and its precursor components formed in the North Atlantic, and AABW and CDW formed in the Southern Ocean. These were introduced in Section 4.3.4 and figure prominently in the global overturning circulation of Chapter 14. The intermediate water components of NADW were already introduced (Section 9.8.3). Here we discuss the NSOW component, then AABW, and end by describing the NADW as a whole, especially in the tropical and South Atlantic where the individual North Atlantic source waters meld into the single water mass that is exported from the Atlantic to the other oceans.

9.8.4.1. Nordic Seas Overflow Waters

The densest new bottom and deep waters in the North Atlantic originate in the Nordic Seas. These are discussed in Section 12.6; reviews can be found in Dickson and Brown (1994) and Hansen and Østerhus (2000). There are three sills in the Greenland-Scotland ridge, all with important dense overflows: (1) the Denmark Strait between Greenland and Iceland with flow into the Irminger Sea, (2) the Iceland-Faroe Ridge with flow into the Iceland Basin, and (3) the Faroe-Shetland Channel. The last includes two routes: the Faroe Bank Channel into the Iceland Basin and the Wyville-Thomson Ridge to the northern Rockall Channel. The dense waters flowing through the three channels are referred to as Denmark Strait Overflow and Iceland-Scotland Overflow Water, the latter containing waters from both sills east of Iceland. We can refer to the overflows collectively as the Nordic Seas Overflows, although this nomenclature is not universal.

The mean transports through the three straits are 3 Sv (DSOW), 0.5–1 Sv (Iceland-Faroe), and 2–2.5 Sv (Faroe-Shetland Channel), for a total of 6 Sv (Figures 9.15a and 12.20). Most of the Faroe-Shetland overflow goes through the Faroe Bank Channel, with just a few tenths of Sverdrups through the Wyville-Thomson Ridge. The overflow waters are separated from the northward-flowing surface waters by the isopycnal σ_θ ~ 27.8 kg/m³ (Figure 9.24a). The overflow layers originate in several water masses within the Nordic Seas, making a single T-S characterization of their properties impossible; temporal variability in the mix of waters in the straits creates variability in the overflow properties (Macrander et al., 2005). DSOW properties are

−0.18°C, 34.88 psu, $\sigma_\theta = 28.02$ kg/m³ to 0.17°C, 34.66 psu, $\sigma_\theta = 27.82$ kg/m³ (Tanhua, Olsson, & Jeansson, 2005). ISOW properties in the Faroe Bank Channel are −0.5–3°C, 34.87–34.90 psu, and $\sigma_\theta = 28.02$ to 27.8 kg/m³ (Hansen & Østerhus, 2000). The upper ocean water masses south of the ridge are separated from those north of the ridge by a strong front; east of Iceland this Iceland-Faroe Front is associated with concentrated eastward flow that feeds the northward Norwegian Atlantic Current.

The overflows at each of the sills are dense and plunge down their respective slopes toward the deep North Atlantic (Figure 9.24). The overflows form eddies as they plunge and turbulently entrain the water masses they pass through. The entrained waters include SPMW, LSW, and ambient deep waters. Therefore the overflow properties change rapidly as their transport increases. Along the zonal section south of the sills (Figure 9.20), the Irminger Sea and Iceland Basin overflows are obvious in the dense layers banked to the west; in Rockall Trough, overflow water is much weaker but still present on the western side near the bottom. DSOW in the Irminger Basin is markedly fresher and more oxygenated than ISOW (see also Figure 9.18), mainly because DSOW entrains newer LSW. Once the DSOW and ISOW plumes equilibrate and begin to move further into the North Atlantic, their maximum densities are reduced to about 27.92 σ_θ (46.1 σ_4; Figure 9.20).

ISOW circulates westward through the Charlie Gibbs Fracture Zone in the Reykjanes Ridge and joins the DSOW in the southern Irminger Sea (Figure 9.15a); the combined NSOW flows cyclonically around Greenland into the Labrador Sea and then out to the south beneath the Labrador Current. At this point, the NSOW is the denser part of the newly forming DWBC (Section 9.6.2). This dense layer crosses under the Gulf Stream relatively easily compared with the LSW (Pickart & Smethie, 1993) and is marked by high oxygen and high CFCs at the western boundary at 24°N

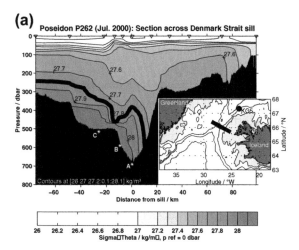

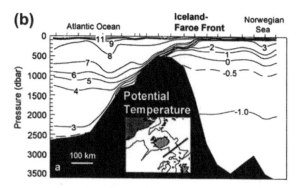

FIGURE 9.24 (a) Potential density in Denmark Strait. The heavy contour marks the upper bound on the overflow layer in the strait. *Source: From Macrander et al. (2005).* (b) Potential temperature (°C) crossing the Iceland-Faroe Ridge. *Source: From Hansen and Østerhus (2000).*

(Figure 9.22). Here the DWBC waters are usually referred to as NADW and the portion associated with NSOW is referred to as Lower North Atlantic Deep Water (LNADW; Section 9.8.4.3).

9.8.4.2. Antarctic Bottom Water

The densest water in most of the Atlantic originates in the Southern Ocean south of the ACC.[6] The very densest Antarctic waters in the Atlantic sector, created by brine rejection in the Weddell Sea, cannot escape northward past the complex topography (Chapter 13; Mantyla & Reid, 1983; Reid, 1994). Nevertheless the water that does escape is often referred to as Antarctic Bottom Water, and we follow this convention.

AABW is the water colder than about 2°C and fresher than about 34.8 psu along the full-Atlantic meridional section (Figures 4.1a and b and 9.17). Potential temperature is <0°C in the south. In the T-S relation (Figure 9.18), AABW is the coldest tail, stretching down to less than 0°C. The northward progress and modification of AABW are severely constrained by deep topography (Figure 14.14). The coldest AABW fills the Argentine Basin in the southwestern South Atlantic. It moves northward through the constricted Vema Channel into the Brazil Basin, where AABW colder than 0°C is present only in the DWBC along the coast up to about 15°S (Figure 9.25). The AABW temperature and salinity increase northward, due to downward diffusion from the overlying warmer, saltier NADW. AABW oxygen also paradoxically increases northward, which is further evidence of downward diffusion from the highly oxygenated NADW.

At the equator, the DWBC carrying AABW splits into an eastward flow that crosses the MAR through the Romanche Fracture Zone, and a northward flow that crosses the equator. The eastward branch is joined by NADW, turns back southward in the eastern tropical Atlantic,

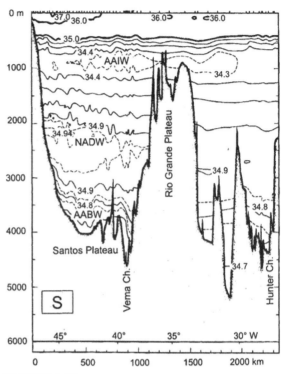

FIGURE 9.25 Salinity at about 28°S in the western South Atlantic, with water masses labeled. *Source: From Hogg, Siedler, and Zenk (1999).*

and fills the abyssal northeastern South Atlantic from the north (Figures 14.14 and 14.15). (The Walvis Ridge blocks direct northward flow into the eastern South Atlantic from the south.) The northward branch of AABW shifts to the western flank of the MAR (Section 9.6). Part of it crosses the ridge through the Vema Fracture Zone at 11°N, where it is one source of the northeastern North Atlantic's abyssal water (van Aken, 2000). Most of the AABW continues northward up to the latitude of Bermuda. At 66°W and 24°N, the AABW is still apparent as water colder than 1.8°C mounded to the east toward the MAR (Figures 9.15 and 9.22).

---

[6] While the NSOWs are denser than the bottom waters formed in the Antarctic, intense entrainment of lighter waters as the NSOW plunges over the sills into the North Atlantic reduces the density of the equilibrated NSOW.

9.8.4.3. *North Atlantic Deep Water*

NADW is the prominent layer of high salinity, high oxygen, and low nutrients between about 1500 and 3500 m depth found through the length of the Atlantic (Figures 4.11, 4.22a, b, and 9.17). We have already examined the North Atlantic sources of NADW in the Nordic Seas, the Labrador Sea, and the Mediterranean Sea. In the subpolar and subtropical North Atlantic, these waters are easily distinguishable (NSOW, LSW, and MW). Most narrowly, the term "NADW" is used where these source water masses become less easily distinguished, beginning in the subtropical North Atlantic's DWBC and in the tropical

Atlantic. However, it is also appropriate to refer to the whole complex as NADW in more generalized water mass studies (e.g., in global overturning schematics such as Figure 14.11); in paleoceanography, the balance of source waters changes dramatically over millennial timescales, so it is useful to refer to the NADW as a whole rather than focus on its individual parts.

NADW is not the only water mass in its depth range. The fresher CDW moves northward into the South Atlantic from the Southern Ocean, as seen in salinity on an isopycnal that lies at about 2500 m (Figure 9.26) and the circulation at 2500 m (Figure 9.14a). However,

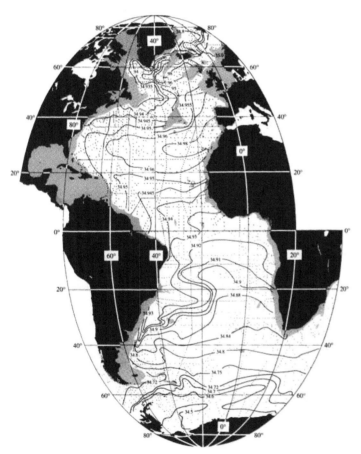

FIGURE 9.26 Salinity on the isopycnal $\sigma_3 = 41.44$ kg/m^3 (referenced to 3000 dbar), which lies at approximately 2500 m depth. *Source: From Reid (1994).*

NADW dominates in terms of net volume transport (Figure 9.16). In the North Atlantic, high salinity in the eastern subtropics in Figure 9.26 is due to downward diffusion of salt from the overlying Mediterranean salt tongue. In the subpolar North Atlantic, the saltier ISOW (>34.98 psu) is seen in the east and the fresher DSOW (~34.94 psu) in the west. The fresher DSOW spreads southward along the western boundary, and also spreads eastward toward the Mid-Atlantic Ridge (MAR) at about 47°N.

In the northeastern North Atlantic, between the high salinity ISOW and high salinity Mediterannean tongue, the lower salinity on the isopycnal in Figure 9.26 is the NEADW (van Aken, 2000). While it is a lateral salinity minimum, NEADW is also a vertical salinity maximum (Figure 9.17) and a vertical oxygen minimum (Figure 9.18c). NEADW is a mixture of local abyssal and intermediate water masses, with high salinity from both the ISOW and Mediterranean salt tongue, and low oxygen from both age/respiration and northward advection of the Mediterranean tongue. This contrasts with the high oxygen ISOW. The underlying deep water (containing modified AABW) and overlying LSW also contribute (van Aken, 2000).

In the subtropical North Atlantic, the southward spread of NADW in the DWBC is marked by high oxygen and high CFCs (Figures 9.22 and 9.15). At the western boundary at 24°N, the NADW includes high oxygen at 2000–5000 m and two striking maxima in CFCs, at 1500 m and 3500 m (Figure 9.22). The upper CFC maximum derives from the Labrador Sea (Bryden et al., 1996; Rhein, Stramma, & Send, 1995). The deeper CFC maximum is coincident with the deep high oxygen layer of the *Lower NADW* (LNADW), which is mostly NSOW.

NADW continues southward in the DWBC, splitting at the equator into eastward and southward flow. In the tropics, it is traditional to distinguish between *Upper* (UNADW), *Middle* (MNADW), and LNADW (Wüst, 1935), which can be seen at the equator on the 25°W section

(Figure 4.11): UNADW is a salinity maximum (about 1700 m), MNADW the upper oxygen maximum (2500 m), and LNADW a separate, deeper oxygen maximum (3500 m). Only the LNADW has a simple correspondence with the upstream sources, with the NSOW (Figures 9.22 and 9.27).

The equatorial UNADW salinity maximum results from low salinity AAIW cutting into the top of the NADW, leaving a salinity maximum at its top, which is deeper and denser than MW. The MNADW oxygen maximum is much deeper than the original high oxygen LSW, which can be seen by comparing the tropical oxygen and CFC maxima (Figures 4.11 and 9.27; Weiss, Bullister, Gammon, & Warner, 1985; Rhein et al., 1995; Andrié et al., 1999; Chapter 3). As at 24°N, equatorial CFCs (in observations taken from 2003–2005) contain two maxima that directly reflect northern North Atlantic sources. The deeper CFC maximum is the same as the Lower NADW oxygen maximum, deriving from NSOW. The upper CFC maximum (1000–1500 m) derives from LSW, and is shallower than the MNADW oxygen maximum, which is depressed to greater depth (>1500 m) because of high consumption of oxygen in the upper layer of the tropical North Atlantic (Weiss et al., 1985) whereas CFCs are biologically inert.

Because CFCs have time-dependent surface sources, they are useful markers of the invasion of high latitude waters (Figure 9.27). The first equatorial CFC observations in the 1980s showed the arrival of LSW as a blob of non-zero CFCs at 1500 m. By the time of the second full set of CFC observations in 2003–2005, this LSW maximum was greatly enlarged and the LNADW (NSOW) was also marked by a CFC maximum. The CFC minimum at about 2700 m is mostly associated with the oxygen minimum between the MNADW and LNADW. The oxygen/CFC minimum results from upwelling of AABW and older LNADW in the eastern tropics (Friedrichs et al., 1994).

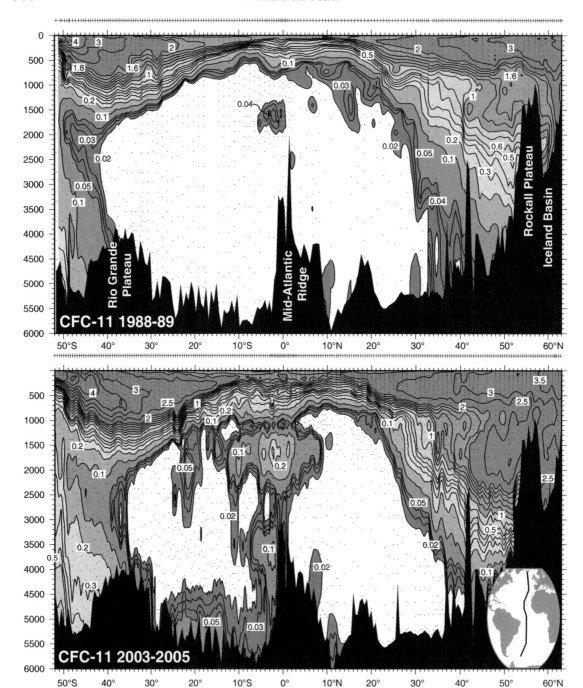

FIGURE 9.27 Chlorofluorocarbon-11 (pmol/kg) along 20–25°W from (a) 1988–1989 and (b) 2003–2005. Section location is shown in the inset map. Sample locations are indicated by small dots on the plots.

In the South Atlantic, the NADW moves southward in the DWBC to 25°S in the Brazil Basin. Here there is an eastward breach of NADW marked at 25°W by higher salinity and higher oxygen, and even non-zero CFC-11 in 2005 (Figures 4.11b, d and 9.27; Tsuchiya et al., 1994). These three maxima are nearly coincident in depth, without the complicated equatorial layering: the NADW here is becoming the more homogenized single layer of high salinity and oxygen that exits the Atlantic. The NADW, whose southernmost boundary is at 35°S, moves eastward across the MAR, which is a formidable mixing barrier. East of the ridge, NADW has markedly lower oxygen and salinity. It gathers in a broad band of about 1000−1500 km width around the southern end of Africa. Part of it moves northward into the Indian Ocean in a broad DWBC underneath and offshore of the Agulhas (Chapter 11). The remainder joins the eastward ACC, where it provides the high salinity core for the LCDW (Chapter 13).

9.9. CLIMATE AND THE ATLANTIC OCEAN

Atlantic climate research tends to be focused on decadal and longer term variability centered on the northern North Atlantic's deep-water formation processes and on sea ice processes in the Nordic Seas and Arctic (Chapter 12).

This is because the mean ventilation age of northern North Atlantic deep waters is on the order of decades or less, with associated measurable variability. However, climate variability at all timescales from interannual to decadal, centennial, and millennial affects all regions of the Atlantic. Trends related to climate change (anthropogenic forcing) have also been documented.

All of the text, figures, and tables relating to Atlantic Ocean climate variability are located in Chapter S15 (Climate Variability and the Oceans) on the supplemental Web site for this textbook. The chapter describes tropical Atlantic climate variability: (1) the *Atlantic Meridional Mode* (AMM), which is a cross-equatorial mode; (2) *Atlantic Niño*, which is a zonal equatorial mode dynamically similar to El-Niño-Southern Oscillation (ENSO) with its tropical Bjerknes feedback (Section 7.9.2); and (3) remote forcing from the Pacific ENSO. Chapter S15 then describes modes of decadal and multidecadal variability in the Atlantic: (1) *North Atlantic Oscillation* (NAO), (2) *East Atlantic Pattern* (EAP), and (3) *Atlantic Multidecadal Oscillation* (AMO). The considerable variability in ocean properties, with an emphasis on salinity variations is described. The section ends with a description of the imprint of climate change on the Atlantic Ocean including difficulties with detection given the large natural variability at all depths in the northern North Atlantic.

10

Pacific Ocean

10.1. INTRODUCTION AND OVERVIEW

The Pacific Ocean is the largest of the three major oceans. It has well-developed wind-driven circulation systems in the subtropics, subpolar North Pacific, and tropics (Sections 10.1–10.7). In the south, the Pacific circulation transitions to the Southern Ocean, which connects it to the other oceans (Chapter 13). The Pacific is also connected at low latitudes to the Indian Ocean through passages in the Indonesian archipelago. It is connected to the Arctic (Chapter 12) through the very shallow Bering Strait.

The Pacific is the freshest of the three major oceans because of small differences in net evaporation/precipitation between the oceans (Chapter 5). Compared with the North Atlantic, this freshness completely inhibits formation of deep waters and weakens formation of intermediate water in the northern North Pacific (Section 10.9). At this global scale, the Pacific is one of the broad regions of deep upwelling that returns deep waters formed elsewhere back to mid-depths or even the surface. Because of its weak thermohaline circulation, the North Pacific upper ocean circulation is mostly associated with wind forcing. Therefore, it can be useful to study the wind-driven circulation first in the context of the North Pacific and equatorial Pacific, followed by study of the other oceans.

The tropical Pacific is the center of action for the interannual climate mode, El Niño-Southern Oscillation (ENSO: Section 10.8), which impacts much of the globe through atmospheric "teleconnections." Important natural climate variability of quasi-decadal timescale is also observed in the Pacific (Section 10.10; Chapter S15 located in the supplemental material found on the textbook Web site http://booksite. academicpress.com/DPO/; "S" denotes supplemental material).

The Pacific Ocean has numerous marginal seas, particularly along its western side; these are described briefly in the online supplement Section S8.10. The complicated passages through the Indonesian archipelago shunt water from the tropical Pacific to the tropical Indian Ocean. The Bering Strait at the northern end of the Bering Sea allows a small leakage of North Pacific water into the Arctic and hence into the Atlantic Ocean. The Okhotsk Sea in the northwestern Pacific is the site for the densest water formation in the North Pacific; this densest Pacific water is only of intermediate depth and is less dense and much smaller in impact than dense water formation in the North Atlantic and Antarctic, which supply the deepest waters of the global ocean, including the Pacific.

The Pacific Ocean's surface circulation (Figures 10.1, 10.2a and Figure S10.1 in the

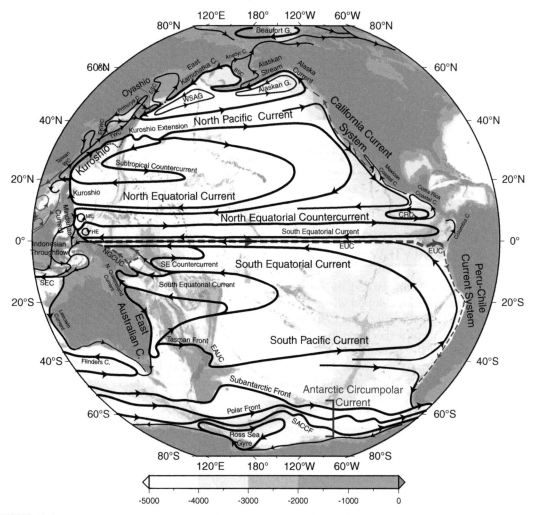

FIGURE 10.1 Pacific Ocean: surface circulation scheme. Major near-surface undercurrents at the equator and along the eastern boundary are also shown (dashed). The South China Sea circulation represents the winter monsoon. Acronyms: SACCF, Southern ACC Front; EAUC, East Auckland Current; NGCUC, New Guinea Coastal Undercurrent; EUC, Equatorial Undercurrent; CRD, Costa Rica Dome; ME, Mindanao Eddy; HE, Halmahera Eddy; TWC, Tsushima Warm Current; EKWC, East Korean Warm Current; WSAG, Western Subarctic Gyre; ESC, East Sakhalin Current; and BSC, Bering Slope Current.

supplementary Web site) includes subtropical gyres in both hemispheres, a subpolar gyre in the North Pacific, and the Antarctic Circumpolar Current (ACC; Chapter 13) in the far south. The western boundary currents for the North and South Pacific's subtropical gyres are the *Kuroshio* and *East Australian Current* (EAC),

respectively. The eastern boundary currents for these subtropical gyres are the *California Current* and the *Peru Current*, respectively. The western boundary current for the North Pacific's subpolar gyre is the *Oyashio/East Kamchatka Current* (EKC). The strongly zonal (east-west) circulation in the equatorial Pacific is described separately (Section 10.7) because of its complexity and dynamics, which differ from mid-latitude wind-driven circulation processes (Section 7.8). The tropical circulation also includes low latitude western boundary currents: the *Mindanao Current* and the *New Guinea Coastal Undercurrent* (NGCUC). The Pacific Ocean's deep circulation (Section 10.6) consists of inflow from the Southern Ocean in a *Deep Western Boundary Current* (DWBC) along the deep plateaus and island chains from New Zealand northward. Much of the deep flow funnels through the Samoan Passage in the South

Pacific and then enters the deep tropical ocean. Deep flow crosses the equator in the west and then follows the western boundary's deep trenches northward, filling in the deep North Pacific. The "end" of the deep circulation is reached in the northeastern Pacific, which has the oldest deep waters of the world, as supported by carbon-14 content (Chapter 4, Figure 4.24b).

The inflowing bottom waters upwell through the length of the Pacific, although most of the upward transport occurs in the South Pacific and tropics. Downward diffusion of heat and

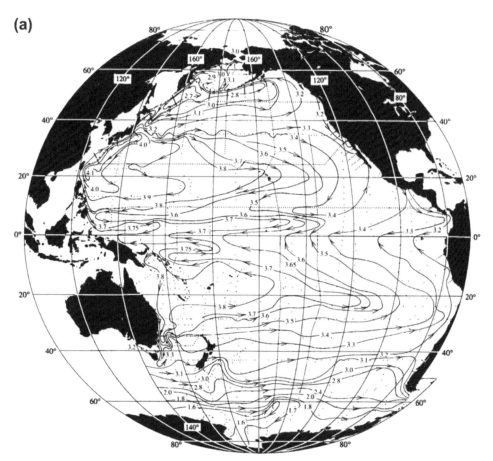

FIGURE 10.2 Adjusted geostrophic streamfunction (steric height, 10 m^2/sec^2) at (a) 0 dbar and (b) 500 dbar. *Source: From Reid (1997).*

(b)

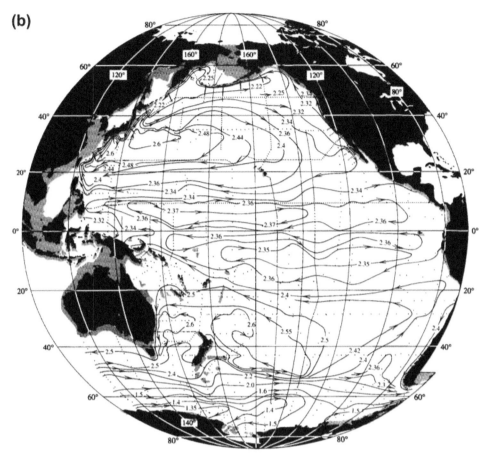

FIGURE 10.2 *(Continued)*.

freshwater modify the water density, and the upwelling deep waters create a relatively homogenous and volumetrically large water mass called the Pacific Deep Water (PDW; or Common Water). This returns back to the Southern Ocean where it joins the Indian Deep Water (which is formed similarly) and the North Atlantic Deep Water (which has an entirely different formation mechanism). Within the Pacific, there is also upwelling from the deep waters to shallower layers, including intermediate and upper ocean layers, with outflow in different parts of these layers in all directions:

through the Indonesian passages to the Indian Ocean, southwestward around Australia, northward through the Bering Strait, and eastward through the Drake Passage (see Chapter 14).

10.2. WIND AND BUOYANCY FORCING

The Pacific's upper ocean gyres and tropical circulation are mainly wind-driven. The mean surface winds are dominated by the westerlies at latitudes poleward of about 30° (north and south) and the easterly trade winds at low latitudes (Figure 5.16a—c and supplementary Figure S10.2a). The resulting Ekman transport convergences and divergences drive Sverdrup transport (Figure 5.17 and supplementary Figure S10.2b) and hence the gyres. The anticyclonic subtropical gyres in Figures 10.1 and 10.2a correspond to the Ekman downwelling regions and equatorward Sverdrup transport. The cyclonic gyres in the subpolar North Pacific and south of the ACC in the Ross Sea correspond to the Ekman upwelling regions and poleward Sverdrup transport. A narrow tropical cyclonic cell, centered at about 5°N, stretches across the width of the Pacific; it includes the Mindanao Current and the North Equatorial Countercurrent (NECC). It is associated with Ekman upwelling beneath the Intertropical Convergence Zone (ITCZ).

The alongshore wind stress component at the eastern boundaries creates Ekman transport divergence that is not represented in wind stress curl maps. There is also non-zero wind stress curl in bands along the boundaries, for instance, upwelling-favorable along the California-Oregon coast. Both mechanisms drive the California and Peru-Chile Current Systems (PCCS).

The Pacific's annual mean buoyancy forcing (Figure 5.15) is dominated by heating/cooling (Figure 5.12). The tropical Pacific has the largest mean heating of any region on the globe, over the upwelling cold tongue in the eastern equatorial region. Bands of ocean heat gain are found along the west coasts of North and South America, in the California Current and Peru-Chile Current upwelling systems. The Kuroshio region in the North Pacific is one of the strongest global air—sea heat loss regions (>125 W/m^2). The equivalent region along the Australian coast, in the EAC, also has significant heat loss (>100 W/m^2).

Net evaporation-precipitation (Figure 5.4a) is directly related to the Pacific's surface salinity pattern. There is net precipitation in the ITCZ (5—10°N). A broader region of net precipitation occurs in the western tropical Pacific under the ascending branch of the Walker circulation. Net precipitation is also found throughout the higher latitudes in both the North and South Pacific. Net evaporation is found in the subtropical gyres under the descending branches of the Hadley circulation.

The air—sea flux maps in Chapter 5 do not represent the brine rejection process that creates dense water when sea ice forms. This process is active in the North Pacific in the Okhotsk and Bering Seas, and in the northern Japan (East) Sea. Okhotsk Sea brine rejection is the densest source of North Pacific Intermediate Water (Section 10.9.2)

10.3. NORTH PACIFIC CIRCULATION

The mid-latitude North Pacific surface circulation (Figures 10.1 and 10.2a; Table S10.1 in the online supplement), with its subtropical and subpolar gyres, is the clearest example seen in all of the oceans of the two-gyre circulation driven by the westerly and trade winds. This is because the North Pacific is almost completely closed to the north and has only a weak thermohaline circulation. The gyres have the familiar east-west asymmetry (with strong western boundary currents and weak meridional flow spread over much of the remainder of the ocean), which is understood in terms of meridional Sverdrup transport (Figure 5.17 and Section 10.2). With increasing depth, the North Pacific gyres weaken and shrink, and the subtropical gyre centers (highest pressure) shift westward and poleward (Section 10.6).

10.3.1. Subtropical Circulation

10.3.1.1. General Description

The North Pacific's subtropical gyre, like all subtropical gyres, is anticyclonic (clockwise in the Northern Hemisphere), associated with Ekman downwelling and equatorward Sverdrup transport (Figure 5.17). Its strong, narrow, northward western boundary current is the *Kuroshio*. After the Kuroshio separates from the western boundary and flows eastward into the North Pacific, the current is referred to as the *Kuroshio Extension*. The broad eastward flow on the northern side of this gyre is called the *North Pacific Current* or the "West Wind Drift." The North Pacific Current also includes the eastward flow of the subpolar gyre; it is also called the Subarctic Current (Sverdrup, Johnson, & Fleming, 1942). The westward flow on the south side of the subtropical gyre is the *North Equatorial Current*, which also includes the westward flow in the elongated tropical cyclonic circulation. The concentrated flow near the eastern boundary is the *California Current System* (CCS), which includes a locally forced eastern boundary current and a poleward undercurrent (Davidson Current); both are forced by coastal upwelling (Section 7.9).

The surface subtropical gyre in the western North Pacific has an overall "C-shape" (Wyrtki, 1975; Hasunuma & Yoshida, 1978). The "C" looks like a large-scale overshoot of the Kuroshio as it becomes the Kuroshio Extension, with a swing back to the west in the *recirculation*, followed by southward flow parallel to the Kuroshio, a turn to the east in the *Subtropical Countercurrent* (STCC) at 20−25°N, and then the westward flow of the North Equatorial Current (NEC) south of 20°N. This C-shape is common to surface flow in all subtropical gyres, but the STCC portion is very shallow; the circulation just 250 dbar below the surface is a simpler, closed anticyclonic gyre.

The broad eastward and westward flows crossing the Pacific include narrow, nearly zonal (east-west) fronts or frontal zones that are narrow (less than 100 km wide). Nomenclature is confusing and contradictory. We adopt Roden's (1975, 1991) terms for the central North Pacific. The *Subarctic Frontal Zone*

(SAFZ; or Subarctic Boundary), centered at about 42°N, is embedded in the North Pacific Current; it roughly separates the subtropical and subpolar gyres, being slightly south of the maximum westerly winds. The *Subtropical Frontal Zone* (or convergence), at about 32°N in the central and eastern Pacific, separates the eastward North Pacific Current from the westward NEC.

With increasing depth, the subtropical gyre shrinks toward the west and toward Japan, and decreases in strength. It disappears around 1500 m depth except in the Kuroshio region (Figures 10.10 and 10.14).

10.3.1.2. *The Kuroshio and Kuroshio Extension*

The Kuroshio (black stream in Japanese, where *shio* means current) arises at the western boundary where the westward flow of the NEC splits at about 15°N into northward and southward boundary currents: the Kuroshio and Mindanao Current, respectively (Figures 10.1 and 10.3). The Kuroshio continues northward, turns to follow the south coast of Japan, then separates and flows out to the mid-subtropical gyre. Maximum surface current speeds in the Kuroshio range between 75 and 250 cm/sec. The width of the current is 80 to 100 km. It has major variability at timescales of weeks to decades.

The Kuroshio velocity decreases with depth (Figure 10.4b). The northward velocity core of the Kuroshio is sometimes flanked on both sides by weak countercurrents (flowing in the opposite direction). Where the Kuroshio begins to leave the western boundary, it passes eastward through Tokara Strait (Figure 10.4a,b), tracks eastward roughly parallel to the south coast of Japan, then passes through gaps in the Izu-Ogasawara (Izu) Ridge, and finally enters the open Pacific at Boso Peninsula (Figure 10.3a). Between Tokara Strait and the Izu Ridge, the Kuroshio exists in one of two (or three) semi-stable states: flowing either nearly directly

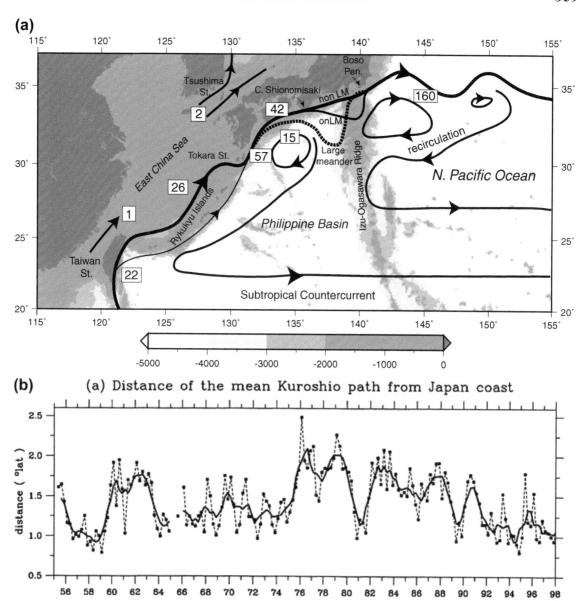

FIGURE 10.3 Kuroshio system in the western North Pacific. (a) Schematic of the large meander (LM), straight (near shore non-large meander) and offshore non-large meander paths (*after Kawabe, 1995*), and recirculation gyre schematics, with transports in Sv (*after Hasunuma and Yoshida, 1978; Qiu and Chen, 2005*). (b) Index of the Kuroshio meander state: distance offshore of the 16°C isotherm at 200 m averaged between 132° and 140°E. ©*American Meteorological Society. Reprinted with permission. Source: From Qiu and Miao (2000).*

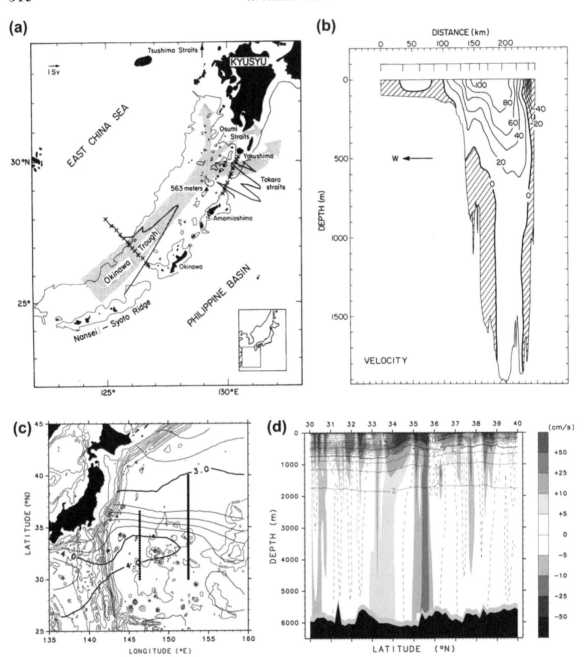

FIGURE 10.4 Kuroshio velocity structure. Vertical sections of (b) northward velocity of the Kuroshio where it is a western boundary current, at 24°N (*Source: From Bingham & Talley, 1991*), and (d) eastward velocity of the Kuroshio Extension at 152° 30′E [red (blue) indicates eastward (westward) flow]. *Source: From Yoshikawa et al. (2004).* Section positions are shown in (a) and (c). Mean temperature at 1000 m is contoured in (c). Figure 10.4d can also be found in the color insert.

along the coast (straight path), or looping far to the south in a meander (large meander path). The Kuroshio remains in one of these states for several years and then switches to the other state (index in Figure 10.3b). The mean eastward-flowing Kuroshio Extension splits close to Shatsky Rise into a southward branch that feeds a westward flow that creates a *recirculation gyre* (*Kuroshio Countercurrent*) and an eastward flow that becomes the North Pacific Current. The recirculation gyre is often split in two by the Izu Ridge, with one gyre west of the ridge and south of Japan, and the other gyre east of the ridge, downstream of the Kuroshio separation point (Figure 10.3a).

Once the Kuroshio crosses the Izu Ridge and enters deep water, its upper ocean structure is like that of the Gulf Stream, with a strong eastward velocity core and marked, but weaker, westward recirculation just to the south. The Kuroshio Extension extends to the ocean bottom in the deepest water downstream of the separation point, with 10 cm/sec velocities even at the bottom (Figure 10.4d). Westward recirculations flank the deep Kuroshio Extension to the bottom.

The volume transport of the Kuroshio increases downstream (Figure 10.3a) from 20 to 25 Sv, where it is a western boundary current east of Taiwan (Johns et al., 2001; Bingham & Talley, 1991), to about 57 Sv east of Tokara Strait but still prior to separation to a maximum of 140 to 160 Sv at 145°E, just to the east of separation. Considerable recirculation causes much of these increases (Imawaki et al., 2001). The transport decreases east of this point, with water lost southward to the recirculation gyre and into the Kuroshio Extension bifurcation fronts (Yoshikawa, Church, Uchida, & White, 2004).

The Kuroshio Extension is highly unstable. It meanders and produces rings when the meanders pinch off. The meanders have somewhat preferred locations, which differs from the Gulf Stream. The first northward meander occurs just downstream of the separation point. This often creates an anticyclonic warm-core ring of about 200 km diameter that moves northward. A second preferred location for northward meandering is at 150°E. Southward meanders, between the northward meanders, form cyclonic cold-core rings south of the Kuroshio Extension. The envelope of paths is several hundred kilometers wide from the separation point out to near 160°E (Shatsky Rise), widening to about 500–600 km with the paths becoming considerably more random (Mizuno & White, 1983; Qiu & Chen, 2005).

10.3.1.3. North Pacific Current and Mid-Latitude Fronts

The *North Pacific Current* is the broad eastward flow of the central and eastern subtropical gyre. The mean speed of the North Pacific Current is small, less than 10 cm/sec. However, synoptic meridional crossings of the North Pacific Current reveal larger geostrophic flows of 20 to 50 cm/sec that reverse direction on the order of every 100 km (at the eddy scale) and are deep-reaching. The difficulty of distinguishing between eddies and permanent flow features obscured observation of the deep penetration of the Kuroshio Extension Front until recently (Figure 10.3d).

The northern and southern "boundaries" of the subtropical gyre can be considered to be the SAFZ (40–44°N) and the Subtropical Front (25–32°N, depending on longitude). In both frontal zones — which are synoptically about 100–200 km wide and often contain at least two sharp fronts — temperature, salinity, and density change rapidly with latitude (Figure S10.3 in the supplementary Web site). The frontal zones are relatively zonal over much of the North Pacific; they veer southward into the CCS in the east.

The SAFZ arises in the western North Pacific from both a branch of the Kuroshio Extension

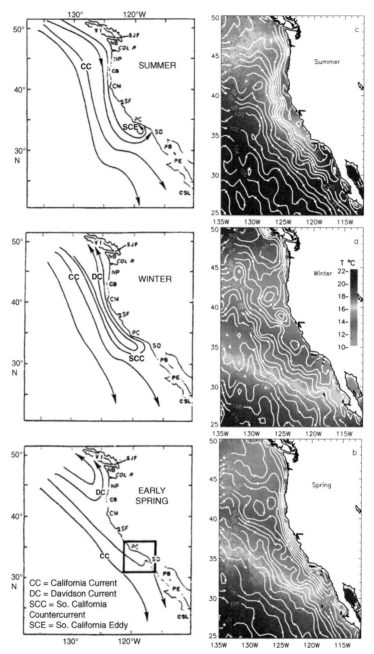

FIGURE 10.5 (a) Schematic of the surface currents in the CCS in different seasons. *Source: From Hickey (1998).* (b) Mean seasonal cycle of satellite-derived surface temperature (color) and altimetric height, showing the geostrophic surface circulation. *Source: From Strub and James (2000, 2009).* This figure can also be found in the color insert.

Front and the STCC. It coincides in the open Pacific with maximum Ekman convergence in the center of the subtropical gyre.

The SAFZ might be partly associated with the separated Oyashio front (Section 10.3.2.2). It coincides approximately with the maximum westerly wind, marking the transition from the Ekman downwelling of the subtropical gyre to the Ekman upwelling of the subpolar gyre. The northern front in the SAFZ is the southern-most limit of the very strong halocline of the subpolar gyre, and the southernmost limit of the shallow temperature minimum in the western subpolar gyre. There is a jump in nutri-ents across the frontal zone to higher values in the subpolar surface waters (Figure S10.3c,d in the supplementary Web site; surface nitrate map in Figure 4.23).

10.3.1.4. California Current System

The CCS stretches from the Strait of Juan de Fuca to the tip of Baja California (Figures 10.1 and 10.5). We describe the CCS in some detail, because it is the principal example in this text of an eastern boundary current system. In-depth overviews of the CCS and its variability can be found in Wooster and Reid (1963), Huyer (1983), Lynn and Simpson (1987), Hickey (1998), and Marchesiello, McWilliams, and Shchepetkin (2003).

The CCS has two regimes: (1) the southward, shallow, narrow, meandering California Current Front, with upwelling zones along the coast, offshore-advecting jets of upwelled water, and a northward undercurrent or inshore surface countercurrent and (2) the broad southward flow of the subtropical gyre. Dynamically, these two components have entirely different origins: (1) southward flow due to locally wind-driven coastal upwelling with a poleward undercur-rent and (2) southward flow that is part of the large-scale subtropical circulation resulting from Ekman downwelling and associated equa-torward Sverdrup transport (Figure 5.17). We discuss only the upwelling system here.

A simplified approach to the dynamics of subtropical eastern boundary systems, based on Ekman transport and upwelling, was provided in Section 7.9. This framework is useful for initial broad understanding, but these systems tend to be far more complex than this, which is evident as soon as we look at satellite images of sea-surface temperature (SST) and ocean color in the CCS (Figure 10.6). The CCS upwelling is forced by the alongshore component of the prevailing westerly winds, which results from their south-ward deflection as they encounter the North American continent (Figure 5.16). The upwelling is apparent off the North American coast from British Columbia to California (50–30°N), as a patchy band of cool surface water within a region 80 to 300 km from shore, strongest from April to August (Figure 10.6a). The upw-elled waters are highly productive, which can be observed with satellite ocean color sensors (Figure 10.6b). The upwelled water does *not* orig-inate from great depth because the ocean is strat-ified. Its source is around 150–200 m depth, but this is deep enough to access the nutrient-enriched waters below the euphotic zone.

The maximum surface velocity of the mean southward California Current is 40–80 cm/sec and its width is 50–100 km. The California Current is in geostrophic balance with the cross-shore pressure gradient force. It decays rapidly with depth and is essentially confined to the top 300 m (Lynn & Simpson, 1987). Thus the CC is much shallower and carries much less transport, on the order of only a few Sver-drups, than a western boundary current such as the Kuroshio. The decrease in geostrophic velocity from the surface to 200 m is evidenced in the upward tilt of isotherms toward the coast (Figure 10.7). The upwelling, surface PGF and upward tilt of the isotherms result from offshore Ekman transport, which has been observed directly by Chereskin (1995) (Figure 7.7).

An idealized steady state requires warming of the upwelled water as it moves offshore. Since the right amount of warming does not generally

(a)   **(b)**

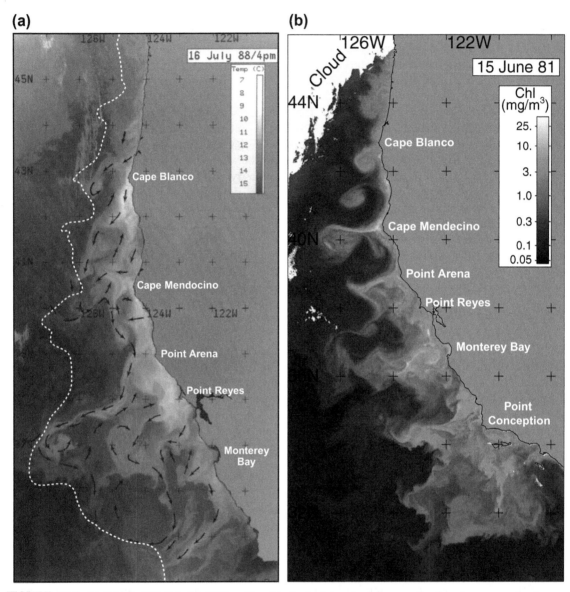

FIGURE 10.6 (a) Satellite SST (July 16, 1988), with subjectively determined flow vectors based on successive images. (b) Surface pigment concentration from the CZCS satellite on June 15, 1981. *Source: From Strub et al. (1991).*

occur at exactly the right time, the actual state is more complicated. The seasonal offshore Ekman transport creates an upwelling front that moves offshore. The California Current's southward core is located at the upwelling front, as seen in Figure 10.7, and moves offshore with the front as it progresses through the upwelling season. The mean location of the California Current is therefore offshore, by about 200–300 km, and not at the coast. This is also evident in the tighter

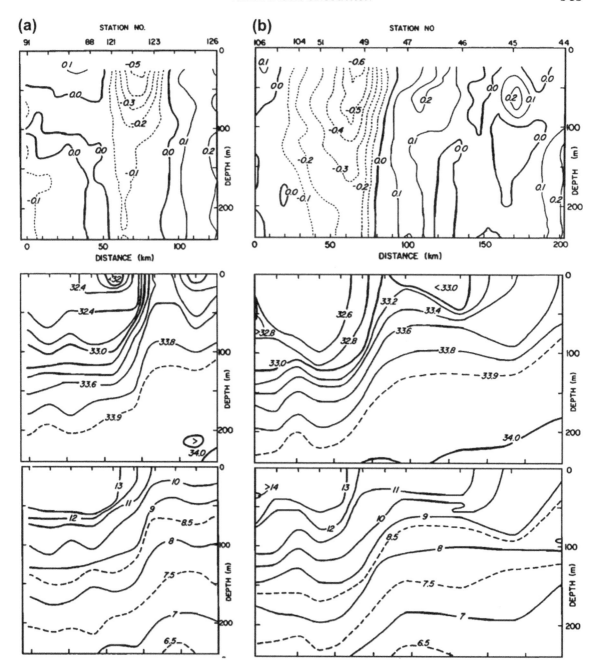

FIGURE 10.7 Sections of (top) velocity (m/sec), (middle) salinity, (bottom) potential temperature (°C) across the CCS at 41.5°N (left) and 40.0°N (right) in June, 1987. The coast is to the right. *Source: From Kosro et al. (1991).*

dynamic height contours in Figure 10.5 and the strong fronts in Figure 10.7.

The mean offshore location of the California Current is also apparent in enhanced dynamic height/sea-surface height variability due to a vigorous eddy field, and in low salinity that reflects the northern source of the surface water (Figure 10.8). Underneath and inshore of the California Current, the mean flow is northward (poleward), centered at the continental shelf break. This is the *California Undercurrent* (CUC). The CUC is approximately 20 km wide and its core lies at about 250 m, although it can extend to more than 1000 m depth. Its maximum speed is more than 10 cm/sec, and its water originates in the warm, saline, low oxygen tropical Pacific. The mean CUC is in geostrophic balance with the offshore pressure gradient force at this depth. The reversal of alongshore geostrophic flow from the southward CC at the surface to the northward CUC requires sloping isopycnals between the two currents. The CUC then weakens below its core. The CUC is thus recognized by a spreading of the isotherms and isopycnals, upward above the undercurrent and downward below it.

During winter, upwelling is weak or inactive. The California Current is far offshore and relatively weak and the coastal flow is northward (the *Inshore Countercurrent* or *Davidson Current*). This poleward flow could be the Sverdrup transport response to the Ekman suction driven by positive wind stress curl in the CCS region; it is overwhelmed by the response to coastal upwelling during the upwelling season (Marchesiello et al., 2003). When upwelling starts up again, an upwelling front appears near the coast as the offshore edge of the Ekman transport. A strong southward California Current jet is associated with the front and moves progressively offshore with time (Figure 10.5; Strub & James, 2000).

The strong seasonal cycle of the wind forcing is quantified with upwelling indices. In Figure 10.9, one index is based on Ekman transport and the

(a)

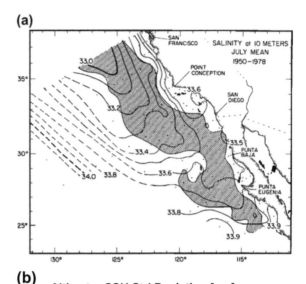

(b)

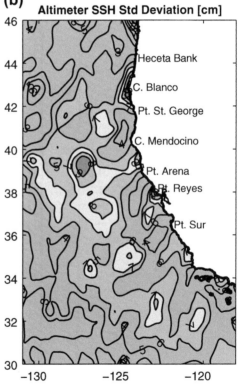

FIGURE 10.8 (a) Mean salinity at 10 m in July (contoured) with dynamic height standard deviations greater than 4 dyn cm in gray. *Source: From Lynn and Simpson (1987).* (b) Sea surface height standard deviation (cm) from satellite altimetry. *©American Meteorological Society. Reprinted with permission. Source: From Marchesiello et al. (2003).*

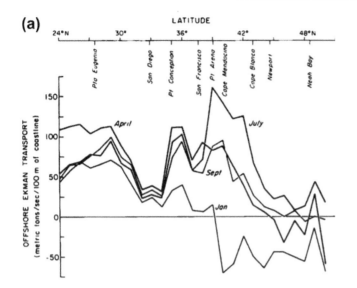

(a)

FIGURE 10.9 (a) Offshore Ekman transport based on long-term mean wind stress. *Source: From Huyer (1983).* (b) Upwelling index based on atmospheric pressure distribution *(from Bakun, 1973),* averaged over 1946–1995. Lower shaded region (positive values or blues in the original figure) is upwelling; upper shaded region (negative values or reds in the original figure) is downwelling. *Source: From Schwing, O'Farrell, Steger, and Baltz (1996).*

(b) **AVERAGE MONTHLY UPWELLING INDEX**

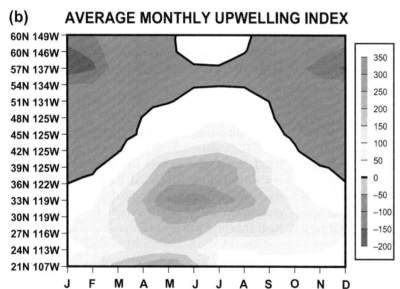

other on the strength of alongshore wind component.[1] Maximum upwelling occurs in late spring and summer (April through July), as evident from enhanced surface chlorophyll content in summer (Figure 10.6b), and is highest near Point Conception (34°N). North of 40°N, the longshore winds actually cause downwelling in winter as the Aleutian Low expands southward; north of 45°N, there is downwelling in the annual mean (Venegas et al., 2008).

[1] Neither index includes the wind stress curl component of the upwelling, although we have already noted that it can be important (Bakun & Nelson, 1991; Pickett & Paduan, 2003).

The quasi-continuous, alongshore mean circulation described in the previous paragraphs is the simplest view of the CCS. However, as seen in satellite images (Figure 10.6), the upwelled water does not move offshore in a "sheet," but rather in jets at recurring locations associated with capes or points in the coastline. The circulation can be either "squirt-like," in which the jet goes out to sea and dies, or meandering, in which the jet goes out and returns. The high mesoscale eddy activity in the CCS (Figure 10.8) may be created by baroclinic instability of the coastal upwelling current. The eddies spawned by this instability move the upwelled cold water offshore and thus maintain the mean balance, which includes Ekman upwelling (Marchesiello et al., 2003). Recent studies of the California Current are beginning to focus on even smaller spatial scales, called the *submesoscale* (order of 1 to 10 km). These are associated with the actual fronts and their instabilities within the mesoscale eddy field (Capet, McWilliams, Molemaker, & Shchepetkin, 2008).

10.3.1.5. *North Equatorial Current*

The NEC is the broad westward flow on the southern side of the subtropical gyre. It is between about 8 and 20°N depending on longitude. The NEC forms gradually in the eastern Pacific from southward flow of the subtropical gyre, including the CCS. At the eastern boundary, it has input from the tropical current system (Costa Rica Dome and NECC).

As the NEC flows westward, some of it moves southward and joins the strong eastward flow of the NECC. When the NEC reaches the western boundary, it bifurcates at about 14°N into a northward portion that becomes the Kuroshio and a southward portion that becomes the Mindanao Current (Section 10.7.4). In the western Pacific, the NEC includes a strong zonal surface salinity front that separates saline water that originates in the subtropical gyre from fresher NECC surface water. The location of this front is similar to the latitude of the NEC bifurcation, and it is also an ecological front that is important for fisheries (Kimura & Tsukamato, 2006). These suggest that the front is a boundary between Ekman upwelling in the tropical NEC/NECC cyclonic gyre and downwelling in the anticyclonic subtropical gyre.

Volume transport of the NEC in the western Pacific is up to 50 Sv in the top 500 m and 80 Sv top to bottom (Kaneko, Takatsuki, Kamiya, & Kawae, 1998; Toole, Millard, Wang, & Pu, 1990).

10.3.1.6. *Depth Dependence of the Subtropical Circulation*

The subtropical gyre shrinks spatially with depth. Like all subtropical gyres, it shrinks toward the most energetic part of its surface flow: westward toward the western boundary, and northward toward the Kuroshio Extension. The Kuroshio Extension extends to the ocean bottom as previously noted.

The gyre shrinkage from the sea surface to about 200 m depth is dramatic (Reid, 1997; represented by Figure 10.2). The boundary between eastward and westward flows shifts from south of 20°N at the sea surface to 25–30°N at 200 m. The C-shape of the western gyre, which includes the STCC, disappears by 200 m. On the other hand, the Kuroshio and Kuroshio Extension do not shift (Figure 10.3d). At 1000–1500 m depth, the anticyclonic subtropical gyre is found entirely in the western North Pacific near the Kuroshio and Kuroshio Extension (Figure 10.10).

Flow in the subtropical regions vacated by the subtropical gyre is very weak. Steric height differences over 1000 km distances are on the order of 1 cm rather than the 10 cm differences within the gyre proper. Dynamically, on isopycnal surfaces that are still within the gyre in the western region, the vacated region is called the shadow zone (Section 7.8.5). Within these regions where there is little direct ventilation from the sea surface, on the eastern and southern flanks of the subtropical gyres, oxygen is depleted to the point where denitrification sets in (Section 10.9.1).

FIGURE 10.10 Steric height ($10\ \mathrm{m}^2/\mathrm{sec}^2$) at 1000 dbar based on hydrographic data and reference geostrophic velocities adjusted to provide absolute circulation at all depths. *Source: From Reid (1997).*

10.3.2. Subpolar Circulation

10.3.2.1. *General Description*

The cyclonic (counterclockwise) subpolar gyre in the North Pacific stretches across the width of the basin and is compressed in the north-south direction between about 42°N (Subarctic Front) and the Aleutian Islands/ Alaskan coast (Figure 10.1). It has a southward western boundary current, the Oyashio/EKC.

A geographic constriction at the southernmost location of the Aleutian Islands (near the date line) separates the subpolar gyre into two portions. The *Western Subarctic Gyre* is centered east of the Kuril Islands, and the *Alaskan Gyre* is centered in the Gulf of Alaska. They are connected through eastward flow along the southern side of the gyre (*Subarctic Current*, which is part of the North Pacific Current, Section 10.3.1.3) and westward flow along the Aleutian Islands (*Alaskan Stream*). Completing the nomenclature for the cyclonic gyre, the *Alaska Current* is the northward eastern boundary current along the coast of Canada and Alaska. An older but exhaustive treatment of this circulation is found in Favorite, Dodimead, and Nasu (1976; Figure S10.4 in the supplementary Web site).

Parts of the subpolar gyre circulation loop through the Bering and Okhotsk Seas (S8.10 located in the Web site supplementary text). Transport of 0.8 Sv from the North Pacific to the Arctic and onward to the Atlantic occurs through the Bering Strait at the northern end of the Bering Sea. Both the Bering and Okhotsk Seas have significant ice cover in winter. As a result, important water mass transformation and modification occur in both seas. The Okhotsk Sea produces the densest water in the subpolar North Pacific, mainly through sea ice processes (Section 10.9.2.1).

The subpolar gyre circulation is forced by Ekman upwelling (suction; Figure 5.16d). The winds throughout the region are westerlies, producing southward Ekman transport. The strongest westerly winds are at about $40°N$. Southward Ekman transport is largest there and decreases with higher latitude to smaller southward transports. This requires upwelling into the Ekman layer, which creates northward mean Sverdrup transport and the cyclonic gyre (Figure 5.17).

The upwelled water in the subpolar gyre comes from just below the Ekman layer. (It cannot come from greater depth because of the strong pycnocline, mainly due to the low salinity surface layer, hence halocline.) The heightened surface nitrate in Figures 4.22 and 4.23 is a result of this upwelling, which greatly enhances biological productivity. Major fisheries including salmon, halibut, saury, and walleyed pollock are found in the subpolar gyre.

Clearly, the Subarctic Front, marking the southern boundary of the subpolar gyre's upwelling, is an important ecosystem boundary.

With increasing depth, the North Pacific's subpolar circulation does not shift location unlike that of the subtropical gyre. It weakens, but its boundary currents reach far down into the water column, even to the bottom. The subpolar gyre is therefore "quasi-barotropic": its surface currents extend to the bottom (barotropic) but weaken (quasi). Near the bottom there are also additional currents associated with the topography and global thermohaline forcing (weak upwelling). The barotropic nature of the gyre is possibly due to geographic restriction, with the Alaskan coast cutting through the region that would be spanned by the gyre if there were no land. On the other hand, similar structure is found in other high latitude cyclonic circulations (North Atlantic subpolar gyre and the Weddell and Ross Sea gyres), suggesting a more general dynamical underpinning.

10.3.2.2. Subpolar Western Boundary Currents

The southward flow in the subpolar western boundary current system includes: (1) the EKC along the Kamchatka peninsula and the northern Kuril Islands and (2) the Oyashio along the southern Kuril Islands and Hokkaido. The division between the two is at Bussol' Strait, which is the deepest strait in the Kuril Island chain. The distinction is drawn because about half of the EKC loops through the Okhotsk Sea, where water properties are greatly modified. This creates a discontinuity in water properties at Bussol' Strait where the Okhotsk Sea waters exit and join the Oyashio.

About 200 km offshore of the Oyashio, there is a northeastward flow called the Subarctic Current (see also Figure S10.4 in the textbook Web site). The Oyashio-Subarctic Current region is very dynamic and includes large (100–200 km diameter), deep-reaching, long-lived anticyclonic eddies with cold, fresh cores (<3°C, <33.5 psu) that are usually found between Hokkaido and Bussol' Strait (online Figure S10.5). The eddies have two different origins: either locally at Bussol' Strait, from water exiting the Okhotsk Sea, or as warm water from the Oyashio intrusions (see next paragraph) that then propagates northeastward between the Oyashio and the Subarctic Current and is modified by the local cold, fresh subpolar water (Yasuda et al., 2001).

The Oyashio separates from the western boundary at the southernmost cape of Hokkaido. After separation, it usually makes two large meanders called the first (coastal) and second (offshore) Oyashio intrusions (Figure S10.6 from the online supplemental material). These are unrelated to the Kuroshio Extension meanders, which are farther south. Water from the coastal Oyashio intrusion can penetrate southward along the Honshu coast, sometimes as far south as the Kuroshio separation point at around 36°N; this cold coastal water is visible in the SST image of Figure 10.11. The location of southernmost penetration is of great interest to Japanese fisheries since the nutrient-rich Oyashio waters support a more biologically productive ecosystem than the nutrient-depleted Kuroshio waters. Therefore the Oyashio penetration latitude is used as a regional climate index.

The Oyashio/EKC is a relatively weak western boundary current. Maximum surface velocities are 20–50 cm/sec. Total Oyashio transport, based on combined direct current observations and hydrographic data east of

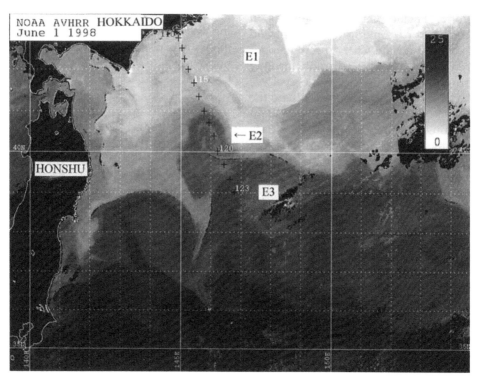

FIGURE 10.11 Oyashio, Kuroshio, and Mixed Water Region east of Japan. Sea surface temperature (NOAA AVHRR satellite infrared image) with temperature scale from 0 to 25°C; E1, E2, and E3 denote anticyclonic eddies. *Source: From Yasuda et al. (2001).*

Hokkaido, ranges from 5 to 20 Sv, with large variability (Kono & Kawasaki, 1997; Yasuda et al., 2001). The EKC transports range from 10 to 25 Sv, relative to various levels of no motion (Talley & Nagata, 1995).[2]

The separated Kuroshio and Oyashio are about 5 degrees of latitude apart (Figure 10.11). The region between them is referred to as the "Transition Region," "Mixed Water Region," or, in older literature, the "Perturbed Area." Water properties in this region are transitional between the Oyashio and Kuroshio properties. Both currents spawn major mesoscale eddy variability, some in the form of "rings," which participate in water mass modification. Sometimes the eddies re-merge with their parent currents, bringing the modified waters back with them.

10.3.2.3. Circulation in the Gulf of Alaska

The North Pacific Current splits as it approaches the North American continent and part turns south into the CCS. The remainder turns north into the Alaska Current, forming the eastern and northern side of the cyclonic Alaskan Gyre in the Gulf of Alaska. Where the coast of Alaska swings southward, at about 143°W, it forms a slanted western boundary along which the swift southwestward Alaskan Stream forms as a western boundary current. The wind field that drives the cyclonic circulation includes intensified Ekman upwelling in the Gulf of Alaska.

Details of the North Pacific Current bifurcation depend on the large-scale wind forcing, which has seasonal variability, and also interannual and decadal variability associated mainly with ENSO and the Pacific Decadal Oscillation (PDO; Sections 10.8 and 10.10; Chapter S15 from the online supplemental material). The position of the North Pacific Current bifurcation is at about 45°N in winter and 50°N in summer (Figure 10.1). The subpolar gyre, including the Alaskan Gyre, intensifies during periods when the atmosphere's Aleutian Low is especially strong such as El Niño years and years of low PDO. When the Aleutian Low and the subpolar gyre are weak, more subpolar water enters the CCS (Van Scoy & Druffel, 1993).

The Alaska Current contains dramatic, large anticyclonic eddies that are permanent, time-dependent components of the circulation. "Sitka Eddies" form west of Sitka, Alaska, at about 57°N and have a diameter of 150–300 km and surface amplitude of 10–20 cm (Tabata, 1982). "Haida Eddies" or "Queen Charlotte Eddies" form west of the Queen Charlotte Islands (Figure S10.7 located in the online supplementary material). The formation sites are related to bottom topography. After formation, these eddies propagate mainly westward into the Gulf of Alaska and are an important means of transporting coastal properties into the interior. Large eddies also populate the Alaskan Stream on the northwest side of the Gulf of Alaska (Crawford, Cherniawsky, & Foreman, 2000).

10.4. SOUTH PACIFIC CIRCULATION

10.4.1. Subtropical Circulation

The South Pacific is dominated by its anticyclonic subtropical gyre, extending from the ACC at about 50°S to the equator (Figures 10.1 and 10.2a; Table S10.2 in the online supplementary material). The gyre is well defined, but its western boundary current is complicated because the western boundary is composed of islands. (Oceanographically, Australia is a large island since it sits entirely within the subtropical gyre latitudes.) Connections with the other Southern Hemisphere oceans occur through the

---

[2] These transport estimates could be low because (1) velocities are often underestimated due to the use of inappropriately shallow levels of no motion and (2) large anticyclonic eddies can pull much of the Oyashio transport offshore, resulting in a weak coastal Oyashio and a stronger offshore component.

complex passages of the Indonesian archipelago and through the Southern Ocean south of Australia and South America.

The main western boundary current is the EAC, which flows southward along the coast of Australia until reaching the northernmost latitude of New Zealand. The EAC then separates and flows eastward to New Zealand, where it re-attaches to the east coast (as a western boundary current called the *East Auckland Current*) and continues a little farther southward. The EAC is very time-dependent and dominated by a series of cyclonic and anticyclonic eddies.

The broad eastward flow on the south side of the subtropical gyre can be called the *South Pacific Current* (SPC), following Stramma, Peterson, and Tomczak (1995), and consistent with usage of "North Pacific Current" and "North Atlantic Current" for the West Wind Drifts in the Northern Hemisphere. The circulation is bounded to the south by the Subantarctic Front, which is the northernmost front of the ACC (Chapter 13).

The northward flow along the coast of South America is the *Peru-Chile Current*. Like the California Current, the Peru-Chile Current is both the northward flow of the subtropical gyre and a full coastal upwelling system (PCCS) with separate eastern boundary current dynamics driven by alongshore winds. The westward flow of the subtropical gyre is the *South Equatorial Current* (SEC). At the sea surface, the SEC is located from about 20°S northward all the way to and across the equator; its structure at low latitudes is described with the tropical circulation in Section 10.7.3.

10.4.1.1. East Australian Current

The EAC is the southward western boundary current along the coast of Australia (Figure 10.12). A thorough description is found in Ridgway and Dunn (2003). The EAC forms from the westward flow of the SEC as it crosses the Coral Sea and reaches the Australian coast. At the sea surface, the SEC bifurcates at about 15°S into the

(a)

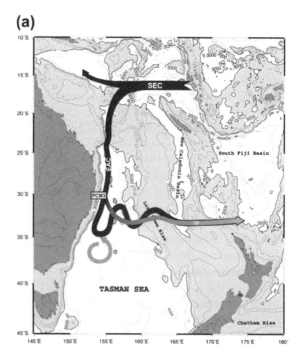

(b)

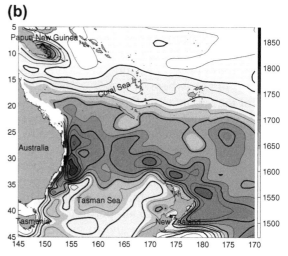

FIGURE 10.12 (a) Schematic of circulation in the western South Pacific (SEC: South Equatorial Current; EAC: East Australian Current; TF: Tasman Front). Eddy shedding from the EAC is depicted in light gray. *Source: From Mata et al. (2006).* (b) Mass transport streamfunction relative to 2000 dbar; contour interval is 25 m². *Source: From Ridgway and Dunn (2003).*

southward EAC and northward flow along Queensland. This bifurcation point moves toward the south with increasing depth, reaching 500 m at about 22°S (Figure 10.12b and Figure S10.8 in the supplementary Web site). The EAC transport intensifies as it flows along the Australian coast, reaching a maximum velocity of around 90 cm/sec at 30°S. It begins to separate from the coast around 31 to 32°S. It reaches its maximum transport of about 35 Sv shortly after separation, at 33°S, where it undergoes a southward meander and retroflection with part of the transport returning northward in a tight recirculation. A mean northward recirculation exists offshore of the EAC between latitudes 33°S and about 24°S, and likely has two separate lobes (Figure 10.12).

Most of the EAC flow that does not recirculate turns eastward into the zonal Tasman Front and crosses the Tasman Sea to the northern cape of New Zealand. Transport in the Tasman Front is estimated at 13 Sv. The EAC flow in the Tasman Front re-attaches to the coastline at New Zealand and forms the East Auckland Current (Roemmich & Sutton, 1998). The East Auckland Current continues southward and finally separates from New Zealand at about 43°S (Figure 10.12), where it meets a northward loop of the Subantarctic Front (ACC).

The remainder of the EAC reaches southward through the Tasman Sea to Tasmania. The location of the southernmost penetration of EAC waters along Tasmania is used as a regional climate index, much like the southward penetration latitude of Oyashio waters along Japan (Section 10.3.2.2). A small portion continues southward past Tasmania and turns westward into the Indian Ocean, connecting the westward flow of the South Pacific and Indian subtropical gyres (Speich et al., 2002; Ridgway & Dunn, 2007).

The EAC separates from the coast at about 32°S and meanders strongly southward and then northward. The meander regularly pinches off into a ring. The EAC undergoes major

retraction and deformation after such eddy shedding, which occurs about every 100 days (Mata, Wijffels, Church, & Tomczak, 2006).

The EAC has long been understood to be particularly rich in eddies (Hamon, 1965; Godfrey et al., 1980). EAC eddies sometimes appear to dominate the mean circulation. Eddy diameters are 200–300 km, and surface speeds are up to 180–200 cm/sec, with lifetimes of up to a year (Boland & Church, 1981). The eddy centers are well mixed to as much as 300 m depth (Nilsson & Cresswell, 1981). In austral winter, the surface water in an eddy may be as much as 2°C warmer than the surrounding water.

Eddy formation sites in the EAC tend to be recurrent, so the eddies appear in the mean dynamic topographies and altimetric height maps (Figure 10.12 and Figure S10.9 from the online supplementary material). Two are found within the recirculation of the EAC along the Australian coast, and three within the Tasman Front and East Auckland Current. The permanence of these eddy sites suggests topographic control (Ridgway & Dunn, 2003).

10.4.1.2. South Pacific Current and Subtropical Front

The eastward flow of the South Pacific subtropical gyre is the SPC (Stramma et al., 1995; Wijffels, Toole, & Davis, 2001). The broad, weak eastward flow of the SPC was long identified with the ACC, but the SPC is dynamically distinct from the ACC. As an analog of the North Pacific Current, we consider the SPC to be all of the eastward flow of the South Pacific's subtropical gyre north of the Subantarctic Front. The SPC flows into the broad, open-ocean part of the northward Peru-Chile Current, and from there to the westward SEC. These three currents constitute the open ocean part of the South Pacific's subtropical gyre. Maximum Sverdrup transport for the subtropical gyre occurs around 30°S and is about 35 Sv (Figure 5.17, Figure S10.2b in the online supplementary materials, and Wijffels et al., 2001).

The SPC forms as eastward outflow from the East Australian and East Auckland Currents. In mid-ocean, it has a somewhat bowed structure, with a slight northward excursion from offshore of the EAC to mid-gyre, around 170°W, then southward to about 140°W and finally, northward in the main Peru-Chile gyre flow. This structure appears to be permanent.

The eastward flow of the SPC bifurcates at the eastern boundary between 40°S and 45°S. The northward flow joins the Peru-Chile Current and the southward flow joins the ACC through Drake Passage.

Within the SPC there is a marked, nearly zonal Subtropical Front, called the Subtropical Convergence in earlier works, including earlier editions of this text. The Subtropical Front is identified by large meridional gradients in temperature and salinity in the upper ocean, with a northward increase of 4°C and 0.5 psu, sometimes over just a few kilometers (Deacon, 1982; Orsi, Whitworth, & Nowlin, 1995). North of the Subtropical Front lies the saline, warm water of the central subtropical gyre; salinities are greater than 34.9 psu just north of the front. South of the Subtropical Front is the fresher, cooler water of the poleward part of the gyre.

Transport of the SPC has not been estimated as such. An estimate for the Subtropical Front alone is less than 5 Sv (Stramma et al., 1995). Otherwise the transport of the broad subtropical gyre has been mainly estimated from the meridional (north-south) component through east-west sections across the gyre, which are described in the next subsection.

10.4.1.3. Northward Flow of the Subtropical Gyre and the Peru-Chile Current System

Northward flow in the subtropical South Pacific consists of the broad subtropical gyre and the swifter, narrow eastern boundary current system along the coast of South America, referred to as the PCCS (Figures 10.1 and 10.13). The northward transport is estimated to be 15 Sv between 180° and the eastern boundary (Wijffels et al., 2001). Within the broad gyre, denser surface waters from the south subduct northward under lighter low latitude waters. This creates the stratified structure of the central South Pacific pycnocline (Section 10.9.1), and the salinity/oxygen layering in the vertical that facilitates identification of various water masses.

At the eastern boundary, the PCCS (Figure 10.13) is a typical eastern boundary current upwelling system (Sections 7.9 and 10.3.1.4), forced by the alongshore component of the large-scale winds and an offshore band of positive wind stress curl. It includes the northward Peru-Chile Current (also called the Peru Current and formerly called the Humboldt Current). The Poleward Undercurrent (also called the Gunther Current) is found along the coast beneath the surface layer, as expected for a typical eastern boundary current system. The PCCS also contains other currents: a poleward Peru-Chile Countercurrent 100–300 km offshore, and an equatorward Peru Coastal Current on the inshore side. The Peru-Chile Current and Peru Coastal Current connect to the equatorial SEC and the cold tongue in the eastern equatorial Pacific (Figure 10.13). The Equatorial Undercurrent (EUC) feeds into the Poleward Undercurrent and Peru-Chile Countercurrent (Strub et al., 1998).

Maximum upwelling, extending southward along the Chilean coast to 45°S, occurs in austral summer. The PCCS upwelling is well known because of the rich fisheries there. Satellite ocean color images (Figure S10.10 in the online supplemental text) vividly show the effects of coastal upwelling, which lifts nutrients to the euphotic zone, resulting in high biological productivity. The permanent upwelling region extends from about 32°S northward to the equator; seasonal upwelling occurs south of this to about 40°S.

Vertical sections across the PCCS at 33°S (Figure 10.13) show the isotherm structure

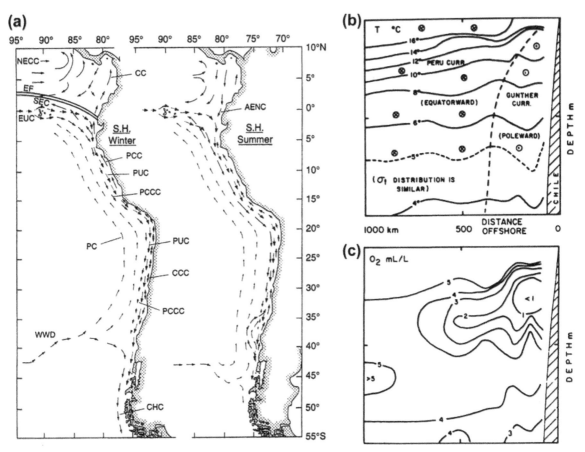

FIGURE 10.13 Peru-Chile Current System. (a) Maps in austral winter and summer. Acronyms: WWD, West Wind Drift; PC, Peru Current; PCCC, Peru-Chile Countercurrent; PUC, Poleward Undercurrent; PCC, Peru Coastal Current; CCC, Chile Coastal Current; and CHC, Cape Horn Current. Also, near the equator: CC, Colombia Current; AENC, Annual El Niño Current; NECC, North Equatorial Countercurrent; SEC, South Equatorial Current; EUC, Equatorial Undercurrent. *Source: From Strub et al. (1998).* (b, c) Eastern South Pacific zonal vertical sections at 33°S: temperature (°C) with meridional current directions and dissolved oxygen (ml/L); companion salinity and phosphate sections appear in Figure S10.11 on the textbook Web site.

typical of a geostrophic eastern boundary current system, including the equatorward Peru-Chile Current above about 500 m and the poleward subsurface Peru-Chile Undercurrent (PCUC) near the coast. The undercurrent is characterized by low oxygen which comes from the tropics and from local high productivity that traps high nutrients and low oxygen just beneath the surface layer (Montecino et al., 2006). High nutrient content, associated with the low oxygen,

helps to create the characteristic high biological productivity of this eastern boundary region.

The PCCS is strongly affected by ENSO (Section 10.8). Collapses of the PCCS fisheries resulting from changing upwelling conditions were among the earliest dramatic evidences for ENSO, which is now known to encompass the entire equatorial Pacific. During normal conditions, the Peru-Chile Current extends to a few degrees south of the equator before turning west into the SEC. The low temperature of the Peru-Chile Current surface waters contrasts with higher equatorial temperatures to the north. During an El Niño (warm phase), the high temperatures extend 5 to 10 degrees farther south than usual and the thermocline deepens by 100 m or so. Upwelling either weakens or simply draws on warmer water from this thicker warm layer, thus causing the surface temperatures to increase. The increase in temperature was thought to kill fish, but recent studies have shown that the fish merely descend below the abnormally warm surface layer. In every austral summer there is a slight warming of the sea surface along with an increase in precipitation. During El Niño years, however, the warming and the rainfall far exceed the norm.

10.4.1.4. South Equatorial Current

The SEC is the broad westward geostrophic flow in the northern limb of the South Pacific's subtropical gyre (Figures 10.1 and 10.12). The SEC forms in the eastern Pacific as the north-ward flow of the subtropical gyre turns west-ward. The narrow eastern boundary current (Peru-Chile Current) also feeds into the SEC close to the equator.

As it reaches the western South Pacific, the SEC carries water into the Coral Sea off north-eastern Australia. The many islands in the region complicate the SEC, including intense zonal jets with large east-west extent (Webb, 2000; Qu & Lindstrom, 2002; Ganachaud, Gour-deau, & Kessler, 2008). When the SEC reaches the Australian coast, it bifurcates into the southward EAC and the northward North Queensland Current. The latter feeds the NGCUC, bringing South Pacific water to the western equatorial Pacific and feeding the EUC (Section 10.7.4).

The SEC also includes the frictional equatorial surface flow (Section 10.7), which is bounded to the north by the powerful eastward NECC. Because the SEC extends across the equator,

whereas the NEC is separated from the equator by the NECC, the South Pacific subtropical gyre is much more directly connected to the equator than is the North Pacific gyre. Subtropical anom-alies in heat or salinity can more easily reach the equator from the South Pacific than from the North Pacific because of this direct SEC connec-tion (Johnson & McPhaden, 1999).

10.5. PACIFIC OCEAN MESOSCALE EDDY VARIABILITY

The ocean circulation focused on in this text is the mean of a highly time-dependent, turbulent flow. Mesoscale eddy variability at timescales of weeks to months is easily detected with instruments such as satellite altimeters, which measure the surface height variability. At depth, eddy variability is measured with moored observations at point locations and using Lagrangian floats that are usually deployed at a single depth.

Surface EKE and horizontal eddy diffusivity in the Pacific are shown in Figures 14.16 and 14.17 and also in Figure S10.12 on the textbook Web site. High EKE is mostly associated with strong mean flows: the Kuroshio Extension (30−40°N), the EAC (25−40°S), the ACC (south of 50°S), and the NECC (5−10°N). Two zonally elongated regions of high eddy energy, at 20°N and 25°S, are associated instead with weak east-ward surface flows. These are the STCCs in both hemispheres; the flow just below the surface, even at 200 dbar, is westward (Figure 10.2). The energy in these unstable mean flows is mainly released through baroclinic instability, creating the high EKE (Stammer, 1998; Qiu, Scott, & Chen, 2008; Section 7.7.5).

High eddy variability in the Pacific in Figures 14.16 and S10.14 also occurs at the locations of recurrent rings, including the Tehuantepec eddies in the eastern tropical Pacific (Section 10.7.6), the Kuroshio rings, the EAC rings, and the rings along the boundaries of the subpolar gyre (Haida & Sitka eddies; eddies in the Oyashio).

10.6. DEPTH DEPENDENCE OF THE PACIFIC OCEAN CIRCULATION AND MERIDIONAL OVERTURN

Below the wind-driven subtropical gyres, and coexisting with the deep-reaching North Pacific subpolar gyre, the Pacific circulation is weak, mostly less than several centimeters per second except in the tropics. Faster currents (>10 cm/sec) occur in the deeper parts of the upper ocean western boundary currents and in the DWBCs, but transports are nevertheless relatively small, of the order of 10 Sv or less.

As we leave the sea surface, the subtropical gyres shrink away from the equator, away from the eastern boundary, and toward the energetic western boundary currents. The Kuroshio gyre shrinkage was described in Section 10.3.1.6. In

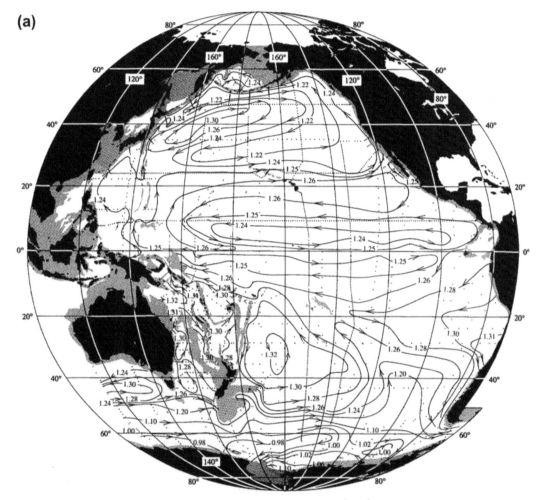

(a)

FIGURE 10.14 Adjusted geostrophic streamfunction (steric height, 10 m²/sec²) at (a) 2000 dbar, (b) 4000 dbar. *Source: From Reid (1997).*

(b)

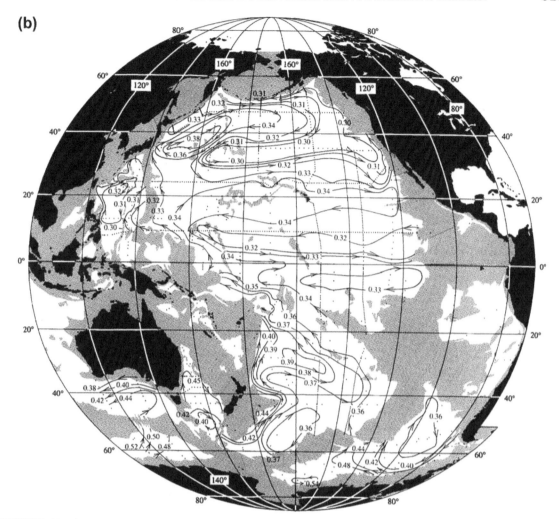

FIGURE 10.14 (*Continued*).

the South Pacific, the subtropical gyre shrinks into the Southwest Pacific Basin, east of New Zealand and the Tonga-Kermadec Ridge.

On the tropical side vacated by these shrinking gyres, the flows are nearly zonal except close to the western and eastern boundaries (Figures 10.2b, 10.10, 10.14 and Figure S10.13 on the textbook Web site). This zonal flow pattern persists down to the tops of the major mid-ocean ridges, roughly between latitudes 20°N and 20°S. Outside the tropics, the deep flow patterns are

influenced by the overlying gyres, the underlying topography, and the DWBCs (Figure 10.14). In the southwest Pacific below 2000 dbar, the circulation is a combination of a northward DWBC and an anticyclonic flow that fills the rest of the basin to the east and north. In the southeast Pacific, in the Bellingshausen Basin, the flow is weak and cyclonic from about 800 dbar to the ocean bottom, with a southward eastern boundary current that carries the thick, low oxygen layer of PDW southward to the Southern Ocean (Shaffer et al., 1995;

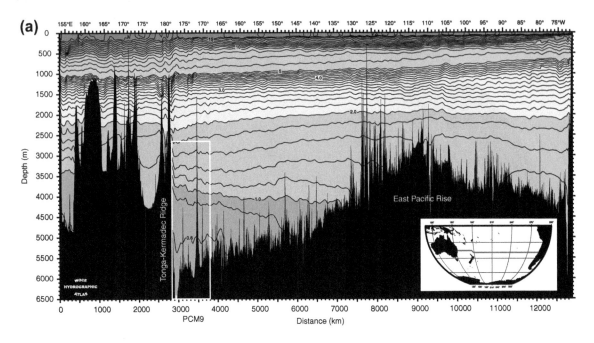

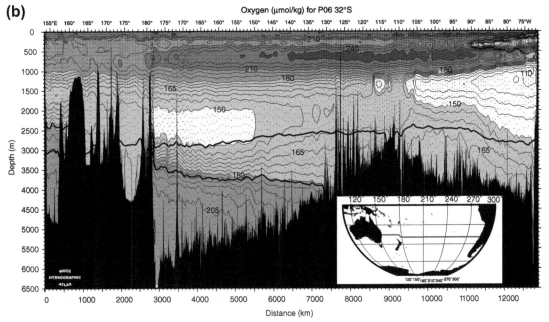

FIGURE 10.15 South Pacific sections at 32°S and DWBC. (a) Potential temperature and (b) oxygen (μmol/kg). Neutral densities 28.00 and 28.10 kg/m³ are superimposed in (a). *Source: From the WOCE Pacific Ocean Atlas, Talley (2007).* (c) Mean northward velocities (cm/sec) from current meters at 32° 30′S northeast of New Zealand in 1991–1992. The array location is within the white box in (a). *Source: From Whitworth et al. (1999).*

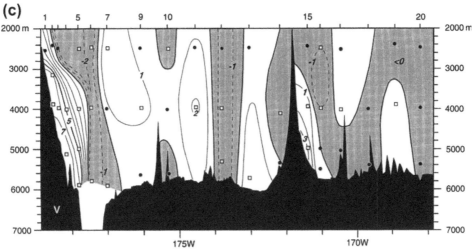

FIGURE 10.15 (*Continued*).

Figure 10.15b; Section 10.9.3). In the deep North Pacific north of about 10°N, the abyssal flow consists of two anticyclonic circulations, one centered south of the Hawaiian Islands, and the other centered at about 45°N (Figure 10.14b). These two gyres are also evident in silica distributions on deep isopycnals (Talley & Joyce, 1992).

The deep flows include well-delineated DWBCs (Section 7.10.3). In the southwest Pacific, the DWBC carries deep and bottom waters from the Southern Ocean northward into the Pacific, as seen in observations at 32°S (Whitworth et al., 1999). Large upward slopes in isotherms within several stations just east of the Tonga-Kermadec Ridge indicate the narrow DWBC, from the bottom up to 1.8°C (~2500 m; Figure 10.15). Northward transport of 16 Sv was measured in a narrow, banked band at the ocean bottom, mostly colder than 1°C. This DWBC continues northward to the tropics. Its most constricted location is at the Samoan Passage at 10°S, 169°W (Figure 10.16). Observed transport of all waters colder than 1.1°C, including those within the passage and banked against the Manahiki Plateau, was 11.7 Sv (Roemmich, Hautala, & Rudnick, 1996). The mean northward transport below 4000 m, within

the Samoan Passage, was 6.0 Sv; velocities were shown in Figure 6.7 (Rudnick, 1997).

The DWBC proceeds northward from the Samoan Passage region and crosses the equator at the deep western boundary (Figure 10.17 and Figure S10.14 on the textbook Web site). Here it splits into two branches, one following the western boundary and the other heading toward the Wake Island Passage (168° 30'E, 18° 20'N). The western boundary branch is observed to carry both Lower Circumpolar Deep Water (LCDW; 1 Sv) and Upper Circumpolar Deep Water (UCDW; 11 Sv). The flow in the Wake Island Passage is up to 10 cm/sec within several hundred meters of the bottom, with a transport of 4 Sv of LCDW (Kawabe, Yanagimoto, Kitagawa, & Kuroda, 2005; Kawabe, Yanagimoto, & Kitagawa, 2006).

North of the Wake Island Passage, the deep flow moves westward to the boundary and then northward to an encounter with the Kuroshio Extension. Further north along the subpolar boundary, abyssal circulation theory indicates that the DWBC should flow southward (even though there is no local source of deep water; Figure 7.16). The western and northern boundaries are complicated by a very deep trench, in

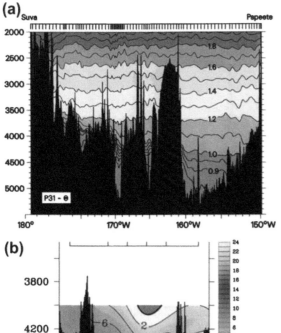

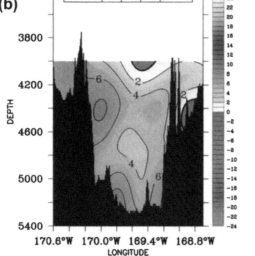

FIGURE 10.16 DWBC in the Samoan Passage. (a) Potential temperature (°C) on WOCE P31 across the passages. (b) Mean northward velocity (cm/sec) through the passage measured by current meters (1992–1994). *Source: From Roemmich et al. (1996).*

which the observed flow is southward/westward at the continental boundary and northward/eastward along the offshore side of the trench (Figure 10.17; Owens & Warren, 2001). The net DWBC transport is small (order 3 Sv) and southward/westward, matching theory. The net meridional overturning in the Pacific consists of northward transport from the

Southern Ocean in the abyssal layers and southward outflow in the deep to intermediate layers (e.g., Figures 10.18, 14.6, and Figure S10.15 on the textbook Web site). Estimates of the northward transport of the deepest water into the South Pacific (LCDW or Antarctic Bottom Water; AABW) range from 7 to 20 Sv, but the large range might simply be due to layer choices. Most of this water upwells into the PDW and returns southward. Most of the upwelling occurs in the South Pacific and tropics; at 24°N in the North Pacific the bottom upwelling cell is much weaker and much more confined to the bottom layers.

10.7. TROPICAL PACIFIC CIRCULATION AND WATER PROPERTIES

10.7.1. Introduction

The Pacific equatorial current system is dominated by strong zonal (east-west) flows with weak meridional (north-south) currents in the ocean interior (Figure 10.2a; Table S10.3 and Figure S10.1 located on the textbook Web site). At the sea surface there are three major zonal currents. Below the surface there is a complex set of reversing zonal flows. At the western boundary, strong meridional currents connect the zonal flows.

The three major zonal surface currents are (1) the westward-flowing NEC between about 8°N and 20°N, (2) the westward SEC from about 3°N to 10°S, and (3) the narrow NECC flowing to the east between them, centered at about 5°N. These were well-known components of the Pacific surface circulation before 1940. The other major equatorial current lies just below the thin surface layer of the SEC and it is the eastward-flowing EUC, which is one of the fastest permanent currents in the world. The eastward *South Equatorial Countercurrent* (SECC) in the western South Pacific between 10 and 12°S is much weaker and more time-dependent than these. Then

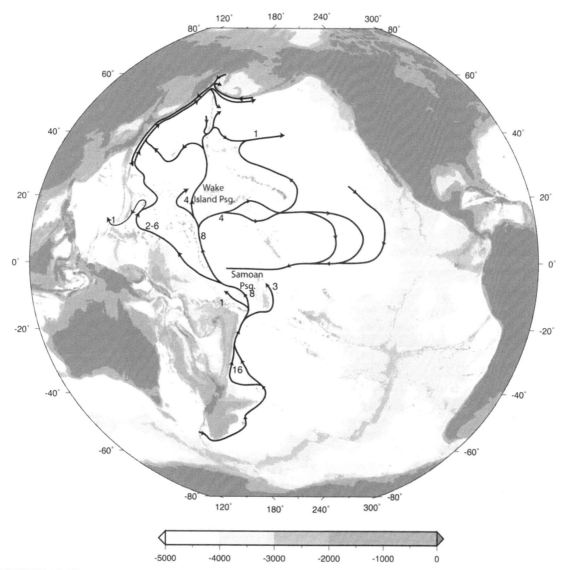

FIGURE 10.17 Abyssal circulation schematics. *After: Owens and Warren (2001), Johnson and Toole (1993), Kato and Kawabe (2009), Komaki and Kawabe (2009), Yanigomoto, Kawabe, and Fujio (2010), Whitworth et al. (1999), and Roemmich, Hautala, and Rudnick (1996).*

a complicated set of subsurface eastward and westward mean flows (Section 10.7.3) is evident.

The low latitude western boundary currents (Section 10.7.4) collect water from the westward SEC and NEC and feed it into the eastward subsurface equatorial flows and into the eastward NECC. The Mindanao Current is the primary equatorward Northern Hemisphere boundary current, connecting the westward flow of the NEC to the eastward flow of the NECC. The NGCUC is the primary equatorward Southern Hemisphere boundary current, connecting the westward flow of the SEC to the eastward flow of the subsurface equatorial currents (EUC, North

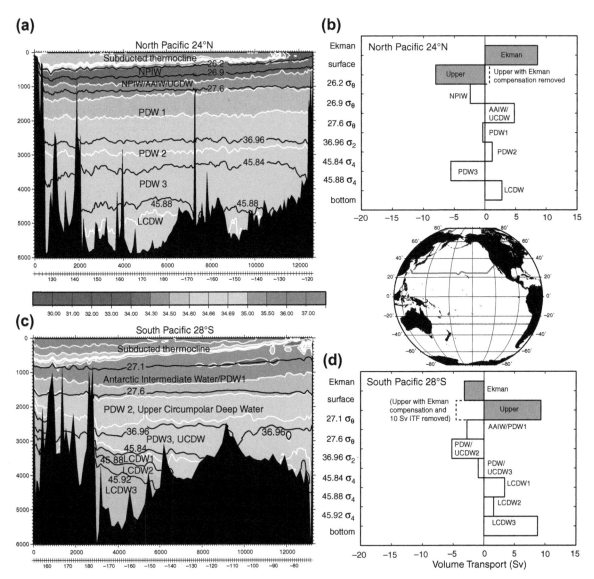

FIGURE 10.18 Salinity and meridional transport in isopycnal layers at 24°N (a, b) and at 28°S (c, d). Inset map shows section locations. The isopycnals (σ_θ, σ_2, σ_4) that define the layers are contoured on the salinity sections. *After Talley (2008).* Overturning transports from Ganachaud (2003) are shown in Figure S10.15 on the textbook Web site.

Subsurface Countercurrent, NSCC; and South Subsurface Countercurrent, SSCC), and also crossing the equator to meet the southward flow of the Mindanao Current and feed the NECC.

Dynamics of the wind-driven equatorial surface currents and the EUC were presented

briefly in Section 7.9.2, and directly on the equator, flow is in the direction of wind stress (in the frictional surface layer) and pressure gradient force. Moving slightly away from the equator, the Coriolis force quickly becomes important; the currents are almost geostrophic and the upper ocean circulation can be considered in terms of the usual Sverdrup dynamics driven by convergence of the wind-driven Ekman layer (Section 7.5).

10.7.2. Tropical Wind and Buoyancy Forcing

The tropical surface current system is driven by the easterly trade winds at the ocean's surface (Figure 5.16, Figure S10.16 located on the textbook Web site, and also the stick plot in Figure 10.20b). The trade winds are part of the atmosphere's Walker and Hadley cells (Section 7.9.2). The trade winds are not uniformly westward; these surface winds converge at the ITCZ north of the equator. (A weak, secondary ITCZ is found in the western South Pacific.) The wind stress curl associated with the ITCZ is positive, creating Ekman suction (Figure 5.16d). This drives cyclonic circulation that is very zonally elongated. This includes westward flow on the northern side (part of the NEC), and eastward flow on the southern side, which is the NECC (Yu, McCreary, Kessler, & Kelly, 2000). The SECC, which appears in the western South Pacific tropics, is driven by a similar mechanism associated with the Southern Hemisphere ITCZ.

Seasonally, the trade winds are stronger in the winter hemisphere (Figure 5.16). The Northern Hemisphere ITCZ lies closer to the equator, at about 5°N in the east, in February than in August. In August, the northern ITCZ shifts northward to 10°N across the whole Pacific. In the western tropical Pacific, there is a seasonal monsoon, which is a reversal in winds in the Northern Hemisphere and equatorial regions. This especially impacts the surface equatorial circulation (Section 10.7.3.1).

Air—sea fluxes of heat and freshwater in the tropical Pacific are important for the global balances of both of these quantities (Figures 5.4, 5.12). The tropical oceans warm due to high solar radiation (Figure 5.11a). The greatest warming is in the equatorial cold tongue in the eastern Pacific (see next section), where lower surface temperatures result in reduced latent and longwave heat losses, hence higher net heating.

The tropical Pacific is also a region of net precipitation. The precipitation is not uniform (Figure 5.4). Beneath the ITCZ of the Northern Hemisphere is a band of net precipitation. The western Pacific is also a region of net precipitation, concentrated in two bands centered at the northern and Southern Hemisphere ITCZs. The eastern tropical Pacific is a region of net evaporation. These patterns are directly related to the Hadley and Walker circulations, with more precipitation where air rises along the ITCZ and in the western tropical Pacific.

The net precipitation in the western tropics creates a low salinity surface layer, with a strong halocline beneath. The mean stratification here consists of a so-called *barrier layer*, in which warm surface temperature extends to greater depth than the fresh surface water. The mixed layer stratification, therefore, is dominated by salinity.

10.7.3. Equatorial Pacific Current Structure

10.7.3.1. Zonal Currents and Associated Mid-Ocean Meridional Flows

Zonal flows dominate meridional flows in the tropics, except at the western boundary. Average upper ocean zonal velocity, temperature, and salinity structure in the central Pacific (154°W) is shown in Figures 10.19 and 10.20 (Wyrtki & Kilonsky, 1984; WK). The deep equatorial currents are described for a nearby longitude (Figure 10.21). These zonal currents are geostrophic except directly on the equator,[3] and are therefore reflected in sloping surface dynamic

---

[3] Geostrophy is valid to within about one-quarter degree of the equator with sufficient temporal averaging. A 12-month set of 43 sections was used for the mean structure in WK.

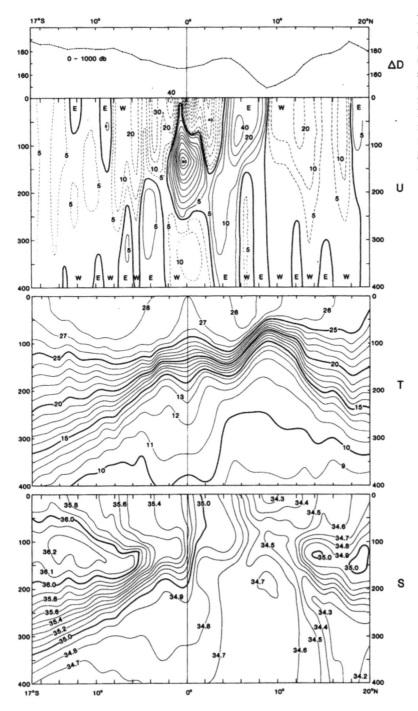

FIGURE 10.19 Mean distributions of surface dynamic height ($\triangle$D dyn cm) relative to 1000 db (dyn cm) and vertical meridional sections of zonal geostrophic flow (U in cm/sec), temperature (T in °C), and salinity (S) between Hawaii and Tahiti, for 12 months from April 1979. ©*American Meteorological Society. Reprinted with permission. Source: From Wyrtki and Kilonsky (1984).*

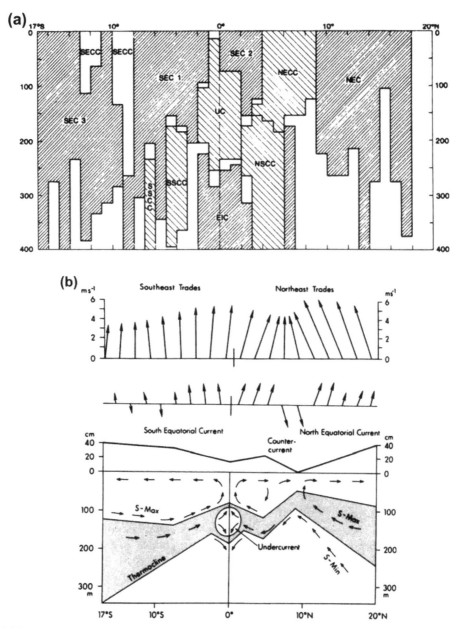

FIGURE 10.20 (a) Schematic of mean areas occupied by zonal currents between Hawaii and Tahiti for 12 months from April 1979. Dark shading indicates westward flow, light shading indicates eastward flow, blank areas have zonal speeds less than 2 cm/sec. Acronyms: NEC, North Equatorial Current; NECC, North Equatorial Countercurrent; SEC, South Equatorial Current (three sections); SECC, South Equatorial Countercurrent; UC, Equatorial Undercurrent (EUC in our notation); EIC, Equatorial Intermediate Current; and NSCC/SSCC, Northern/Southern Subsurface Countercurrents (Tsuchiya jets). (b) Schematic meridional section across the equator showing (top) the mean trade winds, (middle) surface circulation, and (bottom) schematic surface dynamic topography, temperature structure, and meridional circulation below the surface. ("Countercurrent" = "NECC" in our notation.) ©*American Meteorological Society. Reprinted with permission. Source: From Wyrtki and Kilonsky (1984).*

159° W (PEQUOD) 2/82 - 6/83

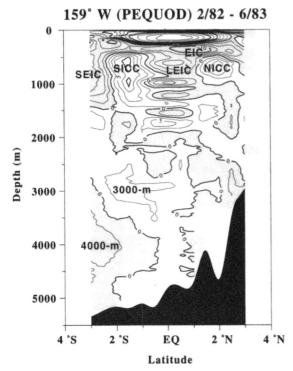

FIGURE 10.21 Zonal velocity (cm/sec) in the equatorial Pacific, averaged from 41 sections of direct current measurements collected in 1982–1983. White is eastward flow, gray is westward. *Source: From Firing, Wijffels, and Hacker (1998).*

The eastward NECC is a strong, permanent current that stretches across the whole width of the Pacific with associated large dynamic height and isotherm slopes in the opposite direction to those of the NEC/SEC. In contrast, the weak, eastward SECC is mostly restricted to the western Pacific with only a weak expression in the central Pacific seen as a slight reversal in surface dynamic height slope (Figure 10.19).

At the surface on the equator, the surface flow is westward (SEC-1). This equatorial SEC is in just a thin layer above the EUC. The equatorial SEC's flow is the downwind, frictional equatorial response to the westward trade winds, in the absence of the Coriolis force and hence an Ekman layer (Section 7.9.2). It can disappear at times since it is driven directly by the wind, and in any case responds quickly to changes in winds; a reversal to eastward occurs regularly during westerly wind bursts at the onset of El Niño (Section 10.8; Hisard & Hénin, 1984).

At the equator, the EUC lies just beneath the SEC. Its maximum velocity core at this central Pacific location lies at 130 m, with average speeds greater than 90 cm/sec.[4] The EUC was considered to be weak during the WK measurement period; it can regularly reach speeds of 120 cm/sec. Despite its thinness in the vertical, its large speeds are reflected in large transport (32.3 ± 3.5 Sv in WK's annual average). The EUC is easily identified in isotherm structure at the equator: the 13–26°C isotherms spread upward above it and downward below it. It has no expression in surface dynamic height since it is not a surface current. This creates the necessary geostrophic vertical shear on

height (ΔD; Figure 10.19), and in isopycnal slopes, which produce the vertical shear of the geostrophic currents. The westward NEC and the southernmost part of the westward SEC (SEC-3) are the primary westward flows of the North and South Pacific's subtropical gyres (Figure 10.1) and extend down through the thermocline. Their dynamic heights slope downward and isotherms tilt upward toward the equator.

[4] The EUC was first discovered in 1951 when researchers from the U.S. Fish and Wildlife Research Service in Honolulu found that their "long-line" deep fishing equipment drifted strongly eastward in spite of the westward surface currents. Their gear traveled eastward at speeds of about 1.5 m/sec, which was about three times that of the westward surface current. A subsequent cruise to investigate this phenomenon was led by Townsend Cromwell; the EUC is also called the "Cromwell Current." Unfortunately Dr. Cromwell died the next year in a plane crash on the way to an oceanographic expedition. See Knauss (1960).

both sides of the equator to create a subsurface eastward flow with westward flows both above it (SEC) and below it (Equatorial Intermediate Current; EIC).

The eastward NSCC and SSCC are just to the north and south of the equator and slightly deeper than the EUC. The NSCC is not always easily distinguishable from the deeper part of the surface-intensified NECC. In the isotherms (Figure 10.19), the SCCs are apparent in the strong upward slopes of the 10 and 11°C isotherms away from the equator. The SCCs were first identified from maps of properties on isopycnals by Tsuchiya (1975). They transport salinity, oxygen, and nutrients characteristic of the western Pacific toward the east. In honor of this first description, the SCCs are often referred to as "Tsuchiya jets".

The westward EIC is a weak but persistent flow along the equator beneath the EUC. Beneath the EIC, the reversing eastward and westward flows between 1000 and 2000 m are referred to as the *equatorial stacked jets* (Figure 10.21). Off the equator, around 700–900 m depth, there are also reversing zonal flows, but in thicker layers with speeds around 15–20 cm/sec. The deepest equatorial flows in Figure 10.21 have small mean speeds, <5 cm/sec, but might be permanent features.[5] Given the local topography at 159°W, which rises to 3000 m in the north, the robust currents are south of the equator. Transports of each of these flows is on the order of several Sverdrups. Farther from the equator, within 15–20° of the equator and above the topography of the mid-ocean ridges (above 3000 m), the intermediate and deep circulation remains dominantly zonal compared with flow at higher latitudes. The zonal nature of the flows is clear in float trajectories at 900 m (Davis, 2005; Figure S10.13 on the textbook Web site), in steric height maps for these mid-depths (Figure 10.2b), and in ocean properties on isopycnals. At 2500 m, flow includes a narrow eastward tongue at

about 2°S and broad flanking westward flows centered at 5–8°N and at 10–15°S (Talley & Johnson, 1994). At the bottom, the westward equatorial flow is possibly fed by broader eastward flow north of the equator (Johnson & Toole, 1993). These complex, zonal deep flows are likely wind-forced (Nakano & Suginohara, 2002).

Returning to the upper ocean, meridional flows in the equatorial Pacific (Figure 10.20b) are associated with the major zonal currents. At the sea surface, the easterly trade winds cause Ekman transport to the north in the Northern Hemisphere and to the south in the Southern Hemisphere. This results in equatorial divergence, which creates *equatorial upwelling*. (There is equatorial downwelling if the winds shift to westerly, as in the western equatorial Pacific at the beginning of an El Niño event.)

The equatorial upwelling is fed by equatorward subsurface flow. The inflow is in the thermocline, based on water properties, including salinity (Figure 10.19 "S" panel). The equatorward inflow can be geostrophic, due to the west-to-east pressure gradient force set up by the westward flow of surface water along the equator to the western boundary. This creates high pressure in the west and low pressure in the east.

10.7.3.2. Zonal Structure of the Equatorial Currents

The equatorial current system extends from at least 143°E (north of Papua, New Guinea) to the Galapagos Islands (90°E) and then eastward to the coast of Ecuador, a distance of approximately 15,000 km. The sea surface is high in the west and slopes down to the east in the equatorial band (Figure 10.2 and Figure S10.1 on the textbook Web site). The west-east difference in surface height is 40–60 cm, with significant interannual variability associated with ENSO; the largest slopes occur during La Niña (Figure 10.22c). Surface dynamic height shows

---

[5] According to Firing (1989), "a 10 year time-series would be ideal for studying annual and interannual variations."

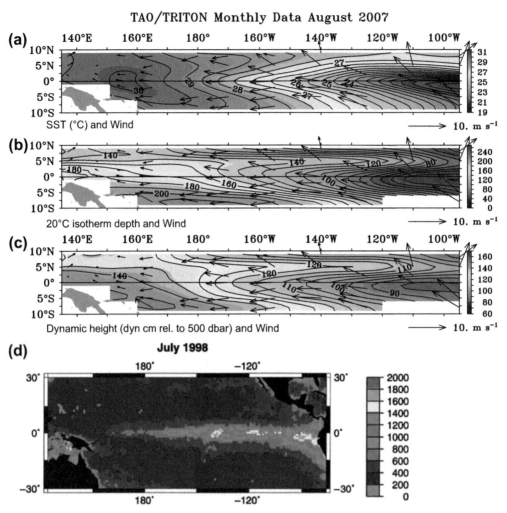

FIGURE 10.22 (a) SST; (b) depth of the 20°C isotherm, which is an indicator of thermocline depth; and (c) dynamic height (dyn cm), with superimposed wind velocity vectors, during a period of a well-developed cold tongue (La Niña; August 2007). *Source: From TAO Project Office (2009a).* (d) Primary productivity (mg C m^{-2} day^{-1}) based on ocean color, during a La Niña (July 1998). *Source: From McClain et al. (2002).*

the same west-east contrast of about 40 dyn cm (Figure S10.17 on the textbook Web site). The equatorial sea-surface height slope is due to the wind-driven westward flow of surface water in the SEC along the equator. This piles warm water up in the west, in the region called the *warm pool*. The westward equatorial flow is also associated with equatorial upwelling in the east. The cold, upwelled surface water in

the east is called the *cold tongue*. These structures are obvious in mean SST (Figures 4.1 and 10.22). Along-equatorial sections of potential temperature, salinity, and potential density show the warmer, lighter water to the west and colder, denser surface water to the east. Surface nutrients have a similar structure, with higher nutrients in the cold tongue and nearly complete depletion in the warm pool (Figure 4.22).

Cold water along the equator has two sources: upwelling in the eastern Pacific due to the westward surface flow (SEC) driven along the equator by the trade winds, and upwelling due to divergent Ekman transport just off the equator, also due to the trade winds, which can occur at all longitudes. Because the warm pool in the western Pacific is so thick, the Ekman divergence component of the upwelling does not bring cold water to the sea surface there.

The pileup of water in the west causes an eastward pressure gradient force along the equator. This pressure gradient force drives the eastward flow of the EUC. The west-to-east pressure gradient force also creates equatorward geostrophic flow that feeds the equatorial upwelling.

The equatorial pycnocline is deep in the west and tilts upward toward the east (Figure 10.23). This upward tilt compensates the downward sea-surface tilt such that the pressure gradient force along the equator beneath the pycnocline is very weak. In fact the equatorial flow beneath the EUC is weakly westward (EIC). The EUC is located within the pycnocline (Figure 10.23c). It shoals toward the east along with the pycnocline (Figure 10.23). It is weak in the western equatorial Pacific, with speeds less than 40 cm/sec. It speeds up east of the date line, and reaches maximum strength around 140°W. This corresponds to longitudes of greater eastward pressure gradient force, evident in surface height and dynamic height. Its transport peaks at about 2.5 Sv in the central Pacific (Leetmaa & Spain, 1981).

At the western boundary, the EUC is fed by the saline NGCUC (Section 10.7.4). As the EUC flows eastward it encounters the Galapagos Islands, located on the equator at 91–89°W. The EUC splits upstream of the islands at about 92°W and flows north and south around the islands. The southern part is stronger; the main core of the EUC core is actually slightly south of the equator from 98°W. East of the Galapagos, part of the EUC penetrates southeast to

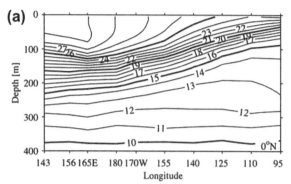

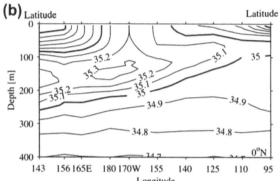

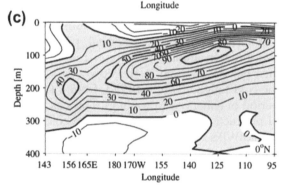

FIGURE 10.23 Mean equatorial (a) potential temperature (°C), (b) salinity, and (c) zonal velocity (cm/sec). Eastward velocities are shaded. *Source: From Johnson et al. (2002).*

5°S and joins the Peru Countercurrent at the surface and the PCUC at the South American coast (Section 10.4.1.3; Lukas, 1986).

Other zonal "asymmetries" are apparent in the other major tropical currents. The eastward NECC shifts northward toward the east. The eastward NSCC and SSCC both shift poleward as well. The SECC is present permanently only in the western Pacific and disappears by the mid-Pacific.

10.7.3.3. Equatorial Upwelling and Biological Productivity

The Pacific equatorial SST structure is strongly influenced by upwelling of cold water from the pycnocline/thermocline. Where the thermocline is shallow, upwelling creates cold surface temperature; where and when the thermocline is deep, upwelling is not as effective in cooling the surface. The cold tongue and warm pool are evident in satellite images in non-El Niño years (Figures 4.1 and 10.24). Coastal upwelling along Ecuador is also evident, joining with the equatorial cold tongue.

Upwelled water is often richer in nutrients than the displaced surface water. Global maps of surface nutrients show a maximum in the Pacific cold tongue (nitrate in Figure 4.24), because of the eastward shoaling of the pycnocline, which is also the nutricline. This nutrient maximum promotes biological production. Biological productivity, measured in amount of carbon produced per area per day, is high in the upwelled water of the cold tongue (Figure 10.22d, from a La Niña period of enhanced upwelling). This calculation of productivity was based on ocean color from the SeaWIFs satellite (Figure S10.318 on the textbook Web site).

10.7.4. Low Latitude Western Boundary Currents

The Mindanao Current is a 200 km wide western boundary current that flows southward along the western boundary of the tropical North Pacific. Dynamically, it is the western boundary current associated with the Sverdrup transport of the elongated tropical cyclonic gyre. The Mindanao Current carries subtropical North Pacific waters toward the equator, including saline water from the subtropical

thermocline and traces of North Pacific Intermediate Water (Bingham & Lukas, 1994).

The Mindanao Current forms near 14°N where the westward-flowing NEC splits, with the northward flow forming the Kuroshio (Figure 10.1 and Figures S10.1 and S10.19 on the textbook Web site). It turns eastward at about 5°N and feeds the NECC. Mindanao Current speeds are typical of western boundary currents, reaching a maximum of 100 cm/sec. Volume transport estimates range from 20 to 40 Sv, consistent with the calculated Sverdrup transport (Wijffels, Firing, & Toole, 1995).

The Mindanao Eddy (ME in Figure 10.1) is a recirculating cyclonic feature at the western boundary of the cyclonic tropical gyre. It forms between the westward NEC and the eastward NECC. Its western side is the Mindanao Current. The Halmahera Eddy (HE in Figure 10.1) is an anticyclonic feature at the western boundary just north of the equator between the eastward NECC and the westward SEC. The Halmahera Eddy mixes waters from the North and South Pacific. The properties of waters that enter the Indonesian Throughflow (ITF) may therefore depend on the activity of this eddy (Kashino et al., 1999). Both eddies are highly dependent on wind forcing.

The NGCUC is the northward western boundary current of the tropical South Pacific. The NGCUC is the northernmost part of the western boundary current that forms from the westward flow of the SEC (Qu & Lindstrom, 2002), which splits at the Australian coast, with the southward flow forming the EAC (Section 10.4.1.1). The split is at 15°S at the sea surface and shifts poleward to 23°S at 800 m. The northward boundary current north of 15°S is referred to as the *North Queensland Current* (NQC). (The northward subsurface flow between 23°S and 15°S is called the *Great Barrier Reef Undercurrent* (GBRUC).) The NQC flows through the Coral Sea, through the Solomon Sea, and then through the Vitiaz Strait between New Guinea and New Britain. Beyond that point it is referred to as the NGCUC. The NGCUC turns north and then east along the equator at about 143°E to feed the EUC. The NGCUC has speeds of 50 cm/sec centered at 200 m depth and a transport of 7 Sv at 2°S, which are equivalent to those of the EUC at the equator.

Lastly, the tropical Pacific and Indian Oceans are connected via the Indonesian Throughflow, through the complex passages of the Indonesian archipelago (Figure 11.11; Section 11.5). Approximately 10–15 Sv flow through the passages, with significant variability, much of it due to ENSO. The Pacific's low latitude western boundary currents are the source of the ITF. The flow through the Makassar Strait originates in the Mindanao Current. South Pacific waters from the NGCUC enter the Halmahera Sea; deeper South Pacific waters from the same source enter through Lifamatola Strait (Hautala, Reid, & Bray, 1996).

10.7.5. Equatorial Property Distributions

Although most of the Pacific water mass description is in Section 10.9, we briefly review the tropical upper ocean distributions in this section because they are so clearly linked to the equatorial current system.

The temperature structure is highly symmetric about the equator (Figure 10.19). The thermocline is most intense a few degrees north and south of the equator, with the isotherms spreading apart north and south of the 10°N and 10°S parallels. At the equator, the spreading of the isotherms marks the core of the EUC. Below the thermocline, between about 5°S and 12°N, there is a marked thermostad (low vertical gradient).

For salinity, there is little symmetry across the equator (Figure 10.19), because the South Pacific is more saline, because the SEC reaches the equator and the NEC does not, and because of the Northern Hemisphere location of the ITCZ.

Salinity maximum layers are subducted equatorward from both the South and North Pacific subtropical evaporation maxima; their core salinities are 36.2 psu and 35.0 psu, respectively. (These are the Subtropical Underwaters, also called Tropical Waters in Johnson & McPhaden, 1999.) Because the SEC extends to the equator, the salinity maximum at the equator comes directly from the South Pacific subtropical gyre. The South and North Pacific salinity maxima are separated laterally by lower salinity arising from the California Current and downward diffusion beneath the rainy ITCZ (Johnson & McPhaden, 1999). The lowest surface salinity is in the NECC, which lies directly below the ITCZ. The subsurface low salinity water entering at about 20°N at 300 m is the North Pacific Intermediate Water (Section 10.9.2.1).

10.7.6. Intraseasonal and Seasonal Variability

The equatorial Pacific includes temporal variability at intraseasonal (20–30 days), seasonal, monthly-to-interannual, interannual (3–7 years), and interdecadal (10–30 years) timescales. The most energetic intraseasonal variations are the *Tropical Instability Waves* (TIWs). Seasonal variability includes response to changes in location and strength of the ITCZs in both hemispheres. Other variability at weekly to interannual periods is associated with Rossby and Kelvin waves (Section 7.7) and at interannual and longer periods, with ENSO and other climate modes (Section 10.8 and Chapter S15 on the textbook Web site).

TIWs are large cusp-like spatial oscillations in SST along the northern edge of the cold tongue (Figure 10.24) (Legeckis, 1977). The oscillations are also apparent in ocean color/chlorophyll (McClain et al., 2002). TIWs have wavelengths of about 1000 km. The TIW pattern propagates westward at an average phase speed of 30 to 50 cm/sec, resulting in a period of about 20 to 30 days. The TIWs are principally

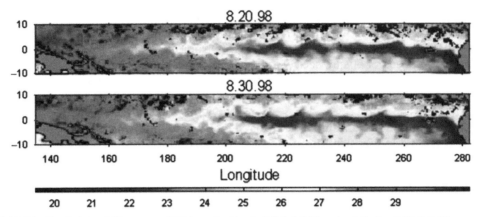

FIGURE 10.24 Tropical instability waves. SST from the Tropical Rainfall Mapping Mission (TRMM) Microwave Imager (TMI) for two successive 10-day periods in August 1998, after establishment of the cold tongue during a La Niña. A more complete time series (June 1–August 30, 1998) is reproduced in Figure S10.20 on the textbook Web site. This figure can also be seen in the color insert. TMI data are produced by Remote Sensing Systems and sponsored by the NASA Earth Science MEASURES DISCOVER Project. Data are available at www.remss.com. *Source: From Remote Sensing Systems (2004).*

due to (barotropic) instability arising from the horizontal shear between the SEC and the NECC (Philander, 1978). TIWs are shallow (100–200 m thick) because the high velocities of the currents that create them are surface-intensified.

TIWs appear in summer (June) when the ITCZ migrates northward and the trade winds accelerate the portion of the SEC that lies north of the equator (Vialard, Menkes, Anderson, & Balmaseda, 2003). In the time series leading to Figure 10.24 (Figure S10.20 on the textbook Web site), the equatorial cold tongue emerges in early June; by June 10 the tongue shows north-south oscillations due to TIWs. Closed anticyclonic vortices are found in the troughs of the waves. Seasonal wind forcing in the tropical Pacific directly affects SST and the surface and upper ocean currents (Figure 10.25). The cold tongue is strongest in August-September, during the period of strongest trade winds, accompanied by warmest temperatures in the warm pool; the west-east contrast is as much as 10°C. By March, both temperature features are much weaker and the west-east contrast is reduced to about 5°C. (The large interannual variability superimposed on the annual cycle in Figure 10.25 is due to ENSO.)

The equatorial part of the SEC, which responds frictionally to the wind stress, varies mostly in phase with the seasonal winds. The EUC, which responds to the west-east pressure gradient set up by the SEC, has a more complicated response that lags the winds. Johnson, Sloyan, Kessler, and McTaggert (2002) provided detailed discussion of the seasonal variability of each of the upper ocean currents, phasing with the winds, and spatial structure. Dramatic seasonal variability occurs just offshore of the Central American mountain chain (Figure 10.26). Trade winds from the Atlantic funnel through three major gaps in the mountains, with wintertime winds reaching 20 m/sec during several 5- to 7-day-long events. The wind jets (Tehuantepec, Papagayo, and Panama) that emerge over the Pacific force dramatic local circulations and upper layer mixing, resulting in cool SST (Chelton, Freilich, & Esbensen, 2000) and ocean color anomalies; the effects are visible even in the global mean wind stress curl map from Chelton et al. (2004; Figure 5.16d).

Monthly Zonal Wind and SST 2°S to 2°N Average

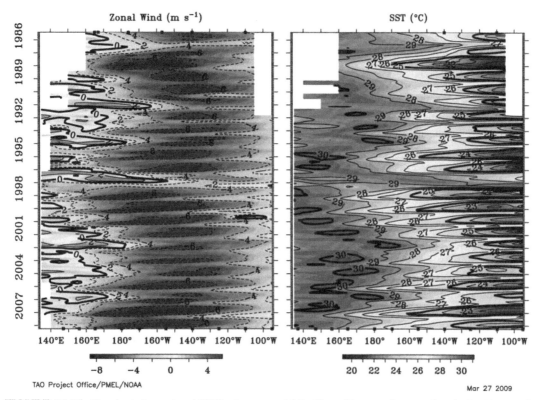

TAO Project Office/PMEL/NOAA

Mar 27 2009

FIGURE 10.25 Zonal wind speed and SST in the equatorial Pacific to illustrate the annual cycle. Positive wind speed is toward the east. Climatological means in February and August and an expanded time series for 2000–2007 are shown in Figure S10.21 on the textbook Web site, to emphasize the seasonal cycle. This figure can also be found in the color insert. *Source: From TAO Project Office (2009a).*

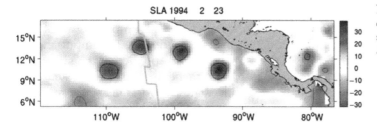

FIGURE 10.26 Tehuantepec eddies evident in sea surface height anomalies from satellite altimetry in February, 1994. *Source: From Palacios and Bograd (2005).*

Anticyclonic eddies are produced by the wind jets, as a combination of eddy shedding from the coastal circulation system (coastally trapped waves) and the strong wind stress curl in the jets. The eddies propagate offshore.

The best known are the Tehuantepec eddies (Figure 10.26 and Figure S10.22 on the textbook Web site). Three to four Tehuantepec eddies and two to three Papagayo eddies form each year between October and July with greater frequency and intensity during El Niño years (Palacios & Bograd, 2005).

10.8. EL NIÑO/ LA NIÑA AND THE SOUTHERN OSCILLATION (ENSO)

El Niño/La Niña is a natural climate variation that is dynamically centered in the tropical Pacific. Its "interannual" timescale is 3 to 7 years for quasi-periodic alternation between the El Niño and La Niña states. The *Southern Oscillation* is an index based on the pressure difference between two tropical South Pacific locations, and is closely related to the El Niño state. Because this index is so closely related to El Niño events, the full climate phenomenon is often referred to as El Niño-Southern Oscillation (ENSO). The ocean and atmosphere are fully coupled in this climate "cycle." The coupling is referred to as the Bjerknes feedback (Section 7.9.2; Bjerknes, 1969).

An El Niño event is marked by an unusual excursion of warm water ($>28°C$) to the east in the equatorial zone, associated with weakened southeast Trade Winds in the east and stronger westerlies in the west. La Niña is the opposite — stronger southeast Trades in the east (and weak westerlies in the far west) with resulting cool water ($<25°C$) extending much further westward along the equator than usual. The alternation between states is not regular since there are many different oceanic and atmospheric phenomena linked in the full system plus random, short-term forcing. Therefore, ENSO predictability is not like that of, say, the tides, which are forced by very regular, predictable progressions in the orbits of the earth, moon, and sun.

El Niño/La Niña events have large and sometime devastating impacts on ocean ecosystems, particularly along the South American coast, but also as far north as the CCS. ENSO impacts air temperature and precipitation on global scales (Figures S10.24 and S10.25 on the textbook Web site), via propagation of large-scale waves through the atmosphere and propagation of Kelvin waves (Section 7.7.6) along the eastern boundary of the Pacific. Precipitation anomalies during a composite El Niño include regions of anomalously low precipitation that are susceptible to drought and fire and high precipitation that are susceptible to flooding. Although it is not located in the tropics, U.S. air temperatures are affected by ENSO; El Niño signatures include anomalous warmth over the northwest and high plains; cool temperatures in the south and Florida; anomalously dry conditions in the northwest, east, and Appalachians; and wet conditions from California through the southeastern U.S.

Early ideas about the cause of El Niño centered on local mechanisms along the South American coast, for instance that the alongshore winds off Peru changed to lessen or stop the coastal upwelling. More intensive studies in the early 1970s, motivated by a major El Niño event in 1972 that resulted in the collapse of the Peru/Ecuador anchovy fishery, showed that El Niño has a much larger geographic scale. Rasmusson & Carpenter's (1982) canonical description of ENSO, based on El Niño's from 1949 to 1980, was the underpinning for an international project (Tropical Ocean Global Atmosphere; TOGA) to study ENSO (1985–1995). TOGA planning was underway when the strong El Niño event of 1982/83 provided additional impetus for the experiment. The importance of ENSO analysis and prediction is such that a massive permanent observing system has been deployed in the tropical Pacific since the 1980s (TAO and TRITON; Section S6.5.6 on the textbook Web site).

Excellent, regularly updated information, including background information on dynamics and impacts, forecasts, and links to many different ENSO products, is available from several different Web sites administered by the National Oceanic and Atmospheric Administration.

10.8.1. ENSO Description

We first recall the "normal" ocean and atmosphere conditions in the tropical Pacific (Section 7.9.2; Figure 10.27b). The easterly trade winds pile up warm equatorial water in the western tropical Pacific and cause upwelling along the equator. This causes the cold tongue in SST in the eastern tropics, and causes the thermocline to be inclined upward from west to east (Section 10.7.3). The warm-to-cold SST difference along the equator maintains the Walker circulation in the atmosphere, thus sustaining this component of the trade winds. This is an equilibrium state of the simple coupled ocean-atmosphere, and the system would remain in this state if it did not include large-scale propagating waves such as Kelvin and Rossby waves.

The exaggerated, strong version of the normal state is the La Niña state (Figure 10.27a). In La Niña, the warm SST shifts slightly more to the west, the thermocline is a bit deeper in the west, the sea surface is higher in the west and lower in the east, and the Walker circulation in the atmosphere is stronger.

In an El Niño state, the trade winds are weaker, because the Walker circulation is weak or reversed, and the thermocline is more level (Figure 10.27c). The cold tongue in the east weakens and disappears, due to both relaxation of the thermocline and eastward movement of warm water from the central and western tropical Pacific. This does not indicate an absence of upwelling but rather that warm water is now occupying the eastern tropical Pacific; the schematic shows easterly trades in the eastern Pacific, but these upwell only warm water from the now thicker and warmer surface layer.

In the time series of SST along the equator (Figure 10.25), warm SST and weaker trade winds mark several El Niño events, with the opposite markers for La Niña events. Time series indicating the occurrence of El Niño and La Niña are constructed in various ways. The first index, the *Southern Oscillation Index* (SOI) is the difference in atmospheric pressure between the western and eastern tropical South Pacific; meteorological stations at Darwin, Australia, and Tahiti are used in the SOI because observations have been made in these locations for a very long time. Every El Niño event is associated with low SOI. However, not every SOI low corresponds to an El Niño. Several indices are based on SSTs averaged spatially over portions of the eastern tropical Pacific because this reflects conditions in the cold tongue (e.g., Oceanic Niño Index in Figure 10.28). A multivariate index based on SST, sea level pressure, surface air temperature, surface wind, and cloudiness is also useful (Wolter & Timlin, 1993; Wolter, 2009). Very long time series have been reconstructed from proxies of temperature measured in coral heads (Cobb, Charles, Cheng,

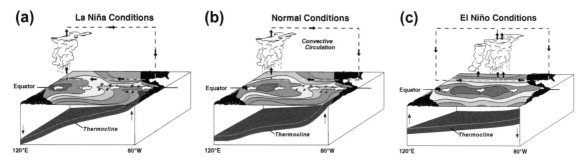

FIGURE 10.27 (a) La Niña, (b) normal, and (c) El Niño conditions. This figure can also be found in the color insert. *Source: From NOAA PMEL (2009b).*

(a)

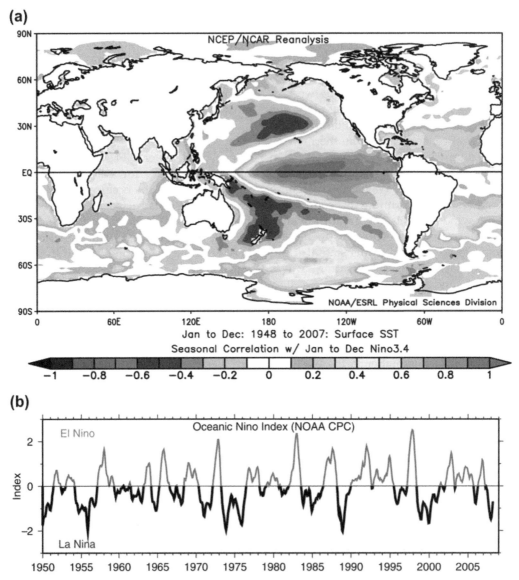

(b)

FIGURE 10.28 (a) Correlation of monthly SST anomalies with the ENSO Nino3.4 index, averaged from 1948 to 2007. The index is positive during the El Niño phase, so the signs shown are representative of this phase. (*Data and graphical interface from NOAA ESRL, 2009b.*) This figure can be found in the color insert. (b) "Oceanic Nino Index" based on SST in the region 5°N to 5°S and 170°W to 120°W. (*Data from Climate Prediction Center Internet Team, 2009*). Gray and black correspond to El Niño and La Niña, respectively. Additional indices representing ENSO and the correlation of monthly sea level pressure anomalies with the ENSO Nino3.4 index are shown in Figure S10.23 on the textbook Web site.

& Edwards, 2003). Long-term reconstructions of tropical Pacific SST show that El Niño events at 2–7 years are ubiquitous, although intensity and duration have varied. Well-documented El Niño events took place in 1941–1942, 1957–1958, 1965–1966, 1972–1973, 1977–1978, 1982–1983, 1997–1998, and 2002–2003. The events of 1982–1983 and 1997–1998 were the largest recorded since the 1880s.

The global reach of ENSO is apparent in correlations of SST and sea level pressure with an ENSO index (Figure 10.28b and Figures S10.23d on the textbook Web site). SST in the equatorial Pacific shows the pattern previously described of anomalously warm eastern equatorial waters during the El Niño phase. The even simpler sea level pressure pattern extends well into the ACC region in an alternating zonal pattern that is similar to that of the Southern Annular Mode (Section 10.10 and Chapter S15 on the textbook Web site).

10.8.2. ENSO Mechanisms

The Bjerknes (1969) feedback is at the heart of ENSO (Section 7.9.2), but it does not describe how each stage of ENSO develops or why there is a transition from one state to another with an "oscillation" timescale of 3 to 7 years. Bjerknes speculated that the transition results from ocean dynamics but could go no further. An oscillation with a period of several years can be produced with a model that includes an eastward-propagating equatorial Kelvin wave that reflects at the eastern boundary, producing westward-propagating Rossby waves (Cane, Münnich, & Zebiak, 1990; Jin, 1996; Van der Vaart, Dijkstra, & Jin, 2000).

The ENSO cycle, based on Rasmusson & Carpenter (1982), Jin (1996), and Van der Waart et al. (2000), is very briefly summarized here. Moving from normal conditions toward a full-blown El Niño, the steps are (1) changes of the trade winds to westerly winds in the western Pacific, often associated with the atmosphere's 30–60 day Madden-Julian oscillation; (2) an oceanic Kelvin wave shooting eastward along the equator in response; (3) resultant warm SST anomalies in the eastern and central equatorial Pacific; and (4) disruption of the Walker circulation through SST feedback on the atmosphere. The "recharge oscillator" that transitions this back toward a La Niña occurs when: (1) the Kelvin wave reflects at the eastern boundary and produces westward-propagating Rossby waves, (2) the Rossby waves move warm water away from the equator which weakens the equatorial SST warm anomaly, (3) the trade winds strengthen a little in response to the somewhat cooler SST, (4) the strengthened trades begin pushing the thermocline back toward a normal state, and (5) Bjerknes feedback then creates a La Niña state.

The adjustment timescale of this nearly free oscillation yields the 3–7 year ENSO timescale. An important property of this system is a delay between the change in thermocline depth in the western Pacific and the SST warming in the eastern Pacific, which can be explained partially by the Kelvin wave propagation (Jin, 1996).

The actual ENSO system is nonlinear and messy. Fedorov et al. (2003) described it as a "slightly damped, swinging pendulum sustained by modest blows at random times." The switch from one state to another, and the intensity and duration of the resulting state, depend on many factors. These include phasing of the shifts relative to the seasonal cycle and also to the occurrence, timing, and intensity of westerly wind bursts in the western tropical Pacific that are associated with the intraseasonal (30–60 day) Madden-Julian Oscillation in the atmosphere. Predictability of onset, intensity, and duration of events is therefore limited.

Because of the widespread economic impacts of ENSO, skillful prediction several months ahead has been a goal for many decades. Two approaches are dynamical modeling and statistical modeling. Dynamical models use a coupled ocean-atmosphere model with initial conditions based on observations. Statistical models use observed parameters such as SST or heat content and winds with a regression method to forecast ENSO several months ahead. Forecasts are generally probabilistic, meaning that an ensemble (large number) of model runs is made with slightly varying initial conditions. Given the randomness of "triggering" mechanisms, Philander and Fedorov (2003) and Fedorov et al. (2003) highly recommended this approach. The International Research Institute for Climate and Society (IRI) at Columbia University currently monitors 15 dynamical and 8 statistical model forecasts (http://iri.columbia.edu/climate/ENSO/currentinfo/SST_table.html).

10.9. PACIFIC OCEAN WATER MASSES

Pacific Ocean water properties, like those of the other oceans, can be considered in four layers (Section 4.1). The upper ocean layer contains the mixed layer and main pycnocline (thermocline/halocline), and is in broad contact with the atmosphere. The intermediate layer contains two low salinity water masses that originate at the sea surface of the subpolar/subantarctic latitudes. The deep layer contains two deep water masses, one from the North Pacific and one from the Southern Ocean. The North Pacific deep water "source" is entirely internal mixing and upwelling of waters from the Southern Ocean, with no contact with the atmosphere. The Southern Ocean deep water source contains a mixture of deep waters from all three oceans (Atlantic, Indian, and Pacific) as well as waters that are locally ventilated in the Southern Ocean. The bottom layer contains the densest water that escapes northward from the Southern Ocean. The distinction between the deep and bottom layers is not sharp, and is usually based on the direction of net meridional transport in the two

layers, with net southward transport in the deep layer and net northward in the bottom layer.

The Pacific Ocean is the freshest of the three main ocean basins. The Atlantic and Indian Oceans are both net evaporative basins, and therefore have high overall salinity. The Pacific evaporation-precipitation balance is nearly neutral which makes the Pacific fresher than the Atlantic and Indian.

The most important distinguishing process for Pacific Ocean water properties is the lack of a surface source of very dense water in the North Pacific. This differs entirely from the Atlantic Ocean. The densest water formed locally is the relatively light North Pacific Intermediate Water. On a global scale, the Pacific Ocean is the low density end-member of the overturning circulation. Its bottom waters, which originate in other oceans, are salty and its upper waters are relatively fresh; cooling to the freezing point, which occurs in the Bering and Okhotsk Seas in the northwest Pacific, cannot increase the surface water density to a high enough value to punch through to the deep and bottom layers.

A potential temperature-salinity (T-S) diagram that represents the major water masses is shown in Figure 10.29. Table S10.4 on the textbook Web site lists the principal water masses and an abbreviated description of the process that initially forms each water mass. The Pacific World Ocean Circulation Experiment (WOCE) Hydrographic Programme Atlas (Talley, 2007) is a comprehensive source of sections, maps, and property plots.

10.9.1. Pacific Ocean Upper Waters

Pacific surface temperature (Figure 4.1) shows the usual tropical maximum with poleward decrease in temperature in both hemispheres. The highest temperatures (>29°C) are in the equatorial warm pool. The lower temperatures of the equatorial cold tongue are also evident. Isotherms in the PCCS and CCS are deformed,

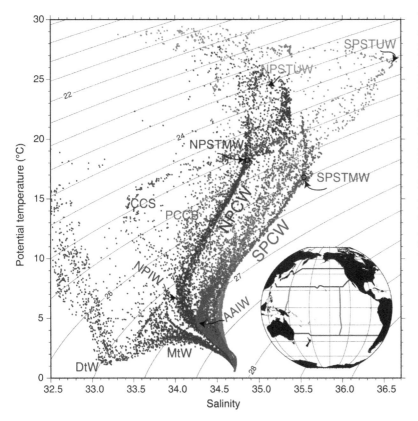

FIGURE 10.29 Potential T-S curves for selected stations (inset map). Acronyms: NPCW, North Pacific Central Water; SPCW, South Pacific Central Water; NPSTUW, North Pacific Subtropical Underwater; SPSTUW, South Pacific Subtropical Underwater; NPSTMW, North Pacific Subtropical Mode Water; SPSTMW, South Pacific Subtropical Mode Water; NPIW, North Pacific Intermediate Water; AAIW, Antarctic Intermediate Water; DtW, Dichothermal Water; MtW, Mesothermal Water; CCS, California Current System waters; and PCCS, Peru-Chile Current System Waters. This figure can also be found in the color insert.

with colder water near the coasts due to equatorward advection and upwelling. The coldest temperatures are in the sea ice areas of the Okhotsk and Bering Seas, and in the Antarctic.

Pacific surface salinity shows the typical maxima in the subtropics, in the major subtropical evaporation centers (Figures 4.14, 5.4). There is a north-south minimum in the tropics beneath the ITCZ at 5–10°N, due to excess precipitation. Salinity is also low at high latitudes due to excess precipitation. The surface salinity in the North Pacific is considerably less than in the North Atlantic, because of the greater runoff and precipitation. In the South Pacific the average surface salinity is higher than in the North Pacific but is lower than in the South Atlantic.

In the subtropics, there are two important processes for creating upper ocean waters: subduction of surface waters equatorward and downward beneath less dense, lower latitude surface waters, and production of thick, well-mixed layers on the warm side of strong current fronts such as the Kuroshio. These result in several recognized subtropical water masses (Table S10.4 on the textbook Web site) as follows.

The waters that make up the thermocline/pycnocline in the subtropics are called *Central Waters* (Figure 10.29), as also found in the Atlantic and Indian Oceans. The pycnocline, or Central Water, is created by subduction and diapycnal mixing (Section 9.8.1). "Central Water" is a T-S relation with a large range of temperatures and salinities, rather than an extremum of some property.

North Pacific Central Water (NPCW) extends from the NECC to about 40°N and is the freshest of the Central Waters of the world's oceans (Figure 4.7). It is separated from the eastern boundary by another, yet fresher water mass that characterizes the CCS. This fresher CCS water is advected southward from the eastern subpolar gyre.

South Pacific Central Water (SPCW) is saltier than NPCW since the South Pacific is saltier overall. SPCW extends from about 10°S southward to the Subantarctic Front at about 55°S. Similar to NPCW, SPCW is separated from the eastern boundary by another, fresher water mass within the PCCS, advected northward from fresher high latitude surface waters.

A second water mass associated with subtropical subduction in both hemispheres is the *Subtropical Underwater* (STUW), or subtropical salinity maximum water. This is identified as a shallow salinity maximum on the equatorward part of the subtropical gyre (Figure 10.30). STUW results from subduction of the very high salinity surface water in the center of each subtropical gyre. STUW is found on every meridional section in the Pacific between 25°S and 25°N. It is very shallow, with its salinity extremum no more than 200 m deep, because the isopycnals that

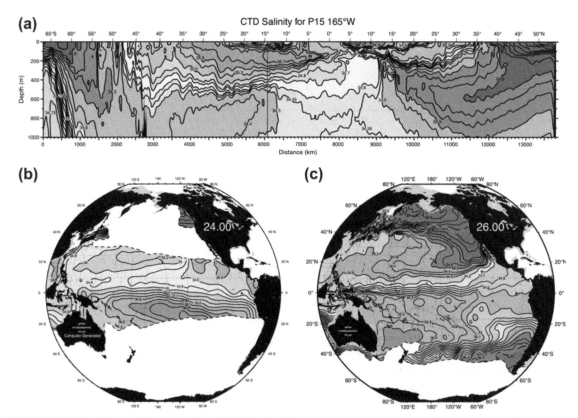

FIGURE 10.30 Salinity: (a) along 165°W (WOCE P15); (b) at neutral density 24.0 kg/m^3, characteristic of STUW; and (c) at neutral density 26.00 kg/m^3, characteristic of SPSTMW. The isopycnals intersect the surface along the dashed contours Gray contours in (c) indicate winter outcrops. *Source: From WOCE Pacific Ocean Atlas, Talley (2007).*

outcrop in the surface salinity maximum water are warm (~26 and 24°C in the South and North Pacific, respectively) and low density (σ_θ ~ 24.0 and 23.5 kg/m^3 in the South and North Pacific, respectively).

The third subtropical water mass that we single out is *Subtropical Mode Water* (STMW; Masuzawa, 1969). "Mode" means relatively large volume on a volumetric potential T-S diagram. Mode Water is a pycnostad embedded in the main pycnocline; it results from subduction of the especially thick winter mixed layers on the warm side of the separated western boundary currents (Kuroshio and EAC; Hanawa & Talley, 2001; Figure S10.26 on the textbook Web site). The STMW in the North Pacific (NPSTMW) is in the temperature range 16–19°C and centered at potential density $\sigma_\theta = 25.2$ kg/m^3 (Figure 10.29 and Figure S10.26b on the textbook Web site). It originates in winter as a thick mixed layer just south of the Kuroshio. The thick layers subduct into the general region of the western subtropical gyre and are evident within the thermocline (Figures 10.31a and S10.26c on the textbook Web site). The temperature of the STMW is highest (>18°C) just south of Japan and decreases toward the east.

In the South Pacific, the South Pacific STMW (SPSTMW) is present north of the Tasman Front and East Auckland Current (Figure 10.31b and Figure S10.26c on the textbook Web site; Roemmich & Cornuelle, 1992). Its core temperature, salinity, and density are 15–17°C (just north of New Zealand) and 17–19°C (region north of 29°S), 35.5 psu, and $\sigma_\theta = 26.0$ kg/m^3 (SPSTMW in Figures 10.29 and 10.30). Thus it has the same temperature range as NPSTMW. It is denser because it is somewhat more saline, because the South Pacific is saltier than the North Pacific. SPSTMW is the weakest of the global STMWs; without a supplementary vertical density gradient calculation, the widening of isopycnals and isotherms on intersecting vertical sections is somewhat difficult to discern (Figure 10.31b).

The North Pacific's subpolar gyre is a region of Ekman upwelling rather than downwelling. Therefore there is no wind-driven subduction. Surface densities increase along the cyclonic path around the gyre; they are higher in the west than in the east, and are highest in the Okhotsk Sea, along Hokkaido and just south of Hokkaido. In the regions of highest surface density, the densest (intermediate) North Pacific waters are formed (Section 10.9.2).

The combination of low surface salinity and upwelling in the subpolar gyre creates a strong halocline. This supports a temperature minimum where the surface water becomes very cold in winter. The temperature minimum is called *Dichothermal Water* and is found in the western subpolar gyre and the adjacent Okhotsk and Bering Seas. Associated with the temperature minimum is very high oxygen saturation in the summertime, due to capping by warm surface water and slight warming of the subsurface T min layer. Below the Dichothermal Water, temperature increases to a maximum and then decreases to the ocean bottom. The temperature maximum layer is called *Mesothermal Water*. The maximum indicates a substantial advective component from the east or the south since otherwise it would acquire the low temperature of the surface layer.

Tropical Pacific water properties were described in Section 10.7.5. Complex vertical structure is created by interleaving of North and South Pacific waters (Figure 10.19). Nearly zonal fronts in salinity occur along the equator (Figure 10.30c). In temperature and density, the equatorial thermocline/pycnocline ascends and intensifies from 150–200 m in the west to less than 50 m in the east (Figure 10.23). The pycnocline inhibits vertical transfer of water properties. In the west, a halocline lies within the upper (warm pool) layer, above the thermocline, so the pycnocline is determined by salinity rather than temperature.

One tropical water mass that is distinguished by a name is the *Equatorial 13°C Water*

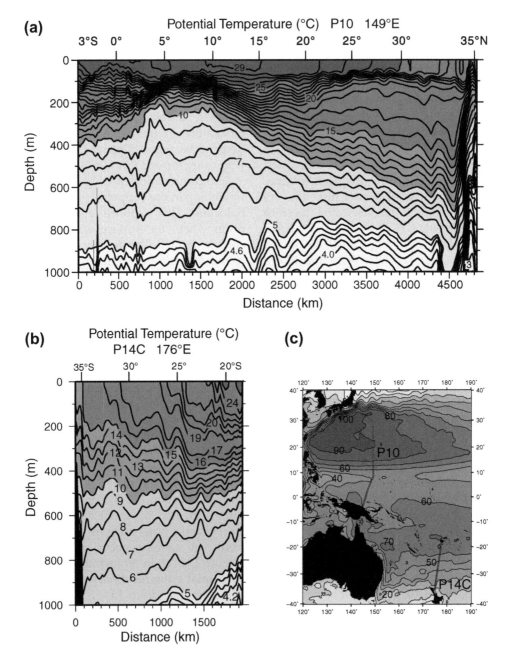

FIGURE 10.31 (a) Potential temperature (°C) along 149°E in the North Pacific. (b) Potential temperature along 170°E in the South Pacific. *Source: From WOCE Pacific Ocean Atlas, Talley (2007).* (c) Station locations superimposed on surface streamfunction. *(Data from Niiler, Maximenko, & McWilliams, 2003.)*

(Montgomery & Stroup, 1962; Tsuchiya, 1981). This is a mode water — a conspicuous thickening of the equatorial layer centered at 13°C, at about 75–300 m depth (Figure 10.19). Water at this temperature is advected eastward across the Pacific from the low latitude western boundary currents. The thickening is possibly linked to the local dynamics of the equatorial currents, as the water mass is associated with the North and South Subsurface Countercurrents (Figure 10.20b).

Finally, two large regions of remarkably low oxygen (<1 μmol/kg) are found in the eastern tropical Pacific, centered at 10°N and 7°S, and most intense near the eastern boundary (Figures 10.32 and 4.20). The most extreme oxygen minima here coincide with well-developed subsurface maxima in nitrite (NO_2; Figure 10.32b). Nitrite normally occurs within or at the base of the euphotic zone (widespread band in the upper 200 m in the figure), as part of the usual nitrification process. The strongly developed subsurface nitrite maxima are a unique feature of *denitrification*. Remarkably, chlorofluorocarbons (CFCs) are non-zero in the oxygen minima (WOCE Pacific Ocean Atlas, Talley, 2007), which means that these waters are ventilated and that oxygen is low because of high biological productivity rather than extreme age.

10.9.2. Intermediate Waters

The intermediate layer of the Pacific is occupied by two low salinity water masses, the *North Pacific Intermediate Water* (NPIW) and the *Antarctic Intermediate Water* (AAIW) (e.g., Figures 4.12b, 14.13 and S10.27 on the textbook Web site). The source waters of both are fresh, cool surface waters at subpolar latitudes. In the subtropics and equatorial Pacific, the overlying water is the higher salinity Central Water, which originates in the high salinity mid-latitude surface waters. Underlying the intermediate waters is higher salinity Circumpolar Deep Water, which obtains its higher salinity

from the North Atlantic. Thus the NPIW and AAIW both appear as vertical salinity minima in the subtropics and tropics.

The NPIW salinity minimum is confined to the subtropical North Pacific. The AAIW salinity minimum, in contrast, is found throughout the subtropical South Pacific, the tropical Pacific, and similar regions of the Atlantic and Indian Oceans. Both NPIW and AAIW are within the ventilated, higher oxygen part of the water column. However, neither have particularly high oxygen content in the Pacific, indicating that residence time is longer than for the overlying Central Waters.

Salinity and oxygen content on isopycnals that represent NPIW and AAIW (Figure 10.33) reflect the low salinity/high oxygen influx from (1) the Okhotsk Sea for the NPIW and (2) the southeast Pacific for the AAIW. These are the source regions of these water masses.

(Salinity at neutral density 27.30 kg/m^3 provides a straightforward example of the importance of diapycnal mixing. Throughout the tropics, salinity is higher and the water is warmer as this is an isopycnal. There is no warm, salty surface outcrop for this isopycnal, so the tropical properties must result from diapycnal mixing.)

10.9.2.1. North Pacific Intermediate Water

NPIW is the densest water that is directly ventilated on a regular basis in the North Pacific. The full NPIW density range is $\sigma_\theta = 26.7$ kg/m^3 to 27.2 kg/m^3 (directly ventilated), to 27.6 kg/m^3 (ventilated through vigorous diapycnal mixing in the Kuril Island straits). The subtropical NPIW salinity minimum has potential density $\sigma_\theta = 26.7$ to 26.8 kg/m^3. On an NPIW isopycnal, the lowest salinity (hence coldest) and highest oxygen, indicating the most recently ventilated water, occur in the Okhotsk Sea and adjacent subpolar gyre (Figure 10.33). The main direct ventilation process for NPIW is brine rejection during sea ice formation in a coastal (latent heat) polynya

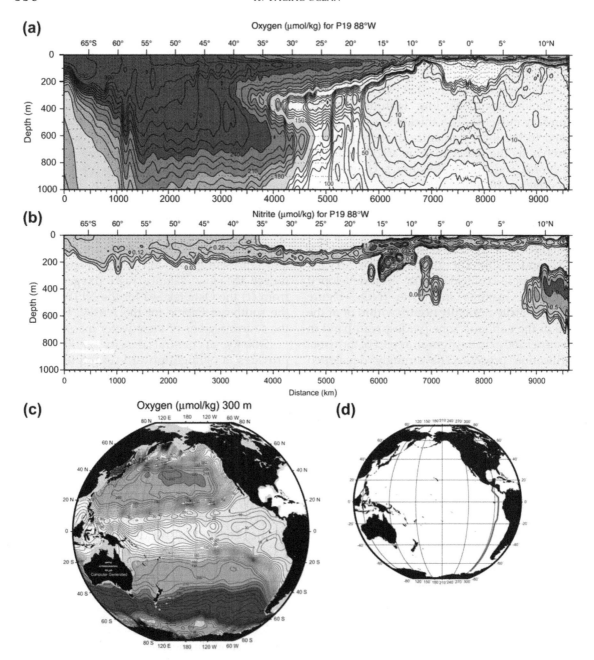

FIGURE 10.32 Tropical oxygen minima and denitrification regions. Eastern Pacific vertical sections of (a) oxygen (μmol/kg) and (b) nitrite (μmol/kg) at 88°W (WOCE P19). (c) Oxygen (μmol/kg) at 300 m depth. (d) P19 station locations. *Source: From WOCE Pacific Ocean Atlas, Talley (2007).*

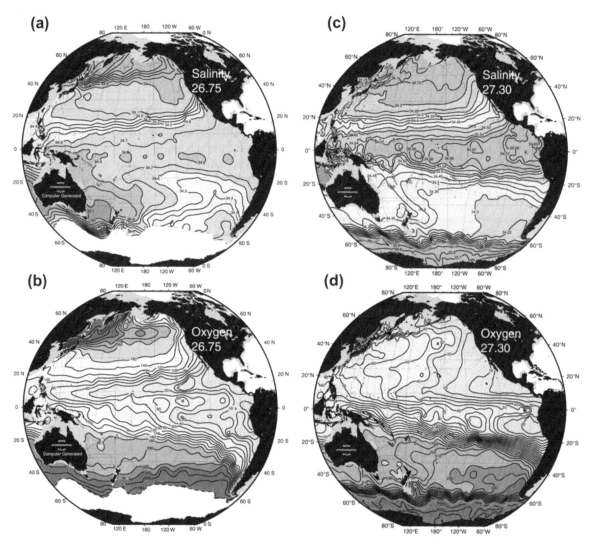

FIGURE 10.33 (a, c) Salinity and (b, d) oxygen (μmol/kg) at neutral densities 26.75 kg/m³ and 27.3 kg/m³, characteristic of NPIW and AAIW, respectively. In the Southern Ocean, white at 26.75 kg/m³ shows the isopycnal outcrops; the gray curve in (c) and (d) is the winter outcrop. Depth of the surfaces is shown in the WOCE Pacific Ocean Atlas. This figure can also be seen in the color insert. *Source: From WOCE Pacific Ocean Atlas, Talley (2007).*

in the northwestern corner of the Okhotsk Sea ("NWP" in Figure 10.34b). Polynyas all along the shelf create brine rejection; the NWP is at the end of the cyclonic circulation, so the water has accumulated the most brine. Historical data suggest that brine rejection can affect densities up to about $\sigma_\theta = 27.1$ kg/m³. See

also the online supplement Section S8.10.6 on the Okhotsk Sea.

A sensible heat polynya maintained by tidal mixing (Figure 3.12b) almost always occurs over Kashevarov Bank ("KBP" in Figure 10.34b). The subsurface temperature maximum is mixed upward, melting the sea ice and fluxing

(a)

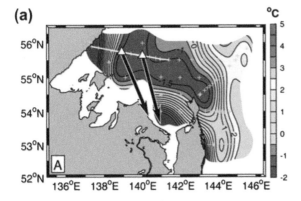

(b)

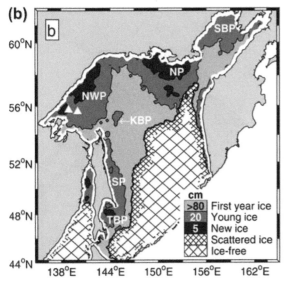

FIGURE 10.34 Dense water formation in the Okhotsk Sea. (a) Bottom potential temperature in September, 1999, and mean velocity vectors at the two moorings. (b) Ice distribution on February 1, 2000, from the SSM/I microwave imager. "NWP" is the northwest polynya where the densest water is formed. Figure 10.34a can also be found in the color insert. *Source: From Shcherbina, Talley, and Rudnick (2003, 2004).*

nutrients to the surface layer; this is a highly productive region biologically. The Okhotsk Sea waters exit back to the northwest Pacific through a deep strait in the Kuril Islands (depth ~ 1500 m). Vigorous tides complete the process of mixing the high oxygen down to the maximum density at the sill, $\sigma_\theta \sim 27.6$ kg/m^3 (Talley, 1991). The renewed waters that exit

into the Oyashio do not have a subsurface salinity minimum; instead, salinity is lowest at the sea surface. The NPIW salinity minimum forms as the renewed Oyashio waters encounter the warmer, saltier, lighter surface waters of the Kuroshio in the transition region between the separated Oyashio and Kuroshio.

The NPIW formation rate based on meridional overturn across 24°N is 2 Sv, which is small compared with the other low salinity intermediate waters. If measured locally, within the subpolar gyre, where most of the newly ventilated water remains, the recycling rate could be higher.

Export of the low salinity NPIW southward into the subtropics balances the net precipitation in the subpolar region and net evaporation in the subtropics. Part of the subpolar freshwater input also exits northward through the Bering Strait, where it eventually becomes part of the North Atlantic Deep Water and is exported to the low latitude North Atlantic (Talley, 2008).

10.9.2.2. Antarctic Intermediate Water

AAIW is the low salinity intermediate layer in all of the Southern Hemisphere oceans north of the ACC (Figure 14.13; Section 13.4.2).

The Pacific AAIW salinity minimum is at a depth of about 700–1000 m through most of the South Pacific. Its potential density is between $\sigma_\theta = 27.05$ and 27.15 kg/m^3 in the southeast Pacific, where it originates in the thick surface layer (Subantarctic Mode Water) just north of the Subantarctic Front. Its potential temperature and salinity in this region are 4–6°C and 34.1–34.5 psu. The salinity minimum is just the top of the AAIW layer. We generally identify the layer down to approximately $\sigma_\theta = 27.5$ kg/m^3 as AAIW, based on properties that indicate an identifiable water mass separate from Circumpolar Deep Water (Section 13.5.2).

AAIW circulates anticyclonically around the South Pacific's subtropical gyre. Tongues of low salinity, high oxygen water on the neutral density surface 27.30 kg/m^3 ($\sigma_\theta = 27.15$ kg/m^3)

originate in the southeast Pacific and stretch northwestward across the South Pacific (Figure 10.32c, d). The AAIW salinity minimum becomes slightly warmer, saltier, and denser along its path. It enters the tropics in the western Pacific, where its density becomes distinctly higher due to higher salinity (mean values of 5.4°C, 34.52 psu, 27.25 kg/m^3 between 15°S and the equator).

The northern boundary of the AAIW is at the Northern Hemisphere tropical-subtropical transition at about 15°N (Figure 14.13); that is, AAIW does not enter the North Pacific subtropical gyre as a salinity minimum. AAIW does extend northward along the eastern boundary to about 35°N, in the "shadow zone" outside the subtropical gyre.

The formation rate of Pacific AAIW is approximately 5−6 Sv based on air−sea fluxes (Cerovecki, Talley, & Mazloff, 2011). A slightly smaller rate of 4 Sv was obtained by Schmitz (1995a), with an additional 10 Sv of AAIW formation for the Atlantic/Indian.

10.9.3. Deep Waters

Two deep waters, distinct from the bottom waters, are identified in the Pacific: Pacific Deep Water and Circumpolar Deep Water. Historically, Sverdrup thought (essentially by analogy with the Atlantic) that a slow southward movement of deep water must occur in the South Pacific. This is the case, but for a different reason than in the Atlantic, which has active deep water formation in the north. PDW, also known as Common Water, originates within the Pacific from upwelled bottom waters and modified UCDW. UCDW originates in the Southern Ocean as a mixture of PDW and Indian Deep Water (IDW; both marked by low oxygen) and deep waters that are formed locally in the Southern Ocean. PDW and UCDW occupy approximately the same density (and depth) range in the Pacific, with UCDW flowing into the Pacific and PDW flowing out. The net transport is southward, hence dominated by PDW (Figure 10.18).

UCDW is described in Section 13.5.3 so is only referred to here where it interacts with PDW.

PDW is one of the major deep waters of the global ocean, with many similarities to IDW (Chapter 11). PDW has no surface sources, unlike North Atlantic Deep Water. PDW is formed entirely internally from upwelling and diffusion. Because PDW is formed internally from waters that flow in from the Southern Ocean, the waters in the PDW are the oldest of the global ocean. PDW is marked by low oxygen, high nutrients, no CFCs, and large Δ^{14}C age (Figures 10.35, 4.12, 4.22, 4.24). The vertical extrema indicating greatest age are centered at 2000−2500 dbar, with the most extreme values in the mid- to high-latitude North Pacific. These signals of age extend southward down the length of the Pacific toward the Southern Ocean. Because the PDW mixes with the younger surrounding waters as it moves south, its age appears to decrease toward the south. These age tracers, especially the low oxygen, mark the presence of PDW in the Southern Ocean. Because it is very old, PDW is well mixed in T-S properties. It includes the highest peak by far in the global volumetric T-S diagram (Figure 4.17), at 1.1−1.2°C, 34.68−34.69 psu (corresponding to $\sigma_4 = 45.87$ kg/m^3). (PDW encompasses a wider range of T-S than this.) For this reason, Montgomery (1958) named it the (Oceanic) Common Water. In sections of Figure 4.12, these T-S properties are found in the North Pacific north of 20°N from about 3500 m to the bottom.

In the North Pacific, north of 40°N, the most extreme PDW is found on the isopycnal in Figure 10.35, as indicated by highest silica and lowest salinity (and also the most negative Δ^{14}C in Figure 4.24b). This is the "new" PDW, which is formed of very old waters. The low salinity is acquired through downward diffusion from above. The high silica in the northern

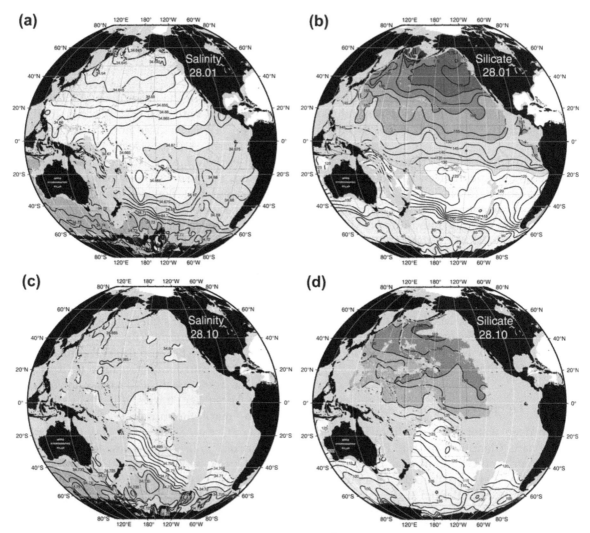

FIGURE 10.35 (a, c) Salinity and (b, d) silicate for PDW/UCDW ($\gamma^N = 28.01$ kg/m^3; $\sigma_2 \sim 36.96$ kg/m^3) and LCDW ($\gamma^N = 28.10$ kg/m^3; $\sigma_4 \sim 45.88$ kg/m^3). Depths of the two surfaces are approximately 2600–2800 m and 3500–5200 m, respectively, north of the ACC. Maps of Δ^{14}C (/mille) and δ^3He (%) at $\gamma^N = 28.01$ kg/m^3 and depth and potential temperature at $\gamma^N = 28.10$ are found in Figures S10.31 and S10.32 on the textbook Web site. *Source: From WOCE Pacific Ocean Atlas, Talley (2007).*

North Pacific, which is also a marker of PDW, comes from both aging of the waters and dissolution from the underlying silica-rich sediments (Talley & Joyce, 1992).

PDW and UCDW are horizontally juxtaposed, especially in the South Pacific. Salinity and silicate on an isopycnal (Figure 10.35a, b)

show the higher salinity/lower silicate UCDW entering in the southeast, and the contrasting low salinity/high silicate PDW moving southward in the west.

There is also southward flow of PDW along the South American boundary, evidenced by the higher silica in Figure 10.35b, but much more obvious in the vertical section of oxygen at 32°S (Figure 10.15b). Salinity in Figure 10.35a does not reflect this southward flow because of the small but noticeable impact of geothermal heating from the East Pacific Rise. The geothermally affected waters are beautifully marked by δ^3He plumes (Talley, 2007). These match the two westward-extending plumes of higher salinity in the tropics in Figure 10.35a. (On an isopycnal, warmer water must be more saline.) The higher salinity at the eastern boundary in the South Pacific is consistent with East Pacific Rise heating, which masks the salinity signature of southward flow.

When it leaves the Pacific and enters the Southern Ocean, PDW joins the IDW, which has a similar density range and is also marked by low oxygen and high nutrients. The layer is then referred to as UCDW, which upwells to the sea surface in the ACC. This upwelled UCDW is the most likely source of the surface waters that are transported northward out of the Southern Ocean (Chapter 14).

10.9.4. Bottom Water (LCDW)

The densest water in the Pacific comes from the Southern Ocean. Its source is a mixture of the deep waters of all three oceans (Atlantic, Indian, and Pacific) that is modified by production of dense waters around the Antarctic continent (Section 13.5.3). In the Pacific and Indian Oceans, it is common to refer to this dense bottom water mass as Lower Circumpolar Deep Water (LCDW). The similar layer in the Atlantic is usually called Antarctic Bottom Water (AABW), which is the nomenclature we use when we discuss this bottom layer globally (Chapter 14).

LCDW is recognized in the Pacific by low temperatures and higher salinity than the overlying PDW (vertical section in Figure 4.12). Its higher oxygen and lower nutrients reflect its somewhat younger age than the very old PDW (Figures 4.12 and 4.22).

At the southern end of the Pacific sections, LCDW is marked by the vertical salinity maximum within the ACC. The higher salinity is a long-distance tracer of North Atlantic Deep Water (Reid & Lynn, 1971). The salinity maximum, which approximately follows an isopycnal, extends northward into the deep Pacific; eventually the maximum salinity is at the ocean bottom. On the 165°W section, this grounding occurs at about 5°S, but in the far eastern Pacific at 88°W, it has already occurred by 45°S (section P19 in the WOCE Pacific Ocean Atlas, Talley, 2007).

LCDW enters the Pacific in the DWBC in the southwest, east of New Zealand (Section 10.6). This inflow is apparent in northward extension of high salinity and low silica in the southwestern Pacific on a deep isopycnal surface characterizing LCDW (Figure 10.35). Some of this signal succeeds in passing through the Samoan Passage at 10°S and crosses into the Northern Hemisphere hugging the western boundary. Silica in particular shows evidence of northward flow all the way along the western boundary to the northern North Pacific.

LCDW properties change to the north as the layer erodes and upwells across isopycnals into the PDW, with downward diapycnal diffusion of heat and freshwater as the source of buoyancy. The upwelling transports were described in Section 10.6, with the budgets suggesting most of the upwelling occurs in the South Pacific and tropics. Evidence of diapycnal diffusion is abundant in the property changes along the LCDW pathway. Salinity on the characteristic LCDW isopycnal decreases to the north, and is lowest in the central North Pacific near the Hawaiian Ridge and in the northwestern North Pacific (with temperature, of course, the mirror image). Similar patterns are apparent on constant depth surfaces and in bottom properties (Figure 14.14).

The bottom water is subject to low levels of geothermal heating that increase its temperature gently, by about 0.05°C from the tropics to the northern North Pacific. This change is consistent with geothermal heating, and affects a bottom layer of about 1000 m thickness (Joyce, Warren, & Talley, 1986). This buoyancy source could be important for the deepest upward flux in the northern North Pacific, where overturn does not extend much higher above the bottom than this (Section 5.6; Figure 10.18).

10.10. DECADAL CLIMATE VARIABILITY AND CLIMATE CHANGE

The Pacific Ocean represents a large fraction of the global ocean's surface and therefore a large potential for coupled atmosphere–ocean feedbacks. The interannual ENSO (Section 10.8), which has maximum amplitude in the tropics, is an excellent example of efficient coupling. The decadal and longer timescale climate modes are characterized by much larger north-south spatial patterns, with extratropical amplitudes that are similar to tropical amplitudes. Outside the tropics, coupling of the ocean and atmosphere is much weaker and so feedbacks are much weaker and harder to discern.

All of the text, figures, and tables relating to climate variability other than ENSO are located in Chapter S15 (Climate Variability and the Oceans) on the textbook Web site. Chapter S15 covers the following modes of decadal climate variability that most directly affect the Pacific: *Pacific Decadal Oscillation* (PDO), *North Pacific Gyre Oscillation* (NPGO), *Pacific North American* teleconnection pattern (PNA), *North Pacific Index* (NPI), *Southern Annular Mode*. It concludes with a discussion of climate change (trends in temperature, salinity, and oxygen).

11.1. INTRODUCTION AND OVERVIEW

The Indian Ocean is the smallest of the three major oceans. It differs from the Atlantic and Pacific Oceans in having no high northern latitudes, extending to only 25°N. The southern boundary of the circulation is the Antarctic Circumpolar Current (ACC), within and north of which the Indian is connected to the Atlantic and Pacific Oceans. The Indian Ocean also has an important low latitude connection to the Pacific Ocean through the Indonesian archipelago. In the north, the Indian Ocean has two large embayments west and east of India: the Arabian Sea and the Bay of Bengal. The deep Indian Ocean is geographically much more complex than the deep Atlantic and Pacific due to its tectonic history (Figure 2.10). Many deep ridges divide the deep circulation that is connected with the Southern Ocean into numerous, complicated pathways.

The Indian Ocean was explored later than the Atlantic and Pacific, with the first truly extensive observations during the International Indian Ocean Expedition (1962 to 1965), whose results are gathered in the Oceanographic Atlas of the International Indian Ocean Expedition (Wyrtki, 1971). During the 1980s and 1990s, international exploration of the Indian Ocean circulation as part of the World Ocean Circulation Experiment (WOCE), major programs in the Arabian and Red Seas, Indonesian Throughflow, Leeuwin Current, and Agulhas Current/Retroflection, along with many national programs, vastly increased the amount of information about all aspects of the circulation and water masses in all regions of the Indian Ocean. While the Indian Ocean remains relatively less explored than, say, the northern North Atlantic, it is now fully integrated in the global observing systems and there are a number of ongoing regional programs.

The principal upper ocean flow regimes of the Indian Ocean are the subtropical gyre of the south Indian Ocean and the monsoonally forced circulation of the tropics and Northern Hemisphere (Figure 11.1). These are separated oceanographically around 10−12°S by a nearly zonal current (*South Equatorial Current*: SEC) carrying fresher Pacific waters westward across the Indian Ocean. The anticyclonic subtropical gyre is similar to those of the other four ocean basins. Differences are that its western boundary current (*Agulhas Current*) overshoots the African coast, hence has a different type of separation from the western boundary, and that its eastern boundary current (*Leeuwin Current*) flows the "wrong way," toward the south. In the tropics and northern Indian Ocean, the circulations are strongly seasonal, forced by the reversing Southwest and Northeast Monsoons. In addition, the Arabian Sea and the Bay of Bengal are thoroughly contrasting oceanographic regimes, with the saline Arabian Sea and its marginal seas (Red Sea and Persian Gulf) dominated by evaporation while the fresher Bay of Bengal is dominated by runoff from all of the major rivers of India, Bangladesh, and Burma (Section S8.8 on the textbook Web site http://booksite.academicpress.com/DPO/; "S" denotes supplementary material). The surface waters of

the tropical Indian Ocean are the warmest of the global open ocean, often exceeding 29°C.

The intermediate and deep flow regimes of the Indian Ocean include a connection to the Southern Ocean that is similar to that of the South Pacific Ocean, with differences largely due to accidents of topography. The main difference

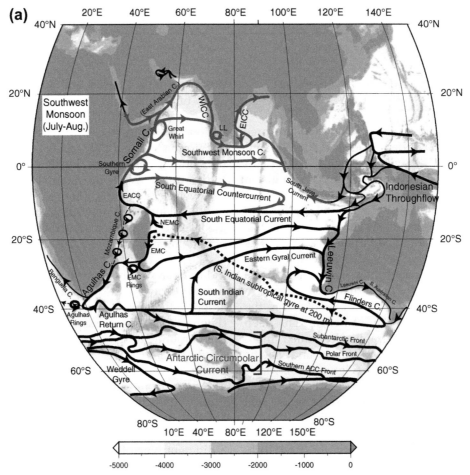

FIGURE 11.1 Indian Ocean schematic surface circulation. Black: mean flows without seasonal reversals. Gray: monsoonally reversing circulation (*after Schott & McCreary, 2001*): (a) Southwest Monsoon (July-August) (b) Northeast Monsoon (January-February). The ACC fronts are taken directly from Orsi, Whitworth, and Nowlin (1995). The subtropical gyre in the Southern Hemisphere just 200 m below the sea surface differs significantly from the surface circulation, as indicated by the dashed curve. Acronyms: EACC, East African Coastal Current; EICC, East Indian Coastal Current; EMC, East Madagascar Current; LH and LL, Lakshadweep high and low; NEC, North Equatorial Current; NEMC, Northeast Madagascar Current; and WICC, West Indian Coastal Current. See also Figure S11.1 from the textbook Web site, which is a surface height map based on Niiler, Maximenko, and McWilliams (2003), with labeled currents, and Figure S11.2, *reproduced from Schott and McCreary (2001)*.

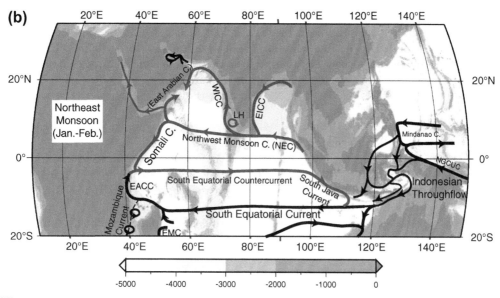

FIGURE 11.1 (*Continued*).

for the Indian Ocean is a limited source of intermediate (deep) water in the Red Sea in the northwest Indian Ocean. This water mass is similar to
the Mediterranean Overflow Water of the
Atlantic. Both are highly saline, hence "dying"
the intermediate and deep waters with high
salinity, but both have low transport and hence
limited impact on deep ventilation rates.

The role of the Indian Ocean in the global
overturning circulation is that of an upwelling
region, like the Pacific Ocean. Near-bottom
waters from the North Atlantic and Antarctic
enter from the south and participate in a complicated upwelling pattern that likely includes
return of Indian Deep Water to the Southern
Ocean, as well as upwelling to near the surface.
Upper ocean waters from the Pacific Ocean that
participate in the global circulation also traverse
the Indian Ocean (*Indonesian Throughflow* or
ITF), enter the Agulhas Current, and finally
enter the Atlantic Ocean.

The principal currents of the Indian Ocean are
shown in Figure 11.1 and also in Figure S11.1 in
the supplement on the textbook Web site, where

they are also listed in Tables S11.1 and S11.2. The
wind forcing, including monsoons, is described
in Section 11.2, followed by the monsoonal and
tropical circulation in Section 11.3. The subtropical circulation, ITF, and Red Sea/Persian Gulf
regimes are described in Sections 11.4–11.6 and
Section S8.10 on the textbook Web site. The intermediate and deep circulations are presented in
Section 11.7. Water masses are described in
Section 11.8 and summarized in Table S11.3 on
the textbook Web site. A few aspects of climate
variability are included in Chapter S15 on the
textbook Web site.

11.2. WIND AND BUOYANCY FORCING

The wind forcing of the Indian Ocean is one
of its most unique features. The mean wind
pattern (Figure 5.16a) of the south Indian Ocean
is like that of the Atlantic and Pacific, with westerly winds at high latitude (Southern Ocean)
and trade winds at low latitudes. The northern

Indian Ocean, however, is dominated by the seasonally reversing monsoons (Figures 5.16b, c and Figures S11.3 and S11.4 on the textbook Web site), which change the ocean circulation seasonally.

11.2.1. Mean Wind Forcing

The mean winds in the Southern Hemisphere result in Ekman downwelling over the broad latitude region from 50°S to 10°S (Figure 5.16d and Figure S11.3a on the textbook Web site). This produces Sverdrup forcing for a standard, anticyclonic subtropical gyre (Figure 5.17 and Figure S11.3b). The gyre forcing is different from that for the South Pacific and South Atlantic, because the southern cape of Africa, at about 35°S, lies well within the major subtropical gyre forcing. The subtropical gyre "runs out" of western boundary before it "runs out" of wind forcing for the gyre. Consequently, the western boundary current, the Agulhas, overshoots the tip of Africa, making it different from the other four subtropical gyre western boundary currents. The wind forcing then continues the subtropical circulation far to the west to the coast of South America, where there is a southward western boundary current (Brazil Current). The actual circulation is much more complex as the Agulhas turns back to the east after it separates from the African coast, shedding large eddies at the retroflection that propagate westward into the South Atlantic rather than continuing westward as a smooth flow to the coast of South America. In any case, the wind forcing ensures that the subtropical gyres of the Indian and South Atlantic are connected.

At the eastern side of the Indian Ocean's subtropical gyre region, there is some connection with the South Pacific's subtropical circulation. East of Tasmania, the subtropical circulation is more part of the South Pacific's circulation, although part of the East Australian Current (EAC) leaks into the Indian Ocean circulation (Section 10.4.1).

The mean winds in the tropical and northern Indian Ocean produce a net upwelling region between the equator and 15–20°S. This is associated with a cyclonic gyre consisting of the westward SEC on the south side, the eastward South Equatorial Countercurrent (SECC) on the north side, and a northward western boundary current (East African Coastal Current; EACC).

The Southwest Monsoon, producing net downwelling and Sverdrup transport forcing for a mean anticyclonic circulation, dominates in the mean winds in the Arabian Sea. The Northeast Monsoon regime, though, is quite different (see next section).

11.2.2. Monsoonal Wind Forcing

The northern and tropical Indian Ocean is subject to monsoonal wind forcing. The word monsoon is derived from the Arabic word "mausim," which means seasons. There is a nearly complete reversal of winds from summer to winter, and the ocean circulation responds accordingly.

Monsoons are the seasonal changes of the large-scale winds (Figure 5.16b, c and online supplementary Figure S11.4 from Schott, Dengler, & Schoenefeldt, 2002, which also includes sea-surface temperature; SST). These arise in response to the change in sign of the large-scale temperature difference between the ocean and land mass. In summertime, the land mass is warm and in winter it is cold. The ocean surface temperature varies a little with seasons, but not nearly as much as the land. So during the summer in the tropics, the large-scale winds blow toward the warm continent, and in winter they blow toward the ocean. A thorough explanation is much more complex and well beyond our oceanographic scope.

Monsoons are named for the prevailing wind direction. In the northern Indian Ocean in summer, the Southwest Monsoon blows from southwest to northeast, from the western Indian

Ocean and Arabian Sea onto India. (The South-west Monsoon winds are a continuation across the equator of the southeast trade winds, which continue throughout the year.) The southwest-erly winds are concentrated in a narrow jet, called the Somali (or Findlater) Jet, which is apparent in the July winds described in Figure S11.4 on the textbook Web site. This is the "wet-season" in India and most of Southeast Asia. In winter (November-March), the North-east Monsoon blows from northeast to south-west, from the continental landmass to the ocean. This is the dry season, with relatively cool conditions.

The Southwest Monsoon winds are stronger than the Northeast Monsoons, so the annual mean wind pattern looks like a weak version of the Southwest Monsoon.

The transitions between Southwest and Northeast Monsoons are relatively quick, taking place in 4−6 week periods in April-June and October-November. During the transitions, the equatorial winds are eastward across the full width of the Indian Ocean.

11.2.3. Buoyancy Forcing

Air−sea fluxes of heat and freshwater are shown in the global maps in Chapter 5 (Figure 5.15). The Indian Ocean has no high northern latitudes that could result in substantial heat loss. Its northernmost reaches, in the Red Sea and Persian Gulf, do experience net cooling and evaporation, and form dense waters. The Red Sea outflow is dense enough to penetrate deep into the water column, but the volume transport of the overturn is small and the saline overflow water mainly results in a salty "dye" for the deep northern Indian waters.

The tropical Indian Ocean is a region of net heating, with largest heating along the coast of Africa, in the Somali Current, associated with upwelling and large permanent eddies. There is net precipitation in the east due to rising air above the Indian Ocean's very warm pool of surface water. These features are reversed in the east-west direction compared with the regions of highest heating and net precipitation in the Pacific and Atlantic Oceans, because the warmest region of the Indian Ocean is the eastern tropics; there is no equatorial cold tongue. This results from the tropical Indian Ocean's strongly seasonal winds as opposed to the prevailing east-erly trade winds of the Pacific and Atlantic.

In the subtropics, the Indian Ocean's surface forcing also differs from that of other subtrop-ical oceans because the eastern boundary regime is dominated by the southward Leeuwin Current rather than an equatorward eastern boundary current. Both the Agulhas and Leeu-win Current regions thus experience net heat loss. The Agulhas region has the highest heat loss of all regions of the Indian Ocean. The high heat loss extends far to the east along the Agulhas Return Current (Section 11.4.2). The subtropics are also a region of net evapora-tion, although the contribution to total buoy-ancy flux is small.

11.3. MONSOONAL AND TROPICAL OCEAN CIRCULATION

The ocean circulation in the tropics and northern Indian Ocean is dominated by the reversing monsoonal wind forcing. Thorough overview and discussion of these circulations is provided by Tomczak and Godfrey (1994) and by Schott and McCreary (2001). Ocean adjustment to strongly variable winds includes generation of large-scale waves such as Rossby and Kelvin waves (Section 7.7.3). Dynamical understanding of the current reversals, produc-tion of undercurrents and eddies, and so forth, requires incorporation of these wave processes. We do not describe these mechanisms here.

The monsoonally forced circulation is north of the SEC front (north of 10−15°S). The SEC flows westward in all seasons and splits at the coast of Madagascar into the Northeast Madagascar

Current (NEMC) and the East Madagascar Current. The latter feeds the Agulhas. The NEMC flows northwestward and reaches the African coast where it splits again, into southward flow through Mozambique Channel and northward flow in the EACC. The northward flow along Madagascar and Africa is expected from the Indian-wide cyclonic forcing south of the equator (Figure 5.17).

The behavior of the EACC as it reaches the equator depends on the monsoon. During the Southwest Monsoon and the buildup to it, the EACC feeds the northward Somali Current, which crosses the equator. This current is notable for its high speeds, measured up to 360 cm/sec. Its transport is about 65 Sv, most of it in the upper 200 m. The continuation of northward flow along the Arabian Peninsula during the Southwest Monsoon is not generally given a name, but Tomczak and Godfrey (1994) and Böhm et al. (1999) referred to it as the East Arabian Current, which we have adopted for Figure 11.1. (At the northeastern termination of the Arabian Peninsula, there is a persistent eastward jet during the Southwest Monsoon called the Ras al Hadd Jet.) The SEC, the Somali Current, and the Southwest Monsoon Current comprise a strong seasonal wind-driven gyre in the northern Indian Ocean.

During the Southwest Monsoon, the mid-ocean circulation from south of the equator to the northern boundary is eastward. The eastward flow between 7°S and Sri Lanka/southern India is called the *Southwest Monsoon Current*. Within both the Arabian Sea and the Bay of Bengal, the circulation is eastward with a tendency to be anticyclonic (Figures 11.1 and 11.2). Both the *West Indian* and *East Indian Coastal Currents* flow eastward.

The western boundary currents during the Southwest Monsoon have remarkably large, recurrent eddy structures (Figure 11.3). As the Somali Current crosses the equator, part of it turns out to the east at 4°N, into the *Southern Gyre*. Another large eddy, the *Great Whirl*, forms at 10°N. Formation of the Great Whirl precedes formation of the Southern Gyre during the transition to the Southwest Monsoon (Schott & McCreary, 2001). There is another smaller recurrent eddy, the Socotra Gyre (or Eddy), at about 12°N. Northeastward flow continues on along the Arabian Peninsula, associated with a major upwelling region off the Oman coast.

During the Northeast Monsoon (November to March), the equatorial current (SECC) continues to flow eastward, but a westward flow, the *Northwest Monsoon Current*, appears from the equator to 8°N, along the south side of Sri Lanka and India. From 8°S to the equator, the South Equatorial Countercurrent flows eastward. The surface circulations in the Arabian Sea and Bay of Bengal reverse. The Somali Current flows southward. The West Indian and East Indian Coastal Currents flow westward. The overall circulation is weaker and more disorganized than during the Southwest Monsoon. The Southwest Monsoon is stronger than the Northeast Monsoon (Figure 11.2), and thus the ocean responds more consistently to the Southwest Monsoon.

During the monsoon transitions in spring and fall, when the equatorial winds are westerlies rather than trades, the equatorial surface circulation reverses. The normal SEC, which is driven by trade winds, is a westward flow. The westerly winds cause the surface currents to flow eastward (Figure 11.4). These flows are called the *Wyrtki Jets*. Surface speeds exceed 100 cm/sec. The Wyrtki Jets are much stronger than the intervening westward flows driven by the trade winds, so the annual mean surface current is also eastward.

The Pacific and Atlantic equatorial circulations have well-defined permanent subsurface Equatorial Undercurrents (EUCs) that flow eastward. Because the Indian Ocean equatorial winds reverse and the equatorial trade winds are rather weak, there is only a weak EUC and only during part of the year. The EUC is found in the thermocline east of 60°E during February-June.

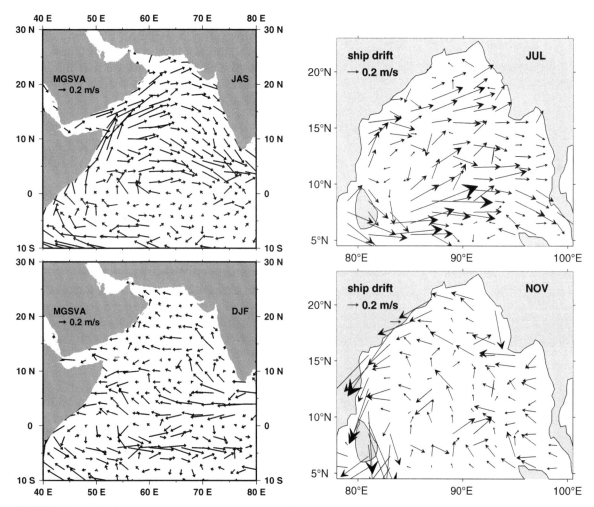

FIGURE 11.2 Surface circulation. Left: Arabian Sea (surface drifters). Right: Bay of Bengal (ship drift). Top: Southwest Monsoon. Bottom: Northeast Monsoon. *Source: From Schott and McCreary (2001).*

During the Southwest Monsoon, the winds are northeastward along the coasts of Somalia and the Arabian Peninsula (Figure 11.2). This causes offshore Ekman transport and upwelling at the coast. The upwelled water along the coasts is cold (~20–24°C) compared with the tropical surface waters (>27°C; Figure 11.5). In addition to the direct coastal upwelling, there is broader scale upwelling since the axis of the Somali Jet lies offshore so Ekman transport increases offshore. The resulting wind stress curl creates upwelling from the coast to the center of the jet. Farther offshore of the Somali Jet, the wind stress curl creates downwelling. (These regions are apparent in the mean wind-stress curl map of Figure 5.16d, since the Southwest Monsoon dominates the Arabian Sea winds.)

The upwelled water during the Southwest Monsoon is rich in nutrients. Ocean productivity changes dramatically in this area with

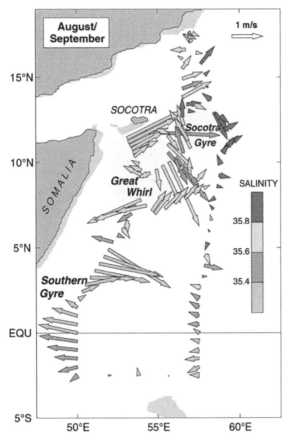

FIGURE 11.3 Somali Current regime during the Southwest Monsoon (August/September, 1995). This figure can also be seen in the color insert. *Source: From Schott and McCreary (2001).*

the reversal of the monsoon. High productivity along the southwest coast of Oman, in the Persian Gulf, along the west coast of India, and the east coast of Somalia is apparent during the monsoon (global maps in Figure 4.29; Arabian Sea maps in Figure S11.5 on the textbook Web site).

Vertical sections across the upwelling system in the Somali Current during the Southwest Monsoon show the uplifted isotherms at the coast, and high nutrients and low oxygen that accompany them (Figure 11.6). Monthly time series of surface temperature and biomass

show minimum temperature in September, coinciding with maximum biomass, at the height of the Southwest Monsoon upwelling season. Maximum temperature and minimum biomass occur in January-March during the Northeast Monsoon.

11.4. SOUTH INDIAN OCEAN SUBTROPICAL CIRCULATION

The subtropical gyre of the Southern Hemisphere Indian Ocean differs from the other ocean basins' subtropical gyres in its connections to the South Atlantic and South Pacific circulations and location relative to the continents. Australia, Tasmania, and New Zealand form an eastern boundary for an Indian Ocean subtropical gyre. The African coast creates only a partial western boundary, allowing an Indian Ocean subtropical gyre only north of the Cape of Good Hope. However, the wind stress pattern for the Southern Hemisphere dictates that the Indian Ocean's gyre extends westward all the way to the western boundary of the South Atlantic. Thus the Indian and South Atlantic gyres are inextricably linked, but with major complications due to the Agulhas Current along the coast of Africa, which is one of the most powerful western boundary currents in the world. The far southeastern boundary of the Indian's subtropical gyre is also not quite complete and there is leakage from the South Pacific's subtropical circulation via a small branch of the EAC. Meanwhile, the eastward flow of the southernmost part of the Indian's subtropical gyre partially continues into the South Pacific. Thus the Southern Ocean connects all three Southern Hemisphere subtropical gyres.

11.4.1. Subtropical Gyre

The wind-forced anticyclonic subtropical gyre of the south Indian Ocean includes the

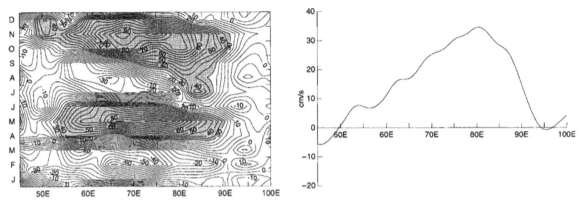

FIGURE 11.4 Mean zonal surface currents at the equator, based on ship drift data. Left: monthly means. Right: annual mean. ©*American Meteorological Society. Reprinted with permission. Source: From Han, McCreary, Anderson, and Mariano (1999).*

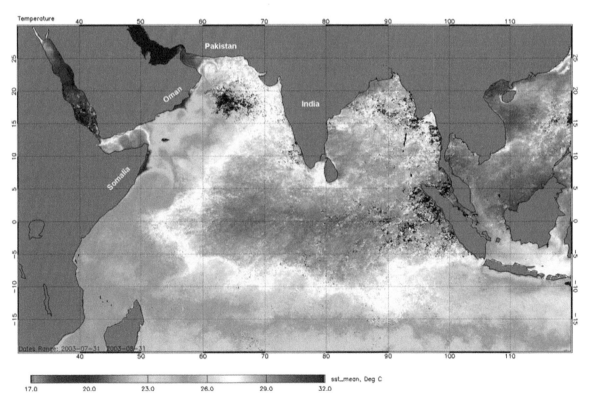

FIGURE 11.5 SST in July 2003 (Southwest Monsoon), from the MODIS satellite. This color can also be seen in the color insert. *Source: From NASA Goddard Earth Sciences (2007a).*

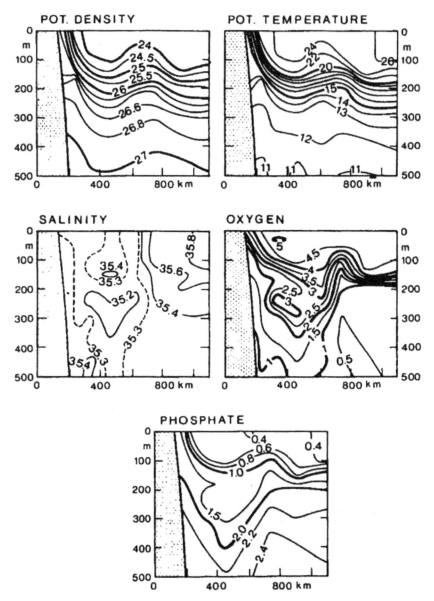

FIGURE 11.6 Sections in the Somali Current upwelling regime at 12°N, 29 August– 1 September 1964. *Source: From Schott and McCreary (2001); after Swallow and Bruce (1966).*

westward flow of the SEC in the north, the eastward flow of the *South Indian Current* in the south, and the southward flow of the *East Madagascar Current* (EMC) and *Agulhas Current* along its western boundary. The northward flow in the eastern part of the gyre is broad in longitude and is sometimes called the *West Australian Current*. The south Indian Ocean does not have a narrow, northward eastern boundary current, unlike the other world ocean subtropical gyres. Instead, the *Leeuwin Current* is a narrow, southward flow along the coast of Australia.

The northern boundary of the subtropical gyre is well defined in the upper ocean by a strong property front within the SEC at about 10–15°S. The SEC carries fresh water from the Indonesian passages westward, resulting in the property front.

The surface geostrophic flow in the subtropical gyre does not seem gyre-like in any depictions (e.g., Figures 11.1, 11.8a, and 14.1; also Stramma & Lutjeharms, 1997). The broad northward flow heads eastward and feeds into the southward Leeuwin Current rather than turning westward to complete the anticyclonic flow. This eastward flow at the sea surface, centered around 17°S, is called the *Eastern Gyral Current* (Wijffels et al., 1996; Domingues et al., 2007); it is analogous to the Subtropical Countercurrent of the North Pacific. A portion of the anticyclonic gyre does reappear south of Australia, in the form of the westward *Flinders Current* (Bye, 1972; Hufford, McCartney, & Donohue, 1997; Middleton & Cirano, 2002); its surface expression turns back to join the Leeuwin Current, but just 200 m below the surface (dashed in Figure 11.1a), the westward flow south of Australia continues on to the northwest across the whole expanse of the south Indian Ocean.

By 200 m depth, the anticyclonic gyre is well formed and conforms to the shape of the net gyre transport and also the wind-forced Sverdrup transport (Figure 5.17). The anticyclonic gyre extends all the way eastward to Tasmania. The center of the gyre at this depth is at 35–36°S, which also, coincidentally, corresponds to the southern tip of Africa. The western boundary current here is the Agulhas. In the northern part of the gyre at this depth, the westward flow of the SEC reaches the coast of Madagascar and splits into southward and northward flows. The southward flow is the EMC, which flows along the coast of Madagascar and then shoots west to the African coast where it forms the Agulhas. The northward flow, the NEMC, also continues west to the coast of Africa where

it splits. A southward portion flows through Mozambique Channel and joins the Agulhas Current. A northward portion joins the EACC to become part of the tropical circulation.

The subtropical gyre shrinks poleward and toward the western boundary with increasing depth, typical of all subtropical gyres (Figures 11.7 and 11.8). A useful measure of the poleward shift is the bifurcation point of the EMC. At 800–900 m, it shifts to about the center of Madagascar. By 1000 m, it shifts to the southern end of Madagascar. By 2000 dbar (Figure 11.14), the anticyclonic circulation is entirely in the western basin, retaining the Agulhas as a strong western boundary current that reaches to the ocean bottom.

11.4.2. Agulhas Current and Retroflection

The Agulhas Current is one of the strongest currents in the global ocean. It is narrow and swift, with synoptic speeds that exceed 250 cm/sec (Figure 11.8). It reaches to the ocean bottom and is narrowest at about 33°S where it is most strongly pinned to the western boundary. The Agulhas location is well indicated by the subtropical gyre Sverdrup transport (Figure 5.17). It forms mainly from southward flow in the EMC and from weaker southward advection through the Mozambique Channel, mainly in the form of large anticyclonic eddies.

The Agulhas Current transport is approximately 70 Sv. On the inshore side of the Agulhas, there is a well-defined, narrow countercurrent/undercurrent that reaches to the ocean bottom. Its surface velocity is also large, exceeding 50 cm/sec (Figure 11.8d). It has a transport of about 15 Sv. The undercurrent is one pathway for transport of North Atlantic Deep Water (NADW) into the Indian Ocean (Bryden & Beal, 2001).

The Agulhas follows the continental shelf to where it ends at about 36°S and then separates from the boundary. It overshoots into the South

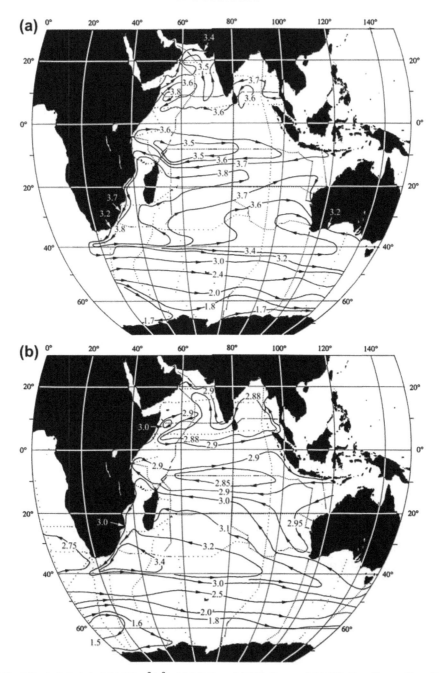

FIGURE 11.7 Adjusted steric height (10 m^2/s^2) at (a) 0 dbar, (b) 200 dbar, and (c) 800 dbar. *Source: From Reid (2003).* The closely related geostrophic streamfunction based on subsurface float observations from Davis (2005) is shown in the supplement Figure S11.6 on the textbook Web site.

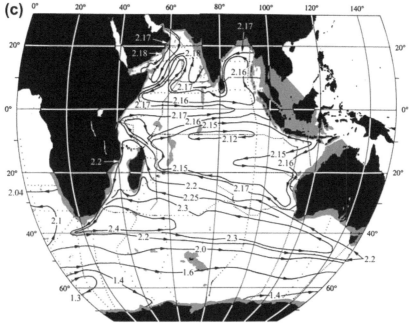

FIGURE 11.7 (*Continued*).

Atlantic and then retroflects back into the Indian Ocean. It sheds large rings at the retroflection that propagate westward into the South Atlantic. The Agulhas Retroflection and Agulhas rings are presented in Chapter 9 because of their impact on South Atlantic circulation and water properties. An infrared satellite image showing the Agulhas Current, its eddies, and retroflection is presented in Figure 9.13 along with a schematic depiction of the Agulhas retroflection and eddy shedding.

Eddy kinetic energy (EKE) is high in the Agulhas retroflection (Figure 11.8c), including a long, narrow band to the east of the retroflection, which follows the Subtropical Front (South Indian Current), also called the Agulhas Return Current in the figure. The peak of EKE variability is west of 27°E, which is the location of the Agulhas Plateau. This band of high EKE is apparent in global maps of EKE (Figure 14.16). Enhanced EKE is also found southwest of Madagascar, in the location where the EMC

separates and flows westward to the African coast.

The location of peak EKE variability varies seasonally. The area of variability is further west in the austral winter and further south in summer. The location of the Agulhas front can be detected from satellite SST data (AVHRR). The Agulhas fronts in SST are farther west in austral winter as well (Quartly & Srokosz, 1993; Figure S11.7 on the textbook Web site).

Quasi-permanent meander sites in the Agulhas Return Current (Subtropical Front) are apparent in the Agulhas front locations, and are included in the Agulhas system schematic (Figure 11.8a). There is a northward meander at 26°E and a second at 32°E. The first one goes around the north side of the Agulhas Plateau. These permanent meander sites resemble those of the Kuroshio (Chapter 10), with similar zonal spacing between meanders.

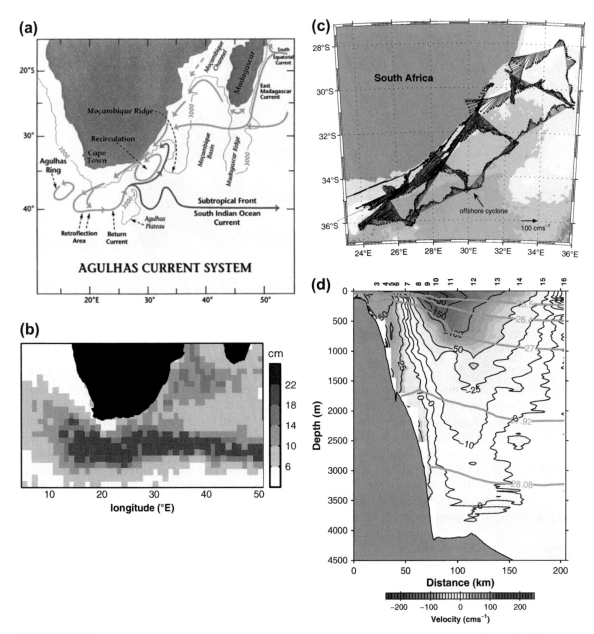

FIGURE 11.8 (a) Schematic of the Agulhas Current system and local topography. *Source: From Schmitz (1995b).* (b) RMS sea surface height variability for eight years from the Topex/Poseidon altimeter. *Source: From Quartly and Srokosz (2003, Satellite observations of the Agulhas Current system, Phil. Trans. Roy. Soc. A., 361, p. 52, Fig. 1b).* (c) Average velocity from 0–75 m depth and (d) velocity section at 36°S, with neutral density surfaces overlaid, from ADCP observations in February–March, 2003. *Source: From Beal, Chereskin, Lenn, and Elipot (2006). Figures c and d are from the ©American Meteorological Society. Reprinted with permission.*

11.4.3. Leeuwin Current

The Leeuwin Current is the eastern boundary current for the south Indian Ocean, and is located off Western Australia. The current was identified prior to 1969, but named only in 1980 following the start of intensive observations (Cresswell & Golding, 1980). It differs from other subtropical eastern boundary currents since it flows poleward rather than equatorward. It is about 50–100 km wide and 2000 km long. It follows along the continental shelf break within 100 km of the coast, from about 22°S, off the Northwest Cape, to the southwestern tip of Australia (Cape Leeuwin) at 35°S. The current then turns eastward toward the Great Australian Bight (Figure 11.9)

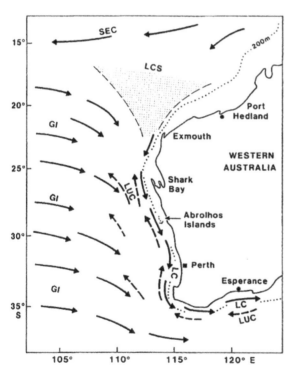

FIGURE 11.9 Leeuwin Current (LC) and Leeuwin Undercurrent (LUC). Other acronyms: SEC, South Equatorial Current; LCS, Leeuwin Current source region; and GI, geostrophic inflow. *Source: From Pearce (1991); from Schott and McCreary (2001).*

where it continues along the shelf break, bringing warm, saline waters into the region. It continues eastward as the South Australian Current, veers southward past Bass Strait between Australia and Tasmania, and then flows southward along the western coast of Tasmania, where the boundary current is historically referred to as the Zeehan Current (Ridgway & Condie, 2004).

The Leeuwin Current has a mean southward speed of 25 cm/sec at the surface, peaking at greater than 50 cm/sec. It opposes the mean wind stress, which is northward (Figures 5.16a, 11.2). Its maximum poleward transport at 33°S is 5 Sv in the upper 250 m (Smith, Huyer, Godfrey, & Church, 1991; Feng, Meyers, Pearce, & Wijffels, 2003). Below it there is an equatorward undercurrent (Leeuwin Undercurrent), with speeds of up to 40 cm/sec, a depth range of 300–800 m, and a transport of 1–2 Sv.

The Leeuwin Current has marked seasonal variability, peaking in surface speed in April-May and in transport in June-July, when the northward alongshore wind stress that opposes the current is weakest (Feng et al., 2003). The Leeuwin Current has elevated levels of mesoscale eddy variability compared with eastern boundary current regions in the other oceans (see the map of surface EKE in Figure 14.16). Both anticyclonic (warm-core) and cyclonic (cold-core) rings are shed by the Leeuwin Current from preferential locations associated with coastline shape (Morrow & Birol, 1998; Fang & Morrow, 2003). The eddies move westward a long distance into the Indian Ocean, preferentially along the band of high eddy energy associated with the Eastern Gyral Current.

In addition to its poleward flow, the Leeuwin Current differs from other eastern boundary currents in that upwelling does not occur on to the shelf. The isotherms off western Australia slope strongly downward from about 200 km offshore to the continental slope (Figure 11.10), in contrast to the situation in the subtropical eastern boundary current regions off the western United

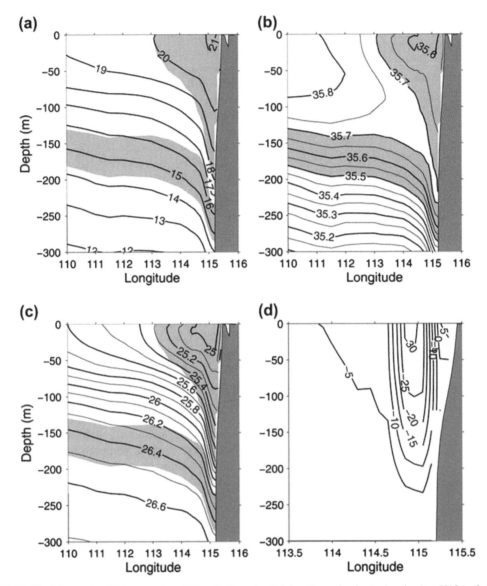

FIGURE 11.10 Mean potential temperature (°C), salinity, potential density, and velocity (cm/sec) at 32°S in the Leeuwin Current. Shading indicates salinity of 35.5–35.7 psu. *Source: From Feng et al. (2003).*

States, South Africa, and South America, where the isotherms slope upward toward the shore and upwelling of cool water occurs.

The Leeuwin Current (in the upper 150 m) in the north is warm and relatively fresh (35.0 psu), and has low dissolved oxygen and high phosphate content. Its northern source waters are both the tropical Indian Ocean and the ITF. As it flows southward, it retains a high temperature core and therefore transports a significant amount of heat to the south. As it moves southward, the Leeuwin Current becomes saltier, reaching about 35.7 psu at 33°S (Figure 11.10), due to inflow of waters from the subtropical gyre (Smith et al., 1991; Domingues et al., 2007). Taken together with its high eddy activity, the overall picture is of inflow of south Indian waters that feed the southward Leeuwin Current, which generates eddies that transport the water back to the west. Thus the Leeuwin Current does not continuously transport ITF waters from the North West Shelf of Australia all the way down to Cape Leeuwin (Domingues et al., 2007).

The Leeuwin Undercurrent, as well as the broad northward West Australia Current, transports South Indian Central Water, Subantarctic Mode Water (SAMW), and Antarctic Intermediate Water (AAIW) northward. These flows are part of the broader anticyclonic subtropical gyre, driven by Sverdrup transport.

The poleward flow of the Leeuwin Current is driven by a southward pressure gradient force associated with the flow from the Pacific through the Indonesian archipelago (Godfrey and Weaver, 1991; Feng, Wijffels, Godfrey, & Meyers, 2005). The downward slope of about 0.3 m in the sea surface along the western coast of Australia from 20 to 32°S is readily apparent in Figure 11.7a and Figure S11.1 in the supplement on the textbook Web site. This overwhelms the local eastern boundary forcing by equatorward winds. Wind variability, however, dominates the seasonality of the Leeuwin Current transport (Smith et al., 1991). Interannual variability of the Leeuwin Current is dominated by El Niño-Southern Oscillation (ENSO) signals that propagate southward along the coast of Australia from the Indonesian archipelago (Feng et al., 2003).

11.5. INDONESIAN THROUGHFLOW

The Indonesian Archipelago is the low latitude connection between the Pacific and Indian Oceans. Flow through the archipelago is referred to as the ITF. The ITF is unidirectional, from the Pacific to the Indian Ocean, since sea-surface pressure (sea level) is higher on the Pacific side. The Indonesian Archipelago has exceedingly complicated geography (Figure 11.11). More than 10 Sv of fresher, high nutrient Pacific waters thread through this complex. The global overturning circulation has transports of this order. The ITF is one of the major upper ocean elements of this global circulation, being part of the movement of 10–15 Sv from the Pacific Ocean, through the Indian Ocean, and back to the Atlantic Ocean (Chapter 14).

Pacific water enters the Indonesian Archipelago mainly through Makassar and Lifamatola Straits. The sources of this Pacific water are discussed briefly in Section 10.7.4. The Makassar Strait is shallower (680 m at Dewakang Sill), but carries most of the transport, at least 9 Sv, which is of North Pacific origin (Gordon, Susanto, & Ffield, 1999). The deeper Lifamatola Strait (1940 m) is the pathway for South Pacific water into the Indonesian Archipelago and for the deeper part of the throughflow into the Indian Ocean. Transport through this strait is at least 2–3 Sv (Gordon, Giulivi, & Ilahude, 2003; Talley & Sprintall, 2005). Some upper layer South Pacific water also passes through the Halmahera Sea. Within the Indonesian Archipelago, the waters are mixed horizontally and vertically. There is also some internal modification through local heating and slight freshening.

The throughflow waters exit the Indonesian Archipelago through three principal routes: Lombok Strait, Ombai Strait (connecting to Savu and Sumba Straits), and Leti Strait (connecting to Timor Passage). The deepest sill for the outflow is 1250 m northeast of Timor, at Leti Strait. All of these outflow straits have been instrumented at some time or another, and the transports through each are 2–5 Sv (Figure 11.11). In addition to various current meter deployments in the principal straits, all of the straits were instrumented with a pair of shallow pressure gauges across the straits in 1995–1999, allowing simultaneous observation of flows

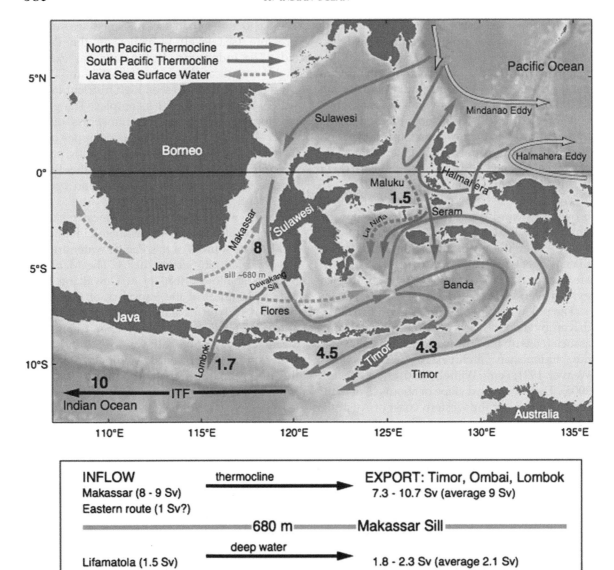

FIGURE 11.11 Indonesian Archipelago and Throughflow with transports (Sv). Lower panel summarizes transport above and below 680 m (Makassar Strait sill depth). This figure can also be seen in the color insert. *Source: From Gordon (2005).*

through the exit straits. Variability is large, and includes an ENSO signal (Hautala et al., 2001). An international array of current meters and pressure gauges is now in place to monitor both the inflows and outflows (Figure 11.11).

After the ITF waters exit the archipelago, they form into a narrow westward flow centered at 12°S, within the SEC. The fresh upper ocean waters are easy to see on any meridional salinity section in the eastern

Indian Ocean (Figure 4.13b). The deeper part of the throughflow is also observable as a salinity minimum at about the same depth and density as the low salinity AAIW that reaches from the south to nearly this latitude. The deeper expression of the ITF is unambiguously of Indonesian Archipelago origin based on its higher nutrient levels, especially in silica (Talley & Sprintall, 2005). The salinity minimum is called the *Indonesian Intermediate Water* (or Banda Sea Intermediate Water in early treatments).

The SEC is a zonal current that carries the throughflow waters westward across the Indian Ocean. Mass balances within the Agulhas Current indicate that the Indonesian waters must join this current and then exit the Indian Ocean. The baroclinic structure of the Agulhas suggests that the excess transport from the throughflow is in the upper ocean. The waters that enter the Agulhas that match the transport through the Indonesian Archipelago are greatly modified within the Indian Ocean and are unlikely to be the same water parcels.

Model studies suggest that dramatic global changes in upper ocean circulation, temperature and salinity, winds, and precipitation would occur if the ITF were cut off for a period (e.g., Schneider, 1998; Song, Vecchi, & Rosati, 2007). Song et al. (2007) show that the eastern tropical Pacific warms and the tropical Indian Ocean cools, reducing the strength of the trade winds and reinforcing the SST changes, and also shifting the precipitation (Figure S11.8 in the online supplementary material). ENSO variability would change. The Pacific Ocean becomes fresher and the Indian Ocean saltier as the fresher Pacific water would no longer be exported to the Indian Ocean. The flow that would normally go through the Archipelago would instead go south along the coast of Australia, leading to marked surface warming southeast of Australia.

11.6. RED SEA AND PERSIAN GULF OUTFLOWS

The Red Sea is one of the two global sources of high salinity intermediate water; the other is the Mediterranean Sea (Chapter 9 and Section S8.10 on the textbook Web site). Despite its relatively low latitude, the Red Sea achieves this distinction because of huge evaporation leading to high salinities even with its relatively high temperatures. Circulation, formation, and properties of the very saline waters within the Red Sea are described in Section S8.10.7 on adjacent seas in Chapter S8 (Figure S8.25).

The pure, newly formed Red Sea Water spills out over the Bab el Mandeb and into the Gulf of Aden (Figure 11.12). Intensive hydrographic and current observations of the Red Sea outflow in the Gulf of Aden document the progress of the highly saline, dense overflow water (Bower, Johns, Fratantoni, & Peters, 2005). The total outflow transport is no greater than 0.4 Sv, but it is extremely saline and dense: 39.7 psu and $\sigma_\theta = 27.5-27.6 \text{ kg/m}^3$. It also has elevated chlorofluorocarbon (CFC) content as a result of its renewal in the Red Sea (Mecking &Warner, 1999). The plume of dense, saline water mixes vigorously as it plunges over the sill. It follows two paths with different mixing characteristics, which are visible in Figure 11.12. The saline water mass, as a whole, turns to the right because of Coriolis force and hugs the southern boundary of the Gulf of Aden where it continues to mix and be diluted. The equilibrated water mass is referred to as either *Red Sea Overflow Water* (RSOW) or *Red Sea Water*, with properties of 38.8–39.2 psu, $\sigma_\theta = 27.0-27.48 \text{ kg/m}^3$ and depth 400–800 m. As it settles into the intermediate layer of the Arabian Sea, the RSOW affects the layer $\sigma_\theta = 27.0-27.6 \text{ kg/m}^3$ at depths of 400–1400m. Its vertical salinity maximum core, at about $\sigma_\theta = 27.3$, is visible on all sections in the western tropical and northern Indian Ocean (Figures 11.13 and 11.19). It is greatly diluted as

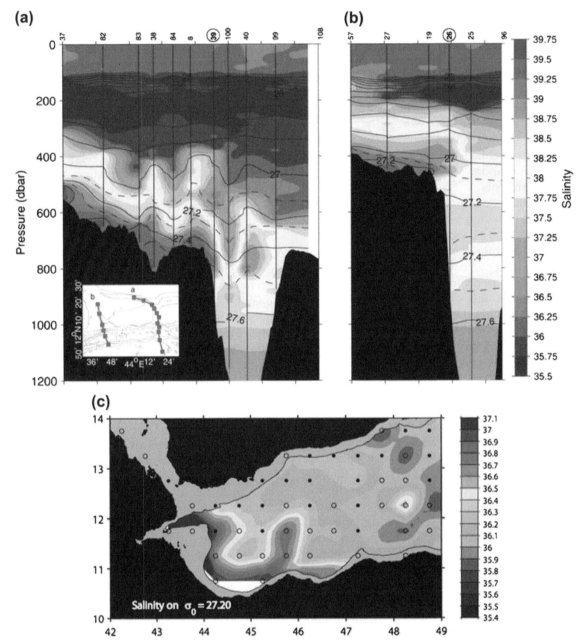

FIGURE 11.12   (a, b) Red Sea Overflow Water: salinity with potential density contours overlaid on sections in the Gulf of Aden in February–March, 2001. North is on the left. *Source: From Bower et al. (2005). ©American Meteorological Society. Reprinted with permission.* (c) Red Sea outflow in the Gulf of Aden: climatological salinity on the isopycnal $\sigma_\theta = 27.20$ kg/m³. *Source: From Bower, Hunt, and Price (2000).* This figure can also be seen in the color insert.

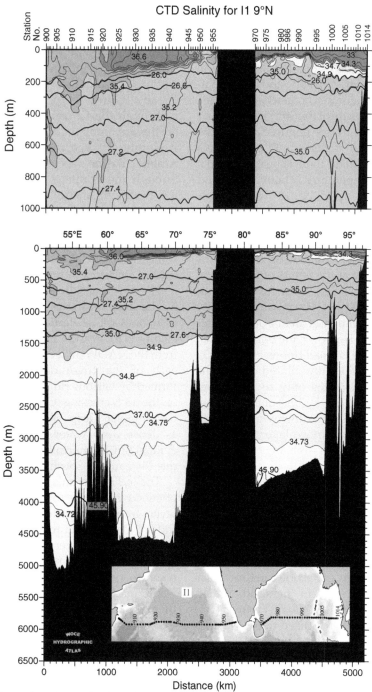

FIGURE 11.13 Salinity along 9°N with selected potential density σ_θ, σ_2, and σ_4 contours and station track overlaid. *After the WOCE Indian Ocean Atlas, Talley 2011).*

it spreads southward, but can be detected in the Mozambique Channel and on into the Agulhas Current (Beal, Ffield, & Gordon, 2000).

In contrast to the Red Sea, the much shallower Persian Gulf contributes its highly saline water to the Arabian Sea at a lower density and hence shallower in the water column. Circulation is into the Gulf on the northern side of the Straits of Hormuz and out on the southern side (Figure S8.25b in the online supplementary materials). Evaporation is in excess of 1.6 m/yr and there is a small annual mean heat loss. Temperature is between 15 and 35°C and salinity is up to 42 psu. Dense water formation ($\sigma_\theta > 29.5$ kg/m^3) occurs in late winter in the southern Persian Gulf where winter temperatures are low and salinities high (<19°C, >41 psu; Swift & Bower, 2003; Johns et al., 2003). Outflow at the Straits of Hormuz in winter has been observed at up to 41 psu at 21°C (averaging 39.5 psu). The potential density of 29 kg/m^3 is much denser than bottom water in the Indian Ocean. However, the outflow transport at about 0.15 Sv is small (Johns et al., 2003). The water is so significantly diluted during the outflow that it contributes only to the upper 200–350 m ($\sigma_\theta \sim 26.4$–26.8 kg/m^3) of the Arabian Sea. In Figure 11.13 this is the downward bulge in high salinity from the surface layer. (The high salinity surface layer in the Arabian Sea is of local evaporative origin.)

11.7. INTERMEDIATE AND DEEP CIRCULATION

The intermediate depth circulation at about 1000 m is dominated by zonal flows in the tropics and the anticyclonic gyre in the south Indian Ocean (Figure 11.7). By 2000 dbar (Figure 11.14) the anticyclonic gyre is restricted to the western Indian Ocean. There is a remnant of the SEC and SECC structure in the tropics, where flows remain basically zonal. Circulation in the Arabian Sea and Bay of Bengal is weak. The tops of the mid-ocean ridges begin to intrude.

By 3500 dbar, the circulation is strongly guided by the topography (see details in Figure 2.10).

Deep Western Boundary Currents (DWBCs) carry Circumpolar Deep Water (CDW) northward into the Indian Ocean, along the deep western boundaries in each of the Indian Ocean's basins. The principal western pathway is through the Crozet and Madagascar Basins, northward along the Madagascar coast, through Amirante Passage, and into the Somali and Arabian Basins. The eastern deep pathway is through the Southeast Indian Ridge at 120°E into the South Australia Basin, and then through gaps east and west of Broken Plateau into the Central Indian and West Australia Basins. The deepest flow entering the Central Indian Basin comes from the West Australia Basin through several fractures in the Ninetyeast Ridge.

The northward deep flows can be recognized by the deep and abyssal water masses that are seen on various vertical sections (Section 11.8 below). NADW and CDW, which has high salinity from the NADW mixed into it, can be seen against each of the (five) deep western boundaries formed by various ridges and undersea plateaus on the 33°S crossing. Cold, dense, fresher Lower Circumpolar Deep Water (LCDW), originating as dense deep water in the Antarctic, is most evident flowing northward against the Mozambique Plateau and Southwest Indian Ridge.

The Indian Ocean's net meridional overturn is obtained from the total transport in isopycnal layers from a complete east-west crossing of the Indian Ocean (Figure 11.15). The direction of transport at any given level is difficult to discern from the circulation maps. The Indian Ocean has no northern deepwater source. It has a small input at intermediate depth from the Red Sea. Therefore there should be net northward inflow in the deep water and outflow in the upper ocean. The zonally integrated transport in isopycnal layers at 33°S is shown in Figure 11.15, based on two independent analyses of the same data set; another analysis, from Talley (2008), is shown in Figure S11.10 on the textbook Web site. (Both analyses shown

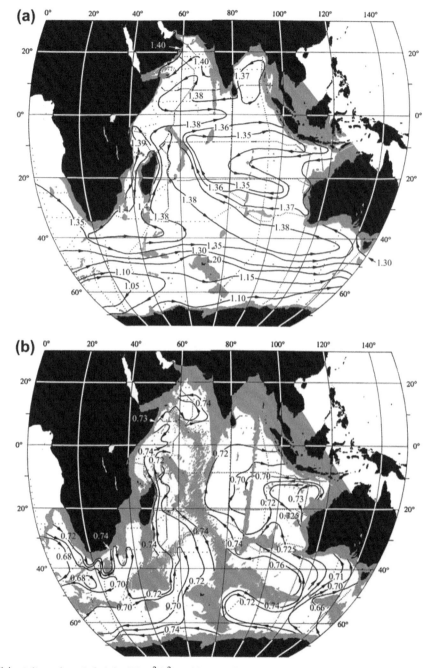

FIGURE 11.14 Adjusted steric height (10 m^2/s^2) at (a) 2000 dbar and (b) 3500 dbar. *Source: From Reid (2003).*

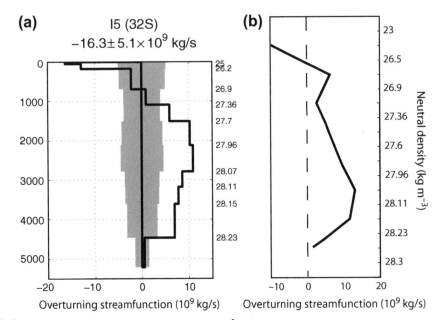

(a) I5 (32S)
$-16.3 \pm 5.1 \times 10^9$ kg/s

Overturning streamfunction (10^9 kg/s)

(b)

Neutral density (kg m^{-3})

Overturning streamfunction (10^9 kg/s)

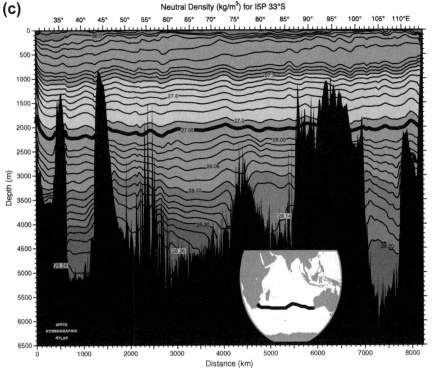

(c) Neutral Density (kg/m^3) for I5P 33°S

Depth (m)

Distance (km)

FIGURE 11.15 (a) and (b) Net northward (meridional) transport (Sv) for the Indian Ocean at 33°S, integrated from the bottom to the top. See also Figure S11.9 in the online supplementary materials. *Source of (a): From Ganachaud, Wunsch, Marotzke, and Toole (2000). Source of (b): From Robbins and Toole (1997).* The right-hand vertical coordinate is neutral density. (c) Neutral density (kg/m^3) at 33°S. Heavy contour is isoneutral surface 27.95 kg/m^3, marking the division between net northward flow below and southward above. *After WOCE Indian Ocean Atlas, Talley (2011).*

here are also integrated from bottom to top, so the actual direction of flow at a given level is the change from below to above the level.) The net meridional overturn from the deep to the intermediate/upper ocean north of 33°S is 11–12 Sv based on the two analyses shown. The transition from northward to southward transport occurs around 2100 dbar, at neutral density 27.96 kg/m^3. (There is also an additional 5 to 10 Sv of southward flow in the upper layers due to the ITF waters moving southward across 33°S.) The required upwelling rate north of the vertical section is approximately 3 to 5 × 10^{-5} cm/sec. This upwelling is an important part of the return of global deep waters to the upper ocean, and is mirrored by similar upwelling in the Pacific Ocean. This upwelling requires a diapycnal diffusivity of 2 to 10 cm^2/sec, which is within the range expected for the global ocean.

11.8. WATER MASSES

We describe water masses from top to bottom in four general layers: upper ocean and thermocline/pycnocline, intermediate layer, deep layer, and bottom layer. These are illustrated in the vertical sections at 33°S, 9°N, 60°E, and 95°E in Figures 11.16, 11.13, 11.19 and 4.13 and 4.22, respectively. Principal water masses of the Indian Ocean are given in Table S11.3 in the online textbook supplement; these generally follow the nomenclature for the Atlantic and Pacific Oceans. A schematic potential temperature-salinity (T-S) diagram with water masses labeled is shown in Figure 11.17. Full water column T-S and potential temperature-oxygen diagrams based on WOCE data are also shown, in Figure 11.18.

11.8.1. Upper Ocean

Surface waters of the Indian Ocean have the nearly zonal distribution of temperature that is typical of all oceans (Figures 4.1 through 4.6). The tropical Indian Ocean represents a westward extension of the Pacific's warm pool. Cooler SSTs are found in the western Indian Ocean, likely due to northward advection in the NEMC and EACC (Section 11.3). There is a notable absence of a cold tongue in the eastern tropics; this differs from the Atlantic and Pacific and is due to the lack of persistent trade winds at the equator in the Indian Ocean.

The western tropical Pacific and the northern Indian Ocean have the warmest SSTs on the globe, together constituting the tropical ocean's warmest pool, remaining between 26 and 30°C. However, the tropical Pacific and Indian Ocean surface heat budgets are entirely different. As argued by Loschnigg and Webster (2000), because the tropical Indian Ocean is nearly cloud-free while the western tropical Pacific is shielded by clouds, the tropical Indian experiences large net heating into northern summer compared with the Pacific (75–100 W m^{-2} compared with 10–20 W m^{-2}). Pacific SST regulation is likely a combination of local balances including cloud feedbacks and atmospheric circulation (Ramanathan & Collins, 1991; Wallace, 1992). The Indian Ocean equilibrium, on the other hand, must be maintained by cross-equatorial ocean heat transports, accomplished by very shallow meridional overturn dominated by the summer monsoon; however, it is unclear whether this overturning cell has significant heat transport (Schott et al., 2002).

Surface salinity in the Southern Hemisphere Indian Ocean includes the usual subtropical salinity maximum due to net evaporation (Figure 4.15). The maximum surface salinity is not as high as in the South Pacific or South Atlantic, and is centered somewhat farther south. In the north, the Arabian Sea and Bay of Bengal have opposite surface salinity characteristics. The Arabian Sea has high surface salinity, up to 36.5 psu, due to evaporation, while in the Bay of Bengal the salinity decreases from about 34 psu at about 5°N to 31 psu or less in the north. The low values in the Bay of Bengal are due to the considerable river runoff. The band of low salinity at about 10°S in the SEC is due to both net precipitation of the Intertropical Convergence Zone and to the ITF, carrying Pacific waters westward (Section 11.5).

The contrast between highly saline *Arabian Sea surface waters* and much fresher *Bay of Bengal surface waters* is apparent in the vertical section of salinity at 9°N (Figure 11.13). Both are best developed in the upper 150 m. Surface temperatures are high in both regions, so the surface density in the Bay of Bengal is much lower than in the Arabian Sea because of the fresh water. Some of this low salinity water is carried past India, especially when the Northeast Monsoon Current is flowing westward. A hint of this low salinity is apparent in Figure 11.13 west of the Indian landmass, at 75°E.

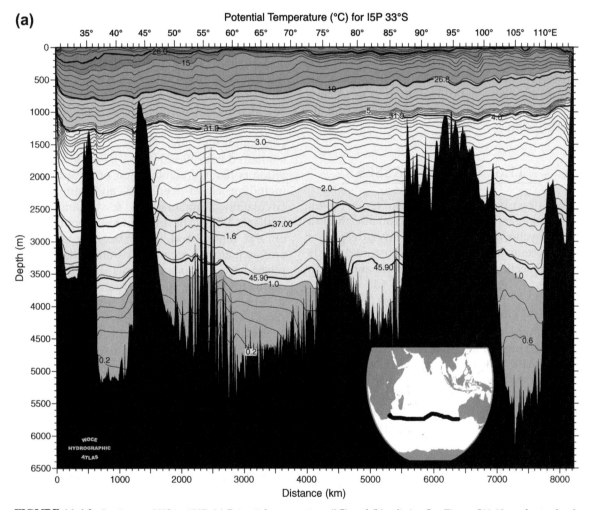

(a)

FIGURE 11.16 Sections at 33°S in 1987. (a) Potential temperature (°C) and (b) salinity. See Figure S11.10 on the textbook Web site for the corresponding oxygen section. Selected isopycnals used for maps in other figures are overlain (bold). Station locations are on the inset maps. *Source: From WOCE Indian Ocean Atlas, Talley (2011); see also Toole & Warren (1993).*

(b)

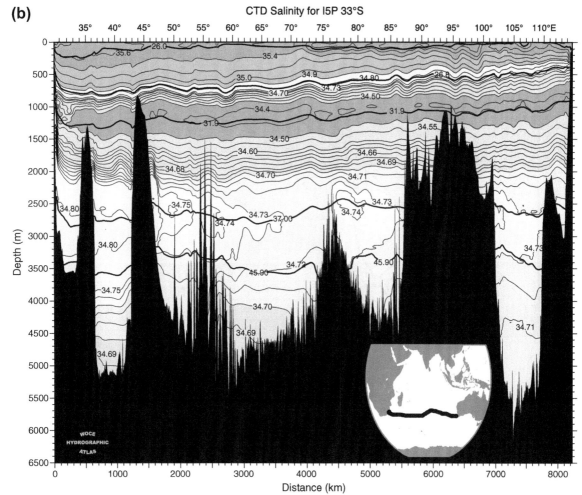

FIGURE 11.16 (*Continued*).

High salinity water (34.9– 35.5 psu) is found in the tropics below the surface layer and north of the SEC front at 10°S. The deeper part of this high salinity layer is referred to as Red Sea Overflow Water because of its high salinity source, while the shallower part attains its high salinity from the Persian Gulf (Beal et al., 2000). The high salinity layer is also found in the Bay of Bengal beneath the thin, fresh surface layer. Although the highest salinity waters are confined north of the SEC, a deep, diluted high salinity layer extends farther south beneath the subtropical gyre in the south Indian Ocean (Figures 11.19 and 4.13b). At 95°E, its core as marked by the 34.73 psu contour is at 2000 m, although it still is not found much farther south than about 15°S. This deep layer is discussed further in the following paragraph.

The upper ocean water mass that marks the division between the tropics and the subtropics,

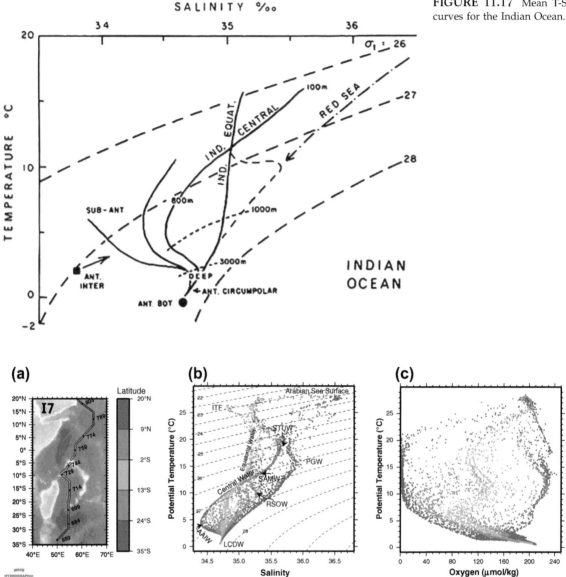

FIGURE 11.17 Mean T-S curves for the Indian Ocean.

FIGURE 11.18 (a) Station locations, (b) potential temperature (°C) — salinity and (c) potential temperature (°C) — oxygen (μmol/kg) for the Indian Ocean along 60°E. This figure can also be seen in the color insert. *After the WOCE Indian Ocean Atlas, Talley (2011).*

at the SEC front at ~10°S, is the *Indonesian Throughflow Water* (Section 11.5). At intermediate depths, the extension of this is the *Indonesian Intermediate Water* (IIW). Both are salinity minima in the north-south direction. The IIW is also a vertical salinity minimum, distinct from the AAIW of the subtropical gyre to its south.

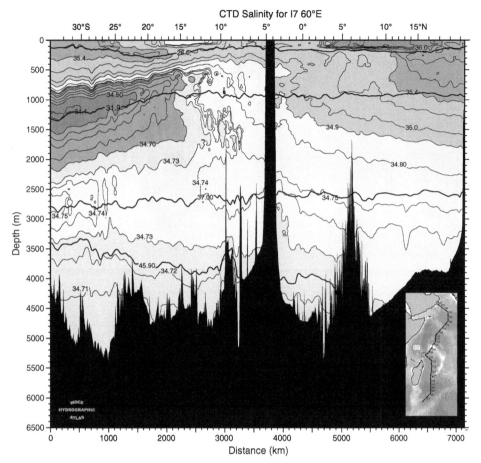

FIGURE 11.19 Salinity along 60°E, with potential density contours used for salinity maps. Station track is overlaid. *After the WOCE Indian Ocean Atlas, Talley (2011).*

In the south Indian subtropical gyre, which extends from the SEC front at 10–12°S southward to the Subantarctic Front (SAF) at about 45°S, the thermocline/pycnocline is typical of all subtropical gyres. Subtropical gyres are ventilated primarily by subduction from the sea surface. The two water masses associated with the subduction, and that have distinctive T-S signatures, are the *Central Water*, which is the main T-S relation of the pycnocline, and the *Subtropical Underwater* (STUW), which is the shallow salinity maximum layer at about

$\sigma_\theta = 26.0 \text{ kg/m}^3$ in the upper part of the Central Water. These water masses are marked in the T-S diagrams (Figures 11.17 and 11.18). The surface water in the center of the gyre at the top of the Central Water is saline, as previously described. The base of the pycnocline (Central Water) can be taken roughly to be the salinity minimum of the AAIW, which lies a little deeper than 1000 m at 40°S and rises up to about 500 m at the SEC front. The STUW is the shallow (<300 m deep), subsurface high salinity layer that extends toward the equator from the center of

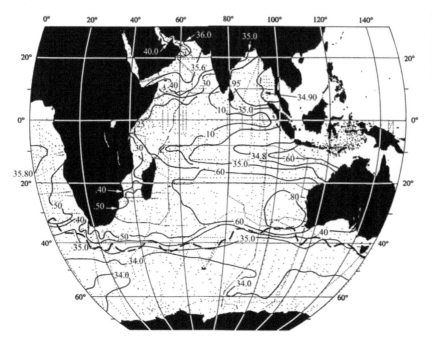

FIGURE 11.20 Salinity at $\sigma_\theta = 26.0$ kg/m³, at a depth of 150–200 m through most of the Indian Ocean. *Source: From Reid (2003).*

the gyre. It is found only north of 25°S, that is, north of the highest surface salinity.

Salinity on the isopycnal $\sigma_\theta = 26.0$ kg/m³ (Figure 11.20) illustrates the high salinity STUW, particularly its high salinity source in the eastern Indian Ocean near Australia. The isopycnal also illustrates the low salinity of the ITF as it crosses the Indian Ocean around 10°S, and the huge difference in salinity between the Arabian Sea and Bay of Bengal.

There are two major mode waters in the upper ocean in the south Indian subtropical gyre. These have no signature in salinity, but are easily identifiable in potential vorticity (inverse isopycnal layer thickness), since they result from thick surface mixed layers in winter. A potential vorticity section at 33°S (Figure 11.21a) illustrates these two water masses, the Indian Ocean *Subtropical Mode Water* (STMW) of the Agulhas and SAMW. The STMW is the weak potential vorticity minimum in the far west around 26.0 kg/m³ (Toole & Warren, 1993; Fine, 1993). SAMW is the major

potential vorticity minimum layer across the whole section, at $\sigma_\theta = 26.5$ to 26.8 kg/m³. The Indian Ocean STMW forms as a thick layer north of the Agulhas Return Current (Figures 11.1 and 11.8). Its potential temperature, salinity and potential density are 17–18°C, 35.6 psu, and $\sigma_\theta = 26.0$ kg/m³. Like the STMW of the EAC, the Indian Ocean STMW is weak compared with the STMWs of the Gulf Stream and Kuroshio, possibly because the air–sea heat loss in the Agulhas region is much lower than in either of the Northern Hemisphere western boundary currents. On the isopycnal map in Figure 11.20, STMW has no signature — it is not a water mass that is identified by a salinity extremum.

Indian Ocean SAMW is a much stronger and much more pervasive mode water than the Indian Ocean's STMW. SAMW forms all along the SAF from the western to the eastern Indian Ocean (McCartney, 1982). In the west (west of 60°E), it has a potential density of about $\sigma_\theta = 26.5$ kg/m³ (14°C, 35.4 psu) and occupies the recirculation region of the Agulhas Return Current and SAF. SAMW is even stronger in the southeast Indian Ocean. This *Southeast Indian Subantarctic Mode Water* (SEISAMW) is the strongest of all of the global SAMWs: mixed layers in the southeast Indian sector of the Southern Ocean are thicker than anywhere else, reaching 700 m in winter (Figure 4.4). The potential temperature, salinity, and potential density of new SEISAMWs are 8–9°C, 34.55

psu, $\sigma_\theta = 26.8-26.9$ kg/m^3 (Hanawa & Talley, 2001).

All of the Indian Ocean SAMWs subduct northward into the subtropical gyre. SEISAMW is the densest water that is directly ventilated in the subtropical gyre and forms the base of the pycnocline and Central Water. It has no extremum in salinity; the underlying AAIW, originating from the Malvinas-Brazil Current confluence in the southwest Atlantic, has lower

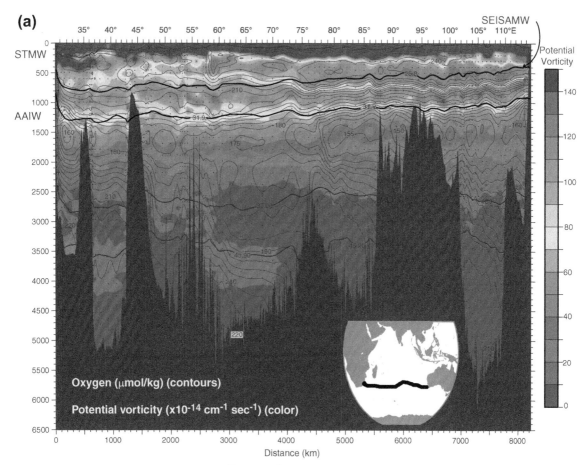

FIGURE 11.21 (a) Potential vorticity $[10^{-14}$ (cm s)$^{-1}]$ (shading), oxygen (light contours), selected isopycnals (dark contours) at 33°S in the Indian Ocean (see supplemental Figures S11.9 and S11.10 on the textbook Web site for additional oxygen and isopycnal contouring). The STMW and SEISAMW potential vorticity minima are labeled, as is the AAIW (salinity minimum). (b) Potential vorticity $[10^{-14}$ (cm s)$^{-1}]$ on the neutral density surface $\gamma^n = 26.88$ kg/m^3, equivalent to $\sigma_\theta = 26.8$ kg/m^3, representative of SEISAMW. *Source: From McCarthy and Talley (1999).*

(b)

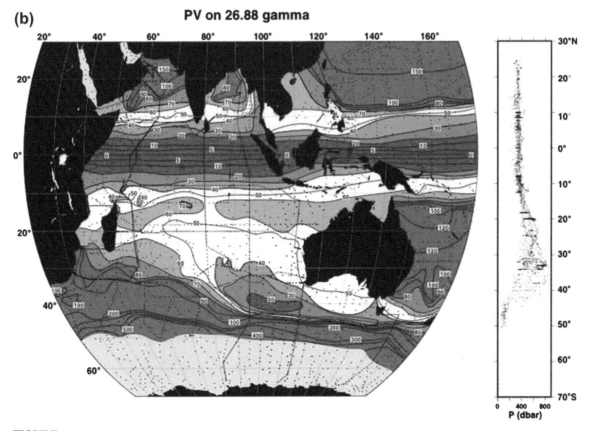

FIGURE 11.21   (*Continued*).

salinity. Potential vorticity on an isopycnal surface representative of SEISAMW shows the formation and gyre subduction regions (Figure 11.21b): low potential vorticity indicates thick layers in the south and central Indian Ocean, with an extension of these low values northward into the subtropical gyre, ending at the southern side of the SEC around 18°S.

As a thick, well-ventilated layer, the SEI-SAMW carries high oxygen waters as well as its thickness. In Figures 4.13d and 11.21a (also supplementary Figure S11.10), it is visible as the high oxygen layer centered at about 500 m, coincident with low potential vorticity. This high oxygen extends all the way northward to the SEC at about 12°S. In the western Indian Ocean, a slight oxygen maximum associated with SEISAMW can be traced along the western boundary all the way into the Arabian Sea.

11.8.2. Intermediate Waters

The two low salinity intermediate waters in the Indian Ocean are AAIW and IIW. IIW has already been discussed in reference to the ITF in Section 11.5 and is mentioned here just for completeness. RSOW is a high salinity intermediate water with its salinity maximum core in the same density range as AAIW, and it was partially described above in Section 11.6. Salinity on an isopycnal at the AAIW and RSOW cores is shown to illustrate the spread

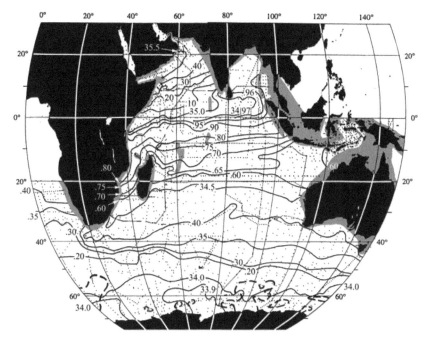

FIGURE 11.22 Salinity at $\sigma_1 = 31.87$ kg/m^3 (equivalent to $\sigma_\theta = 27.3$ to 27.4 kg/m^3), at a depth of 900−1200 m in most of the Indian Ocean. *Source: From Reid (2003).*

of both water masses (Figure 11.22). IIW also affects the same isopycnal.

AAIW is a global, Southern Hemisphere water mass characterized by a salinity minimum in the vertical at densities of $\sigma_\theta = 27.0$−27.3 kg/m^3 and at about 500−1000 m depth (Figure 14.13). Within the Indian Ocean, AAIW can be readily recognized as the low salinity layer (salinity minimum) below the thermocline throughout the subtropical Indian Ocean south of the SEC at about 12°S. The greatly eroded salinity minimum extends into the tropics along the western boundary (in the EACC and Somali Current) and is found along the equator and into the western Arabian Sea. The main part of the AAIW, in the subtropical gyre, is at about 1100 m just north of the SAF and shoals with the subtropical gyre's isopycnals to about 500 m at about 15°S. Its salinity minimum core south of 25°S has a mean potential temperature, salinity, and potential density of 4.7°C, 34.39 psu, $\sigma_\theta = 27.2$ kg/m^3 ($\sigma_1 = 31.8$ kg/m^3).

AAIW in the Indian Ocean comes from the southwestern Atlantic Ocean, where the cold, fresh waters of the Malvinas (Falkland) Current loop far to the north and encounter the subtropical waters of the South Atlantic. The southeast Pacific's SAMW and AAIW carried in this current are then submerged beneath the new South Atlantic SAMWs, producing a different type of AAIW that is denser and of higher potential vorticity and lower oxygen than the southeast Pacific AAIW. This Atlantic AAIW fills the Atlantic and Indian Ocean subtropical gyres. That AAIW in the Indian Ocean does not originate there, which is apparent from global salinity, oxygen, and potential vorticity maps on the AAIW isopycnals; the lowest salinity, highest oxygen, and lowest potential vorticity are from the Malvinas Current region. The southeastern Indian Ocean AAIW has much higher potential vorticity than the Pacific and Atlantic AAIWs, indicating that the Indian Ocean AAIW is the most eroded and hence most distant from its surface source (Figure 11.21a; Talley, 1996).

AAIW from the south freely circulates up to about 20°S. North of this there is a large gradient in depth and properties with the salinity minimum shoaling and eroded to lower density and higher salinity to the north, displaced by IIW and RSOW. This shift is clear in the vertical section at 60°E (Figure 11.19). This latitude is the northern boundary of the subtropical gyre at this depth and density, and is readily apparent in the circulation maps at 800 and 900 m (Figure 11.7). A potential vorticity map for AAIW shows an especially striking subtropical-tropical boundary with well-mixed potential vorticity within the subtropical gyre and nearly zonal contours north of the boundary (McCarthy & Talley, 1999).

In terms of global meridional overturn, the AAIW layer in the Indian Ocean paradoxically has southward transport, although its low salinities are advected northward around the subtropical gyre. However, there is more volume transport upwelling from the deep water into the AAIW layer and moving south than there is actual AAIW moving north (Figure 11.15 and Figure S11.9 on the textbook Web site).

RSOW (or Red Sea Water, depending on the author) is the salinity maximum core at $\sigma_\theta = 27.2-27.4$ kg/m^3 in the Arabian Sea and western Indian Ocean (Figures 11.13, 11.19, and map in 11.22). RSOW results from overflow of 0.4 Sv of highly saline Red Sea Water that has a density of $\sigma_\theta = 27.6$ kg/m^3 as it flows over the sill at Bab el Mandeb into the Gulf of Aden (Section 11.6). High salinity fills the Arabian Sea on the RSOW isopycnal, spreading eastward to the eastern boundary at 5°N and southward along the western boundary toward the Agulhas (Beal et al., 2000).

The high salinity within the Arabian Sea extends downward across isopycnals to much greater depths than the RSOW salinity maximum. CFCs are present in the RSOW depth range, but are essentially absent below 1500 m

in the Arabian Sea, indicating that whatever process diffuses high salinity downward is slow (Mecking & Warner, 1999). This deeper high salinity is described in the next section.

11.8.3. Deep and Bottom Waters

There are no surface sources of deep or bottom water in the Indian Ocean, even though the densities of new Red Sea Water and Persian Gulf Water are high enough to match the bottom density. Both overflows have small volumes and mix and settle out at intermediate and shallow depths, respectively. Based on mass budgets, the deepest Indian Ocean waters upwell to the deep, intermediate, and thermocline layers. Therefore, water parcels in the deepest layers come from the ocean surface in the Atlantic and Southern Oceans. The water mass entering the Indian Ocean from the south is the CDW. In the western Indian Ocean, NADW also enters directly from the South Atlantic without passing through the ACC.

Despite the lack of Indian Ocean surface ventilation for the deep and bottom waters, we distinguish a deep water of Indian Ocean origin (the *Indian Deep Water*; IDW). This is deep water that is "formed" within the Indian Ocean by diffusion and upwelling rather than by surface ventilation. Its low oxygen and high nutrient content reflect high age as it advects back to the Southern Ocean. Here IDW joins the fresher Pacific Deep Water, which is also marked by low oxygen and high nutrients, and together they upwell to the surface in the Southern Ocean as Upper Circumpolar Deep Water. The circuit of the bottom and deep waters through the Indian (and Pacific) Ocean is thus an important part of the global overturning circulation.

On any given isopycnal, or at any given depth in the deep Indian Ocean, we might find both CDW and IDW, so distinguishing between them is a highly regional exercise. One way to distinguish between the deep and bottom layers is in terms of net meridional transport (Section

11.7; Figure 11.15 and Figure S11.9 on the textbook Web site). Waters below about 2000 m depth ($\sigma_2 \sim 37.0$ kg/m^3 or neutral density 27.96 kg/m^3) have net northward transport, and waters above have net southward transport. We could, for instance, consider the southward and northward layers to be the deep and bottom layers, respectively. However, this masks important modification in the "bottom" layer, and much of what we define as "Indian Deep Water" would then occur in the bottom layer.

In terms of water masses, we will consider the deep waters to be the layer containing the high salinity core of CDW/NADW and a high salinity core of IDW, and the bottom waters to be the colder, fresher bottom layer. This latter is also CDW, and is referred to as such in most water mass descriptions of the Indian Ocean. Here, as for the Pacific and Southern Ocean descriptions (Chapters 10 and 13), we call these deepest waters *Lower Circumpolar Deep Water* (also known as Antarctic Bottom Water).

In the deep water layer, there are both southern and northern source salinity maxima (e.g., 2500–3000 m depth in the salinity section at 60°E in Figure 11.19). These are (1) CDW, with the high salinity of NADW, found south of 25°S and (2) IDW, from the north, in which the elevated salinity is created by downward diffusion that accompanies the deep upwelling in the northwest Indian Ocean (Arabian Sea). These two saline deep waters affect a representative isopycnal ($\sigma_2 = 37.0$ kg/m^3 in Figure 11.23). The Arabian Sea's high salinity is clearly separated from the CDW/NADW. The southern CDW/NADW salinity maximum has high oxygen and low silica as well (Reid, 2003; WOCE Indian Ocean Atlas in Talley, 2010), and potential vorticity also transitions abruptly at about 25°S (McCarthy & Talley, 1999). The separation between the southern CDW and northern IDW high salinity layers is even more marked in the eastern Indian Ocean (Figure 4.13b).

The bottom layer of the Indian Ocean has net northward transport. The bottom waters are greatly modified as they circulate northward into the Indian Ocean as a result of diapycnal mixing, acquisition of silica from the bottom sediments, and aging that reduces oxygen and increases nutrients. The amount of activity of each of these depends strongly on the deep basin. As a result, it is not useful to distinguish between CDW and IDW by depth or density ranges unless looking carefully at a specific region.

The principal bottom water mass is LCDW, also called Antarctic Bottom Water in Southern Ocean and global contexts (Section 13.5). LCDW is formed as dense water around Antarctica, although the variety that extends northward into the Indian Ocean is not the densest Antarctic water. The northward circulation pathways of LCDW, including DWBCs, are described in Section 11.7.

At 33°S, this deep, cold, fresh, dense, high oxygen water mass (<1°C, <34.71 psu, $\sigma_4 > 45.96$ kg/m^3, >210 μmol/kg) is found in the deep basins that connect to the Southern Ocean (Agulhas region, and the Mozambique, Crozet, and Perth Basins). The densest, coldest waters are not present in the Madagascar and Central Indian Basins since they are not open to the south. Bottom waters that make it to the Arabian Basin in the northwest and Bay of Bengal in the northeast have densities of $\sigma_4 > 45.88$ kg/m^3 and 45.94 kg/m^3, respectively, and their potential temperatures are 1.4°C and 0.8°C, respectively. The Central Indian Basin is connected to the southern source waters via the West Australia Basin through several gaps in the Ninetyeast Ridge; therefore its bottom waters are warmer and less dense (1.0°C, 45.92 kg/m^3) than in the West Australia Basin.

LCDW upwells into the IDW. Observations of its transformation and overturning transport calculations using WOCE data can be found in several sources (Johnson et al., 1998; Warren & Johnson, 2002). Of the 12 Sv or so that upwell out of the bottom layer (Section 11.7), about 4

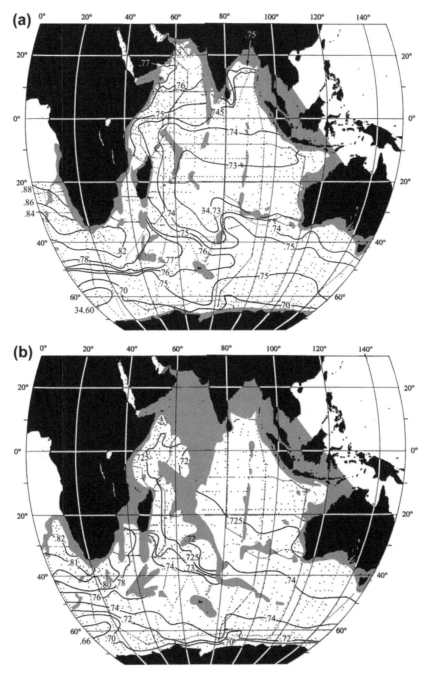

FIGURE 11.23 Salinity maps. (a) At $\sigma_2 = 37.0$ kg/m^3, at about 2600 m depth, representative of the deep waters. (b) At $\sigma_4 = 45.89$ kg/m^3, at about 3500 m depth, representative of the bottom waters. *Source: From Reid (2003).*

Sv progress northward in the westernmost Indian Ocean basin (Mascarene Basin) and less than 2 Sv make it through Amirante Passage into the Somali Basin. All of this upwells. By implication, the remaining ~8 Sv proceeds into the central and eastern Indian Ocean. Of this, 2 Sv crosses into the Central Indian Basin from the West Australian Basin and upwells. In contrast, southward transport in the western Indian Ocean of deep waters, including IDW, appears to account for almost all of the upwelled water from all of the Indian Ocean.

11.9. CLIMATE AND THE INDIAN OCEAN

Climate variability at interannual to decadal timescales has been documented in the Indian Ocean. Because of its importance to agriculture, interannual and longer term variability in the monsoon has been of special interest. Although the air–sea coupling process that creates ENSO is centered in the tropical Pacific, ENSO dominates interannual climate variability in the Indian Ocean. Beyond its response to ENSO, the tropical Indian Ocean has internal interannual variability. A tropical *Indian Ocean dipole mode* has been described, whose simplest index is the east-west difference in tropical SST. In the Southern Hemisphere, the Indian Ocean is affected by the decadal Southern Annular Mode (Antarctic Oscillation).

The text, figures, and tables relating to climate variability are included in Chapter S15 (Climate Variability and the Oceans) on the textbook Web site. It covers the following modes of climate variability that most directly affect the Indian Ocean: ENSO effects in the Indian Ocean, the Indian Ocean dipole mode, the Southern Annular Mode, and climate change (trends in temperature, salinity and circulation).

Arctic Ocean and Nordic Seas

12.1. INTRODUCTION

The Arctic Ocean is a mediterranean sea surrounded by the North American, European, and Asian continents (Figures 2.11 and 12.1). It is connected to the Atlantic Ocean on both sides of Greenland and to the Pacific Ocean through the shallow Bering Strait. The *Nordic Seas* is the region south of Svalbard and north of Iceland. This region is central for transformation and production of some of the densest waters in the global ocean, creating the densest part of the North Atlantic Deep Water (Chapter 9), and is a high latitude connection of the fresher North Pacific waters to the saltier North Atlantic waters. The Arctic's sea ice cover is a vital component of global climate because of its high albedo (high solar reflectivity; Section 5.4). The Arctic's sea ice cover is sensitive to climate change. Because of important climate changes and initiation of difficult hydrographic time series in this ice-covered region beginning in the 1990s, there is a large and growing body of information about circulation, water masses, and ice cover in the Arctic. In addition to numerous journal publications, we note the volume edited by Hurdle (1986) in which the useful term "Nordic Seas" was first introduced, a recent compendium from the Arctic-Subarctic Ocean Fluxes study (Dickson, Meincke, & Rhines, 2008), and an upcoming volume from the Arctic Climate System Study (ACSYS;

Lemke, Fichefet, & Dick, in preparation). Rudels' (2001) review is a good overview of the materials presented in this chapter.

The Arctic Ocean is divided into the Canadian Basin (depth about 3800 m), and the Eurasian Basin (depth about 4200 m; Section 2.11 and Figure 2.11). These basins are separated by the Lomonosov Ridge, which extends from Greenland past the North Pole to Siberia. The maximum sill depth is about 1870 m (Björk et al., 2007). The Eurasian Basin is subdivided into the Nansen and Amundsen Basins; the Canadian Basin is subdivided into the Makarov and Canada Basins. Broad continental shelves of 50 to 100 m depth characterize the Arctic margin north of Eurasia and the Alaskan coast, occupying about 53% of the area of the Arctic Ocean (north of Fram Strait) but containing less than 2% of the total volume of water (Jakobsson, 2002).

The deepest connection of the Arctic Ocean with the other oceans is to the Nordic Seas through Fram Strait, which lies between Greenland and Spitsbergen with a sill depth of 2600 m (Section 12.2). The sill depth north and east of Svalbard, separating it from Franz Josef Land and Novaya Zemlya, is only about 200 m (Coachman & Aagaard, 1974). The Bering Strait connection to the Bering Sea and the Pacific Ocean is narrow and has a sill depth of only 45 m, but the transport, especially its freshwater content, from the Pacific into the Arctic is significant, on the order of 1 Sv. There are also connections from the Arctic to the North Atlantic through the Canadian Archipelago by several channels, principally Nares Strait (sill depth 250 m) and Lancaster Sound (sill depth 130 m), which lead to Baffin Bay and then to the Atlantic.

The Nordic Seas, between Fram Strait and the Greenland-Scotland ridge, include the Norwegian, Greenland, and Iceland Seas. These commonly used oceanographic names are loosely linked to the formal topographic names (e.g., Perry, 1986). The Greenland-Scotland ridge is comprised of three main sections (Hansen & Østerhus, 2000): Denmark Strait between Greenland and Iceland (sill depth 620 m), the Iceland-Faroe Ridge (sill depth 480 m), and the Faroe-Shetland ridge (sill depth of 840 m in Faroe Bank Channel). Within the Nordic Seas, the Greenland Sea is separated from the Norwegian Sea by Mohns Ridge, and from the Iceland Sea by the Jan Mayen fracture zone. The Norwegian and Iceland Seas are separated by Aegir Ridge. Each of these seas has a somewhat separate circulation and water mass structure, which is discussed in Section 12.2. West of Greenland, the Arctic and Atlantic connect through Baffin Bay and Davis Strait and through (or past) Hudson Bay. These regions are discussed in Section 12.3. The remainder of this chapter is devoted to the Arctic Ocean circulation (Section 12.4), water mass structure (Section 12.5), sea ice (Section 12.7), and climate variability (Section S15.4 in Chapter S15 on the textbook Web site http://booksite.academic press.com/DPO/; "S" denotes supplemental material).

The surface circulation is shown schematically in Figure 12.1, which is referred to throughout this chapter. An overall schematic of the surface circulation and water mass formation in the Arctic and Nordic Seas is shown in Figure 12.2, relevant to conditions in previous decades when deep water was still actively forming in the Greenland Sea. The schematic is

still useful even though Nordic Seas convection is currently to intermediate depth only. Both Figures 12.1 and 12.2 show inflows from the Atlantic and Pacific and surface outflow back to the Atlantic. The Arctic surface circulation is divided into principally cyclonic circulation in the Nordic Seas and Eurasian Basin, and principally anticyclonic circulation in the Canadian Basin (Beaufort Gyre). The Transpolar Drift (TPD) is the major cross-polar circulation between these two systems. Figure 12.2 also shows the overturn by open ocean convection in the Nordic Seas and by shelf brine rejection in the Arctic, and denser outflow back into the North Atlantic.

12.2. THE NORDIC SEAS

The Nordic Seas are comprised of the Greenland and Norwegian Seas, the Iceland Basin between Iceland and Jan Mayen, and the Boreas Basin between Greenland and Svalbard. The densest water renewal in the Northern Hemisphere is in the Greenland Sea. The Greenland Sea produces denser waters than the Arctic because it is closer to the high salinity inflow from the Atlantic Ocean; winter cooling of this more saline water produces denser waters than in the fresher Arctic. Dense Arctic waters also flow into the Greenland Sea and are an important part of the mixture of waters that ultimately overflows the sills into the North Atlantic (Aagaard, Swift, & Carmack, 1985).

The Nordic Seas waters that overflow the Greenland-Scotland ridge to become the dense core of North Atlantic Deep Water (NADW) are not the deep waters of the Nordic Seas, which lie below the sill depth. Therefore, the issue of whether Nordic Seas deep water renewal extends to the ocean bottom, which has not occurred since the 1980s, is not as important for NADW formation as the processes that determine properties at sill depth. These also include the properties of the deepest water,

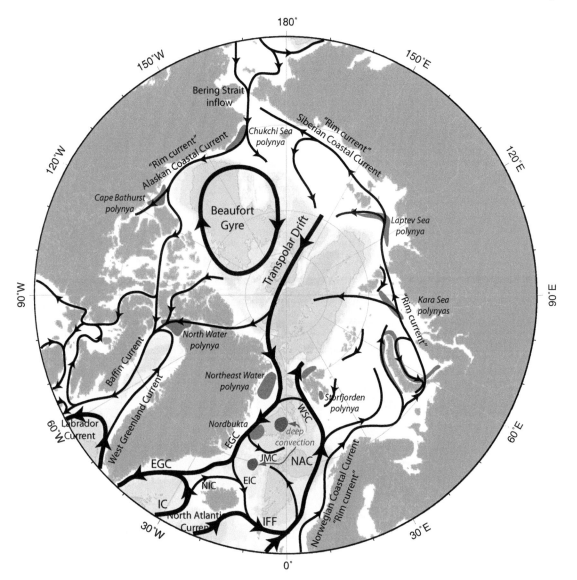

FIGURE 12.1 Schematic surface circulation of the Arctic and Nordic Seas, including some of the major polynyas (gray shading) and the Greenland Sea and Iceland Sea deep convection sites (dark gray). Topography as in Figure 2.11, where place names can be found. Heavy lines indicate the principal circulation components, generally with larger transports than those depicted with finer lines. Acronyms: EGC, East Greenland Current; EIC, East Iceland Current; IC, Irminger Current; IFF, Iceland-Faroe Front; JMC, Jan Mayen Current; NAC, Norwegian Atlantic Current; and NIC, North Irminger Current. *(After Rudels, 2001; Loeng et al., 2005; Rudels et al., 2010; Østerhus & Gammelsrød, 1999; and Straneo & Saucier, 2008. & Polynya locations from IAPP (2010) and Martin (2001)).*

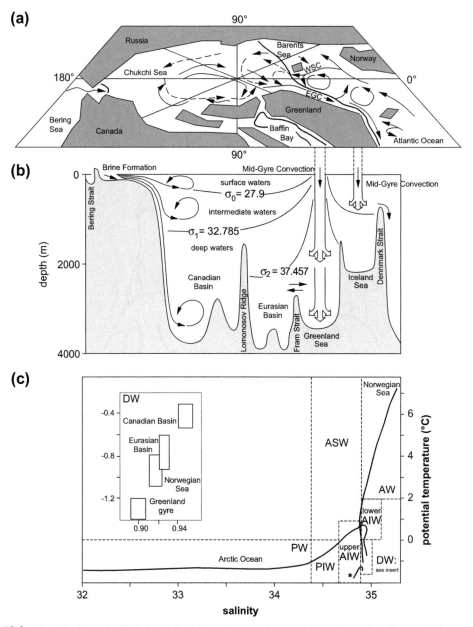

FIGURE 12.2 Overall schematic of (a) circulation, (b) water mass layers and transformation sites, and (c) water masses in potential temperature-salinity. Deep convection in the Greenland Sea in (b) has been replaced by mid-depth convection since the 1980s. Acronyms in (a): EGC, East Greenland Current; WSC, West Spitsbergen Current. Acronyms in (c): AW, Atlantic Water; AIW, Arctic Intermediate Water; ASW, Arctic Surface Water; DW, Deep Water; PIW, Polar Intermediate Water; PW, Polar Water. *Source: From Aagaard, Swift, & Carmack (1985); amended by Schlichtholz and Houssais (2002).*

since they affect the overall stratification of the Nordic Seas.

In the next subsections, circulation, water masses, and deep-water formation are briefly described.

12.2.1. Nordic Seas Circulation

The overall circulation of the Nordic Seas is cyclonic (Figure 12.1 and Figure S12.1 on the textbook Web site). Exchange with the North Atlantic is in the upper ocean, above the ridges that stretch between Greenland and Scotland. Warm, saline waters from the North Atlantic enter in the east, in the *Norwegian Atlantic Current*, which is a continuation of part of the North Atlantic Current (Chapter 9). The North Atlantic Current enters the Norwegian Sea in two branches: an eastern (near-coastal) branch along the coast of Ireland that reaches and passes over the Wyville-Thompson Ridge between the Shetland and Faroe Islands ("Faroe-Shetland Ridge" in Figure 2.11), and a western (mid-ocean) branch that reaches the east coast of Iceland, then turns eastward along the Iceland-Faroe Ridge where it forms a strong current/front, and finally joins the Norwegian Atlantic Current.

The southward-flowing western boundary current of the Nordic Seas is the *East Greenland Current* (EGC). The EGC enters the Nordic Seas from the Arctic through Fram Strait. (This is the main export route for sea ice from the Arctic Ocean.) At about 72°N, part of the EGC continues southward along the coast of Greenland and part splits off to the east into the *Jan Mayen Current*. This bifurcation is likely due to the bathymetry, which steers circulation throughout the water column. The Jan Mayen Current is important for dense water formation in the Greenland Sea. The speeds in the Norwegian Atlantic Current and EGC are up to 30 cm/sec, but the average is more like 20 cm/sec.

The Norwegian Atlantic Current flows northward along the coast of Norway to the Arctic Ocean. It includes a separate coastal current, called the Norwegian Atlantic Coastal Current. As it rounds the northern side of Norway, a branch of the Norwegian Atlantic Current splits off to the east into the Barents Sea, following the coast. The rest of the Norwegian Atlantic Current continues toward Spitsbergen/Svalbard and splits again, a portion flowing northward through Fram Strait as the *West Spitsbergen Current* and the remainder turning southward and joining the EGC. Upper ocean flow from the Arctic into the Nordic Seas occurs through Fram Strait as the EGC. Within the Nordic Seas, there are several gyral circulations, each associated with topographic features that split the boundary currents.

Subsurface waters exit the Nordic Seas southward as overflows over each of the three sills between Greenland and Scotland. The water masses that dominate in the outflows depend on the sill depths, with intermediate water exiting at Denmark Strait and the Iceland-Faroe Ridge, and the densest overflow waters (but still at intermediate depth) through the deeper Faroe Bank Channel. Deep water from the Arctic Ocean also enters the Nordic Seas through Fram Strait (2500 m depth).

12.2.2. Nordic Seas Water Masses

The water masses of the Nordic Seas are complicated and changing in time (Section S15.4 on the textbook Web site) because of the local nature of intermediate to deep convection responding to variations in local air-sea fluxes, and because of the Nordic Seas' location between the northern North Atlantic and Arctic Oceans, both with variable surface waters. Here we describe the dominant Nordic Seas water masses, following Aagaard et al. (1985), Rudels (2001), and Jones (2001). These include two surface waters, three intermediate waters, and three deep waters (listed in Table S12.1 in the textbook Web site).

The two major surface waters are the warm, saline *Atlantic Water* (AW) and the cold, fresh *Polar Surface Water*. AW inflow enters in the Norwegian Atlantic Current (Figure 12.1), at 7 to 9°C and about 35.2 psu. There is a strong pycnocline at about 400 m, which separates the upper layer from the underlying Norwegian Sea Deep Water (described at the end of this section). The AW cools and freshens as it moves northward in the Norwegian Atlantic Current. By the time it reaches Spitsbergen, the surface layer is 1 to 3°C with a salinity of about 35.0 psu. Because the upper layer of the Norwegian Atlantic Current is so warm, the eastern Norwegian Sea is usually ice-free in winter (Figure 12.20a). The warmth of this current is critical for the relatively mild climate of Scandinavia.

The Polar Surface Water is relatively fresh (<34 psu) and close to freezing (<−1.5°C). Polar Surface Water enters the Nordic Seas from the Arctic through Fram Strait in the EGC. In Figure 12.3, this is the very cold, fresh surface layer in the top 200 m on the west side of the section. By the time this water reaches the middle of the Greenland Sea (Figure 12.4 at 73.5°N), the layer is thinner (top 100 m) and warmer. The presence of very cold, relatively fresh surface water, with a strong halocline, is typical of ice-covered regions, which include both the upstream Arctic and also the EGC region locally. The presence of Polar Surface Water results in much colder upper waters in the Greenland Sea than in the Norwegian Sea (Figure 12.3). Within the Greenland gyre, offshore of the EGC, upper ocean temperature and salinity are less stratified, resulting from local convection that mixes the water column to intermediate depths.

We describe three intermediate waters in the Nordic Seas. One is the shallow, subsurface, warm, and saline layer (~150 m, >2°C, 35 psu) in the EGC (on the western side of both panels in Figure 12.4). This is a remnant of AW that has been cooled, densified, and capped

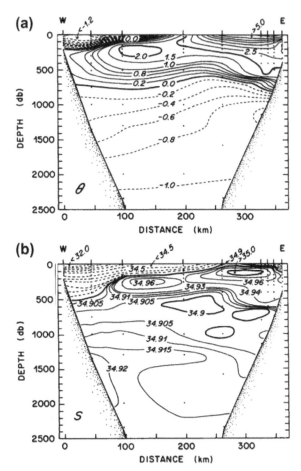

FIGURE 12.3 (a) Potential temperature (°C) and (b) salinity in the Fram Strait in 1980. See Figure 2.11 for location of the strait. *Source: From Mauritzen (1996).*

(covered) at the top by the Polar Surface Water; its sources are both modified AW from the Arctic and recirculation with modification within the Nordic Seas. This remnant is sometimes called the recirculating AW.

A second intermediate water, *Arctic Intermediate Water* (AIW), is the cold, fresher layer (−1.2°C, 34.88 psu), centered at truly intermediate depths (~800 m). Through much of the Nordic Seas, AIW is a salinity minimum layer, lying below the salinity maximum AW. AIW is supplied from the Arctic Ocean through Fram

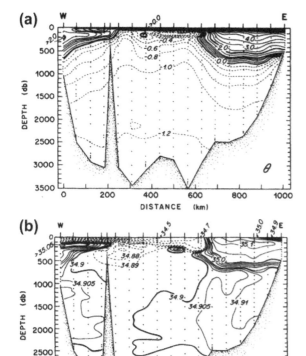

FIGURE 12.4 (a) Potential temperature (°C) and (b) salinity across the southern Greenland Sea at 73.5°N in 1985. *Source: From Mauritzen (1996).*

Strait and is modified by deep convection in the Greenland Sea; production (transformation) of AIW has continued to the present although production of the densest Greenland Sea Deep Water ceased in the early 1990s. In the Greenland gyre where AIW is formed, it is a salinity extremum only during non-winter months when it is capped by a warmer surface layer.

The third intermediate layer in the Nordic Seas is called *upper Polar Deep Water* (uPDW). uPDW enters the Nordic Seas through Fram Strait from the Arctic. In the Nordic Seas, uPDW is found in the EGC, more prominently at Fram Strait than farther south. It is characterized by cold temperatures (0°C declining to −0.5°C) and salinities of 34.85 to 34.9 psu.

At least three distinct deep waters are found in the Nordic Seas: *Greenland Sea Deep Water, Norwegian Sea Deep Water,* and *Arctic Ocean Deep Water.* Greenland Sea Deep Water is the bottom layer colder than −1.2°C and fresher than 34.896 psu in Figure 12.4. It is formed by very intermittent deep convection within the Greenland gyre. Convection also occurs in the Boreas Basin in the northern Greenland Sea, creating dense water similar to Greenland Sea Deep Water. This densest layer has not been formed in recent decades and is shrinking. In the past several decades, convection has been limited to 1700–2000 m depth; the water formed there primarily encompasses AIW.

Arctic Ocean Deep Water is the saline deep water in the Nordic Seas (S > 34.92 psu); its high salinity comes from brine rejection in the shelf seas of the Arctic (Section 12.5). It is composed of deep waters from both the Eurasian and Canadian Basins. It flows southward through Fram Strait into the Nordic Seas as a deep western boundary current. Its core of high salinity hugs the Greenland coast between 1500 and 2000 m (Figure 12.3).

Norwegian Sea Deep Water is a mixture of Arctic Ocean Deep Water and Greenland Sea Deep Water. It does not have a separate convective or brine rejection source. Norwegian Sea Deep Water is also found in the eastern and northern Greenland Sea, where it forms a barrier to the passage of the colder Greenland Sea Deep Water into the Arctic.

12.2.3. Vertical Convection in the Nordic Seas and Dense Water Formation

Historically, the deep-water renewal with highest density in the Northern Hemisphere has been in the Greenland Sea (and its neighboring Boreas Basin). This and the intermediate waters of the Nordic Seas contribute to the densest part of the NADW after they flow over the Greenland-Scotland ridge complex and plunge to the bottom layer of the northern North Atlantic (Chapter 9). (Because of the sill depth, the densest Nordic Seas waters do not cross the ridge.) Dense water renewal is apparent in the high oxygen content of the deep waters of both the Norwegian and Greenland Seas (260–325 μmol/kg or 6–7.5 ml/L), reflecting a short residence time of about 40 years. Formation of deep waters in the Nordic Seas occurs as open ocean convection, which can be either simple mixed layer deepening of the existing waters in winter, or deep penetrative plume convection that pushes through the existing, mid-depth stratification (Ronski & Budéus, 2005a). Dense water formation through brine rejection is not an important factor in the Nordic Seas, unlike the Arctic, likely due to the lack of extensive shallow continental shelves that would allow the water column to become brine-enriched.

From data collected in the first half of the twentieth century, winter cooling resulted in overturning from the surface to the ocean bottom. Deepest convection occurs in the Greenland Sea. However, top-to-bottom convection became very rare after the mid-1980s, so much so that the vertical stratification of the Greenland Sea has changed from a one-layer to a two-layer structure (Ronski & Budéus, 2005b; see Figure S12.2 on the textbook Web site).

Deep vertical convection cells or chimneys (Section 7.10.1) renew the dense Nordic Seas waters in the northern Greenland Sea. (In convection regions, chimneys have scales on the order of 50 km, while convective plumes within the chimneys have scales on the order of 1 km.) There are at least two other convection regions as well, in the Boreas Basin, which is just north of the Greenland Sea, closer to Fram Strait, and in the Iceland Sea (Swift & Aagaard, 1981), which both contribute to the important dense intermediate waters. We concentrate here on the Greenland Sea chimney as it is well defined and observed.

The chimney-formation region of the Greenland Sea is well defined east of the EGC, north of the Jan Mayen Current, and west of Spitsbergen (Clarke, Swift, Reid, & Koltermann, 1990). The deep circulation here is cyclonic and topographically steered, which then partially steers the upper ocean circulation and chimney location. There is often an ice tongue, called the Odden, stretching around the southern part of this cyclonic circulation, along the Jan Mayen Current (Figures 12.1 and 12.5a). The open water inshore of the ice tongue is called the Nordbukta. This general region is referred to as Odden-Nordbukta. The Odden is a region of active ice formation. The Nordbukta is a partial polynya (Section 3.9.6) kept open by deep mixing that brings warmer subsurface water to the surface and by offshore winds; it has characteristics of both latent and sensible heat polynyas, and is not always open (Comiso, Wadhams, Pedersen, & Gersten, 2001). The relationship between the presence of sea ice in the Odden and deep convection is unclear; although one might expect brine rejection to contribute to buoyancy loss and convection, the presence of sea ice might inhibit deep convection, which is the deep water renewal mechanism here.

Formation of deep vertical convection cells or chimneys has been directly observed near the Odden-Nordbukta (Figure 12.5; Morawitz et al., 1996; Wadhams, Holfort, Hansen, & Wilkinson, 2002; Wadhams et al., 2004). Using acoustic tomography (Section S6.6.1 on the textbook Web site) and moored measurements in winter 1988–1989, the development of the winter mixed layer and its temperature were observed. Truly well mixed layers were not seen, most likely because the horizontal resolution of tomography is chimney-scale and not plume-scale, but the deepening of the chimney was clear. The column of near-freezing water extended to almost 1500 m in late March (vernal

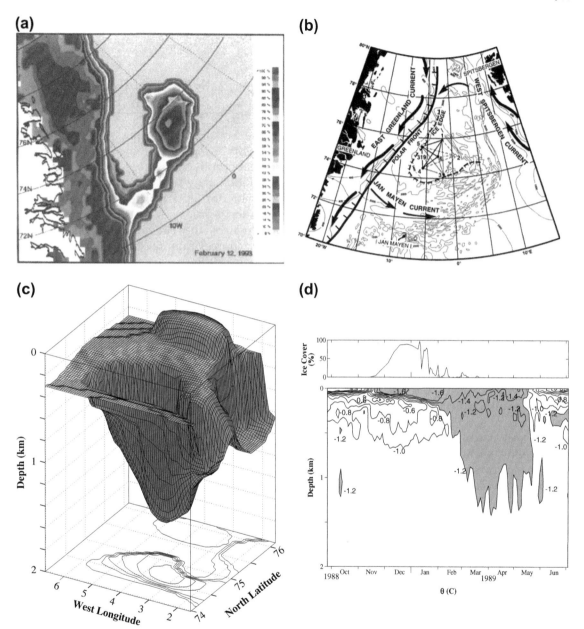

FIGURE 12.5 (a) The Odden ice tongue off the east coast of Greenland, February 12, 1993. *Source: From Wadhams et al. (1996).* (b) Greenland Sea chimney region with 1988–1989 tomographic array location. (c) Mixed layer depth (with contours on bottom plane). Source: From Morawitz et al. (1996). (d) Potential temperature (°C, contour intervals of 0.2°C) time series at the array. *Source: From Morawitz, Cornuelle, and Worcester (1996). Figures b, c, and d are © by the American Meteorological Society. Reprinted with permission.* See also Figure S12.3 in the online supplement.

equinox). More traditional wintertime, ship-based observations in 2001 also showed "deep" convection in the Greenland Sea, to 1800 m, which extended through the temperature minimum layer (1000–1500 m) and into the underlying temperature maximum layer (Wadhams et al., 2002; Figure S12.3 on the textbook Web site).

In both of these experiments, the convection had the two-layer vertical structure of recent decades, without penetration to the ocean bottom, hence not renewing the cold, bottom layer of (now older) Greenland Sea Deep Water (e.g., Ronski & Budéus, 2005b; Figure S12.2 seen on the textbook Web site).

What mechanisms other than deep convection might ventilate the deepest waters in the Greenland Sea? Other possibilities include *double diffusion* (Section 7.4.3.2; Carmack & Aagaard, 1973) and *thermobaricity* (Section 3.5.5) during deep plume convection (Clarke et al., 1990; Ronski & Budéus, 2005a). Double diffusion in the Greenland Sea is of the diffusive variety, with cold, fresh water overlying warmer, saltier water. The thermobaric effect resulting from simply shifting the colder upper ocean water parcels down by several hundred meters into the warmer underlying water could cause overturn sufficient to extend plumes to the bottom because the equation of state is nonlinear (cold water being more compressible than warm).

Dense water production in the northern Nordic Seas is also due to ice formation and brine rejection, specifically in a recurrent, wind-forced (latent heat) polynya in the Storfjorden on the southern side of Svalbard (Haarpaintner, Gascard, & Haugan, 2001). The polynya occurs between fast ice attached to the coast and the offshore pack ice. Brine rejection there enhances the shelf salinity by more than 1 psu. The resulting plume of dense water cascading off the shelf contributes to the deep waters of the Norwegian Sea, and also modifies the AW flowing

northward into the Arctic through the adjacent Fram Strait.

12.3. BAFFIN BAY AND HUDSON BAY

The Labrador Sea, lying west of Greenland and within the geographic North Atlantic, is an important source of intermediate depth ventilation that feeds into the NADW. Since the Labrador Sea is part of the subpolar North Atlantic, its processes are considered in Chapter 9. However, the Labrador Sea has important Arctic sources from the Canadian Archipelago, through Hudson and Baffin Bays, which connect to the Labrador Sea through Hudson and Davis Straits, respectively. Most of the North Pacific input to the Arctic through the Bering Strait reaches the North Atlantic through these bays. Surface flow is in only one direction, from the Arctic to the Labrador Sea; however, there is flow into Baffin Bay and Hudson Bay from the Labrador Sea. The freshwater export through Davis Strait, which includes considerable sea ice, is an important factor in conditions for deep convection in the Labrador Sea; with greater freshwater flux, the Labrador Sea is more likely to be "capped," and not convecting as efficiently.

Hudson Bay (Figure 12.6) is an extensive shallow body of water, averaging only about 90 m in depth, with maximum depths of about 200 m. Hudson Bay is ice-covered in winter and ice-free in summer. Hudson Bay contributes a significant amount, 50%, of the freshwater transport of the Labrador Current, based on observations in Hudson Strait (Straneo & Saucier, 2008). Hudson Bay has substantial river freshwater input, from many (42) rivers, each of moderate flow (Déry, Stieglitz, McKenna, & Wood, 2005). There is considerable seasonal river runoff from the south and east sides, giving rise to a marked horizontal stratification and an estuarine-type circulation. In summer,

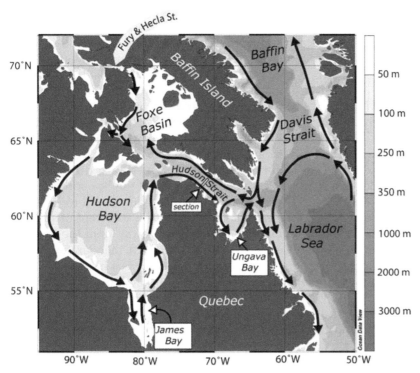

FIGURE 12.6 Schematic circulation in Hudson Bay and, peripherally, Baffin Bay. *Source: From Straneo and Saucier (2008).*

the upper water properties range from 1 to 9°C and S = 25 to 32 psu while the deeper water properties range from −1.6 to 0°C and 32 to 33.4 psu. The low salinities are generally in the south and east, near the main sources of runoff and consistent with a general anticlockwise circulation in the upper layer. A few observations taken in winter through the ice indicate upper salinities from 28 psu in the southeast to 33 psu in the north, with temperatures everywhere at the freezing point appropriate to the salinity. The implication is that the waters are vertically mixed each year; the high dissolved oxygen values of 200 to 350 μmol/kg in the deepest water are consistent with this condition.

Baffin Bay, with a maximum depth of 2400 m, is separated from the Labrador Sea (and hence from the Atlantic) by the sill in the Davis Strait, which is about 640 m deep (Rudels, 1986). Sill depths between the Arctic and Baffin Bay are

120−150 m (Jones et al., 2003). Baffin Bay's temperature and salinity structure include a cold, fresh surface layer to about 200 m, a temperature and salinity maximum at about 700 m (>0.5°C, 34.5 psu), and cold, fresher bottom waters (<−0.4°C, 34.25−34.5 psu; Rudels, 1986). Winter convection within Baffin Bay is likely limited to 200 m, and therefore does not produce either the temperature maximum or cold bottom waters. The temperature maximum signature comes from the Labrador Sea, via the West Greenland Current (Chapter 9). However, much of the water in the temperature maximum layer, and most of the deep and bottom water, come from the Arctic through Nares Strait (Bailey, 1957; Rudels, 1986). As the annual inflow to Baffin Bay is relatively small, the bay is a deep hole compared with the inlet sills, and deep water formation is minimal. Its deep water has a long residence time, reflected in depleted oxygen content and elevated nutrients, and denitrification occurs in the deep waters (Jones et al., 2003).

12.4. ARCTIC OCEAN: CIRCULATION AND ICE DRIFT

The Arctic Ocean's surface circulation is dominantly cyclonic (counterclockwise) on the Eurasian side and anticyclonic (clockwise) in the *Beaufort Gyre* in the Canadian Basin (Figure 12.1 and Section 12.4.2). A major current, the TPD, flows directly across the Arctic between these two circulations, from the Bering Strait side to the Fram Strait. Inflows to the Arctic are from the Nordic Seas, via the Norwegian Atlantic Current that splits into the *West Spitsbergen Current* (on the west side of Spitsbergen) and flows into the Barents Sea, and from the Pacific, via Bering Strait. There is some flow from the Labrador Sea into Baffin and Hudson Bays, but this does not continue onward into the Arctic proper. The intermediate and deep circulations (Figure 12.10 and Section 12.4.3) resemble each other and are cyclonic throughout. They are strongly topographically controlled.

Much of what is known about surface circulation is based on ice drift, but there are some differences between the two. Geostrophic calculations and water mass tracking also provide information on the surface flows. Ice drift is important since the large amounts of ice that exit the Arctic into the Nordic Seas affect the salinity structure of the region and the albedo (surface reflectivity) of the high northern latitudes, which in turn affect Earth's climate.

12.4.1. Ice Drift and Wind Forcing

The oldest records of Arctic ice movement were based on ships held in the ice, such as the *Fram* (Figure 12.7) and the *Sedov*, and from movements of camps on the ice. Modern ice drift is obtained from microwave satellite imagery and from buoys deployed on the ice (International Arctic Buoy Program; Figures 12.8 and 12.9). These various sources yield a consistent picture of the surface-layer movement. Some of the ice drift features oppose local upper ocean circulation (Section 12.4.2). The mean ice drift includes an anticyclonic (clockwise) circulation in the Canadian Basin (Beaufort Gyre) leading out to the TPD, with westward drift along the Alaskan sector as part of the Beaufort Gyre. Ice drifts southward from Baffin Bay through Davis Strait into the Labrador Sea. Except in summer, there is mean ice drift away from the Eurasian coast toward the TPD. In the Eurasian Basin, ice flows from the Laptev Sea into the TPD and subsequently the Fram Strait (coincidentally the track of the ship *Fram*, Figure 12.7). The TPD feeds into strong southward flow (ice export) through Fram Strait and the anticyclonic Beaufort Gyre. In the Eurasian Basin, ice also flows from the Kara Sea around the northern tip of Novaya Zemlya into the Barents Sea and then into the Norwegian Sea.

Ice drift speeds are of the order of 1 to 4 cm/ sec, equivalent to 300 to 1200 km/yr; in comparison, the Arctic Ocean is about 4000 km across. The speed and distance may be compared to the 3 years taken by the *Fram* to drift from the Laptev Sea to Spitsbergen, and the 2.5 years for the *Sedov* to drift about 3000 km. The movement is not steady, but has frequent variations of speed and direction. There is a definite seasonal change in the ice movements. The weakest ice drift is in summer. Large variations in ice drift are associated with the phase of the Arctic Oscillation and the Atlantic Multidecadal Oscillation (Chapter S15 on the textbook Web site).

The ice motion is related to both wind driving (e.g., Ekman response) and advection by non-Ekman surface currents, including the geostrophic flow. Ice buoy vectors and the mean sea level pressure (SLP) associated with the wind forcing are shown in Figure 12.9, and

(a)

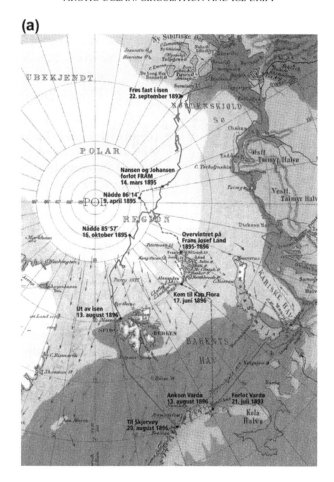

(b)

FIGURE 12.7   (a) Track of the *Fram* (1893–1896). (b) The ship was intentionally frozen into the ice in 1893 and drifted with the pack until 1896. © www.frammuseum.no. *Source: From Frammuseet (2003).*

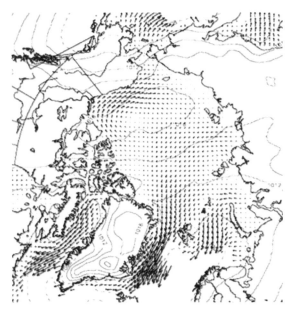

FIGURE 12.8 Annual mean Arctic sea ice motion from 1979–2003 from Special Sensor Microwave Imager (SSM/I) passive microwave satellite data (extended from *Emery, Fowler, & Maslanik, 1997*; data from *NSIDC, 2008a*). Monthly means are shown in Figure S12.4 seen on the textbook Web site.

also in the overlying contours in Figure 12.8. SLP is dominated in the Siberian/Canadian sector by the Beaufort high, which is an extension of the Siberian High. This high-pressure zone forces the anticyclonic Beaufort Gyre. Mean geostrophic winds over the pole are from the Eurasian to the Canadian/Greenland side, roughly in the direction of the TPD. The SLP ridge over Greenland in winter creates strong northerly winds through Fram Strait and southward along the coast of Greenland, roughly paralleling the major ice export path (see also supplementary Figure S12.5 from Bitz, Fyfe, & Flato, 2002 on the textbook Web site). The low pressure over the Nordic and Barents Seas is a northward extension of the Iceland Low, and forces cyclonic circulation in these seas. In summer, the SLP contrasts are much smaller, the winds much weaker, and

the atmospheric low is centered over the North Pole. The Beaufort high is pushed much closer to the Canadian/Siberian sides.

12.4.2. Upper Layer Circulation

The upper ocean circulation pattern (Figure 12.1 and Table S12.2 on the textbook Web site) is cyclonic in the Eurasian Basin and around the rim of the Arctic above the shelves. A large-scale anticyclonic circulation (Beaufort Gyre) occurs in the Canadian Basin. Inflows come from the Nordic Seas and from the Bering Sea (Pacific). Outflows occur through Fram Strait to the Nordic Seas in the EGC and through the Canadian Archipelago to Baffin Bay and the Labrador Sea.

The major currents (heavy curves in Figure 12.1) are the:

1. Inflowing *Norwegian Atlantic Current*, which splits into the northward-flowing West Spitsbergen Current and eastward flow into the shallow Barents Sea. The latter joins near coastal inflow from the Norwegian Coastal Current (Figure 12.1).
2. TPD that flows across the pole from the Alaskan and eastern Asian coasts toward Greenland and the Fram Strait, forming the EGC.
3. Anticyclonic *Beaufort Gyre*, which is driven by the mean high-pressure system above the Beaufort Sea. It is a superficial feature. The intermediate and deep circulations are cyclonic (Figure 12.10).

Shown in thinner curves in Figure 12.1 are weaker, but nevertheless critical, flows. These include the *Bering Strait inflow* from the Pacific, which has much smaller transport than from the Nordic Seas. A cyclonic *rim current* connects the shelf seas and feeds dense water formation on the shelves (Rudels, Friedrich, & Quadfasel, 1999). Each portion of this current has a separate name (see Rudels, 2001; Rudels et al., 2010). The rim current is found from the Norwegian Sea to

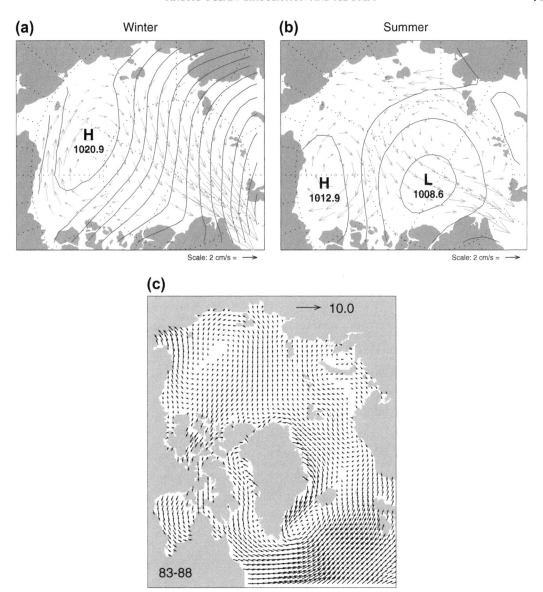

FIGURE 12.9 Mean sea level pressure (1979–1998) with mean ice buoy velocities for (a) winter (January-March) and (b) summer (July-September). ©*American Meteorological Society. Reprinted with permission. Source: From Rigor, Wallace, and Colony (2002).* (c) Mean wind vectors from ECMWF for 1983–1988. *Source: From Zhang and Hunke (2001).* Mean sea level pressure maps from Bitz et al. (2002) are also shown in Figure S12.5 seen on the textbook Web site.

the Barents Sea, and around the Arctic to the Kara and Laptev Seas. It branches off into the interior Arctic at each of the major island groups in each of these seas, joining the TPD toward Greenland. Each of the island groups in the Barents and Kara Seas includes cyclonic flow between the island and the coastal rim current and anticyclonic flow around the island group.

(a)

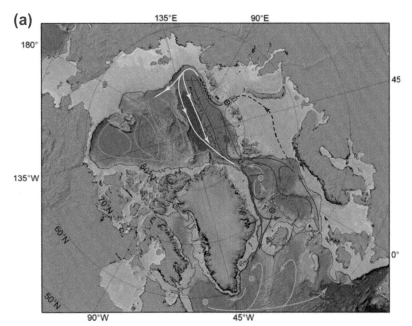

(b)

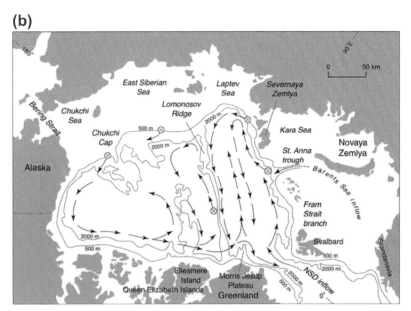

FIGURE 12.10 Circulation schematics. (a) Subsurface Atlantic and intermediate layers of the Arctic Ocean and the Nordic Seas. Convection sites in the Greenland and Iceland Seas, and in the Irminger and Labrador Seas are also shown (light blue), as is a collection point for brine-rejected waters from the Barents Sea. *Source: From Rudels et al. (2010).* This figure can also be found in the color insert. (b) Deep circulation; circled crosses indicate entry sites from dense shelf waters, and the Lomonosov Ridge overflow site. *Source: From Rudels (2001).*

The rim current continues into the Canadian Basin, picks up the Bering Strait inflow, and onward as the Alaskan Coastal Current, transporting Bering Strait water eastward to the Canadian archipelago (Jones, Anderson, & Swift, 1998; Rudels, 2001).

Water enters the Canadian archipelago along several different routes. The most important are the western routes feeding through Lancaster Sound, a central route through Jones Sound, and an eastern route through Nares Strait (Figure 12.1).

The circulation differs somewhat from ice drift, especially in the Makarov Basin. The TPD and anticyclonic Beaufort Gyre (Canadian Basin) are evident in both circulation and ice drift. However, the rim current is not apparent in ice drift. Similarly, cyclonic flow in Baffin Bay and the Labrador Sea is not apparent in ice drift, which is dominated by southward export.

12.4.3. Intermediate and Deep Circulation

Circulation in the intermediate layer, including the subsurface, warm AW layer (Sections 12.3 and 12.5) and the intermediate layer of the Arctic is shown in Figure 12.10a, representing flows between 200 and 900 m depth. The large-scale circulation is cyclonic. Cyclonic cells are embedded in this overall cyclonic circulation, with separate cyclonic cells in each of the major basins (Nansen, Amundsen, Makarov, and Canada). This circulation has many similarities to the surface flow and ice drift (Figures 12.1, 12.8, and 12.9), but the anticyclonic Beaufort Gyre has completely disappeared, replaced by cyclonic flow throughout the Canadian Basin. Major sources of water masses at this level are also indicated in Figure 12.10, including brine-rejected waters from the Siberian shelves that flow out into the deeper Arctic, and deep convection sites in the Nordic Seas (Section 12.2), and in the Irminger and Labrador Seas.

The deep circulation patterns (Figure 12.10b) are nearly identical to the intermediate circulation; that is, the Arctic circulation is nearly barotropic. Because of topography, the deep flow cannot connect across the Lomonosov Ridge, so the continuous cyclonic rim current at mid-depths is absent at depth.

Deep water enters and exits the Arctic from the Nordic Seas through Fram Strait, with the boundaries to the right of the flows (northward flow on the east side and southward flow on the west side). The overall flow in both the Eurasian and Canadian Basins is cyclonic, with the Lomonosov Ridge acting as a barrier. The sources of deep water within the Arctic are the brine-rejected waters from the continental shelves; injection points from the shelves to the deep ocean are denoted in Figure 12.10b by crossed circles. Also indicated in this figure is the saddle in the Lomonosov Ridge, where an intensive experiment in 2005 showed incursions of waters from Makarov Basin over to the Eurasian side of the ridge (Björk et al., 2007). (Figure 12.10b implies the opposite direction, which was the generally accepted concept prior to this experiment.)

12.5. ARCTIC OCEAN WATER MASSES

The Arctic Ocean can be described in terms of three main layers (Figures 12.11 and 12.12; Table S12.3 on the textbook Web site): (1) Polar Surface Water from the sea surface to about 200 m depth, (2) intermediate waters, including AW, from about 200 to 800 m (0°C isotherm), and (3) various deep/bottom waters below this to the bottom. Within the main water mass classifications, the details can be complex. We mainly follow Swift and Aagaard (1981), Aagaard et al. (1985), Rudels (2001), and Loeng et al. (2005); the latter two are reviews.

There are two external oceanic sources for the Arctic Ocean waters: the Atlantic via the Nordic Seas in the Norwegian Atlantic Current, and the Pacific via Bering Strait. These inflow waters can be identified far into the Arctic.[1] In addition, there is significant freshwater input, mainly

---

[1] Water mass properties in the Arctic persist a long distance from their sources, reflecting lower turbulence and hence lower mixing than in other major ocean basins, due to the ice cover that isolates the ocean from direct wind forcing and waves.

from river runoff. Because of their low density, the Bering Strait and river inputs enter the near surface layer (Section 12.5.1), while the AW enters an intermediate layer (Section 12.5.2).

Sea ice formation is the mechanism for Arctic water mass transformation. Through brine rejection over the broad continental shelves, dense shelf waters are created. High production occurs in recurrent latent heat polynyas in the Laptev, Barents, and Kara Seas (Figure 12.1). As these brine-rejected waters leave the shelves, they mix mostly into the pycnocline, but they are also the source of the deeper waters, depending on their initial density and vigor of mixing.

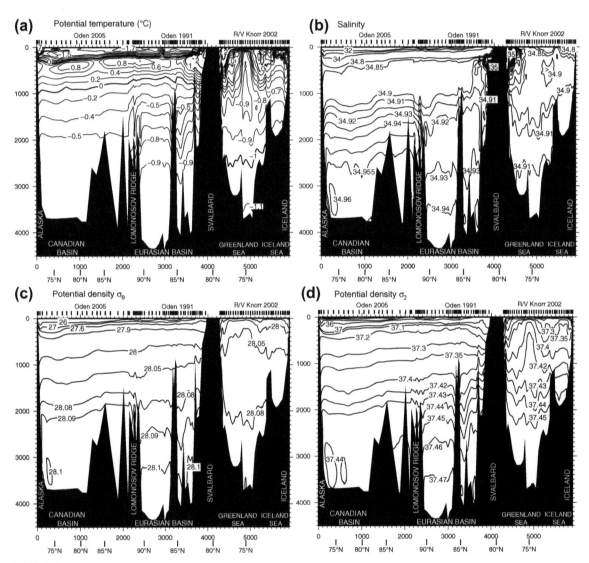

FIGURE 12.11   Arctic Ocean and Nordic Seas: (a) potential temperature (°C), (b) salinity, (c) potential density referenced to the sea surface, (d) potential density referenced to 2000 dbar. (e) station locations. Oxygen and CFC-11 are shown in Figure 12.16. Data sets were collected between 2000 and 2005. *After Aagaard et al. (1985).*

(e)

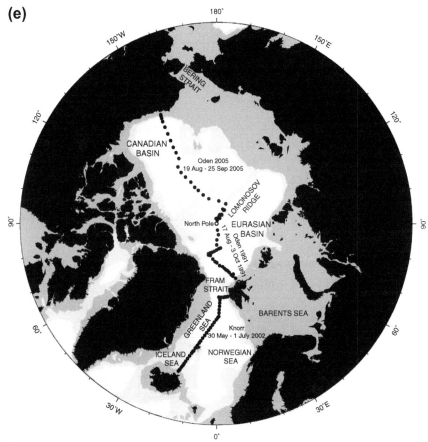

FIGURE 12.11 (Continued).

Also because of brine rejection and the freshness of sea ice, ice formation and melt over the open Arctic freshen the surface layer, contributing to a strong underlying halocline. River outflows also contribute to this fresh surface layer. This salinity structure can then stably support vertical temperature inversions, just as in the Southern Ocean (Chapter 13) and the northern North Pacific (Chapter 10).

Laterally, there is an important demarcation between the Eurasian and Canadian Basins. In the upper ocean, this arises from the different properties of the separate Atlantic and Pacific inflows. In the deep water, the Lomonosov Ridge blocks communication.

12.5.1. Surface and Near-Surface Waters

The surface layer, down to about 200 m, is comprised of the Polar Mixed Layer (PML), a shallow temperature maximum layer in some regions (Canadian Basin), and the halocline. It includes significant inputs from Bering Strait (summer and winter Bering Strait Waters), from river runoff, and from brine-rejected shelf waters. Following Rudels (2001), this whole complex is called the Polar Surface Water (Figure 12.12; Table S12.3 seen on the textbook Web site).

The PML exists across the whole Arctic; it extends from the surface to between 25 and 50 m depth. Its salinity is strongly influenced

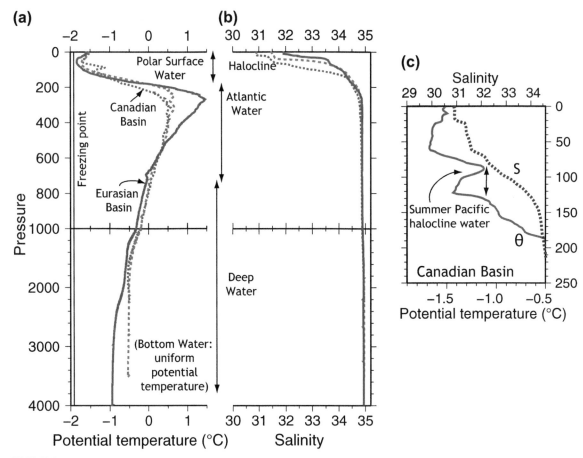

FIGURE 12.12 Arctic Ocean: (a) Potential temperature and (b) salinity profiles for the Canadian (dashed) and Eurasian Basins (solid). Station locations are shown in Figure 12.17a: dashed profiles are stations CaB and MaB and the solid profile is NaB. (c) Expanded potential temperature and salinity profile in the Canadian Basin (CaB in Figure 12.17a). *After Steele et al. (2004).*

by the freezing or melting of ice and has a wide range from 28 to 33.5 psu. The temperature is also controlled by melting and freezing, which involves considerable heat transfer at constant temperature (the freezing point). As a consequence, the temperature remains close to the freezing point, from −1.5°C at a salinity of 28 psu to −1.8°C at a salinity of 33.5 psu. Seasonal variations in water properties are largely limited to this layer and range up to 2 psu in salinity and 0.2°C in temperature.

In the Eurasian Basin, temperature is nearly constant (isothermal), near the freezing point, through the shallow halocline (solid in Figure 12.12, which includes warmer water at the surface since this is a summer observation). The halocline depth is 25–100 m. Because it is nearly isothermal, the halocline cannot be a simple vertical mixture of the PML and AW. Rather, it includes shelf waters from the Eurasian Shelf (Coachman & Aagaard, 1974; Aagaard, Coachman, & Carmack, 1981). The considerable Siberian river runoff flows into the cold, low salinity surface layer. Ice formation creates saline shelf waters at the freezing point. These mix together and continue out into the Arctic Ocean in the 25 to 100 m layer, creating the isothermal halocline. Major canyons along the shelf feed the saline AW onto the shelf; the vertical mixing process is similar to an estuary in which fresh river water flows over saline seawater (Section 8.8).

Below 100 m in the Eurasian Basin, there is a thermocline with temperature increasing downward to the temperature maximum of the intermediate Atlantic layer (AW) that enters from the Nordic Seas.

The brine-rejected shelf waters in the Eurasian sector are relatively saline compared with other brine-rejected waters in the Arctic because the saline, warm AW (Section 12.5.2) is a source. These shelf waters can reach a sufficiently high density to ventilate the deep water in the Eurasian sector. Shelf waters from the Barents and Kara Seas are especially implicated (Aagaard et al., 1981).

In the Canadian Basin, the Polar Surface Water below the mixed layer includes summer and winter Bering Strait waters and Alaskan Coastal Water (ACW), as well as brine-rejected shelf water components (Figure 12.12c). These multiple sources create more complicated vertical and horizontal structures than in the Eurasian Basin. The ACW and summer Bering Strait Water (sBSW) are warm and create a temperature maximum at 50 to 100 m depth beneath the PML (labeled "summer Pacific halocline water" in Figure 12.12c). The temperature maximum is supported by a strong halocline. Below this, there is a temperature minimum at about 150 m depth, due to winter Bering Sea Water. Below this, the temperature increases downward to the maximum in the AW (see next section).

Circulation and temperature in the upper temperature maximum layer (ACW and sBSW) are shown in Figure 12.13. The warmest temperature maxima are in the Beaufort Gyre, and are

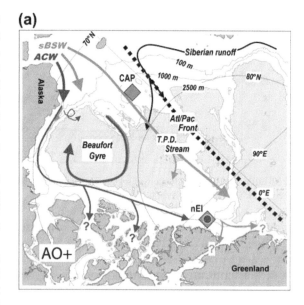

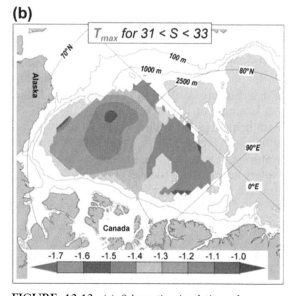

FIGURE 12.13 (a) Schematic circulation of summer Bering Strait Water (blue) and Alaskan Coastal Water (red) during the positive phase of the Arctic Oscillation (Chapter S15 on the textbook Web site). (b) Temperature (°C) of the shallow temperature maximum layer, which lies between 50 and 100 m depth, in the Canadian Basin. This figure can also be seen in the color insert. *Source: From Steele et al. (2004).*

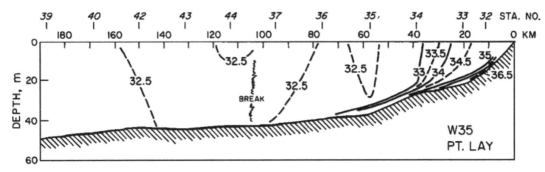

FIGURE 12.14 Salinity along a section in the Chukchi Sea (March 1982), including a high salinity bottom layer created by brine rejection. *Source: From Aagaard et al. (1985).*

due to ACW. The cooler temperature maxima are in the sBSW. ACW enters the Arctic from the eastern coastal side of the Chuckchi Sea, and Bering Strait Water enters from the center and western side. ACW joins an eastward coastal circulation and also forms eddies that move into the central Beaufort Sea (loops in Figure 12.13a). Bering Strait Water stays more in the center of the Arctic and joins the TPD.

Brine rejection on the shelves in the Canadian Basin produces waters that enter the halocline (Polar Surface Water) in the Canadian Basin. An example of late winter salinity distribution with brine-rejected waters in the Chukchi Sea is shown in Figure 12.14. Because the ambient water is not saline, these new brine-rejected waters are not salty (dense) enough to penetrate through the Atlantic layer and do not contribute to the Canadian Basin Deep Water (CBDW; Section 12.5.3).

12.5.2. Atlantic Water

Below the cold Polar Surface Water, the Arctic Ocean is characterized throughout by the AW temperature maximum at a depth of 200 to 900 m (Figures 12.11, 12.12, 12.15, and 12.17). In the Nordic Seas, the AW is a surface water mass, with maximum temperature at the sea surface. Where it enters the West Spitsbergen Current in Fram Strait, it becomes a subsurface maximum with cold, fresh Polar Surface Water

riding over the top. Some of the AW branches back into the Nordic Seas in the EGC. The remainder flows around the Arctic cyclonically, mostly as a "rim" current along the continental shelf break (Figure 12.1; Rudels et al., 1999). This circulation is not in the same direction as the surface circulation or ice drift.

Along its cyclonic path, both the temperature and salinity of the AW decrease (Figures 12.12 and 12.15 and Figure S15.12 in Chapter S15 seen on the textbook Web site). At Fram Strait, the AW temperature is around 3°C and its salinity is greater than 35.0 psu. In the Arctic Ocean, AW temperature decreases gradually to 0.4°C and its salinity to 34.80−34.9 psu. Its core shifts downward from the surface (from 200 m in the Fram Strait to 500 m in the Canadian Basin) and becomes more dense. These changes are due to mixing with waters above and below, and with cold shelf waters that advect in from the side (Aagaard et al., 1981; Rudels et al., 1999).

12.5.3. Deep and Bottom Water

Deep Water extends downward from the lower 0°C isotherm, at about 800 m depth, to the bottom (Figures 12.11 and 12.12; Table S12.3 on the textbook Web site). Deep Water comprises about 60% of the total water volume of the Arctic Ocean (Aagaard et al., 1985). The densest water in the Arctic is produced within

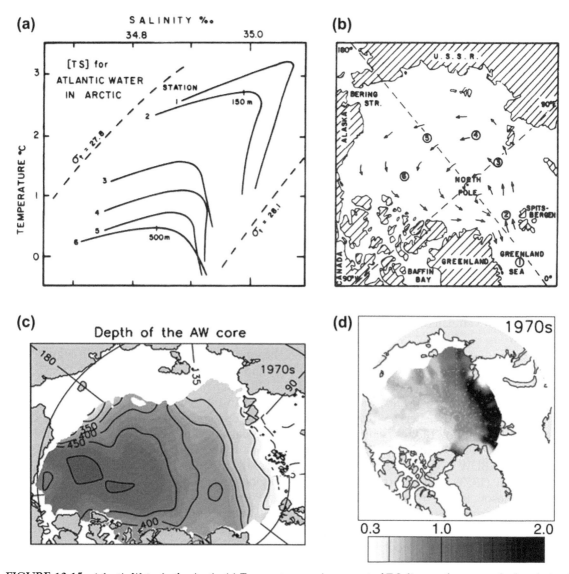

FIGURE 12.15 Atlantic Water in the Arctic. (a) Temperature maximum part of T-S diagram for core method analysis of flow direction for Atlantic Water, (b) circulation inferred from successive erosion of core shown in (a) stations 1 to 6, (c) depth, and (d) potential temperature (°C) of the Atlantic Water temperature maximum in the 1970s. (c, d) are ©*American Meteorological Society. Reprinted with permission; Polyakov et al. (2004) and Polyakov et al. (2010).*

the Arctic. Because relatively shallow sills separate the Arctic from the Atlantic and Pacific, most of the deep water cannot flow out into either the Atlantic or Pacific, nor can deep waters from either of these regions enter. Deep-water production that fills this isolated deep layer must therefore be balanced by upwelling. Consequently, Arctic deep waters are relatively uniform in temperature and salinity, including in the vertical.

We recognize three deep waters, following Jones (2001). First is uPDW, which is found throughout the Arctic and is exported into the Nordic Seas through the Fram Strait. This layer lies below the AW and above the Lomonosov Ridge at about 1700 m, so there is open communication with all regions of the Arctic. uPDW is not obviously marked in potential temperature and salinity: with increasing depth through the UPDW, potential temperature decreases and salinity increases. However, uPDW can be readily distinguished in oxygen, silicate, and chlorofluorocarbons (CFCs). The Arctic Ocean is relatively well ventilated above the ridge depth of the Lomonosov Ridge. On the Canadian Basin side, the waters below ridge depth have lower oxygen, higher silicate, and low CFCs (Figure 12.16). Above this deep layer and below the warm AW, the water column is well ventilated, including an oxygen maximum.

The other two major deep waters are separated by the Lomonosov Ridge, between the Canadian and Eurasian Basins. Deep water in the Eurasian Basin is called *Eurasian Basin Deep Water* (EBDW). Deep water on the Canadian side of the Lomonosov Ridge is called *Canadian Basin Deep Water* (CBDW) and has different properties from EBDW (Figures 12.12 and 12.17). Worthington (1953) deduced the existence of the Lomonosov Ridge from the difference in deep water properties in the Canadian and Eurasian sectors. Deep waters from the Nordic Seas also enter the Arctic through the Fram Strait, so both the relatively cold, fresh, dense Greenland Sea Deep Water and slightly warmer, saltier Norwegian Sea Deep Water are found in the Eurasian Basin (e.g., Aagaard et al., 1985).

The EBDW and CBDW can be split vertically into deep and bottom waters. The bottom water layer is recognized by uniform properties in the vertical, hence it is nearly adiabatic (see the end of this section).

In potential temperature, the progression is from coldest deep waters in the Nordic Seas, to slightly warmer ($\sim -0.95^\circ$C) in the Eurasian Basin. (In Figure 12.17, stations in the Makarov and Amundsen Basins are included; these are sub-basins of the Canadian and Eurasian Basins, respectively, as can be seen from similarities of the Makarov-Canadian and Amundsen-Eurasian properties.) Potential temperature is much more uniform in the deep water than in situ temperature, with the deep temperature minimum erased when the effect of adiabatic compression is taken into account. The bottom-most layer, which can be more than 1000 m thick, is adiabatic (see the end of this section). On the other hand, there is a small but remarkable potential temperature minimum in both the EBDW and CBDW, associated with a characteristic smooth upward curve in potential temperature-salinity (T-S) space (Figures 12.17b and 12.18b). This minimum does not result from choice of reference pressure. This "hook" in the T-S relation, which is due to geothermal heating, is even more apparent in Figure 12.18 (Timmermans, Garrett, & Carmack, 2003).

In salinity, the freshest bottom waters are found in the Nordic Seas, with higher salinity in the Eurasian Basin and highest in the Canadian Basin. In any given region, the vertical variation in the deep water is smaller than the overall difference in salinity between these regions.

Potential density variation between the regions is dominated by potential temperature. Thus the cold, fresher Nordic Seas Deep Waters are denser than the EBDW, and the CBDW is the least dense. To compare the density of the bottom waters, it is important to use a deep pressure reference level. Relative to 4000 dbar, the potential density progression is from densest in the Nordic Seas to least dense in the Canadian Basin. However, relative to 0 dbar, the Eurasian Basin waters are the densest.

EBDW is ventilated from the Eurasian continental shelves around the Arctic through brine rejection, which contributes salt. The densest shelf water is formed in the Barents and Kara

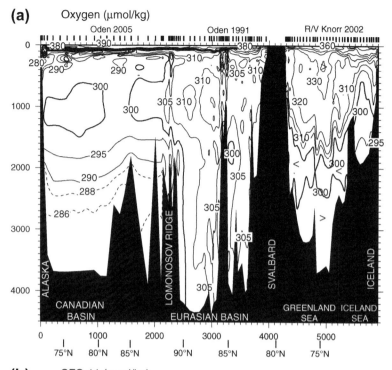

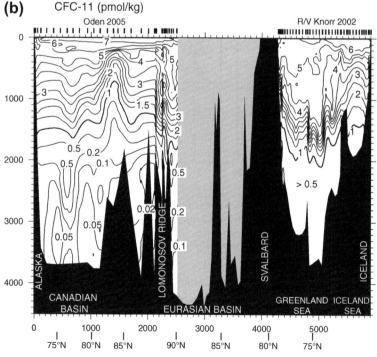

FIGURE 12.16 Vertical section across the Arctic and Nordic Seas. The section extends from the Chukchi Sea north of Bering Strait to the North Pole to Svalbard and Iceland (on the right). Corresponding sections of potential temperature, salinity and potential density were shown in Figure 12.11, along with a station location map. (a) Oxygen (μmol/kg), and (b) CFC-11 (pmol/kg). Station locations are shown in Figure 12.11e. Vertical sections from the Canadian Basin (*Swift et al., 1997*) and the Eurasian Basin (*Schauer et al., 2002*) are shown in Figure S12.6 on the textbook Web site.

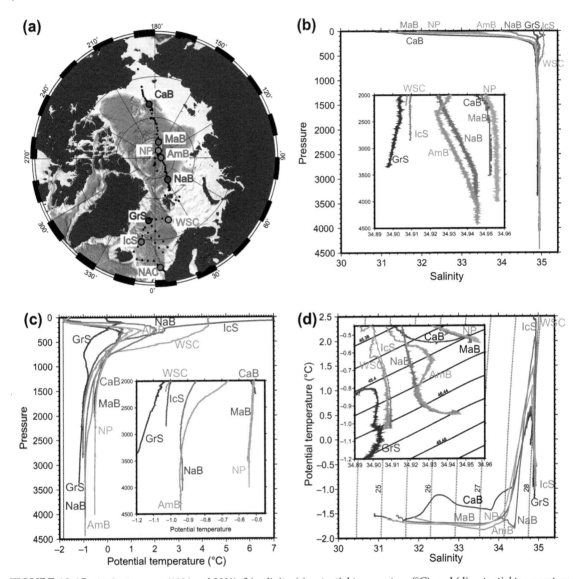

FIGURE 12.17 (a) Station map (1994 and 2001), (b) salinity, (c) potential temperature (°C), and (d) potential temperature-salinity. Acronyms: CaB, Canada Basin; MaB, Makarov Basin; NP, North Pole; AmB, Amundsen Basin; NaB, Nansen Basin; WSC, West Spitsbergen Current; GrS, Greenland Sea; IcS, Iceland Sea; and NAC, Norwegian Atlantic Current. This figure can also be found in the color insert. *Expanded from Timmermans and Garrett (2006).*

Seas. About 10% of the EBDW can be accounted for by brine rejection, with the remainder being the original Nordic Seas Deep Waters that enter through the Fram Strait (Östlund, Possnert, & Swift, 1987).

In the Canada Basin (a sub-basin of the Canadian Basin — see Figure 2.11), the brine-rejected shelf waters are not dense enough to renew the deep water. The bottom waters of the Canada Basin have a mean age

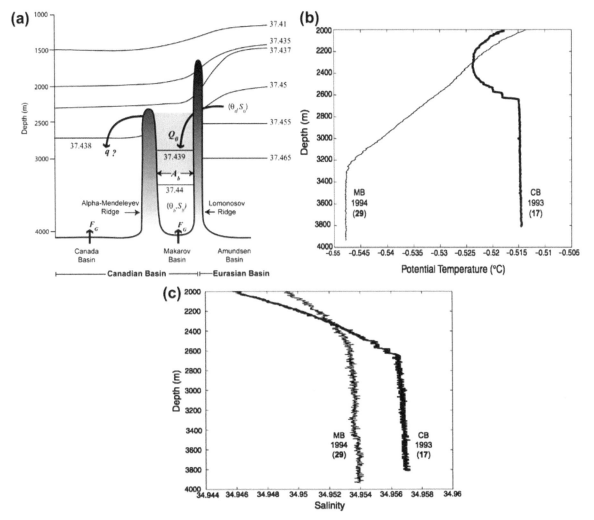

FIGURE 12.18 (a) Schematic of bottom water connections, including approximate sill depths. Potential density is relative to 2000 dbar. Arrows indicate mass fluxes overflowing into the basins and bottom arrows indicate geothermal heat flux. (b) Potential temperature (°C) and (c) salinity at stations in the Makarov and Canada Basins (MB and CB, respectively). The Makarov station "MB" is MaB in Figure 12.17a. ©*American Meteorological Society. Reprinted with permission. Source: From Timmermans and Garrett (2006).*

of about 450 years compared with 250 years in the Eurasian Basin, based on ^{14}C and the low levels of anthropogenic tracers (Macdonald, Carmack, & Wallace, 1993; Schlosser et al., 1997).

For the deep waters just above the Lomonosov sill depth, the connection is mainly from the Canadian Basin side to the Amundsen Basin side. That is, the Amundsen Basin bottom water is identical to the Eurasian Basin bottom water, but there is a remarkable transition to the warmer, more saline Canadian Basin water above about 2000 m (Figure 12.17; Björk et al., 2007).

Adiabatic (vertically uniform) bottom layers of about 500 m thickness are apparent in the Amundsen and Nansen Basin potential temperature and salinity profiles in Figure 12.17, and also at the North Pole station, which is on the Makarov Basin side of the Lomonosov Ridge. The Canada Basin has an even thicker adiabatic bottom layer, from 2600 m to the bottom (Figure 12.18). (The "Canada Basin" profile in Figure 12.17 is not deep enough to capture this layer.) The existence of adiabatic bottom layers indicates that water crossed a sill to fill the deep basin, which is then uniform in properties below sill depth. The sill depth can be inferred from the "break" at the top of the adiabatic bottom layer.

In the Amundsen and Canada Basins, the adiabatic bottom layer is warmer than the overlying water, with a remarkably smooth curve connecting the temperature minimum with the adiabatic bottom layer. This temperature structure is due to geothermal heating from below (Timmermans et al., 2003). The absence of this structure in the Makarov Basin indicates ongoing replenishment (spillover) of cool water from the Amundsen Basin. Based on the deep properties in each of the basins, there is likely some small flow of Amundsen Basin deep water to the Makarov Basin, while any significant deep flow from the Makarov Basin to the Canada Basin can be ruled out (Timmermans & Garrett, 2006).

12.6. ARCTIC OCEAN TRANSPORTS AND BUDGETS

The Arctic Ocean/Nordic Seas is a globally important region of heat loss and dense water mass production. Cooling of the AW flowing through the Nordic Seas is the only major air—sea heat exchange that occurs at high latitudes. This Nordic Seas heat loss is responsible for the existence of northward heat transport through the full length of the Atlantic Ocean, including the South Atlantic, compared with the equatorially symmetric Pacific Ocean, assuming that the Gulf Stream and Kuroshio regions can be considered equivalent in terms of their roles in heat loss in the Atlantic and Pacific (Sections 5.5, 9.7, and 14.3).

The Arctic/Nordic Seas region is also important for the global freshwater budget because of its connection between the Pacific and Atlantic Oceans, its net runoff and precipitation, and its ice export to the Atlantic Ocean. The Arctic freshwater export to the North Atlantic becomes part of the newly formed NADW, and is an important control on the salinity of that water mass. Both as part of natural climate cycles and as a response to anthropogenic change, the Arctic's ice cover varies. This changes the albedo of the Northern Hemisphere and can be an important part of climate feedback.

And not least of all, the Nordic Seas/Arctic Ocean, together with the Labrador Sea, comprise the main Northern Hemisphere region where upper ocean waters are converted to dense waters, thus providing the downward "limb" of the part of the global overturning circulation associated with the North Atlantic. This transformation, from upper ocean to deep ocean water, results from the large heat loss. The transformation process shifts freshwater from the surface layer down to the deep water layer, so the dense water is the primary means of exporting freshwater southward out of the Arctic and into the mid-latitude Atlantic.

What is the current picture of the production of dense waters in the Nordic Seas/Arctic Ocean? Warm AW enters the Nordic Seas via several routes and gathers in the Norwegian Atlantic Current (Figures 12.1 and 12.19). Part recirculates in the Nordic Seas and part proceeds northward into the Arctic. It is joined there by Bering Strait Water and surface water from rain and runoff. This whole upper layer cools further in the Arctic and is a source of the dense EBDW; part of the AW simply becomes more dense. Most of this modified AW and Arctic Ocean Deep

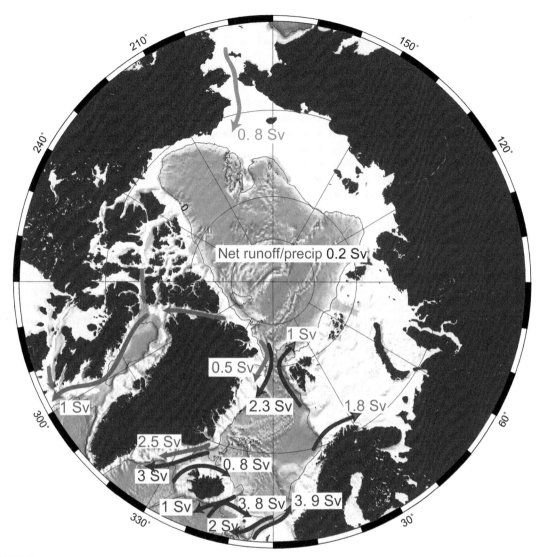

FIGURE 12.19 Volume transport budget. Red and orange are upper ocean inflows. Green is upper ocean outflow. Blue is intermediate/deep outflow. Transports are listed in Sverdrups. See Figure S12.7 on the textbook Web site for the color version.

Water returns to the Nordic Seas via the EGC, and a smaller part returns southward west of Greenland.

In the Nordic Seas, this returned, modified Arctic water joins the locally circulating AW and surface waters. A further densification step occurs, mostly through deep convection in the Greenland Sea (and Boreas Basin, which adjoins it to the north). Buoyancy loss and deep mixing also occur in the Iceland Sea (Figures 12.1 and 12.10), contributing overall to the new AIW layer. Brine rejection, specifically in Storfjorden on the south side of Svalbard (Figure 12.1), is also a densification process in the Nordic Seas. The net result is production of AIW (in the current decades), and, in earlier decades, Greenland Sea Deep Water. The portions of these that can overflow the relatively shallow sills into the North Atlantic then become part of the NADW.

Transports within this overturning system, consisting of both the Nordic Seas and Arctic, are as follows (Figure 12.19). The Norwegian Atlantic Current transports 8.5 Sv of AW northward into the Nordic Seas. The Bering Strait funnels 0.8 Sv from the Pacific Ocean into the Arctic Ocean (Roach et al., 1995). There is approximately 0.2 Sv of runoff and precipitation within the Arctic and Nordic Seas. The net input is therefore 9.5 Sv. Outflows across the Greenland-Scotland ridge include 6 Sv of denser water beneath the Atlantic inflows and 3.5 Sv of lighter water from the Arctic Ocean west and east of Greenland (via Davis Strait and the EGC, respectively). Of the dense overflows, 3 Sv is in the Denmark Strait, 1 Sv is over the Iceland-Faroe Ridge, and about 2 Sv is through the Faroe-Shetland Channel. Therefore, within the overall system, 6 Sv is converted to denser water from the 9.5 Sv of lighter inflow (Figure 12.19; following Jones, 2001 and Rudels et al., 1999).

For the Arctic portion alone, the inflow consists of 1.8 Sv into the Barents Sea, 1–1.5 Sv in the West Spitsbergen Current into the Arctic, 0.8 Sv through the Bering Strait, and 0.1–0.2 Sv of runoff. The net input to the Arctic is thus 3.7–4.3 Sv. Outflow from the Arctic includes 1 Sv west of Greenland to the Labrador Sea, and 2.8–3.3 Sv through the Fram Strait into the Nordic Seas. Of this Fram Strait transport, 0.5 Sv is Polar Surface Water and the remainder is denser water — modified AW (~1 Sv) and uPDW/EBDW (~1.3 Sv). Here "modified Atlantic Water" is the AW core that has been modified within the Arctic Ocean, becoming denser ($\sigma_\theta > 27.97$), colder, and fresher, and

returning southward. This is the northernmost transformation pathway leading to NADW production, with a net conversion to 2.2–2.8 Sv of denser water.

To reach the total conversion of 6 Sv of dense overflow waters, this already denser water from the Arctic joins the Nordic Seas water and all are further transformed to the net 6 Sv of waters denser than the Greenland-Scotland AW inflows. Thus about half of the transformation that feeds NADW is from waters that remain within the Nordic Seas and do not circulate through the Arctic; the other half is initially transformed to denser water during a circuit through the Arctic.

Residence times are estimated from volume transports and layer volumes (Section 4.7). Using their complete volume budget, Aagaard and Greisman (1975) estimated that the surface water is substantially replaced in 3 to 10 years, the deep water in 20 to 25 years, and the bottom water in the Eurasian Basin in about 150 years.

12.7. SEA ICE IN THE ARCTIC

The properties of sea ice were introduced in Section 3.9, with a discussion of how salt water freezes and the accompanying brine-rejection process. We also introduced the concept of polynyas, which are regions of open water within ice-covered regions. Here we specifically describe Arctic sea ice and its seasonal cycle. Photographs of sea ice in the Beaufort Sea are shown in Figure S12.8 seen on the textbook Web site.

12.7.1. Distribution of Arctic Sea Ice

Sea ice covers most of the Arctic. Year-round (multi-year) sea ice is found in some parts of the Arctic, although the coverage is declining (Chapter S15). Even in late winter, there are regions that are almost always ice-free

(a) **(b)**

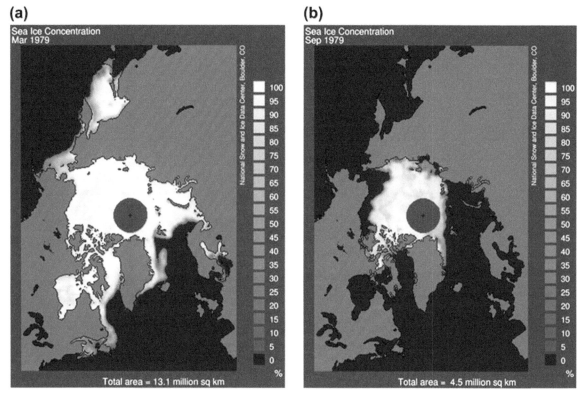

FIGURE 12.20 Ice concentration in 1979 in: (a) late winter (March) and (b) late summer (September). *Source: From NSIDC (2009a).*

(Figure 12.20a). These include the eastern Nordic Seas and part of the Barents Sea shelf, where warm Atlantic waters flow northward in the Norwegian Atlantic Current. Multi-year ice is found throughout the Canadian Basin and Greenland side of the Arctic (Figure 12.21). First-year ice, by definition, is the ice in regions of open water in late summer. Comparison of the late winter and late summer panels in Figure 12.20 gives an idea of where first-year ice occurs: in the Barents and Kara Seas on the Eurasian side, and periphery of the Chukchi and Beaufort Seas on the Canadian side. Ice in the Labrador Sea and Hudson Bay is also first-year ice.

Multi-year ice is found in the central Arctic, particularly in the Canadian Basin (e.g., late summer 1979 coverage in Figure 12.20b). The oldest ice (>4 years) borders the Canadian Archipelago (Figure 12.21). As sea ice cover has been declining, all of the shelf regions and Canadian archipelago areas have become more ice free in late summer, and at some point there will no longer be multi-year ice in the Arctic (Section S15.4 on the textbook Web site).

In addition to categorizing sea ice by its age, Arctic ice may be divided into three categories that are closely related to age: Polar Cap Ice, Pack Ice, and Fast Ice. The most extensive is the *Polar Cap Ice*. It is always present and covers about 70% of the Arctic Ocean, extending from the pole to approximately the 1000 m isobath. Cap Ice is very hummocky and is, on average, several years old. In winter, the average ice

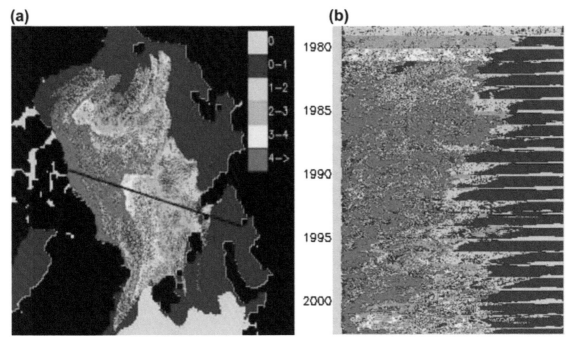

FIGURE 12.21 Arctic ice ages: (a) 2004 and (b) cross-section of ice age classes (right) as a function of time (Hovmöller diagram), extending along the transect across the Arctic from the Canadian Archipelago to the Kara Sea shown in (a). This figure can also be seen in the color insert. *Source: Extended from Fowler et al. (2004).*

thickness is 3 to 3.5 m but hummocks increase the height locally up to 10 m above sea level. (In the ridging process, two ice floes meet and deform vertically to form a ridge, with one-third of the ridge going up and two-thirds of the ridge going down. In rafting, two ice floes also meet, but one floe rises up and over the other.) Some of this Cap Ice melts in the summer and the average thickness decreases to about 2.5 m. Leads and polynyas, which are open water spaces, may form. In the autumn these freeze over and the ice in them is squeezed into ridges or is rafted. Polar Cap Ice is only penetrable by the heaviest icebreakers.

The occasional ice islands, which have fairly uniform ice thickness that is considerably greater than the regular Cap Ice, originate from glaciers on northern Ellesmere Island.

Pack Ice lies outside the Polar Cap. It consists of a smaller fraction of multi-year and more first-year ice than Cap Ice. It is lighter than Cap Ice and up to a few meters thick. It covers about 25% of the Arctic area, extending inshore of the 1000 m isobath. Its area varies somewhat from year to year. Seasonally, its areal extent is least in September and greatest in May. Some of it melts in summer and some is added to the Cap by rafting. Pack Ice is advected southward in the EGC and the Baffin and Labrador Currents. While icebreakers can penetrate Pack Ice, it impedes navigation in the northern parts of the Canadian Archipelago, along the east coast of Greenland, in Baffin Bay and the Labrador Sea, and in the Bering Sea.

The edge of the Pack Ice is the *marginal ice zone*. In this region, which can be tens to hundreds of kilometers wide, the sea ice is loose and broken. Surface waves provide energy to break up the ice. As the waves enter the marginal ice zone, they are scattered by the ice floes and their energy is attenuated. Upwelling, eddies, and jets occur along the ice edge. Higher levels of biological productivity are found in the marginal ice zone compared with surrounding waters.

Lastly, *Fast Ice* forms from the shore out to the Pack and consists of first-year ice that forms each winter. This ice is "fast" or anchored to the shore and extends to about the 20 m isobath. In the winter it develops to a thickness of 1–2 m, but it breaks up and melts completely in summer. When it breaks away from the shore, it may have beach material frozen into it and this may be carried some distance before being dropped as the ice melts, giving rise to "erratic" material in the bottom deposits.

The general circulation of the Cap and Pack Ice is similar to that of the Polar Surface Water (Section 12.4). This moves the ice around and exports it from the Arctic. Although Polar Cap Ice is always present, it is not always the same ice in a given location. Up to one-third of the total Cap and Pack Ice is carried away through Fram Strait in the EGC each year, while other ice is added from the Pack Ice. Ice export through Fram Strait and down the coast of Greenland is at a rate of about 3 km/day. The ice exports through Fram and Davis Straits are major factors in the Arctic freshwater budget. The volume of freshwater exported as ice is approximately equal to the total continental runoff into the Arctic basin.

12.7.2. Build-Up and Break-Up of Arctic Sea Ice; Polynyas

To give some idea of the variation in ice conditions with latitude, we present brief accounts of the build-up and break-up of sea ice from about 48°N to about 80°N in the Canadian north. In the Gulf of St. Lawrence (46–51°N), there is only first-year ice. Ice forms first in the inner area (river), then along the north shore, and becomes a hazard to shipping in the main Gulf by January by covering most

of the area by the end of February with ice to 0.6 m thickness. Break-up starts in mid-March and ships can move freely by mid-April along mid-Gulf over the deep Laurentian Channel. All the ice melts by summer. In severe winters, build-up and break-up can be two months earlier or later.

In Baffin Bay/Davis Strait (63–78°N), there is mostly first-year ice of 1.5–2 m thickness, with some older ice of up to 3 m thickness entering from the north through Smith Sound. In this region, ice cover is more common than open water. Baffin Bay and the west side of Davis Strait are largely covered by ice until mid-May, mostly clear by mid-August except off Baffin Island; ice then starts to develop from the north by late October. Interannual variations are considerable, with some areas clearing by mid-June in a good year but freeze-up starting as early as the end of August in a "bad" year. A large, recurrent polynya, called the "North Water Polynya," is found in the north end of Baffin Bay (Figures 12.1, 12.22); it is generally clear of persistent ice throughout the winter and is a source of dense water for Baffin Bay.

In the Canadian Archipelago, ice cover may break up but floes remain present throughout the year, and icebreakers are needed for surface supply to northern outposts there. First-year ice develops to 2.4 m thickness and multi-year ice to 4.5 m. Some clearing does take place in Lancaster Sound (74°N, leading west from Baffin Bay) and in passages further west by July, but floes continue to be present.

The western Arctic (120°W to the Bering Strait) is largely an open sea area north from the Canada/Alaska coast, which is at about 70°N, with a slow eastward current (Alaska Coastal Current). Multi-year ice (Arctic Pack) of up to 4.5 m thickness is general over the open sea south to 72°N, while fast ice develops to 2 m thickness along the coast. Open water is usually found near the coast from mid-August to mid-September and can even extend to 73°N, but in extreme years

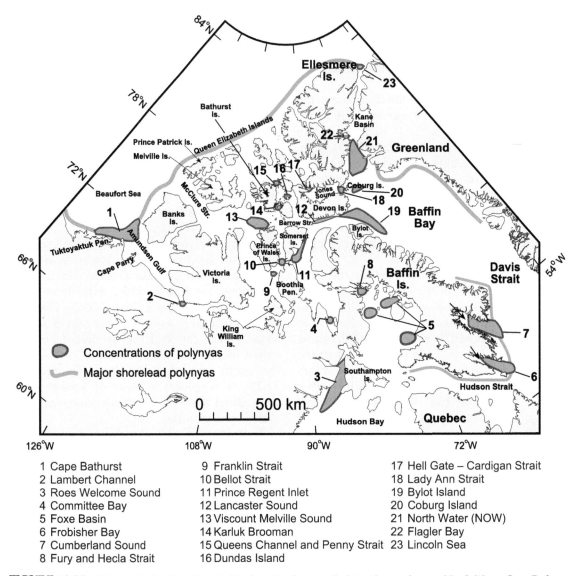

1 Cape Bathurst	9 Franklin Strait	17 Hell Gate – Cardigan Strait
2 Lambert Channel	10 Bellot Strait	18 Lady Ann Strait
3 Roes Welcome Sound	11 Prince Regent Inlet	19 Bylot Island
4 Committee Bay	12 Lancaster Sound	20 Coburg Island
5 Foxe Basin	13 Viscount Melville Sound	21 North Water (NOW)
6 Frobisher Bay	14 Karluk Brooman	22 Flagler Bay
7 Cumberland Sound	15 Queens Channel and Penny Strait	23 Lincoln Sea
8 Fury and Hecla Strait	16 Dundas Island	

FIGURE 12.22 Polynyas in the Canadian Archipelago. Predominantly latent heat polynyas: North Water, Cape Bathurst. Tidally mixed polynyas: Committee Bay, Dundas Island, Lambert Channel and possibly Queens Channel, Bellot Strait, Fury, and Hecla Strait. Polynyas in the Barents, Kara, and Laptev Seas are illustrated in Figures S12.9–S12.11 on the textbook Web site. *Source: From Hannah et al. (2009).*

the Arctic Pack may extend to the coast in August. Ship movements along the coast are generally limited to September.

In the open North Pacific, ice does not occur, but it is formed in the adjacent seas to the north and west, that is, the Bering Sea, the Sea of Okhotsk, and the northern Sea of Japan. In the Bering Sea, pack ice extends in winter to about 58°N but clears completely in the summer, retreating north through the Bering Strait to

70–72°N. Likewise in the Okhotsk and Japan Seas, sea ice is seasonal, disappearing completely each summer.

Polynyas (Section 3.9.6) are found all along the Arctic margins and throughout the Canadian archipelago (Figures 12.1, 12.22). Several satellite images of Arctic polynyas are included in Figures S12.9–S12.11 in the textbook Web site). The wind-forced latent heat polynyas are especially important for dense water formation because of the continual production of sea ice and hence brine within them; much of the AIW is formed in the Siberian shelf polynyas (Martin & Cavalieri, 1989; Smith et al., 1990). Latent heat polynyas depicted in Figure 12.1 include the North Water and Northeast Water around northern Greenland, the Laptev Sea and Cape Bathurst polynyas, and the polynyas around Svalbard, Franz Josef Land, Nova Zemlya, and Severnaya Zemlya. The Storfjorden polynya (Svalbard) is the source of very dense water for the Nordic Seas. The Laptev Sea flaw polynya is the region of highest ice production in the Arctic. The many polynyas of the Canadian archipelago are shown in Figure 12.22; of these, several are kept open through tidal forcing that mixes warmer subsurface waters upward (sensible heat polynyas) while others are wind-forced (Hannah, DuPont, & Dunphy, 2009).

12.7.3. Arctic Icebergs

Icebergs differ from sea ice in that they originate on land, have no salt content, have a density of about 900 kg/m^3 (which is less than that for pure ice because there are gas bubbles in icebergs), and have much greater vertical dimensions. They are a more serious hazard to shipping than sea ice because of their large mass. In the North Atlantic, the chief source of icebergs is calving from the glaciers of west Greenland, with a much smaller number from the western side of Baffin Bay. The total number formed each year is estimated at as many as 40,000. Icebergs vary considerably in dimensions (height

above sea level/length), from 1.5 m/5 m for "growlers," 1–5 m/10 m for "bergy bits," 5–15 m/15–60 m for small bergs, and 50–100 m/ 120–220 m for large bergs. The ratio of volume below sea level to that above is close to 7 to 1, but the ratio of maximum depth below sea level to height above it is less than this, depending on the shape of the iceberg.

Icebergs have a large draft, so their movements are chiefly determined by ocean currents. (Pack ice motions are much more determined by wind stress.) Icebergs have an average life of 2–3 years. They may travel up to 4000 km from their origin in west Greenland. From there they move northward in the WGC, across Baffin Bay, and then south in the Baffin Island and Labrador Currents at about 15 km/day, many becoming grounded on the shelf. A small proportion passes into the North Atlantic off Newfoundland where they are usually a few tens of meters high. The tallest recorded in this region was 80 m high and the longest was about 500 m. The main season for icebergs in the Grand Banks region is from March to July. Since its inception in 1914 following the Titanic disaster in 1912, the U.S. Coast Guard International Ice Patrol has provided information about icebergs coming south in the Labrador Current to the Grand Banks region. The regular annual surveys of both ice and oceanographic conditions, as well as basic descriptions and understanding of conditions in this region, provide a century of information about the ice, circulation, and water masses of the Labrador Sea and Newfoundland regions (Chapter 9).

12.8. CLIMATE VARIATIONS AND THE ARCTIC

The Arctic Ocean and Nordic Seas are central to Northern Hemisphere climate variability. Four modes of climate variability/change are frequently used for describing variability in this region: the *Arctic Oscillation* (also called the

Northern Annular Mode), the *North Atlantic Oscillation*, the *Atlantic Multidecadal Oscillation*, and global change driven by anthropogenic forcing. Anthropogenic climate change scenarios show the largest temperature changes in the Arctic. Arctic sea ice extent and volume have been decreasing and the ice has been becoming younger and thinner since the late 1970s. Climate feedbacks involving Arctic sea ice cover are central to understanding and forecasting climate change. Upper ocean temperature structure, which is affected by sea ice and salinity as well as by circulation and air—sea fluxes, is an important factor for understanding sea ice.

All of the remaining text, figures, and tables relating to Arctic climate variability are found in Section S15.4 in Chapter S15 (Climate Variability and the Oceans) on the textbook Web site. The following Arctic Ocean topics are covered in Section S15.4: (1) Arctic Oscillation or Northern Annular Mode, (2) Atlantic Multidecadal Oscillation, (3) variations in Arctic sea ice cover, and (4) variations in Nordic Seas and AW properties, including discussion of long-term trends that might reflect anthropogenically forced climate change. The North Atlantic Oscillation is discussed in Section S15.1.

13

Southern Ocean

13.1. INTRODUCTION

The "Southern Ocean" is the broad ocean region surrounding Antarctica (Figures 13.1 and 2.12). It is not a formal geographic region in the sense of the Pacific, Atlantic, or Indian Oceans or the many marginal seas, as it is not surrounded by continental land masses. However, the concept of a Southern Ocean is important because the latitude range of the Drake Passage between South America and the Antarctic Peninsula has no north-south boundaries (except in the deep water). As a result, the strong Antarctic Circumpolar Current (ACC) flows continuously eastward, encircling Antarctica without wrapping back to the west; it dominates the Southern Ocean's large-scale circulation. There is no western boundary at the Drake Passage latitudes to support western boundary currents and wind-driven gyres in the upper ocean, although deep topography does provide barriers for western boundary currents in the deep and abyssal waters (Section 7.10.3). The ACC is the ocean's closest analog to the major wind systems, the westerlies and easterlies, since the atmosphere also has no boundaries. However, adding to the complexity of the ACC, its strongest currents lie mostly north or south of the Drake Passage, where there are western boundaries (South America to the north and the Antarctic Peninsula to the south), with just a brief sojourn within the actual Drake Passage. The coastline of Antarctica, south of the ACC, includes two major indentations: the Weddell and Ross Seas. These do have western boundaries and thus support regional wind-driven gyres with western boundary currents.

The Southern Ocean is bounded to the south by the Antarctic continent. Its northern "boundary" is not well defined. The Antarctic Treaty Limit at 60°S could be taken as a political northern limit of the Southern Ocean. However, the Southern Ocean oceanographic regime extends well north of 60°S. If the presence of the ACC is used to define the Southern Ocean, then its northernmost boundary is at about 38°S, which is the northernmost excursion of the ACC (Figure 13.1; Section 13.3). The most inclusive definition in recent use extends the region up to 30°S to fully encompass all Southern Ocean phenomena northward to the Subtropical Front in each ocean (Chapters 9–11). We do not insist on one definition of the Southern Ocean. The processes described in this chapter are associated mainly with the ACC and regions to its south, and also include the connections between the ACC and the ocean basins to its north.

The narrowest constriction north of Antarctica is the Drake Passage, between South America and the Antarctic Peninsula. The complicated bathymetry here and to the east, in the Scotia Arc, presents the greatest latitudinal blockage for the flow of the ACC. Two wider

437

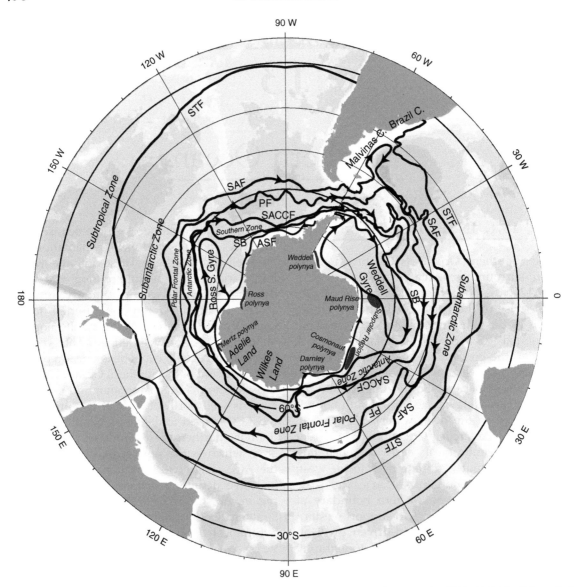

FIGURE 13.1 The Southern Ocean geography, principal fronts, and oceanographic zones (see Table 13.1). The Subtropical Front (STF) is the oceanographic northern boundary for the region. The eastward Antarctic Circumpolar Current (ACC) includes these fronts: Subantarctic Front (SAF), Polar Front (PF), Southern ACC Front (SACCF), Southern Boundary (SB). Front locations from Orsi et al. (1995). The westward Antarctic Slope Front (ASF) (thin) follows the continental slope. Circulation of the ocean basins north of the SAF is not represented; see the maps in Chapters 9, 10 and 11. Major polynyas (dark gray patches) are labeled; all polynyas are shown in Figure 13.20.

constrictions are set by southern Africa and Australia. In all three constrictive regions, the southward-flowing subtropical western boundary currents (Brazil Current, Agulhas, and East Australian Current) interact with the Southern Ocean circulation. Mid-ocean ridges cross through the Southern Ocean, resulting in strong steering of the ACC through gaps in the ridges. Several large undersea plateaus (Kerguelen, Campbell, and Falkland) deflect the ACC. In the latitudes of Drake Passage, deep topography that can allow meridional geostrophic flow occurs in the Drake Passage-Scotia Arc region, Kerguelen Plateau, and Macquarie Ridge south of New Zealand (Warren, 1990; Section 13.5).

Because the ACC connects the three major ocean basins and because it is a deep-reaching current, it is the vehicle for most flow between the oceans. (There is a small transport of about 1 Sv from the Pacific to the Atlantic through the Arctic Ocean, and a transport of 10—15 Sv between the Pacific and Indian Oceans through the Indonesian passages, but these are weaker than the more than 100 Sv in the ACC.) The unique character of deep-water masses originating in each of the oceans is present in the ACC. The waters mingle, upwell, are transformed into both denser and lighter waters, and then re-emerge to enter the ocean basins to the north of the ACC.

Because of its high southern latitude and sea ice formation, the Southern Ocean produces its own very dense deep and bottom waters, mostly along the coast of Antarctica. These dense waters fill the deepest part of the oceans to the north.

13.2. FORCING

The annual mean wind forcing for the Southern Ocean is dominated by westerlies in the latitude band 40—60°S and easterlies closer to Antarctica, south of 60°S (Figure 13.2a). The westerlies are not zonally uniform. They are maximum in the Indian Ocean sector, centered at about 50°S. The westerlies also have a significant southward component, especially in the eastern Indian Ocean, south of Australia. The westerlies drive northward Ekman transport in the latitudes of the *Subantarctic Front* (SAF) and *Polar Front* (PF), which are part of the ACC (Section 13.3). The net transport northward across the (circumpolar) SAF is significant, on the order of 30 Sv, which must be fed by upwelled water from the south.

The wind stress curl is associated with Ekman upwelling and downwelling (Section 7.5.4). The zero wind stress curl, associated with zero Ekman upwelling, occurs at the maximum wind stress, which is around 50°S. Upwelling (positive values in Figure 13.2a) occurs south of this, with highest Ekman upwelling rates closer to the continent. Ekman downwelling occurs north of the westerly wind maximum and is strongest in the eastern Atlantic and throughout the Indian Ocean sector.

Close to the Antarctic continent, the winds are easterlies, and can be very strong as a result of the continental forcing (katabatic winds, which also include a northward component). The easterlies drive Ekman transport toward the continent, inducing downwelling at the boundary. This results in a mounding of the sea surface and deepening of the pycnocline next to the continent, which results in the westward geostrophic flow found near the continent at most locations.

Surface buoyancy forcing is the sum of air—sea heat and freshwater fluxes. These two separate components are shown globally in Figures 5.4 and 5.12 and S5.8, which can be seen on the textbook Web site http://booksite.academicpress.com/DPO/; "S" denotes supplemental material. The net buoyancy flux, converted to equivalent heat flux units of W/m^2 as in Figure 5.15, is shown for the Southern Ocean in Figure 13.2b. The net buoyancy flux south of about 45°S is positive, meaning that the surface waters become less dense. (This

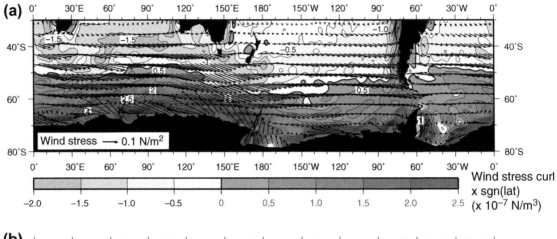

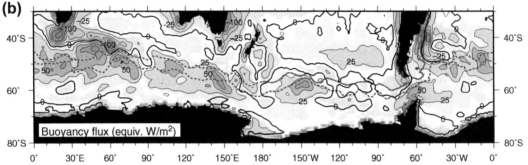

FIGURE 13.2 (a) Annual average wind stress (N/m²) (vectors) and wind stress curl (× 10⁻⁷ N/m³) (shading) multiplied by −1 in the Southern Hemisphere so that positive values (dark grays) indicate Ekman upwelling, from the NCEP reanalysis 1968–1996 (Kalnay et al., 1996). (b) Annual mean air–sea buoyancy flux, converted to equivalent heat flux (W/m²), based on Large and Yeager (2009) air–sea fluxes. Positive values indicate that the ocean is becoming less dense. Contour interval is 25 W/m² (grid-scale contouring along the Antarctic coast has been removed). Dashed contours are the Subantarctic and Polar Fronts from Orsi et al. (1995).

map is clearly missing the buoyancy loss and hence density gain in the coastal polynyas, which are not represented in these products, and which produce the deep and bottom waters of the Antarctic.) This is the only large region of the world ocean where freshwater fluxes are a significant contribution to the net air–sea flux, but heat fluxes here are of similar magnitude and warm the ocean. How can such a cold, high latitude region be warming on average? The upwelling of very cold water and its subsequent northward Ekman transport appear to control the air–sea fluxes such that

the slightly warmer maritime air equilibrates the cooler water. The highest buoyancy/heat gain occurs along the SAF and PF of the ACC, where westerly winds and hence northward Ekman transport are high.

In the Southern Ocean, the regions of highest buoyancy loss, due almost entirely to heat loss (Figures 5.15 and S5.8 in the supplemental material on the textbook Web site), are in the western boundary current regions (Agulhas, East Australian Current, and Brazil Current), and in the Leeuwin Current (west coast of Australia). Annual mean heat losses exceed 100 W/m² in these regions. A zonal band of buoyancy (heat) loss extends along the Agulhas Return Current, stretching southeastward from Africa more than half way across the Indian Ocean, with values exceeding 25 to 50 W/m^2. In the Pacific Ocean, there is also a quasi-zonal band of buoyancy loss in a similar position north of the SAF. The highest buoyancy loss regions are associated with southward mean flows, which bring warmer waters into cooler regimes.

13.3. SOUTHERN OCEAN FRONTS AND ZONES

Because of the open zonal passage and the nearly zonal ACC (Section 13.4), isopleths of all properties in the Southern Ocean are nearly zonal (east-west) to great depth. Near-surface potential temperature, salinity, and geopotential anomaly (Figures 13.3 and 13.7) illustrate this zonal nature, particularly in the latitude range of Drake Passage. South of the ACC, the surface properties and circulation are organized by the cyclonic gyres (clockwise in the Southern Hemisphere) in the Weddell and Ross Seas, and are not as zonal as in the ACC latitude band.

The nearly zonal isopleths of properties in the ACC are organized into three major fronts separating four broad zones in which isopleths are more widely spaced (Figure 13.1). Within the fronts, the currents are strong and eastward. In the zones between the fronts, the flow is dominated by eddies and can be in any direction.

The fronts encircling Antarctica as part of the ACC are the SAF, the PF, and the *Southern ACC Front* (SACCF; Figure 13.1 and Table 13.1). South of the SACCF, Orsi, Whitworth, and Nowlin (1995) define the *Southern Boundary* (SB), which is the southern edge of the low oxygen layer of the *Upper Circumpolar Deep Water* (UCDW); it is not a dynamical front (Section 13.5.3 below). Separate from, and south of the ACC, the *Antarctic Slope Front* (ASF) is found at most locations along the continental slope, with westward flow and separating very dense shelf water from offshore water (Jacobs, 1991; Whitworth, Orsi, Kim, & Nowlin, 1998). On the shelves, especially where they are broad, and close to the coast, the westward flow of the *Antarctic Coastal Current* (ACoC) is found.

The fronts separate the zones: *Subantarctic Zone* (SAZ; north of the SAF), *Polar Frontal Zone* (PFZ; between the SAF and the PF), *Antarctic Zone* (AZ; between the PF and SACCF), the *Southern Zone* (SZ; between the SACCF and SB), and the *Subpolar Region* (south of the SB; Orsi et al., 1995). The Subpolar Region (or *Subpolar Zone*; SPZ) includes the Weddell and Ross Sea gyres. On the continental shelf south of the ASF, dense shelf water is found; this can be considered the Continental Zone.

This classification scheme, from Orsi et al. (1995), supersedes a commonly used older scheme that did not include the SACCF and SB, but instead identified a Continental Water Boundary and a *Continental Zone* (CZ). The older schemes were only appropriate for the Drake Passage region.

The fronts and zones, as well as the typical meridional (north-south) circulation and water masses, are summarized schematically in Figure 13.4. The water masses are discussed in Section 13.5.

13.3.1. Fronts

Mean positions of the ACC fronts from Orsi et al. (1995) as well as the ASF are shown in Figure 13.1. In Figure 13.5 the ACC fronts are shown in two regions that have been very well mapped and described. The strong impact of topography on the paths of the fronts is apparent. In the Drake Passage and southwest Atlantic (Figure 13.5a), the SAF follows the boundary closely; it is an actual western boundary current along the coast of South America, where it is called the Malvinas (or Falkland) Current. The other ACC fronts flow

(a)

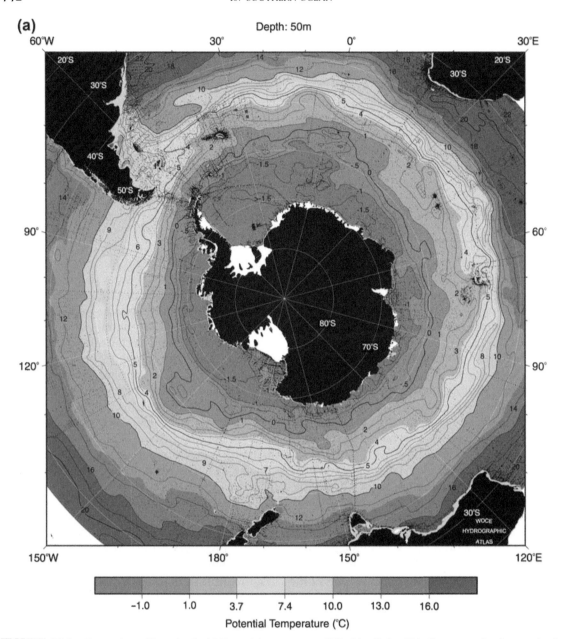

FIGURE 13.3 Properties at 50 m depth. (a) Potential temperature (°C), (b) salinity. This figure can also be seen in the color insert. *Source: From WOCE Southern Ocean Atlas, Orsi and Whitworth (2005).*

(b)

Depth: 50m

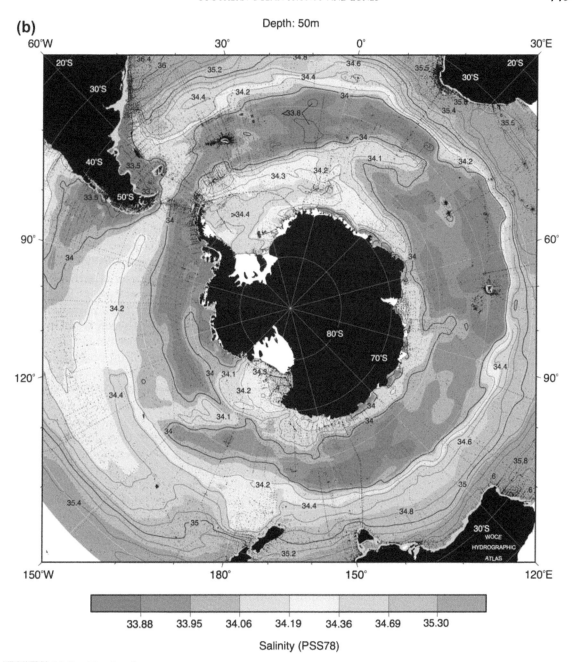

FIGURE 13.3 (*Continued*).

TABLE 13.1 Fronts and Zones of the Antarctic Circumpolar Current and Southern Ocean

Feature	Acronym	Short Description
Subantarctic Front	SAF	Northernmost ACC front
Polar Front	PF	Central ACC front
Southern ACC Front	SACCF	Southernmost dynamical ACC front
Southern Boundary	SB	Mostly along the continental shelf, but also including the Weddell Sea front
Antarctic Slope Front	ASF	Continental slope front, south of the ACC
Antarctic Coastal Current	ACoC	Westward coastal flow
Subantarctic Zone	SAZ	North of the SAF
Polar Frontal Zone	PFZ	Between the SAF and PF
Antarctic Zone	AZ	Between the PF and SACCF
Southern Zone	SZ	Between the SACCF and SB
Subpolar Region	SPZ	Between the ASF and SB
Continental Zone	CZ	South of the Antarctic Slope Front

Following Orsi et al., (1995) and Whitworth et al., (1998).

through passages in the many island chains and loop around following the topography. The SAF and PF merge together along the northern edge of the Falkland Plateau. This merger of ACC fronts is not uncommon; the distinction between the SACCF and SB is also not always strong.

Likewise, in some regions the fronts are split into multiple stable fronts, as observed between Tasmania and Antarctica (Figure 13.5b). Again these fronts are strongly influenced by topography, in this case the Southeast Indian Ridge and Macquarie Ridge.

The ACC fronts are sharpest in or just below the surface layer. They are associated with strongly sloping isopycnals in the water column below, over much wider latitude ranges than the surface fronts (Figure 13.6). These underlying zones of steeply sloping isopycnals are referred to using the surface front nomenclature (SAF, PF, etc.). Most of the eastward flow of the ACC is carried in these fronts.

The ACC fronts and the ASF are also associated with transitions in water properties, as depicted in potential temperature-salinity (T-S) profiles crossing the ACC (Figure 13.7). These transitions are often the practical means of identifying the ACC fronts, especially when using large data sets for which detailed examination of each crossing of the ACC is impractical, or for which velocity measurements are not available. Such "proxy" markers of the fronts include: (1) the existence of a particular water property (e.g., temperature, temperature gradient, salinity, oxygen) at a particular depth and (2) the transition between water property regimes typical of the zones between the fronts. These markers are based on observations that link the strongest eastward currents with subsurface temperature and salinity structure. The markers are not completely robust; they may vary from region to region and the fronts split into many different time-dependent fronts in some regions, or merge in others (e.g., Figure 13.5). But the markers are a useful starting point for finding the fronts in many regions.

A vertical section south of Tasmania that crosses all the ACC fronts is used to illustrate the indicators of the fronts (Figure 13.6).

The SAF is the northern edge of the ACC. It was first identified in the region south of Australia, and it exists in all other sectors of the Southern Ocean (Emery, 1977; Orsi et al., 1995). The SAF has large eastward flow, reflected in steeply sloping isopycnals at all depths. At most locations the SAF is the southernmost limit of the low salinity intermediate layer, Antarctic Intermediate Water (AAIW), and of the thick layer of surface water, Subantarctic Mode Water (SAMW; Section 13.5). Numerous other indicators of the SAF have also been used, many

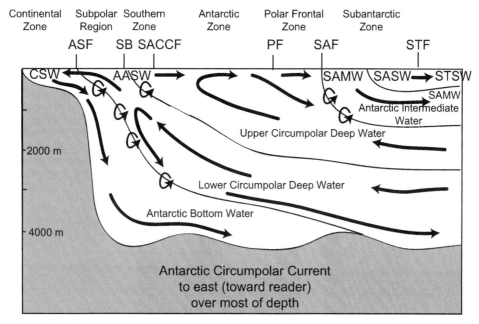

FIGURE 13.4 A schematic meridional section in the Southern Ocean showing the water masses, meridional circulation, fronts, and most zones. Acronyms: Continental Shelf Water (CSW), Antarctic Surface Water (AASW), Subantarctic Mode Water (SAMW), Subantarctic Surface Water (SASW), Subtropical Surface Water (STSW), Antarctic Slope Front (ASF), Southern Boundary (SB), Southern ACC Front (SACCF), Polar Front (PF), Subantarctic Front (SAF), and Subtropical Front (STF). *After Speer, Rintoul, and Sloyan (2000).*

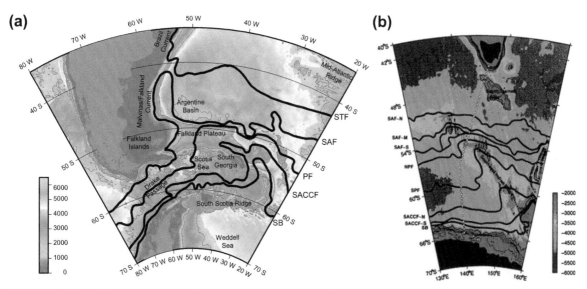

FIGURE 13.5 (a) Drake Passage and southwest Atlantic fronts. (Fronts from Orsi et al., 1995; bathymetry (m) from Smith & Sandwell, 1997.) (b) Fronts south of Australia (Tasmania). N, M, and S refer to northern, middle, and southern branches of the given fronts. *Source: From Sokolov and Rintoul (2002).*

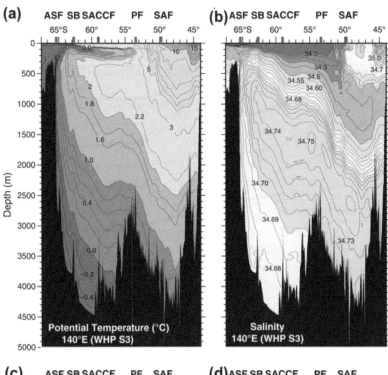

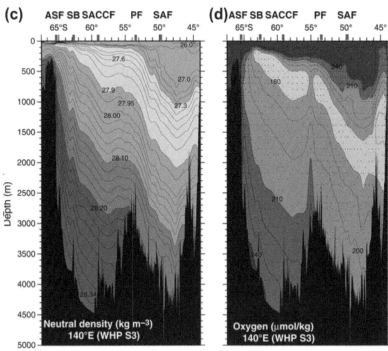

FIGURE 13.6 (a) Potential temperature (°C), (b) salinity, (c) neutral density (kg m^{-3}), and (d) oxygen (μmol/kg) along 140°E from Antarctica to Tasmania (WOCE Hydrographic Programme Atlas section S3, from Talley, 2007). Fronts: Subantarctic Front (SAF), Polar Front (PF), Southern ACC Front (SACCF), Southern Boundary (SB), and Antarctic Slope Front (ASF). Location of section is shown by station dots in Figure 13.5b.

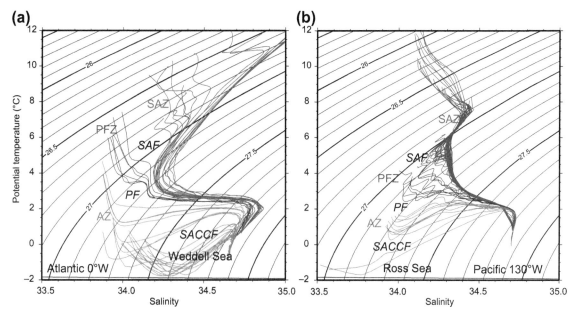

FIGURE 13.7 Potential temperature-salinity relations: (a) Atlantic Ocean (Greenwich meridian) and (b) Pacific Ocean (130°W), encompassing the fronts and zones of the ACC (Table 13.1). Contours are potential density σ_θ (kg/m^3). Line near bottom is the freezing point. Acronyms as in Table 13.1: SAZ (Subantarctic Zone), SAF (Subantarctic Front), PFZ (Polar Frontal Zone), PF (Polar Front), AZ (Antarctic Zone), SACCF (Southern ACC Front).

associated with large horizontal (north-south) changes in properties in the upper ocean (see list in Belkin & Gordon, 1996). The SAF can be identified at many locations by the occurrence of the 4 or 5°C isotherm at 200 m depth, or with a maximum horizontal gradient between the 3 and 5°C isotherms (Sievers & Emery, 1978). All of these indicators are present in Figure 13.6.

The SAF is farthest north (39–40°S) in the western South Atlantic, just off the coast of Argentina (Figures 13.1 and 13.5a). It shifts southward as it progresses toward the east and is farthest south (about 58°S) when it reaches the eastern South Pacific and the Drake Passage. At the eastern end of Drake Passage, the SAF is close to the northern boundary at 55°S. As it leaves Drake Passage, the SAF hugs the western boundary, shooting northward to about 39°S to regain its northernmost

position. Along this coast, the SAF is a true western boundary current — the Malvinas Current.

The PF is within the ACC and is also a strong eastward flow. The PF is identified by water properties as the northern edge of the shallow temperature minimum (Section 13.5). Again, there are numerous indicators (see summary in Belkin & Gordon, 1996). For instance, in most regions it can be identified as the northernmost location of the 2°C isotherm surrounding the temperature minimum layer (Botnikov, 1963; Joyce, Zenk, & Toole, 1978; Orsi et al., 1995), or as the location where the shallow temperature minimum begins to steeply descend toward the north. The PF is found, on average, at about 50°S in the Atlantic and Indian Oceans and at about 60°S in the Pacific, reaching its southernmost location of about 63°S west of the Drake Passage (Figures 13.1 and 13.5).

The SACCF, introduced by Orsi et al. (1995), is a major front with a large current near the southern side of the ACC. Practical indicators of the SACCF, at least in the southwest Atlantic, include potential temperature less than 0°C in the temperature minimum at depths shallower than 150 m, or potential temperature greater than 1.8°C in the potential temperature maximum at depths greater than 500 m (Meredith et al., 2003). It is distinct from the SB, described next, in that the SACCF is a strong dynamical feature, whereas the SB marks the southern edge of the ACC in terms of water properties.

The SB is the southern boundary of the oxygen minimum that characterizes the UCDW (Section 13.5.3). Of the major ACC water masses, only Lower Circumpolar Deep Water (LCDW) is found south of this. The SB is also the northern boundary of the very cold, nearly isothermal water mass found near Antarctica. The SB is circumpolar in extent. It was first observed and defined as the Continental Water Boundary from observations in Drake Passage (Sievers & Emery, 1978). Since this circumpolar boundary is not close to the Antarctic continent in large regions such as the Weddell Sea, the more general "Southern Boundary" was proposed (Orsi et al., 1995) and is used here. It is located at the continental shelf only along the west side of the Antarctic Peninsula.

The ASF lies along the continental slope at many locations around Antarctica (Whitworth et al., 1998). Flow in the ASF is westward. It is mainly characterized by a pycnocline that angles downward toward the continental slope, due both to Ekman downwelling driven by the easterly winds, and to downward penetration of dense shelf waters. The ASF separates very cold, dense waters on the continental shelf from the offshore waters of the SZ, which include *Antarctic Surface Water* (ASW) and upwelled Lower Circumpolar Deep Water (LCDW). Figure 13.17 shows a good example of the front, which is "V-shaped." The ASF is absent along the western side of the Antarctic Peninsula, where the ACC comes close to the continental slope and isopycnals slope upward rather than downward.

The ACoC is a westward coastal current that lies on top of the continental shelf, within the dense shelf water. It is not a water mass boundary. It is sometimes nearly identical to the ASF, especially where the continental shelf is narrow. On the other hand, in some places, notably along the western side of the Antarctic Peninsula, the only westward flow is in the ACoC (Klinck et al., 2004).

13.3.2. Zones

The SAZ is the region north of the SAF. At most longitudes in the SAZ, salinity decreases downward to a minimum value at 500 m or deeper, and then increases below this. This salinity minimum is known as Antarctic Intermediate Water (AAIW; Section 13.5.2). The higher salinity surface waters above the salinity minimum are characteristic of the evaporative subtropical gyres. Close to the SAF, the SAZ is also characterized by a thick, near-surface layer of nearly uniform properties, known as Subantarctic Mode Water (SAMW; Section 13.5). The northern boundary of the SAZ can be taken to be the Subtropical Front, located at about 30°S in each ocean (Figure 13.1).

Despite the name "Subantarctic," the SAZ is the poleward part of the subtropical circulation regime in the Pacific, Atlantic, and Indian Oceans, in which surface flow is dominantly eastward. A difference between the Southern Ocean subtropical regimes and the two Northern Hemisphere subtropical gyres is that part of the eastward flow in the SAZ in the Pacific leaks through the Drake Passage into the South Atlantic's SAZ. Also, the SAZ is continuous from the Atlantic to the Indian Ocean, because the eastward part of the subtropical circulation connects these two oceans.

The PFZ (Gordon, Georgi, & Taylor, 1977) is between the PF and the SAF. This zone varies dramatically in both width and shape. There are places where the SAF and PF merge, particularly in the southwestern Atlantic, and there is no PFZ at all. Within the PFZ, there is a dramatic transition in T-S characteristics from the almost isothermal T-S curve of the ASW that is south of the PF to the much warmer and more saline conditions north of the SAF (Figure 13.7). The T-S relation in the PFZ is complicated due to interleaving. (Interleaving is apparent as zigzagging between the T-S profiles of the AZ and the T-S profiles of the SAZ.)

The PFZ is occupied by strong eddies that form as northward meanders of the PF (Savchenko, Emery, & Vladimirov, 1978) or southward meanders of the SAF (Figure 13.18). The cold PF eddies can move northward to become linked with the SAF, and thus carry water from south of the PF across to the north part of the PFZ, and vice versa, contributing to the meridional exchange of heat between the north and the south (Section 13.6).

The AZ is south of the PF and north of the SACCF. It is characterized by a thin surface layer of cold ASW (Section 13.5.1) with low salinity from summer melting of sea ice. In non-winter profiles, there is a subsurface temperature minimum in the upper 200 m, with temperatures from −1.5 to 2°C.

The SPZ lies between the SACCF and SB. Within the SPZ, the low oxygen UCDW upwells to the surface and is converted to very cold ASW.

The SZ lies between the SB and the Antarctic Shelf Front. In some sectors, this is a very broad region, encompassing most of the Weddell and Ross Sea gyres. In other regions, it is extremely narrow, such as where the SB impinges on the continental shelf west of the Antarctic Peninsula.

In the CZ, south of the Antarctic Shelf Front, there is a very cold water mass (<0°C). In winter the layer is nearly isothermal and extends to great depth (>500 m). Its density is controlled by salinity. In some locations, these continental shelf waters are the source of the very dense deep and bottom waters known as Antarctic Bottom Water (AABW; Section 13.5.4).

13.4. SOUTHERN OCEAN CIRCULATION AND TRANSPORTS

The Southern Ocean circulation is dominated by the strong, deep, eastward-flowing current known as the Antarctic Circumpolar Current (ACC), which runs completely around the globe (Figures 13.1, 13.8, 14.2). The ACC was once known as the "West Wind Drift" because it is partially driven by the strong westerly winds in the region, that is, winds from the west causing the ocean to flow to the east. The westerly wind in the Southern Ocean was notorious in sailing ship days and, together with the eastward current, made it difficult for such vessels to round Cape Horn from the Atlantic to the Pacific. The wind stress, combined with the Coriolis force, also contributes a northward Ekman component to the surface current. This affects the formation of sharp fronts (Section 13.3) and convergences. The northward Ekman transport is an important part of the meridional overturn of the Southern Ocean. Below this wind-driven surface layer, the density structure appears to be in geostrophic balance with the circulation.

The ACC is not purely zonal. The ACC as a whole is farthest north just off the coast of Argentina in the southwest Atlantic (northern edge at 38°S), and farthest south just west of the Drake Passage in the southeast Pacific (northern edge at 58°S). The 2000 km southward spiral of the ACC from the western Atlantic to the eastern Pacific has important consequences for the water masses of the Southern Ocean (Section 13.5).

South of the ACC are two cyclonic "subpolar" gyres, one in the Weddell Sea and the other in the Ross Sea. These gyres result in

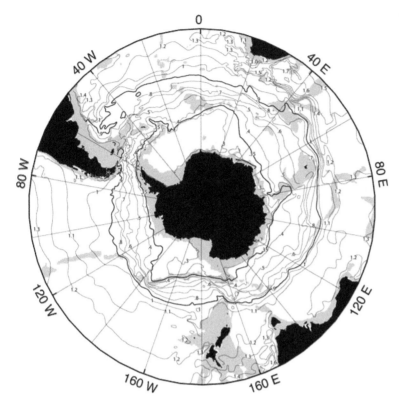

FIGURE 13.8 Geopotential height anomaly at 50 dbar relative to 1000 dbar, in dynamic meters (10 J kg^{-1}). *Source: From Orsi et al. (1995).*

westward flow along the Antarctic coast, as seen in the surface steric height maps for the Atlantic, Pacific, and Indian Oceans (Figure 13.8 and Figures 9.2, 10.2, 11.7). A nearly continuous circumpolar westward flow driven by easterly winds was hypothesized by Deacon (1937); these two gyres and westward flow along the continental shelf break in the Indian Ocean result in such a picture, but with no apparent westward flow in Drake Passage (Figure 13.9).

13.4.1. Antarctic Circumpolar Current

Early concepts of the ACC were that it is a broad current of uniform velocity. It is now clear that the ACC is composed of a series of narrow jets that provide the overall large eastward transport of the ACC (Section 13.3). The narrow jets are confined within the broader

envelope of the ACC defined by the southern- and northernmost streamlines that are continuous all the way around Antarctica (Figure 13.8; see also Figure 14.2). In its circuit around the continent, the ACC is severely obstructed in the narrow Drake Passage (Figure 13.9), followed downstream by a major northward excursion along the western boundary of South America (Malvinas/Falkland Current). In the Australasian sector, the bottom topography of Campbell Plateau (New Zealand) also constricts the ACC, again accompanied by a northward excursion of the ACC with the plateau acting as a western boundary. The ACC path is also affected by mid-ocean ridges as mentioned in Section 13.1.

The ACC jets extend to the ocean bottom, with bottom velocities in the same direction as the jets (Section 13.4.3). This means that

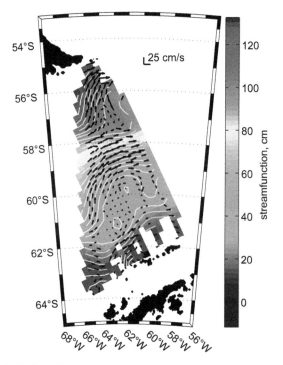

FIGURE 13.9 Mean currents in the Drake Passage, averaged over 30–300 m depth, from 128 ADCP crossings over 5 years. Strong currents from north to south are the Subantarctic Front (56°S), the Polar Front (59°S), and the Southern ACC Front (62°S). This figure can also be found in the color insert. *After Lenn, Chereskin, and Sprintall (2008).*

current profiler (ADCP) velocities and expendable bathythermograph (XBT) temperature profiles (Figure 13.9). A long time series is also continuing between Tasmania and Antarctica, and other estimates have been made in the central South Atlantic and Indian Oceans.

Mean surface current speeds for the whole of the ACC are about 20 cm/sec. However, as noted previously, most of the flow is carried in the fronts. The three jets in the Drake Passage were first clearly identified from data collected in 1976 (Nowlin, Whitworth, & Pillsbury, 1977; Figure 13.5). From surface drifters throughout the Southern Ocean, the highest speeds are in the SAF, with means from 30 to 70 cm/sec; the PF is nearly as energetic with mean speeds of 30 to 50 cm/sec (Hofmann, 1985). Within the Drake Passage, near-surface SAF and PF speeds range up to about 50 cm/sec; the SACCF speeds are somewhat lower (Figure 13.9).

Most ACC transport measurements have been made in the Drake Passage because here the ACC is clearly limited to the north and south. The frontal structure observed in the Drake Passage is the canonical structure described in 13.3.1. Even just east of the Passage, the PF splits in two where the ACC encounters the Falkland Plateau (Arhan, Naveira Garabato, Heywood, & Stevens, 2002). In the other intensely observed region of the ACC, south of Tasmania, the SAF, PF, and SACCF are each normally two or more separate fronts (Sokolov & Rintoul, 2002).

Transport estimates for the ACC in the Drake Passage from 1933 through 1988 were published by Peterson (1988). An early credible estimate of 110 Sv (see Sverdrup, Johnson, & Fleming, 1942) is in the range of present estimates. The first modern observations were made in the 1970s during the International Southern Ocean Study, using a combination of current meters, geostrophic calculations, and pressure gauges. Means of 124 Sv (range 110 to 138 Sv), 139 Sv (range 28 to 290 Sv), and 134 Sv (range 98 to 154 Sv) were estimated using different sets of

transport estimates based on temperature and salinity measurements and the geostrophic method with a "depth of no motion" are too low. Direct current measurements are required for total transports, at a minimum to provide a reference velocity for geostrophic velocity calculations.

Velocity and transport measurements of the ACC have been made at a number of locations. Because the Drake Passage is constricted and relatively easy to access, the most comprehensive observations have been made there, starting in 1933 and continuing to the present (see summary in Peterson, 1988). Recent monitoring programs include annual hydrographic sections and monthly sections of acoustic Doppler measurements covering different lengths of time (respectively, Nowlin et al., 1977; Bryden & Pillsbury, 1977; Whitworth & Peterson, 1985).

Using six repeated hydrographic sections from 1993 to 2000 in the Drake Passage, Cunningham, Alderson, King, and Brandon (2003) reported a baroclinic eastward transport of 107.3 Sv ± 10.4 Sv relative to no motion at 3000 m. Most of this transport is in the SAF (53 ± 10 Sv) and PF (57.5 ± 5.7 Sv). Along a vertical section just east of the Drake Passage, between the Falklands and South Georgia Island, Arhan et al. (2002) found a mean eastward transport of 129 ± 21 Sv concentrated in the SAF (52 ± 6 Sv), and in two branches of the PF — one located over the sill of the Falkland Plateau (44 ± 9 Sv) and the other in the northwestern Georgia Basin (45 ± 9 Sv).

In the Australian sector, between Tasmania and Antarctica, six repeated hydrographic sections yielded a mean transport of 147 Sv relative to the ocean bottom (Rintoul, Hughes, & Olbers, 2001). This is larger than the Drake Passage transport. Eastward transport south of Australia includes a contribution on the order of 10 Sv that enters the Pacific and flows back into the Indian Ocean north of Australia through the Indonesian Passages (Chapters 10, 11).

The assumption of zero velocity at the ocean bottom for referencing geostrophic transports in the ACC is a useful starting point since flow in the fronts is in the same direction from top to bottom. However, since bottom velocities are on the order of 4 to 10 cm/sec based on direct current measurements, such zero-at-the-bottom-referenced transports can have large errors (Donohue, Firing, & Chen, 2001). Global and Southern Ocean inverse models, which use geostrophic velocities calculated from hydrographic data constrained so that transports through closed sections must balance, provide independent estimates of net transport of the ACC. Macdonald and Wunsch (1996) obtained 142 Sv through the Drake Passage and 153 Sv between Australia and

Antarctica. Sloyan and Rintoul (2001) obtained 135 Sv for the Drake Passage and the section between Africa and Antarctica, and 147 Sv between Tasmania and Antarctica, similar to the bottom reference level results.

13.4.2. Weddell and Ross Sea Gyres

The cyclonic circulations south of the ACC in the Weddell and Ross Sea gyres are important sites for formation of the densest waters in the Antarctic and hence the global ocean. The Weddell gyre is separated from the ACC by the Weddell gyre front, which is identical with the SACCF and is nearly co-located with the SB (Section 13.3.1; Orsi et al., 1995). Within the Weddell Sea, the flow is cyclonic (Figure 13.8). The track of the *Endurance* from 1914 to 1916, led by Sir Ernest Shackleton, illustrates the cyclonic flow as the ship became frozen into the pack ice (Figure 13.10).

The Weddell Gyre extends far to the east, to the longitude of Africa, to 20°E at about 54°S (Orsi, Nowlin, & Whitworth, 1993). Its northern boundary is the Scotia Ridge in the west and

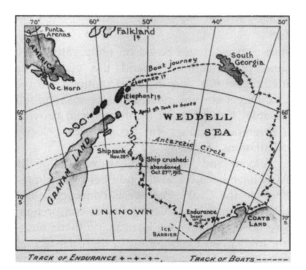

FIGURE 13.10 Track of the *Endurance* (1914—1916). *Source: From Stone (1914); © Royal Geographical Society.*

then it approximately follows the 4000 and 5000 m bottom contour. There may be two separate cyclonic gyres contained within the full Weddell Gyre, centered at 30°W and 10°E. Southward flow into the Weddell Gyre carries water from the ACC and, therefore, the oceans north of the ACC.

The Weddell Gyre has a western boundary current that flows northward along the Antarctic Peninsula. It carries new dense waters from the Weddell shelves.

The net transport of the Weddell Gyre had been estimated to be greater than 20 Sv based on an absolute velocity analysis (Reid, 1994) or 15 Sv relative to 3000 dbar (Orsi et al., 1995). Direct current measurements in the early 1990s suggest 30 to 50 Sv (Schröder & Fahrbach, 1999).

The southern Weddell Sea is occupied by the Filchner-Ronne ice shelf (Figure 13.11a). The eastern portion is the Filchner and the western portion is the Ronne, separated by Berkner Island. Beneath the ice shelf there is a sub-ice shelf cavity of seawater. This is mixed vigorously by tides, and is a factor in modifying water masses of the southern Weddell Sea (Makinson & Nicholls, 1999). Although the Weddell gyre is cyclonic, flow in the cavity appears to be anticyclonic, with ocean waters (new dense shelf water, Section 13.5.4) entering in the west, modified under the ice shelf, and emerging colder and fresher in the east. The outflow in the east from under the Filchner Ice Shelf is a major source of the dense shelf water that becomes Weddell Sea Bottom Water (Jenkins & Holland, 2002).

The Ross Sea gyre is in the Pacific sector of the Southern Ocean. Its northern edge is strongly associated with topography (like that of the Weddell gyre), following the Pacific-Antarctic Ridge. Its transport is on the order of 20 Sv based on absolute geostrophic velocities (Reid, 1997), or 10 Sv relative to 3000 m (Orsi et al., 1995), and it, too, has a northward western boundary current along Victoria Land, carrying dense shelf waters.

The Ross Sea ice shelf is the largest ice shelf in the world (Figure 13.11b). The sub-ice cavity beneath the ice shelf is an important site for

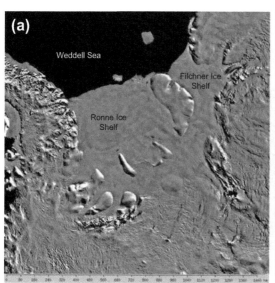 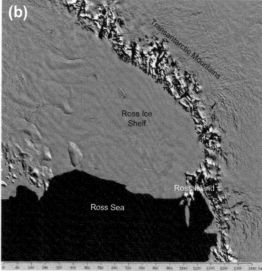

FIGURE 13.11 (a) The Filchner-Ronne Ice Shelf in the southern Weddell Sea. (b) The Ross Ice Shelf in the southern Ross Sea. *Source: From Scambos et al. (2007) database.*

dense shelf water formation and modification. Inflow is from the east and outflow is to the west and north.

There are numerous other ice shelves around Antarctica. The NSIDC (2009c) Web site is an excellent source of information.

13.4.3. Mid-Depth to Bottom Circulation

The eastward ACC extends from the surface through mid-depths to near the ocean bottom, as seen in the global circulation maps in

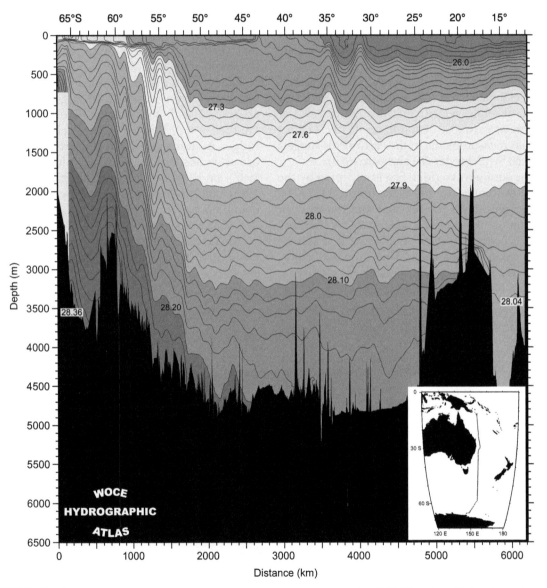

FIGURE 13.12 Neutral density section in the western Pacific into the Tasman Sea (WOCE section P11, location on insert). *Source: From WOCE Pacific Ocean Atlas, Talley (2007).*

Chapter 14 (Figures 14.1 through 14.4), and in the Pacific and Indian Oceans at 900 m from floats (Figures S10.13 and S11.6 from Davis, 2005 as seen on the textbook Web site). Whether specific portions of the currents reach the bottom depends on bottom depth and topography, but the ACC and the gyres are continuous to at least 3000 dbar (Reid, 1994, 1997, 2003). Below this depth, mid-ocean ridges begin to impede the continuous eastward progress of the ACC. At 3500 dbar there are no continuous streamlines through the Drake Passage, thus separating the Pacific and Atlantic below this depth.

By 4000 dbar, the circulation is broken into regional deep gyres confined within the deep basins. The gyres are cyclonic. Deep Western Boundary Currents (DWBCs) are evident as part of these deep gyres, especially along the eastern coasts of South America (into the South Atlantic) and New Zealand (into the South Pacific). The DWBCs carry dense AABW northward away from the continent. Because of the generally cyclonic deep flow, currents along the coast of Antarctica below about 3000 dbar are westward. This is an important route connecting dense Antarctic shelf waters from one formation region to another.

The Weddell and Ross Sea gyres also extend to the ocean bottom (Figure 14.4b). The Weddell gyre is evident to at least 5000 dbar, and the Ross Sea gyre down to at least 4000 dbar, both within the confines of their deep basins.

The geostrophic shear of the ACC is large, as reflected in the downward slope of isopycnals toward the north across the current (Figure 13.12). Surface currents decrease from about 50 cm/sec at the sea surface to 4–10 cm/sec at the bottom (direct current observations in the Pacific by Donohue et al., 2001). The ACC fronts, which are identified using potential temperature and salinity (not shown), are evident to the ocean bottom embedded within the general slope of the ACC isopycnals. The rise of mid-depth and abyssal isopycnals to the upper ocean south of the ACC is an important factor in creation of very dense AABWs from the deep waters that enter the ACC from the oceans to the north.

The top-to-bottom extent of the ACC is important to note since its dynamical balance is presumed to be between surface westerly wind stress and bottom stress associated with the topography. This differs from the dynamics of the wind-driven gyres of the rest of the oceans because there is no meridional boundary to support a western boundary current at the latitude of Drake Passage.

13.5. SOUTHERN OCEAN WATER MASSES

Water masses in the Southern Ocean can be considered in four layers: surface/upper ocean waters, intermediate waters, deep waters, and bottom waters (Figure 13.4; Table S13.2 in the online supplement on the textbook Web site). There are differing conventions for naming water masses in the Southern Ocean; we follow Whitworth et al. (1998) and Orsi, Johnson, and Bullister (1999). These are mostly identified by salinity, potential temperature, and potential density, although one of the deep-water masses (UCDW) is often identified by an oxygen extremum. The surface waters are of local origin. The one intermediate water of the Southern Ocean, AAIW, originates as a fresh, relatively dense surface layer in the Drake Passage region. The deep waters mainly originate from the Atlantic, Pacific, and Indian Oceans, and mingle and mix in the ACC to become CDW. It upwells south of the ACC where a portion becomes the source of the bottom waters around Antarctica. Some of these dense Antarctic waters also modify the CDW.

A potential T-S diagram (Figure 13.13) shows a typical station from the Atlantic from each of the zones of the ACC. The zones can be categorized by the water masses within them. The AZ, south of the PF, contains ASW, Upper and

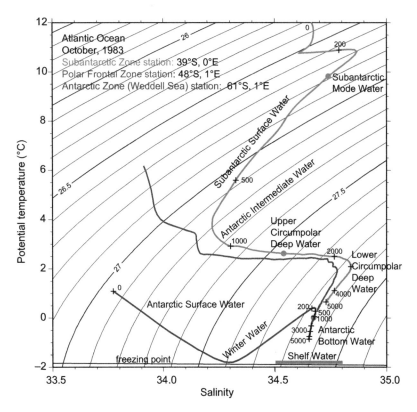

FIGURE   13.13 Potential temperature-salinity curve of Southern Ocean waters in the Atlantic sector showing the different water masses.

Lower CDW, Weddell Sea Deep Water, Ross Sea Deep Water, and the Bottom Waters. The PFZ, between the PZ and SAFs, contains the same water masses, but at greater depth. The SAZ contains Subantarctic Surface Water, SAMW, AAIW, and the deep and bottom waters.

13.5.1. Surface Waters

13.5.1.1. Subantarctic Surface Water and Subantarctic Mode Water

The Subantarctic Surface Water occupies up to 500 m of the upper water column north of the SAF. It has a temperature of 4 to 10°C in the winter and up to 14°C in summer, and a salinity from 33.9 to 34 psu in winter and as low as 33 psu in summer as ice melts. The lowest temperatures and salinities are found in the Pacific sector and the highest in the Atlantic

sector. Temperature and salinity of the surface water increase toward the north. A high salinity surface layer, to a depth of 150 to 450 m, is present in all sectors. This is the surface layer of the three (Atlantic, Indian, and Pacific) subtropical gyres, which are dominated by evaporation.

Within the Subantarctic Surface Water, just north of the SAF, there are very thick mixed layers in wintertime. These are known as SAMW (McCartney, 1977, 1982; Hanawa & Talley, 2001). In the central and eastern Indian Ocean, these mixed layers can reach to more than 500 m depth over a large region (Figure 4.4 and Figure S13.1 found on the textbook Web site). In the South Pacific, the winter mixed layer depths are not quite as extreme, but they are, nevertheless, greater than about 300 m thick at most longitudes. In the South Atlantic, the thick winter mixed layers are more modest, on the order of 200 m thick. These thick mixed layers are capped in summertime by surface warming, and are advected either eastward along the SAF or subducted northward into the subtropical gyres of the three oceans. Where they subduct, the thick surface mixed layers become thick layers within the permanent pycnocline. Given that these thick layers move around the circulation at speeds typical of the gyres, the volume transport of SAMW within the gyres is higher than that of thinner layers that are also subducted into the gyres. Therefore, the SAMW layer supplies a relatively large amount of surface water to the subtropical pycnocline. This may explain why the SAMW can be identified within the subtropical gyres by high oxygen content.

The meridional (south-north) sections of neutral density in Figures 13.6c and 13.12 show the local type of SAMW just north of the SAF. SAMW temperatures are warmest (>14°C) east of South America, where the SAF is farthest north. They decrease toward the east as the SAF moves southward. At the longitude of Australia, SAMW temperatures are 8−9°C. The very thick winter mixed layers in this region are the densest waters that outcrop in the Indian Ocean subtropical gyre, and they become the primary source of high oxygen to the base of the Indian Ocean pycnocline. For this reason, it is useful to apply a special name to the SAMW of this region: the *Southeast Indian SAMW* or *SEISAMW*.

In the South Pacific, the SAF continues to shift southward and the SAMW temperatures continue to decline toward the east to a minimum of about 4°C just west of Drake Passage. This is the coldest, densest (and also freshest) SAMW. This is nearly identical with the salinity minimum of the AAIW (Section 13.5.2). This southeast Pacific SAMW and the AAIW are the densest waters that outcrop in the South Pacific's subtropical gyre. Therefore the portion of SAMW and AAIW that subducts

northward forms the base of the permanent pycnocline in the subtropical gyre.

13.5.1.2. *Antarctic Surface Water*

The surface layer south of the SAF is referred to as Antarctic Surface Water (ASW). ASW is very cold and fresh because of cooling and freezing in winter and ice melt in summer. ASW extends to the base of the mixed layer in winter. In summer, the ASW consists of a warm, fresh surface layer of less than 50 m thickness, overlying a cold, fresh layer (temperature minimum) that is the remnant of the cold winter surface layer. The temperature minimum is sometimes referred to as "Winter Water." The warm surface layer is cooled to freezing in winter, erasing this vertical temperature structure.

The subsurface temperature maximum layer below the ASW lies below the influence of winter freezing. This warmer water is the CDW.

Because there is easy exchange of surface waters across the SB and the SACCF, Whitworth et al. (1998) argued that *Continental Shelf Waters* that are less dense than the CDW should be considered as part of the ASW. Over the continental shelf there is sometimes no CDW temperature maximum underlying the ASW. They suggest using the density of the nearby CDW to define the ASW on the shelf.

The open ocean ASW layer is 100- to 250-m thick. Its salinity ranges from 33 to 34.5 psu. The ASW temperature is between −1.9 and 1°C in winter and between −1 and 4°C in summer. The seasonal cycle of sea ice formation and melting limits the range of winter-summer temperature variation. A considerable proportion of the heat inflow during summer is necessary to melt the ice, leaving only a small part to raise the temperature of the water.

South of the SACCF, the ASW is a true surface layer, with the temperature minimum in summer located at about 50 m depth at all longitudes. Because it is tightly associated with winter sea ice formation, the temperature of the ASW temperature minimum (e.g., Winter

Water) is nearly uniform throughout the Southern Ocean south of the SACCF. ASW water mass variations are therefore controlled by salinity variations. As reviewed in Whitworth et al. (1998), on the continental shelves ASW can sometimes reach to the bottom, or even as deep as 600 m on the slope in the Weddell Sea.

Between the SACCF and the PF, the temperature minimum of the ASW increases in depth and the temperature rises toward the north, likely due to greater absorption of heat during summer. The bottom of the ASW layer is no more than 250 m deep. The PF is identified in most places as the northernmost location of the ASW's temperature minimum.

13.5.1.3. Continental Shelf Water

South of the ASF, sitting on the continental shelf, is a thick, very cold, nearly isothermal layer. This layer is very close to the freezing point in winter. It can be characterized in places by an increase of salinity with depth. The salinity stratification is most likely due to brine rejection from the ice formation, which creates denser, saltier water that settles at the bottom on the shelf. Antarctic continental shelves are quite deep (400–500 m) because of the large mass of ice on Antarctica that depresses the entire continent and its shelves.

Shelf Water is defined by Whitworth et al. (1998) to be water as dense as AABW (neutral density greater than 28.27 kg m^{-3}) but near the freezing point, hence colder than $-1.7°C$. Waters above shelf water but denser than ASW are called *Modified CDW* (Section 13.5.3).

Because shelf water is close to the freezing point, variations in properties depend on salinity. The saltiest continental shelf water is the source of the dense bottom waters (Section 13.5.4).

13.5.2. Antarctic Intermediate Water

Throughout the subtropical gyres of the Southern Hemisphere and the tropics of the Pacific and Atlantic, there is a low salinity layer at about 500 to 1500 m depth (Figures 13.6, 13.7, 13.13, 13.14). This is known as AAIW. It is found north of the SAF, which is identified by the presence of AAIW on its northern side at almost all longitudes (Section 13.3.2). In the Pacific, the AAIW spreads north to about 10–20°N where it meets the *North Pacific Intermediate Water* (NPIW) with lower density and salinity (Chapter 10). In the Atlantic, the AAIW also spreads to about 15–20°N where it meets the *Mediterranean Water* (MW) with its much higher salinity (Chapter 9). A weak signature of AAIW can also be found in the Gulf Stream (Tsuchiya, 1989). In the Indian Ocean, AAIW is found to about 10°S where it meets the fresh intermediate water originating from the *Indonesian Throughflow* (ITF; Banda Sea Intermediate Water) (e.g. Talley & Sprintall, 2005).

In the T-S diagram for the Atlantic between the SAF in the south and Iceland in the north (Figure 13.14), the salinity minimum of the AAIW can be identified easily (see also Chapter 9). Its temperature is 4–5°C and its potential density is about $\sigma_\theta = 27.3$ kg/m^3, which characterizes AAIW in the Atlantic and Indian Oceans. Throughout the Pacific Ocean, its potential density is lower, around $\sigma_\theta = 27.1$ kg/m^3.

The AAIW temperature and salinity change toward the north in the Atlantic, although its density remains relatively constant. The AAIW salinity minimum is coldest and freshest at the SAF. With increasing latitude, its salinity and temperature increase. In the subtropical North Atlantic (most of the profiles north of 15°N in Figure 13.14), the AAIW salinity minimum disappears, replaced by the salinity maximum of the *Mediterranean Overflow Water*.

AAIW has relatively high oxygen content in the southeast Pacific and southwest Atlantic of 250–300 µmol/kg since it has only recently left the surface in those regions. Oxygen on the AAIW isopycnal indicates that it is in the surface layer just west of the Drake Passage and

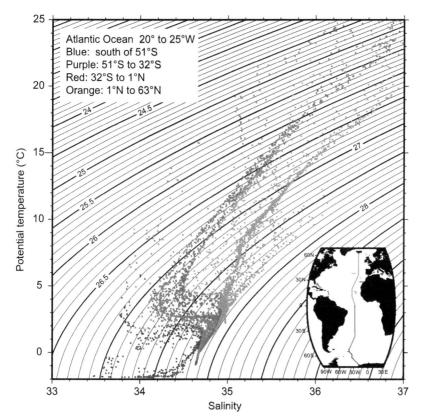

FIGURE   13.14 Potential temperature-salinity diagram in the Weddell Sea and Atlantic Ocean. This figure can also be found in the color insert.

southern coast of Chile (Talley, 1999). In the southeast Pacific, part of this low salinity surface layer subducts northward and becomes the AAIW of the Pacific Ocean.

AAIW in the Atlantic and Indian Oceans is a modified version of this Pacific AAIW. New Pacific AAIW is advected by the SAF through the Drake Passage and into the Malvinas loop east of South America. During the transit from the Pacific, the AAIW properties change some-what to higher density and lower temperature. As the AAIW rounds the loop, it plunges down-ward to just beneath the thermocline in the subtropical South Atlantic. From here it spreads eastward along the SAF, and then northward into the South Atlantic's subtropical gyre. Part of this Atlantic AAIW continues eastward into the Indian Ocean and is advected northward into the Indian subtropical gyre. In the Indian Ocean, AAIW is a long way from its surface origin, and it does not have especially large oxygen content.

There is a long-standing controversy about the origin of AAIW. The traditional view is that it is formed by the sinking of ASW across the SAF, at all longitudes around Antarctica, as a natural result of northward Ekman transport of the ASW. The opposing view of a more local-ized source of the salinity minimum in the southeast Pacific and the Drake Passage, as described previously, may be supported by the distribution of oxygen, salinity, and potential vorticity (inverse layer thickness) on the AAIW isopycnals (Talley, 1999).

The traditional view of circumpolar formation is possibly appropriate for the waters directly beneath the AAIW salinity minimum. From this perspective, the AAIW should be defined as the salinity minimum and the layer below the minimum that differs from CDW (see the next section). In the T-S diagram (Figure 13.14), the AAIW would then be defined to include the salinity minimum and the part of the nearly isothermal layer beneath it that lies above the oxygen minimum of the UCDW (see the next section). The division between AAIW and UCDW occurs at about $\sigma_\theta = 27.5 \text{ kg/m}^3$. The densest outcrop on the north side of the SAF sets the salinity minimum that defines the top of the layer. The remainder of the so-defined AAIW layer then comes from surface waters in the PFZ that cross the SAF.

13.5.3. Circumpolar Deep Water

CDW is the very thick layer that extends from just below the ASW (south of the SAF) or the AAIW (north of the SAF) to just above the dense bottom waters that are created on the Antarctic shelves. CDW is partially derived from the Deep Water of each of the ocean basins: North Atlantic Deep Water (NADW), Pacific Deep Water (PDW), and Indian Deep Water (IDW). These northern deep waters enter the ACC where they mix together. CDW upwells across the ACC into the upper ocean in the AZs and PFZs where it is transformed into the Antarctic water masses (Figure 13.4). Shelf water formed around Antarctica that is not dense enough to become bottom water becomes part of the CDW. Weddell Sea Deep Water is a major source of such renewal of CDW. CDW thus has an important component of locally formed Antarctic waters.

CDW is usually divided into Upper CDW and Lower CDW (UCDW and LCDW). There are differing conventions on how to make this division. We identify UCDW as an oxygen minimum layer and LCDW as the salinity maximum layer, following Whitworth et al. (1998), Orsi et al. (1999), and Rintoul et al. (2001). We also define the bottom of the CDW as the isopycnal that is completely circumpolar in the Southern Ocean, connecting through Drake Passage, following Whitworth et al. (1998) and Orsi et al. (1999). These definitions differ from previous editions of this text.

In the AZ (south of the PF), UCDW includes the temperature maximum layer at 1.5 to 2.5°C that lies at 200–600 m, below the ASW. The oxygen minimum layer (oxygen <180 µmol/kg) in the AZ is nearly coincident with the temperature maximum. The oxygen minimum is a very large-scale feature that comes from the deep waters north of the ACC, whereas the temperature maximum is found only south of the PF where the sea surface is near the freezing point. Therefore, the oxygen minimum is the most useful way to identify UCDW. Of the three deep waters that form CDW, the PDW and IDW have low oxygen (Chapters 10 and 11). Their contribution to CDW creates the UCDW oxygen minimum. North of the SAF, the UCDW oxygen minimum lies at about 1500 m, centered at a potential density of about $\sigma_\theta = 27.6 \text{ kg/m}^3$ and potential temperature of about 2.5°C. The oxygen minimum slopes upward across the SAF, following the upward slope of the isopycnals. Oxygen is higher and potential temperature is lower in the UCDW within and south of the ACC, due to mixing with colder, newer surface waters in this region.

UCDW also has high nutrient concentrations. Where the UCDW upwells to just below the surface layer in the AZ, it supplies nutrients to the surface layer. This is one of the reasons for prolific phytoplankton (plant) growth and consequently, zooplankton in this region. Zooplankton is a food source for larger animals in the sea, which drew the major whaling industry to the Southern Ocean.

LCDW includes the vertical salinity maximum that comes from NADW (Chapter 9; Reid & Lynn, 1971; Reid, 1994). The lower boundary of

LCDW is the neutral density 28.27 kg m^{-3} (approximately $\sigma_4 = 46.06$ kg m^{-3}), which roughly corresponds to older definitions of 0°C as the top of the AABW (Section 13.5). In the AZ, LCDW lies at 400–700 m. In the SAZ, north of the SAF, LCDW is found at 2500–3000 m in the Atlantic but reaches to the ocean bottom in the Pacific and most of the Indian Oceans.

The definition of the boundary between LCDW and AABW is somewhat arbitrary. As a result, LCDW so defined is the bottom water for most of the world ocean outside the Southern Ocean, except in the northern North Atlantic where the densest water originates in the Nordic Seas (Chapter 9). In many contexts, this LCDW is referred to as AABW, but we retain the more restrictive definition here. In Chapters 9 and 14 we refer to the whole LCDW complex as AABW.

Maps of properties in the LCDW are shown in Figure 13.15. Potential temperature at the core of LCDW is 1.3–1.8°C and potential density is around $\sigma_\theta = 27.8$ kg/m^3. Salinity in the LCDW salinity maximum is highest in the Atlantic sector, around 34.8 to 34.9 psu. In the Indian Ocean its maximum salinity is around 34.75 psu and in the Pacific around 34.72 psu. This eastward salinity decrease comes from the lower salinity IDW and PDW that join the ACC in their sectors. Lower salinity deep waters south of the ACC also reduce its salinity. (The salinity of LCDW is lower in the AZ than in the PFZ.)

The NADW salinity maximum in the Atlantic that yields the LCDW salinity maximum was first observed in 1821, but was only later recognized as originating in the North Atlantic by Merz and Wüst (1922). In the western South Atlantic, the NADW salinity maximum even has a slight potential temperature maximum at about 3°C, just below the slightly colder AAIW (vertical section in Figure 4.11a and T-S diagram of Figure 13.14). This slight temperature maximum completely disappears in the SAZ and ACC and is not a characteristic of LCDW. LCDW flows northward from the SAZ into the eastern South Atlantic; its salinity maximum is less extreme than that of the NADW in the west and includes no potential temperature maximum.

LCDW also flows northward into the Indian and Pacific Oceans, where its presence is indicated by high salinity. The high salinity core remains above the bottom in the Indian Ocean but lies on the bottom in the Pacific north of about 10–20°S, depending on longitude.

Some recent authors refer to the LCDW salinity maximum core as NADW throughout the Southern Ocean and well northward into the Indian and Pacific basins. This ignores the important inputs from the Antarctic, Pacific, and Indian regions, so we prefer the CDW nomenclature as used by Southern Ocean specialists.

Specifically, the densest LCDW fills a much greater region of the world oceans than the densest NADW. The global impact of AABW/LCDW is shown in Chapter 14 (Figures 14.14 and 14.15).

13.5.4. Antarctic Bottom Water

AABW is water in the Southern Ocean that is denser than CDW and warmer than the freezing point (Orsi et al., 1999; Whitworth et al., 1998). As described in Section 13.5.3, CDW is defined as being truly circumpolar, hence extending through Drake Passage. The isopycnal that divides AABW and CDW is therefore neutral density 28.27 kg m^{-3}. Potential temperature and salinity on this neutral surface are shown in Figure 13.16. This neutral surface covers the entire ACC region and extends northward in the western South Atlantic, and into two basins in the western Indian Ocean. Otherwise it is confined to the southern regions by the major ridges of the Southern Ocean. The coldest water at this and higher densities is at the freezing point on the continental shelves of Antarctica; this water is Continental Shelf Water and is considered separate from AABW.

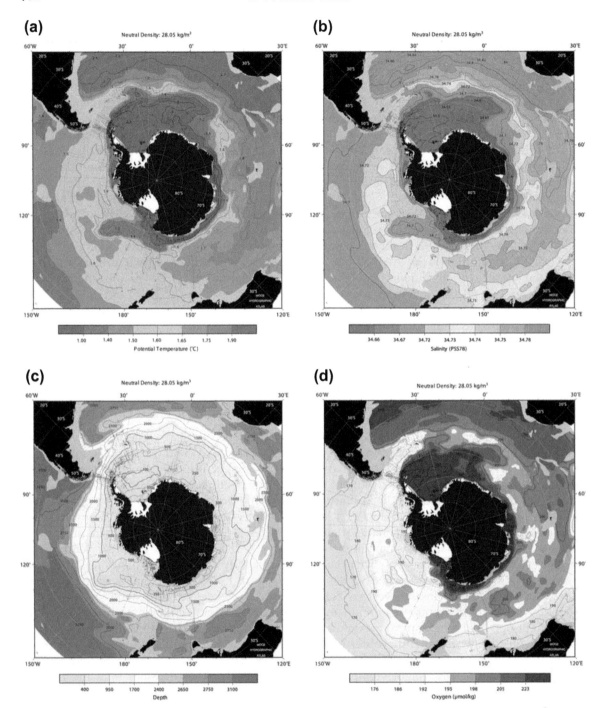

FIGURE 13.15 Properties along a Lower Circumpolar Deep Water isopycnal (neutral density 28.05 kg m^{-3}), corresponding roughly to the salinity maximum core. (a) Potential temperature (°C), (b) salinity, (c) depth (m), (d) oxygen (μmol/kg). This figure can also be found in the color insert. *Source: From WOCE Southern Ocean Atlas, Orsi and Whitworth (2005).*

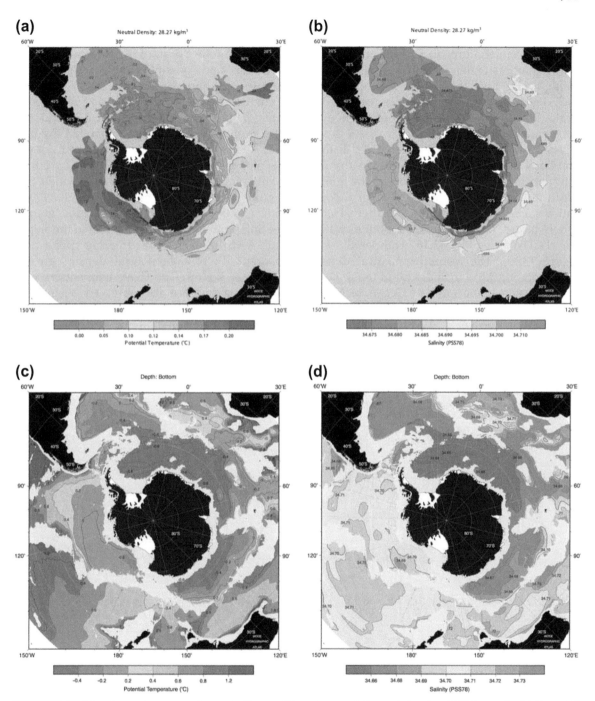

FIGURE 13.16 Properties on an Antarctic Bottom Water isopycnal (neutral density 28.27 kg m^{-3}). (a) Potential temperature and (b) salinity. Bottom properties (depths greater than 3500 m): (c) potential temperature (°C) and (d) salinity. This figure can also be found in the color insert. *Source: From WOCE Southern Ocean Atlas, Orsi and Whitworth (2005).*

An older definition of AABW is all southern deep water that is colder than 0°C. The bottom potential temperature map in Figure 13.16 shows that this region is more restricted in the South Pacific than that of the neutral density 28.27 kg m^{-3}, and does not quite reach to the Drake Passage. We therefore adopt the neutral density definition.

The rather arbitrary neutral density distinction between AABW and CDW means that the southern-origin bottom waters of the global ocean are AABW only in the Southern Ocean and a small distance into the Southern Hemisphere basins. North of this, the bottom waters are LCDW (Figure 13.16 compared with Figure 13.15). The restrictive neutral density definition of AABW includes all of the regional bottom waters in the Southern Ocean, including Weddell Sea, Adélie, and Ross Sea Bottom Waters (Whitworth et al., 1998), as well as Weddell Sea Deep Water, which is colder than 0°C.

AABW is formed in polynyas along the continental margins of the Weddell Sea, the Ross Sea, the Adélie coast of Antarctica south of Australia, and possibly also in Prydz Bay (Tamura, Ohshima, & Nihashi, 2008). AABW is a mixture of the near-freezing, dense Continental Shelf Water (Section 13.5.1.3) and the offshore CDW, which are separated by the ASF. As the very dense shelf water spills down the slope, it mixes with CDW to produce AABW. In an example from the Weddell Sea (Figure 13.17 from Whitworth et al., 1998), both Continental Shelf Water close to the freezing point (on the shelf and down the slope) and AABW above the freezing point are apparent. Both have neutral density greater than 28.27 kg/m^3. The CDW temperature and salinity maxima are also observed offshore in the figure. The V-shaped ASF is also evident, reflecting geostrophic shear with westward flow along the shelf break.

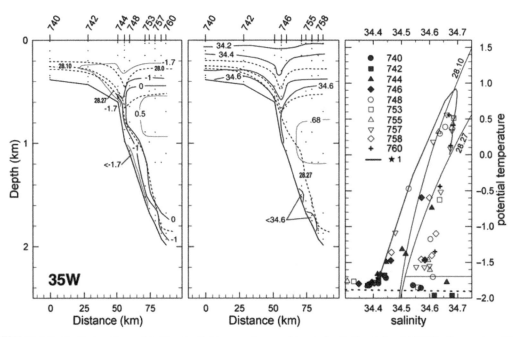

FIGURE 13.17 Vertical sections of (left) potential temperature and (center) salinity at about 35°W in the western Weddell Sea. (Right) Potential temperature versus salinity. Dashed contours in (left) and (center) are neutral density. Near-horizontal dashed line in right panel is the freezing point at 0 dbar. *Source: From Whitworth et al. (1998).*

AABW formed in the Weddell Sea is freshest and coldest (34.53–34.67 psu, -0.9 to $0°C$), along the Adélie coast is intermediate in properties (34.45–34.69 psu, -0.5 to $0°C$), and in the Ross Sea is warmest and saltiest (34.7–34.72 psu, $-0.3°$ to $0°C$; Rintoul, 1998). Volumetrically, most AABW is of Weddell Sea origin (66%), with Adélie Land contributing an intermediate amount (25%), and the Ross Sea the smallest amount (7%; also Rintoul, 1998). The fresh Weddell Sea AABW and the salty Ross Sea AABW are clear on the neutral surface in Figure 13.16. Because the Adélie Land AABW is intermediate in properties, it is not as obvious.

Within the Weddell Sea, the water masses involved in dense water formation are the ASW, CDW (also called "Warm Deep Water"), Shelf Water, Weddell Sea Deep Water, and Weddell Sea Bottom Water. Weddell Sea Deep Water and Weddell Sea Bottom Water are defined by potential temperature between $0°$ and $-0.7°C$ and potential temperature less than $-0.7°C$, respectively. Weddell Sea Deep Water is a very thick water mass, occupying depths of about 1500 to 4000 m. It has no particular property extremum. It is formed within the Weddell Sea in a manner similar to Weddell Sea Bottom Water.

Weddell Sea Bottom Water is formed through two processes: (1) mixing of ASW, UCDW (known in the Weddell Sea literature as Warm Deep Water), and Shelf Water formed on the western shelf of the Weddell Sea and (2) mixing of Ice Shelf Water (western Shelf Water that is modified under the ice shelves) with Weddell Sea Deep Water and UCDW. Western Shelf Water is at the freezing point of almost $-2.0°C$, which is possible because its pressure is about 400 dbar (station 742 in Figure 13.17). Its salinity increases from 34.4 to 34.8 psu from east to west along the shelf, enriched by brine rejection during sea ice formation along its cyclonic circulation. Potential density reaches $\sigma_\theta = 27.96 \text{ kg/m}^3$ (neutral density of 28.75 kg/m^3), among the highest in the Southern Ocean. (Values to $\sigma_\theta = 28.1 \text{ kg/m}^3$ are found in the Ross Sea where the shelf waters are saltier than in the Weddell Sea.)

Only AABW from the Weddell Sea can escape northward from the Antarctic region through a deep gap in the South Scotia Ridge. This AABW enters the Scotia Sea, flows westward to Drake Passage, and then eastward with the ACC. As it crosses the ACC, it spreads northward into the western South Atlantic, reaching northward to the Brazil Basin. It also spreads northward in the Indian Ocean into the Mozambique and Crozet basins (Figure 13.16).

13.5.5. Overturning Budgets

The meridional overturning cell of the Southern Ocean is shown schematically in Figure 13.4. Ekman transport in the surface layer is northward. UCDW and LCDW move southward into the Southern Ocean and upwell. Buoyancy loss due to cooling and salinification through brine rejection create the dense Continental Shelf Waters. These mix with LCDW and create Modified CDW and AABW, which are the dense waters that move northward out of the Southern Ocean to fill the basins to the north.

UCDW experiences a buoyancy gain (becoming lighter) through freshwater and some heat input (Speer, Rintoul, & Sloyan, 2000). UCDW is incorporated in ASW and moves northward along with the northward Ekman transport. This northward transport is incorporated in the denser part of AAIW.

Estimates of the overturning rates vary. The northward Ekman transport across the SAF is between 20 and 30 Sv based on various wind products. An AABW formation rate of about 10 Sv is estimated by Orsi et al. (1999) based on transient tracers. Various estimates of the net northward transport of the denser part of LCDW and of AABW northward out of the Antarctic are 22–27 Sv, 32 Sv, 48 Sv, and 50 Sv (Talley et al., 2003; Macdonald & Wunsch, 1996; Schmitz, 1995a; Sloyan & Rintoul, 2001, respectively). Taking these together with the Orsi et al. (1999) estimate for AABW formation, the LCDW formation rate in the Antarctic is at least equal to the formation rate of AABW, and may be much larger. Southward transport in the UCDW and possibly LCDW must balance the sum of northward Ekman and dense water transports.

The dynamics of the Southern Ocean over-turning are beyond the scope of this text. However, we do note that net southward geostrophic transport in the upper ocean is not possible across the latitude band of the Drake Passage, since such transport requires a west-east pressure gradient that must be supported by a meridional boundary (Warren, 1990). Above the depth of the undersea topography, there is no such boundary.[1] Yet it is in precisely this depth range that UCDW must cross to the south. Speer et al. (2000) and others proposed that this occurs through eddies. The eddy field of the Southern Ocean is described next.

13.6. EDDIES IN THE SOUTHERN OCEAN

Eddies are present in all regions of the global ocean (Section 14.4), but have a special role in the Southern Ocean because of the lack of an upper ocean north-south boundary at the lati-tude of the Drake Passage. By "eddies," we mean features with horizontal scales of at least several kilometers, up to about 200 km, which are departures from the time mean velocity or properties such as temperature. (We do not mean purely closed elliptical features in the total flow or property contours, although the depar-ture from the mean sometimes has this sort of shape.) In some Southern Ocean literature, there is also reference to "standing eddies," which are departures from the zonal (west-east) mean, but which have no time dependence; these can have much larger spatial scale than the temporal eddies. Most eddies arise from instabilities of the ocean currents. Strong currents, such as the fronts of the ACC, are especially unstable and therefore have highly energetic eddy fields.

The wind-driven gyres in all other ocean basins transport properties like heat, freshwater, and chemicals. These gyres consist of largely upper ocean currents forced by Ekman conver-gence and divergence, closed by a western boundary current (Section 7.8). No similar wind-driven gyre can be present across the lati-tude band of the Drake Passage since there is no meridional boundary. We know from property distributions that major exchange does occur. One mechanism for exchange is the eddy field. Consequently, evaluation of the eddy field is central to understanding the ACC. In this respect, the Southern Ocean in this latitude range is analogous to the mid-latitude atmo-sphere, where eddies play a dominant role in the dynamics.

Heat transport at the latitude of the Drake Passage is southward and is carried by eddies rather than the mean flow (deSzoeke & Levine, 1981). This result was originally based on inference from the mean heat transport and esti-mated air—sea heat flux in the Southern Ocean, but has been substantiated by eddy-resolving studies in recent years.

There have been few in situ regional studies of eddy variability in the ACC due to its remote-ness. Long time series of velocity and tempera-ture have been collected only in the Drake Passage and south of Australia, with results extrapolated to other regions. Wide geographic information but at limited depths is available from subsurface floats, surface drifters, and altimetry.

---

[1] More precisely, there is no meridional boundary above the *density* that occurs at the sill in the Drake Passage latitude range; this sill is actually located at Macquarie Ridge south of New Zealand and not in the Drake Passage. The other shallow region in the Drake Passage latitude range is at Kerguelen Plateau in the central Indian Ocean.

Most of the eddy variability of the ACC is at the mesoscale, with space and timescales of about 90 km and 1 month (Gille, 1996). This mesoscale variability is largely associated with meanders of the SAF and PF, presumably due to their instabilities. A snapshot of the eddy field in the southeast Pacific, from altimetry, is shown in Figure 13.18. The climatological positions of the SAF and PF are superimposed. The largest anomalies are the order of 20 to 30 cm; these are either meanders of the fronts or cutoff eddies from the fronts. (Superposition of the mean field is required to determine which it is.)

Eddy activity is often depicted using *eddy kinetic energy* (EKE), which is proportional to the mean squared velocity anomaly (e.g., total velocity with the time mean subtracted). Global maps of EKE have been based in recent years on Lagrangian surface drifters and subsurface floats and on geostrophic velocity anomalies calculated from sea surface height measured by satellite altimeters (Figure 14.16).

A circumpolar band of high EKE follows the ACC, mostly due to eddies of the SAF and PF, which are vigorous, unstable eastward currents. The EKE band for the ACC is most easily defined in the Pacific Ocean, where it jumps northward as the ACC passes New Zealand (Campbell Plateau) and then shifts smoothly southward toward the Drake Passage. In the Atlantic Ocean, the band of high EKE along the western boundary also includes eddies of the Brazil Current extending eastward from South America, and eddies of the Agulhas Retroflection (Chapters 9 and 11). In the Indian Ocean, the Agulhas front extending eastward and shifting southward merges with the ACC so that it is difficult from EKE alone to determine where the high EKE of the ACC begins.

Cyclonic eddies in the Australian sector of the ACC have been studied in situ. The cyclonic

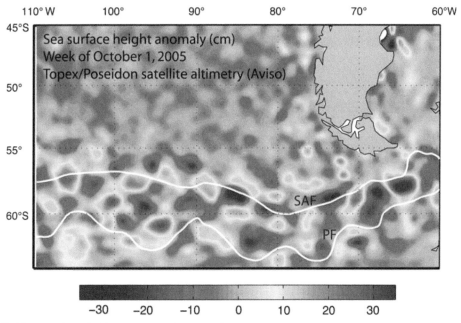

FIGURE 13.18 Snapshot of eddies in the southeast Pacific and Drake Passage: sea surface height anomalies (cm) for the week of October 1, 2005 from Topex/Poseidon altimetry (Aviso product). The climatological Subantarctic Front (SAF) and Polar Front (PF) are marked.

eddies are spawned by meanders of the PF and SAF. The single eddy surveyed in Savchenko et al. (1978) originated south of the PF and had a cold core (Figure S13.2 on the textbook Web site). Morrow, Donguy, Chaigneau, and Rintoul (2004) paired in situ observations and satellite altimetry to study a large ensemble of long-lived cyclonic eddies generated by meanders of the SAF. They concluded that these eddies play an important role in cooling and freshening the region north of the SAF where mode waters are formed, equivalent to that of Ekman transport.

13.7. SEA ICE IN THE SOUTHERN OCEAN

13.7.1. Sea Ice Cover

Sea ice in the Southern Ocean has a major impact on Southern Hemisphere albedo and on water properties, including deep and bottom water formation in the Southern Hemisphere. Southern Ocean sea ice covers an enormous area at its maximum extent in late winter, but, unlike the Arctic Ocean, almost all of the sea ice is lost each year (Figure 13.19). Therefore much of the sea ice in the Southern Ocean is "first-year ice." The exceptions are in the western Weddell Sea and along the Ross Sea Ice Shelf where ice cover usually persists throughout the year.

In winter, the pack ice extends out 65 to 60°S. Icebergs may be found between 50 and 40°S. The relatively zonal distribution of the sea ice edge is probably due to the zonal character of the currents in the Southern Ocean.

Tabular icebergs in the Southern Ocean originate from the ever-evolving ice shelves, described in part in Section 13.4.2. A map of all of the shelves is shown in Figure S13.3 on the textbook Web site. Shelf ice is very thick

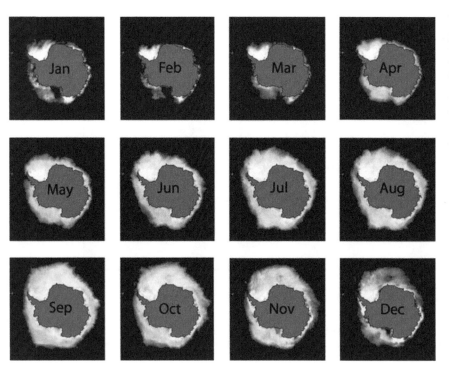

FIGURE 13.19 Annual progression of sea ice concentration in 1991, computed from the Special Sensor Microwave Imager (SSM/I) carried on the Defense Meteorological Satellite Program satellites. *Source: From Cavalieri, Parkinson, Gloersen, and Zwally (1996, 2008).*

and extensive: the Ross Ice Shelf is 35 to 90 m above sea level with corresponding depth below, extending 700 km out to the Pacific. Shelf ice is the extension of glaciers from the Antarctic continent out on to the sea where the ice floats until bergs break off. These tabular bergs may be 80 to 100 km long and tens of kilometers wide. In late 1987, the biggest berg recorded broke off from the Ross Ice Shelf. It was 208 km long, 53 km wide, and 250 m thick — it was claimed to provide enough fresh water if melted to satisfy the needs of Los Angeles or New Zealand for 1000 years.

In March 2002, the northernmost of the ice shelves, the Larsen B, with an area similar to that of the state of Rhode Island, broke up due to warming of the Antarctic Peninsula. The unexpectedly high speed of the break-up has been attributed to the presence of meltwater ponds on top of the ice shelf; these filled the crevasses with water and allowed them to extend to the bottom of the ice shelf, thus creating faster break-up than if the crevasses had been filled with air (Scambos, Hulbe, Fahnestock, & Bohlander, 2000).

When floating ice shelves break off and melt, there is no change in sea level because the ice is already displacing the water before it melts. However, when the break-up includes continental ice, or if the break-up contributes to increased flow of land-fast glacial ice to the sea, then it does cause sea level rise.

Regions of low ice cover, or polynyas, occur in the Antarctic as well as in the Arctic (Section 12.7.2; Section 3.9). Much information about them has been obtained from satellite observations as well as from ships (Comiso & Gordon, 1987). Latent heat polynyas are found in many locations around the coastline and ice shelf edge. The resulting brine rejection produces dense shelf water, some of which is dense enough to create AABW. Three polynya regions are most productive of AABW: the southern Weddell Sea (68%), the Ross Sea (8%), and Adélie Land (24%; Rintoul, 1998; Barber & Massom,

2007). Sea ice production is large in latent heat polynyas, so a map of this production (Figure 13.20) is a good indication of the location of the polynyas and hence of dense water formation, although the relationship between sea ice production and dense water formation is not one-to-one. Of the many polynyas displayed in the East Antarctic region, the productive Mertz glacier region is the main Adélie Land source of dense water (Williams et al., 2010). The Darnley polynya on the west side of Prydz Bay is another potential source of dense water that is just beginning to be explored (Tamura et al., 2008).

Sensible heat polynyas in the Weddell sector of the Antarctic have been observed in the Cosmonaut Sea area (43°E, 66°S) and over Maud Rise (2°E, 64°S; Comiso & Gordon, 1987). While these are not "ice factories" in the sense of latent heat polynyas, they may be locations of open ocean deep convection. The Maud Rise polynya ("Weddell polynya") was very large in 1974 and persisted through three winters. This was an unusual event, having not recurred as of 2008; it has been linked to feedbacks with the Southern Annular Mode, which amplified the existing forcing due to upwelling over the rise (Gordon, Visbeck, & Comiso, 2007).

Significant year-to-year changes in sea ice cover occur in the Southern Ocean. These are linked to climate change at interannual to decadal timescales, including El Niño-Southern Oscillation, the various circumpolar modes of decadal variability that have been determined, and variations in the Southern Annular Mode (Section 13.8).

13.7.2. Sea Ice Motion

The motion of the Southern Ocean ice cover is related to the winds and, less importantly, to the general circulation. The sea ice has been tracked with the passive microwave satellite SSM/I sensor; daily through long-term average

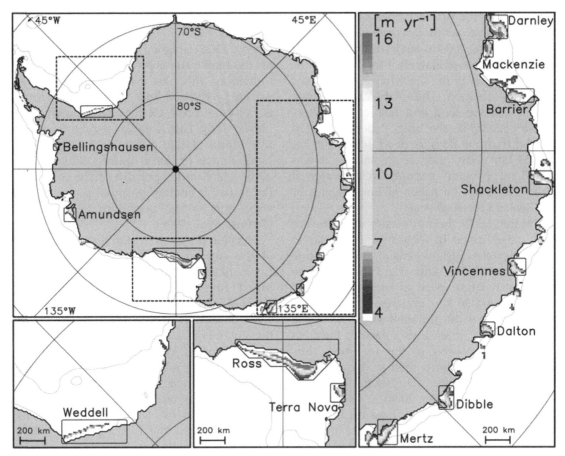

FIGURE 13.20   Antarctic latent heat polynyas: sea ice production, averaged over 1992–2001. This figure can also be found in the color insert. *Source: From Tamura et al. (2008).*

data sets are available from the National Snow and Ice Data Center (Fowler, 2003). The mean annual ice motion is shown in Figure 13.21. The ice drifts generally westward next to the continent; this matches the cyclonic general circulation in the Weddell and Ross Sea gyres. Eastward ice motion in the ACC matches both the wind forcing and mean circulation there. Northward ice motion occurs in wide regions of the Ross and Weddell Sea gyres, as well as in a wide region between 90 and 150°E, north of Prydz Bay; katabatic winds blowing off the Antarctic continent are a factor in this

northward motion. These large-scale patterns of ice motion in the annual mean persist throughout the year, based on monthly mean maps using the same data set.

13.8. CLIMATE VARIABILITY IN THE SOUTHERN OCEAN

Climate variability in the Southern Ocean is still being characterized because of the shortness of good time series. It is dominated by the circumpolar *Southern Annular Mode*. El

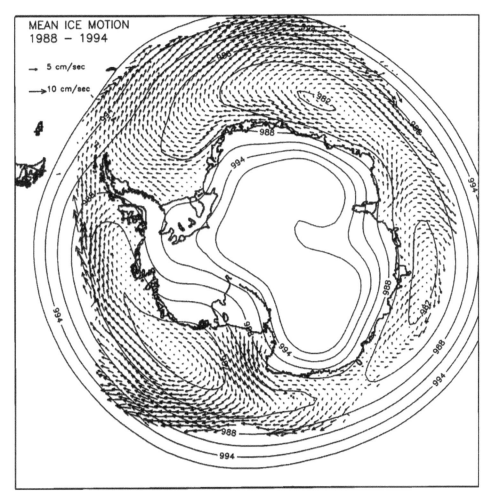

FIGURE 13.21   Mean ice motion for 1988–1994 with the mean atmospheric pressure superimposed. *Source: From Emery, Fowler, and Maslanik (1997).*

Niño-Southern Oscillation (Chapter 10) has an impact on Southern Ocean climate modes, especially at interannual timescales. Longer timescales may be partially linked to anthropogenic change.

All of the remaining text, figures, and tables relating to these Southern Ocean climate variability topics are located in Chapter S15 (Climate Variability and the Oceans) on the textbook Web site.

14

Global Circulation and Water Properties

In this chapter we summarize the circulation and water properties at a global scale, synthesizing the regional elements from the individual ocean basins (Chapters 9 through 13), and present some evolving views of the global overturning circulation. For courses providing just a limited introduction to the ocean's circulation and water properties, it might suffice to use materials from Chapter 4 and this chapter, with highlights from the forcing fields in Chapter 5 and introductory materials in the basin Chapters 9 through 13.

The surface circulation systems (Section 14.1.1) have been observed for centuries in all of their complexity, and are the best mapped part of the circulation because of ease of access. These circulations impact navigation, pollutant dispersal, the upper ocean's productive euphotic zone, and continental shelves and coastal zones. As the interface with the atmosphere, the surface layer and circulation are directly involved in ocean-atmosphere feedbacks that affect both the mean states of the ocean and atmosphere and also seasonal to climate scale variability.

Just a few hundred meters below the sea surface, some parts of the circulation change dramatically as the wind-driven gyres contract and weaken. At intermediate and abyssal depths (Section 14.1.2), the circulation is dominated by the deep penetration of the most vigorous surface currents, and by circulation associated with large-scale buoyancy forcing and weak diapycnal processes that can change the density of the water internally (Section 14.5).

The large-scale circulations include very weak vertical velocities that connect these deeper layers with each other and with the upper ocean, referred to as the overturning circulation (Section 14.2). The overturning circulation includes shallow cells that cycle water within the warmest, lowest density parts of the ocean, which can be important for poleward heat export from the tropics and subtropics. The deeper overturning circulations, connecting intermediate and deep waters to the sea surface, are generally much more global in scope than the wind-driven, upper ocean circulation systems. The grandest scale overturning circulations are those associated with North Atlantic Deep Water (NADW) formation in the northern North Atlantic and Nordic Seas, and with dense water formation in the Southern Ocean. A weaker, smaller scale overturning circulation is associated with North Pacific Intermediate Water (NPIW) formation in the North Pacific.

The drivers for the ocean circulation, its variability, and its mixing are the winds and air–sea-ice buoyancy fluxes; the tides are an additional source of energy for turbulent dissipation, which is central to the overturning circulation. In Chapter 5 we presented all of these forcing fields. In Section 14.3 we revisit the ocean's heat and freshwater transports with an emphasis on their relation to elements of the ocean circulation.

Time dependence characterizes fluid flows at all timescales. While this text mostly emphasizes the large-scale, time-averaged circulation, each basin chapter also introduces regions of persistent local eddy variability. Here we summarize the global distribution of eddy variability and associated eddy diffusivity (Section 14.5). A brief overview of climate variability and climate change in the global ocean is provided in Section 14.6, but the main materials are presented in the supplemental material as Chapter S15 on the text Web site http://booksite.academicpress.com/DPO/; "S" denotes supplemental material.

14.1. GLOBAL CIRCULATION

14.1.1. Upper Ocean Circulation Systems

The global surface circulation is shown schematically in Figure 14.1. This surface circulation derives most of its characteristic shape and strength from the basin-scale wind forcing in each ocean (Section 5.8, Figures 5.16 and 5.17). The anticyclonic subtropical gyres in each of the five ocean basins are evident, with their poleward western boundary currents: the Gulf Stream and North Atlantic Current (NAC), Kuroshio, Brazil Current, East Australian Current (EAC), and Agulhas Current systems. Each anticyclonic gyre has its eastern boundary current regime: Canary, California, Benguela, Peru-Chile, and Leeuwin Current systems, respectively. The eastern boundary currents

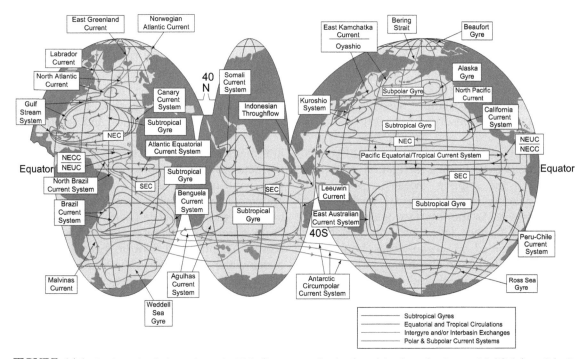

FIGURE 14.1 Surface circulation schematic. This figure can also be found in the color insert. *Modified from Schmitz (1996b).*

flow equatorward with the exception of the Leeuwin Current, which flows poleward.

The higher latitude cyclonic circulations with their equatorward western boundary currents are evident in the Arctic and Nordic Seas, North Atlantic, North Pacific including the marginal seas, and the Weddell and Ross Seas. The respective boundary currents are the East Greenland (EGC) and Labrador Currents, the East Kamchatka Current (EKC) and Oyashio, and the boundary currents of the Weddell and Ross Sea gyres.

In the tropics, the quasi-zonal tropical circulation systems are apparent, including equatorial countercurrents, equatorial currents, and low-latitude western boundary currents. Large-scale tropical cyclonic circulation systems include the zonally elongated North Equatorial Current and Countercurrent "gyres" at 5–10°N in the Pacific and Atlantic, the Angola Dome (South Atlantic), and Costa Rica Dome (North Pacific).

While all circulation is time dependent to some extent, tropical circulation variability is particularly strong relative to the mean, with fast responses to changing winds yielding strong seasonal and interannual variability. Of the major western boundary currents, only the Somali Current system in the northwestern Indian Ocean and the circulation in the Bay of Bengal change direction completely (seasonally), responding quickly to the reversing monsoonal winds because of the narrow width of their basins, which reduces the response time to the changing winds.

The ocean circulations are connected to each other. The North Pacific is connected to the North Atlantic with a small transport (<1 Sv) through the Bering Strait, through the Arctic, and then southward both west and east of Greenland. The tropical Pacific feeds water into the Indian Oceans through the Indonesian passages with a modest transport (~10 Sv). The three major oceans south of South America (Drake Passage), Africa, and Australia/New Zealand are connected through the Antarctic Circumpolar Current (ACC), with a large transport (>100 Sv). (The Bering Strait and Indonesian Throughflow (ITF) outflows from the Pacific are supplied from the Southern Ocean as well.)

There are also many connections with marginal seas that affect water properties within the open ocean. Many of these connections are shown in Figure 14.1; they are mostly discussed in the ocean basin chapters.

Surface circulation mapped directly from data has improved dramatically in recent years because of more complete surface drifter data sets and satellite altimetry (see Chapter S16 on the textbook Web site). A drifter-based surface dynamic topography map, which reflects the surface geostrophic circulation, is shown in Figure 14.2a (Maximenko et al., 2009). Globally, the highest dynamic height is in the subtropical North Pacific, which is around 70 cm higher than the highest dynamic topography of the subtropical Atlantic. The lowest dynamic height surrounds Antarctica, south of the ACC. Relatively low dynamic height is found in the subpolar North Atlantic and North Pacific.

Surface velocity (Figure 14.2b) highlights include the high velocities and narrow boundary currents, the zonal tropical circulation systems, and the ACC. This total surface velocity field includes both geostrophic and Ekman components. In the geostrophic flow field, depicted by contours of surface dynamic height (Figure 14.2a), complete closed subtropical gyres are missing or distorted in some regions, especially the North Atlantic. But when the total including the Ekman component is considered, the surface circulation appears more gyre-like, and more similar to the 200 m geostrophic flow (Figure 14.3). The streamlines for total surface velocity also indicate regions of convergence, where the streamlines terminate in mid-gyre, and divergence where the streamlines originate in mid-gyre. Convergence and divergence are associated only with the Ekman velocity, since geostrophic flow is non-divergent by definition (Section 7.6).

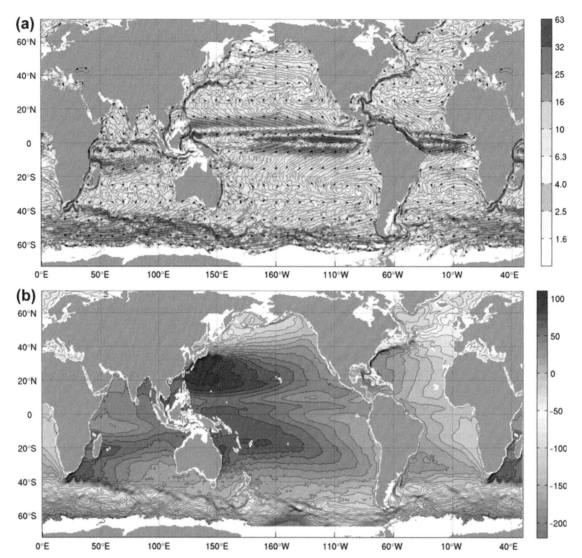

FIGURE 14.2 (a) Surface dynamic topography (dyn cm), with 10 cm contour intervals, and (b) surface velocity streamlines, including both geostrophic and Ekman components; color is the mean speed in cm/sec. This figure can also be found in the color insert. *Source: From Maximenko et al. (2009).*

Just below the surface layer, even at 200 dbar, all five wind-driven subtropical gyres are tighter (more localized) than at the surface (Figure 14.3). At this depth there is no Ekman flow, so the total mean velocity is represented by the absolute dynamic topographies in the basin chapters, which the relative dynamic topography in Figure 14.3 strongly resembles. Compared with the sea surface, the subtropical gyres are shifted toward their strong western boundary currents and extensions, that is, toward the west and the poles.

At 1000 dbar relative to 2000 dbar, the anticyclonic gyres retreat even more toward their

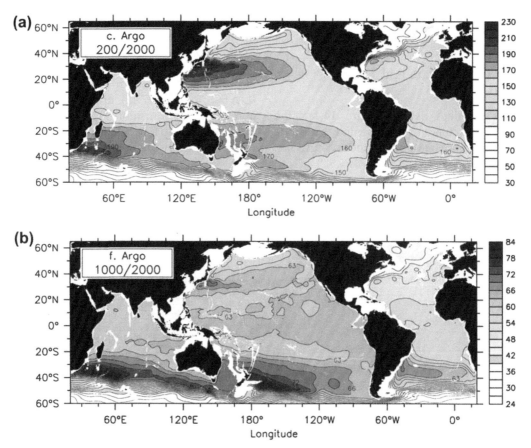

FIGURE 14.3 Steric height (dyn cm) relative to 2000 dbar at (a) 200 dbar and (b) 1000 dbar, using mean temperature and salinity from five years of float profiles (2004–2008). *Source: From Roemmich and Gilson (2009).*

western boundary current extensions (Figure 14.3b).[1] The contrast between maximum dynamic height of the Pacific and Atlantic remains, with the Pacific higher than the Atlantic. The Southern Hemisphere gyres are much more exaggerated at 1000 dbar than in the upper ocean, while the Kuroshio and Gulf Stream gyres are weaker. In Figure 14.3b, the Gulf Stream gyre even appears to have disappeared in this relative velocity calculation, in favor of general

northeastward flow into the subpolar gyre, but the absolute geostrophic streamfunction even as deep as 2500 dbar in Chapter 9 (Figure 9.14a) retains a closed anticyclonic Gulf Stream gyre.

The subpolar circulations and ACC are more barotropic than the subtropical circulations, with little change in position from the sea surface to the ocean bottom. This marked shift in behavior from the subtropics to the subpolar regions is most likely due to a reduction in

---

[1] The absolute geostrophic streamfunctions at 1000 dbar in the basin chapters differ somewhat from the relative geostrophic flow at 1000 dbar relative to 2000 dbar in Figure 14.3b, because the 1000 and 2000 dbar circulations are both weak. Therefore the non-zero flow field at 2000 dbar is important to include when computing the 1000 dbar flow relative to 2000 dbar.

stratification, which then allows much deeper penetration of surface signals.

14.1.2. Intermediate and Deep Circulation

At intermediate depth (Figure 14.4a), the geostrophic circulation, represented by steric height, retains the western boundary currents, their recirculations, and the ACC of the upper ocean. It also retains a strongly zonal character in the tropics (where flows are not well resolved in the set of studies used in the figure). Importantly, Deep Western Boundary Currents (DWBCs; Section 7.10) appear by this depth, and there is a transition to the structure of the open-ocean deep flows that are affected by topography, especially the mid-ocean ridges.

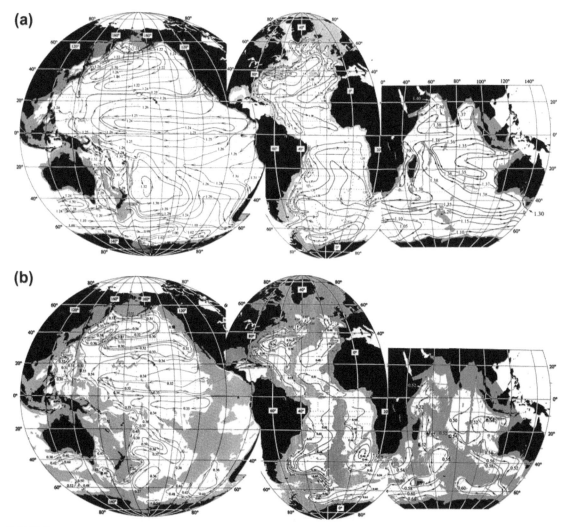

FIGURE 14.4   Streamlines for the (a) mid-depth circulation at 2000 dbar and (b) deep circulation at 4000 dbar. (Adjusted steric height, representing the absolute geostrophic flow.) *Source: From Reid (1994, 1997, and 2003).*

Unlike the surface currents, deep currents that are not a deep expression of a surface current have only generic names. They are mostly identified by location.

In more detail, each of the five subtropical anticyclonic gyres retains some expression at 2000 dbar, including a western boundary current that separates from the coast and flows eastward, and a very compact anticyclonic circulation on the equatorward, offshore side. The Gulf Stream, Brazil Current, Agulhas, and a deep version of the South Pacific's anticyclonic gyre (somewhat east of New Zealand at this depth due to topography) are present. The Kuroshio gyre is shifted north of its surface location in Figure 14.4a, but in Chapter 10 we noted that the Kuroshio Extension does extend, weakly, to the ocean bottom at the same location as its surface core (Figure 10.3).

The high latitude cyclonic circulations evident at the sea surface are also present in the northern North Atlantic, North Pacific, and south of the ACC in the Weddell and Ross Seas, continuing the near-barotropic character previously noted.

Major circulation features that appear at 2000 dbar, but not at the sea surface, include the DWBCs and poleward flows along the eastern boundaries. The DWBCs at 2000 dbar are southward in the Atlantic and mostly northward in the Pacific and Indian Oceans. The Atlantic DWBC carries NADW from the northern North Atlantic southward to about 40°S, where it enters the ACC system. At this depth, unlike at 4000 dbar, there is no DWBC in the subtropical South Pacific, which has a deep-reaching subtropical gyre. However, a northward DWBC does form within the tropical Pacific and can be traced northward along the western boundary to past the northern boundary of Japan, where it encounters a southward DWBC. These flows are more clearly defined at 4000 dbar. In the Indian Ocean, the northward DWBCs are also more easily identified at 4000 dbar (next), but include northward flow along the east coast of Madagascar and a hint of

northward flow in mid-basin that follows the Central Indian Ridge.

At 2000 dbar, all three oceans export water southward into the Southern Ocean. Part of this southward transport is gathered in broad poleward flows near the eastern boundaries, and is evident in water properties in the South Pacific and South Atlantic (see the basin chapters, 9–13). In the Indian Ocean, poleward flow is evident in water properties west of Australia, but it does not continue southward to the ACC. In the South Atlantic, southward flow of NADW is most vigorous in the DWBC along the western boundary. For a dynamically complete description, we note that the 2000 dbar flows near the eastern boundaries of the North Atlantic and North Pacific are also poleward, indicating that basin-scale cyclonic circulation at this depth is ubiquitous.

The circulation at 4000 dbar (Figure 14.4b) is greatly affected by topography. Here DWBCs carry deep and bottom water northward from the Antarctic into each of the three oceans. The northward DWBC in the South Atlantic, carrying Antarctic Bottom Water (AABW), shifts eastward at the equator, becoming an eastern boundary current along the mid-Atlantic Ridge (Chapter 9); the DWBC at the continental western boundary in the North Atlantic is southward, carrying the deepest components of NADW. The northward DWBCs in the Indian Ocean follow the ridge systems, east of Madagascar and northward into the Arabian Sea, east of the Southeast Indian Ridge and Central Indian Ridge, and east of Broken Plateau into the western Australia Basin. In the Pacific, the principal northward DWBC is east of New Zealand, flowing through Samoan Passage into the tropics, crossing the equator and then northward along the western boundary and also through the Wake Island Passage and along the Izu-Ogasawara Ridge.

In the far northern North Pacific, the DWBC is southward. This is counterintuitive if one assumes (incorrectly) that DWBCs must carry water away from their sinking sources. There is no surface source of deep water in the northern North Pacific. This DWBC thus best illustrates the dynamics of the deep circulation, which is driven by upwelling from the bottom and deep water layers into the overlying intermediate and shallow layers. According to the Stommel and Arons (1960a,b) solution, upwelling stretches the deep water column thereby creating poleward flow in the deep layers in mid-ocean through potential vorticity conservation (Section 7.10.3). This poleward flow is counterintuitive, since the main flow in the basins is toward the high latitude sources of deep water. In this theoretical framework, the DWBCs are a consequence of closing the mass balance for this upwelling. They are not simply drains of dense water.

14.2. GLOBAL MASS TRANSPORTS AND OVERTURNING CIRCULATION

The overturning circulation in each ocean is described in Chapters 9–13. We present here a global picture of the volume/mass overturns. Their role in global heat and freshwater transport is summarized in Section 14.3.

The global overturning circulation is complex and three-dimensional, with dominant paths that we attempt to depict in a simplified manner. The student is cautioned that these pathways are not isolated tubes flowing through the ocean. Many transport pathways depicted schematically as narrow "ribbons" are broad and thick flows, covering large regions. There is much circulation and mixing from one "path" to another. Throughout, and especially in the upper ocean, the pathways can be caught up in the much stronger wind-driven circulation.

Historically, there has been an emphasis on the "meridional" overturning circulation (MOC), which is important for the latitudinal redistribution of heat, freshwater, and other properties. However, some important elements of the global overturning circulation and these redistributions are not meridional. Inter-basin transports between the Pacific, Indian, and Atlantic Oceans are crucial for the global mean ocean state. Even at the ocean gyre scale, the mean state is maintained by some zonal components, such as between western boundary air-sea heat loss regions and eastern boundary air-sea heat gain regions.

Calculation and depiction of the MOC usually includes computing the meridional mass transports across each coast-to-coast, zonal transect in isopycnal (or depth) layers (Section 14.2.1), computing the upwelling and downwelling transports between the layers in closed geographic regions bounded by two transects (Section 14.2.2), and computing the overturning streamfunction to visualize the overturn in two dimensions (Section 14.2.3). Schematics of the overturn (14.2.4) are often constructed to assist interpretation, but are obviously not essential to the calculations.

More generally, this methodology applies to any closed region, and could be used for zonal transports across meridional sections, or even transports into an open ocean region enclosed by station data.

14.2.1. Mass Transports in Layers into Closed Regions

The overturning circulation is calculated by first defining closed geographical regions within which net mass must be (nearly) conserved.[2] For example, a closed region can be defined by two coast-to-coast, top-to-bottom transects (labeled "N" and "S" in Figure 14.5);

---

[2] The mass balance is not exactly zero because there is a very small exchange of freshwater with the atmosphere. When time dependence is considered, there can also be a time-dependent storage or deficit of mass within some regions; this also is proportionally small when considering large regions.

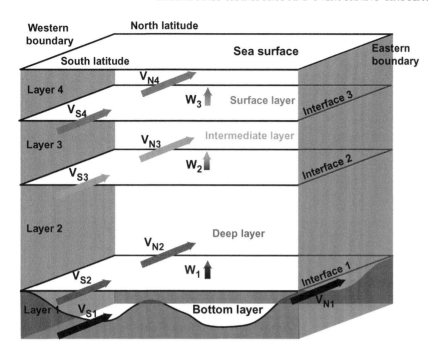

FIGURE 14.5 Meridional overturning circulation transport calculation: example for four layers. The mass transports for each layer "i" through the southern and northern boundaries of each layer are V_{Si} and V_{Ni}. The vertical transport across each interface is W_i. Arrow directions are those for positive sign; the actual transports can be of any magnitude and sign. The sum of the four transports (two horizontal and two vertical) into a given closed layer must be 0 Sv. The small amount of transport across the sea surface due to evaporation and precipitation is not depicted.

in the mean, the same amount of water must move out of the region through one section as moves in through the other (Section 5.1). For data analysis, the latitudes are those of ocean transects along which the data were collected. For models, any latitudes can be chosen and often many are used, thinking ahead to the overturning streamfunction calculation described in Section 14.2.3.

To look at overturn, the closed region is next divided vertically into layers (c.f. i = 1, 2, 3, 4 in Figure 14.5). The boundaries between the layers can be defined in different ways. Exact choices depend on the purposes of the calculation. The usual choices are isopycnals (or neutral density surfaces), constant depths, and sometimes even isotherms (of potential temperature). Isopycnal (isoneutral) surfaces are usually the most informative, because they are directly related to the air–sea buoyancy fluxes and diapycnal diffusion that transform waters from one layer to another.

The net mass transports in the layers along the southern and northern boundaries of the region are then calculated (V_{Si} and V_{Ni} in Figure 14.5). For hydrographic sections, the transports are usually based on geostrophic velocities from the sections, plus Ekman transports perpendicular to the sections. Calculating the geostrophic velocities from hydrographic station data is not trivial since reference velocities are required (Section 7.6); the overall mass conservation constraint is one of the important inputs for determining the best set of reference velocities.

Three different global calculations are superimposed in Figure 14.6a (two based on data and one on a global ocean model) and a fourth is shown in Figure 14.6b; a fifth is represented in Figure 14.9b and c by its overturning streamfunction (Lumpkin & Speer, 2007). The most robust elements of the overturning circulation are common to all of these calculations.

For all five analyses in Figures 14.6 and 14.9, mass transports were first computed for a large number of isopycnal layers. These were combined into three or four larger layers representing: the upper ocean above the main pycnocline; a deep

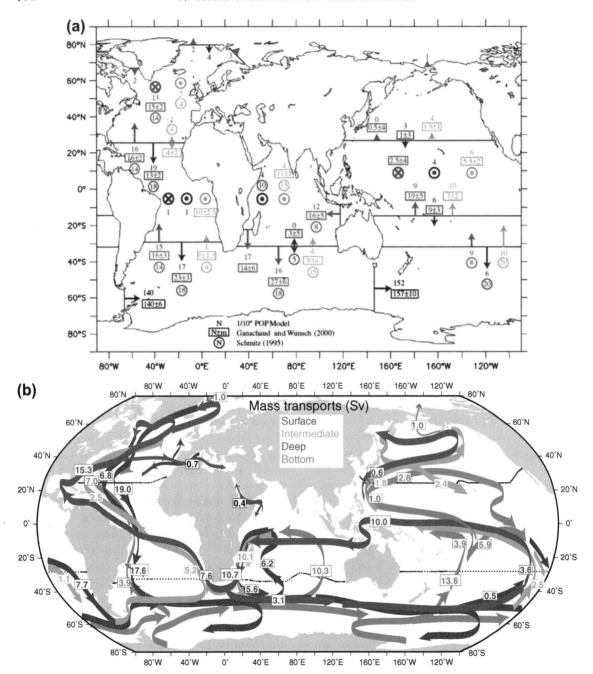

FIGURE 14.6 Net transports (Sv) in isopycnal layers across closed hydrographic sections (1 Sv $= 1 \times 10^6$ m^3/sec). (a) Three calculations from different sources are superimposed, each using three isopycnal layers (see header). Circles between sections indicate upwelling (arrow head) and downwelling (arrow tail) into and out of the layer defined by the circle color. This figure can also be found in the color insert. *Source: From Maltrud and McClean (2005)*, combining results from their POP model run, *Ganachaud and Wunsch (2000)*, and *Schmitz (1995)*. (b) Fourth calculation based on velocities from *Reid (1994, 1997, 2003)*, with ribbons indicating flow direction and oveturn locations schematically. *Source: From Talley (2008).*

layer that includes North Atlantic, Pacific, and Indian Deep Waters; and a bottom layer that is mainly dense Antarctic water (Lower Circumpolar Deep Water, LCDW or AABW).

For the upper ocean layer, robust results are (a) net northward mass transport through the entire Atlantic (also including intermediate water in Figure 14.6b), (b) southward transport out of the Indian Ocean, (c) northward transport into the Pacific, (d) westward transport from the Pacific into the Indian Ocean through the Indonesian passages (ITF), and (e) northward transport out of the Pacific into the Atlantic through the Arctic (Bering Strait). Along these pathways, there is also weak upwelling into the warm water path from deeper layers within the Pacific and Indian Oceans.

Deep water in Figures 14.6 and 14.9 is transported southward through the length of the Atlantic and southward out of the Pacific. These are the NADW and Pacific Deep Water (PDW), respectively. Deep transport in the Indian Ocean in Figure 14.6a is small and northward. When the deep Indian layer is subdivided, as in Figure 14.6b, the thinner layers have nearly balancing northward and southward transports of about 6 Sv; these are NADW moving northward and Indian Deep Water (IDW) moving southward at a slightly lower density.

Bottom water moves northward from the Antarctic into all three oceans. Figure 14.6b shows the northward penetration of bottom water into the subtropical North Atlantic as well, while the thicker layer used in Figure 14.6a subsumes this Antarctic water in the bottom part of the southward-moving NADW.

The layer transports differ from one section to another within each map. Therefore there is convergence or divergence between the sections. This results in upwelling and downwelling, as described next.

The weak overturn of the North Pacific is also depicted in Figure 14.6b. This cell transports approximately 2 Sv of warm water northward, and slightly denser NPIW southward.

14.2.2. Upwelling and Downwelling

Returning to the method (Figure 14.5), we next calculate the vertical (diapycnal) transport through the interfaces between layers within the closed regions. The transports and velocities for each layer i are

$$M_{Ti} = V_{Ni} - V_{Si} + W_i - W_{i-1} = 0 \quad (14.1a)$$

$$W_i = -V_{Ni} + V_{Si} + W_{i-1} \quad (14.1b)$$

$$w_i = W_i/A_i \quad (14.1c)$$

in which the vertical transport through the bottom ("W_0") is zero. A_i is the area of each interface and w_i is the average vertical velocity through the interface. The upwelling or transport W_i , in units of Sverdrups, across the top interface of each layer is calculated from the divergence of the horizontal transports in the closed region plus the upwelling transport across the bottom interface (Eq. 14.1b). We start with the bottom layer, which has no flow through its bottom boundary, labeled i = 1 in Figure 14.5. The total transport M_{T1} into the closed region must be 0 by continuity (Eq. 14.1a). This yields the upwelling or downwelling transport W_1 across the upper interface of the bottom layer since there is no transport through the ocean bottom. The average upwelling velocity w_1, in m/sec, across this interface is this transport divided by the surface area A_1 of that interface.

Next move upward through each of the layers and find the sum of the transports through the side boundaries and through the bottom interface (V_{Ni}, V_{Si}, and W_{i-1}). This yields the net transport W_i through the upper interface of this box. Continue this for all layers up to the surface. Because the overall velocity calculation should have been carried out with mass conservation (including Ekman transport in the uppermost layer), there should be no net upwelling or downwelling across the sea surface, which is the upper interface of the topmost box.

For example, if there is a net flow of 2 Sv northward into the southern side of the bottom box and a net flow of 1 Sv northward out of the northern side of the box, then there must be a net loss of 1 Sv within the closed bottom region. Therefore, 1 Sv must upwell across its top interface. The average upwelling velocity, with an interface surface area of, say, 10^{13}m^2, would be 10^{-5}cm/sec.

Now, switching back to results for the actual global ocean (Figure 14.6), we find: (a) the net lateral (meridional) transport requires net downwelling from the surface to the deep water in the northern North Atlantic, (b) there is also downwelling from the upper ocean and deep water to bottom water in the Antarctic, and (c) there is diapycnal upwelling of bottom water in all three oceans in the low latitude region between about 30°S and 24°N (the locations of the zonal hydrographic sections). While most of this upwelling is into just the overlying deep water layer, some reaches the upper ocean in the Indian and Pacific Oceans.

The "downwelling" process in the North Atlantic is localized deep convection in the Greenland and Labrador Seas followed by entrainment of surrounding waters; this is the production of NADW (Chapter 9). The "downwelling" in the Antarctic is localized brine rejection combined with entrainment that increases its volume tenfold, mainly along continental shelves and near ice shelves; this is the production of AABW (Chapter 13).

"Upwelling" in low latitudes is associated with eddy diffusion, driven by deep turbulence, which has large geographical heterogeneity (see Sections 7.3.2 and 14.5; Figure 14.7). In the Indian and Pacific Oceans, this upwelling produces the Indian Deep Waters (IDW) and the PDW from upwelled AABW. In the Atlantic Ocean, the upwelled AABW joins the NADW.

In greater detail, in the Indian Ocean there is upwelling from the bottom to the deep water, from the deep water to the intermediate layer, and even a small amount of upwelling to the thermocline layer. The South Pacific has similar

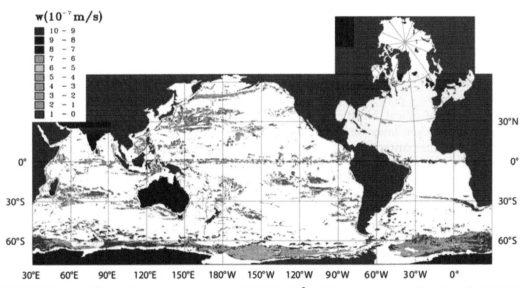

FIGURE 14.7 Modeled upwelling across the isopycnal 27.625 kg/m^3, which represents upwelling from the NADW layer. This figure can also be found in the color insert. *Source: From Kuhlbrodt et al. (2007); adapted from Döös and Coward (1997).*

processes, with net inflow in the bottom layer and outflow in all layers above it, hence with net upwelling into each of the layers, although the quantities and vertical distribution differ from the Indian Ocean.

In the Southern Ocean, using a finer division of layers than in Figure 14.6, there is also diapycnal upwelling from the deep water to the upper ocean. The upwelling source waters are largely the IDW and PDW that enter the Southern Ocean at a slightly lower density and shallower depth than the NADW. All three northern source deep waters (NADW, PDW, and IDW) physically upwell here to near the sea surface, mostly adiabatically along the steeply sloped isopycnals. The actual Southern Ocean diapycnal "upwelling" can occur mostly near the sea surface where air–sea buoyancy flux can directly transform the upwelled water. The air–sea buoyancy flux map of Figure 5.15 shows the requisite (small) net heating and net precipitation along the ACC that create lighter surface waters. Part of the adiabatically upwelled water also experiences cooling and brine rejection, hence buoyancy loss, and sinks to make the deep and bottom waters in the Antarctic.

The actual location of diapycnal upwelling (buoyancy gain) is likely very complex. Observations and budgets are as yet relatively sparse so we do not have a detailed picture from observations. Much more detail is available from general circulation models than from data, and suggest very localized processes. For the isopycnal layer associated with NADW, much of the diapycnal upwelling occurs in the Southern Ocean south of the ACC, but there is also enhancement along the equator and in other regions associated with the circulation and with complex topography (Figure 14.7).

14.2.3. Meridional Overturning Streamfunction

A final quantitative step in depicting the overturning circulation is to compute the meridional

overturning transport streamfunction for each ocean and for the globe. The overturning streamfunction is one of the basic diagnostic outputs for ocean models used to study climate, as in the Coupled Model Intercomparison Project. In Chapter 7, we introduced the concept of a streamfunction for geostrophic flow in the horizontal plane (Eq. 7.23f): the velocity is parallel to the streamfunction and its magnitude is equal to the derivative of the streamfunction in the direction perpendicular to the flow. Therefore, the geostrophic streamfunction is the horizontal integral of the geostrophic velocity field.

The overturning transport streamfunction is conceptually similar. It is calculated and plotted in a vertical plane, with a single horizontal direction. For the MOC, this horizontal direction is north-south. At any given latitude, the overturning streamfunction Ψ is the vertical integral of the mass transport, summed from the bottom of the ocean (bottom layer) to the surface:

$$\Psi_i = \sum_{i=1}^{N} V_i$$

$$\Psi(z) = \int_0^z \int_{x_{west}}^{x_{east}} v(x', z')dx' \, dz' \qquad (14.2a,b,c)$$

$$\Psi(\rho) = \int_o^\rho \int_{x_{west}}^{x_{east}} v(x', \rho')dx' \, d\rho'$$

The transport streamfunction Ψ has units of transport (Sv). The discrete sum form (14.2a) is the calculation that is actually carried out in N layers that can be defined in depth or density (or any other pseudo-vertical coordinate). Upper case V_i is the transport through the section in each layer, that is, the integral of velocities in that layer, however the layer is defined. For the more mathematical reader/student, two integral forms are given in Eq. (14.2b,c), to make explicit the difference between integrating in depth or in density. Lower case v is the velocity (in m/sec) perpendicular to the transect, which proceeds from one coast, at x_{east}, to the other coast, at x_{west}.

To calculate the overturning streamfunction Ψ (14.2a), it is preferable to subdivide the water column at each latitude into a large number of layers, many more than the 3 or 4 depicted in Figures 14.5 and 14.6. Again, the optimal layers are isopycnal or isoneutral, rather than defined in depth, although most published overturning streamfunctions are computed with depth layers.

The overturning streamfunction is calculated at each available latitude (very few for hydrographic data; many for an ocean model). The transport streamfunctions for all latitudes can then be contoured as a function of latitude and vertical coordinate (Figures 14.8 and 14.9). If the layers are defined by isopycnals or neutral density surfaces, the streamfunction can be projected back to depth coordinates by choosing the average depth of the isopycnals at each latitude.

The depiction of the overturn in each separate ocean from a global ocean model (Figure 14.8) is representative of most published calculations, although actual numerical values of the overturn (in Sv) differ. These individual ocean overturns were described in Chapters 9–13.

1. The Atlantic has an NADW cell with sinking in the north and an AABW cell with inflow from the south and upwelling into the NADW layer.
2. The combined Pacific/Indian overturn includes inflow of bottom waters (mostly AABW) that upwell into the deep water and thermocline layers, mostly in the Southern Hemisphere and tropics.
3. Meridional overturns in the Northern Hemispheres of the Pacific and Indian are weak. The 2 Sv overturn of NPIW is apparent, as is the weak, deeper overturn of Red Sea Water (RSW) in the Indian Ocean.

The global overturning streamfunction is constructed by summing the layer transports for all oceans at each latitude (Figure 14.9). The major features, found in all recent calculations, are (a) shallow subtropical-tropical overturning cells with sinking at about 30° latitude and rising in the tropics; (b) the large deep cell due to NADW with sinking in the north, occupying most of Northern Hemisphere water column and extending southward to about 35°S; (c) the deep cell centered in the Southern Hemisphere at 3000 m, with northward bottom flow (AABW) and southward deep flow, mainly as PDW and IDW; and (d) a top-to-bottom overturn in the south next to Antarctica that forms AABW. These are all principally connected to diapycnal processes, with downwelling and upwelling limbs.

When the globally averaged overturning streamfunction is calculated in depth layers (Figure 14.9a,c) rather than isopycnal layers (Figure 14.9b), the Southern Ocean also includes a strong surface-intensified overturning cell with sinking around 50°S and upwelling between 35 and 50°S. This is called the "Deacon cell." It mostly disappears when the overturning is calculated in isopycnal layers, that is, it is not associated with a large amount of diapycnal flux. The Deacon cell is mostly due to (adiabatic) flow along isopycnals that change depth and latitude: in the Southern Ocean there is a large component of northward flow that returns southward at the same density but at greater depth (Döös & Webb, 1994; Kuhlbrodt et al., 2007).

There is also well-documented diapycnal upwelling in the Southern Ocean (Chapter 13). The part of the Deacon cell that is mostly due to depth-averaging is the downwelling limb between 50 and 40°S. The "diabatic" Deacon cell (Speer, Rintoul, & Sloyan, 2000), which involves diapycnal transport, includes northward Ekman transport in the surface layer across the ACC, fed by upwelling from deep waters that mostly rise to the surface adiabatically (along isopycnals) as they move southward across the ACC. Buoyancy gain at the sea surface in the ACC vicinity then allows these waters to move northward, decreasing in density. They are mostly deposited into the Subantarctic Mode Water (SAMW) layer just

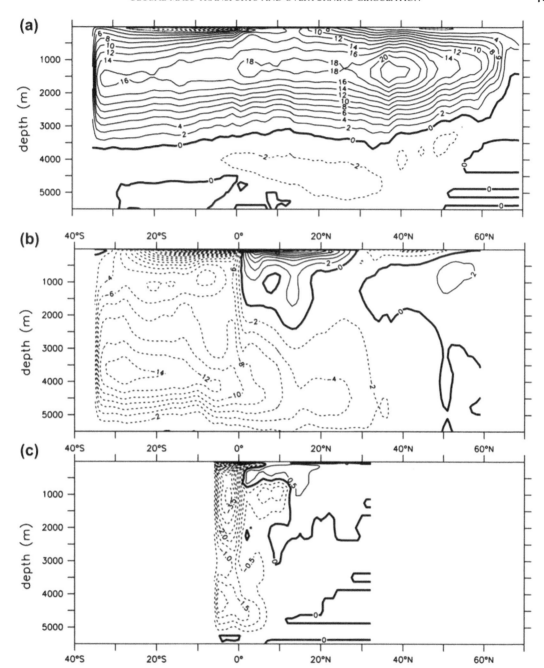

FIGURE 14.8 Meridional overturning streamfunction (Sv) from a high resolution general circulation model for the (a) Atlantic, (b) Pacific plus Indian, and (c) Indian north of the ITF. The Southern Ocean is not included. *Source: From Maltrud and McClean (2005).*

north of the ACC and then move into the Southern Hemisphere gyre circulations to the north.

14.2.4. Overturning Circulation Schematics

Here we use schematics to summarize the elements of the global overturn, based on the preceding transport, upwelling, and streamfunction calculations. All such schematics are incomplete since they cannot represent the complexities of the large-scale circulation or eddying and time-dependent paths. Therefore they should always be interpreted cautiously. Richardson (2008) presented a good history of such overturning schematics from the earliest in the nineteenth century to the present.

The widely known popularized depiction of the global NADW cell, often referred to as the "great ocean conveyor," is shown in Figure 14.10 (after Broecker, 1987, 1991, which were based on Gordon, 1986). Although this diagram has deep deficiencies in terms of representing the actual global overturn, it is nevertheless useful for public education: it is simple and it is global. It nicely illustrates the Atlantic-Pacific/Indian asymmetry in deep water formation, with sinking somewhere near the northern North Atlantic but not in the other two oceans. This particular simplification captures only a part of the global overturn because the important multiple roles of the Southern Ocean were intentionally excluded; other overly simplified descriptions do not include the essential roles of the Indian and Pacific Oceans.

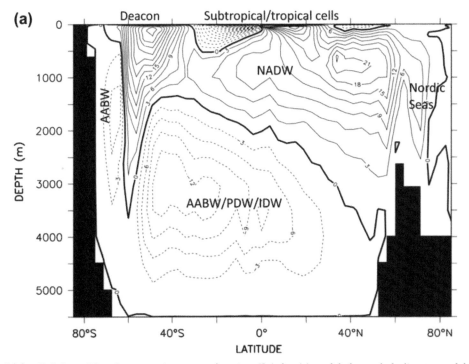

FIGURE 14.9 Global meridional overturning streamfunction (Sv) for (a) a global coupled climate model with high resolution in latitude. *Source: After Kuhlbrodt et al. (2007).* (b, c) For hydrographic section data at several latitudes, plotted as a function of neutral density and pressure; contour intervals are 2 Sv. The white contours are typical winter mixed layer densities; gray contours indicate bathymetric features (ocean ridge crests). *Source: After Lumpkin and Speer (2007).*

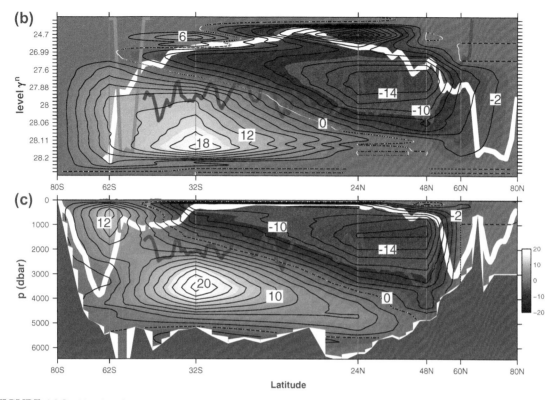

FIGURE 14.9 (*Continued*).

The next three simplified schematics illustrate the elements that we consider essential for a comprehensive teaching presentation of the global overturn. A number of global overturn schematics capture most of the aspects; in particular we note Gordon (1991), Schmitz (1995), Lumpkin and Speer (2007), and Kuhlbrodt et al. (2007).

The global overturning can be divided into two major, connected global cells, one with dense water formation around the North Atlantic and the other with dense water formation around Antarctica. These are the NADW and AABW cells, respectively. These two cells are interconnected, especially in the Southern Ocean, complicating any simple representation of the overturn. A third, weak overturning cell is found in the North Pacific, forming a small amount of intermediate water (NPIW); it is mostly unconnected to the NADW/AABW cells, but is included because it contrasts the weakness of dense water formation in this high-latitude region with that in the high-latitude regions of the Atlantic and Southern Ocean.

Essential features for the global NADW cell are as follows.[3] The NADW cell begins with

---

[3] We ignore along-isopycnal exchange of deep waters between oceans. We also have to ignore the finer steps of the large-scale upwelling process, which could better be likened to hundreds of steps on different staircases in a building of many floors, rather than a single leap from one very thick layer to another.

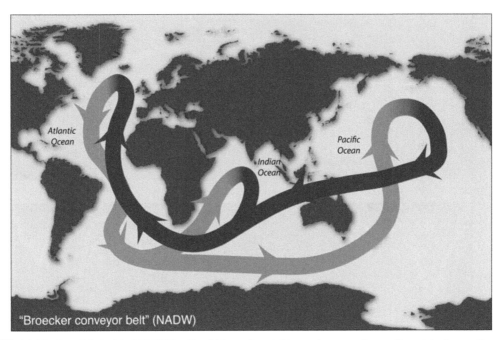

FIGURE 14.10 Simplified global NADW cell, which retains sinking only somewhere adjacent to the northern North Atlantic and upwelling only in the Indian and Pacific Oceans. See text for usefulness of, and also issues with, this popularization of the global circulation, which does not include any Southern Ocean processes. *Source: After Broecker (1987).*

warm water entering the Atlantic from the Indian, via the Agulhas, and from the Pacific, via Drake Passage. This upper ocean water moves northward through the entire length of the Atlantic (becoming first lighter and then denser), and then sinks at several sites in the northern North Atlantic (Nordic Seas, Labrador Sea, and Mediterranean Sea). These denser waters flow south and exit the Atlantic as NADW. Bottom water (AABW) also enters the Atlantic from the south. It upwells into the bottom of the NADW layer in a diffusive process.

The warm Indian Ocean source water for NADW includes water from (1) the Pacific via the ITF, (2) the southeastern Indian Ocean south of Australia (sourced from the Southern Ocean and also somewhat from the Pacific), and (3) upwelling from the underlying deep water layers within the Indian Ocean. The ITF waters in the

Pacific Ocean originate from upwelling from deep and intermediate waters within the Pacific (South Pacific and tropics) and from the upper ocean in the southeastern Pacific. The upper ocean source waters from Drake Passage arise in the southeastern Pacific (SAMW and some Antarctic Intermediate Water).

Now following the NADW as it leaves the Atlantic, part enters the Indian Ocean directly around the southern tip of Africa, contributing to the IDW. Most enters the ACC, where it upwells. Here it becomes the source for deep water formation around Antarctica. This is the main connection of the NADW and AABW cells.

The AABW cell begins with this NADW upwelling to near the sea surface around Antarctica, where it is subjected to brine rejection in polynyas (Chapter 13). The densest waters thus formed sink; the part that escapes northward across topography and into the main ocean basins is referred to as AABW (although the densest bottom waters are confined to the Southern Ocean). This AABW moves northward at the bottoms of the Atlantic, Indian, and Pacific Oceans. In all three oceans, AABW upwells into the local deep water, that is, into the NADW, IDW, and PDW. Because there are no volumetrically important surface sources of dense water in the northern Indian and Pacific Oceans, this upwelled AABW is the primary volumetric source of the IDW and PDW, whereas AABW is only a minor component of NADW.

The IDW and PDW (which can be traced by low oxygen because they are composed of old, upwelled waters) flow southward into the Southern Ocean *above* the NADW layer because they are less dense than NADW. Here, like NADW, they upwell to the sea surface. However, they upwell farther north than NADW because they are less dense. The up-welled IDW/PDW in the Antarctic feeds two cells: (1) northward flux of surface water across the ACC that joins the upper ocean circulation, accomplished initially by Ekman transport; and (2) the dense AABW formation, which then recycles this mass back through the deep water routes, along with the NADW. The first of these is a major source of the upper ocean waters that then feed northward to the NADW formation region, again connecting the AABW and NADW cells.

The vertical pathways connecting NADW, AABW, and also, importantly, IDW and PDW, are illustrated in Figure 14.11c, which is a collapsed, two-dimensional version of Figure 14.11a and b. If we tried to sketch the NADW and AABW cells directly from a global meridi-onal overturning streamfunction (Figure 14.9a), they would appear to be completely separate. This is incorrect as the global average is missing the important basin-specific roles of the Indian and Pacific upwelling and diffu-sive formation of IDW and PDW, which are high-volume water masses with large associ-ated meridional and upwelling transports.

14.3. HEAT AND FRESHWATER TRANSPORTS AND OCEAN CIRCULATION

The global circulation redistributes heat and freshwater within and between the ocean basins. In Chapter 5, the heat and freshwater budgets, air—sea fluxes, and meridional trans-ports were described. Here we briefly describe the components of the circulation that redis-tribute heat and freshwater.

Globally averaged, heat is transported merid-ionally by the ocean from the tropics to higher latitudes. The largest heat gains are in the tropics, with heat gain also in upwelling regions such as the eastern boundary currents. Individ-ually, the Pacific and Indian Oceans move heat poleward. The Atlantic Ocean transports heat northward throughout its length to balance the combined Gulf Stream and Nordic Seas heat loss regions.

The meridional heat transports are mostly associated with the upper ocean circulations, which are wind driven. The shallow tropical cells carry heat from the tropics to the subtropics. The subtropical gyres then carry the heat toward the enhanced heat-loss regions of their western boundary currents. The some-what cooled water returns southward, sub-ducted into the upper part of the subtropical gyres. This leads to a net poleward heat trans-port in all five anticyclonic subtropical gyres (Figure S14.2 seen on the textbook Web site). The cyclonic subpolar gyres of the North Pacific and North Atlantic also transport heat pole-ward, with warmer surface inflow in the east cooling to form the colder, denser waters in the northern and western parts of both gyres (NPIW and Labrador Sea Water/NADW).

In the subtropical North and South Pacific and Indian Oceans, this upper ocean gyre process accounts for almost all of the net poleward heat transport. However, in the North Atlantic, the Gulf Stream gyre accounts for only part of the northward heat transport (about 0.4 PW of 1.2 PW total in the calculation in Talley, 2003). In the South Atlantic, the net heat transport is northward (~0.4 PW), toward the equator, even though the upper ocean gyre carries heat southward (~0.1 PW). The formation of NADW, associated with heat loss in the northern North Atlantic and Nordic Seas, accounts for the remaining northward heat transport, due to northward volume transport of warm upper ocean water and southward

return of the new, cold NADW (Figure 5.12 and Figures S5.9 and S14.3 from the textbook Web site).

Freshwater is transported by the oceans from regions of net precipitation and runoff to regions of net evaporation. The tropical cells export freshwater poleward from the rainy Intertropical Convergence Zone toward the net evaporation centers (Figure 5.4a). On the poleward side of the evaporation centers, the subtropical gyres transport freshwater equatorward (salty water poleward in the western boundary currents, and freshened subducted water toward the evaporation centers).

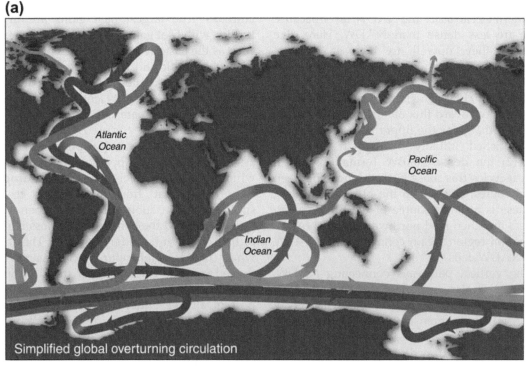

FIGURE 14.11 Global overturning circulation schematics. (a) The NADW and AABW global cells and the NPIW cell. (b) Overturn from a Southern Ocean perspective. *Source: After Gordon (1991), Schmitz (1996b), and Lumpkin and Speer (2007).* (c) Two-dimensional schematic of the interconnected NADW, IDW, PDW, and AABW cells. The schematics do not accurately depict locations of sinking or the broad geographic scale of upwelling. Colors: surface water (purple), intermediate and Southern Ocean mode water (red), PDW/IDW/UCDW (orange), NADW (green), AABW (blue). See Figure S14.1 on the textbook Web site for a complete set of diagrams. This figure can also be found in the color insert. *Source: From Talley (2011).*

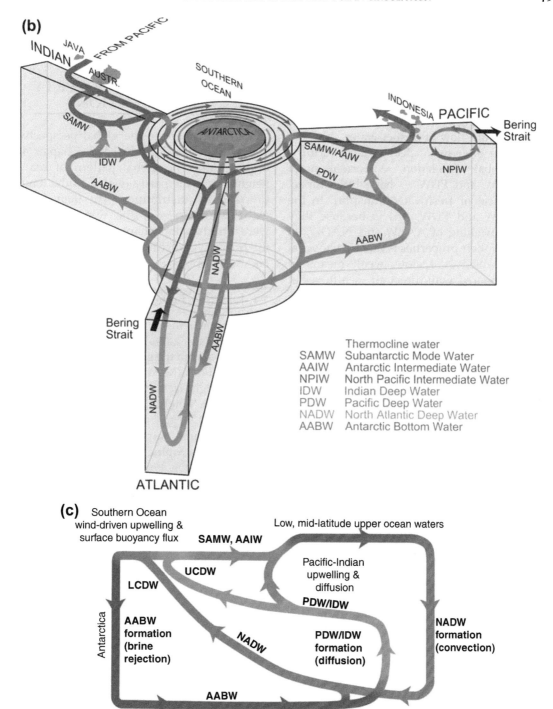

FIGURE 14.11 (*Continued*).

Of the deeper overturning cells, only the NADW and NPIW overturns carry a significant amount of freshwater equatorward. Both of these cells consist of saltier poleward flow of surface water that is freshened and joined by higher latitude fresh water (Nordic Seas, Arctic and Bering Strait input for the NADW), with southward flow of fresher water.

The other three major deep water overturns in the global circulation — formation of AABW, IDW, and PDW — have little impact on either heat or freshwater transport. In the case of IDW and PDW, this is because they form by upwelling of AABW and NADW, so alteration of their properties is due to diapycnal diffusion, which is a slow, weak means of change compared with direct air—sea fluxes at surface outcrops. In the case of AABW, even though there is direct atmospheric forcing, the heat and freshwater transports are small because the source water is already cold, so it can be cooled only slightly more and can only be freshened by a limited amount and still remain dense enough to sink.[4]

Heat and freshwater are also transported in the global overturning circulation by the ITF, moving 10 to 15 Sv from the Pacific to the Indian Ocean, and by flow through the Bering Strait, moving less than 1 Sv of low salinity water (32.5 psu) from the Pacific to the Atlantic. The ITF loop exports heat and freshwater from the Pacific because the ITF is warmer and fresher than the compensating inflow into the Pacific from the Southern Ocean. In the Indian, the ITF imports heat and freshwater input because the ITF is warmer and fresher than the Agulhas outflow that drains the ITF water. Bering Strait exports freshwater from the Pacific to the Atlantic, because the flow through the strait, at 32.5 psu, is fresher than the

volumetrically compensating inflow from the Southern Ocean.

14.4. GLOBAL PROPERTY DISTRIBUTIONS

We return here to the global perspective of water properties introduced in Chapter 4. We first describe the global pattern of sea level height since it is partly related to the temperature/salinity distribution. We then focus on global summaries of the water masses that were introduced in Chapters 9–13, here relating the property structures to the global circulation and overturning.

14.4.1. Sea Level

The ocean's mean surface height distribution (relative to the global mean surface height) can be inferred from the global dynamic topography of Figure 14.2a (Maximenko et al., 2009). Actual surface height (relative to the geoid) is close to the dynamic height divided by $g = 981$ cm s^{-2} (Eq. 7.28).

The dynamic topography also yields the surface geostrophic circulation. Using the global map, we compare the corresponding large-scale features in each ocean. For instance, the surface height difference from west to east across the North Pacific subtropical gyre is about 70 cm. In contrast, the west-to-east drop across the North Atlantic subtropical gyre is about 40 cm. There is a similar contrast between the South Pacific and South Atlantic subtropical gyres of about 70 cm versus 40 cm difference. This means that there is more equatorward volume transport in the Pacific subtropical gyres than in the Atlantic gyres.

---

[4] Although the upwelled Antarctic surface waters incorporate a large amount of freshwater in the Antarctic, the newly forming AABW can only be freshened a limited amount and still retain a density that is high enough to allow sinking. The remaining freshwater stays in the upper ocean and is exported in the Southern Ocean's upper ocean overturns, contributing to Antarctic Intermediate Water (Talley, 2008).

The simplest reason is that the Sverdrup transport in the Pacific is proportionally higher than in the Atlantic because the Pacific is that much wider and the winds, and hence Ekman pumping, are similar.

Looking at the global scale, the very low surface height in the Southern Ocean contrasts with the rest of the world ocean. The high gradient between the low Southern Ocean pressure and high pressure just to its north marks the eastward geostrophic flow of the ACC, which is principally wind-driven.

Separate from, and somewhat masked by these wind-driven gyre differences, a remarkable global feature is the overall higher surface height in the Pacific compared with the Atlantic. This is associated with the relatively lower density of the Pacific compared with the Atlantic, which is associated with the lower mean salinity of the Pacific.

14.4.2. Water Mass Distributions

Water masses in the upper ocean, at intermediate depth (below the pycnocline), in the deep ocean (2000–4000 m), and near the bottom are presented here mostly using schematics; maps and sections were shown in Chapters 9-13. Only a subset of the water masses introduced in previous chapters are included, but these are representative of most of the processes that determine the property structures.

The upper ocean water masses are represented here by mode waters, reviewed in more detail in Hanawa and Talley (2001; Figure 14.12). (Unrepresented by this schematic are the upper ocean water masses associated with subduction — the Central Water and Subtropical Underwater of the main pycnocline of each ocean basin.) All mode waters are associated with strong fronts, most of which are

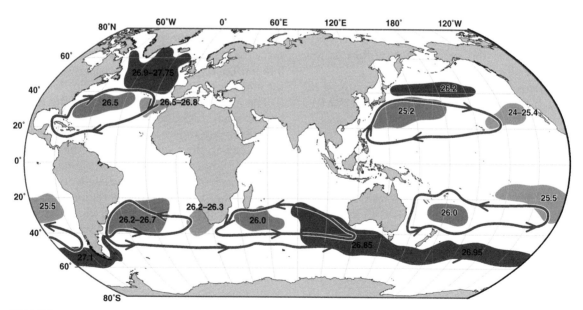

FIGURE 14.12 Mode Water distributions, with typical potential densities and schematic subtropical gyre, and ACC circulations. *Source: After Hanawa and Talley (2001).* Medium grays are STMWs in each subtropical gyre. Light grays are eastern STMWs in each subtropical gyre. Dark grays are SPMW (North Atlantic), Central Mode Water (North Pacific), and SAMW (Southern Ocean).

well-known strong currents, such as the Gulf Stream, Subantarctic Front, and so forth. These fronts have strongly sloping isopycnals that favor lower stratification and hence deeper mixing on the warm side of the front.

Subtropical Mode Waters (STMWs) are associated with each subtropical western boundary current. STMWs fill a large portion of the western subtropical gyres. Each STMW has a temperature of around 16 to 19°C; the ubiquity of this temperature is due to the similarity of western boundary current separation latitudes and the surface temperature distribution in each subtropical gyre (Figure 4.1). However, the potential densities of the STMWs differ greatly because of the salinity differences between the oceans. The North Atlantic is the saltiest, so it has the densest STMW, and so forth.

STMW formation mechanisms include deep winter mixed layer outcrops close to the strong fronts, preconditioned by the isopycnal slopes associated with the fronts, and cross-frontal transports driven by wind or the dynamics of the front. Each STMW is subducted into the interior of its subtropical gyre and becomes isolated from the sea surface within a few hundred kilometers of the front.

Eastern STMWs (lighter grays Figure 14.12) in each ocean basin are less dense than STMWs in most oceans, and are the least dense class of mode waters shown in the map. They are found where the fronts that define the gyres swing southward, except in the North Atlantic, where the Azores Current is the relevant front, and head directly for the Strait of Gibraltar.

Subpolar Mode Water (SPMW) in the North Atlantic is associated with the northeastward flow of the NAC and the cyclonic Irminger and Labrador Sea circulations. Within the NAC separation region, SPMW functions like an STMW, and subducts southward into the subtropical North Atlantic circulation. SPMW in the northeastern North Atlantic is associated with the northeastward branches of the NAC that enter the Norwegian Sea. SPMW in the

northwestern North Atlantic ultimately becomes the new Labrador Sea Water (LSW) that sinks and spreads away from the Labrador Sea (McCartney & Talley, 1982). Formation of SPMW is like that of STMW: deep winter mixed layers on the warm side of the strong fronts.

Central Mode Water in the North Pacific is associated with the eastward flow of the North Pacific Current's subarctic front, which is more or less a continuation of the Oyashio, and lies north of the Kuroshio front. Again, this front favors deep mixed layers on its warm side.

SAMW is the series of mode waters along the northern side of the Subantarctic Front (McCartney, 1977, 1982). Like STMWs, these are associated with deep winter mixed layers within several hundred kilometers or less of the strong front. The SAMWs subduct northward into the subtropical gyres where they become an important part of the pycnocline. The best-developed (thickest, lowest stratification) SAMWs are found in the eastern Indian Ocean and across the Pacific where winter mixed layers are thickest. After subducting northward into the pycnocline, these SAMWs carry tracers associated with large surface ventilation (high oxygen, chlorofluorocarbons; CFCs) far into the Indian and South Pacific.

The major intermediate waters of the global ocean, each identified by a vertical salinity extremum, are shown in Figure 14.13. The greens and blues are low salinity intermediate waters and the oranges are high salinity intermediate waters. Each of these intermediate waters forms predominantly in a localized region (locations marked on the map) and then is advected by the circulation. Each is associated with the global overturning circulation, in that formation involves a conversion of surface waters to densities that reach to intermediate depths, below the pycnocline. The overall reach of each intermediate water is greater than indicated by the location of its vertical extremum, which is simply an imperfect marker of the spread of water from a surface source. For instance, most of the

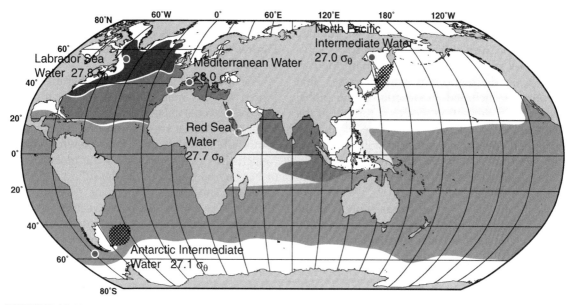

FIGURE 14.13 Low- and high-salinity intermediate waters. AAIW (dark green), NPIW (light green), LSW (dark blue), MW (orange in Atlantic), RSW (orange in Indian). Light blue in Pacific: overlap of AAIW and NPIW. Light blue in Indian: overlap of AAIW and RSW. Cross-hatching: mixing sites that are particularly significant for the water mass. Red dots indicate the primary formation site of each water mass; fainter dots mark the straits connecting the Mediterranean and Red Seas to the open ocean. The approximate potential density of formation is listed. This figure can also be found in the color insert. *Source: After Talley (2008).*

Okhotsk Sea water that provides the NPIW salinity minimum in the subtropical North Pacific resides in the subpolar gyre; however, it is not a vertical salinity extremum, and is therefore not as easily identified.

The three major low salinity intermediate waters — LSW, NPIW, and AAIW — result from relatively fresh, cold, dense water at subpolar latitudes that sinks beneath the warmer, saltier, lighter subtropical waters. The formation mechanism differs for each of these water masses: LSW, deep convection and sinking in the Labrador Sea; NPIW, brine rejection and sinking in the Okhotsk Sea followed by strong mixing in the Kuril Island passages; AAIW, deep mixed layers and underlying fresh subantarctic water sinking in the Drake Passage region and subducted smoothly northward into the Pacific and subducted with large mixing northward into the Atlantic/Indian.

The temperatures of these three low salinity intermediate waters are similar: 3 to 5°C. Their densities differ greatly because the overall salinity of their respective oceans differs. NPIW forms in the relatively fresh North Pacific and is the freshest and least dense of the intermediate waters, while LSW forms in the salty North Atlantic and is the saltiest and most dense.

The high salinity intermediate waters, Mediterranean Water (MW) and RSW, result from high evaporation and winter cooling in the Mediterranean/mid-east region. The resulting deep convection is strongly localized within the Mediterranean and Red Seas. The dense waters flow out over sills through narrow straits to join their respective ocean circulations, both sinking to intermediate depth and entraining large amounts of ambient ocean water as they equilibrate at depth in the open ocean. Once equilibrated, they are much warmer than the low salinity intermediate waters, of the order 12 to 15°C, but they are dense because of their high salinity.

NADW and AABW (or LCDW) are the two large-scale dense/bottom water masses that are always included in describing the global overturning circulation. Both are formed at the sea surface by buoyancy loss due to air–sea fluxes. The overturn associated with NADW is estimated at 15 to 20 Sv. The overturning estimates associated with AABW range from 12 to more than 25 Sv.

IDW and PDW are not represented in the maps included here. Both are formed by upwelling of bottom water (AABW) and admixture of NADW within their respective oceans. Downward diffusion from above modifies their properties relative to their deep source waters.

They have little or no surface source, and therefore the water mass decompositions shown in Figures 14.14 and 14.15 focus on NADW and AABW.

The sources of NADW and AABW are represented by the map showing the location of a deep isopycnal in Figure 14.14a. At this density, these two sources are separated by topography south of Nova Scotia and Newfoundland (which thus under-represents the global reach of NADW as described in the following paragraphs). The dense source of Nordic Seas Overflow water is deep convection east of Greenland. The multiple, distributed sources of AABW are brine rejection due to sea ice formation in polynyas within the Weddell and Ross Seas and at several locations along the Antarctic coast.

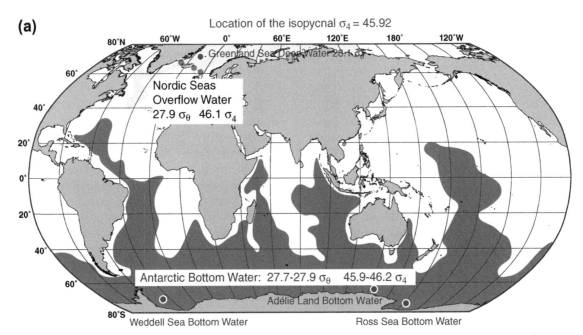

FIGURE 14.14 Deep and bottom waters. (a) Distribution of waters that are denser than $\sigma_4 = 45.92$ kg/m³. This is approximately the shallowest isopycnal along which the Nordic Seas dense waters are physically separated from the Antarctic's dense waters. At lower densities, both sources are active, but the waters are intermingled. Large dots indicate the primary formation site of each water mass; fainter dots mark the straits connecting the Nordic Seas to the open ocean. The approximate potential density of formation is listed. *Source: After Talley (1999).* (b) Potential temperature (°C), and (c) salinity at the ocean bottom, for depths greater than 3500 m. *Source: After Mantyla and Reid (1983).*

(b)

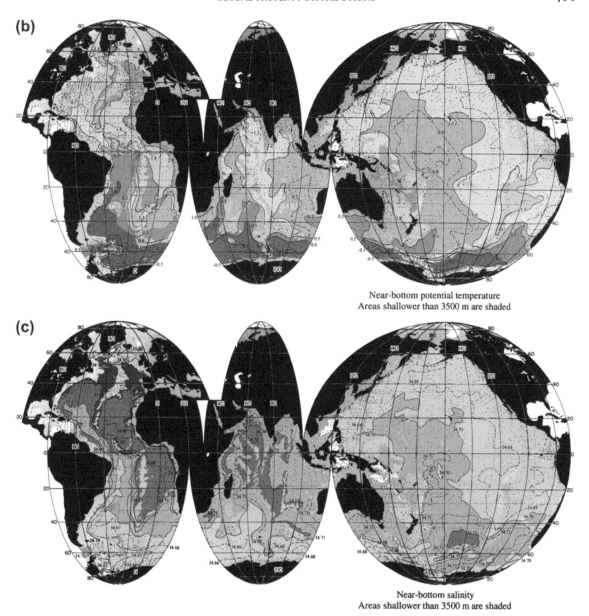

Near-bottom potential temperature
Areas shallower than 3500 m are shaded

(c)

Near-bottom salinity
Areas shallower than 3500 m are shaded

FIGURE 14.14 (*Continued*).

The global maps of bottom potential temperature and salinity in Figure 14.14b,c show the contrasting warm, saline Nordic Seas and cold, fresh AABW properties. The NADW occupies most of the bottom of the North Atlantic and eastern South Atlantic, as also seen in the water mass decomposition considered next (Figure 14.19). The colder AABW occupies the Southern Ocean, the western South Atlantic, and dominates in the Indian and Pacific Oceans.

(a)

Fraction of NADW at γ^N=28.06 kg/m³ (2500-3000 m)

(b)

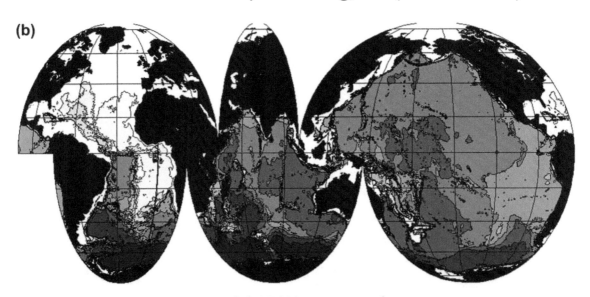

Fraction of AABW at ocean bottom

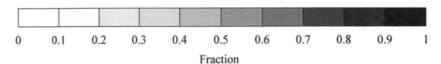

Fraction

FIGURE 14.15 Fractions of NADW and AABW. (a) Fraction of NADW on the isoneutral surface $\gamma^N = 28.06$ kg/m³ ($\sigma_4 \sim$ 45.84 kg/m³, at a depth of 2500–3000 m north of the ACC; G. Johnson, personal communication, 2009). (b) Fraction of AABW in the bottom water (with the remaining fraction being mostly NADW). *Source: From Johnson (2008).* Both maps are from an OMP analysis using as inputs the properties of NADW at a location just south of Greenland, downstream from the Nordic Seas Overflows, and of AABW in the Weddell Sea. The complete figures are reproduced on the textbook Web site as Figures S14.4 and S14.5.

The effect of vertical mixing is apparent in these global maps. The western Indian Ocean has higher bottom salinity than the entering AABW, which fills the bottommost layer. This is due to downward mixing from the overlying higher salinity RSW. In the Pacific Ocean, the bottom salinity distribution also indicates downward mixing from overlying waters: in the southwestern Pacific, a deep vertical salinity maximum remains from the NADW influence in the circumpolar deep waters, and this creates higher salinity at the bottom. Farther north in the Pacific, the bottom waters are fresher and warmer, which is partially due to the elimination of the densest bottom waters through upwelling (diapycnal mixing) and downward mixing of overlying fresher water. Full explanation requires detailed consideration of properties on deep isopycnal surfaces.

The global reach of NADW is demonstrated using the fraction of NADW (compared with AABW) on an isopycnal surface that typifies the high salinity core in the ACC (Figure 14.15 and Figure S14.4 seen on the textbook Web site). The maps were computed by G. Johnson (personal communication, 2009) using his (Johnson, 2008) application of optimum multi-parameter analysis (OMP) (Section 6.7.3) to global water masses. Reid and Lynn (1971) were the first to show and describe the global pattern of salinity on this NADW isopycnal. Salinity on nearly the same isopycnal surface in the Southern Ocean is shown in Figure 13.17.

Water with a large fraction of NADW reaches southward from its source in the northern North Atlantic, down along the western side of the South Atlantic, and then eastward at 20−30°S. There is a transition to a lower NADW fraction (i.e., salinity) between 30 to 40°S, which indicates the onset of much more AABW. A tongue of higher NADW fraction (i.e., salinity) spreads eastward south of Africa and then in patches along the core of the ACC eastward into the Pacific Ocean. The pattern of somewhat higher fractions (salinity) then extends northward

into the Pacific Ocean in the DWBC just east of New Zealand. Along the entire path, the NADW fraction (higher salinity) decreases downstream as the AABW fraction increases.

The global reach of AABW is also demonstrated with OMP analysis applied at the ocean bottom (Figure 14.15b and S14.5 from the textbook Web site). The pattern is similar to that on the deep isopycnal (Figure 14.15a). AABW dominates the world ocean's bottom water, but with a respectable NADW fraction of about 0.3. AABW is mostly blocked from crossing the mid-Atlantic and Walvis Ridges in the South Atlantic, so NADW dominates the eastern South Atlantic as well as the North Atlantic.

In both the deep and bottom water maps, AABW covers significantly more of the ocean than NADW. Johnson (2008) estimated that two-thirds of the deep and bottom water arises from AABW and one-third from NADW. If the overturning rates for NADW and AABW of 19 and 28 Sv from Figure 14.5 are used together with Johnson's (2008) calculations of the volumes of NADW and AABW, a residence time of about 500 years for both water masses is obtained. If, however, the overturning rates for the two water masses are more like 17 Sv each, as summarized in Johnson (2008), the residence times differ, with NADW around 500 years and AABW around 870 years. Uncertainties in these values are large. As both water masses are in similar deep/bottom water environments, differences in residence times would rely on differences in how they are affected by geographically heterogeneous diapycnal diffusion.

14.5. EDDY VARIABILITY AND DIFFUSIVITY

This introductory textbook is mostly written from the point of view of a mean circulation and simple departures from the mean, including some seasonal and climate variability, and energetic, recurring time-dependent features such as Gulf Stream or Agulhas rings. However, all regions of the ocean have some level of eddy variability, which is defined as the departure of instantaneous velocities or sea surface/isopycnal heights from the mean. Variability can range from random noise, to wavelike disturbances, to closed, coherent features. Eddy variability is responsible for stirring in the nearly horizontal (along-isopycnal) direction, and is thus critically important to along-isopycnal eddy diffusivity (Sections 7.2.4 and 7.3.2).

Warm- and cold-core rings generated by the meandering of major currents are large, closed features that one might typically associate with "eddies." On the other hand, eddy variability in the central parts of the gyres may look more like spectral noise. Some kinds of eddies extend from the sea surface to the bottom, while others are concentrated in the surface layer, and others can be embedded entirely within subsurface layers.

Horizontal, eddy length scales are kilometers to thousands of kilometers. Timescales are typically weeks to months, but can sometimes be many years for coherent vortices such as Meddies (Chapter 9). This is considered to be the ocean's *mesoscale*. These are the length and timescales associated with planetary waves such as Rossby and Kelvin waves. The most important length scale for this variability is the Rossby deformation radius, which depends on latitude and vertical stratification (Section 7.7.4, Figure S7.30 on the textbook Web site). A new category of shorter *submesoscale* variability, mostly associated with the surface layer, is now being vigorously analyzed through theory, modeling, and observations. Because this layer is so shallow, the horizontal spatial scales are small, on the order of kilometers. (This can be thought of as related to an internal Rossby deformation radius using a vertical length scale of about 100 m rather than 1000 m.)

Internal waves and tides (Sections 7.5.1, 8.4, and 8.6) have even shorter timescales. This high frequency variability is critically important

for ocean turbulence and hence diapycnal diffusivity and mixing (Sections 7.3 and 7.4). We therefore present some recent global results for near-inertial and tidal variability.

14.5.1. Eddy Energy and Lateral Eddy Diffusivity Distributions

The basic physics concepts of kinetic and potential energy were described in terms relevant to the ocean in Section 7.7.5. *Eddy kinetic energy* (EKE) is calculated using departures of the instantaneous (synoptic) velocity from the mean velocity, regardless of how the mean is defined (leaving some ambiguity that should be carefully described in any given study). EKE maps are almost always based on lateral currents and not on the vertical velocities, which are much smaller (but important for the diapycnal eddy diffusivity described in the following section). EKE for surface currents was first derived from ship drift observations, but is now much more easily constructed from surface drifter velocities and from surface velocities derived from altimetric surface heights. EKE maps for deeper levels are calculated from Lagrangian float observations; moored current meter arrays are also used to calculated eddy energy locally. Eddy potential energy is calculated using departures of instantaneous sea surface height and isopycnal heights from their mean values; currently satellite altimetry data are valuable for this, and in situ Argo profiling float data set will also be valuable after many more years of data are collected.

Surface EKE (Figure 14.16 and Figure S14.6 seen on the textbook Web site) is mostly related to the mean current speeds (Figure 14.2b, Section 14.1). Eddy energy has several sources, including current instabilities (Section 7.7.5). Mean flows with strong velocity shear in both the horizontal and vertical tend to be the most unstable. Horizontal shear generates barotropic instabilities that draw energy from the shear. Vertical shear in geostrophic flow is associated

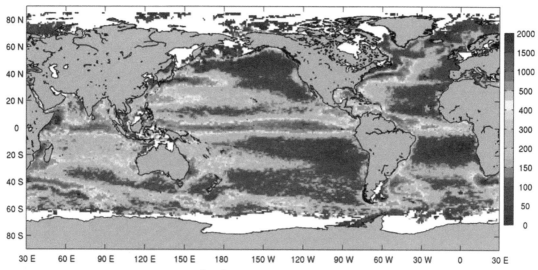

FIGURE 14.16 Eddy kinetic energy (cm^2 s^{-2}) from surface drifters. *Source: From NOAA AOML PHOD (2009)*. A complementary figure based on satellite altimetry *(from Ducet, Le Traon, & Reverdin, 2000)* is reproduced in Figure S14.6c on the textbook Web site. This figure can also be found in the color insert.

with sloped isopycnals; the eddy energy is generated through release of the potential energy of the sloping isopycnals through baroclinic instability. Overall, satellite altimetry analysis has reinforced the importance of flow instabilities, and especially of baroclinic instability, in all regions (Stammer, 1998).

On the other hand, mid-ocean eddies away from strong currents can be generated by mechanisms other than flow instability, that is, through direct wind forcing. An easily visible example in Figure 14.16 is the high EKE band just west of Central America; these are eddies in the Gulf of Tehuantepec, forced by very strong winds through the adjacent mountain passes (Figures 5.16 and 10.21).

Although it is mostly related to the currents, the EKE distribution differs in important ways from the speed distribution (Figure 14.2b and Figure S14.6 seen on the textbook Web site). The strongest eddies, such as Agulhas rings, propagate away from the mean flow that created them, accounting for broader EKE maxima compared with the speed (mean kinetic

energy) maxima. The most striking large-scale difference of EKE from the mean speed distribution is the high EKE in bands around 20 to 30° latitude, especially in the Pacific and Indian Oceans, but also with a signature in the Atlantic. These regions have low mean surface velocities, yet the enhanced EKE stands out starkly in the global mean. These regions have shallow eastward surface flow with underlying westward flow (the Subtropical Countercurrents in the Pacific, Eastern Gyral Current in the Indian, and the Azores Current and Subtropical Countercurrents in both hemispheres in the Atlantic). This vertical shear is associated with tilted isopycnals and enhanced baroclinic instability (e.g., Palastanga, van Leeuwen, Schouten, & deRuijter, 2007; Qiu, Scott, & Chen, 2008).

Horizontal eddy diffusivity can be calculated from the eddy variability measured by Lagrangian drifters. The highest values of surface eddy diffusivity might exceed 2×10^4 m^2/sec (2×10^8 cm^2/sec) (Figure 14.17a and Figure S14.7 on the textbook Web site). The pattern of surface diffusivity corresponds roughly to the

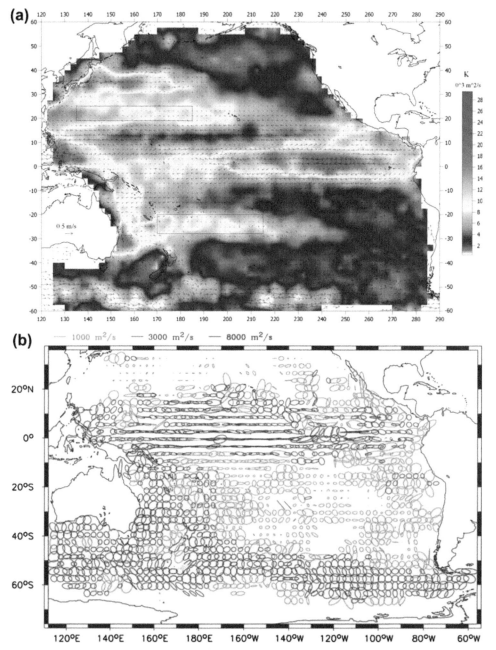

FIGURE 14.17 (a) Horizontal eddy diffusivity (m²/sec) at the sea surface (color) with mean velocity vectors, based on surface drifter observations. *Source: From Zhurbas and Oh (2004).* (b) Eddy diffusivity ellipses at 900 m based on subsurface float velocities. Colors indicate different scales (see figure headers). *Source: From Davis (2005).* The Atlantic surface map and Indian 900 m map from the same sources are reproduced in Chapter S14 (Figures S14.7 and S14.8) on the textbook Web site. Both Figures 14.7a and 14.7b can also be found in the color insert.

EKE pattern. The eddy diffusivity is not exactly proportional to EKE because timescales for diffusion depend on location and on the underlying processes that create the variability (Lumpkin, Treguier, & Speer, 2002; Shuckburgh, Jones, Marshall, & Hill, 2009). The map in Figure 14.17a shows a scalar diffusivity, calculated using a modified version of Davis' (1991) method; the full horizontal diffusivity is a tensor, hence can have different values in the zonal and meridional directions, since velocity variations can be in any direction relative to the mean velocity.

Below the sea surface, out of reach of satellites and surface drifters, eddy statistics are more difficult to compile.[5] Horizontal eddy diffusivities based on subsurface floats at 900 m (Figure 14.17b) have maximum values of about 0.8×10^4 m^2/sec, which are robustly less than those at the sea surface despite differences in computation methods, including the use of ellipses that show the directional difference in diffusivity. As at the sea surface, the 900 m eddy diffusivity is high mostly where currents are strong. It is also mostly isotropic (the ellipses are "round") except in the tropics, where it is highly directional, with much larger values in the east-west direction, matching the strongly zonal direction of the tropical currents (Davis, 2005).

14.5.2. Observed Scales, Speeds, and Coherence of Eddy Variability

This is a very brief introduction to the large amount of work describing the ocean's eddy variability. A simple time-space display (Hovmöller diagram) of sea-surface height (SSH) anomalies at mid-latitude in each ocean

(Figure 14.18) reveals generally westward propagation of the dominant features, which is typical behavior for Rossby waves (Section 7.7.3). Similar patterns are found at almost all latitudes, except near the equator where eastward-propagating Kelvin waves are also found, and in strong eastward flows such as the ACC that advect the variability to the east (a Doppler shift).

Phase speeds calculated from SSH plots like those of Figure 14.18 yield robustly westward propagation, close to the speeds of first-mode baroclinic Rossby waves (Figure 14.19; Chelton & Schlax, 1996; Stammer, 1997). However, the difference from simple Rossby wave speeds is important: the observed speeds are almost twice as fast at mid-latitudes. The non-Rossby-wavelike behavior of the variability is likely due to nonlinear interactions with other modes (e.g., Wunsch, 2009), and to the prevalence of coherent eddies that propagate westward (Figure 14.21). Such coherent eddies are nonlinear by definition.

Frequency and wavenumber spectra (Section 6.5.3) provide statistical information about variability observed from satellites, current meters, and so forth. The directional wavenumber spectrum from satellite altimetry in Figure 14.20b again shows that most energy propagates westward (solid curve) rather than eastward (dashed). In the frequency spectrum, the annual cycle is the most energetic signal (peak indicated by dashed vertical line in the left panel), because this is the strongest external forcing frequency, associated with seasonal changes. Other than this peak, the frequency spectrum is relatively smooth.

The spectra in Figure 14.20 are nearly flat at lower frequencies and wavenumbers (longer

---

[5] The best EKE estimates are made at long-term current meter moorings, which tend to be deployed in dynamically interesting regions such as the Gulf Stream, with little sampling elsewhere. Subsurface floats provide information at their target depths. Acoustically tracked floats provide the best statistics since their locations are observed nearly continuously, but they are not global. Profiling floats that are tracked when they surface, approximately every 10 days, can provide global statistics, although global maps are not yet available.

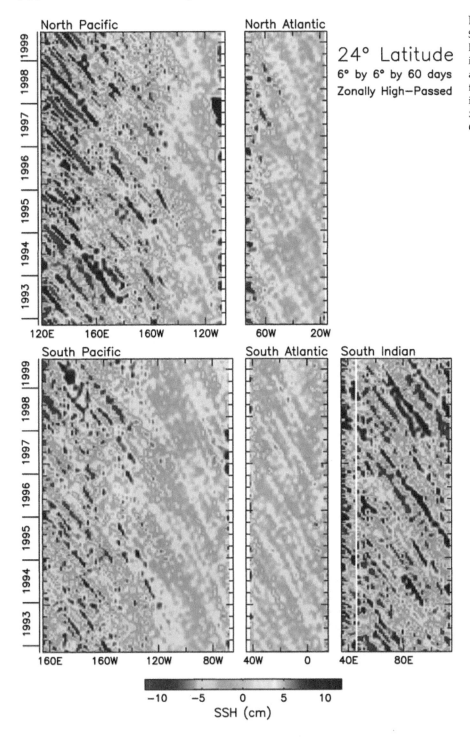

24° Latitude
6° by 6° by 60 days
Zonally High–Passed

FIGURE   14.18
Surface-height anomalies at 24 degrees latitude in each ocean, from a satellite altimeter. This figure can also be found in the color insert. *Source: From Fu and Chelton (2001).*

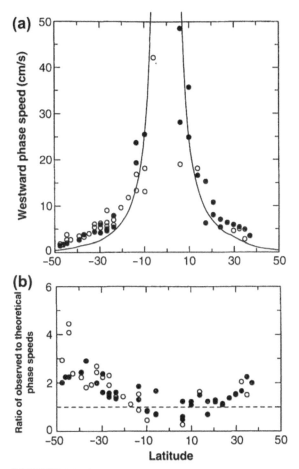

FIGURE 14.19 (a) Westward phase speeds (cm/sec) in the Pacific Ocean, calculated from the visually most-dominant SSH anomalies from satellite altimetry. The underlying curves are the fastest first-mode baroclinic Rossby waves speeds at each latitude. (b) The ratio of observed and theoretical phase speeds, showing that the observed phase speeds are generally faster than theorized. *Source: From Chelton and Schlax (1996).*

and wavenumber; this can be called a "cutoff" frequency or wavenumber. The cutoff marks a shift in the physical processes that dominate the white versus the red parts of the spectra. The spectral slopes are consistent with generation of energy through baroclinic instability (Section 7.7.5) rather than external changes in forcing (mainly wind; Stammer, 1997).

One explanation for the lack of detailed correspondence between observations of westward propagation and the obvious initial explanation of Rossby waves is that a great deal of energy is actually contained in coherent eddies, which differ from Rossby wave behavior in many ways, but retain the westward propagation of Rossby waves. In each ocean basin chapter (9–13), we described some of the major eddy (ring) formation and propagation locations. These phenomena often have names ("Halmahera eddy," "Gulf Stream rings," "Agulhas rings," "Queen Charlotte Eddy," "Brazil Current Rings," etc.) because they occur so frequently in a given location and so greatly dominate variability and often transport of properties near those locations. There have been indications of more widespread coherent eddies in Lagrangian data sets (e.g. Shoosmith, Richardson, Bower, & Rossby, 2005).

Coherent vortices (eddies) have now been shown, from satellite altimetric data, to exist in large regions of the oceans (Figure 14.21; Chelton, Schlax, Samelson, & de Szoeke, 2007). These maps of vortices resemble, to some extent, the EKE maps of Figure 14.16. The major bands of eddies are in the western boundary currents/extensions, in large bands at mid-latitudes where flow is eastward (subtropical countercurrents, Azores Current), and in the ACC. The eddies mostly propagate westward; in the eastward flow of the ACC, they are advected eastward. These eddies are strongest (highest SSH) in the western boundary current extensions, but their populations are highest in the ACC and mid-latitude bands (subtropical

time and space scales), and steeply sloped at higher frequencies and wavenumbers. (The flat parts are called "white" because white noise includes all frequencies with roughly the same energy; the sloped parts are called "red" because they slope up to higher energy at lower frequency.) The spectra transition from flat to sloped at a relatively well-defined frequency

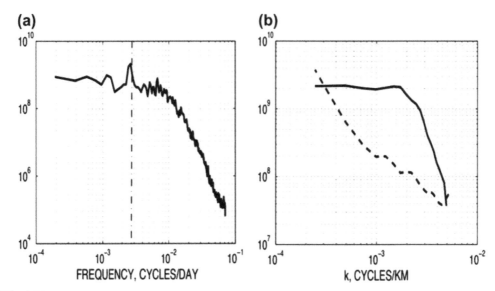

FIGURE 14.20   (a) Frequency and (b) wavenumber spectra of SSH in the eastern subtropical North Pacific, using 15 years of satellite altimetry observations. The dashed line in (a) is the annual frequency. In the wavenumber panel, solid is westward propagating, and dashed is eastward propagating energy. *Source: From Wunsch (2009).*

countercurrents). The eddy diameters are generally larger than the Rossby deformation radius (Figure S7.30 seen on the textbook Web site), exceeding 200 km in the regions most dominated by eddies, and dropping off to about 100 km at higher latitudes (Chelton et al., 2007).

14.5.3. Diapycnal Diffusion and Near-Inertial Motion

In the ocean's overturning circulation, density is altered along the circulation path (Section 7.10). Surface water becomes dense enough to sink to great depth mostly in well-defined small regions. The dense waters eventually return to lower density, due to downward diapycnal eddy diffusion of buoyancy (Figure S7.40 on the textbook Web site). In the ocean's interior, this is a weak and slow process, associated with turbulence generated by internal wave breaking (Section 7.3.2). Based on the observed ocean stratification and simple models, the globally averaged diapycnal eddy diffusivity is

about 10^{-4} m^2/sec (Munk, 1966; see Sections 7.3.2 and 7.10.2). Within the main pycnocline the diapycnal eddy diffusivity is much smaller, of the order 10^{-5} m^2/sec (e.g., Gregg, 1987; Ledwell, Watson, & Law, 1993). On the other hand, diapycnal diffusivity as observed near the ocean bottom, and up into the water column over regions of very rough topography, is higher than the Munk value (Figure 7.2).

If isopycnals are raised up to near the sea surface through mechanical forcing (e.g., Ekman suction due to wind stress curl) or due to sloping associated with strongly sheared geostrophic currents, then much higher diffusivities are available to waters on those isopycnals because of the much higher levels of near-surface turbulence due to wind and air–sea flux forcing. This uplift occurs in shallow cells in the tropics and along the eastern boundaries, where upwelling is strong, and also in the Southern Ocean and subpolar North Pacific, where open-ocean isopycnals rise up to the sea surface.

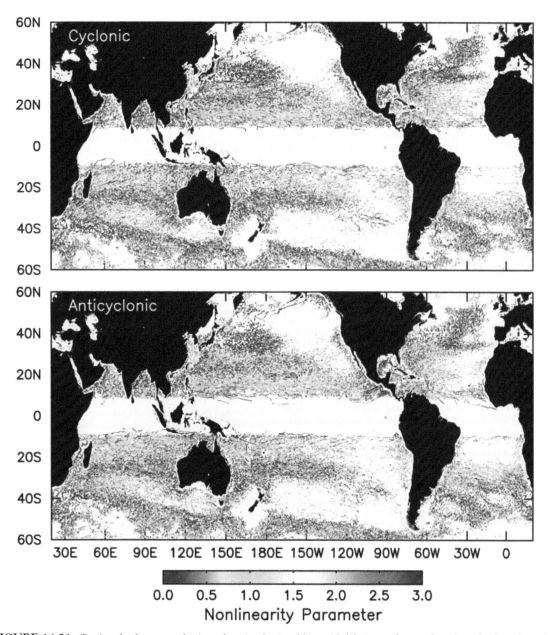

FIGURE 14.21 Tracks of coherent cyclonic and anticyclonic eddies with lifetimes of more than 4 weeks, based on altimetric SSH, color coded by a "nonlinearity parameter," which is the ratio of velocity within the eddy compared with the eddy propagation speed. White areas indicate no eddies or trajectories within 10 degrees latitude of the equator. This figure can also be found in the color insert. *Source: From Chelton et al. (2007).*

Wind-forced near-inertial motion (Figure 7.4) in the surface layer is expected to result in higher diapycnal diffusivities there; the geographical distribution of this motion should be an indication of geographic variations in surface layer diffusivity. The near-inertial motion has been mapped globally from the drifter data set (Figure 14.22a). Mean speeds of the near-inertial motions are 10 cm/sec, ranging up to much higher than 20 cm/sec beneath the atmosphere's mid-latitude storm tracks and in the eastern tropical Pacific and western tropical Atlantic. Observed inertial current radii are 10–30 km (Chaigneau, Pizarro, & Rojas, 2008).

Energy spectra from surface drifter velocities show the prevalence of inertial energy at all latitudes (Figure 14.22b). This is important because it demonstrates that the wind energy is indeed concentrating in the inertial band, and therefore energy for the surface layer's turbulent mixing is largely near-inertial. Thus the map of inertial energy in Figure 14.22a reflects the spatially varying capability of the surface ocean to mix. The inertial frequency depends on latitude, going from 0 at the equator to almost 2 cycles per day at 70° latitude (solid curve in the figure). There is also energy at low frequencies at all latitudes, which is largely due to geostrophic motion, and energy at the semidiurnal period (2 cycles per day; vertical yellow bands in the figure) mainly due to tides.

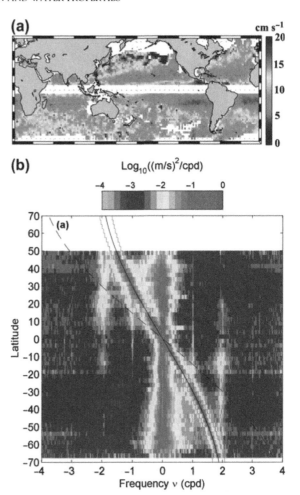

FIGURE 14.22   Near-inertial motion. (a) Average inertial current speeds (cm/sec), based on surface drifters. *Source: From Chaigneau et al. (2008).* (b) Rotary power spectra in 2.5 degree latitude bins in the Pacific Ocean. The solid curve is the inertial frequency at each latitude; the dashed curve is twice the inertial frequency. Negative frequencies rotate counterclockwise and positive frequencies rotate clockwise. *Source: From Elipot and Lumpkin (2008).* Both Figures 14.22a and 14.22b can also be found in the color insert.

14.6. CLIMATE AND THE GLOBAL OCEAN

In present usage, *climate variability* refers to natural climate variability and *climate change* refers to anthropogenically forced variations in climate. The latter is also referred to as "global change." We include climate in an oceanography text not necessarily because of ocean-atmosphere feedbacks, which might be weak in all but the tropical modes, but because climate variability and change affect ocean variability in properties and circulation.

All of the remaining text, figures, and tables relating to climate variability have been moved to Chapter S15 (Climate Variability and the

Oceans) on the supplemental Web site for the text-book, which also includes climate variability materials from each of the basin chapters. In the supplement, we present figures and a table of the modes of the interannual, decadal, and longer term climate variability that appear to have the most imprint on the ocean. We also discuss changes in ocean properties (temperature, salinity, oxygen) and to some extent circulation, as they relate to climate variability and climate change.

References

Aagaard, K., Coachman, L.K., Carmack, E., 1981. On the halocline of the Arctic Ocean. Deep-Sea Res. 28, 529–545.

Aagaard, K., Greisman, P., 1975. Toward new mass and heat budgets for the Arctic Ocean. J. Geophys. Res. 80, 3821–3827.

Aagaard, K., Swift, J.H., Carmack, E.C., 1985. Thermohaline circulation in the arctic mediterranean seas. J. Geophys. Res. 90, 4833–4846.

Ambar, I., Fiuza, A.F.G., 1994. Some features of the Portugal Current System: A poleward slope undercurrent, an upwelling-related summer southward flow and an autumn-winter poleward coastal surface current. In: Katsaros, K.B., Fiuza, A.F.G., Ambar, I. (Eds.), Proceedings of the Second International Conference on Air-Sea Interaction and on Meteorology and Oceanography of the Coastal Zone. American Meteorological Society, pp. 286–287.

Andrié, C., Ternon, J.-F., Bourlès, B., Gouriou, Y., Oudot, C., 1999. Tracer distributions and deep circulation in the western tropical Atlantic during CITHER 1 and ETAMBOT cruises, 1993–1996. J. Geophys. Res. 104, 21195–21215.

Ångström, A., 1920. Applications of heat radiation measurements to the problems of the evaporation from lakes and the heat convection at their surfaces. Geogr. Ann. 2, 237–252.

Annamalai, H., Xie, S.P., McCreary, J.P., Murtugudde, R., 2005. Impact of Indian Ocean sea surface temperature on developing El Niño. J. Climat. 18, 302–319.

Anselme, B., 1998. Sea ice fields and atmospheric phenomena in Eurasiatic arctic seas as seen from the NOAA-12 satellite. Int. J. Rem. Sensing 19, 307–316.

Antonov, J.I., Locarnini, R.A., Boyer, T.P., Mishonov, A.V., Garcia, H.E., 2006. World Ocean Atlas 2005, vol 2: Salinity. In: Levitus, S. (Ed.), NOAA Atlas NESDIS 62. U.S. Government Printing Office, Washington, D.C., 182 pp.

Aoki, S., Bindoff, N.L., Church, J.A., 2005. Interdecadal water mass changes in the Southern Ocean between 30E and 160E. Geophys. Res. Lett. 32, doi:10.1029/2004GL022220.

Arhan, M., Naveira Garabato, A., Heywood, K.J., Stevens, D.P., 2002. The Antarctic Circumpolar Current between the Falkland Islands and South Georgia. J. Phys. Oceanogr. 32, 1914–1931.

Armi, L., 1978. Some evidence for boundary mixing in the deep ocean. J. Geophys. Res. 83, 1971–1979.

Armi, L., Farmer, D.M., 1988. The flow of Mediterranean Water through the Strait of Gibraltar. Progr. Oceanogr. 21, 1–105 (also Farmer and Armi, 1988).

Assaf, G., Gerard, R., Gordon, A., 1971. Some mechanisms of oceanic mixing revealed in aerial photographs. J. Geophys. Res. 76, 6550–6572.

Bailey, W.D., 1957. Oceanographic features of the Canadian Archipelago. J. Fish. Res. Bd. Can. 14, 731–769.

Bainbridge, A.E., Broecker, W.S., Spencer, D.W., Craig, H., Weiss, R.F., Ostlund, H.G., 1981–1987. The Geochemical Ocean Sections Study (7 volumes). National Science Foundation, Washington, D.C.

Bakun, A., 1973. Coastal upwelling indices, west coast of North America, 1946–71. U.S. Dept. of Commerce, NOAA Tech. Rep. NMFS SSRF-671, 103 pp.

Bakun, A., Nelson, C.S., 1991. The seasonal cycle of wind-stress curl in subtropical eastern boundary current regions. J. Phys. Oceanogr. 21, 1815–1834.

Balmforth, N.J., Llewellyn-Smith, S., Hendershott, M., Garrett, C., 2005. Geophysical Fluid Dynamics/WHOI 2004 Program of Study: Tides. WHOI Technical Report, WHOI-2005-08, 327 pp. http://hdl.handle.net/1912/98 (accessed 10.01.09).

Bane, J.M., 1994. The Gulf Stream System: An observational perspective. Chapter 6. In: S.K., Majumdar, S.K., Miller, E.W., Forbes, G.S., Schmalz, R.F., Panah, A.A. (Eds.), The Oceans: Physical-Chemical Dynamics and Human Impact. The Pennsylvania Academy of Science, pp. 99–107.

Barber, D.G., Massom, R.A., 2007. Chapter 1: The role of sea ice in Arctic and Antarctic polynyas. In: Polynyas: Windows to the World, Elsevier Oceanography Ser, vol. 74. Elsevier, Amsterdam, pp. 1–54.

Baringer, M., Larsen, J., 2001. Sixteen years of Florida Current transport at 27° N. Geophys. Res. Lett. 28, 3179–3182.

Barnett, T.P., Pierce, D.W., AchutaRao, K., Gleckler, P., Santer, B., Gregory, J., et al., 2005. Penetration of human-induced warming into the World's Oceans. Science 309, 284–287.

Barnston, A.G., Livezey, R.E., 1987. Classification, seasonality and persistence of low-frequency atmospheric circulation patterns. Mon. Weather Rev. 115, 1083–1126.

Beal, L.M., Chereskin, T.K., Lenn, Y.D., Elipot, S., 2006. The sources and mixing characteristics of the Agulhas Current. J. Phys. Oceanogr. 36, 2060–2074.

Beal, L.M., Ffield, A., Gordon, A.L., 2000. Spreading of Red Sea overflow waters in the Indian Ocean. J. Geophys. Res. 105, 8549–8564.

Beal, L.M., Hummon, J.M., Williams, E., Brown, O.B., Baringer, W., Kearns, E.J., 2008. Five years of Florida Current structure and transport from the Royal Caribbean Cruise Ship Explorer of the Seas. J. Geophys. Res. 113, C06001. doi:10.1029/2007JC004154.

Beardsley, R.C., Boicourt, W.C., 1981. On estuarine and continental-shelf circulation in the Middle Atlantic Bight. In: Warren, B.A., Wunsch, C. (Eds.), Evolution of Physical Oceanography. MIT Press, Cambridge, MA, pp. 198–223.

Becker, J.J., Sandwell, D.T., 2008. Global estimates of seafloor slope from single-beam ship soundings. J. Geophys. Res. 113, C05028. doi:10.1029/2006JC003879.

Becker, J.J., Sandwell, D.T., Smith, W.H.F., Braud, J., Binder, B., Depner, J., et al., 2009. Global bathymetry and elevation data at 30 arc seconds resolution: SRTM30_PLUS. Mar. Geod. 32, 4355–4371.

Belkin, I.M., 2004. Propagation of the "Great Salinity Anomaly" of the 1990s around the northern North Atlantic. Geophys. Res. Lett. 31, L08306. doi:10.1029/2003GL019334.

Belkin, I.M., Gordon, A.L., 1996. Southern Ocean fronts from the Greenwich meridian to Tasmania. J. Geophys. Res. 101, 3675–3696.

Bendat, J.S., Piersol, A.G., 1986. Random Data: Analysis and Measurement Procedures, second ed. Wiley, New York, 566 pp.

Bennett, A.F., 1976. Poleward heat fluxes in southern hemisphere oceans. J. Phys. Oceanogr. 8, 785–789.

Bernard, E.N., Robinson, A.R. (Eds.), 2009. The Sea, vol. 15: Tsunamis. Harvard University Press, Cambridge, MA, 462 pp.

Bevington, P.R., Robinson, D.K., 2003. Data Reduction and Error Analysis for the Physical Sciences. McGraw Hill, Dubuque, IA, 320 pp.

Biastoch, A., Böning, C.W., Lutjeharms, J.R.E., 2008. Agulhas leakage dynamics affects decadal variability in the Atlantic overturning circulation. Nature 456, 489–492.

Bindoff, N.L., McDougall, T.J., 2000. Decadal changes along an Indian Ocean section at 32 degrees S and their interpretation. J. Phys. Oceanogr. 30, 1207–1222.

Bindoff, N.L., Willebrand, J., Artale, V., Cazenave, A., Gregory, J., Gulev, S., et al., 2007. Observations: Oceanic climate change and sea level. In: Solomon, S., Qin, D., Manning, M., Chen, Z., Marquis, M., Averyt, K.B. (Eds.), Climate Change 2007: The Physical Science Basis. Contribution of Working Group I to the Fourth Assessment Report of the Intergovernmental Panel on Climate Change. Cambridge University Press, Cambridge, UK and New York.

Bingham, F.M., Lukas, R., 1994. The southward intrusion of North Pacific Intermediate Water along the Mindanao Coast. J. Phys. Oceanogr. 24, 141–154.

Bingham, F.M., Talley, L.D., 1991. Estimates of Kuroshio transport using an inverse method. Deep-Sea Res. 38 (Suppl.), S21–S43.

Bishop, J.K.B., 1999. Transmissometer measurement of POC. Deep-Sea Res. I 46, 353–369.

Bitz, C.M., Fyfe, J.C., Flato, G.M., 2002. Sea ice response to wind forcing from AMIP models. J. Clim. 15, 522–536.

Bjerknes, J., 1969. Atmospheric teleconnections from the equatorial Pacific. Mon. Weather Rev. 97, 163–172.

Bjerknes, V., Bjerknes, J., Solberg, H., Bergeron, T., 1933. Physikalische Hydrodynamik mit Anwendung auf die Dynamische Meterologie, 3. Springer, Berlin.

Björk, G., Jakobsson, M., Rudels, B., Swift, J.H., Anderson, L., Darby, D.A., et al., 2007. Bathymetry and deep-water exchange across the central Lomonosov Ridge at 88–89°N. Deep-Sea Res. I 54, 1197–1208.

Boccaletti, G., Ferrari, R., Fox-Kemper, B., 2007. Mixed layer instabilities and restratification. J. Phys. Oceanogr. 37, 2228–2250.

Boebel, O., Davis, R.E., Ollitrault, M., Peterson, R.G., Richardson, P.L., Schmid, C., Zenk, W., 1999. The intermediate depth circulation of the western South Atlantic. Geophys. Res. Lett. 26, 3329–3332. doi:10.1029/1999GL002355.

Böhm, E., Morrison, J.M., Manghnani, V., Kim, H.-S., Flagg, C.N., 1999. The Ras al Hadd Jet: Remotely sensed and acoustic Doppler current profiler observations in 1994–1995. Deep-Sea Res. II 46, 1531–1549.

Boland, F.M., Church, J.A., 1981. The East Australian Current 1978. Deep-Sea Res. 28A, 937–957.

Botnikov, V.N., 1963. Geographical position of the Antarctic Convergence zone in the Southern Ocean. Sov. Antarct. Exped. Inf. Bull., Engl. Transl. 4 (41), 324–327.

Bourlès, B., Lumpkin, R., McPhaden, M.J., Hernandez, F., Nobre, P., Campos, E., et al., 2008. The Pirata program: History, accomplishments, and future directions. B. Am. Meteorol. Soc. 89, 1111–1125.

Bowden, K.F., 1983. Physical Oceanography of Coastal Waters. Ellis Horwood Series in Marine Science. Halsted Press, New York, 302 pp.

Bower, A.S., Armi, L., Ambar, I., 1997. Lagrangian observations of Meddy formation during a Mediterranean undercurrent seeding experiment. J. Phys. Oceanogr. 27, 2545–2575.

Bower, A.S., Hunt, H.D., Price, J.F., 2000. Character and dynamics of the Red Sea and Persian Gulf outflows. J. Geophys. Res. 105, 6387–6414.

Bower, A.S., Johns, W.E., Fratantoni, D.M., Peters, H., 2005. Equilibration and circulation of Red Sea Outflow Water in the western Gulf of Aden. J. Phys. Oceanogr. 35, 1963–1985.

Bower, A.S., LeCann, B., Rossby, T., Zenk, W., Gould, J., Speer, K., et al., 2002. Directly measured mid-depth circulation in the northeastern North Atlantic Ocean. Nature 410, 603–607.

Bower, A.S., Lozier, M.S., Gary, S.F., Böning, C.W., 2009. Interior pathways of the North Atlantic meridional overturning circulation. Nature 459, 243–247.

Boyer, T.P., Antonov, J.I., Levitus, S., Locarnini, R., 2005. Linear trends of salinity for the world ocean: 1955–1998. Geophys. Res. Lett. 32, L01604. doi:1029/2004GL021791.

Brambilla, E., Talley, L.D., 2008. Subpolar Mode Water in the northeastern Atlantic: 1. Averaged properties and mean circulation. J. Geophys. Res. 113, C04025. doi_10.1029/2006JC004062.

Bray, N., Ochoa, J., Kinder, T., 1995. The role of the interface in exchange through the Strait of Gibraltar. J. Geophys. Res. 100, 10755–10776.

Bretherton, F.P., Davis, R.E., Fandry, C.B., 1976. A technique for objective analysis and design of oceanographic experiments applied to MODE-73. Deep-Sea Res. 23, 559–582.

Brink, K.H., 1991. Coastal-trapped waves and wind-driven currents over the continental shelf. Annu. Rev. Fluid Mech. 23, 389–412.

Brink, K.H., 2005. Coastal physical processes overview. In: Robinson, A.F., Brink, K.H. (Eds.), The Sea, vol. 13: The Global Coastal Ocean: Multiscale Interdisciplinary Processes. Harvard University Press, Cambridge, MA, pp. 37–60.

Brink, K.H., Robinson, A.R. (Eds.), 1998. The Sea, vol. 10: The Global Coastal Ocean: Processes and Methods. Harvard University Press, Cambridge, MA, 628 pp.

Broecker, W.S., 1974. "NO," a conservative water-mass tracer. Earth Planet. Sci. Lett. 23, 100–107.

Broecker, W.S., 1987. The biggest chill. Nat. Hist 97, 74–82.

Broecker, W.S., 1991. The great ocean conveyor. Oceanography 4, 79–89.

Broecker, W.S., 1998. Paleocean circulation during the last deglaciation: A bipolar seesaw? Paleoceanography 13, 119–121.

Broecker, W.S., Clark, E., Hajdas, I., Bonani, G., 2004. Glacial ventilation rates for the deep Pacific Ocean. Paleoceanography 19, PA2002. doi: 10.1029/2003PA000974.

Broecker, W.S., Peng, T., 1982. Tracers in the Sea. Lamont-Doherty Geological Observatory. Columbia University, 690 pp.

Bryan, K., 1963. A numerical investigation of a nonlinear model of a wind-driven ocean. J. Atm. Sci. 20, 594–606.

Bryden, H.L., Beal, L.M., 2001. Role of the Agulhas Current in Indian Ocean circulation and associated heat and freshwater fluxes. Deep-Sea Res. I 48, 1821–1845.

Bryden, H.L., Candela, J., Kinder, T.H., 1994. Exchange through the Strait of Gibraltar. Progr. Oceanogr. 33, 201–248.

Bryden, H.L., Griffiths, M.J., Lavin, A.M., Millard, R.C., Parrilla, G., Smethie, W.M., 1996. Decadal changes in water mass characteristics at 24°N in the subtropical North Atlantic Ocean. J. Clim. 9, 3162–3186.

Bryden, H.L., Imawaki, S., 2001. Ocean heat transport. In: Siedler, G., Church, J. (Eds.), Ocean Circulation and Climate, International Geophysics Series. Academic Press, San Diego, CA, pp. 455–474.

Bryden, H.L., Johns, W.E., Saunders, P.M., 2005a. Deep western boundary current east of Abaco: Mean structure and transport. J. Mar. Res. 63, 35–57.

Bryden, H.L., Longworth, H.R., Cunningham, S.A., 2005b. Slowing of the Atlantic meridional overturning circulation at 26.5°N. Nature 438, 655–657.

Bryden, H.L., Pillsbury, R.D., 1977. Variability of deep flow in the Drake Passage from year-long current measurements. J. Phys. Oceanogr. 7, 803–810.

Bye, J.A.T., 1972. Ocean circulation south of Australia. In: Hayes, D.E. (Ed.), Antarctic Oceanology I, The Australian-New Zealand Sector, Antarctic Research Series, 19. AGU, Washington, D.C, pp. 95–100.

CalCOFI ADCP, 2008. CalCOFI ADCP. Scripps Institution of Oceanography. http://adcp.ucsd.edu/calcofi/ (accessed 10.18.18).

Cameron, W.M., Pritchard, D.W., 1963. Estuaries. In: Hill, M.N. (Ed.), The Sea, vol. 2: Ideas and Observations. Wiley-Interscience, New York, pp. 306–324.

Candela, J., Tanahara, S., Crepon, M., Barnier, B., Sheinbaum, J., 2003. Yucatan Channel flow: Observations versus CLIPPER ATL6 and MERCATOR PAM models. J. Geophys. Res. 108, 3385. doi:10.1029/2003JC001961.

Candela, J., 2001. Mediterranean Water and global circulation. In: Siedler, G., Church, J. (Eds.), Ocean Circulation and Climate, International Geophysics Series. Academic Press, San Diego, CA, pp. 419–430.

Cane, M.A., Münnich, M., Zebiak, S.F., 1990. A study of self-excited oscillations of the tropical ocean-atmosphere system. Part I: Linear analysis. J. Atmos. Sci. 47, 1562–1577.

Capet, X., McWilliams, J.C., Molemaker, M.J., Shchepetkin, A.F., 2008. Mesoscale to submesoscale transition in the California Current System. Part III: Energy balance and flux. J. Phys. Oceanogr. 38, 2256–2269.

Carmack, E., Aagaard, K., 1973. On the deep water of the Greenland Sea. Deep-Sea Res. 20, 687–715.

Cartwright, D.E., 1999. Tides: A Scientific History. Cambridge University Press, Cambridge, UK, 292 pp.

Castro, S.L., Wick, G.A., Emery, W.J., 2003. Further refinements to models for the bulk-skin sea surface temperature difference. J. Geophys. Res. 108, 3377–3395.

Cavalieri, D., Parkinson, C., Gloersen, P., Zwally, H.J., 1996, updated 2008. Sea ice concentrations from Nimbus-7 SMMR and DMSP SSM/I passive microwave data, 1991. Boulder, Colorado USA: National Snow and Ice Data Center. Digital media. http://nsidc.org/data/nsidc-0051.html (accessed 11.11.08).

Cayan, D.R., 1992. Latent and sensible heat flux anomalies over the northern oceans: Driving the sea surface temperature. J. Phys. Oceanogr. 22, 859–881.

CDIP, 2009. The Coastal Data Information Program, Scripps Institution of Oceanography. http://cdip.ucsd.edu/ (accessed 5.15.09).

Cerovecki, I., Talley, L.D., Mazloff, M., 2011. Transformation and formation rates of Subantarctic Mode Water based on air–sea fluxes, in preparation.

Cetina, P., Candela, J., Sheinbaum, J., Ochoa, J., Badan, A., 2006. Circulation along the Mexican Caribbean coast. J. Geophys. Res. 111, C08021. doi:10.1029/2005JC003056.

Chaigneau, A., Pizarro, O., Rojas, W., 2008. Global climatology of near-inertial current characteristics from Lagrangian observations. Geophys. Res. Lett. 35, L13603. doi:10.1029/2008GL034060.

Chang, P., Ji, L., Li, H., 1997. A decadal climate variation in the tropical Atlantic Ocean from thermodynamic air-sea interactions. Nature 385, 516–518.

Chapman, B., 2004. Initial visions of paradise: Antebellum U.S. government documents on the South Pacific. J. Gov. Inform. 30, 727–750.

Chatfield, C., 2004. The Analysis of Time Series: An Introduction, sixth ed. Chapman and Hall/CRC Press, Boca Raton, FL, 333 pp.

Chelton, D.B., deSzoeke, R.A., Schlax, M.G., El Naggar, K., Siwertz, N., 1998. Geographical variability of the first baroclinic Rossby radius of deformation. J. Phys. Oceanogr. 28, 433–460.

Chelton, D.B., Freilich, M.H., Esbensen, S.K., 2000. Satellite observations of the wind jets off the Pacific coast of Central America. Part II: Regional relationships and dynamical considerations. Mon. Weather Rev. 128, 2019–2043.

Chelton, D.B., Ries, J.C., Haines, B.J., Fu, L.L., Callahan, P.S., Cazenave, A., 2001. Satellite altimetry. In: Fu, L.-L.,

Cazenave, A. (Eds.), Satellite Altimetry and Earth Sciences. Academic Press, San Diego, CA, 463 pp.

Chelton, D.B., Schlax, M.G., 1996. Global observations of oceanic Rossby waves. Science 272, 234–238.

Chelton, D.B., Schlax, M.G., Freilich, M.H., Milliff, R.F., 2004. Satellite measurements reveal persistent small-scale features in ocean winds. Science 303, 978–983.

Chelton, D.B., Schlax, M.G., Samelson, R.M., de Szoeke, R.A., 2007. Global observations of large oceanic eddies. Geophys. Res. Lett. 34, L15606. doi:10.1029/2007GL030812.

Chen, C., Beardsley, R.C., 2002. Cross-frontal water exchange on Georges Bank: Some results from an U.S. GLOBEC/Georges Bank Program Model Study. J. Oceanogr. 58, 403–420.

Chen, C.T., Millero, F.J., 1977. Speed of sound in seawater at high pressures. J. Acoust. Soc. Am. 62, 1129–1135.

Chereskin, T.K., 1995. Direct evidence for an Ekman balance in the California Current. J. Geophys. Res. 100, 18261–18269.

Chereskin, T.K., Trunnell, M., 1996. Correlation scales, objective mapping, and absolute geostrophic flow in the California Current. J. Geophys. Res. 101, 22619–22629.

Chiang, J.C.H., Vimont, D.J., 2004. Analogous Pacific and Atlantic meridional modes of tropical atmosphere–ocean variability. J. Clim. 17, 4143–4158.

CIESM, 2001. CIESM Round table session on Mediterranean water mass acronyms. 36th CIESM Congress, Monte Carlo, 26 September 2001. https://www.ciesm.org/catalog/WaterMassAcronyms.pdf (accessed 6.5.09).

Clark, C.O., Webster, P.J., Cole, J.E., 2003. Interdecadal variability of the relationship between the Indian Ocean zonal mode and East African coastal rainfall anomalies. J. Clim. 16, 548–554.

Clarke, R.A., Swift, J.H., Reid, J.L., Koltermann, K.P., 1990. The formation of Greenland Sea Deep Water: Double diffusion or deep convection? Deep-Sea Res. Part A 37, 1385–1424.

Climate Prediction Center Internet Team, 2006. AAO, AO, NAO, PNA. NOAA National Weather Service Climate Prediction Center. http://www.cpc.noaa.gov/products/precip/CWlink/daily_ao_index/teleconnections.shtml (accessed 4.20.09).

Climate Prediction Center Internet Team, 2009. Cold and warm episodes by season. NOAA National Weather Service Climate Prediction Center. http://www.cpc.noaa.gov/products/analysis_monitoring/ensostuff/ensoyears.shtml (accessed 3.26.09).

Coachman, L.K., Aagaard, K., 1974. Physical oceanography of arctic and subarctic seas. In: Herman, Y. (Ed.), Marine Geology and Oceanography of the Arctic Seas. Springer-Verlag, New York, pp. 1–81.

Cobb, K.M., Charles, C.D., Cheng, H., Edwards, R.L., 2003. El Niño/Southern Oscillation and tropical Pacific climate during the last millennium. Nature 424, 271–276.

Cole, S.T., Rudnick, D.L., Hodges, B.A., Martin, J.P., 2009. Observations of tidal internal wave beams at Kauai Channel, Hawaii. J. Phys. Oceanogr. 39, 421–436.

Comiso, J., Wadhams, P., Pedersen, L., Gersten, R., 2001. Seasonal and interannual variability of the Odden ice tongue and a study of environmental effects. J. Geophys. Res. 106, 9093–9116.

Comiso, J.C., Gordon, A.L., 1987. Recurring polynyas over the Cosmonaut Sea and the Maud Rise. J. Geophys. Res. 92, 2819–2833.

Conkright, M.E., Levitus, S., Boyer, T.P., 1994. World Ocean Atlas 1994 vol 1: Nutrients. NOAA Atlas NESDIS 1. U.S. Department of Commerce, Washington, D.C., 150 pp.

Cornillon, P., 1986. The effect of the New England Seamounts on Gulf Stream meandering as observed from satellite IR imagery. J. Phys. Oceanogr. 16, 386–389.

Cox, R.A., McCartney, M.J., Culkin, F., 1970. The specific gravity/salinity/temperature relationship in natural seawater. Deep-Sea Res. 17, 679–689.

Cox, R.A., Smith, N.D., 1959. The specific heat of seawater. Philos. Trans. Roy. Soc. London A 252, 51–62.

Craik, A.D.D., 2005. George Gabriel Stokes on water wave theory. Ann. Rev. Fluid Mech. 37, 23–42.

Crawford, W., 2002. Physical characteristics of Haida Eddies. J. Oceanogr. 58, 703–713.

Crawford, W., Cherniawsky, J., Foreman, M., 2000. Multi-year meanders and eddies in the Alaskan Stream as observed by TOPEX/Poseidon altimeter. Geophys. Res. Lett. 27, 1025–1028.

Cresswell, G.R., Golding, T.J., 1980. Observations of a south-flowing current in the southeastern Indian Ocean. Deep-Sea Res. 27A, 449–466.

Cunningham, S.A., Alderson, S.G., King, B.A., Brandon, M.A., 2003. Transport and variability of the Antarctic Circumpolar Current in Drake Passage. J. Geophys. Res. 108 (C5), 8084. doi:10.1029/2001JC001147.

Cunningham, S.A., Kanzow, T., Rayner, D., Baringer, M.O., Johns, W.E., Marotzke, J., et al., 2007. Temporal variability of the Atlantic meridional overturning circulation at 26.5°N. Science 317, 935–938.

Curray, J.R., Emmel, F.J., Moore, D.G., 2003. The Bengal Fan: Morphology, geometry, stratigraphy, history and processes. Mar. Petr. Geol. 19, 1191–1223.

Curry, R., Dickson, B., Yashayaev, I., 2003. A change in the freshwater balance of the Atlantic Ocean over the past four decades. Nature 426, 826–829.

Curry, R.G., McCartney, M.S., 2001. Ocean gyre circulation changes associated with the North Atlantic Oscillation. J. Phys. Oceanogr. 31, 3374–3400.

Cushman-Roisin, B., 1994. Introduction to Geophysical Fluid Dynamics. Prentice Hall, Englewood Cliffs, NJ, 320 pp.

da Silva, A.M., Young, A.C., Levitus, S., 1994. Atlas of surface marine data, vol. 1: Algorithms and Procedures.

NOAA Atlas NESDIS 6. U.S. Department of Commerce, NOAA, NESDIS.

Dai, A., Trenberth, K.E., 2002. Estimates of freshwater discharge from continents: Latitudinal and seasonal variations. J. Hydromet. 3, 660–687.

d'Asaro, E.A., Eriksen, C.C., Levine, M.D., Paulson, C.A., Niiler, P., Van Meurs, P., 1995. Upper-ocean inertial currents forced by a strong storm. Part 1: Data and comparisons with linear theory. J. Phys. Oceanogr. 25, 2909–2936.

Davis, R.E., 1976. Predictability of sea surface temperature and sea level pressure anomalies over the North Pacific. J. Phys. Oceanogr. 6, 249–266.

Davis, R.E., 1991. Observing the general circulation with floats. Deep-Sea Res. 38 (Suppl.), S531–S571.

Davis, R.E., 2005. Intermediate-depth circulation of the Indian and South Pacific Oceans measured by autonomous floats. J. Phys. Oceanogr. 35, 683–707.

Davis, R.E., deSzoeke, R., Niiler, P., 1981. Variability in the upper ocean during MILE. Part II: Modeling the mixed layer response. Deep-Sea Res. 28A, 1453–1475.

Deacon, G.E.R., 1933. A general account of the hydrology of the South Atlantic Ocean. Disc. Rep. 7, 171–238.

Deacon, G.E.R., 1937. The hydrology of the Southern Ocean. Disc. Rep. 15, 3–122.

Deacon, G.E.R., 1982. Physical and biological zonation in the Southern Ocean. Deep-Sea Res. 29, 1–15.

deBoyer Montégut, C., Madec, G., Fischer, A.S., Lazar, A., Iudicone, D., 2004. Mixed layer depth over the global ocean: An examination of profile data and a profile-based climatology. J. Geophys. Res. 109, C12003. doi:10.1029/2004JC002378.

Defant, A., 1936. Die Troposphäre des Atlantischen Ozeans. In Wissenschaftliche Ergebnisse der Deutschen Atlantischen Expedition auf dem Forschungs- und Vermessungsschiff "Meteor" 1925-1927, 6(1), 289-411 (in German).

Defant, A., 1961. Physical Oceanography, vol. 1. Pergamon Press, New York, 729 pp.

Del Grosso, V.A., 1974. New equation for the speed of sound in natural waters (with comparisons to other equations. J. Acoust. Soc. Am. 56, 1084–1091.

Delworth, T.L., Mann, M.E., 2000. Observed and simulated multidecadal variability in the northern hemisphere. Clim. Dynam. 16, 661–676.

Dengler, M., Quadfasel, D., Schott, F., Fischer, J., 2002. Abyssal circulation in the Somali Basin. Deep-Sea Res. II 49, 1297–1322.

Dengler, M., Schott, F.A., Eden, C., Brandt, P., Fischer, J., Zantopp, R.J., 2004. Break-up of the Atlantic deep western boundary current into eddies at 8°S. Nature 432, 1018–1020.

Déry, S.J., Stieglitz, M., McKenna, E.C., Wood, E.F., 2005. Characteristics and trends of river discharge into Hudson, James, and Ungava Bays, 1964–2000. J. Clim. 18, 2540–2557.

Deser, C., Holland, M., Reverdin, G., Timlin, M., 2002. Decadal variations in Labrador Sea ice cover and North Atlantic sea surface temperatures. J. Geophys. Res. 107, 3035. doi:10.1029/2000JC000683.

Deser, C., Phillips, A.S., Hurrell, J.W., 2004. Pacific interdecadal climate variability: Linkages between the tropics and the North Pacific during boreal winter since 1900. J. Clim. 17, 3109–3124.

deSzoeke, R.A., Levine, M.D., 1981. The advective flux of heat by mean geostrophic motions in the Southern Ocean. Deep-Sea Res. 28, 1057–1085.

Deutsch, C., Emerson, S., Thompson, L., 2005. Fingerprints of climate change in North Pacific oxygen. Geophys. Res. Lett. 32, L16604. doi:10.1029/2005GL023190.

Dickson, R., Brown, J., 1994. The production of North Atlantic Deep Water: Sources, rates, and pathways. J. Geophys. Res. 99, 12319–12341.

Dickson, R., Lazier, J., Meincke, J., Rhines, P., Swift, J., 1996. Long-term coordinated changes in the convective activity of the North Atlantic. Progr. Oceanogr. 38, 241–295.

Dickson, R., Yashayaev, I., Meincke, J., Turrell, B., Dye, S., Holfort, J., 2002. Rapid freshening of the deep North Atlantic Ocean over the past four decades. Nature 416, 832–837.

Dickson, R.R., Curry, R., Yashayaev, I., 2003. Recent changes in the North Atlantic. Phil. Trans. Roy. Soc. A 361, 1917–1933.

Dickson, R.R., Meincke, J., Malmberg, S.-A., Lee, A.J., 1988. The "Great Salinity Anomaly" in the northern North Atlantic 1968–1982. Progr. Oceanogr. 20, 103–151.

Dickson, R.R., Meincke, J., Rhines, P. (Eds.), 2008. Arctic-Subarctic Ocean Fluxes: Defining the Role of the Northern Seas in Climate. Springer, The Netherlands, 736 pp.

Dietrich, G., 1963. Allgemeine Meereskunde. Gebruder Borntraeger Verlagsbuchhandlung, Berlin-Stuttgart. (English translation: General Oceanography. Wiley-Interscience, New York), 492 pp.

DiLorenzo, E., Schneider, N., Cobb, K.M., Franks, P.J.S., Chhak, K., Miller, A.J., et al., 2008. North Pacific Gyre Oscillation links ocean climate and ecosystem change. Geophys. Res. Lett. 35, L08607. doi:10.1029/2007GL032838.

Dittmar, W., 1884. Report of researches into the composition of ocean water collected by HMS Challenger during the years 1873–76. Voyage of the H.M.S.Challenger: Physics and chemistry, 1, part 1. Longmans & Co., London.

Dmitrenko, I.A., Tyshko, K.N., Kirillov, S.A., Eicken, H., Hölemann, J.A., Kassens, H., 2005. Impact of flaw polynyas on the hydrography of the Laptev Sea. Global Planet. Change 48, 9–27.

Dmitrenko, I.A., Wegner, C., Kassens, H., Kirillov, S.A., Krumpen, T., Heinemann, G., et al., 2010. Observations of supercooling and frazil ice formation in the Laptev Sea coastal polynya. J. Geophys. Res. 115, C05015. doi:10.1029/2009JC005798.

Domingues, C.M., Church, J.A., White, N.J., Gleckler, P.J., Wijffels, S.E., Barker, P.M., et al., 2008. Improved estimates of upper-ocean warming and multi-decadal sea-level rise. Nature 453, 1090–1093.

Domingues, C.M., Maltrud, M.E., Wijffels, S.E., Church, J.A., Tomczak, M., 2007. Simulated Lagrangian pathways between the Leeuwin Current System and the upper-ocean circulation of the southeast Indian Ocean. Deep-Sea Res. II 54, 797–817.

Dong, S., Gille, S.T., Sprintall, J., Talley, L., 2008. Southern Ocean mixed-layer depth from Argo float profiles. J. Geophys. Res. 113, C06013. doi:10.1029/2006JC004051.

Donohue, K.A., Firing, E., Chen, S., 2001. Absolute geostrophic velocity within the Subantarctic Front in the Pacific Ocean. J. Geophys. Res. 106, 19869–19882.

Döös, K., Coward, A., 1997. The Southern Ocean as the major upwelling zone of North Atlantic Deep Water. International WOCE Newsletter, 27, 3–4. http://woce.nodc.noaa.gov/wdiu/wocedocs/newsltr/ (accessed 09.01.09).

Döös, K., Webb, D.J., 1994. The Deacon cell and the other meridional cells of the Southern Ocean. J. Phys. Oceanogr. 24, 429–442.

Doron, P., Bertuccioli, L., Katz, J., Osborn, T.R., 2001. Turbulence characteristics and dissipation estimates in the coastal ocean bottom boundary layer from PIV data. J. Phys. Oceanogr. 31, 2108–2134.

Doronin, Y.P., Khesin, D.E., 1975. Sea Ice (Trans., 1977). Amerind Publishing Company, New Delhi, India, 323 pp.

Ducet, N., Le Traon, P.Y., Reverdin, G., 2000. Global high-resolution mapping of ocean circulation from TOPEX/Poseidon and ERS-1 and -2. J. Geophys. Res. 105, 19477–19498.

Durack, P.J., Wijffels, S.E., 2010. Fifty-year trends in global ocean salinities and their relationship to broad-scale warming. J. Clim. 23, 4342–4362.

Dyer, K.R., 1997. Estuaries: A Physical Introduction, second ed. Wiley, New York, 195 pp.

Egbert, G.D., Ray, R., 2001. Estimates of M2 tidal energy dissipation from TOPEX/Poseidon altimeter data. J. Geophys. Res. 106, 22475–22502.

Ekman, V.W., 1905. On the influence of the Earth's rotation on ocean currents. Arch. Math. Astron. Phys. 2 (11), 1–53.

Ekman, V.W., 1923. Über Horizontalzirkulation bei winderzeugten Meeresströmungen. Ark. Math. Astron Fys. 17, 1–74 (in German).

Elipot, S., Lumpkin, R., 2008. Spectral description of oceanic near-surface variability. Geophys. Res. Lett. 35, L05606. doi:10.1029/2007GL032874.

Emery, W.J., 1977. Antarctic polar frontal zone from Australia to the Drake Passage. J. Phys. Oceanogr. 7, 811–822.

Emery, W.J., Fowler, C.W., Maslanik, J.A., 1997. Satellite derived Arctic and Antarctic sea ice motions: 1988–1994. Geophys. Res. Lett. 24, 897–900.

Emery, W.J., Meincke, J., 1986. Global water masses: summary and review. Oceanol. Acta 9, 383–391.

Emery, W.J., Thomson, R.E., 2001. Data Analysis Methods in Physical Oceanography, second ed. Elsevier, Amsterdam, 638 pp.

Enfield, D., Mestas-Nuñez, A., Trimble, P., 2001. The Atlantic Multidecadal Oscillation and its relation to rainfall and river flows in the continental U.S. Geophys. Res. Lett. 28, 2077–2080.

Eriksen, C.C., 1982. Geostrophic equatorial deep jets. J. Mar. Res. 40 (Suppl.), 143–157.

Fang, F., Morrow, R., 2003. Evolution, movement and decay of warm-core Leeuwin Current eddies. Deep-Sea Res. II 50, 2245–2261.

Farmer, D.M., Freeland, H.J., 1983. The physical oceanography of fjords. Progr. Oceanogr. 12, 147–219.

Favorite, F., Dodimead, A.J., Nasu, K., 1976. Oceanography of the subarctic Pacific region, 1960–71. International North Pacific Fisheries Commission, Vancouver, Canada, 33, 187 pp.

Fedorov, A.V., Harper, S.L., Philander, S.G., Winter, B., Wittenberg, A., 2003. How predictable is El Niño? B. Am. Meteorol. Soc. 84, 911–919.

Feely, R.A., Sabine, C.L., Lee, K., Berelson, W., Kleypas, J., Fabry, V.J., et al., 2004. Impact of anthropogenic CO_2 on the $CaCO_3$ system in the oceans. Science 305, 362–366.

Feng, M., Meyers, G., Pearce, A., Wijffels, S., 2003. Annual and interannual variations of the Leeuwin Current at 32°S. J. Geophys. Res. 108, C11. doi:10.1029/2002JC001763.

Feng, M., Wijffels, S., Godfrey, S., Meyers, G., 2005. Do eddies play a role in the momentum balance of the Leeuwin Current? J. Phys. Oceanogr. 35, 964–975.

Field, J.G., Shillington, F.A., 2006. Variability of the Benguela Current System. In: Robinson, A.R., Brink, K.H. (Eds.), The Sea, vol. 14B: The Global Coastal Ocean: Interdisciplinary Regional Studies and Syntheses. Harvard University Press, Boston, MA, pp. 835–864.

Fine, R.A., 1993. Circulation of Antarctic Intermediate Water in the South Indian Ocean. Deep-Sea Res. I 40, 2021–2042.

Fine, R.A., Lukas, R., Bingham, F.M., Warner, M.J., Gammon, R.H., 1994. The western equatorial Pacific: A water mass crossroads. J. Geophys. Res. 99, 25063–25080.

Fine, R.A., Maillet, K.A., Sullivan, K.F., Willey, D., 2001. Circulation and ventilation flux of the Pacific Ocean. J. Geophys. Res. 106, 22159–22178.

Firing, E., 1989. Mean zonal currents below 1500 m near the equator, 159°W. J. Geophys. Res. 94, 2023–2028.

Firing, E., Wijffels, S.E., Hacker, P., 1998. Equatorial sub-thermocline currrents across the Pacific. J. Geophys. Res. 103, 21413–21423.

Flatau, M.K., Talley, L.D., Niiler, P.P., 2003. The North Atlantic Oscillation, surface current velocities, and SST changes in the subpolar North Atlantic. J. Clim. 16, 2355–2369.

Fofonoff, N.P., 1954. Steady flow in a frictionless homogeneous ocean. J. Mar. Res. 13, 254–262.

Fofonoff, N.P., 1977. Computation of potential temperature of seawater for an arbitrary reference pressure. Deep-Sea Res. 24, 489–491.

Fofonoff, N.P., 1985. Physical properties of seawater: A new salinity scale and equation of state for seawater. J. Geophys. Res. 90, 3332–3342.

Forch, C., Knudsen, M., Sorensen, S.P., 1902. Berichte über die Konstantenbestimmungen zur Aufstellung der hydrographischen Tabellen. Kgl. Dan. Vidensk. Selsk. Skr., 6, Raekke, Naturvidensk. Mat., Afel. XII. 1, 151 (in German).

Forchhammer, G., 1865. On the composition of seawater in the different parts of the ocean. Philos. Trans. Roy. Soc. Lond. 155, 203–262.

Foster, T.D., 1972. An analysis of the cabbeling instability in seawater. J. Phys. Oceanogr. 2, 294–301.

Fowler, C., 2003, updated 2007. Polar Pathfinder daily 25 km EASE-Grid sea ice motion vectors. National Snow and Ice Data Center. http://nsidc.org/data/nsidc-0116.html. ftp://sidads.colorado.edu/pub/DATASETS/ice_motion/browse (accessed 11.01.08).

Fowler, C., Emery, W.J., Maslanik, J., 2004. Satellite-derived evolution of Arctic sea ice age: October 1978 to March 2003. IEEE Remote Sensing Lett. 1, 71–74.

Frammuseet, 2003. Picture archive — 1st Fram voyage. http://www.fram.museum.no (accessed 3.17.09).

Fratantoni, D.M., 2001. North Atlantic surface circulation during the 1990s observed with satellite-tracked drifters. J. Geophys. Res. 106, 22067–22093.

Fratantoni, D.M., Johns, W.E., Townsend, T.L., Hurlburt, H.E., 2000. Low-latitude circulation and mass transport pathways in a model of the tropical Atlantic Ocean. J. Phys. Oceanogr., 301944–301966.

Fratantoni, D.M., Zantopp, R.J., Johns, W.E., Miller, J.L., 1997. Updated bathymetry of the Anegada-Jungfern Passage complex and implications for Atlantic inflow to the abyssal Caribbean Sea. J. Mar. Res. 55, 847–860.

Friedrichs, M., McCartney, M., Hall, M., 1994. Hemispheric asymmetry of deep water transport modes in the western Atlantic. J. Geophys. Res. 99, 25165–25179.

Fu, L.-L., Chelton, D.B., 2001. In: Fu, L.-L., Cazenave, A. (Eds.), International Geophysics Series. Large-scale ocean circulation. In Satellite Altimetry and Earth Sciences: A Handbook of Techniques and Applications, 69. Academic Press, San Diego, CA, pp. 133–170.

Fuglister, F.C., 1960. Atlantic Ocean Atlas of temperature and salinity profiles and data from the IGY of 1957–1958. Woods Hole Oceanographic Institution Atlas Series 1, 209 pp.

Fyfe, J.C., 2006. Southern Ocean warming due to human influence. Geophys. Res. Lett. 33, L19701. doi:10.1029/2006GL027247.

Ganachaud, A., 2003. Large-scale mass transports, water mass formation, and diffusivities estimated from World Ocean Circulation Experiment (WOCE) hydrographic data. J. Geophys. Res. 108, 3213. doi: 10.1029/2002JC002565.

Ganachaud, A., Gourdeau, L., Kessler, W., 2008. Bifurcation of the subtropical south equatorial current against New Caledonia in December 2004 from a hydrographic inverse box model. J. Phys. Oceanogr. 38, 2072–2084.

Ganachaud, A., Wunsch, C., 2000. Improved estimates of global ocean circulation, heat transport and mixing from hydrographic data. Nature 408, 453–457.

Ganachaud, A., Wunsch, C., 2003. Large-scale ocean heat and freshwater transports during the World Ocean Circulation Experiment. J. Clim. 16, 696–705.

Ganachaud, A., Wunsch, C., Marotzke, J., Toole, J., 2000. Meridional overturning and large-scale circulation of the Indian Ocean. J. Geophys. Res. 105, 26117–26134.

Gardner, W.D., 2009. Visibility in the ocean and the effects of mixing. Quarterdeck 5(1), Spring 1997. http://oceanz.tamu.edu/~pdgroup/Qdeck/gardner-5.1.html (accessed 2.18.09).

Gardner, W.D., Mishonov, A.V., Richardson, M.J., 2006. Global POC concentrations from in-situ and satellite data. Deep-Sea Res. II 53, 718–740.

Garrett, C., 1972. Tidal resonance in the Bay of Fundy and Gulf of Maine. Nature 238, 441–443.

Garrett, C.J., Munk, W., 1972. Space–time scales of internal waves. Geophys. Fluid Dyn. 2, 255–264.

Garrett, C.J., Munk, W., 1975. Space–time scales of internal waves: A progress report. J. Geophys. Res. 80, 291–297.

Garrison, T., 2001. Essentials of Oceanography, second ed. Brooks/Cole, Pacific Grove, CA, 361 pp.

Gent, P.R., McWilliams, J.C., 1990. Isopycnal mixing in ocean circulation models. J. Phys. Oceanogr. 20, 150–155.

Giarolla, E., Nobre, P., Malaguti, M., Pezzi, L.P., 2005. The Atlantic Equatorial Undercurrent: PIRATA observations and simulations with GFDL modular ocean model at CPTEC. Geophys. Res. Lett. 32, L10617. doi:10.1029/2004GL022206.

Gill, A.E., 1982. Atmospheric-Ocean Dynamics. Academic Press, New York, 662 pp.

Gill, A.E., Niiler, P., 1973. The theory of seasonal variability in the ocean. Deep-Sea Res. 20, 141–177.

Gille, S.T., 1996. Scales of spatial and temporal variability in the Southern Ocean. J. Geophys. Res. 101, 8759–8773.

Gille, S.T., 2002. Warming of the Southern Ocean since the 1950s. Science 295, 1275–1277.

Gille, S.T., 2003. Float observations of the Southern Ocean. Part I: Estimating mean fields, bottom velocities, and topographic steering. J. Phys. Oceanogr. 33, 1167–1181.

Gille, S.T., 2005. MAE 127: Statistical methods for environmental sciences and engineering. http://www-pord.ucsd.edu/~sgille/mae127/index.html (accessed 4.14.09).

Giosan, L., Filip, F., Constatinescu, S., 2009. Was the Black Sea catastrophically flooded in the early Holocene? Quaternary Sci. Rev. 28, 1–6.

Girton, J.B., Pratt, L.J., Sutherland, D.A., Price, J.F., 2006. Is the Faroe Bank Channel overflow hydraulically controlled? J. Phys. Oceanogr. 36, 2340–2349.

Gladyshev, S., Talley, L., Kantakov, G., Khen, G., Wakatsuchi, M., 2003. Distribution, formation and seasonal variability of Okhotsk Sea Intermediate Water. J. Geophys. Res. 108, 3186. doi:10.1029/2001JC000877.

Godfrey, J., Cresswell, G., Golding, T., Pearce, A., Boyd, R., 1980. The separation of the East Australian Current. J. Phys. Oceanogr. 10, 430–440.

Godfrey, J.S., Weaver, A.J., 1991. Is the Leeuwin Current driven by Pacific heating and winds? Progr. Oceanogr. 27, 225–272.

Gonzalez, F.I., 1999. Tsunami! Sci. Am., 56–63. May 1999.

Gordon, A., 1991. The role of thermohaline circulation in global climate change. In: Lamont–Doherty Geological Observatory 1990 & 1991 Report. Lamont-Doherty Geological Observatory of Columbia University, Palisades, New York, pp. 44–51.

Gordon, A.L., 1986. Interocean exchange of thermocline water. J. Geophys. Res. 91, 5037–5046.

Gordon, A.L., 2003. Oceanography: The brawniest retroflection. Nature 421, 904–905.

Gordon, A.L., 2005. Oceanography of the Indonesian Seas and their throughflow. Oceanography 18, 14–27.

Gordon, A.L., Bosley, K.T., 1991. Cyclonic gyre in the tropical South Atlantic. Deep-Sea Res. 38 (Suppl.), S323–S343.

Gordon, A.L., Georgi, D.T., Taylor, H.W., 1977. Antarctic polar front zone in the western Scotia Sea — Summer 1975. J. Phys. Oceanogr. 7, 309–328.

Gordon, A.L., Giulivi, C.F., Ilahude, A.G., 2003. Deep topographic barriers within the Indonesian seas. Deep-Sea Res. II 50, 2205–2228.

Gordon, A.L., Susanto, R.D., Ffield, A., 1999. Throughflow within Makassar Strait. Geophys. Res. Lett. 26, 3321–3328.

Gordon, A.L., Visbeck, M., Comiso, J.C., 2007. A possible link between the Weddell polynya and the Southern Annular Mode. J. Clim. 20, 2558–2571.

Gregg, M.C., 1987. Diapycnal mixing in the thermocline: A review. J. Geophys. Res. 94, 5249–5286.

Grist, J.P., Josey, S.A., 2003. Inverse analysis adjustment of the SOC air-sea flux climatology using ocean heat transport constraints. J. Clim. 20, 3274–3295.

Grötzner, A., Latif, M., Barnett, T.P., 1998. A decadal climate cycle in the North Atlantic Ocean as simulated by the ECHO coupled GCM. J. Clim. 11, 831–847.

Gruber, N., Sarmiento, J.L., 1997. Global patterns of marine nitrogen fixation and denitrification. Glob. Biogeochem. Cyc. 11, 235–266.

Guza, R., Thornton, E., 1982. Swash oscillations on a natural beach. J. Geophys. Res. 87, 483–491.

Haarpaintner, J., Gascard, J.-C., Haugan, P.M., 2001. Ice production and brine formation in Storfjorden, Svalbard. J. Geophys. Res. 106, 14001–14013.

Häkkinin, S., Rhines, P.B., 2004. Decline of subpolar North Atlantic circulation during the 1990s. Science 304, 555–559.

Hall, A., Visbeck, M., 2002. Synchronous variability in the southern hemisphere atmosphere, sea ice and ocean resulting from the annular mode. J. Clim. 15, 3043–3057.

Hamon, B.V., 1965. The East Australian Current, 1960–1964. Deep-Sea Res. 12, 899–921.

Han, W., McCreary, J.P., Anderson, D.L.T., Mariano, A.J., 1999. Dynamics of the eastern surface jets in the equatorial Indian Ocean. J. Phys. Oceanogr. 29, 2191–2209.

Hanawa, K., Talley, L.D., 2001. Mode Waters. In: Siedler, G., Church, J. (Eds.), Ocean Circulation and Climate. International Geophysics Series. Academic Press, San Diego, CA, pp. 373–386.

Hannah, C.G., Dupont, F., Dunphy, M., 2009. Polynyas and tidal currents in the Canadian Arctic Archipelago. Arctic 62, 83–95.

Hansen, B., Østerhus, S., 2000. North Atlantic-Nordic Seas exchanges. Progr. Oceanogr. 45, 109–208.

Hansen, B., Østerhus, S., Turrell, W.R., Jónsson, S., Valdimarsson, H., Hátún, H., et al., 2008. The inflow of Atlantic water, heat, and salt to the Nordic Seas across the Greenland-Scotland Ridge. In: Dickson, R.R., Meincke, J., Rhines, P. (Eds.), Arctic-Subarctic Ocean Fluxes: Defining the Role of the Northern Seas in Climate. Springer, The Netherlands, pp. 15–44.

Hardisty, J., 2007. Estuaries: Monitoring and Modeling the Physical System. Blackwell Publishing, Maiden, MA, 157 pp.

Hasunuma, K., Yoshida, K., 1978. Splitting of the subtropical gyre in the western North Pacific. J. Oceanogr. Soc. Japan 34, 160–172.

Hautala, S., Reid, J., Bray, N., 1996. The distribution and mixing of Pacific water masses in the Indonesian Seas. J. Geophys. Res. 101, 12375–12389.

Hautala, S.L., Sprintall, J., Potemra, J., Chong, J.C.C., Pandoe, W., Bray, N., et al., 2001. Velocity structure and transport of the Indonesian throughflow in the major straits restricting flow into the Indian Ocean. J. Geophys. Res. 106, 19527–19546.

Hecht, M.W., Hasumi, H. (Eds.) 2008. Ocean Modeling in an Eddying Regime. AGU Geophys. Monogr. Ser. 177, 350 pp.

Heezen, B.C., Ericson, D.B., Ewing, M., 1954. Further evidence for a turbidity current following the 1929 Grand Banks earthquake. Deep-Sea Res. 1, 193–202.

Helland-Hansen, B., 1916. Nogen hydrografiske metoder. In: Forhandlinger ved de 16 Skandinaviske Naturforskeremote, pp. 357–359 (in Norwegian).

Helland-Hansen, B., 1934. The Sognefjord section. Oceanographic Observations in the northernmost part of the North Sea and the southern part of the Norwegian Sea. J. Johnstone Mem. Vol., Liverpool, 257 pp.

Herbers, T.H.C., Elgar, S., Sarap, N.A., Guza, R.T., 2002. Nonlinear dispersion of surface gravity waves in shallow water. J. Phys. Oceanogr. 32, 1181–1193.

Hickey, B.M., 1998. Coastal oceanography of western North America from the tip of Baja California to Vancouver Island. In: Robinson, A.R., Brink, K.H. (Eds.), The Sea, vol. 11, The Global Coastal Ocean: Regional Studies and Syntheses. John Wiley and Sons, New York, pp. 345–394.

Hisard, P., Hénin, C., 1984. Zonal pressure gradient, velocity and transport in the Atlantic Equatorial Undercurrent from FOCAL cruises (July 1982–February 1984). Geophys. Res. Lett. 11, 761–764.

Hofmann, E.E., 1985. The large-scale horizontal structure of the Antarctic Circumpolar Current from FGGE drifters. J. Geophys. Res. 90, 7087–7097.

Hogg, N.G., 1983. A note on the deep circulation of the western North Atlantic: Its nature and causes. Deep-Sea Res. 30, 945–961.

Hogg, N.G., Owens, W.B., 1999. Direct measurement of the deep circulation within the Brazil Basin. Deep-Sea Res. II 46, 335–353.

Hogg, N.G., Pickart, R.S., Hendry, R.M., Smethie Jr., W.M., 1986. The northern recirculation gyre of the Gulf Stream. Deep-Sea Res. 33, 1139–1165.

Hogg, N.G., Siedler, G., Zenk, W., 1999. Circulation and variability at the southern boundary of the Brazil Basin. J. Phys. Oceanogr. 29, 145–157.

Holliday, N.P., Meyer, A., Bacon, S., Alderson, S.G., de Cuevas, B., 2007. Retroflection of part of the east Greenland current at Cape Farewell. Geophys. Res. Lett. 34, L07609. doi:10.1029/2006GL029085.

Holte, J., Gilson, J., Talley, L., Roemmich, D., 2010. Argo mixed layers. Scripps Institution of Oceanography, UCSD. http://mixedlayer.ucsd.edu (accessed 2.24.10).

Holte, J., Talley, L., 2009. A new algorithm for finding mixed layer depths with applications to Argo data and Subantarctic Mode Water formation. J. Atmos. Ocean. Tech. 26, 1920–1939.

Horrillo, J., Knight, W., Kowalik, Z., 2008. Kuril Islands tsunami of November 2006: 2. Impact at Crescent City by local enhancement. J. Geophys. Res. 113, C01021. doi:10.1029/2007JC004404.

Hosoda, S., Suga, T., Shikama, N., Mizuno, K., 2009. Surface and subsurface layer salinity change in the global ocean using Argo float data. J. Oceanogr. 65, 579–586.

Hu, J., Kawamura, H., Hong, H., Qi, Y., 2000. A review on the currents in the South China Sea: seasonal circulation, South China Sea Warm Current and Kuroshio intrusion. J. Oceanogr. 56, 607–624.

Hu, S., Townsend, D.W., Chen, C., Cowles, G., Beardsley, R.C., Ji, R., et al., 2008. Tidal pumping and nutrient fluxes on Georges Bank: A process-oriented modeling study. J. Marine Syst. 74, 528–544.

Hufford, G.E., McCartney, M.S., Donohue, K.A., 1997. Northern boundary currents and adjacent recirculations off southwestern Australia. Geophys. Res. Lett. 24 (22), 2797–2800. doi_10.1029/97GL02278.

Hughes, S.L., Holliday, N.P., Beszczynska-Möller, A. (Eds.), 2008. ICES Report on Ocean Climate 2007. ICES Cooperative Research Report No. 291, p. 64. http://www.noc.soton.ac.uk/ooc/ICES_WGOH/iroc.php (accessed 7.1.09).

Hurdle, B.G. (Ed.), 1986. The Nordic Seas. Springer-Verlag, New York, 777 pp.

Hurlburt, H.E., Thompson, J.D., 1973. Coastal upwelling on a β-plane. J. Phys. Oceanogr. 19, 16–32.

Hurrell, J., 2009. Climate Indices. NAO Index Data provided by the Climate Analysis Section, NCAR, Boulder, Colorado (Hurrell, 1995). http://www.cgd.ucar.edu/cas/jhurrell/nao.stat.winter.html (accessed 6.23.09).

Hurrell, J.W., Kushnir, Y., Ottersen, G., Visbeck, M., 2003. An overview of the North Atlantic Oscillation. In: The North Atlantic Oscillation: Climate Significance and Environmental Impact. Geophys. Monogr. Ser. 134, 1–35.

Huyer, A., 1983. Coastal upwelling in the California Current System. Progr. Oceanogr. 12, 259–284.

IAPP, 2010. International Arctic Polynya Programme. Arctic Ocean Sciences Board. http://aosb.arcticportal.org/iapp/iapp.html (accessed 11/26/10).

Ihara, C., Kushnir, Y., Cane, M.A., 2008. Warming trend of the Indian Ocean SST and Indian Ocean dipole from 1880 to 2004. J. Clim. 21, 2035–2046.

Imawaki, S., Uchida, H., Ichikawa, H., Fukasawa, M., Umatani, S., 2001. Satellite altimeter monitoring the Kuroshio transport south of Japan. Geophys. Res. Lett. 28, 17–20.

IOC, SCOR, IAPSO, 2010. The international thermodynamic equation of seawater — 2010: Calculation and use of thermodynamic properties. Intergovernmental Oceanographic Commission, Manuals and Guides No. 56. UNESCO (English), 196 pp.

IPCC, et al., 2001. Climate Change 2001: The Scientific Basis. In: Houghton, J.T., Ding, Y., Griggs, D.J., Noguer, M., van der Linden, P.J., Dai, X. (Eds.), Contribution of Working Group I to the Third Assessment Report of the Intergovernmental Panel on Climate Change. Cambridge University Press, Cambridge, UK and New York, 881 pp.

IPCC, 2007. Summary for Policymakers. In: Solomon, S., Qin, D., Manning, M., Chen, Z., Marquis, M., Averyt, K.B., et al. (Eds.), Climate Change 2007: The Physical Science Basis. Contribution of Working Group I to the Fourth Assessment Report of the Intergovernmental Panel on Climate Change. Cambridge University Press, Cambridge, UK, New York.

ISCCP, 2007. ISCCP and other cloud data, maps, and plots available on-line. NASA Goddard Institute for Space Studies. http://isccp.giss.nasa.gov/products/onlineData.html (accessed 10.16.10).

Iselin, C.O'D. 1936. A study of the circulation of the western North Atlantic. Papers in Physical Oceanography and Meteorology, 4(4), 10 pp. MIT and Woods Hole Oceanographic Institution.

Iselin C.O'D, 1939. The influence of vertical and lateral turbulence on the characteristics of the waters at mid-depths. Trans. Am. Geophys. Union 20, 414–417.

Ishii, M., Kimoto, M., Sakamoto, K., Iwasaki, S.I., 2006. Steric sea level changes estimated from historical ocean subsurface temperature and salinity analyses. J. Oceanogr. 62, 155–170.

Ivers, W.D., 1975. The deep circulation in the northern Atlantic with special reference to the Labrador Sea. Ph.D. Thesis, University of California at San Diego, 179 pp.

Jackett, D.R., McDougall, T.J., 1997. A neutral density variable for the world's oceans. J. Phys. Oceanogr. 27, 237–263.

Jacobs, S.S., 1991. On the nature and significance of the Antarctic Slope Front. Mar. Chem. 35, 9–24.

Jakobsen, F., 1995. The major inflow to the Baltic Sea during January 1993. J. Marine Syst. 6, 227–240.

Jakobsson, M., 2002. Hypsometry and volume of the Arctic Ocean and its constituent seas. Geochem. Geophys. Geosys. 3 (5), 1028. doi:10.1029/2001GC000302.

Jenkins, A., Holland, D., 2002. A model study of ocean circulation beneath Filchner-Ronne Ice Shelf, Antarctica: Implications for bottom water formation. Geophys. Res. Lett. 29, 8. doi:10.1029/2001GL014589.

Jenkins, W.J., 1998. Studying thermocline ventilation and circulation using tritium and 3He. J. Geophys. Res. 103, 15817–15831.

Jerlov, N.G., 1976. Marine Optics. Elsevier, Amsterdam, 231 pp.

Jin, F.F., 1996. Tropical ocean-atmosphere interaction, the Pacific cold tongue, and the El Niño-Southern Oscillation. Science 274, 76–78.

JISAO, 2004. Arctic Oscillation (AO) time series, 1899 — June 2002. JISAO. http://www.jisao.washington.edu/ao/ (accessed 3.18.10).

Jochum, M., Malanotte-Rizzoli, P., Busalacchi, A., 2004. Tropical instability waves in the Atlantic Ocean. Ocean Model. 7, 145–163.

Johannessen, O.M., Shalina, E.V., Miles, M.W., 1999. Satellite evidence for an arctic sea ice cover in transformation. Science 286, 1937–1939.

Johns Hopkins APL Ocean Remote Sensing, 1996. Sea surface temperature imagery. http://fermi.jhuapl.edu/avhrr/sst.html (accessed 6.10.09).

Johns, W., Lee, T., Schott, F., Zantopp, R., Evans, R., 1990. The North Brazil Current retroflection: seasonal structure and eddy variability. J. Geophys. Res. 95, 22103–22120.

Johns, W.E., Beal, L.M., Baringer, M.O., Molina, J.R., Cunningham, S.A., Kanzow, T., et al., 2008. Variability of shallow and deep western boundary currents off the Bahamas during 2004–05: Results from the 26°N RAPID–MOC Array. J. Phys. Oceanogr. 38, 605–623.

Johns, W.E., Jacobs, G.A., Kindle, J.C., Murray, S.P., Carron, M., 1999. Arabian Marginal Seas and Gulfs: Report of a Workshop held at Stennis Space Center, Miss. 11–13 May, 1999. University of Miami RSMAS. Technical Report 2000–01.

Johns, W.E., Lee, T.N., Beardsley, R.C., Candela, J., Limeburner, R., Castro, B., 1998. Annual cycle and variability of the North Brazil Current. J. Phys. Oceanogr. 28, 103–128.

Johns, W.E., Lee, T.N., Zhang, D., Zantopp, R., Liu, C.-T., Yang, Y., 2001. The Kuroshio east of Taiwan: Moored transport observations from the WOCE PCM-1 array. J. Phys. Oceanogr. 31, 1031–1053.

Johns, W.E., Shay, T.J., Bane, J.M., Watts, D.R., 1995. Gulf Stream structure, transport, and recirculation near 68°W. J. Geophys. Res. 100, 817–838.

Johns, W.E., Townsend, T.L., Fratantoni, D.M., Wilson, W.D., 2002. On the Atlantic inflow to the Caribbean Sea. Deep-Sea Res. I 49, 211–243.

Johns, W.E., Yao, F., Olsen, D.B., Josey, S.A., Grist, J.P., Smeed, D.A., 2003. Observations of seasonal exchange through the Straits of Hormuz and the inferred heat and freshwater budgets of the Persian Gulf. J. Geophys. Res. 108 (C12), 3391. doi:10.1029/2003JC001881.

Johnson, G.C., 2008. Quantifying Antarctic Bottom Water and North Atlantic Deep Water volumes. J. Geophys. Res. 113, C05027. doi:10.1029/2007JC004477.

Johnson, G.C., Gruber, N., 2007. Decadal water mass variations along 20°W in the northeastern Atlantic Ocean. Progr. Oceanogr. 73, 277–295.

Johnson, G.C., McPhaden, M.J., 1999. Interior pycnocline flow from the subtropical to the equatorial Pacific Ocean. J. Phys. Oceanogr. 29, 3073–3089.

Johnson, G.C., Musgrave, D.L., Warren, B.A., Ffield, A., Olson, D.B., 1998. Flow of bottom and deep water in the Amirante Passage and Mascarene Basin. J. Geophys. Res. 103, 30973–30984.

Johnson, G.C., Sloyan, B.M., Kessler, W.S., McTaggart, K.E., 2002. Direct measurements of upper ocean currents and water properties across the tropical Pacific during the 1990s. Progr. Oceanogr. 52, 31–61.

Johnson, G.C., Toole, J.M., 1993. Flow of deep and bottom waters in the Pacific at 10°N. Deep-Sea Res. I 40, 371–394.

Jones, E.P., 2001. Circulation in the Arctic Ocean. Polar Res. 20, 139–146.

Jones, E.P., Anderson, L.G., Swift, J.H., 1998. Distribution of Atlantic and Pacific waters in the upper Arctic Ocean: Implications for circulation. Geophys. Res. Lett. 25, 765–768.

Jones, E.P., Swift, J.H., Anderson, L.G., Lipizer, M., Civitarese, G., Falkner, K.K., et al., 2003. Tracing Pacific water in the North Atlantic Ocean. J. Geophys. Res. 108, 3116. doi:10.1029/2001JC001141.

Josey, S.A., Kent, E.C., Taylor, P.K., 1999. New insights into the ocean heat budget closure problem from analysis of the SOC air-sea flux climatology. J. Clim. 12, 2856–2880.

Josey, S.A., Marsh, R., 2005. Surface freshwater flux variability and recent freshening of the North Atlantic in the eastern subpolar gyre. J. Geophys. Res. 110, C05008. doi:10.1029/2004JC002521.

Joyce, T.M., Hernandez-Guerra, A., Smethie, W.M., 2001. Zonal circulation in the NW Atlantic and Caribbean from a meridional World Ocean Circulation Experiment hydrographic section at 66°W. J. Geophys. Res. 106, 22095–22113.

Joyce, T.M., Warren, B.A., Talley, L.D., 1986. The geothermal heating of the abyssal subarctic Pacific Ocean. Deep-Sea Res. 33, 1003–1015.

Joyce, T.M., Zenk, W., Toole, J.M., 1978. The anatomy of the Antarctic Polar Front in the Drake Passage. J. Geophys. Res. 83, 6093–6114.

Juliano, M.F., Alvés, M.L.G.R., 2007. The Subtropical Front/Current systems of Azores and St. Helena. J. Phys. Oceanogr. 37, 2573–2598.

Kalnay, E., Kanamitsu, M., Kistler, R., Collins, W., Deaven, D., Gandin, L., et al., 1996. The NCEP-NCAR 40-year reanalysis project. Bull. Am. Meteorol. Soc. 77, 437–471.

Kaneko, I., Takatsuki, Y., Kamiya, H., Kawae, S., 1998. Water property and current distributions along the WHP-P9 section (137°−142°E) in the western North Pacific. J. Geophys. Res. 103, 12959−12984.

Kanzow, T., Send, U., McCartney, M., 2008. On the variability of the deep meridional transports in the tropical North Atlantic. Deep-Sea Res. I 55, 1601−1623.

Kaplan, A., Cane, M., Kushnir, Y., Clement, A., Blumenthal, M., Rajagopalan, B., 1998. Analyses of global sea surface temperature 1856−1991. J. Geophys. Res. 103, 18567−18589.

Kara, A.B., Rochford, P.A., Hurlburt, H.E., 2003. Mixed layer depth variability over the global ocean. J. Geophys. Res. 108, 3079. doi:10.1029/2000JC000736.

Karbe, L., 1987. Hot brines and the deep sea environment. In: Edwards, A.J., Head, S.M. (Eds.), Red Sea. Pergamon Press, Oxford, UK, 441 pp.

Karstensen, J., 2006. OMP (Optimum Multiparameter) analysis — USER GROUP. OMP User group. http://www.ldeo.columbia.edu/~jkarsten/omp_std/ (accessed 4.24.09).

Karstensen, J., Stramma, L., Visbeck, M., 2008. Oxygen minimum zones in the eastern tropical Atlantic and Pacific oceans. Progr. Oceanogr. 77, 331−350.

Kashino, Y., Watanabe, H., Herunadi, B., Aoyama, M., Hartoyo, D., 1999. Current variability at the Pacific entrance of the Indonesian Throughflow. J. Geophys. Res. 104, 11021−11035.

Kato, F., Kawabe, M., 2009. Volume transport and distribution of deep circulation at 156°W in the North Pacific. Deep-Sea Res. I 56, 2077−2087.

Kawabe, M., 1995. Variations of current path, velocity, and volume transport of the Kuroshio in relation with the large meander. J. Phys. Oceanogr. 25, 3103−3117.

Kawabe, M., Yanagimoto, D., Kitagawa, S., 2006. Variations of deep western boundary currents in the Melanesian Basin in the western North Pacific. Deep-Sea Res. I 53, 942−959.

Kawabe, M., Yanagimoto, D., Kitagawa, S., Kuroda, Y., 2005. Variations of the deep circulation currents in the Wake Island Passage. Deep-Sea Res. I 52, 1121−1137.

Kawai, H., 1972. Hydrography of the Kuroshio Extension. In: Stommel, H., Yoshida, K. (Eds.), Kuroshio: Physical Aspects of the Japan Current. University of Washington Press, Seattle and London, pp. 235−352.

Kawano, T., Doi, T., Uchida, H., Kouketsu, S., Fukasawa, M., Kawai, Y., et al., 2010. Heat content change in the Pacific Ocean between 1990s and 2000s. Deep-Sea Res. II 57, 1141−1151.

Kawano, T., Fukasawa, M., Kouketsu, S., Uchida, H., Doi, T., et al., 2006. Bottom water warming along the pathway of Lower Circumpolar Deep Water in the Pacific Ocean. Geophys. Res. Lett. 33, L23613. doi:10.1029/2006GL027933.

Kelley, D.E., Fernando, H.J.S., Gargett, A.E., Tanny, J., Özsoy, E., 2003. The diffusive regime of double-diffusive convection. Progr. Oceanogr. 56, 461−481.

Kennett, J.P., 1982. Marine Geology. Prentice Hall, Englewood Cliffs, NJ, 813 pp.

Kern, S., 2008. Polynya area in the Kara Sea, Arctic, obtained with microwave radiometry for 1979−2003. IEEE T. Geosc. Remote Sens. Lett. 5, 171−175.

Kessler, W.S., 2006. The circulation of the eastern tropical Pacific: A review. Progr. Oceanogr. 69, 181−217.

Kessler, W.S.. The Central American mountain-gap winds and their effects on the ocean. http://faculty.washington.edu/kessler/t-peckers/t-peckers.html (accessed 3.27.09).

Key, R.M., 2001. Ocean Process Tracers: Radiocarbon. In: Steele, J., Thorpe, S., Turekian, K. (Eds.), Encyclopedia of Ocean Sciences. Academic Press, Ltd., London, pp. 2338−2353.

Kieke, D., Rhein, M., Stramma, L., Smethie, W.M., LeBel, D.A., Zenk, W., 2006. Changes in the CFC inventories and formation rates of Upper Labrador Sea Water, 1997−2001. J. Phys. Oceanogr. 36, 64−86.

Killworth, P.D., 1979. On chimney formation in the ocean. J. Phys. Oceanogr. 9, 531−554.

Killworth, P.D., 1983. Deep convection in the world ocean. Rev. Geophys. 21, 1−26.

Kimura, S., Tsukamoto, K., 2006. The salinity front in the North Equatorial Current: A landmark for the spawning migration of the Japanese eel (Anguilla japonica) related to the stock recruitment. Deep-Sea Res. II 53, 315−325.

Kinder, T.H., Coachman, L.K., Galt, J.A., 1975. The Bering Slope Current System. J. Phys. Oceanogr. 5, 231−244.

King, M.D., Menzel, W.P., Kaufman, Y.J., Tanre, D., Bo-Cai, G., Platnick, S., et al., 2003. Cloud and aerosol properties, precipitable water, and profiles of temperature and water vapor from MODIS. IEEE T. Geosc. Remote Sens. 41, 442−458.

Klein, B., Roether, W., Civitarese, G., Gacic, M., Manca, B.B., d'Alcalá, M.R., 2000. Is the Adriatic returning to dominate the production of Eastern Mediterranean Deep Water? Geophys. Res. Lett. 27, 3377−3380.

Klein, B., Roether, W., Manca, B.B., Bregant, D., Beitzel, V., Kovacevic, V., et al., 1999. The large deep water transient in the Eastern Mediterranean. Deep-Sea Res. I 46, 371−414.

Klein, S.A., Soden, B.J., Lau, N.C., 1999. Remote sea surface temperature variations during ENSO: Evidence for a tropical atmospheric bridge. J. Clim. 12, 917−932.

Klinck, J.M., Hofmann, E.E., Beardsley, R.C., Salihoglu, B., Howard, S., 2004. Water mass properties and circulation on the west Antarctic Peninsula continental shelf in austral fall and winter 2001. Deep-Sea Res. II 51, 1925−1946.

Knauss, J.A., 1960. Measurements of the Cromwell Current. Deep-Sea Res. 6, 265–286.

Knauss, J.A., 1997. Introduction to Physical Oceanography, second ed. Waveland Press, Long Grove, IL, 309 pp.

Knudsen, M. (Ed.), 1901. Hydrographical Tables. G.E.C. Goad, Copenhagen, 63 pp.

Kobayashi, T., Suga, T., 2006. The Indian Ocean HydroBase: A high-quality climatological dataset for the Indian Ocean. Progr. Oceanogr. 68, 75–114.

Koltermann, K.P., Gouretski., V., Jancke, K., 2011. Hydrographic Atlas of the World Ocean Circulation Experiment (WOCE). Vol 3: Atlantic Ocean. In: Sparrow, M., Chapman, P., Gould, J. (Eds.). International WOCE Project Office, Southampton, UK in press.

Komaki, K., Kawabe, M., 2009. Deep-circulation current through the Main Gap of the Emperor Seamounts Chain in the North Pacific. Deep-Sea Res. I 56, 305–313.

Komar, P.D., 1998. Beach Processes and Sedimentation. Prentice Hall, Upper Saddle River, NJ, 544 pp.

Komar, P.D., Holman, R.A., 1986. Coastal processes and the development of shoreline erosion. Annu. Rev. Earth Pl. Sc. 14, 237–265.

Kono, T., Kawasaki, Y., 1997. Modification of the western subarctic water by exchange with the Okhotsk Sea. Deep-Sea Res. I 44, 689–711.

Kosro, P.M., Huyer, A., Ramp, S.R., Smith, R.L., Chavez, F.P., Cowles, T.J., et al., 1991. The structure of the transition zone between coastal waters and the open ocean off northern California, winter and spring 1987. J. Geophys. Res. 96, 14707–14730.

Kossina, E., 1921. Die Tiefen des Weltmeeres, Veröffentil. des Inst. für Meereskunde, Neue Folge, Heft 9, E.S. Mittler und Sohn, Berlin (in German).

Kraus, E.B., Turner, J.S., 1967. A one-dimensional model of the seasonal thermocline, II. The general theory and its consequences. Tellus 19, 98–105.

Krishnamurthy, V., Kirtman, B.P., 2003. Variability of the Indian Ocean: relation to monsoon and ENSO.Q. J. Roy. Meteorol. Soc. 129, 1623–1646.

Krummel, O., 1882. Bemerkungen über die Meeresstromungen und Temperaturen in der Falklandsee, Archiv der deutschen Seewarte, V(2), 25 pp. (in German).

Kuhlbrodt, T., Griesel, A., Montoya, M., Levermann, A., Hofmann, M., Rahmstorf, S., 2007. On the driving processes of the Atlantic meridional overturning circulation. Rev. Geophys. 45, RG2001. doi:10.1029/2004RG000166.

Kunze, E., Firing, E., Hummon, J.M., Chereskin, T.K., Thurnherr, A.M., 2006. Global abyssal mixing inferred from lowered ADCP shear and CTD strain profiles. J. Phys. Oceanogr. 36, 1553–1576.

Kuo, H.-H., Veronis, G., 1973. The use of oxygen as a test for an abyssal circulation model. Deep-Sea Res. 20, 871–888.

Kushnir, Y., Seager, R., Miller, J., Chiang, J.C.H., 2002. A simple coupled model of tropical Atlantic decadal climate variability. Geophys. Res. Lett. 29, 2133. doi:10.1029/2002GL015874.

Kwon, Y.-O., Riser, S.C., 2004. North Atlantic Subtropical Mode Water: A history of ocean-atmosphere interaction 1961–2000. Geophys. Res. Lett. 31, L19307. doi:10.1029/2004GL021116.

Langmuir, I., 1938. Surface motion of water induced by wind. Science 87, 119–123.

Laplace, P.S., 1790. Mémoire sur le flux et reflux de la mer. Mém. Acad. Sci. Paris, 45–181 (in French).

Large, W.G., McWilliams, J.C., Doney, S.C., 1994. Oceanic vertical mixing: A review and a model with a non-local K-profile boundary layer parameterization. Rev. Geophys. 32, 363–403.

Large, W.G., Yeager, S.G., 2009. The global climatology of an interannually varying air-sea flux data set. Clim. Dynam. 33, 341–364.

Lavender, K.L., Davis, R.E., Owens, W.B., 2000. Mid-depth recirculation observed in the interior Labrador and Irminger seas by direct velocity measurements. Nature 407, 66–69.

Lavín, M.F., Marinone, S.G., 2003. An overview of the physical oceanography of the Gulf of California. In: Velasco Fuentes, O.U., Sheinbaum, J., Ochoa de la Torre, J.L. (Eds.), Nonlinear Processes in Geophysical Fluid Dynamics. Kluwer Academic Publishers, Dordrecht, Holland, pp. 173–204.

Le Traon, P.Y., 1990. A method for optimal analysis of fields with spatially variable mean. J. Geophys. Res. 95, 13543–13547.

Leaman, K., Johns, E., Rossby, T., 1989. The average distribution of volume transport and potential vorticity with temperature at three sections across the Gulf Stream. J. Phys. Oceanogr. 19, 36–51.

Ledwell, J.R., Watson, A.J., Law, C.S., 1993. Evidence for slow mixing across the pycnocline from an open–ocean tracer–release experiment. Nature 364, 701–703.

Ledwell, J.R., Watson, A.J., Law, C.S., 1998. Mixing of a tracer in the pycnocline. J. Geophys. Res. 103, 21499–21529.

Lee, Z., Weidemann, A., Kindle, J., Arnone, R., Carder, K.L., Davis, C., 2007. Euphotic zone depth: Its derivation and implication to ocean-color remote sensing. J. Geophys. Res. 112, C03009. doi:10.1029/2006JC003802.

Leetmaa, A., Spain, P.F., 1981. Results from a velocity transect along the equator from 125 to 159°W. J. Phys. Oceanogr. 11, 1030–1033.

Legeckis, R., 1977. Long waves in the eastern Equatorial Pacific Ocean: A view from a geostationary satellite. Science 197, 1179–1181.

Legeckis, R., Brown, C.W., Chang, P.S., 2002. Geostationary satellites reveal motions of ocean surface fronts. J. Marine Syst. 37, 3–15.

Legeckis, R., Reverdin, G., 1987. Long waves in the equatorial Atlantic Ocean during 1983. J. Geophys. Res. 92, 2835–2842.

Lemke, P., Fichefet, T., Dick, C., 2011. Arctic Climate Change — The ACSYS decade and beyond. Springer Atmospheric and Oceanographic Sciences Library, in preparation since 2005.

Lenn, Y.-D., Chereskin, T.K., Sprintall, J., Firing, E., 2008. Mean jets, mesoscale variability and eddy momentum fluxes in the surface layer of the Antarctic Circumpolar Current in Drake Passage. J. Mar. Res. 65, 27–58.

Lentini, C.A.D., Goni, G.J., Olson, D.B., 2006. Investigation of Brazil Current rings in the confluence region. J. Geophys. Res. 111, C06013. doi:10.1029/2005JC002988.

Lentz, S.J., 1995. Sensitivity of the inner-shelf circulation to the form of the eddy viscosity profile. J. Phys. Oceanogr. 25, 19–28.

Leppäranta, M., Myrberg, K., 2009. Physical Oceanography of the Baltic Sea. Springer, Berlin, 378 pp. with online version.

Lerczak, J.A., 2000. Internal waves on the southern California shelf. Ph.D. Thesis, University of California, San Diego, 253 pp.

Levine, M.D., 2002. A modification of the Garrett–Munk internal wave spectrum. J. Phys. Oceanogr. 32, 3166–3181.

Levitus, S., 1982. Climatological Atlas of the World Ocean. NOAA Professional Paper 13. NOAA, Rockville, MD, 173 pp.

Levitus, S., 1988. Ekman volume fluxes for the world ocean and individual ocean basins. J. Phys. Oceanogr. 18, 271–279.

Levitus, S., Antonov, J.I., Boyer, T.P., 2005. Warming of the world Ocean, 1955–2003. Geophys. Res. Lett. 32, L02604. doi:10.1029/2004GL021592.

Levitus, S., Boyer, T.P., 1994. World Ocean Atlas 1994 Volume 4: Temperature. NOAA Atlas NESDIS 4. U.S. Department of Commerce, Washington, D.C., 117 pp.

Levitus, S., Boyer, T.P., Antonov, J., 1994a. World Ocean Atlas Volume 5: Interannual variability of upper ocean thermal structure. NOAA/NESDIS. Tech. Rpt. OSTI ID:137204.

Levitus, S., Burgett, R., Boyer, T.P., 1994b. World Ocean Atlas 1994 Volume 3: Salinity. NOAA Atlas NESDIS 3. U.S. Department of Commerce, Washington, D.C., 99 pp.

Lewis, E.L., 1980. The practical salinity scale 1978 and its antecedents. IEEE J. Oceanic Eng OE-5, 3–8.

Lewis, E.L., Fofonoff, N.P., 1979. A practical salinity scale. J. Phys. Oceanogr. 9, 446.

Lewis, E.L., Perkin, R.G., 1978. Salinity: Its definition and calculation. J. Geophys. Res. 83, 466–478.

Lewis, M.R., Kuring, N., Yentsch, C., 1988. Global patterns of ocean transparency: Implications for the new production of the open ocean. J. Geophys. Res. 93, 6847–6856.

Libes, S., 2009. Introduction to Marine Biogeochemistry, Second Edition. Elsevier, Amsterdam, 909 pp.

Lien, R.-C., Gregg, M.C., 2001. Observations of turbulence in a tidal beam and across a coastal ridge. J Geophys. Res. 106, 4575–4591.

Lighthill, J., 1978. Waves in Fluids. Cambridge University Press, New York and London, 504 pp.

Liu, W.T., Katsaros, K.B., 2001. Air-sea fluxes from satellite data. In: Siedler, G., Church, J. (Eds.), Ocean Circulation and Climate, International Geophysics Series. Academic Press, pp. 173–180.

Loeng, H., Brander, K., Carmack, E., Denisenko, S., Drinkwater, K., et al., 2005. Chapter 9 Marine Systems. In: Symon, C., Arris, L., Heal, B. (Eds.), Arctic Climate Impact Assessment — Scientific Report. Cambridge University Press, UK, 1046 pp.

Lorenz, E., 1956. Empirical orthogonal functions and statistical weather prediction. Scientific Report No. 1. Air Force Cambridge Research Center, Air Research and Development Command, Cambridge, MA, 49 pp.

Loschnigg, J., Webster, P.J., 2000. A coupled ocean–atmosphere system of SST modulation for the Indian Ocean. J. Clim. 13, 3342–3360.

Lozier, M.S., Owens, W.B., Curry, R.G., 1995. The climatology of the North Atlantic. Progr. Oceanogr. 36, 1–44.

Lukas, R., 1986. The termination of the Equatorial Undercurrent in the eastern Pacific. Progr. Oceanogr. 16, 63–90.

Lukas, R., Yamagata, T., McCreary, J.P., 1996. Pacific low-latitude western boundary currents and the Indonesian throughflow. J. Geophys. Res. 101, 12209–12216.

Lumpkin, R., Speer, K., 2007. Global ocean meridional overturning. J. Phys. Oceanogr. 37, 2550–2562.

Lumpkin, R., Treguier, A.M., Speer, K., 2002. Lagrangian eddy scales in the northern Atlantic Ocean. J. Phys. Oceanogr. 32, 2425–2440.

Luyten, J.R., Pedlosky, J., Stommel, H., 1983. The ventilated thermocline. J. Phys. Oceanogr. 13, 292–309.

Lynn, R.J., Reid, J.L., 1968. Characteristics and circulation of deep and abyssal waters. Deep-Sea Res. 15, 577–598.

Lynn, R.J., Simpson, J.J., 1987. The California Current system: The seasonal variability of its physical characteristics. J. Geophys. Res. 92, 12947–12966.

Maamaatuaiahutapu, K., Garçon, V., Provost, C., Boulahdid, M., Osiroff, A., 1992. Brazil-Malvinas confluence: Water mass composition. J. Geophys. Res. 97, 9493–9505.

Maamaatuaiahutapu, K., Garçon, V., Provost, C., Mercier, H., 1998. Transports of the Brazil and Malvinas Currents at their confluence. J. Mar. Res. 56, 417–438.

Macdonald, A.M., Suga, R., Curry, R.G., 2001. An isopycnally averaged North Pacific climatology. J. Atmos. Ocean Tech. 18, 394–420.

Macdonald, A.M., Wunsch, C., 1996. An estimate of global ocean circulation and heat fluxes. Nature 382, 436–439.

Macdonald, R.W., Carmack, E.C., Wallace, D.W.R., 1993. Tritium and radiocarbon dating of Canada Basin deep waters. Science 259, 103–104.

Mackas, D.L., Denman, K.L., Bennett, A.F., 1987. Least-square multiple tracer analysis of water mass composition. J. Geophys. Res. 92, 2907–2918.

Mackas, D.L., Strub, P.T., Thomas, A., Montecino, V., 2006. Eastern ocean boundaries pan-regional overview. In: Robinson, A.R., Brink, K.H. (Eds.), The Sea, vol. 14A: The Global Coastal Ocean: Interdisciplinary Regional Studies and Syntheses. Harvard University Press, Boston, MA, pp. 21–60.

Mackenzie, K.V., 1981. Nine-term equation for the sound speed in the oceans. J. Acoust. Soc. Am. 70, 807–812.

Macrander, A., Send, U., Valdimarsson, H., Jónsson, S., Käse, R.H., 2005. Interannual changes in the overflow from the Nordic Seas into the Atlantic Ocean through Denmark Strait. Geophys. Res. Lett. 32, L06606. doi:10.1029/2004GL021463.

Madden, R., Julian, P., 1994. Observations of the 40-50 day tropical oscillation: A review. Mon. Weather Rev. 122, 814–837.

Makinson, K., Nicholls, K.W., 1999. Modeling tidal currents beneath Filchner-Ronne Ice Shelf and on the adjacent continental shelf: their effect on mixing and transport. J. Geophys. Res. 104, 13449–13466.

Malanotte-Rizzoli, P., Manca, B.B., Salvatore Marullo, Ribera d' Alcalá, M., Roether, W., Theocharis, A., et al., 2003. The Levantine Intermediate Water Experiment (LIWEX) Group: Levantine basin — A laboratory for multiple water mass formation processes. J. Geophys. Res. 108 (C9), 8101. doi:10.1029/2002JC001643.

Malmgren, F., 1927. On the properties of sea-ice. Norwegian North Polar Expedition with the Maud, 1918–1925. Sci. Res. 1 (5), 67 pp.

Maltrud, M.E., McClean, J.L., 2005. An eddy resolving global 1/10° ocean simulation. Ocean Model. 8, 31–54.

Mantua, N.J., Hare, S.R., Zhang, Y., Wallace, J.M., Francis, R.C., 1997. A Pacific interdecadal climate oscillation with impacts on salmon production. B. Am. Meteor. Soc. 78, 1069–1079.

Mantyla, A.W., Reid, J.L., 1983. Abyssal characteristics of the World Ocean waters. Deep-Sea Res. 30, 805–833.

Marchesiello, P., McWilliams, J.C., Shchepetkin, A., 2003. Equilibrium structure and dynamics of the California Current System. J. Phys. Oceanogr. 33, 753–783.

Mariano, A.J., Ryan, E.H., Perkins, B.D., Smithers, S., 1995. The Mariano Global Surface Velocity Analysis 1.0. USCG Report CG-D-34–95, p. 55. http://oceancurrents.rsmas.miami.edu/index.html (accessed 3.4.09).

Marshall, G., 2003. Trends in the Southern Annular Mode from observations and reanalyses. J. Clim. 16, 4134–4143.

Marshall, J., Schott, F., 1999. Open-ocean convection: observations, theory, and models. Rev. Geophys. 37, 1–64.

Martin, S., 2001. Polynyas. In: Steele, J.H., Turkeian, K.K., Thorpe, S.A. (Eds.), Encyclopedia of Ocean Sciences. Academic Press, pp. 2243–2247.

Martin, S., Cavalieri, D.J., 1989. Contributions of the Siberian shelf polynyas to the Arctic Ocean intermediate and deep water. J. Geophys. Res. 94, 12725–12738.

Martinson, D.G., Steele, M., 2001. Future of the Arctic sea ice cover: Implications of an Antarctic analog. Geophys. Res. Lett. 28, 307–310.

Maslanik, J., Serreze, M., Agnew, T., 1999. On the record reduction in 1998 western Arctic sea-ice cover. Geophys. Res. Lett. 26, 1905–1908.

Masuzawa, J., 1969. Subtropical Mode Water. Deep-Sea Res. 16, 453–472.

Mata, M.M., Wijffels, S.E., Church, J.A., Tomczak, M., 2006. Eddy shedding and energy conversions in the East Australian Current. J. Geophys. Res. 111, C09034. doi:10.1029/2006JC003592.

Mauritzen, C., 1996. Production of dense overflow waters feeding the North Atlantic across the Greenland-Scotland Ridge. Part 1: Evidence for a revised circulation scheme. Deep-Sea Res. I 43, 769–806.

Maury, M.F., 1855. The Physical Geography of the Sea. Harper and Brothers, New York, 304 pp.

Maximenko, N., Niiler, P., Rio, M.H., Melnichenko, O., Centurioni, L., Chambers, D., et al., 2009. Mean dynamic topography of the ocean derived from satellite and drifting buoy data using three different techniques. J. Atmos. Ocean. Tech. 26, 1910–1919.

McCarthy, M.C., Talley, L.D., 1999. Three-dimensional isoneutral potential vorticity structure in the Indian Ocean. J. Geophys. Res. 104, 13251–13268.

McCartney, M., Curry, R., 1993. Transequatorial flow of Antarctic Bottom Water in the western Atlantic Ocean: Abyssal geostrophy at the equator. J. Phys. Oceanogr. 23, 1264–127.

McCartney, M.S., 1977. Subantarctic Mode Water. In: Angel, M.V. (Ed.), A Voyage of Discovery: George Deacon 70th Anniversary Volume, supplement to Deep-Sea Res., pp. 103–119.

McCartney, M.S., 1982. The subtropical circulation of Mode Waters. J. Mar. Res. 40 (Suppl.), 427–464.

McCartney, M.S., Talley, L.D., 1982. The Subpolar Mode Water of the North Atlantic Ocean. J. Phys. Oceanogr. 12, 1169–1188.

McClain, C., Christian, J.R., Signorini, S.R., Lewis, M.R., Asanuma, I., Turk, D., et al., 2002. Satellite ocean-color observations of the tropical Pacific Ocean. Deep-Sea Res. II 49, 2533–2560.

McClain, C., Hooker, S., Feldman, G., Bontempi, P., 2006. Satellite data for ocean biology, biogeochemistry, and climate research. Eos Trans. AGU 87 (34), 337–343.

McDonagh, E.L., Bryden, H.L., King, B.A., Sanders, R.J., Cunningham, S.A., Marsh, R., 2005. Decadal changes in the south Indian Ocean thermocline. J. Clim. 18, 1575–1590.

McDougall, T.J., 1987a. Neutral surfaces. J. Phys. Oceanogr. 17, 1950–1964.

McDougall, T.J., 1987b. Thermobaricity, cabbeling, and water-mass conversion. J. Geophys. Res. 92, 5448–5464.

McDougall, T.J., Jackett, D.R., Millero, F.J., 2010. An algorithm for estimating Absolute Salinity in the global ocean. Submitted to Ocean Science, a preliminary version is available at Ocean Sci. Discuss. 6, 215–242. http://www.ocean-sci-discuss.net/6/215/2009/osd-6-215-2009-print.pdf the computer software is available from http://www.TEOS-10.org.

McPhaden, M.J., Busalacchi, A.J., Cheney, R., Donguy, J.-R., Gage, K.S., Halpern, D., et al., 1998. The Tropical Ocean-Global Atmosphere observing system: A decade of progress. J. Geophys. Res. 103, 14169–14240.

MEDOC Group, 1970. Observations of formation of deep-water in the Mediterranean Sea, 1969. Nature 227, 1037–1040.

Mecking, S., Warner, M.J., 1999. Ventilation of Red Sea Water with respect to chlorofluorocarbons. J. Geophys. Res. 104, 11087–11097.

Meehl, G.A., Stocker, T.F., Collins, W.D., Friedlingstein, P., Gaye, A.T., Gregory, J.M., et al., 2007. Global climate projections. In: Solomon, S., Qin, D., Manning, M., Chen, Z., Marquis, M., Averyt, K.B., et al. (Eds.), Climate Change 2007: The Physical Science Basis. Contribution of Working Group I to the Fourth Assessment Report of the Intergovernmental Panel on Climate Change. Cambridge University Press, Cambridge, UK and New York.

Mei, C.C., Stiassnie, M., Yue, D.K.-P., 2005. Theory and Applications of Ocean Surface Waves: Part I, Linear Aspects; Part II, Nonlinear Aspects. World Scientific, New Jersey and London, 1136 pp.

Meinen, C.S., Watts, D.R., 1997. Further evidence that the sound-speed algorithm of Del Grosso is more accurate than that of Chen and Millero. J. Acoust. Soc. Am. 102, 2058–2062.

Meinen, C.S., Watts, D.R., 2000. Vertical structure and transport on a transect across the North Atlantic Current near 42°N: Time series and mean. J. Geophys. Res. 105, 21869–21891.

Menard, H.W., Smith, S.M., 1966. Hypsometry of ocean basin provinces. J. Geophys. Res. 71, 4305–4325.

Meredith, J.P., Watkins, J.L., Murphy, E.J., Ward, P., Bone, D.G., Thorpe, S.E., et al., 2003. Southern ACC Front to the northeast of South Georgia: Pathways, characteristics and fluxes. J. Geophys. Res. (C5), 108. doi:10.1029/2001JC001227.

Meredith, M.P., Hogg, A.M., 2006. Circumpolar response of Southern Ocean eddy activity to a change in the Southern Annular Mode. Geophys. Res. Lett. 33, L16608. doi:10.1029/2006GL026499.

Merz, A., Wüst, G., 1922. Die Atlantische Vertikal Zirkulation. Z. Ges. Erdkunde Berlin 1, 1–34 (in German).

Merz, A., Wüst, G., 1923. Die Atlantische Vertikal Zirkulation. 3 Beitrag. Zeitschr. D.G.F.E. Berlin (in German).

Middleton, J.F., Cirano, M., 2002. A northern boundary current along Australia's southern shelves: The Flinders Current. J. Geophys. Res. 107. doi:10.1029/2000JC000701.

Millero, F.J., 1967. High precision magnetic float densimeter. Rev. Sci. Instrum. 38, 1441–1444.

Millero, F.J., 1978. Freezing point of seawater, Eighth report of the Joint Panel of Oceanographic Tables and Standards. Appendix 6. UNESCO Tech. Papers.

Millero, F.J., Feistel, R., Wright, D.G., McDougall, T.J., 2008. The composition of Standard Seawater and the definition of the reference-composition salinity scale. Deep-Sea Res. I 55, 50–72.

Millero, F.J., Perron, G., Desnoyers, J.E., 1973. The heat capacity of seawater solutions from 5 to 35°C and from 0.5 to 22% chlorinity. J. Geophys. Res. 78, 4499–4507.

Millero, F.J., Poisson, A., 1980. International one-atmosphere equation of state of seawater. Deep-Sea Res. 28, 625–629.

Millot, C., 1991. Mesoscale and seasonal variabilities of the circulation in the western Mediterranean. Dynam. Atmos. Oceans 15, 179–214.

Millot, C., Taupier-Letage, I., 2005. Circulation in the Mediterranean Sea. In: Saliot, E.A. (Ed.), The Handbook of Environmental Chemistry, vol. 5. Part K. Springer-Verlag, Berlin Heidelberg, pp. 29–66.

Mills, E.L., 1994. Bringing oceanography into the Canadian university classroom. Scientia Canadensis. Can. J. Hist. Sci. Tech. Med. 18, 3–21.

Mittelstaedt, E., 1991. The ocean boundary along the northwest African coast: Circulation and oceanographic properties at the sea surface. Progr. Oceanogr. 26, 307–355.

Mizuno, K., White, W.B., 1983. Annual and interannual variability in the Kuroshio current system. J. Phys. Oceanogr. 13, 1847–1867.

Mobley, C.D., 1995. Optical properties of water. In: Bass, M., Van Stryland, E.W., Williams, D.R., Wolfe, W.L. (Eds.), Handbook of Optics, Vol. 1, Fundamentals, Techniques, and Design. McGraw-Hill 43.1–43.56.

Molinari, R.L., Fine, R.A., Wilson, W.D., Curry, R.G., Abell, J., McCartney, M.S., 1998. The arrival of recently formed Labrador Sea Water in the Deep Western Boundary Current at 26.5°N. Geophys. Res. Lett. 25, 2249–2252.

Monismith, S.G., 2007. Hydrodynamics of coral reefs. Annu. Rev. Fluid Mech. 39, 37–55.

Montecino, V., Strub, P.T., Chavez, F., Thomas, A., Tarazona, J., Baumgartner, T., 2006. Biophysical interactions off western South-America. In: Robinson, A.R., Brink, K.H. (Eds.), The Sea, vol. 14A: The Global Coastal Ocean: Interdisciplinary Regional Studies and Syntheses. Harvard University Press, Boston, MA, pp. 329–390.

Montgomery, R.B., 1938. Circulation in the upper layers of the Southern North Atlantic deduced with the use of isentropic analysis. Papers Phys. Oceanogr. and Met. 6, MIT and Woods Hole Oceanographic Institution, 55 pp.

Montgomery, R.B., 1958. Water characteristics of Atlantic Ocean and of world ocean. Deep-Sea Res. 5, 134–148.

Montgomery, R.B., Stroup, E.D., 1962. Equatorial waters and currents at 150°W in July-August 1952. Johns Hopkins Oceanographic Study, No.1, 68 pp.

Morawitz, W.M.L., Cornuelle, B.D., Worcester, P.F., 1996. A case study in three-dimensional inverse methods: combining hydrographic, acoustic, and moored thermistor data in the Greenland Sea. J. Atm. Oceanic Tech. 13, 659–679.

Morawitz, W.M.L., Sutton, P.J., Worcester, P.F., Cornuelle, B.D., Lynch, J.F., Pawlowicz, R., 1996. Three-dimensional observations of a deep convective chimney in the Greenland Sea during winter 1988/1989. J. Phys. Oceanogr. 26, 2316–2343.

Morel, A., Antoine, D., 1994. Heating rate within the upper ocean in relation to its bio-optical state. J. Phys. Oceanogr. 24, 1652–1665.

Morrow, R., Birol, F., 1998. Variability in the southeast Indian ocean from altimetry: Forcing mechanisms for the Leeuwin Current. J. Geophys. Res. 103, 18529–18544.

Morrow, R., Donguy, J.-R., Chaigneau, A., Rintoul, S.R., 2004. Cold-core anomalies at the subantarctic front, south of Tasmania. Deep-Sea Res. I, 1417–1440.

Moum, J.N., Farmer, D.M., Smyth, W.D., Armi, L., Vagle, S., 2003. Structure and generation of turbulence at interfaces strained by internal solitary waves propagating shoreward over the continental shelf. J. Phys. Oceanogr. 33, 2093–2112.

Müller, R.D., Sdrolias, M., Gaina, C., Roest, W.R., 2008. Age, spreading rates and spreading symmetry of the world's ocean crust. Geochem. Geophys. Geosyst. 9, Q04006. http://www.ngdc.noaa.gov/mgg/ocean_age/. doi:10/1029/2007GC001743(accessed 2.01.09).

Munk, W., 1966. Abyssal recipes. Deep-Sea Res. 13, 707–730.

Munk, W., 1981. Internal waves and small-scale processes. In: Warren, B.A., Wunsch, C. (Eds.), Evolution of Physical Oceanography. The MIT Press, Boston, MA, pp. 264–290.

Munk, W., Wunsch, C., 1982. Observing the ocean in the 1990's: a scheme for large-scale monitoring. Philos. Trans. Roy. Soc. A 307, 439–464.

Murray, J.W., Jannasch, H.W., Honjo, S., Anderson, R.F., Reeburgh, W.S., Top, Z., et al., 1989. Unexpected changes in the oxic/anoxic interface in the Black Sea. Nature 338, 411–413.

Naimie, C.E., Blain, C.A., Lynch, D.R., 2001. Seasonal mean circulation in the Yellow Sea — A model-generated climatology. Continental Shelf Res. 21, 667–695.

Nakano, H., Suginohara, N., 2002. Importance of the eastern Indian Ocean for the abyssal Pacific. J. Geophys. Res. 107, 3219. doi:10.1029/2001JC001065.

Nansen, F., 1922. In: Nacht und Eis. F.U. Brodhaus, Leipzig, Germany, 355 pp. (in German).

NASA, 2009a. Ocean color from space: global seasonal change. NASA Goddard Earth Sciences Data and Information Services Center. http://daac.gsfc.nasa.gov/oceancolor/scifocus/space/ocdst_global_seasonal_change.shtml (accessed 2.18.09).

NASA, 2009b. Ocean Color Web. NASA Goddard Space Flight Center. http://oceancolor.gsfc.nasa.gov/ (accessed 2.18.09).

NASA Earth Observatory, 2010. Global maps. NASA Goddard Space Flight Center. http://earthobservatory.nasa.gov/GlobalMaps/ (accessed 12.13.10).

NASA Goddard Earth Sciences, 2007a. An assessment of the Indian Ocean, Monsoon, and Somali Current using NASA's AIRS, MODIS, and QuikSCAT data. NASA Goddard Earth Sciences Data Information Services Center. http://daac.gsfc.nasa.gov/oceancolor/scifocus/modis/IndianMonsoon.shtml (accessed 7.1.08).

NASA Goddard Earth Sciences, 2007b. Sedimentia. NASA Goddard Earth Sciences Ocean Color. http://disc.gsfc.nasa.gov/oceancolor/scifocus/oceanColor/sedimentia.shtml (accessed 4.3.09).

NASA Goddard Earth Sciences, 2008. Ocean color: classic CZCS scenes, Chapter 4. NASA Goddard Earth Sciences Data Information Services Center. http://disc.gsfc.nasa.gov/oceancolor/scifocus/classic_scenes/04_classics_arabian.shtml (accessed 1.9.09).

NASA Visible Earth, 2006. Visible Earth: Sun glint in the Mediterranean Sea. NASA Goddard Space Flight Center. http://visibleearth.nasa.gov/view_rec.php?id=732 (accessed 10.01.08).

NASA Visible Earth, 2008. Eddies off the Queen Charlotte Islands. NASA Goddard Space Flight Center. http://visibleearth.nasa.gov/view_rec.php?id=2886 (accessed 3.26.09).

National Data Buoy Center, 2006. How are estimates of wind-seas and swell made from NDBC wave data? NOAA/NDBC. http://www.ndbc.noaa.gov/windsea.shtml (accessed 3.28.09).

National Data Buoy Center, 2009. NDBC Web Site. NOAA/NDBC. http://www.ndbc.noaa.gov/ (accessed 5.15.09).

National Research Council, 2010. Ocean acidification: A national strategy to meet the challenges of a changing ocean. National Academies Press, Washington D.C., 152 pp.

Naval Postgraduate School, 2003. Basic concepts in physical oceanography: tides. Navy Operational Ocean Circulation and Tide Models. Department of Oceanography, Naval Postgraduate School. http://www.oc.nps.edu/nom/day1/partc.html (accessed 3.30.09).

Neilson, B.J., Kuo, A., Brubaker, J., 1989. Estuarine Circulation. Humana Press, Clifton, N.J, 377 pp.

New, A.L., Jia, Y., Coulibaly, M., Dengg, J., 2001. On the role of the Azores Current in the ventilation of the North Atlantic Ocean. Progr. Oceanogr. 48, 163–194.

Niiler, P.P., Maximenko, N.A., McWilliams, J.C., 2003. Dynamically balanced absolute sea level of the global ocean derived from near-surface velocity observations. Geophys. Res. Lett. 30, 22. doi:10.1029/2003GL018628.

Nilsson, C.S., Cresswell, G.R., 1981. The formation and evolution of East Australian Current warm-core eddies. Progr. Oceanogr. 9, 133–183.

NOAA, 2008. Tides and Water Levels. NOAA Ocean Service Education. http://oceanservice.noaa.gov/education/kits/tides/welcome.html (accessed 3.29.09).

NOAA, 2009. Arctic Change: Climate indicators — Arctic Oscillation. NOAA Arctic. http://www.arctic.noaa.gov/detect/climate-ao.shtml (accessed 3.17.09).

NOAA AOML PHOD, 2009. The Global Drifter Program. NOAA AOML. http://www.aoml.noaa.gov/phod/dac/gdp.html (accessed 9.09).

NOAA CO-OPS, 2010. Tides and Currents. NOAA/National Ocean Service. http://co-ops.nos.noaa.gov/index.shtml (accessed 10.26.10).

NOAA CPC, 2005. Madden/Julian Oscillation (MJO). NOAA/National Weather Service. http://www.cpc.ncep.noaa.gov/products/precip/CWlink/MJO/mjo.shtml (accessed 12.28.09).

NOAA ESRL, 2009. Linear correlations in atmospheric seasonal/monthly averages. NOAA Earth System Research Laboratory Physical Sciences Division. http://www.cdc.noaa.gov/data/correlation/ (accessed 10.30.09).

NOAA ESRL, 2010. PSD Map Room Climate Products Outgoing Longwave Radiation (OLR). NOAA ESRL PSD. http://www.cdc.noaa.gov/map/clim/olr.shtml (accessed 12.14.10).

NOAA National Weather Service, 2005. Hydrometeorological Prediction Center (HPC) Home Page. National Weather Service. http://www.hpc.ncep.noaa.gov/ (accessed 1.3.05).

NOAA NESDIS, 2009. Ocean Products Page. NOAA/NESDIS/OSDPD. http://www.osdpd.noaa.gov/PSB/EPS/SST/SST.html (accessed 2.18.09).

NOAA NGDC, 2008. Global Relief Data — ETOPO. NOAA National Geophysical Data Center. http://www.ngdc.noaa.gov/mgg/global/global.html (accessed 9.24.08).

NOAA PMEL TAO Project Office, 2009a. The TAO project. TAO Project Office, NOAA Pacific Marine Environmental Laboratory. http://www.pmel.noaa.gov/tao/ (accessed 6.1.09).

NOAA PMEL TAO Project Office, 2009b. El Niño theme page: access to distributed information on El Niño. NOAA Pacific Marine Environmental Laboratory. http://www.pmel.noaa.gov/tao/elnino/nino-home.html (accessed 3.26.09).

NOAA PMEL, 2009c. Global tropical moored array. NOAA Pacific Marine Environmental Laboratory. http://www.pmel.noaa.gov/tao/global/global.html (accessed 5.20.09).

NOAA PMEL, 2009d. Impacts of El Niño and benefits of El Niño prediction. NOAA Pacific Marine Environmental Laboratory. http://www.pmel.noaa.gov/tao/elnino/impacts.html (accessed 3.26.09).

NOAA Wavewatch III, , 2009. NCEP MMAB operational wave models. NOAA/NWS Environmental Modeling Center/Marine Modeling and Analysis Branch. http://polar.ncep.noaa.gov/waves/index2.shtml (accessed 5.14.09).

NODC, 2005a. World Ocean Atlas 2005 (WOA05). NOAA National Oceanographic Data Center. http://www.nodc.noaa.gov/OC5/WOA05/pr_woa05.html (accessed 4.28.09).

NODC, 2005b. World Ocean Database 2005 (WOD05). NOAA National Oceanographic Data Center. http://www.nodc.noaa.gov/OC5/WOD05/pr_wod05.html (accessed 4.28.09).

NODC, 2009. Data sets and products, National Oceanographic Data Center Ocean Climate Laboratory. http://www.nodc.noaa.gov/OC5/indprod.html (accessed 12.15.09).

Nowlin, W.D., Whitworth, T., Pillsbury, R.D., 1977. Structure and transport of the Antarctic Circumpolar Current at Drake Passage from short-term measurements. J. Phys. Oceanogr. 7, 788–802.

NSIDC, 2007. Arctic sea ice news fall 2007. National Snow and Ice Data Center. http://nsidc.org/arcticseaicenews/2007.html (accessed 3.17.09).

NSIDC, 2008a. Polar Pathfinder Daily 25 km EASE-Grid Sea Ice Motion Vectors. National Snow and Ice Data Center. http://nsidc.org/data/docs/daac/nsidc0116_icemotion.gd.html (accessed 02.01.09).

NSIDC, 2008b. Arctic sea ice down to second-lowest extent; likely record-low volume. National Snow and Ice Data Center. http://nsidc.org/news/press/20081002_seaice_pressrelease.html (accessed 3.17.09).

NSIDC, 2009a. Cryospheric climate indicators. National Snow and Ice Data Center. http://nsidc.org/data/seaice_index/archives/index.html (accessed 2.25.09).

NSIDC, 2009b. Arctic climatology and meteorology primer. National Snow and Ice Data Center. http://nsidc.org/arcticmet/ (accessed 3.1.09).

NSIDC, 2009c. Images of Antarctic Ice Shelves. National Snow and Ice Data Center. http://nsidc.org/data/iceshelves_images/index.html (accessed 3.5.09).

NWS Internet Services Team, 2008. ENSO temperature and precipitation composites. http://www.cpc.noaa.gov/products/precip/CWlink/ENSO/composites/EC_LNP_index.shtml (accessed 3.27.09).

O'Connor, B.M., Fine, R.A., Olson, D.B., 2005. A global comparison of subtropical underwater formation rates. Deep-Sea Res. I. 52, 1569–1590.

Ochoa, J., Bray, N.A., 1991. Water mass exchange in the Gulf of Cadiz. Deep-Sea Res. 38 (Suppl.), S465–S503.

ODV, 2009. Ocean Data View. Alfred Wegener Institute. http://odv.awi.de/en/home/ (accessed 4.28.09).

Officer, C.B., 1976. Physical Oceanography of Estuaries (and Associated Coastal Waters). Wiley, New York. 465 pp.

Oguz, T., Tugrul, S., Kideys, A.E., Ediger, V., Kubilay, N., 2006. Physical and biogeochemical characteristics of the Black Sea. In: Robinson, A.R., Brink, K.H. (Eds.), The Sea, Vol., 14A: The Global Coastal Ocean: Interdisciplinary Regional Studies and Syntheses. Harvard University Press, Boston, MA, pp. 1333–1372.

Olbers, D.J.M., Wenzel, M., Willebrand, J., 1985. The inference of North Atlantic circulation patterns from climatological hydrographic data. Rev. Geophys. 23, 313–356.

Olsen, S.M., Hansen, B., Quadfasel, D., Østerhus, S., 2008. Observed and modeled stability of overflow across the Greenland-Scotland ridge. Nature 455, 519–523.

Olson, D., Schmitt, R., Kennelly, M., Joyce, T., 1985. A two-layer diagnostic model of the long-term physical evolution of warm-core ring 82B. J. Geophys. Res. 90, 8813–8822.

Oort, A.H., Vonder Haar, T.H., 1976. On the observed annual cycle in the ocean-atmosphere heat balance over the northern hemisphere. J. Phys. Oceanogr. 6, 781–800.

Open University, 1999. Waves, Tides and Shallow-Water Processes, second ed. Butterworth-Heinemann, Burlington, MA, 228 pp.

Orsi, A.H., Johnson, G.C., Bullister, J.L., 1999. Circulation, mixing, and production of Antarctic Bottom Water. Progr. Oceanogr. 43, 55–109.

Orsi, A.H., Nowlin, W.D., Whitworth, T., 1993. On the circulation and stratification of the Weddell Gyre. Deep-Sea Res. I. 40, 169–203.

Orsi, A., Whitworth, T., 2005. Hydrographic Atlas of the World Ocean Circulation Experiment (WOCE). Volume 1: Southern Ocean. In: Sparrow, M., Chapman, P., Gould, J. (Eds.). International WOCE Project Office, Southampton, UK ISBN 0-904175-49-9. http://woceatlas.tamu.edu/ (accessed 4.20.09).

Orsi, A.H., Whitworth, T., Nowlin, W.D., 1995. On the meridional extent and fronts of the Antarctic Circumpolar Current. Deep-Sea Res. I. 42, 641–673.

Osborn, T.R., Cox, C.S., 1972. Oceanic fine structure. Geophys. Astrophys. Fluid Dynam. 3, 321–345.

Osborne, J., Swift, J.H., 2009. Java OceanAtlas. http://odf.ucsd.edu/joa/ (accessed 4.20.09).

Østerhus, S., Gammelsrød, T., 1999. The abyss of the Nordic Seas is warming. J. Clim. 12, 3297–3304.

Östlund, H., Possnert, G., G., Swift, J., 1987. Ventilation rate of the deep Arctic Ocean from carbon 14 data. J. Geophys. Res. 92, 3769–3777.

Owens, W.B., 1991. A statistical description of the mean circulation and eddy variability in the northwestern Atlantic using SOFAR floats. Progr. Oceanogr. 28, 257–303.

Owens, W.B., Warren, B.A., 2001. Deep circulation in the northwest corner of the Pacific Ocean. Deep-Sea Res. I 48, 959–993.

Özsoy, E., Hecht, A., Ünlüata, Ü., Brenner, S., Oguz, T., Bishop, J., et al., 1991. A review of the Levantine Basin circulation and its variability during 1985–1988. Dynam. Atmos. Oceans 15, 421–456.

Özsoy, E., Ünlüata, U., 1998. The Black Sea. In: Robinson, A.R., Brink, K.H. (Eds.), The Sea, Vol. 11: The Global Coastal Ocean: Regional Studies and Syntheses. Harvard University Press, Boston, MA, pp. 889–914.

Palacios, D.M., Bograd, S.J., 2005. A census of Tehuantepec and Papagayo eddies in the northeastern tropical Pacific. Geophys. Res. Lett. 32, L23606. doi: 10.1029/2005GL024324.

Palastanga, V., van Leeuwen, P.J., Schouten, M.W., deRuijter, P.M., 2007. Flow structure and variability in the subtropical Indian ocean: instability of the South Indian Ocean Countercurrent. J. Geophys. Res. 112, C01001. doi:10.1029/2005JC003395.

Parker, C.E., 1971. Gulf Stream rings in the Sargasso Sea. Deep-Sea Res. 18, 981–993.

Paulson, C.A., Simpson, J.J., 1977. Irradiance measurements in the upper ocean. J. Phys. Oceanogr. 7, 952–956.

Payne, R.E., 1972. Albedo of the sea surface. J. Atmos. Sci. 29, 959–970.

Pearce, A.F., 1991. Eastern boundary currents of the southern hemisphere. J. Roy. Soc. Western Austral. 74, 35–45.

Pedlosky, J., 1987. Geophysical Fluid Dynamics, second ed. Springer-Verlag, New York, 732 pp.

Pedlosky, J., 2003. Waves in the Ocean and Atmosphere. Springer-Verlag, Berlin, 260 pp.

Peeters, F.J.C., Acheson, R., Brummer, G.-J.A., de Ruijter, W.P.M., Schneider, R.R., Ganssen, G.M., et al., 2004. Vigorous exchange between the Indian and Atlantic oceans at the end of the past five glacial periods. Nature 430, 661–665.

Pelegrí, J.L., Arístegui, J., Cana, L., González-Dávila, M., Hernández-Guerra, A., Hernández-León, S., et al., 2005. Coupling between the open ocean and the coastal upwelling region off northwest Africa: Water recirculation and offshore pumping of organic matter. J. Marine Syst. 54, 3–37.

Perry, R.K., 1986. Bathymetry. In: Hurdle, B. (Ed.), The Nordic Seas. Springer-Verlag, New York, pp. 211–236.

Peterson, R.G., 1988. On the transport of the Antarctic Circumpolar Current through Drake Passage and its relation to wind. J. Geophys. Res. 93, 13993–14004.

Peterson, R.G., 1992. The boundary currents in the western Argentine Basin. Deep-Sea Res. 39, 623–644.

Peterson, R.G., Stramma, L., Kortum, G., 1996. Early concepts and charts of ocean circulation. Progr. Oceanogr. 37, 1–115.

Philander, S.G.H., 1978. Instabilities of zonal equatorial currents: II. J. Geophys. Res. 83, 3679–3682.

Philander, S.G.H., Fedorov, A., 2003. Is El Niño sporadic or cyclic? Annu. Rev. Earth Pl. Sci. 31, 579–594.

Phillips, O.M., 1977. The Dynamics of the Upper Ocean. Cambridge University Press, Cambridge, UK, 336 pp.

Pickard, G.L., 1961. Oceanographic features of inlets in the British Columbia mainland coast. J. Fish. Res. Bd. Can. 18, 907–999.

Pickard, G.L., Donguy, J.R., Hénin, C., Rougerie, F., 1977. A review of the physical oceanography of the Great Barrier Reef and western Coral Sea. Australian Institute of Marine Science, 2. Australian Government Publishing Service, Canberra, 134 pp.

Pickard, G.L., Stanton, B.R., 1980. Pacific fjords — A review of their water characteristics. In: Freeland, H.J., Farmer, D.M., Levings, C.D. (Eds.), Fjord Oceanography. Plenum Press, pp. 1–51.

Pickart, R.S., McKee, T.K., Torres, D.J., Harrington, S.A., 1999. Mean structure and interannual variability of the slopewater system south of Newfoundland. J. Phys. Oceanogr. 29, 2541–2558.

Pickart, R.S., Smethie, W.M., 1993. How does the Deep Western Boundary Current cross the Gulf Stream? J. Phys. Oceanogr. 23, 2602–2616.

Pickart, R.S., Torres, D.J., Clarke, R.A., 2002. Hydrography of the Labrador Sea during active convection. J. Phys. Oceanogr. 32, 428–457.

Pickett, M.H., Paduan, J.D., 2003. Ekman transport and pumping in the California Current based on the U.S. Navy's high-resolution atmospheric model (COAMPS). J. Geophys. Res. 108, 3327. doi:10.1029/2003JC001902.

Polton, J.A., Smith, J.A., MacKinnon, J.A., Tejada-Martínez, A.E., 2008. Rapid generation of high-frequency internal waves beneath a wind and wave forced oceanic surface mixed layer. Geophys. Res. Lett. 35, L13602. doi:10.1029/2008GL033856.

Polyakov, I.V., 22 co-authors, 2005. One more step toward a warmer Arctic. Geophys. Res. Lett. 32, L17605. doi:10.1029/2005GL023740.

Polyakov, I.V., 17 co-authors, 2010. Arctic Ocean warming contributes to reduced polar ice cap. J. Phys. Oceanogr 40, 2743–2756.

Polyakov, I.V., Alekseev, G.V., Timokhov, L.A., Bhatt, U.S., Colony, R.L., Simmons, H.L., et al., 2004. Variability of the intermediate Atlantic Water of the Arctic Ocean over the last 100 years. J. Clim. 17, 4485–4497.

Polzin, K.L., Toole, J.M., Ledwell, J.R., Schmitt, R.W., 1997. Spatial variability of turbulent mixing in the abyssal ocean. Science 276, 93–96.

Pond, S., Pickard, G.L., 1983. Introductory Dynamical Oceanography, second ed. Pergamon Press, Oxford, 329 pp.

Poole, R., Tomczak, M., 1999. Optimum multiparameter analysis of the water mass structure in the Atlantic Ocean thermocline. Deep-Sea Res. I 46, 1895–1921.

Potter, R.A., Lozier, M.S., 2004. On the warming and salinification of the Mediterranean outflow waters in the North Atlantic. Geophys. Res. Lett. 31, L01202. doi:10.1029/2003GL018161.

Press, W.H., Flannery, B.P., Teukolsky, S.A., Vetterline, W.T., 1986. Numerical Recipes. Cambridge University Press, Cambridge, UK, 818 pp.

Price, J.F., Baringer, M.O., 1994. Outflows and deep water production by marginal seas. Progr. Oceanogr. 33, 161–200.

Price, J.F., Weller, R.A., Pinkel, R., 1986. Diurnal cycling: Observations and models of the upper ocean response to diurnal heating, cooling and wind mixing. J. Geophys. Res. 91, 8411–8427.

Pritchard, D.W., 1989. Estuarine classification — a help or a hindrance. In: Neilson, B.J., Kuo, A., Brubaker, J. (Eds.), Estuarine Circulation. Humana Press, Clifton, N.J, pp. 1–38.

Proshutinsky, A.Y., Johnson, M.A., 1997. Two circulation regimes of the wind-driven Arctic Ocean. J. Geophys. Res. 102, 12493–12514.

Provost, C., Escoffier, C., Maamaatuaiahutapu, K., Kartavtseff, A., Garçon, V., 1999. Subtropical mode waters in the South Atlantic Ocean. J. Geophys. Res. 104, 21033–21049.

Pugh, D.T., 1987. Tides, Surges, and Mean Sea-level. J. Wiley, Chichester, UK, 472 pp.

Purkey, S.G., Johnson, G.C., 2010. Antarctic bottom water warming between the 1990s and 2000s: Contributions to global heat and sea level rise budgets. J. Clim. 23, 6336–6351.

Qiu, B., Chen, S., 2005. Variability of the Kuroshio Extension jet, recirculation gyre, and mesoscale eddies on decadal time scales. J. Phys. Oceanogr. 35, 2090–2103.

Qiu, B., Huang, R.X., 1995. Ventilation of the North Atlantic and North Pacific: Subduction versus obduction. J. Phys. Oceanogr. 25, 2374–2390.

Qiu, B., Miao, W., 2000. Kuroshio path variations south of Japan: Bimodality as a self-sustained internal oscillation. J. Phys. Oceanogr. 30, 2124–2137.

Qiu, B., Scott, R.B., Chen, S., 2008. Length scales of eddy generation and nonlinear evolution of the seasonally modulated South Pacific Subtropical Countercurrent. J. Phys. Oceanogr. 38, 1515–1528.

Qu, T., Lindstrom, E., 2002. A climatological interpretation of the circulation in the western South Pacific. J. Phys. Oceanogr. 32, 2492–2508.

Quartly, G.D., Srokosz, M.A., 1993. Seasonal variations in the region of the Agulhas retroflection: Studies with Geosat and FRAM. J. Phys. Oceanogr. 23, 2107–2124.

Quartly, G.D., Srokosz, M.A., 2003. Satellite observations of the Agulhas Current system. Philos. Trans. Roy. Soc. A 361, 51–56.

Ralph, E.A., Niiler, P.P., 1999. Wind-driven currents in the tropical Pacific. J. Phys. Oceanogr. 29, 2121–2129.

Ramanathan, V., Collins, W., 1991. Thermodynamic regulation of ocean warming by cirrus clouds deduced from observations of the 1987 El Nino. Nature 351, 27–32.

Rasmusson, E.M., Carpenter, T.H., 1982. Variations in tropical sea surface temperature and surface wind fields associated with the outer Oscillation/El Niño. Mon. Weather Rev. 110, 354–384.

Ray, R.D., 1999. A global ocean tide model from TOPEX/POSEIDON altimetry: GOT99.s. NASA/TM -1999–209478, 58 pp.

Redfield, A.C., 1934. On the proportion of organic derivatives in sea water and their relation to the composition of plankton. James Johnstone Memorial Volume, Liverpool, UK, pp. 176–192.

Redi, M.H., 1982. Oceanic isopycnal mixing by coordinate rotation. J. Phys. Oceanogr. 12, 1154–1158.

Reed, R.K., 1995. On geostrophic reference levels in the Bering Sea basin. J. Oceanogr. 51, 489–498.

Reid, J.L., 1973. The shallow salinity minima of the Pacific Ocean. Deep-Sea Res. 20, 51–68.

Reid, J.L., 1989. On the total geostrophic circulation of the South Atlantic Ocean: Flow patterns, tracers and transports. Progr. Oceanogr. 23, 149–244.

Reid, J.L., 1994. On the total geostrophic circulation of the North Atlantic Ocean: Flow patterns, tracers and transports. Progr. Oceanogr. 33, 1–92.

Reid, J.L., 1997. On the total geostrophic circulation of the Pacific Ocean: Flow patterns, tracers and transports. Progr. Oceanogr. 39, 263–352.

Reid, J.L., 2003. On the total geostrophic circulation of the Indian Ocean: Flow patterns, tracers and transports. Progr. Oceanogr. 56, 137–186.

Reid, J.L., Lynn, R.J., 1971. On the influence of the Norwegian-Greenland and Weddell seas upon the bottom waters of the Indian and Pacific Oceans. Deep-Sea Res. 18, 1063–1088.

Remote Sensing Systems, 2004. TMI sea surface temperatures (SST). <http://www.ssmi.com/rss_research/tmi_sst_pacific_equatorial_current.html> (accessed 3.27.09).

Reverdin, G., Durand, F., Mortensen, J., Schott, F., Valdimarsson, H., 2002. Recent changes in the surface salinity of the North Atlantic subpolar gyre. J. Geophys. Res. 107, 8010. doi:10.1029/2001JC001010.

Rhein, M., Stramma, L., Send, U., 1995. The Atlantic Deep Western Boundary Current: Water masses and transports near the equator. J. Geophys. Res. 100, 2441–2457.

Richardson, P.L., 1980a. Benjamin Franklin and Timothy Folger's First Printed Chart of the Gulf Stream. Science 207, 643–645.

Richardson, P.L., 1980b. Gulf Stream ring trajectories. J. Phys. Oceanogr. 10, 90–104.

Richardson, P.L., 1983. Gulf Stream rings. In: Robinson, A.R. (Ed.), Eddies in Marine Science. Springer-Verlag, Berlin, pp. 19–45.

Richardson, P.L., 2005. Caribbean Current and eddies as observed by surface drifters. Deep-Sea Res. II 52, 429–463.

Richardson, P.L., 2007. Agulhas leakage into the Atlantic estimated with subsurface floats and surface drifters. Deep-Sea Res. I 54, 1361–1389.

Richardson, P.L., 2008. On the history of meridional overturning circulation schematic diagrams. Progr. Oceanogr. 76, 466–486.

Richardson, P.L., Bower, A.S., Zenk, W., 2000. A census of Meddies tracked by floats. Progr. Oceanogr. 45, 209–250.

Richardson, P.L., Cheney, R.E., Worthington, L.V., 1978. A census of Gulf Stream rings Spring 1975. J. Geophys. Res. 83, 6136–6144.

Richardson, P.L., Hufford, G., Limeburner, R., Brown, W., 1994. North Brazil Current retroflection eddies. J. Geophys. Res. 99, 5081–5093.

Richardson, P.L., Lutjeharms, J.R.E., Boebel, O., 2003. Introduction to the "Interocean exchange around Africa." Deep-Sea Res. II 50, 1–12.

Richardson, P.L., Mooney, K., 1975. The Mediterranean outflow — a simple advection-diffusion model. J. Phys. Oceanogr. 5, 476–482.

Richardson, P.L., Reverdin, G., 1987. Seasonal cycle of velocity in the Atlantic North Equatorial Countercurrent as measured by surface drifters, current meters, and ship drifts. J. Geophys. Res. 92, 3691–3708.

Ridgway, K.R., Condie, S.A., 2004. The 5500-km-long boundary flow off western and southern Australia. J. Geophys. Res. 109, C04017. doi: 10.1029/2003JC001921.

Ridgway, K.R., Dunn, J.R., 2003. Mesoscale structure of the mean East Australian Current System and its relationship with topography. Progr. Oceanogr. 56, 189–222.

Ridgway, K.R., Dunn, J.R., 2007. Observational evidence for a Southern Hemisphere oceanic supergyre. Geophys. Res. Lett. 34 doi:10.1029/2007GL030392.

Rigor, I.G., Wallace, J.M., Colony, R.L., 2002. Response of sea ice to the Arctic Oscillation. J. Clim. 15, 2648–2663.

Rintoul, S.R., 1998. On the origin and influence of Adelie Land Bottom Water. In: Jacobs, S., Weiss, R. (Eds.), 1998. Ocean, Ice, and Atmosphere: Interactions at the Antarctic Continental Margin. Antarctic Research Series 75, American Geophysical Union, Washington, pp. 151–171.

Rintoul, S.R., Hughes, C.W., Olbers, D., 2001. The Antarctic Circumpolar Current System. In: Siedler, G., Church, J. (Eds.), Ocean Circulation and Climate, International Geophysics Series. Academic Press, San Diego, CA, pp. 271–302.

Risien, C.M., Chelton, D.B., 2008. A global climatology of surface wind and wind stress fields from eight years of QuikSCAT scatterometer data. J. Phys. Oceanogr. 38, 2379–2413.

Roach, A.T., Aagaard, K., Pease, C.H., Salo, S.A., Weingartner, T., Pavlov, V., et al., 1995. Direct measurements of transport and water properties through the Bering Strait. J. Geophys. Res. 100, 18443–18458.

Robbins, P.E., Toole, J.M., 1997. The dissolved silica budget as a constraint on the meridional overturning circulation in the Indian Ocean. Deep-Sea Res. I 44, 879–906.

Robinson, A.R., Brink, K.H. (Eds.), 1998. The Sea, Vol. 11: The Global Coastal Ocean: Regional Studies and Syntheses. Harvard University Press, Cambridge, MA, 1090 pp.

Robinson, A.R., Brink, K.H. (Eds.), 2005. The Sea, Vol. 13: The Global Coastal Ocean: Multiscale Interdisciplinary Studies. Harvard University Press, Cambridge, MA, 1033 pp.

Robinson, A.R., Brink, K.H. (Eds.), 2006. The Sea, Vol. 14A: The Global Coastal Ocean: Interdisciplinary Regional Studies and Syntheses. Harvard University Press, Cambridge, MA, 840 pp.

Robinson, A.R., Golnaraghi, M., Leslie, W.G., Artegiani, A., Hecht, A., Lazzoni, E., et al., 1991. The eastern Mediterranean general circulation: Features, structure and variability. Dynam. Atmos. Oceans 15, 215–240.

Robinson, I.S., 2004. Measuring the Oceans from Space: The Principles and Methods of Satellite Oceanography. Springer-Verlag, Chichester, UK, 669 pp.

Rochford, D.J., 1961. Hydrology of the Indian Ocean. 1. The water masses in intermediate depths of the southeast Indian Ocean. Aust. J. Mar. Fresh. Res. 12, 129–149.

Roden, G.I., 1975. On North Pacific temperature, salinity, sound velocity and density fronts and their relation to the wind and energy flux fields. J. Phys. Oceanogr. 5, 557–571.

Roden, G.I., 1991. Subarctic-subtropical transition zone of the North Pacific: Large-scale aspects and mesoscale structure. In: Wetherall, J.A. (Ed.), Biology, Oceanography and Fisheries of the North Pacific Transition Zone and the Subarctic Frontal Zone. NOAA Technical Report, 105, pp. 1–38.

Rodhe, J., 1998. The Baltic and North Seas: A process-oriented review of the physical oceanography. In: Robinson, A.R., Brink, K.H. (Eds.), The Sea, Vol. 11: The Global Coastal Ocean: Regional Studies and Syntheses. Harvard University Press, Boston, MA, pp. 699–732.

Rodhe, J., Tett, P., Wulff, F., 2006. The Baltic and North Seas: A regional review of some important physical-chemical-biological interaction processes. In: Robinson, A.R., Brink, K.H. (Eds.), The Sea, Vol. 14A: The Global Coastal Ocean: Interdisciplinary Regional Studies and Syntheses. Harvard University Press, Boston, MA, pp. 1033–1076.

Roemmich, D., Cornuelle, B., 1992. The Subtropical Mode Waters of the South Pacific Ocean. J. Phys. Oceanogr. 22, 1178–1187.

Roemmich, D., Gilson, J., Davis, R., Sutton, P., Wijffels, S., Riser, S., 2007. Decadal spin-up of the South Pacific subtropical gyre. J. Phys. Oceanogr. 37, 162–173.

Roemmich, D., Hautala, S., Rudnick, D., 1996. Northward abyssal transport through the Samoan passage and adjacent regions. J. Geophys. Res. 101, 14039–14055.

Roemmich, D., Sutton, P., 1998. The mean and variability of ocean circulation past northern New Zealand: Determining the representativeness of hydrographic climatologies. J. Geophys. Res. 103, 13041–13054.

Roemmich, D.L., 1983. Optimal estimation of hydrographic station data and derived fields. J. Phys. Oceanogr. 13, 1544–1549.

Roemmich, D.L., Gilson, J., 2009. The 2004-2008 mean and annual cycle of temperature, salinity, and steric height in the global ocean from the Argo Program. Progr. Oceanogr. 82, 81–100.

Ronski, S., Budéus, G., 2005a. How to identify winter convection in the Greenland Sea from hydrographic summer data. J. Geophys. Res. 110, C11010. doi:10.1029/2003JC002156.

Ronski, S., Budéus, G., 2005b. Time series of winter convection in the Greenland Sea. J. Geophys. Res. 110, C04015. doi:10.1029/2004JC002318.

Ross, D. A., 1983. The Red Sea. In Estuaries and Enclosed Seas, Ed. B.H. Ketchum. Ecosystems of the World, 26, Elsevier, 293–307.

Rossby, T., 1996. The North Atlantic Current and surrounding waters: at the crossroads. Rev. Geophys. 34 (4), 463–481.

Rossby, T., 1999. On gyre interactions. Deep-Sea Res. II 46, 139–164.

Rothrock, D., Yu, Y., Maykut, G., 1999. Thinning of the Arctic sea-ice cover. Geophys. Res. Lett. 26, 3469–3472.

Rowe, E., Mariano, A.J., Ryan, E.H., 2010. The Antilles Current. Ocean Surface Currents. University of Miami, RSMAS, CIMAS. <http://oceancurrents.rsmas.miami.edu/atlantic/antilles.html> (accessed 1.10.10).

Rudels, B., 1986. The outflow of polar water through the Arctic archipelago and the oceanographic conditions in Baffin Bay. Polar Res. 4, 161–180.

Rudels, B., 2001. Arctic Basin circulation. In: Steele, J.H., Thorpe, S.A., Turekian, K.K. (Eds.), Encyclopedia of Ocean Sciences. Elsevier Science Ltd., Oxford, UK, pp. 177–187.

Rudels, B., Anderson, L.G., Eriksson, P., Fahrbach, E., Jakobsson, M., Jones, E.P., et al., 2011. ACSYS Chapter 4: Observations in the Ocean. In: Lemke, P., Fichefet, T., Dick, C. (Eds.), Arctic Climate Change — The ACSYS Decade and Beyond. Springer-Verlag, Berlin in press.

Rudels, B., Bjork, G., Nilsson, J., Winsor, P., Lake, I., Nohr, C., 2005. Interaction between waters from the Arctic Ocean the Nordic Seas north of Fram Strait and along the East Greenland Current: results from the Arctic Ocean-20 Oden expedition. J. Marine Sys. 55, 1–30.

Rudels, B., Friedrich, H.J., Quadfasel, D., 1999. The Arctic circumpolar boundary current. Deep-Sea Res. II 46, 1023–1062.

Rudnick, D.L., 1996. Intensive surveys of the Azores Front 2. Inferring the geostrophic and vertical velocity fields. J. Geophys. Res. 101, 16291–16303.

Rudnick, D.L., 1997. Direct velocity measurements in the Samoan Passage. J. Geophys. Res. 102, 3293–3302.

Rudnick, D.L., 2008. SIO 221B: Analysis of physical oceanographic data. <http://chowder.ucsd.edu/Rudnick/SIO_221B.html> (accessed 4.14.09).

Rudnick, D.L., Boyd, T.J., Brainard, R.E., Carter, G.S., Egbert, G.D., Gregg, M.C., et al., 2003. From tides to mixing along the Hawaiian Ridge. Science 301, 355–357.

Rydevik, D., 2004. A picture of the 2004 tsunami in Ao Nang, Thailand. Wikipedia OTRS system. <http://en.wikipedia.org/wiki/File:2004-tsunami.jpg#file> (accessed 9.22.05).

Sabine, C.L., Feely, R.A., Gruber, N., Key, R.M., Lee, K., Bullister, J.L., et al., 2004. The oceanic sink for anthropogenic CO_2. Science 305, 367–371.

Saji, N.H., Goswami, B.N., Vinayachandran, P.N., Yamagata, T., 1999. A dipole mode in the tropical Indian Ocean. Nature 401, 360–363.

Salmon, R., 1998. Lectures on Geophysical Fluid Dynamics. Oxford University Press, New York, 378 pp.

Sandström, J., 1908. Dynamische Versuche mit Meerwasser. Annalen der Hydrographie und Maritimen Meteorologie, 6–23 (in German).

Sankey, T., 1973. The formation of deep water in the Northwestern Mediterranean. Progr. Oceanogr. 6, 159–179.

Savchenko, V.G., Emery, W.J., Vladimirov, O.A., 1978. A cyclonic eddy in the Antarctic Circumpolar Current south of Australia: results of Soviet-American observations aboard the R/V Professor Zubov. J. Phys. Oceanogr. 8, 825–837.

Scambos, T., Haran, T., Fahnestock, M., Painter, T., Bohlander., J., 2007. MODIS-based Mosaic of Antarctica (MOA) data sets: Continent-wide surface morphology and snow grain size. Remote Sens. Environ. 111, 242–257.

Scambos, T., Hulbe, C., Fahnestock, M., Bohlander, J., 2000. The link between climate warming and break-up of ice shelves in the Antarctic Peninsula. J. Glaciol. 46, 516–530.

Schauer, U., Rudels, B., Jones, E.P., Anderson, L.G., Muench, R.D., Björk, G., et al., 2002. Confluence and redistribution of Atlantic water in the Nansen, Amundsen and Makarov basins. Ann. Geophys. 20, 257–273.

Schlichtholz, P., Houssais, M.-H., 2002. An overview of the theta-S correlations in Fram Strait based on the MIZEX 84 data. Oceanologia 44, 243–272.

Schlitzer, R., Roether, W., Oster, H., Junghans, H.-G., Hausmann, M., Johannsen, H., et al., 1991. Chlorofluoromethane and oxygen in the Eastern Mediterranean. Deep-Sea Res. 38, 1531–1551.

Schlosser, P., co-authors, 1997. The first trans-Arctic 14C section: Comparison of the mean ages of the deep waters in the Eurasian and Canadian basins of the Arctic Ocean. Nucl. Instrum. Methods 123B, 431–437.

Schmitt, R.W., 1981. Form of the temperature-salinity relationship in the Central Water: Evidence for double-diffusive mixing. J. Phys. Oceanogr. 11, 1015–1026.

Schmitt, R.W., Perkins, H., Boyd, J.D., Stalcup, M.C., 1987. C-SALT: An investigation of the thermohaline staircase in the western tropical North Atlantic. Deep-Sea Res. 34, 1655–1665.

Schmitz, W.J., 1995. On the interbasin-scale thermohaline circulation. Rev. Geophys. 33, 151–173.

Schmitz, W.J., 1996a. On the eddy field in the Agulhas Retroflection, with some global considerations. J. Geophys. Res. 101, 16259–16271.

Schmitz, W.J., 1996b. On the World Ocean Circulation: Volume I: Some global features/North Atlantic circulation. Woods Hole Oceanographic Institution Technical Report, WHOI-96-03, Woods Hole, MA, 141 pp.

Schneider, N., 1998. The Indonesian throughflow and the global climate system. J. Clim. 11, 676–689.

Schneider, N., Cornuelle, B.D., 2005. The forcing of the Pacific Decadal Oscillation. J. Clim. 18, 4355–4373.

Schott, F.A., Brandt, P., 2007. Circulation and deep water export of the subpolar North Atlantic during the 1990s. Geophys. Monogr. Ser. 173, 91–118. doi:10.1029/173GM08.

Schott, F.A., Dengler, M., Brandt, P., Affler, K., Fischer, J., Bourlès, B., et al., 2003. The zonal currents and transports at 35°W in the tropical Atlantic. Geophys. Res. Lett. 30, 1349. doi:10.1029/2002GL016849.

Schott, F.A., Dengler, M., Schoenefeldt, R., 2002. The shallow overturning circulation of the Indian Ocean. Progr. Oceanogr. 53, 57–103.

Schott, F.A., McCreary Jr., J., 2001. The monsoon circulation of the Indian Ocean. Progr. Oceanogr. 51, 1–123.

Schott, F.A., McCreary, J.P., Johnson, G.A., 2004. Shallow overturning circulations of the tropical-subtropical oceans. In Earth's Climate: The Ocean-Atmosphere Interaction. AGU Geophy. Monogr. Ser. 147, 261–304.

Schott, F.A., Stramma, L., Giese, B.S., Zantopp, R., 2009. Labrador Sea convection and subpolar North Atlantic Deep Water export in the SODA assimilation model. Deep-Sea Res. I 56, 926–938.

Schott, F.A., Zantopp, R., Stramma, L., Dengler, M., Fischer, J., Wibaux, M., 2004. Circulation and deep water export at the western exit of the subpolar North Atlantic. J. Phys. Oceanogr. 34, 817–843.

Schröder, M., Fahrbach, E., 1999. On the structure and the transport of the eastern Weddell Gyre. Deep-Sea Res. 46, 501–527.

Schwing, F.B., O'Farrell, M., Steger, J., Baltz, K., K., 1996. Coastal Upwelling Indices, West coast of North America, 1946–1995. U.S. Dept. of Commerce, NOAA Tech. Memo. NOAA-TM-NMFS-SWFC-231, 207.

Sclater, J., Parsons, B., Jaupart, C., 1981. Oceans and continents: similarities and differences in the mechanisms of heat loss. J. Geophys. Res. 86, 11535–11552.

SeaWiFS Project, 2009. SeaWiFS captures El Nino-La Nina transitions in the equatorial Pacific. NASA Goddard Space Flight Center. <http://oceancolor.gsfc.nasa.gov/SeaWiFS/BACKGROUND/Gallery/pac_elnino.jpg> (accessed 3.26.09).

Seibold, E., Berger, W.H., 1982. The Sea Floor. Springer Verlag, Berlin, 356 pp.

Sekine, Y., 1999. Anomalous southward intrusions of the Oyashio east of Japan 2. Two-layer numerical model. J. Geophys. Res. 104, 3049–3058.

Serreze, M.C., Holland, M.M., Stroeve, J., 2007. Perspectives on the Arctic's shrinking sea-ice cover. Science 16, 1533–1536.

Shaffer, G., Salinas, S., Pizarro, O., Vega, A., Hormazabal, S., 1995. Currents in the deep ocean off Chile (30°S). Deep-Sea Res. I 42, 425–436.

Shcherbina, A., Talley, L.D., Rudnick, D.L., 2003. Direct observations of brine rejection at the source of North Pacific Intermediate Water in the Okhotsk Sea. Science 302, 1952–1955.

Shcherbina, A., Talley, L.D., Rudnick, D.L., 2004. Dense water formation on the northwestern shelf of the Okhotsk Sea: 1. Direct observations of brine rejection. J Geophys. Res. 109, C09S08. doi:10.1029/2003JC002196.

Shell, K.M., Frouin, R., Nakamoto, S., Somerville, R.C.J., 2003. Atmospheric response to solar radiation absorbed by phytoplankton. J. Geophys. Res. 108 (D15), 4445. doi:10.1029/2003JD003440.

Shillington, F.A., 1998. The Benguela upwelling system off southwestern Africa. In: Robinson, A.R., Brink, K.H. (Eds.), The Sea, Vol. 11: The Global Coastal Ocean: Regional Studies and Syntheses. Harvard University Press, Boston, MA, pp. 583–604.

Shimada, K., Kamoshida, T., Itoh, M., Nishino, S., Carmack, E.C., McLaughlin, F., et al., 2006. Pacific Ocean Inflow: influence on catastrophic reduction of sea ice cover in the Arctic Ocean. Geophys. Res. Lett. 33, L08605. doi; 10.1029/2005GL025624.

Shinoda, T., Kiladis, G.N., Roundy, P.E., 2009. Statistical representation of equatorial waves and tropical instability waves in the Pacific Ocean. Atmos. Res. 94, 37–44.

Shoosmith, D.R., Richardson, P.L., Bower, A.S., Rossby, H.T., 2005. Discrete eddies in the northern North Atlantic as observed by looping RAFOS floats. Deep-Sea Res. II 52, 627–650.

Shuckburgh, E., Jones, H., Marshall, J., Hill, C., 2009. Understanding the regional variability of eddy diffusivity in the Pacific sector of the Southern Ocean. J. Phys. Oceanogr. 39, 2011–2023.

Siedler, G., Church, J., Gould, J., 2001. Ocean Circulation and Climate: Observing and Modelling the Global Ocean. AP International Geophysics Series Vol. 77, 715 pp.

Siedler, G., Kuhl, A., Zenk, W., 1987. The Madeira Mode Water. J. Phys. Oceanogr. 17, 1561–1570.

Siedler, G., Zanbenberg, N., Onken, R., Morlière, A., 1992. Seasonal changes in the tropical Atlantic circulation: Observation and simulation of the Guinea Dome. J. Geophys. Res. 97, 703–715.

Sievers, H., Emery, W., 1978. Variability of the Antarctic Polar Frontal Zone in the Drake Passage — Summer 1976–1977. J. Geophys. Res. 83, 3010–3022.

Simpson, J.H., 1998. Tidal processes in shelf seas. In: Brink, K.H., Robinson, A.R. (Eds.), The Sea, Vol. 10: The Global Coastal Ocean: Processes and Methods. Harvard University Press, Boston, MA, pp. 113−150.

Simpson, J.J., Koblinsky, C.J., Peláez, J., Haury, L.R., Wiesenhahn, D., 1986. Temperature — plant pigment — optical relations in a recurrent offshore mesoscale eddy near Point Conception, California. J. Geophys. Res. 91, 12919−12936.

SIO, 2008. SRTM30_plus, Satellite Geodesy, Scripps Institution of Oceanography. University of California San Diego. <http://topex.ucsd.edu/WWW_html/srtm30_plus.html> (accessed 9.24.08).

Sloyan, B.M., Rintoul, S.R., 2001. The Southern Ocean limb of the global deep overturning circulation. J. Phys. Oceanogr. 31, 143−173.

Smethie, W.M., Fine, R.A., 2001. Rates of North Atlantic Deep Water formation calculated from chlorofluorocarbon inventories. Deep-Sea Res. I 48, 189−215.

Smith, J.A., 2001. Observations and theories of Langmuir circulation: a story of mixing. In: Lumley, J.L. (Ed.), Fluid Mechanics and the Environment: Dynamical Approaches. Springer, New York, pp. 295−314.

Smith, R.D., Maltrud, M.E., Bryan, F.O., Hecht, M.W., 2000. Numerical simulation of the North Atlantic Ocean at 1/10°. J. Phys. Oceanogr. 30, 1532−1561.

Smith, R.L., Huyer, A., Godfrey, J.S., Church, J.A., 1991. The Leeuwin Current off Western Australia, 1986−1987. J. Phys. Oceanogr. 21, 323−345.

Smith, S.D., 1988. Coefficients for sea surface wind stress, heat flux, and wind profiles as a function of wind speed and temperature. J. Geophys. Res. 93, 15467−15472.

Smith, S.D., Muench, R.D., Pease, C.H., 1990. Polynyas and leads: An overview of physical processes and environment. J. Geophys. Res. 95, 9461−9479.

Smith, W.H.F., Sandwell, D.T., 1997. Global seafloor topography from satellite altimetry and ship depth soundings. Science 277, 1957−1962.

Smith, W.H.F., Scharroo, R., Titov, V.V., Arcas, D., Arbic, B.K., 2005. Satellite altimeters measure tsunami. Oceanography 18, 10−12.

Sofianos, S.S., Johns, W.E., 2003. An Oceanic General Circulation Model (OGCM) investigation of the Red Sea circulation: 2. Three-dimensional circulation in the Red Sea. J. Geoph. Res. 108, 3066. doi: 10,1029/200IJC001185.

Sokolov, S., Rintoul, S.R., 2002. Structure of Southern Ocean fronts at 140°E. J. Marine Syst. 37, 151−184.

Song, Q., Vecchi, G.A., Rosati, A.J., 2007. The role of the Indonesian Throughflow in the Indo-Pacific climate variability in the GFDL coupled climate model. J. Clim. 20, 2434−2451.

Sosik, H., 2003. Patterns and scales of variability in the optical properties of Georges Bank waters, with special reference to phytoplankton biomass and production. H. Sosik, Woods Hole Oceanographic Institution. <http://www.whoi.edu/science/B/sosiklab/gbgom.htm> (accessed 3.29.08).

Spadone, A., Provost, C., 2009. Variations in the Malvinas Current volume transport since October 1992. J. Geophys. Res. 114, C02002. doi:10.1029/2008JC004882.

Spall, M.A., Richardson, P.L., Price, J., 1993. Advection and eddy mixing in the Mediterranean salt tongue. J. Marine Res. 51, 797−818.

Speer, K., Rintoul, S.R., Sloyan, B., 2000. The diabatic Deacon cell. J. Phys. Oceanogr. 30, 3212−3222.

Speich, S., Blanke, B., de Vries, P., Drijfhout, S., Doos, K., Ganachaud, A., et al., 2002. Tasman leakage: A new route in the global ocean conveyor belt. Geophys. Res. Lett. 29, 1416. doi:10.1029/2001GL014586.

Spiess, F., 1928. Die Meteor Fahrt: Forschungen und Erlebnisse der Deutschen Atlantischen Expedition, 1925−1927. Verlag von Dietrich Reimer, Berlin, 376 pp. (in German, English translation Emery, W.J., Amerind Publishing Co. Pvt. Ltd., New Delhi, 1985).

Stabeno, P.J., Reed, R.K., 1995. Circulation in the Bering Sea basin observed by satellite-tracked drifters: 1986−1993. J. Phys. Oceanogr. 24, 848−854.

Stammer, D., 1997. Global characteristics of ocean variability estimated from regional TOPEX/POSEIDON altimeter measurements. J. Phys. Oceanogr. 27, 1743−1769.

Stammer, D., 1998. On eddy characteristics, eddy transports, and mean flow properties. J. Phys. Oceanogr. 28, 727−739.

Stammer, D., Wunsch, C., 1999. Temporal changes in eddy energy of the oceans. Deep-Sea Res. 46, 77−108.

Stammer, D., Wunsch, C., Ponte, R.M., 2000. De-aliasing of global high frequency barotropic motions in altimeter observations. Geophys. Res. Lett. 27, 1175−1178.

Steele, M., Morison, J., Ermold, W., Rigor, I., Ortmeyer, M., Shimada, K., 2004. Circulation of summer Pacific halocline water in the Arctic Ocean. J. Geophys. Res. 109, C02027. doi:10.1029/2003JC002009.

Steger, J.M., Carton, J.A., 1991. Long waves and eddies in the tropical Atlantic Ocean: 1984−1990. J. Geophys. Res. 96, 15161−15171.

Stewart, R.H., 2008. Introduction to Physical Oceanography. Open-source textbook. <http://oceanworld.tamu.edu/ocean410/ocng410_text_book.html> (accessed 3.28.09).

Stocker, T.F., Marchal, O., 2000. Abrupt climate change in the computer: Is it real? Proc. Natl. Acad. Sci. USA 97l, 1362−1365.

Stommel, H., 1948. The westward intensification of wind-driven currents. Trans. Am. Geophys. Union 29, 202−206.

Stommel, H.M., 1958. The abyssal circulation. Deep-Sea Res. 5, 80−82.

Stommel, H.M., 1961. Thermohaline convection with two stable regimes of flow. Tellus 13, 224−230.

Stommel, H.M., 1965. The Gulf Stream: A Physical and Dynamical Description, second ed. University of California Press, Berkeley, and Cambridge University Press, London, 248 pp.

Stommel, H.M., 1979. Determination of water mass properties of water pumped down from the Ekman layer to the geostrophic flow below. Proc. Nat. Acad. Sci. USA 76, 3051–3055.

Stommel, H.M., Arons, A., 1960a. On the abyssal circulation of the World Ocean — I. Stationary planetary flow patterns on a sphere. Deep-Sea Res. 6, 140–154.

Stommel, H.M., Arons, A., 1960b. On the abyssal circulation of the World Ocean — II. An idealized model of the circulation pattern and amplitude in oceanic basins. Deep-Sea Res. 6, 217–233.

Stommel, H.M., Arons, A., Faller, A., 1958. Some examples of stationary planetary flow patterns in bounded basins. Tellus 10, 179–187.

Stommel, H.M., Niiler, P.P., Anati, D., 1978. Dynamic topography and recirculation of the North Atlantic. J. Marine Res. 36, 449–468.

Stone, B., 1914. Map of track of the 'Endurance' in Weddell Sea. Royal Geographical Society. <http://images.rgs.org/imageDetails.aspx?barcode=27820> (accessed 10.15.06).

Stramma, L., Cornillon, P., Woller, R.A., Price, J.F., Briscoe, M.G., 1986. Large diurnal sea surface temperature variability: satellite and in situ measurements. J. Phys. Oceanogr. 16, 827–837.

Stramma, L., Ikeda, Y., Peterson, R.G., 1990. Geostrophic transport in the Brazil Current region north of 20°S. Deep-Sea Res. 37, 1875–1886.

Stramma, L., Johnson, G.C., Sprintall, J., Mohrholz, V., 2008. Expanding oxygen minimum zones in the tropical oceans. Science 320, 655–658.

Stramma, L., Kieke, D., Rhein, M., Schott, F., Yashayaev, I., Koltermann, K.P., 2004. Deep water changes at the western boundary of the subpolar North Atlantic during 1996 to 2001. Deep-Sea Res. I 51, 1033–1056.

Stramma, L., Lutjeharms, J.R.E., 1997. The flow field of the subtropical gyre of the South Indian Ocean. J. Geophys. Res. 102, 5513–5530.

Stramma, L., Peterson, R.G., 1990. The South Atlantic Current. J. Phys. Oceanogr. 20, 846–859.

Stramma, L., Peterson, R.G., Tomczak, M., 1995. The South Pacific Current. J. Phys. Oceanogr. 25, 77–91.

Stramma, L., Schott, F., 1999. The mean flow field of the tropical Atlantic Ocean. Deep-Sea Res. II 46, 279–303.

Stramski, D., Reynolds, R.A., Babin, M., Kaczmarek, S., Lewis, M.R., Röttgers, R., et al., 2008. Relationships between the surface concentration of particulate organic carbon and optical properties in the eastern South Pacific and eastern Atlantic Oceans. Biogeosciences 5, 171–201.

Straneo, F., Saucier, F., 2008. The outflow from Hudson Strait and its contribution to the Labrador Current. Deep-Sea Res. I 55, 926–946.

Strub, P.T., James, C., 2000. Altimeter-derived variability of surface velocities in the California Current System: 2. Seasonal circulation and eddy statistics. Deep-Sea Res. II 47, 831–870.

Strub, P.T., James, C., 2009. Altimeter-derived circulation in the California Current. College of Oceanic and Atmospheric Sciences. Oregon State University. <http://www.coas.oregonstate.edu/research/po/research/strub/index.html> (accessed 4.2.09).

Strub, P.T., Kosro, P.M., Huyer, A., Brink, K.H., Hayward, T.L., Niiler, P.P., et al., 1991. The nature of cold filaments in the California current system. J. Geophys. Res. 96, 14743–14769.

Strub, P.T., Mesias, J.M., Montecino, V., Ruttlant, J., Salinas, S., 1998. Coastal ocean circulation off western South America. In: Robinson, A.R., Brink, K.H. (Eds.), The Sea, Vol. 11: The Global Coastal Ocean — Regional Studies and Syntheses. Wiley, New York, pp. 273–313.

Sundby, S., Drinkwater, K., 2007. On the mechanisms behind salinity anomaly signals of the northern North Atlantic. Progr. Oceanogr. 73, 190–202.

Sutton, R.T., Jewson, S.P., Rowell, D.P., 2000. The elements of climate variability in the tropical Atlantic region. J. Clim. 13, 3261–3284.

Sverdrup, H.U., 1947. Wind-driven currents in a baroclinic ocean. Proc. Nat. Acad. Sci. USA 33, 318–326.

Sverdrup, H.U., Johnson, M.W., Fleming, R.H., 1942. The Oceans: Their Physics, Chemistry and General Biology. Prentice Hall Inc., Englewood Cliffs, NJ, 1057 pp.

Swallow, J.C., Bruce, J.C., 1966. Current measurements off the Somali coast during the southwest monsoon of 1964. Deep-Sea Res. 13, 861–888.

Swallow, J.C., Worthington, L.V., 1961. An observation of a deep countercurrent in the western North Atlantic. Deep-Sea Res. 8, 1–19.

Swift, J.H., 1986. The Arctic Waters. In: Hurdle, B. (Ed.), The Nordic Seas. Springer-Verlag, New York, pp. 129–154.

Swift, J.H., Aagaard, K., 1981. Seasonal transitions and water mass formation in the Iceland and Greenland seas. Deep-Sea Res. 28, 1107–1129.

Swift, J.H., Aagaard, K., Timokhov, L., Nikiforov., E.G., 2005. Long-term variability of Arctic Ocean Waters: Evidence from a reanalysis of the EWG data set. J. Geophys. Res. 110, C03012. doi:10.1029/2004JC002312.

Swift, J.H., Jones, E.P., Aagaard, K., Carmack, E.C., Hingston, M., Macdonald, R.W., et al., 1997. Waters of the Makarov and Canada basins. Deep-Sea Res. II 44, 1503–1529.

Swift, S.A., Bower, A.S., 2003. Formation and circulation of dense water in the Persian/Arabian Gulf. J. Geophys. Res. 108 (C10) doi:10.1029/2002JC001360.

TAO Project Office, 2009a. TAO/TRITON data display and delivery. NOAA Pacific Marine Environmental Laboratory. <http://www.pmel.noaa.gov/tao/disdel/disdel.html> (accessed 3.27.09).

TAO Project Office, 2009b. TAO Climatologies. NOAA Pacific Marine Environmental Laboratory. <http://www.pmel.noaa.gov/tao/clim/clim.html> (accessed 7.5.09).

Tabata, S., 1982. The anti-cyclonic, baroclinic eddy off Sitka Alaska, in the northeast Pacific Ocean. J. Phys. Oceanogr. 12, 1260−1282.

Talley, L.D., 1991. An Okhotsk Sea anomaly: Implication for ventilation in the North Pacific. Deep-Sea Res. 38 (Suppl.), S171−S190.

Talley, L.D., 1993. Distribution and formation of North Pacific Intermediate Water. J. Phys. Oceanogr. 23, 517−537.

Talley, L.D., 1996a. Antarctic Intermediate Water in the South Atlantic. In: Wefer, G., Berger, W.H., Siedler, G., Webb, D. (Eds.), The South Atlantic: Present and Past Circulation. Springer-Verlag, New York, pp. 219−238.

Talley, L.D., 1996b. North Atlantic circulation and variability, reviewed for the CNLS conference. Physica D 98, 625−646.

Talley, L.D., 1999. Some aspects of ocean heat transport by the shallow, intermediate and deep overturning circulations. In: Clark, P.U., Webb, R.S., Keigwin, L.D. (Eds.), Mechanisms of Global Climate Change at Millennial Time Scales, Geophys. Mono. Ser. 112, American Geophysical Union, pp. 1−22.

Talley, L.D., 2003. Shallow, intermediate, and deep overturning components of the global heat budget. J. Phys. Oceanogr. 33, 530−560.

Talley, L.D., 2007. Hydrographic Atlas of the World Ocean Circulation Experiment (WOCE). Volume 2: Pacific Ocean. In: Sparrow, M., Chapman, P., Gould, J. (Eds.). International WOCE Project Office, Southampton, UK ISBN 0-904175-54-5.

Talley, L.D., 2008. Freshwater transport estimates and the global overturning circulation: Shallow, deep and throughflow components. Progr. Oceanogr. 78, 257−303. doi:10.1016/j.pocean.2008.05.001.

Talley, L.D., 2011a. Hydrographic Atlas of the World Ocean Circulation Experiment (WOCE). vol 3: Indian Ocean. In: Sparrow, M., Chapman, P., Gould, J. (Eds.). International WOCE Project Office, Southampton, U.K. <http://www-pord.ucsd.edu/whp_atlas/indian_index.htm> Online version (accessed 4.20.09).

Talley, L.D., 2011b. Schematics of the global overturning circulation. In preparation.

Talley, L.D., Johnson, G.C., 1994. Deep, zonal subequatorial jets. Science 263, 1125−1128.

Talley, L.D., Joyce, T.M., 1992. The double silica maximum in the North Pacific. J. Geophys. Res. 97, 5465−5480.

Talley, L.D., McCartney, M.S., 1982. Distribution and circulation of Labrador Sea Water. J. Phys. Oceanogr. 12, 1189−1205.

Talley, L.D., Min, D.-H., Lobanov, V.B., Luchin, V.A., Ponomarev, V.I., Salyuk, A.N., et al., 2006. Japan/East Sea water masses and their relation to the sea's circulation. Oceanography 19, 33−49.

Talley, L.D., Nagata, Y., 1995. The Okhotsk Sea and Oyashio Region. PICES Scientific Report, 2. North Pacific Marine Science Organization (PICES), Sidney, B.C., Canada, 227 pp.

Talley, L.D., Sprintall, J., 2005. Deep expression of the Indonesian Throughflow: Indonesian Intermediate Water in the South Equatorial Current. J. Geophys. Res. 110, C10009. doi:10.1029/2004JC002826.

Talley, L.D., Tishchenko, P., Luchin, V., Nedashkovskiy, A., Sagalaev, S., Kang, D.-J., et al., 2004. Atlas of Japan (East) Sea hydrographic properties in summer, 1999. Progr. Oceanogr. 61, 277−348.

Talley, L.D., Yun, J.-Y., 2001. The role of cabbeling and double diffusion in setting the density of the North Pacific Intermediate Water salinity minimum. J. Phys. Oceanogr. 31, 1538−1549.

Tamura, T., Ohshima, K.I., Nihashi, S., 2008. Mapping of sea ice production for Antarctic coastal polynyas. Geophys. Res. Lett. 35, L07606. do:10.1029/2007GL032903.

Tanhua, T., Olsson, K.A., Jeansson, E., 2005. Formation of Demark Strait overflow water and its hydro-chemical composition. J. Marine Sys. 57, 264−288.

Taylor, P.K. (Ed.), 2000. Intercomparison and validation of ocean-atmosphere energy flux fields — Final report of the Joint WCRP/SCOR Working Group on Air-Sea Fluxes. WCRP-112, WMO-TD-1036, 306 pp.

Teague, W.J., Ko, D.S., Jacobs, G.A., Perkins, H.T., Book, J.W., Smith, S.R., et al., 2006. Currents through the Korea/Tsushima Strait. Oceanography 19, 50−63.

Thompson, D.W.J., Solomon, S., 2002. Interpretation of recent southern hemisphere climate change. Science 296, 895−899.

Thompson, D.W.J., Wallace, J.M., 1998. The Arctic- Oscillation signature in the wintertime geopotential height and temperature fields. Geophys. Res. Lett. 25, 1297−1300.

Thompson, D.W.J., Wallace, J.M., 2000. Annular modes in the extratropical circulation. Part I: Month-to-month variability. J. Clim. 13, 1000−1016.

Thompson, S.L., Warren, S.G., 1982. Parameterization of outgoing infrared radiation derived from detailed radiative calculations. J. Atmos. Sci. 39, 2667−2680.

Thomson, J., Elgar, S., Herbers, T.H.C., 2005. Reflection and tunneling of ocean waves observed at a submarine canyon. Geophys. Res. Lett. 32, L10602. doi:10.1029/2005GL022834.

Thorpe, S.A., 2004. Langmuir circulation. Annu. Rev. Fluid Mech. 36, 55−79. doi:10.1146/annurev.fluid.36.052203. 071431.

Thoulet, J., Chevallier, A., 1889. Sur la chaleur spécifique de l'eau de mer a divers degres de dilution et de concentration. C.R. Acad. Sci. 108, 794−796 (in French).

Thurman, H.V., Trujillo, A.P., 2002. Essentials of Oceanography, 7th ed. Prentice Hall, NJ, 524 pp.

Timmermans, M.L., Garrett, C., 2006. Evolution of the deep water in the Canadian Basin in the Arctic Ocean. J. Phys. Oceanogr. 36, 866−874.

Timmermans, M.L., Garrett, C., Carmack, E., 2003. The thermohaline structure and evolution of the deep waters in the Canada Basin, Arctic Ocean. Deep-Sea Res. I 50, 1305−1321.

Titov, V., Rabinovich, A.B., Mofjeld, H.O., Thomson, R.E., Gonzalez, F.I., 2005. The global reach of the 26 December 2004 Sumatra tsunami. Science 309, 2045−2048.

Tomczak, M., 1981. A multiparameter extension of temperature/salinity diagram techniques for the analysis of non-isopycnal mixing. Progr. Oceanogr. 10, 147−171.

Tomczak, M., 2000, 2002. Shelf and Coastal Oceanography. Open-source textbook. <http://www.es.flinders.edu.au/~mattom/ShelfCoast/index.html> (accessed 3.28.09).

Tomczak, M., Godfrey, J.S., 1994. Regional Oceanography: An Introduction. Pergamon Press, Oxford, UK, 422 pp.

Tomczak, M., Godfrey, J.S., 2003. Regional Oceanography: An Introduction, second ed. Daya Publications, Delhi, 390 pp., ISBN: 8170353068. (Online, open source version at. <http://www.es.flinders.edu.au/~mattom/regoc/pdfversion.html>.

Tomczak, M., Large, D., 1989. Optimum multiparameter analysis of mixing in the thermocline of the eastern Indian Ocean. J. Geophys. Res. 94, 16141−16149.

Toole, J.M., Millard, R.C., Wang, Z., Pu, S., 1990. Observations of the Pacific North Equatorial Current bifurcation at the Philippine coast. J. Phys. Oceanogr. 20, 307−318.

Toole, J.M., Warren, B.A., 1993. A hydrographic section across the subtropical South Indian Ocean. Deep-Sea Res. I 40, 1973−2019.

Tourre, Y.M., White, W.B., 1995. ENSO Signals in global upper-ocean temperature. J. Phys. Oceanogr. 25, 1317−1332.

Tourre, Y.M., White, W.B., 1997. Evolution of the ENSO signal over the Indo-Pacific domain. J. Phys. Oceanogr. 27, 683−696.

Treguier, A.M., 2006. Ocean models. In: Chassignet, E.P., Verron, J. (Eds.), Ocean Weather Forecasting: An Integrated view of Oceanography. Springer, The Netherlands.

Trenberth, K.E., Caron, J.M., 2001. Estimates of meridional atmosphere and ocean heat transports. J. Clim. 14, 3433−3443.

Trenberth, K.E., Hurrell, J.W., 1994. Decadal atmosphere-ocean variations in the Pacific. Clim. Dyn. 9, 303−319.

Trenberth, K.E., Jones, P.D., Ambenje, P., Bojariu, R., Easterling, D., Klein Tank, A., et al., 2007. Observations: Surface and Atmospheric Climate Change. In: Solomon, S., Qin, D., Manning, M., Chen, Z., Marquis, M., Averyt, K.B., et al. (Eds.), Climate Change 2007: The Physical Science Basis. Contribution of Working Group I to the Fourth Assessment Report of the Intergovernmental Panel Climate Change. Cambridge University Press, Cambridge, UK and New York.

Tsuchiya, M., 1975. Subsurface countercurrents in the eastern equatorial Pacific Ocean. J. Mar. Res. 33 (Suppl), 145−175.

Tsuchiya, M., 1981. The origin of the Pacific equatorial 13°C water. J. Phys. Oceanogr. 11, 794−812.

Tsuchiya, M., 1989. Circulation of the Antarctic Intermediate Water in the North Atlantic Ocean. J. Mar. Res. 47, 747−755.

Tsuchiya, M., Talley, L.D., McCartney, M.S., 1992. An eastern Atlantic section from Iceland southward across the equator. Deep-Sea Res. 39, 1885−1917.

Tsuchiya, M., Talley, L.D., McCartney, M.S., 1994. Water mass distributions in the western Atlantic: A section from South Georgia Island (54S) northward across the equator. J. Mar. Res. 52, 55−81.

Tully, J.P., 1949. Oceanography and prediction of pulp-mill pollution in Alberni Inlet. Fish. Res. Bd. Can., Bulletin 83, 169 pp.

UCT Oceanography Department, 2009. Monthly sea surface temperature (SST) composites. Marine remote sensing unit at the Department of Oceanography. University of Cape Town. <http://www.sea.uct.ac.za/projects/remsense/index.php> (accessed 6.9.09).

UNESCO, 1981. The Practical Salinity Scale 1978 and the International Equation of State of Seawater 1980. Tech. Paper Mar., Sci. 36, 25 pp.

UNESCO, 1983. Algorithms for computation of fundamental properties of seawater. Tech. Paper Mar., Sci. 44, 53 pp.

UNESCO, 1987. International oceanographic tables. Tech. Paper Mar., Sci. 40, 196 pp.

Urick, R.J., 1983. Principles of Underwater Sound, 3rd ed. McGraw-Hill, New York, 423 pp.

Vallis, G.K., 2006. Atmospheric and Oceanic Fluid Dynamics: Fundamentals and Large-scale Circulation. Cambridge University Press, Cambridge, UK, 745 pp.

van Aken, H.M., 2000. The hydrography of the mid-latitude northeast Atlantic Ocean I: The deep water masses. Deep-Sea Res. I 47, 757−788.

van Aken, H.M., van Veldhoven, A.K., Veth, C., de Ruijter, W.P.M., van Leeuwen, P.J., Drijfhout, S.S., et al., 2003. Observations of a young Agulhas ring, Astrid, during MARE in March 2000. Deep-Sea Res. II 50, 167−195.

Van der Vaart, P.C.F., Dijkstra, H.A., Jin, F.F., 2000. The Pacific cold tongue and the ENSO mode: A unified theory within the Zebiak-Cane model. J. Atmos. Sci. 57, 967–988.

Van Dorn, W.G., 1993. Oceanography and Seamanship, second ed. Dodd, Mead Publishers, New York, 440 pp.

VanScoy, K.A., Druffel, E.R.M., 1993. Ventilation and transport of thermocline and intermediate waters in the northeast Pacific during recent El Ninos. J. Geophys. Res. 98, 18083–18088.

Vellinga, M., Wood, R.A., 2002. Global climatic impacts of a collapse of the Atlantic thermohaline circulation. Climatic Change 43, 251–267.

Venegas, R.M., Strub, P.T., Beier, E., Letelier, R., Thomas, A.C., Cowles, T., et al., 2008. Satellite-derived variability in chlorophyll, wind stress, sea surface height, and temperature in the northern California Current System. J. Geophys. Res. 113, C03015. doi:10.1029/2007JC004481.

Veronis, G., 1966. Wind-driven ocean circulation–part II. Numerical solution of the nonlinear problem. Deep-Sea Res. 13, 30–55.

Vialard, J., Menkes, C., Anderson, D.L.T., Balmaseda, M.A., 2003. Sensitivity of Pacific Ocean tropical instability waves to initial conditions. J. Phys. Oceanogr. 33, 105–121.

Visbeck, M., 2002. The ocean's role in climate variability. Science 297, 2223–2224.

Visbeck, M., Chassignet, E.P., Curry, R.G., Delworth, T.L., Dickson, R.R., Krahmann, G., 2003. The ocean's response to North Atlantic Oscillation variability. In: The North Atlantic Oscillation: Climate significance and environmental impact. Geophys. Monogr. Ser. 134, 113–146.

Vivier, F., Provost, C., 1999. Direct velocity measurements in the Malvinas Current. J. Geophys. Res. 104, 21083–21103.

Von Storch, H., Zwiers, F.W., 1999. Statistical Analysis in Climate Research. Cambridge University Press, Cambridge, UK, 496 pp.

Wacongne, S., 1990. On the difference in strength between Atlantic and Pacific undercurrents. J. Phys. Oceanogr. 20, 792–800.

Wacogne, S., Piton, B., 1992. The near-surface circulation in the northeastern corner of the South Atlantic Ocean. Deep-Sea Res. A 39, 1273–1298.

Wadhams, P., Budéus, G., Wilkinson, J.P., Løyning, T., Pavlov, V., 2004. The multi-year development of long-lived convective chimneys in the Greenland Sea. Geophys. Res. Lett. 31, L06306. doi:10.1029/2003GL019017.

Wadhams, P., Comiso, J., Prussen, E., Wells, S.T., Brandon, M., Aldworth, E., et al., 1996. The development of the Odden ice tongue in the Greenland Sea during winter 1993 from remote sensing and field observations. J. Geophys. Res. 101, 18213–18235.

Wadhams, P., Holfort, J., Hansen, E., Wilkinson, J.P., 2002. A deep convective chimney in the winter Greenland Sea. Geophys. Res. Lett. 29, 10. doi:10.1029/2001GL014306.

Walin, G., 1982. On the relation between sea-surface heat flow and thermal circulation in the ocean. Tellus 34, 187–195.

Wallace, J.M., 1992. Effect of deep convection on the regulation of tropical sea surface temperature. Nature 357, 230–231.

Wallace, W.J., 1974. The development of the chlorinity/salinity concept in oceanography. Elsevier Oceanography Series 7, 227 pp.

Wang, C., 2002. Atlantic climate variability and its associated atmospheric circulation cells. J. Clim. 15, 1516–1536.

Warren, B.A., 1981. Deep circulation of the world ocean. In: Warren, B.A., Wunsch, C. (Eds.), Evolution of Physical Oceanography. MIT Press, Cambridge, MA, pp. 6–41.

Warren, B.A., 1990. Suppression of deep oxygen concentrations by Drake Passage. Deep-Sea Res. 37, 1899–1907.

Warren, B.A., Johnson, G.C., 2002. The overflows across the Ninetyeast Ridge. Deep-Sea Res. II, 1423–1439.

WCRP (World Climate Research Programme), 1998. CLIVAR Initial Implementation Plan. WCRP-103, WMO/TD No. 869, ICPO No. 14, 367 pp.

Webb, D.J., 2000. Evidence for shallow zonal jets in the South Equatorial Current region of the southwest Pacific. J. Phys. Oceanogr. 30, 706–720.

Webster, P.J., Magana, V.O., Palmer, T.N., Shukla, J., Tomas, R.A., Yanai, M., et al., 1998. Monsoons: Processes, predictability, and the prospects for prediction. J. Geophys. Res. 103, 14451–14510.

Webster, P.J., Moore, A.M., Loschnigg, J.P., Leben, R.R., 1999. Coupled ocean-atmosphere dynamics in the Indian Ocean during 1997-98. Nature 401, 356–360.

Weiss, R.F., Bullister, J.L., Gammon, R.H., Warner, M.J., 1985. Atmospheric chlorofluoromethanes in the deep equatorial Atlantic. Nature 314, 608–610.

Weller, R., Dean, J.P., Marra, J., Price, J., Francis, E.A., Boardman, D.C., 1985. Three-dimensional flow in the upper ocean. Science 118, 1–22.

Whitworth, T., Orsi, A.H., Kim, S.-J., Nowlin, W.D., 1998. Water masses and mixing near the Antarctic slope front. In: Jacobs, S.S., Weiss, R.F. (Eds.), Ocean, Ice, and Atmosphere: Interactions at the Antarctic Continental Margins. Antarctic Research Series 75, American Geophysical Union, Washington, pp. 1–27.

Whitworth, T., Peterson, R., 1985. The volume transport of the Antarctic Circumpolar Current from bottom pressure measurements. J. Phys. Oceanogr. 15, 810–816.

Whitworth, T., Warren, B.A., Nowlin Jr., W.D., Rutz, S.B., Pillsbury, R.D., Moore, M.I., 1999. On the deep western-boundary current in the Southwest Pacific Basin. Progr. Oceanogr. 43, 1–54.

Wick, G.A., 1995. Evaluation of the variability and predictability of the bulk-skin sea surface temperature difference with application to satellite-measured sea surface temperature. Ph.D. Thesis, University of Colorado, Boulder, CO, 146 pp.

Wick, G.A., Emery, W.J., Kantha, L., Schluessel, P., 1996. The behavior of the bulk — skin temperature difference at varying wind speeds. J. Phys. Oceanogr. 26, 1969–1988.

Wijffels, S., Bray, N., Hautala, S., Meyers, G., Morawitz, W.M.L., 1996. The WOCE Indonesian Throughflow repeat hydrography sections: I10 and IR6. International WOCE Newsletter 24, 25–28.

Wijffels, S., Firing, E., Toole, J., 1995. The mean structure and variability of the Mindanao Current at 8°N. J. Geophys. Res. 100, 18421–18436.

Wijffels, S.E., 2001. Ocean transport of fresh water. In: Siedler, G., Church, J. (Eds.), Ocean Circulation and Climate. International Geophysics Series. Academic Press, pp. 475–488.

Wijffels, S.E., Schmitt, R.W., Bryden, H.L., Stigebrandt, A., 1992. Transport of fresh water by the oceans. J. Phys. Oceanogr. 22, 155–162.

Wijffels, S.E., Toole, J.M., Davis, R., 2001. Revisiting the South Pacific subtropical circulation: A synthesis of World Ocean Circulation Experiment observations along 32°S. J. Geophys. Res. 106, 19481–19513.

Wilks, D.S., 2005. Statistical Methods in the Atmospheric Sciences. In: International Geophysics Series, second ed., vol. 91. Academic Press, 648 pp.

Williams, G.D., Aoki, S., Jacobs, S.S., Rintoul, S.R., Tamura, T., Bindoff, N.L., 2010. Antarctic Bottom Water from the Adelie and George V Land coast, East Antarctica (140–149°E). J. Geophys. Res. 115, C04027. doi:10.1029/2009JC005812.

Williams, R.G., 1991. The role of the mixed layer in setting the potential vorticity of the main thermocline. J. Phys. Oceanogr. 21, 1803–1814.

Williams, W.J., Carmack, E.C., Ingram, R.G., 2007. Chapter 2 Physical oceanography of polynyas, in Polynyas: Windows to the World. Elsevier Oceanogr. Ser. 74, 55–85.

Willis, J., Roemmich, D.L., Cornuelle, B., 2004. Interannual variability in upper-ocean heat content, temperature and thermosteric expansion on global scales. J. Geophys. Res. 109, C12036. doi:10.1029/2003JC002260.

Wilson, C., 2002. Newton and celestial mechanics. In: Cohen, I.B., Smith, G.E. (Eds.), The Cambridge Companion to Newton. Cambridge University Press, Cambridge, UK.

Wilson, T.R.S., 1975. Salinity and the major elements in seawater. Ch. 6. In: Riley, J.P., Skirrow, G. (Eds.), Chemical Oceanography, Vol. 1 (second ed.). Academic Press, San Diego, CA, pp. 365–413.

Winsor, P., Rodhe, J., Omstedt, A., 2001. Baltic Sea ocean climate: An analysis of 100 years of hydrographic data with focus on freshwater budget. Clim. Res. 18, 5–15.

Winther, N.G., Johannessen, J.A., 2006. North Sea circulation: Atlantic inflow and its destination. J. Geophys. Res. 111, C12018. doi:10.1029/2005JC003310.

Witte, E., 1902. Zur Theorie der Stromkabbelungen. Gaea, 38, 484–487(in German).

Wolanski, E. (Ed.), 2001. Oceanographic Processes of Coral Reefs: Physical and Biological Links in the Great Barrier Reef. CRC Press, Boca Raton, Florida, 356 pp.

Wolfram, 2009. Wolfram Demonstrations Project and Wolfram MathWorld. Wolfram Research Inc. <http://demonstrations.wolfram.com/> and <http://mathworld.wolfram.com> (accessed 4.3.09).

Wolter, K., 2009. Multivariate ENSO Index (MEI). NOAA Earth System Research Laboratory. http://www.cdc.noaa.gov/people/klaus.wolter/MEI/ (accessed 3.26.09).

Wolter, K., Timlin, M.S., 1993. Monitoring ENSO in COADS with a seasonally adjusted principal component index. Proc. of the 17th Climate Diagnostics Workshop, Norman, OK, NOAA/NMC/CAC, NSSL, Oklahoma Climate Survey, CIMMS and the School of Meteor., University of Oklahoma, pp. 52–57.

Wong, A.P.S., Bindoff, N.L., Church, J.A., 2001. Freshwater and heat changes in the North and South Pacific Oceans between the 1960s and 1985–94. J. Clim. 14, 1613–1633.

Woodgate, R.A., Aagaard, K., 2005. Revising the Bering Strait freshwater flux into the Arctic Ocean. Geophys. Res. Lett. 32, L02602. doi:10.1029/2004GL021747.

Woods, J.D., 1985. The World Ocean Circulation Experiment. Nature 314, 501–511.

Wooster, W.S., Reid, J.L., 1963. Eastern boundary currents. In: Hill, M.N. (Ed.), The Sea, Vol. 2: Ideas and Observations. Wiley-Interscience, New York, pp. 253–280.

World Meteorological Organization, 2005a. Natural hazards. WMO. <http://www.wmo.int/pages/themes/hazards/index_en.html> (accessed 9.22.05).

World Meteorological Organization, 2005b. Our World: International Weather. WMO. <http://www.wmo.int/pages/about/wmo50/e/world/weather_pages/chronicle_e.html> (accessed 3.28.09).

Worthington, L.V., 1953. Oceanographic results of project Skijump 1 and Skijump 2 in the Polar Sea 1951–1952. Eos T. Am. Geophys. Union 34, 543.

Worthington, L.V., 1959. The 18° Water in the Sargasso Sea. Deep-Sea Res. 5, 297–305.

Worthington, L.V., 1976. On the North Atlantic circulation. Oceanographic Studies. The Johns Hopkins University, Baltimore, Maryland, 110 pp.

Worthington, L.V., 1981. The water masses of the world ocean: some results of a fine-scale census. In: Warren, B.A., Wunsch, C. (Eds.), Evolution of Physical Oceanography. MIT Press, Cambridge, MA.

Worthington, L.V., Wright, W.R., 1970. North Atlantic Ocean Atlas of potential temperature and salinity in the deep water. Woods Hole Oceanographic Institution Atlas Series, 2, 24 pp and 58 plates.

Wu, Y., Tang, C.L., Sathyendranath, S., Platt, T., 2007. The impact of bio-optical heating on the properties of the upper ocean: A sensitivity study using a 3-D circulation model for the Labrador Sea. Deep-Sea Res. II 54, 2630–2642.

Wunsch, C., 1996. The Ocean Circulation Inverse Problem. Cambridge University Press, New York, 458 pp.

Wunsch, C., 2009. The oceanic variability spectrum and transport trends. Atmosphere-Ocean 47, 281–291.

Wunsch, C., Ferrari, R., 2004. Vertical mixing, energy, and the general circulation of the oceans. Annu. Rev. Fluid Mech. 36, 281–314.

Wüst, G., 1935. Schichtung und Zirkulation des Atlantischen Ozeans. Die Stratosphäre. In Wissenschaftliche Ergebnisse der Deutschen Atlantischen Expedition auf dem Forschungs- und Vermessungsschiff "Meteor" 1925–1927,6 1st Part, 2, 109–288 (in German).

Wüst, G., 1957. Wissenschaftliche Ergebnisse der deutschen atlantischen Expedition "Meteor", vol. 6. Walter de Gruyter, Berlin, part 2, pp. 1–208 (in German).

Wüst, G., 1961. On the vertical circulation of the Mediterranean Sea. J. Geophys. Res. 66, 3261–3271.

Wyrtki, K., 1971. Oceanographic Atlas of the International Indian Ocean Expedition. National Science Foundation Publication. OCE/NSF 86-00-001, Washington, D.C, 531 pp.

Wyrtki, K., 1973. An equatorial jet in the Indian Ocean. Science 181, 262–264.

Wyrtki, K., 1975. Fluctuations of the dynamic topography in the Pacific Ocean. J. Phys. Oceanogr. 5, 450–459.

Wyrtki, K., Kilonsky, B., 1984. Mean water and current structure during the Hawaii-to-Tahiti shuttle experiment. J. Phys. Oceanogr. 14, 242–254.

Xue, H., Chai, F., Pettigrew, N., Xu, D., Shi, M., Xu, J., 2004. Kuroshio intrusion and the circulation in the South China Sea. J. Geophys. Res. 109, C02017. doi:10.1029/2002JC001724.

Yanigomoto, D., Kawabe, M., Fujio, S., 2010. Direct velocity measurements of deep circulation southwest of the Shatsky Rise in the western North Pacific. Deep-Sea Res. I 57, 328–337.

Yashayaev, I., 2007. Hydrographic changes in the Labrador Sea, 1960-2005. Progr. Oceanogr. 73, 242–276.

Yasuda, I., Hiroe, Y., Komatsu, K., Kawasaki, K., Joyce, T.M., Bahr, F., et al., 2001. Hydrographic structure and transport of the Oyashio south of Hokkaido and the formation of North Pacific Intermediate Water. J. Geophys. Res. 106, 6931–6942.

Yates, M.L., Guza, R.T., O'Reilly, W.C., Seymour, R.J., 2009. Overview of seasonal sand level changes on southern California beaches. Shore Beach 77 (1), 39–46.

Yoshikawa, Y., Church, J.A., Uchida, H., White, N.J., 2004. Near bottom currents and their relation to the transport in the Kuroshio Extension. Geophys. Res. Lett. 31, L16309. doi:10.1029/2004GL020068.

Yu, Z., McCreary, J.P., Kessler, W.S., Kelly, K.A., 2000. Influence of equatorial dynamics on the Pacific North Equatorial Countercurrent. J. Phys. Oceanogr. 30, 3179–3190.

Yuan, X., 2004. ENSO-related impacts on Antarctic sea ice: a synthesis of phenomenon and mechanisms. Antarct. Sci. 16, 415–425. doi:10/1017/S0954102004002238.

Yun, J.-Y., Talley, L.D., 2003. Cabbeling and the density of the North Pacific Intermediate Water quantified by an inverse method. J. Geophys. Res. 108, 3118. doi:10.1029/2002JC001482.

Zamudio, L., Hurlburt, H.E., Metzger, E.J., Morey, S.L., O'Brien, J.J., Tilburg, C.E., et al., 2006. Interannual variability of Tehuantepec eddies. J. Geophys. Res. 111, C05001. doi:10.1029/2005JC003182.

Zaucker, F., Broecker, W.S., 1992. The influence of atmospheric moisture transport on the fresh water balance of the Atlantic drainage basin: General circulation model simulations and observations. J. Geophys. Res. 97, 2765–2773.

Zemba, J.C., 1991. The structure and transport of the Brazil Current between 27° and 36° South. Ph.D. Thesis, Massachusetts Institute of Technology and Woods Hole Oceanographic Institution, 160 pp.

Zenk, W., 1975. On the Mediterranean outflow west of Gibraltar. "Meteor" Forschungsergebnisse A16, 23–34.

Zhang, R., Vallis, G.K., 2006. Impact of Great Salinity Anomalies on the low-frequency variability of the North Atlantic Climate. J. Clim. 19, 470–482.

Zhang, Y., Hunke, E., 2001. Recent Arctic change simulated with a coupled ice-ocean model. J. Geophys. Res. 106, 4369–4390.

Zhong, A., Hendon, H.H., Alves, O., 2005. Indian Ocean variability and its association with ENSO in a global coupled model. J. Clim. 18, 3634–3649.

Zhurbas, V., Oh, I.S., 2004. Drifter-derived maps of lateral diffusivity in the Pacific and Atlantic Oceans in relation to surface circulation patterns. J. Geophys. Res. 109, C05015. doi:10.1029/2003JC002241.

Index

Color Plates

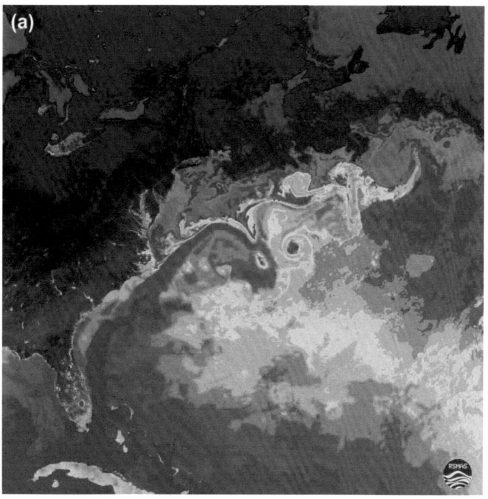

FIGURE 1.1 (a) Sea surface temperature from a satellite advanced very high resolution radiometer (AVHRR) instrument (Otis Brown, personal communication, 2009).

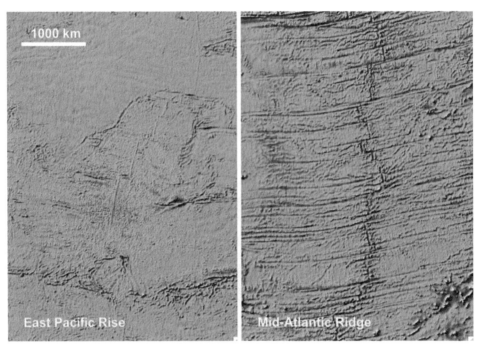

FIGURE 2.4 Seafloor topography for a portion of (a) the fast-spreading EPR and (b) the slow-spreading MAR. Note the ridge at the EPR spreading center and rift valley at the MAR spreading center. (Sandwell, personal communication, 2009.)

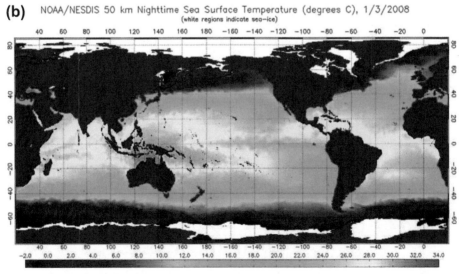

FIGURE 4.1 (b) Satellite infrared sea surface temperature (°C; nighttime only), averaged to 50 km and 1 week, for January 3, 2008. White is sea ice. (See Figure S4.1 from the online supplementary material for an austral winter image from July 3, 2008). *Source: From NOAA NESDIS (2009b).*

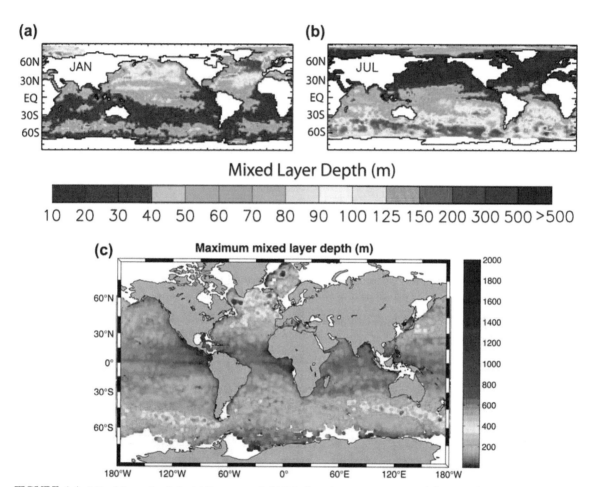

FIGURE 4.4 Mixed layer depth in (a) January and (b) July, based on a temperature difference of 0.2°C from the near-surface temperature. *Source: From deBoyer Montégut et al. (2004).* (c) Averaged maximum mixed layer depth, using the 5 deepest mixed layers in 1° × 1° bins from the Argo profiling float data set (2000–2009) and fitting the mixed layer structure as in Holte and Talley (2009).

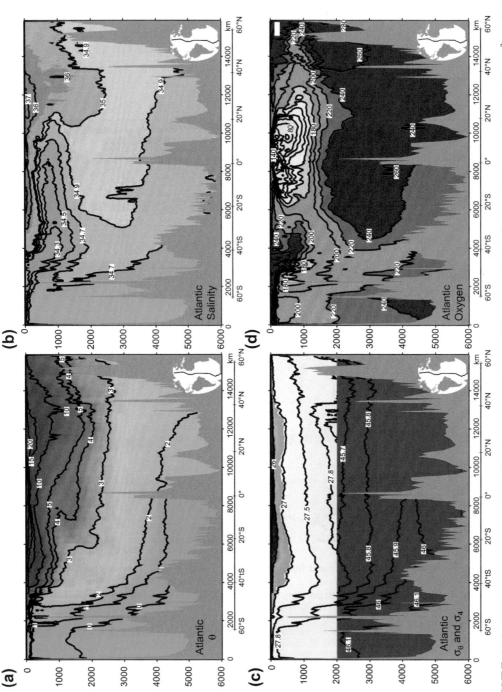

FIGURE 4.11 (a) Potential temperature (°C), (b) salinity (psu), (c) potential density σ_θ (top) and potential density σ_4 (bottom) (kg m^{-3}), and (d) oxygen (μmol/kg) in the Atlantic Ocean at longitude 20° to 25°W. Data from the World Ocean Circulation Experiment.

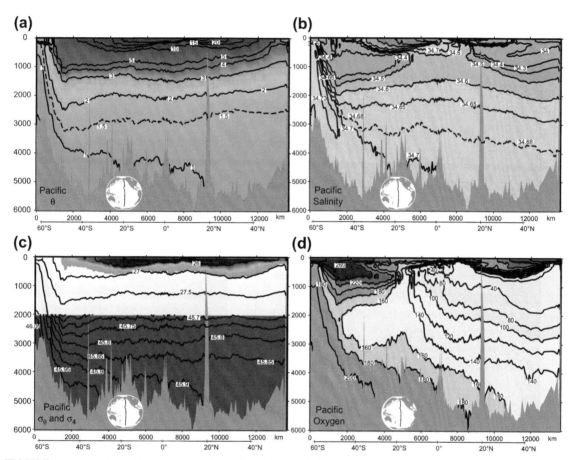

FIGURE 4.12 (a) Potential temperature (°C), (b) salinity (psu), (c) potential density σ_θ (top) and potential density σ_4 (bottom; kg m^{-3}), and (d) oxygen (μmol/kg) in the Pacific Ocean at longitude 150°W. Data from the World Ocean Circulation Experiment.

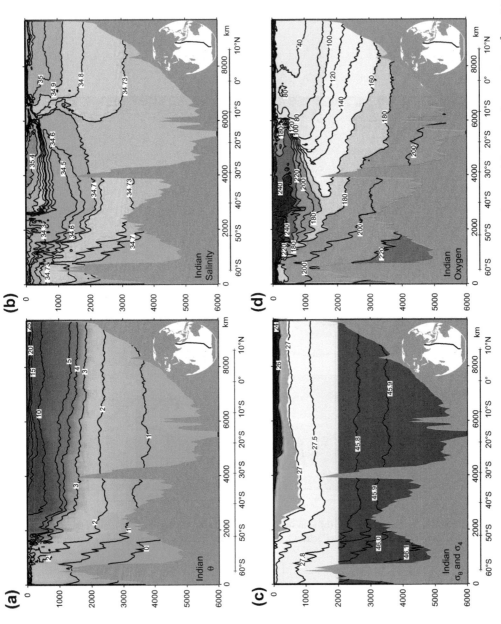

FIGURE 4.13 (a) Potential temperature (°C), (b) salinity (psu), (c) potential density σ_θ (top) and potential density σ_4 (bottom; kg m^{-3}), and (d) oxygen (µmol/kg) in the Indian Ocean at longitude 95°E. Data from the World Ocean Circulation Experiment.

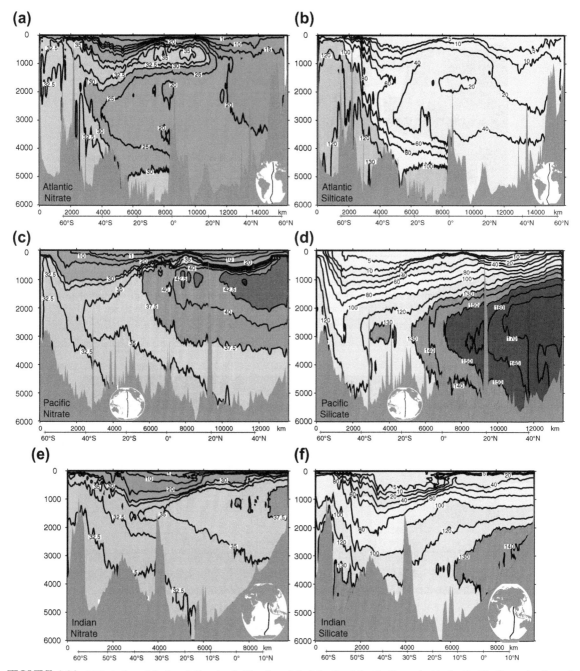

FIGURE 4.22 Nitrate (μmol/kg) and dissolved silica (μmol/kg) for the Atlantic Ocean (a, b), the Pacific Ocean (c, d), and the Indian Ocean (e, f). Note that the horizontal axes for each ocean differ. Data from the World Ocean Circulation Experiment.

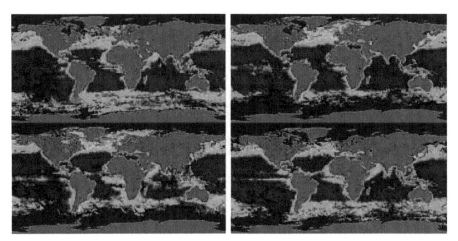

FIGURE 4.28 Global images of chlorophyll derived from the Coastal Zone Color Scanner (CZCS). Global phytoplankton concentrations change seasonally, as revealed by these three-month "climatological" composites for all months between November 1978–June 1986 during which the CZCS collected data: January–March (upper left), April–June (upper right), July–September (lower left), and October–December (lower right). Note the "blooming" of phytoplankton over the entire North Atlantic with the advent of Northern Hemisphere spring, and seasonal increases in equatorial phytoplankton concentrations in both Atlantic and Pacific Oceans and off the western coasts of Africa and Peru. See Figure S4.2 from the online supplementary material for maps showing the similarity between particulate organic carbon (POC) and chlorophyll. *Source: From NASA (2009a).*

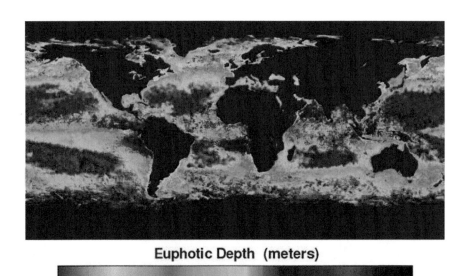

FIGURE 4.29 Euphotic zone depth (m) from the Aqua MODIS satellite, 9 km resolution, monthly composite for September 2007. (Black over oceans is cloud cover that could not be removed in the monthly composite.) See Figure S4.3 from the online supplementary material for the related map of photosynthetically available radiation (PAR). *Source: From NASA (2009b).*

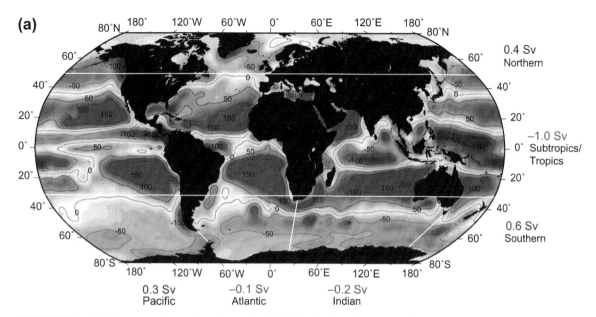

FIGURE 5.4 (a) Net evaporation and precipitation (E−P) (cm/yr) based on climatological annual mean data (1979–2005) from the National Center for Environmental Prediction. Net precipitation is negative (blue), net evaporation is positive (red). Overlain: freshwater transport divergences (Sverdrups or 1×10^9 kg/sec) based on ocean velocity and salinity observations. *Source: After Talley (2008).*

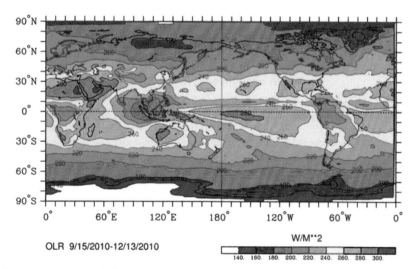

OLR 9/15/2010-12/13/2010

W/M**2

140. 160. 180. 200. 220. 240. 260. 280. 300.

FIGURE 5.9 Outgoing Longwave Radiation (OLR) for Sept. 15–Dec. 13, 2010. *Source: From NOAA ESRL (2010).*

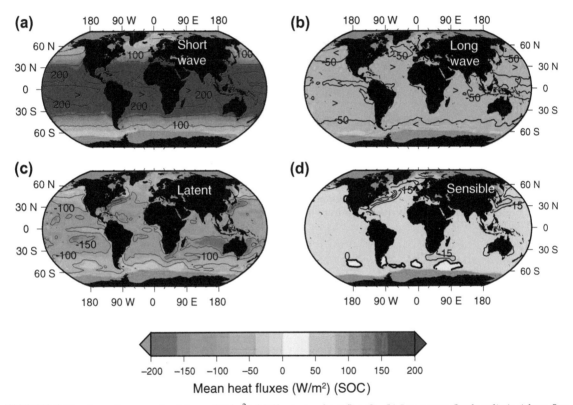

FIGURE 5.11 Annual average heat fluxes (W/m²). (a) Shortwave heat flux Q_s. (b) Longwave (back radiation) heat flux Q_b. (c) Evaporative (latent) heat flux Q_e. (d) Sensible heat flux Q_h. Positive (yellows and reds): heat gain by the sea. Negative (blues): heat loss by the sea. Contour intervals are 50 W/m² in (a) and (c), 25 W/m² in (b), and 15 W/m² in (d). Data are from the National Oceanography Centre, Southampton (NOCS) climatology (Grist and Josey, 2003).

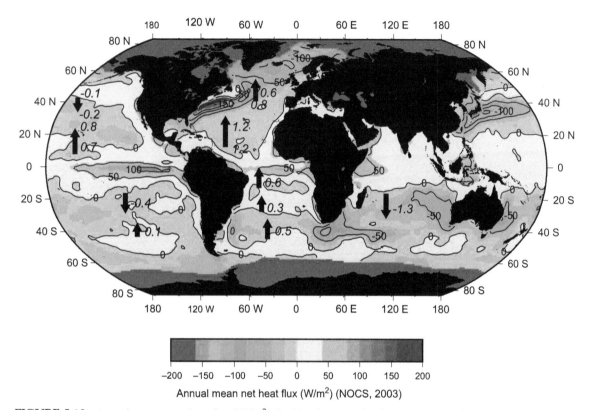

FIGURE 5.12 Annual average net heat flux (W/m²). Positive: heat gain by the sea. Negative: heat loss by the sea. Data are from the NOCS climatology (Grist and Josey, 2003). Superimposed numbers and arrows are the meridional heat transports (PW) calculated from ocean velocities and temperatures, based on Bryden and Imawaki (2001) and Talley (2003). Positive transports are northward. The online supplement to Chapter 5 (Figure S5.8) includes another version of the annual mean heat flux, from Large and Yeager (2009).

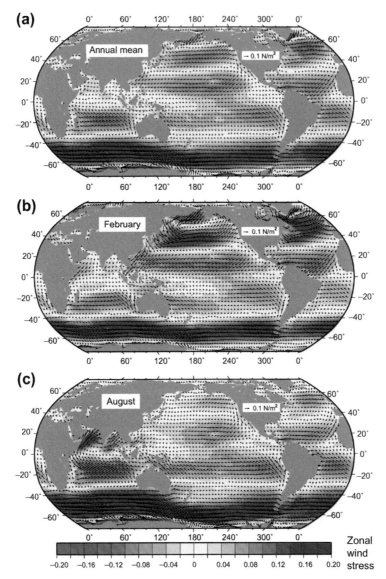

FIGURE 5.16 Mean wind stress (arrows) and zonal wind stress (color shading) (N/m²): (a) annual mean, (b) February, and (c) August, from the NCEP reanalysis 1968–1996 (Kalnay et al., 1996).

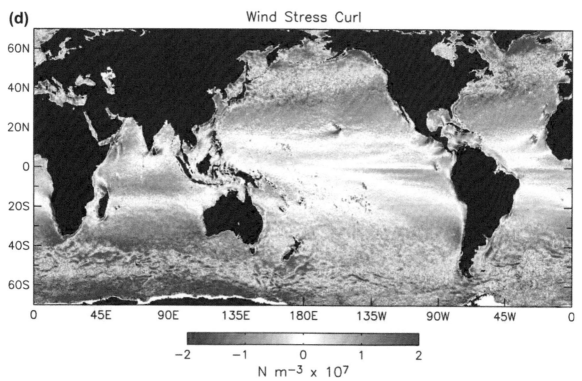

(d) Wind Stress Curl

$N\ m^{-3} \times 10^7$

FIGURE 5.16 (d) Mean wind stress curl based on 25 km resolution QuikSCAT satellite winds (1999–2003). Downward Ekman pumping (Chapter 7) is negative (blues) in the Northern Hemisphere and positive (reds) in the Southern Hemisphere. *Source: From Chelton et al. (2004).*

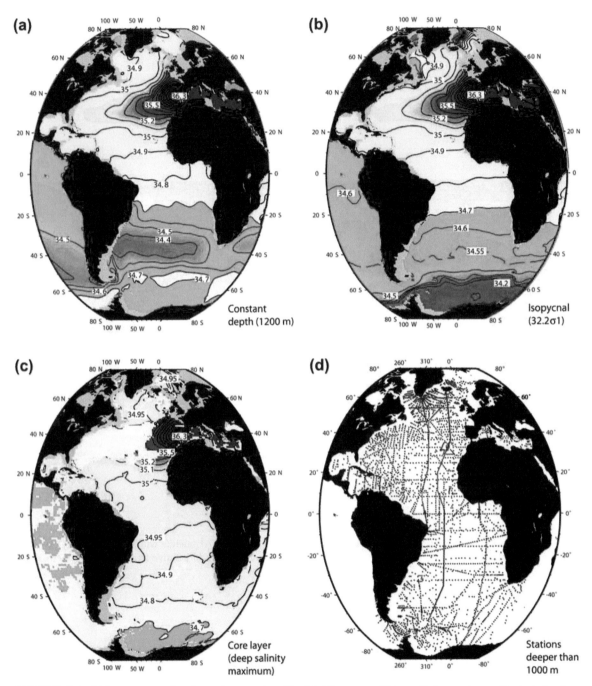

FIGURE 6.4 Different types of surfaces for mapping. The Mediterranean Water salinity maximum illustrated using: (a) a standard depth surface (1200 m); (b) an isopycnal surface (potential density $\sigma_1 = 32.2$ kg/m^3 relative to 1000 dbar, $\sigma_\theta \sim 26.62$ kg/m^3 relative to 0 dbar, and neutral density ~ 26.76 kg/m^3); (c) at the salinity maximum of the Mediterranean Water and North Atlantic Deep Water (white areas are where there is no deep salinity maximum); and (d) data locations used to construct these maps.

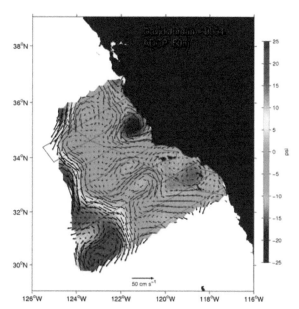

FIGURE 6.5 Objective mapping of velocity data, combining density and ADCP velocity measurements. California Current: absolute surface streamfunction and velocity vectors in April, 1999, using the method from Chereskin and Trunnell (1996). *Source: From Calcofi ADCP (2008).*

(a)

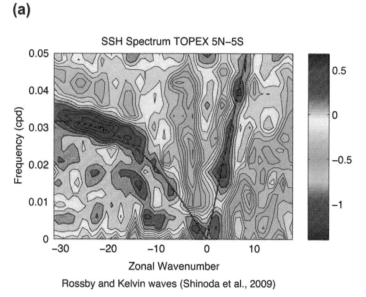

Rossby and Kelvin waves (Shinoda et al., 2009)

FIGURE 6.10 Examples of frequency-wavenumber spectra. (a) Equatorial waves (Kelvin and Rossby) from SSH anomalies, compared with theoretical dispersion relations (curves). *Source: From Shinoda et al. (2009)*

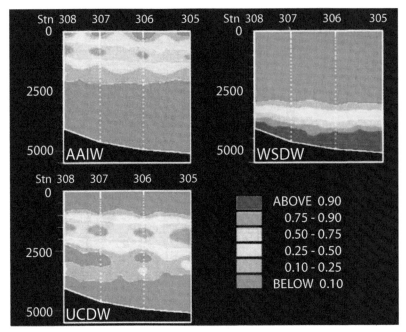

FIGURE 6.17 Example of optimum multiparameter (OMP) water mass analysis. Southwestern Atlantic about 36°S, showing the fraction of three different water masses. Antarctic Intermediate Water, AAIW; Upper Circumpolar Deep Water, UCDW; and Weddell Sea Deep Water, WSDW. *Source: From Maamaatuaiahutapu et al. (1992).*

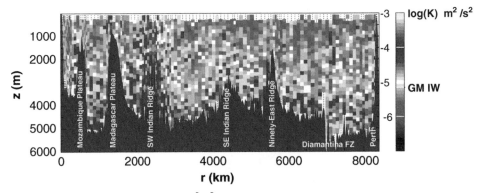

FIGURE 7.2 Observed diapycnal diffusivity (m²/s²) along 32°S in the Indian Ocean, which is representative of other ocean transects of diffusivity. See Figure S7.4 for diffusivity profiles. © *American Meteorological Society. Reprinted with permission. Source: From Kunze et al. (2006).*

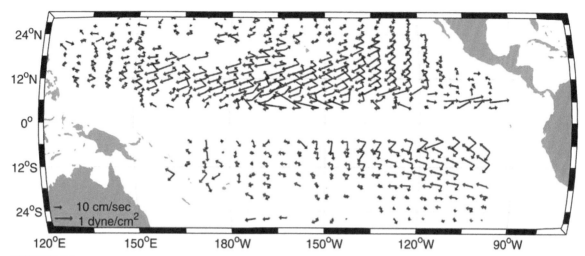

FIGURE 7.8 Ekman response. Average wind vectors (blue) and average ageostrophic current at 15 m depth (red). The current is calculated from 7 years of surface drifters drogued at 15 m, with the geostrophic current based on average density data from Levitus et al. (1994a) removed. (No arrows were plotted within 5 degrees of the equator because the Coriolis force is small there.) © *American Meteorological Society. Reprinted with permission. Source: From Ralph and Niiler (1999).*

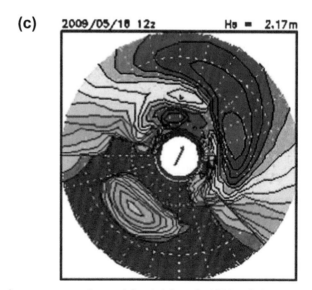

FIGURE 8.2 (c) Directional wave spectrum (spectral density) from the NE Pacific (station 46006, 40°53′ N 137°27′ W, May 16, 2009. In (c), wave periods are from about 25 sec at the center of the ring to 4 sec at the outer ring. Blue is low energy, purple is high. Direction of the waves is the same as direction relative to the center of the circle. Gray arrow in center indicates wind direction. "Hs" indicates significant wave height. *Source: From NOAA Wavewatch III (2009).*

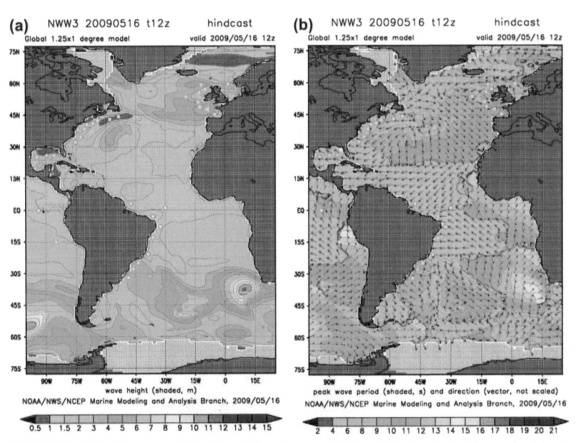

FIGURE 8.3 (a) Significant wave height (m) and (b) peak wave period (s) and direction (vectors) for one day (May 16, 2009). *Source: From NOAA Wavewatch III (2009).*

(c)

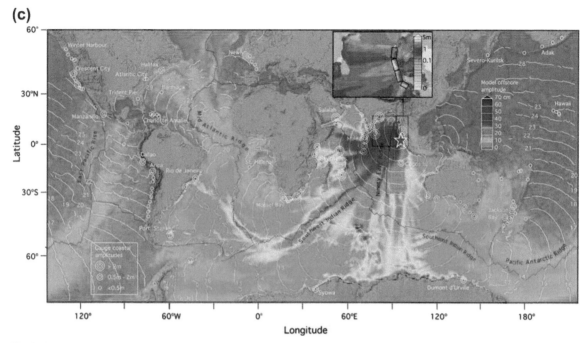

FIGURE 8.7 Sumatra Tsunami (December 26, 2004). (c) Global reach: simulated maximum sea-surface height and arrival time (hours after earthquake) of wave front. *Source: From Titov et al. (2005).*

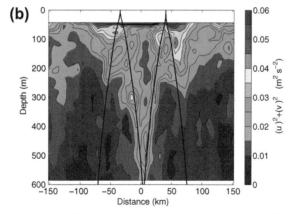

FIGURE 8.11 (b) Velocity variance (variability) observed along a section crossing the Hawaiian Ridge, which is located just below the bottom of the figure at 0 km; the black rays are the (group velocity) paths expected for an internal wave with frequency equal to the M_2 tide; distance (m) is from the center of the ridge. *Source: From Cole, Rudnick, Hodges, & Martin (2009).*

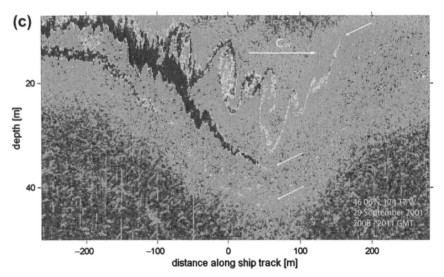

FIGURE 8.11 (c) Breaking internal solitary wave, over the continental shelf off Oregon. The image shows acoustic backscatter: reds indicate more scatter and are related to higher turbulence levels. © American Meteorological Society. Reprinted with permission. *Source: From Moum et al. (2003).*

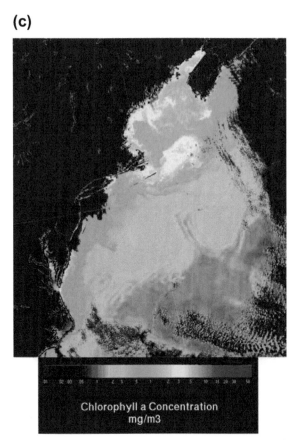

FIGURE 8.15 Tidal effects on Georges Bank. (c) Chlorophyll a concentration (mg/m^3) on October 8, 1997, from the SeaWiFS satellite. *Source: From Sosik (2003).*

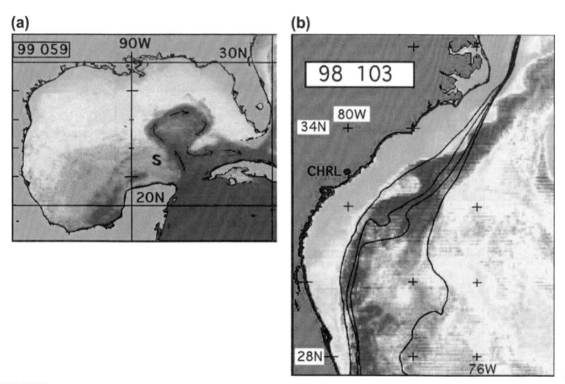

FIGURE 9.4 Sea surface temperature from the GOES satellite. (a) Gulf of Mexico showing the Loop Current beginning to form an eddy. (b) Gulf Stream, showing meander at the Charleston Bump and downstream shingling. Black contours are isobaths (100, 500, 700, 1000 m). *Source: From Legeckis, Brown and Chang (2002).*

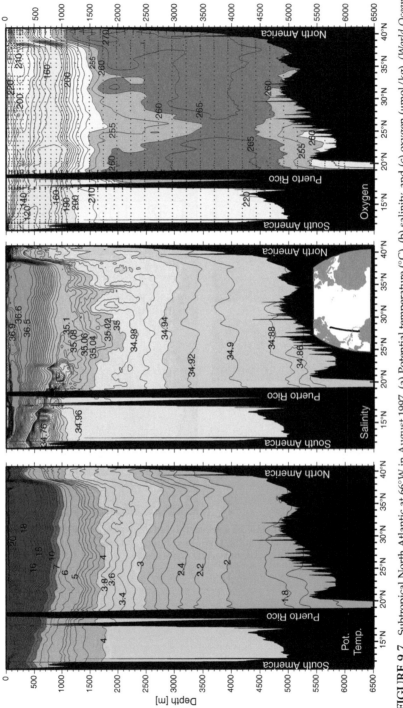

FIGURE 9.7 Subtropical North Atlantic at 66°W in August 1997. (a) Potential temperature (°C), (b) salinity, and (c) oxygen (μmol/kg). (*World Ocean Circulation Experiment section A22.*)

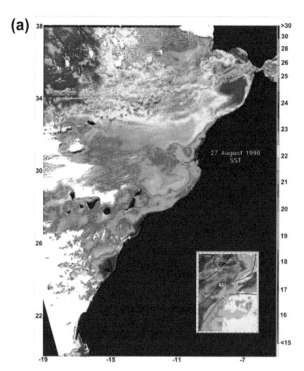

FIGURE 9.8 Canary Current System. (a) SST (satellite AVHRR image) on August 27, 1998. *Source: From Pelegrí et al. (2005).*

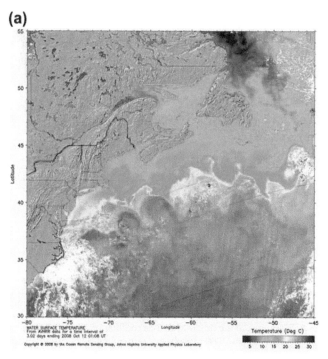

FIGURE 9.9 North Atlantic Current and Labrador Current at the Grand Banks. (a) SST (AVHRR) on October 12, 2008, showing cold Labrador Current moving southward along the edge of the Grand Banks. *Source: From Johns Hopkins APL Ocean Remote Sensing (1996).*

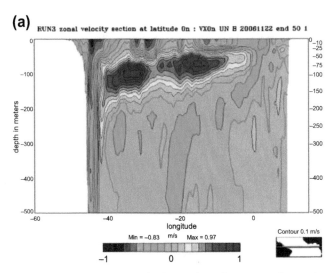

FIGURE 9.11 Tropical current structures. (a) Eastward velocity along the equator, from a data assimilation. *Source: From Bourlès et al. (2008).*

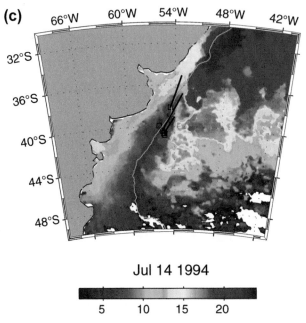

FIGURE 9.12 (c) Infrared satellite image of the Brazil-Malvinas confluence. Black lines are current vectors at moorings, at approximately 200 m depth. Light curve is the 1000 m isobath. *Source: From Vivier and Provost (1999).*

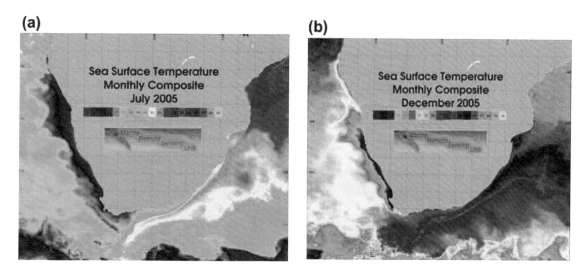

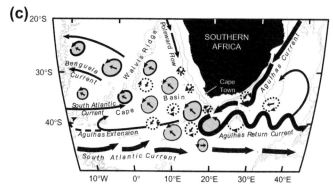

FIGURE 9.13 Benguela Current and Agulhas retroflection. (a, b) AVHRR SST monthly composite for July (winter) and December (summer) 2005. *Source: From UCT Oceanography Department (2009).* (c) Schematic of Agulhas retroflection and eddies, with flow directions in the intermediate water layer. Gray-shaded rings are the Agulhas anticyclones. Dashed rings are cyclones that are generated in the Agulhas. *Source: From Richardson (2007).*

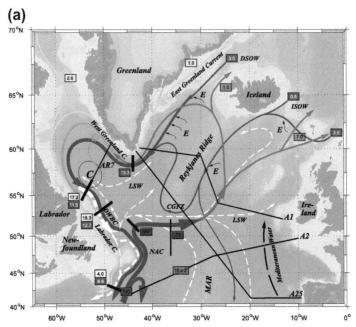

FIGURE 9.15 Schematics of deep circulation. (a) NSOW (blue), LSW (white dashed), and upper ocean (red, orange, and yellow) in the northern North Atlantic. *Source: From Schott and Brandt (2007).*

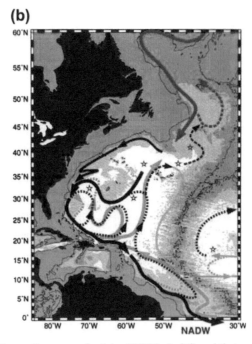

FIGURE 9.15 (b) Deep circulation pathways emphasizing DWBCs (solid) and their recirculations (dashed). Red: NSOW. Brown: NADW. Blue: AABW. (M.S. McCartney, personal communication, 2009.)

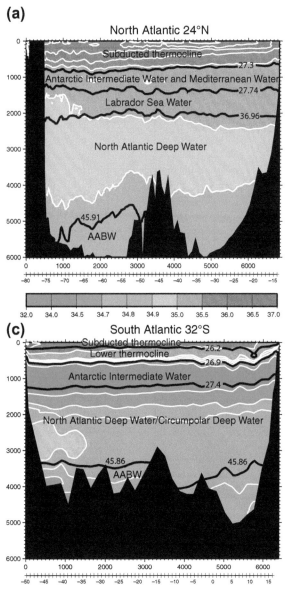

FIGURE 9.16 Salinity (color and white contours) and isopycnals (black contours) at (a) 24°N in 1981 and (c) 32°S in 1959/1972. *After Talley (2008), based on Reid (1994) velocities.*

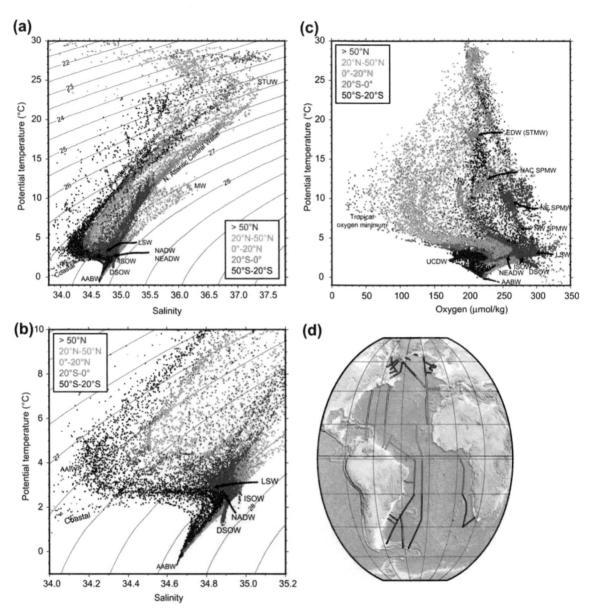

FIGURE 9.18 Potential temperature (°C) versus salinity for (a) full water column, and (b) water colder than 10°C. (c) Potential temperature versus oxygen for full water column. (d) Station location map. Colors indicate latitude range. Contours are potential density referenced to 0 dbar. Data are from the World Ocean Circulation Experiment (1988–1997).

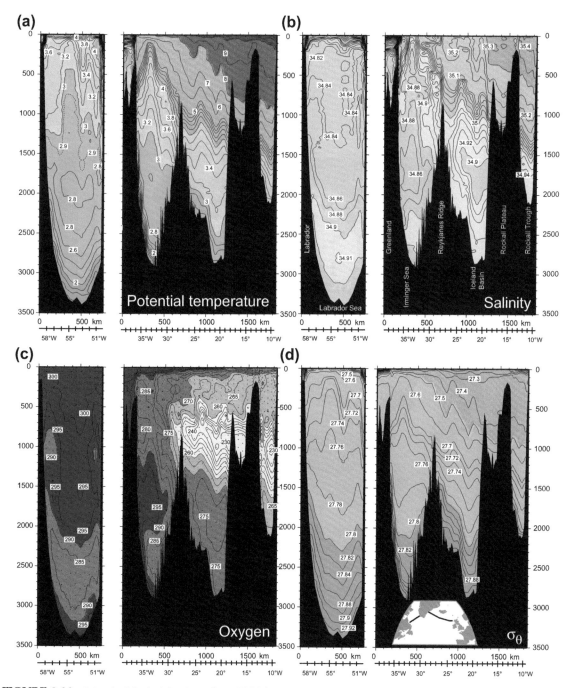

FIGURE 9.20 Subpolar North Atlantic at about 55°N from May to June, 1997. (a) Potential temperature (°C), (b) salinity, (c) oxygen, and (d) potential density (σ_θ) in the Labrador Sea (left side) and from Greenland to Ireland (right side). (World Ocean Circulation Experiment sections AR7W and A24)

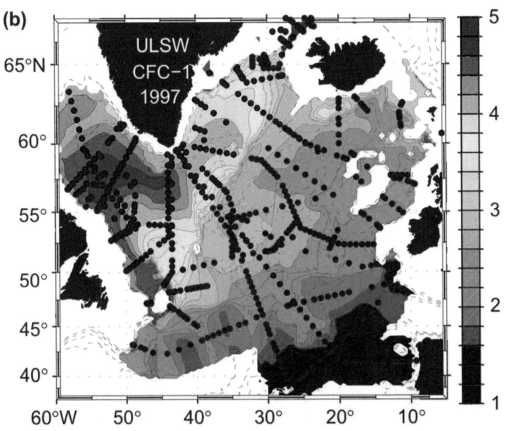

FIGURE 9.21 Labrador Sea Water. (b) Chlorofluorocarbon-11 (pmol/kg) in the upper LSW layer, at $\sigma_\theta \sim 27.71$ kg/m^3. *Source: From Schott et al. (2009) and from Kieke et al. (2006).*

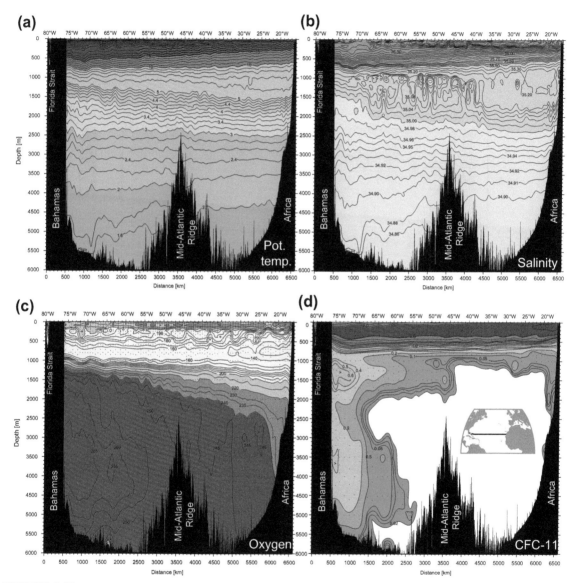

FIGURE 9.22 Subtropical North Atlantic at 24°N from July to August 1992. (a) Potential temperature (°C), (b) salinity, (c) oxygen (μmol/kg), and (d) CFC-11 (pmol/kg) at 24°N. (World Ocean Circulation Experiment section A05). *Adapted: From WOCE Atlantic Ocean Atlas, Jancke, Gouretski, and Koltermann (2011).*

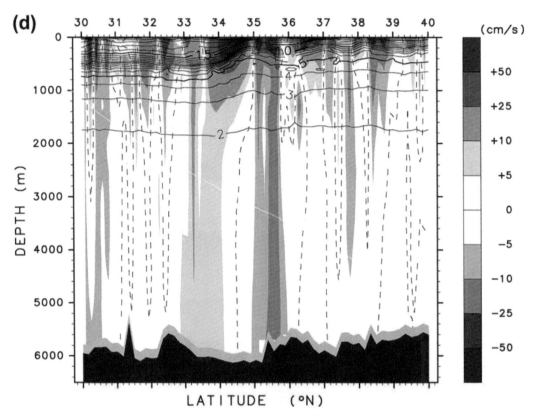

FIGURE 10.4 Kuroshio velocity structure. (d) Eastward velocity of the Kuroshio Extension at 152° 30′E [red (blue) indicates eastward (westward) flow]. *Source: From Yoshikawa et al. (2004).*

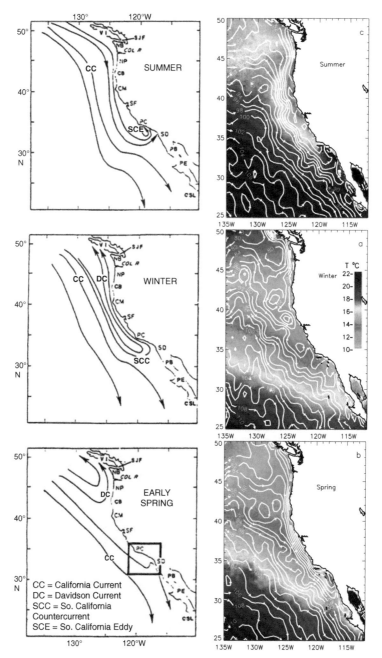

FIGURE 10.5 (a) Schematic of the surface currents in the CCS in different seasons. *Source: From Hickey (1998).* (b) Mean seasonal cycle of satellite-derived surface temperature (color) and altimetric height, showing the geostrophic surface circulation. *Source: From Strub and James (2000, 2009).*

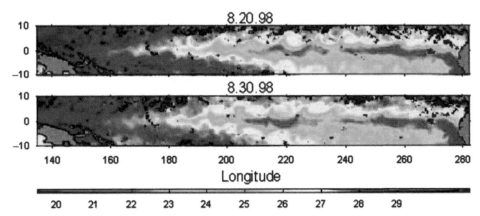

FIGURE 10.24 Tropical instability waves. SST from the Tropical Rainfall Mapping Mission (TRMM) Microwave Imager (TMI) for two successive 10-day periods in August 1998, after establishment of the cold tongue during a La Niña. A more complete time series (June 1–August 30, 1998) is reproduced in Figure S10.20 on the textbook Web site. *Source: From Remote Sensing Systems (2004).*

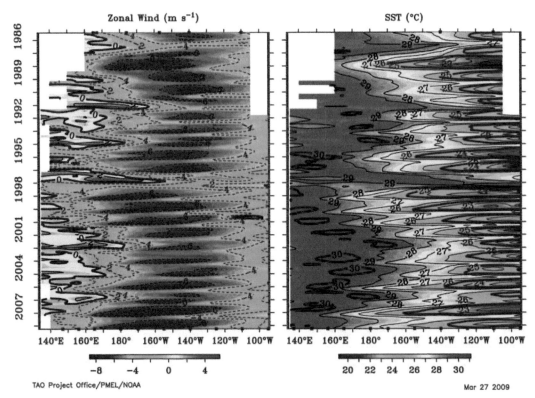

FIGURE 10.25 Zonal wind speed and SST in the equatorial Pacific to illustrate the annual cycle. Positive wind speed is toward the east. Climatological means in February and August and an expanded time series for 2000–2007 are shown in Figure S10.21 on the textbook Web site, to emphasize the seasonal cycle. *Source: From TAO Project Office (2009a).*

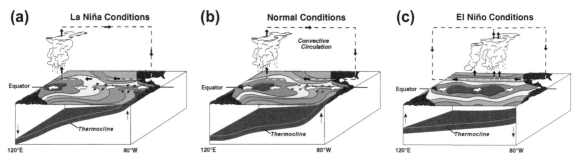

FIGURE 10.27 (a) La Niña, (b) normal, and (c) El Niño conditions. *Source: From NOAA PMEL (2009b).*

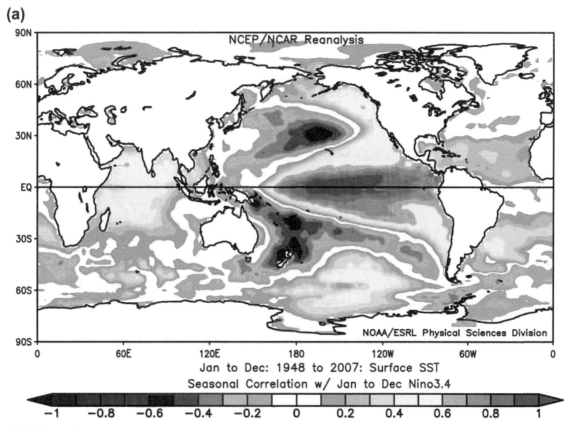

FIGURE 10.28 (a) Correlation of monthly SST anomalies with the ENSO Nino3.4 index, averaged from 1948 to 2007. The index is positive during the El Niño phase, so the signs shown are representative of this phase. (*Data and graphical interface from NOAA ESRL, 2009b.*)

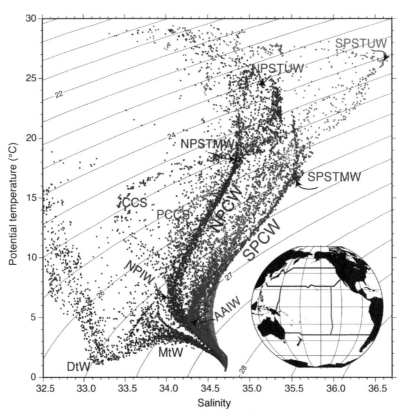

FIGURE 10.29 Potential T-S curves for selected stations (inset map). Acronyms: NPCW, North Pacific Central Water; SPCW, South Pacific Central Water; NPSTUW, North Pacific Subtropical Underwater; SPSTUW, South Pacific Subtropical Underwater; NPSTMW, North Pacific Subtropical Mode Water; SPSTMW, South Pacific Subtropical Mode Water; NPIW, North Pacific Intermediate Water; AAIW, Antarctic Intermediate Water; DtW, Dichothermal Water; MtW, Mesothermal Water; CCS, California Current System waters; and PCCS, Peru-Chile Current System Waters.

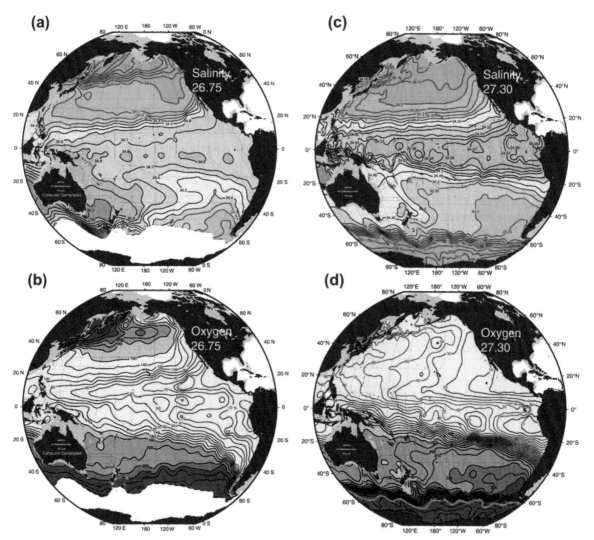

FIGURE 10.33 (a, c) Salinity and (b, d) oxygen (μmol/kg) at neutral densities 26.75 kg/m^3 and 27.3 kg/m^3, characteristic of NPIW and AAIW, respectively. In the Southern Ocean, white at 26.75 kg/m^3 shows the isopycnal outcrops; the gray curve in (c) and (d) is the winter outcrop. Depth of the surfaces is shown in the WOCE Pacific Ocean Atlas. *Source: From WOCE Pacific Ocean Atlas, Talley (2007).*

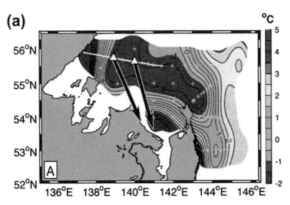

FIGURE 10.34 Dense water formation in the Okhotsk Sea. (a) Bottom potential temperature in September, 1999, and mean velocity vectors at the two moorings. *Source: From Shcherbina, Talley, and Rudnick (2003, 2004).*

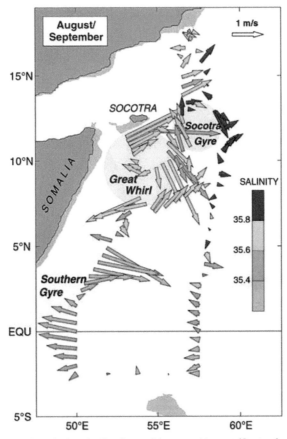

FIGURE 11.3 Somali Current regime during the Southwest Monsoon (August/September, 1995). *Source: From Schott and McCreary (2001).*

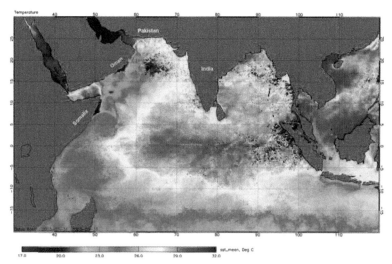

FIGURE 11.5 SST in July 2003 (Southwest Monsoon), from the MODIS satellite. *Source: From NASA Goddard Earth Sciences (2007a).*

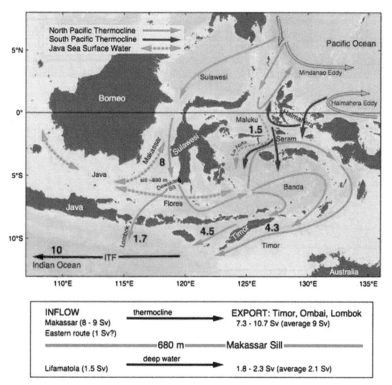

FIGURE 11.11 Indonesian Archipelago and Throughflow with transports (Sv). Lower panel summarizes transport above and below 680 m (Makassar Strait sill depth). *Source: From Gordon (2005).*

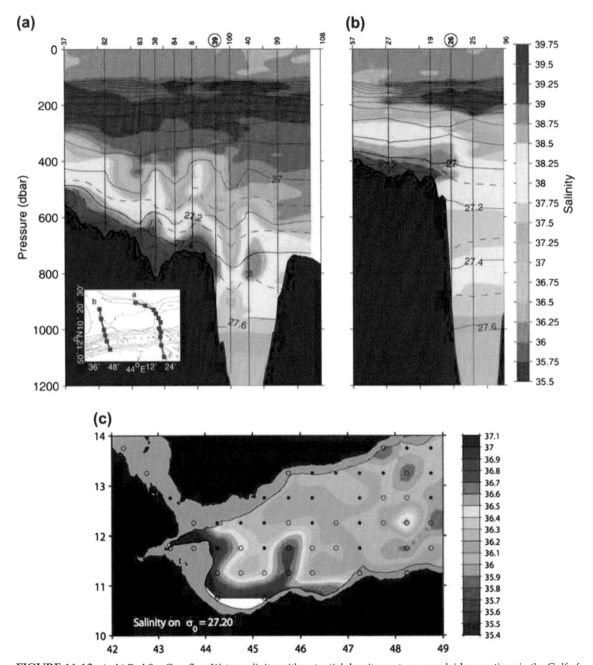

FIGURE 11.12 (a, b) Red Sea Overflow Water: salinity with potential density contours overlaid on sections in the Gulf of Aden in February–March, 2001. North is on the left. *Source: From Bower et al. (2005). © American Meteorological Society. Reprinted with permission.* (c) Red Sea outflow in the Gulf of Aden: climatological salinity on the isopycnal $\sigma_\theta = 27.20$ kg/m^3. *Source: From Bower, Hunt, and Price (2000).*

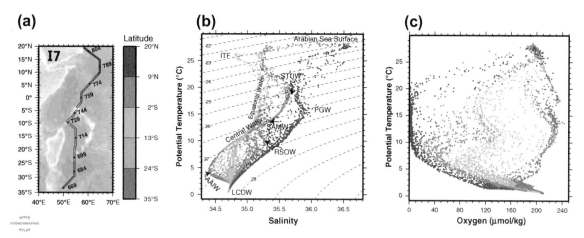

FIGURE 11.18 (a) Station locations, (b) potential temperature (°C) — salinity and (c) potential temperature (°C) — oxygen (μmol/kg) for the Indian Ocean along 60°E. *After the WOCE Indian Ocean Atlas, Talley (2011).*

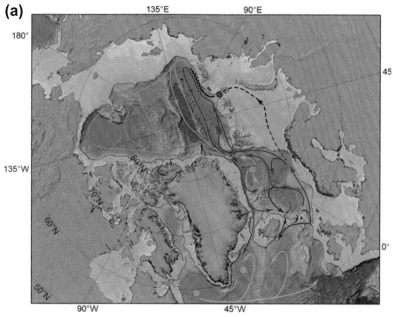

FIGURE 12.10 Circulation schematics. (a) Subsurface Atlantic and intermediate layers of the Arctic Ocean and the Nordic Seas. Convection sites in the Greenland and Iceland Seas, and in the Irminger and Labrador Seas are also shown (light blue), as is a collection point for brine-rejected waters from the Barents Sea. *Source: From Rudels et al. (2010).*

(a)

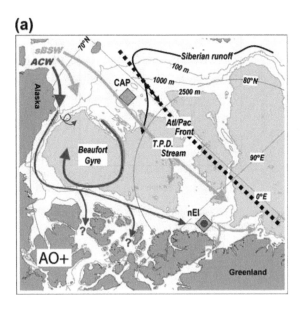

(b)

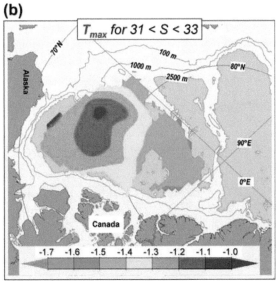

FIGURE 12.13 (a) Schematic circulation of summer Bering Strait Water (blue) and Alaskan Coastal Water (red) during the positive phase of the Arctic Oscillation (Chapter S15 on the textbook Web site). (b) Temperature (°C) of the shallow temperature maximum layer, which lies between 50 and 100 m depth, in the Canadian Basin. *Source: From Steele et al. (2004).*

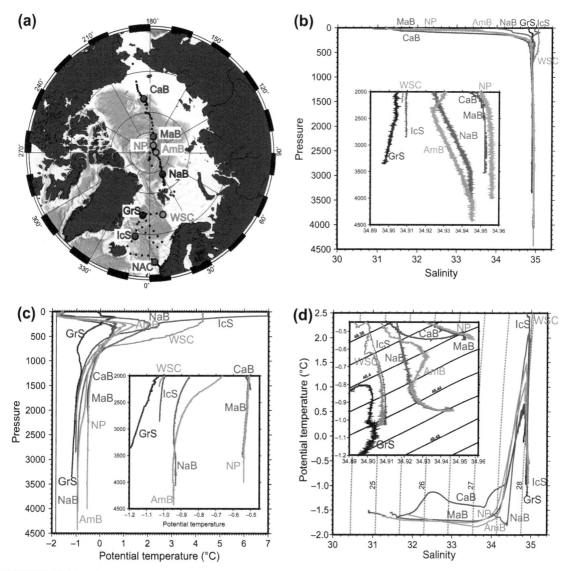

FIGURE 12.17 (a) Station map (1994 and 2001), (b) salinity, (c) potential temperature (°C), and (d) potential temperature-salinity. Acronyms: CaB, Canada Basin; MaB, Makarov Basin; NP, North Pole; AmB, Amundsen Basin; NaB, Nansen Basin; WSC, West Spitsbergen Current; GrS, Greenland Sea; IcS, Iceland Sea; and NAC, Norwegian Atlantic Current. *Expanded from Timmermans and Garrett (2006).*

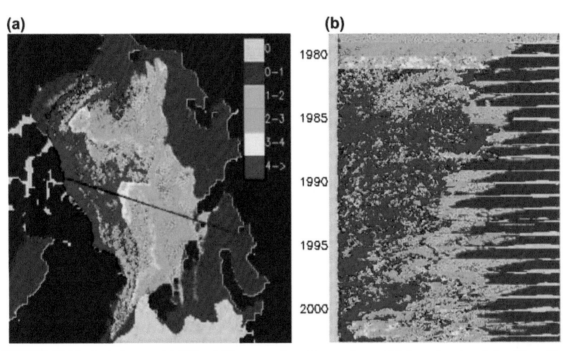

FIGURE 12.21 Arctic ice ages: (a) 2004 and (b) cross-section of ice age classes (right) as a function of time (Hovmöller diagram), extending along the transect across the Arctic from the Canadian Archipelago to the Kara Sea shown in (a). *Source: Extended from Fowler et al. (2004).*

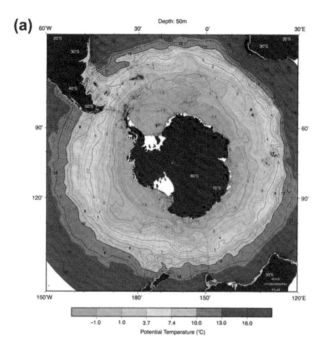

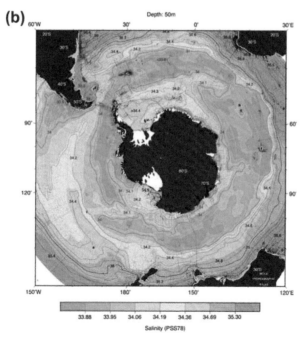

FIGURE 13.3 Properties at 50 m depth. (a) Potential temperature (°C), (b) salinity. *Source: From WOCE Southern Ocean Atlas, Orsi and Whitworth (2005).*

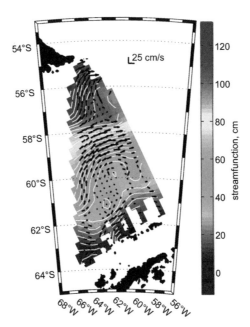

FIGURE 13.9 Mean currents in the Drake Passage, averaged over 30–300 m depth, from 128 ADCP crossings over 5 years. Strong currents from north to south are the Subantarctic Front (56°S), the Polar Front (59°S), and the Southern ACC Front (62°S). *After Lenn, Chereskin, and Sprintall (2008).*

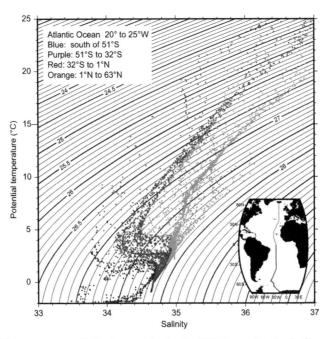

FIGURE 13.14 Potential temperature-salinity diagram in the Weddell Sea and Atlantic Ocean.

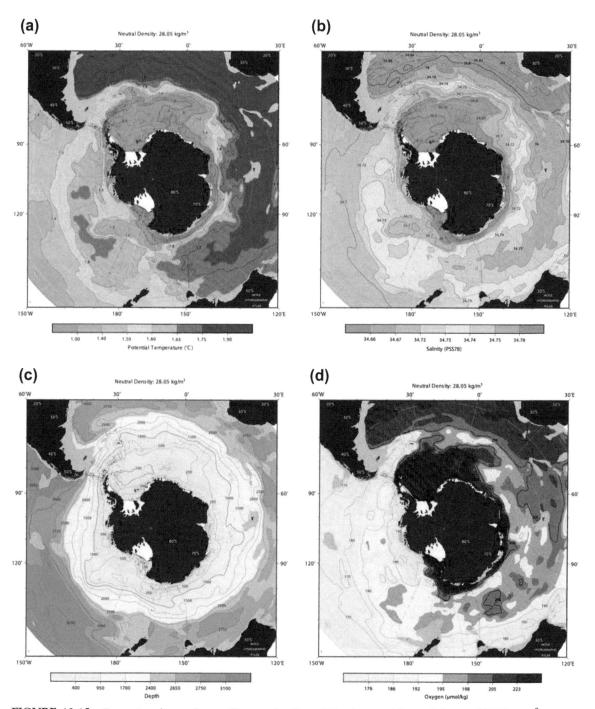

FIGURE 13.15 Properties along a Lower Circumpolar Deep Water isopycnal (neutral density 28.05 kg m^{-3}), corresponding roughly to the salinity maximum core. (a) Potential temperature (°C), (b) salinity, (c) depth (m), (d) oxygen (μmol/kg). *Source: From WOCE Southern Ocean Atlas, Orsi and Whitworth (2005).*

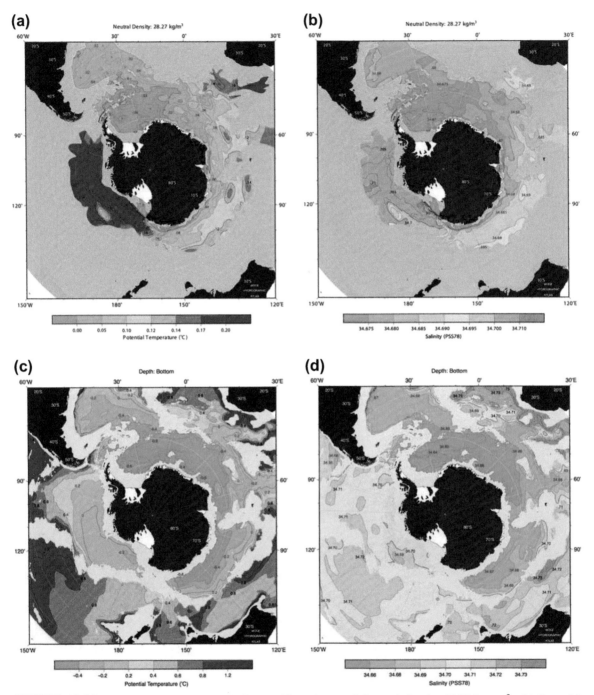

FIGURE 13.16 Properties on an Antarctic Bottom Water isopycnal (neutral density 28.27 kg m^{-3}). (a) Potential temperature and (b) salinity. Bottom properties (depths greater than 3500 m): (c) potential temperature and (d) salinity. *Source: From WOCE Southern Ocean Atlas, Orsi and Whitworth (2005).*

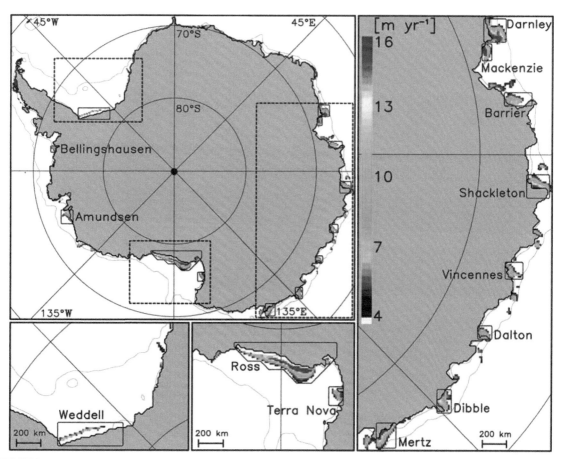

FIGURE 13.20 Antarctic latent heat polynyas: sea ice production, averaged over 1992–2001. *Source: From Tamura et al. (2008).*

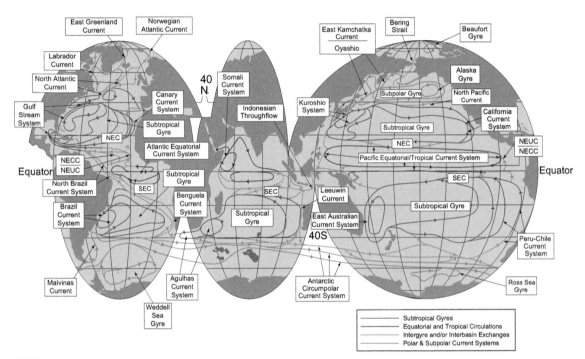

FIGURE 14.1 Surface circulation schematic. *Modified from Schmitz (1996b).*

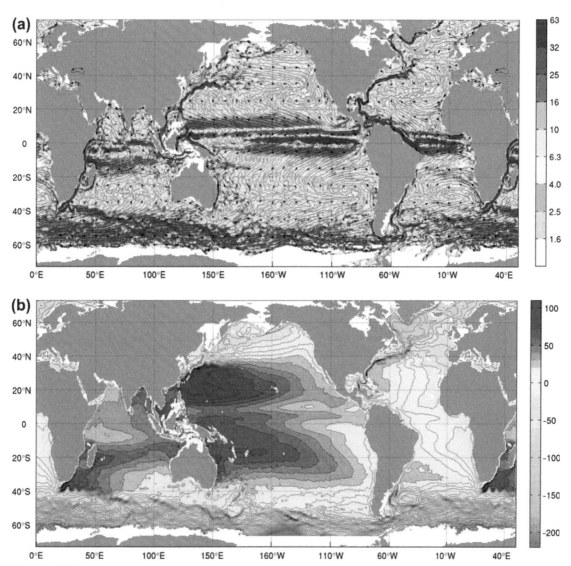

FIGURE 14.2 (a) Surface dynamic topography (dyn cm), with 10 cm contour intervals, and (b) surface velocity streamlines, including both geostrophic and Ekman components; color is the mean speed in cm/sec. *Source: From Maximenko et al. (2009).*

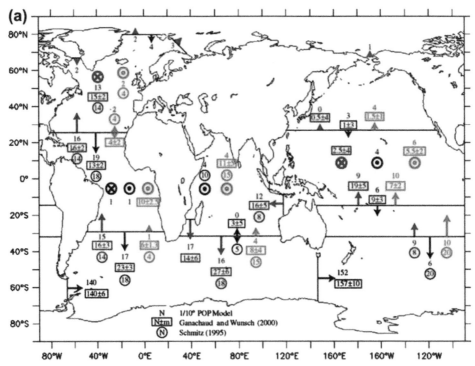

FIGURE 14.6 Net transports (Sv) in isopycnal layers across closed hydrographic sections (1 Sv = 1×10^6 m^3/sec). (a) Three calculations from different sources are superimposed, each using three isopycnal layers (see header). Circles between sections indicate upwelling (arrow head) and downwelling (arrow tail) into and out of the layer defined by the circle color. *Source: From Maltrud and McClean (2005), combining results from their POP model run, Ganachaud and Wunsch (2000), and Schmitz (1995).*

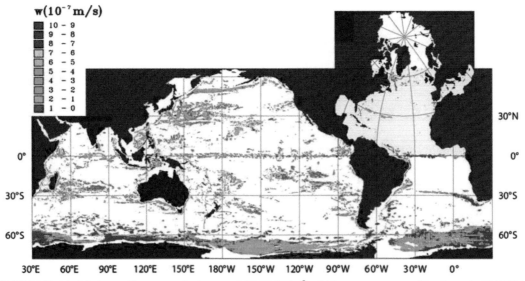

FIGURE 14.7 Modeled upwelling across the isopycnal 27.625 kg/m^3, which represents upwelling from the NADW layer. *Source: From Kuhlbrodt et al. (2007); adapted from Döös and Coward (1997).*

(a)

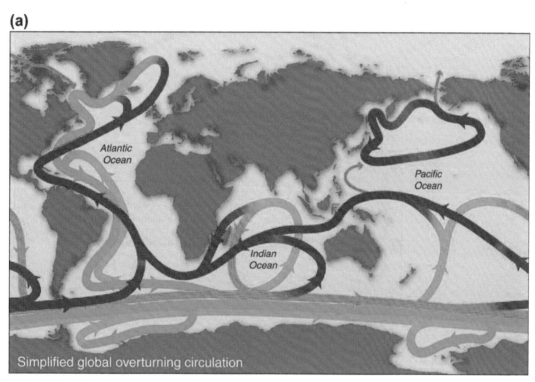

FIGURE 14.11 Global overturning circulation schematics. (a) The NADW and AABW global cells and the NPIW cell.

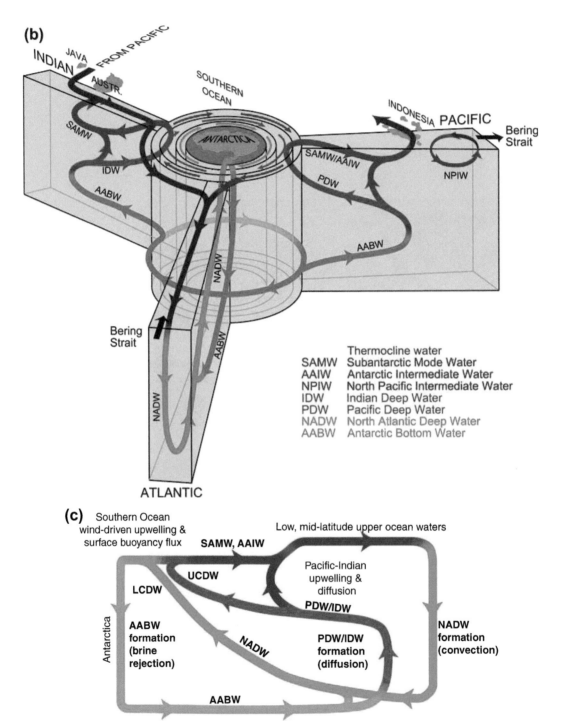

FIGURE 14.11 (b) Overturn from a Southern Ocean perspective. *Source: After Gordon (1991), Schmitz (1996b), and Lumpkin and Speer (2007).* (c) Two-dimensional schematic of the interconnected NADW, IDW, PDW, and AABW cells. The schematics do not accurately depict locations of sinking or the broad geographic scale of upwelling. Colors: surface water (purple), intermediate and Southern Ocean mode water (red), PDW/IDW/UCDW (orange), NADW (green), AABW (blue). See Figure S14.1 on the textbook Web site for a complete set of diagrams. *Source: From Talley (2011).*

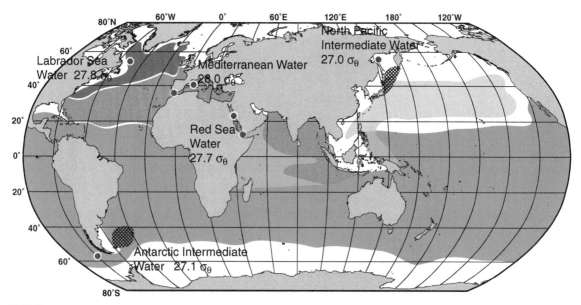

FIGURE 14.13 Low- and high-salinity intermediate waters. AAIW (dark green), NPIW (light green), LSW (dark blue), MW (orange in Atlantic), RSW (orange in Indian). Light blue in Pacific: overlap of AAIW and NPIW. Light blue in Indian: overlap of AAIW and RSW. Cross-hatching: mixing sites that are particularly significant for the water mass. Red dots indicate the primary formation site of each water mass; fainter dots mark the straits connecting the Mediterranean and Red Seas to the open ocean. The approximate potential density of formation is listed. *Source: After Talley (2008).*

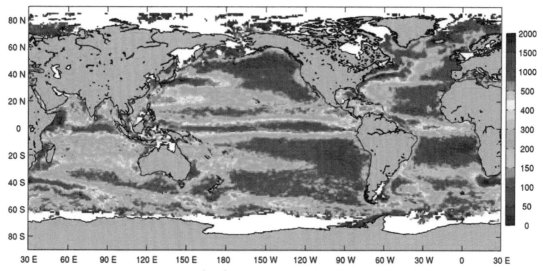

FIGURE 14.16 Eddy kinetic energy (cm^2 s^{-2}) from surface drifters. *Source: From NOAA AOML PHOD (2009).* A complementary figure based on satellite altimetry (*from Ducet, Le Traon, & Reverdin, 2000*) is reproduced in Figure S14.6c on the textbook Web site.

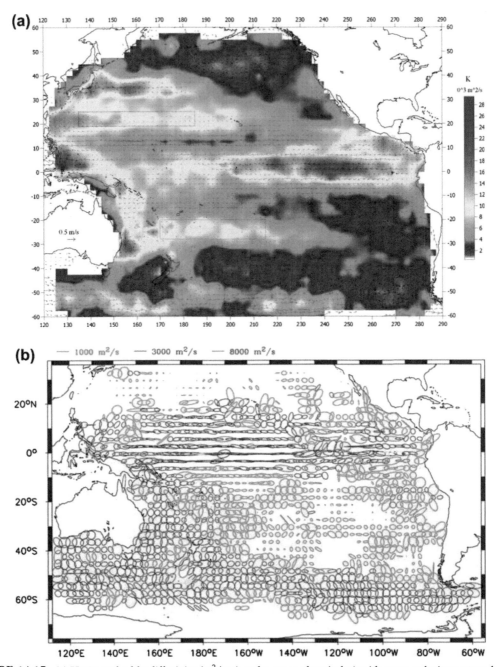

FIGURE 14.17 (a) Horizontal eddy diffusivity (m^2/sec) at the sea surface (color) with mean velocity vectors, based on surface drifter observations. *Source: From Zhurbas and Oh (2004).* (b) Eddy diffusivity ellipses at 900 m based on subsurface float velocities. Colors indicate different scales (see figure headers). *Source: From Davis (2005).* The Atlantic surface map and Indian 900 m map from the same sources are reproduced in Chapter S14 (Figures S14.7 and S14.8) on the textbook Web site.

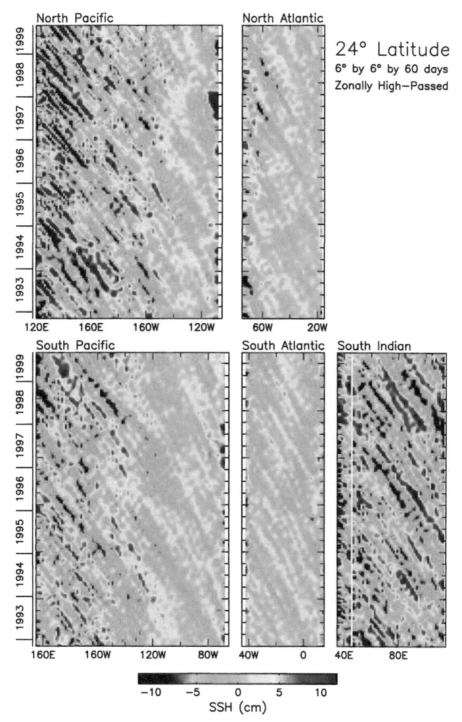

FIGURE 14.18 Surface-height anomalies at 24 degrees latitude in each ocean, from a satellite altimeter. *Source: From Fu and Chelton (2001).*

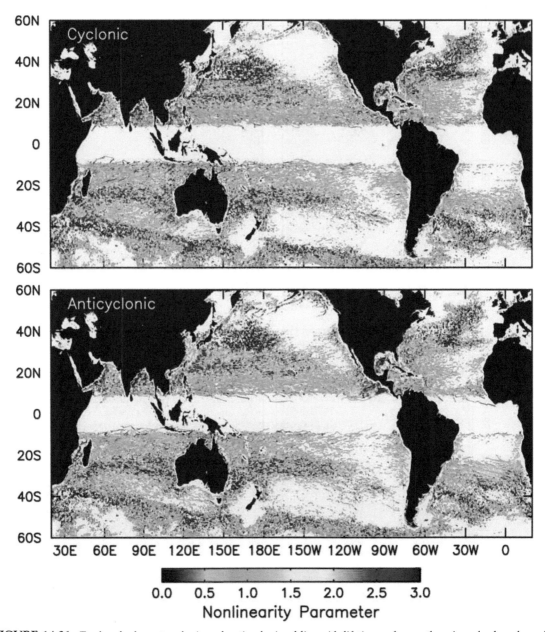

FIGURE 14.21 Tracks of coherent cyclonic and anticyclonic eddies with lifetimes of more than 4 weeks, based on altimetric SSH, color coded by a "nonlinearity parameter," which is the ratio of velocity within the eddy compared with the eddy propagation speed. White areas indicate no eddies or trajectories within 10 degrees latitude of the equator. *Source: From Chelton et al. (2007).*

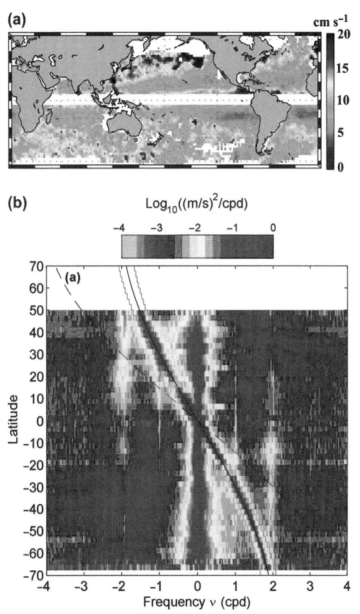

FIGURE 14.22 Near-inertial motion. (a) Average inertial current speeds (cm/sec), based on surface drifters. *Source: From Chaigneau et al. (2008).* (b) Rotary power spectra in 2.5 degree latitude bins in the Pacific Ocean. The solid curve is the inertial frequency at each latitude; the dashed curve is twice the inertial frequency. Negative frequencies rotate counterclockwise and positive frequencies rotate clockwise. *Source: From Elipot and Lumpkin (2008).*

Printed and bound by CPI Group (UK) Ltd, Croydon, CR0 4YY

03/10/2024

01040332-0001